Five Membered Bioactive N and O-Heterocycles:

Models and Medical Applications

Shrikaant Kulkarni
Victorian Institute of Technology, Australia

Hemantkumar Akolkar
Rayat Shikshan Sanstha's Abasaheb Marathe Arts and New Commerce Science College, India

Bapurao B. Shingate
Dr. Babasaheb Ambedkar Marathwada University, India

Vijay M. Khedkar
Vishwakarma University, India

IGI Global
Scientific Publishing
Publishing Tomorrow's Research Today

Vice President of Editorial	Melissa Wagner
Managing Editor of Acquisitions	Mikaela Felty
Managing Editor of Book Development	Jocelynn Hessler
Production Manager	Mike Brehm
Cover Design	Phillip Shickler

Published in the United States of America by
IGI Global Scientific Publishing
701 East Chocolate Avenue
Hershey, PA, 17033, USA
Tel: 717-533-8845
Fax: 717-533-8661
Website: https://www.igi-global.com E-mail: cust@igi-global.com

Library of Congress Cataloging-in-Publication Data

Names: Kulkarni, Shrikaant, editor. | Akolkar, Hemantkumar, 1986- editor |
Shingate, Bapurao, 1975- editor. | Khedkar, Vijay, 1983- editor.
Title: Five membered bioactive N and O-heterocycles : models and medical
applications / edited by Shrikaant Kulkarni, Hemantkumar Akolkar,
Bapurao Shingate, Vijay Khedkar.
Description: Hershey, PA : IGI Global Scientific Publishing, [2025] |
Includes bibliographical references and index. | Summary: "This book
explores the comprehensive review of syntheses and pharmaceutical
applications of five-membered bioactive heterocycles containing two or
more heteroatoms. This book will help medicinal chemists to develop a new
synthetic methodology that enables quick access to a wide range of
functionalized heterocyclic compounds"-- Provided by publisher.
Identifiers: LCCN 2024049769 (print) | LCCN 2024049770 (ebook) | ISBN
9798369372678 (h/c) | ISBN 9798369372685 (s/c) | ISBN 9798369372692
(eISBN)
Subjects: LCSH: Heterocyclic compounds--Synthesis. | Bioactive
compounds--Synthesis. | Medicinal chemistry.
Classification: LCC QD400.5.S95 F57 2025 (print) | LCC QD400.5.S95
(ebook) | DDC 547/.595--dc23/eng/20250117
LC record available at https://lccn.loc.gov/2024049769
LC ebook record available at https://lccn.loc.gov/2024049770

British Cataloguing in Publication Data
A Cataloguing in Publication record for this book is available from the British Library.

Table of Contents

Detailed Table of Contents

Chapter 1
Saurabh Chetia, Mizoram University, India
Lalremruati, Mizoram University, India
Brilliant N. Marak, Mizoram University, India
Ved Prakash Singh, Mizoram University, India

Oxazoles, a class of five-membered aromatic heterocycles containing nitrogen and oxygen atoms, are of significant importance across various fields, including medicine and agriculture. Renowned for their diverse biological activities, including antibacterial, antiviral, anticancer, and anti-inflammatory properties, oxazoles have garnered significant attention in research. This chapter offers a comprehensive overview of the chemical properties, synthesis methods, and reactivity of oxazoles, alongside their pharmacological applications. It covers both conventional and forefront synthetic pathways of oxazole and its derivatives. The reactivity section explores the unique behavior of oxazoles in cycloadditions, reductions, and nucleophilic and electrophilic substitution reactions. Additionally, the chapter highlights the occurrence of oxazole derivatives in natural products and their potential medicinal applications.

Chapter 2
Aditi Boruah, Gauhati University, India
Ariful Islam, Gauhati University, India
Pranjal K. Baruah, Gauhati University, India

This chapter discusses about bioactive pyrazoles, which have garnered substantial attention in medicinal chemistry and drug development. Pyrazoles are five-membered heterocyclic compounds characterized by a nitrogen-rich structure, imparting them with unique physicochemical and biological and making them useful for treating various diseases. The chapter explores the synthesis, structural diversity, and functional modifications of pyrazoles that enhance their bioactivity. Detailed discussions highlight the mechanisms of action and therapeutic potential of pyrazole derivatives in various pharmacological areas, including anti-inflammatory, anticancer, antimicrobial, antiviral, neuroprotective, and other applications. The chapter also looks at the latest research and advancements in pyrazole chemistry, including new ways to make them and studies on how their structure affects their activity. By exploring the many uses of bioactive pyrazoles, this chapter shows their importance in contemporary medicinal chemistry in developing new and better medicines.

Chapter 3

Popat Mohite, AETs St. John Institute of Pharmacy and Research, Palghar, India

Savita Tauro, AETs St. John Institute of Pharmacy and Research, Palghar, India

Aarati Pawar, AETs St. John Institute of Pharmacy and Research, Palghar, India

Isoxazole is a versatile five-membered heterocyclic compound that has attracted significant attention due to its diverse medicinal chemistry and material science applications. This chapter provides a comprehensive overview of synthetic methodologies for preparing isoxazole derivatives, including traditional approaches such as cyclization reactions and modern techniques like microwave-assisted synthesis. The advantages and limitations of each method, highlighting the influence of structural variations on synthesis efficiency, are discussed. Furthermore, the biological activities associated with isoxazole compounds, showcasing their roles as potential pharmacological agents including anti-inflammatory, antibacterial, antifungal, and anticancer properties, are explored. By integrating synthetic strategies with biological evaluation, this chapter aims to illustrate the significance of isoxazole derivatives in drug discovery and development and their potential applications in therapeutic settings.

Chapter 4

Bishal Bhattacharyya, Dibrugarh University, India

Ramyata Priyam Borah, Dibrugarh University, India

Dipankar Nath, Dibrugarh University, India

Parishma Gogoi, Dibrugarh University, India

Diganta Sarma, Dibrugarh University, India

Cu catalysed azide alkyne cycloaddition reactions, known as click reactions, result in the regioselective formation of 1,4 disubstituted 1,2,3-triazoles. 1,2,3-triazoles are five membered nitrogen containing heterocycles that have gained much importance in recent times due to their various applications such as anti-cancer, anti- TB etc. Several greener approaches have been developed by different researchers to prepare 1,2,3-triazoles using benign methodologies. Greener solvents like Deep Eutectic

Solvents (DES) and Ionic Liquids (IL) can serve as good alternative for volatile organic solvents used during azide alkyne cycloaddition reactions

Chapter 5

Mubarak Hanif Shaikh, Radhabai Kale Mahila Mahavidyalaya, Ahmednagar, India
Amol A. Nagargoje, K.M.C. College of Arts, Science, and Commerce, Khopoli, India
Dattatraya N. Pansare, Deogiri College, India
Bapurao B. Shingate, Dr. Babasaheb Ambedkar Marathwada University, India

Click chemistry is not a single specific reaction, but was meant to mimic nature, which also generates substances by joining small modular units. The 1,3-dipolar azide, alkyne cycloaddition (CuAAC) reaction catalyzed by copper, as nearly quantitative and easy to execute has emerged as the leading example of "click chemistry". Given the importance of the triazole scaffold in medicinal chemistry, its synthesis has attracted the attention of the drug discovery and development community. This book chapter will summarizes the major synthetic methods currently used for the preparation of triazole and pharmacological significance such as antifungal, antibacterial, antitubercular, anticancer, anti-inflammatory, antioxidant and many more properties will discussed. Furthermore, this book chapter will comprise the literature from 2020 to till date

Chapter 6

Indazoles Chemistry and Biological Activities: Synthesis, Properties, and

Navnath Tulshiram Hatvate, Institute of Chemical Technology, Mumbai, India
Khushbu Bagul, Institute of Chemical Technology, Marathwada, India
Nandini Anilsingh Gour, Institute of Chemical Technology, Marathwada, India
Kalyani S. Sonawane, Institute of Chemical Technology, Marathwada, India

In organic and medicinal chemistry research, indazole is an essential nitrogen-containing heterocyclic unit that is also a helpful precursor molecule for the synthesis of several kinds of heterocycles. The diverse tautomeric forms and unique chemical properties make it a versatile scaffold in medicinal chemistry. Indazole's relevance in the pharmaceutical industry is underscored by its presence in currently marketed drugs and investigational compounds, highlighting its therapeutic potential. In addition, the present ring structure has already been explored for diverse biological activity, including anti-microbial, anti-viral, anti-protozoal, anti-cancer, anti-inflammatory, analgesic, antipyretic, anti-oxidant, anti-convulsant, anti-depressant, anti-emetic, anti-diabetic, neuroprotective, antihypertensive, and anti-arrhythmic properties. This chapter comprehensively reviews indazole's synthesis, properties, and biological applications, along with an update on recent patents and ongoing clinical trials.

Chapter 7

> *Paran Jyoti Borpatra, Institute of Science, Banaras Hindu University, India*
>
> *Mintu Maan Dutta, Arya Vidyapeeth College, India*

Heterocyclic compounds play a crucial role in medicinal chemistry, serving as key components in the development of pharmacologically active molecules. The therapeutic promise of many synthesized drugs can be attributed to their heterocyclic scaffolds, wherein even minor modifications in the heterocyclic structure can significantly impact the drug's efficacy. Among these, benzimidazoles are particularly significant. These class of compounds comprises a combination of the aromatic benzene ring and an imidazole ring. A significant natural form of benzimidazole found in nature is N-ribosyl-dimethyl benzimidazole, which plays a crucial role in coordinating to the cobalt metal in vitamin B12. Extensive biochemical and pharmacological research has demonstrated that benzimidazoles are highly effective against various strains of microorganisms. Furthermore, they have exhibit a broad spectrum of biological activities, including anti-inflammatory, anticancer, antihistamine, antimicrobial, antifungal, antioxidant, antidiabetic and antiviral activities.

Chapter 8

> *G. K. Prashanth, Sir M. Visvesvaraya Institute of Technology, India*
>
> *Srilatha Rao, Nitte Meenakshi Institte of Technology, India*
>
> *H. S. Lalithamba, Siddaganga Institute of Technology, India*
>
> *K. V. Rashmi, Sir M. Visvesvaraya Institute of Technology, India*
>
> *N. P. Bhagya, Sai Vidya Institute of Technology, India*
>
> *Mithun Kumar Ghosh, Medi-Caps University, India*
>
> *Manoj Gadewar, KR Mangalam University, India*
>
> *Nirmala R. Darekar, Department of Chemistry, Radhabai Kale Mahila Mahavidyalaya, India*

Benzoxazoles are heterocyclic compounds featuring a benzene ring fused to an oxazole ring. They have attracted significant attention in medicinal chemistry due to their diverse biological activities. Many pharmacological compounds have a core structure called a heterocyclic scaffold, which is essential to their therapeutic actions. Modest alterations to this fundamental structure may result in notable variations in the medication's mechanism of action. Benzoxazole and its derivatives have shown substantial and noteworthy therapeutic effects. This chapter explores the synthesis, structural characteristics, and various medical applications of benzoxazoles, highlighting their roles as antibacterial, antifungal, antiviral, anticancer, anti-inflammatory, and analgesic agents.

Chapter 9

Amol Arjun Nagargoje, K.M.C. College, Khopoli, India
Sharad Pandit Panchgalle, K.M.C. College, Khopoli, India
Mubarak Hanif Shaikh, Radhabai Kale Mahila Mahavidyalaya,
    Ahmednagar, India
Bapurao B. Shingate, Dr. Babasaheb Ambedkar Marathwada
    University, India

Benzotriazole represents a vital scaffold in the design of new pharmacologically active compounds. Its diverse biological activities and the potential for structural modification make it a promising candidate for future drug development endeavours. Continued research into its synthesis and SAR will further enhance its therapeutic potential. This chapter aims to provide a comprehensive overview of benzotriazole, focusing on its synthesis, biological activities, and potential applications in drug design. By exploring the historical development, modern synthetic methods, and the diverse biological activities of benzotriazole and its derivatives, we will highlight its significance in medicinal chemistry. The environmental and safety aspects of benzotriazole are also discussed, emphasizing the importance of sustainable and safe practices in its use and development.

Chapter 10

H. S. Lalithamba, Siddaganga Institute of Technology, Tumakuru, India
G. K. Prashanth, Sir M. Visvesvaraya Institute of Technology, India
Aisha Siddekha, Government First Grade College, India
M. Ramya, REVA University, Bangalore, India
G. Nagendra, REVA University, Bangalore, India

The present chapter Tetrazoles: synthetic strategies and Biomedical applications reveals different synthetic protocols for the preparation of tetrazoles. A standard aside, nitrile condensation route to metal-catalyzed condensation route, including the greener approaches and ionic liquid mediated approaches were discussed further these tetrazoles are considered on bio isosteres of carboxylic and found various applications in pharma and medicine and exhibit catalytic activities in organic

Synthesis. Various methods limit the role of NaN_3 and emissions the alternate protocol for the construction of tetrazole. Tetrazole alone and in conjunction with the organic groups including heterocycles exhibit various pharmacological activities that were covered in the chapter. Finally, the chapter allows a reader to choose the best possible pathway to construct the tetrazole ring and provides all necessary references.

Chapter 11

Kartik Sanghavi, RK University, India
Bonny Patel, RK University, India
Vijay Khedkar, Vishvkarma University, India
Khushal M. Kapadiya, RK University, India

The structural resemblance between the fused imidazopyridine heterocyclic ring system (a purine system) has prompted biological investigations to assess their potential therapeutic significance. They are known to play a crucial role in numerous disease conditions. In recent years, new preparative methods for the synthesis of imidazopyridines using various catalysts or non-catalytic systems have been described. In the present chapter, we summarise the recent approaches adopted for the synthesis of functionalized imidazo-pyridines over the last two decades along with their clinical advancement and applications. The key points adopted here including, traditional cyclo-condensation, reaction with nitro olefins, reaction with alkynes, 3-CCR (3-Component Condensation Reaction) based MCRs and miscellaneous aspects of imidazo-pyridine along with its biological importance have been presented and discussed. This chapter will provide new initiatives to the chemists towards the synthesis of imidazo-pyridines and possible medicinal applications with the reported methods.

Chapter 12

Shalini Jaiswal, Amity University, Greater Noida, India

In the literature, several Drugs substituted with isoxazole ring shows various biological activities. The various drugs available in the market like sulfamethoxazole A, muscimol-B, ibotenic acid C, parecoxib-D, and leounomide contain the core structure of isoxazole. The use of water and phase transfer catalyst combined acted as a green attribute for the current approach. The current methodology has several benefits, such as a single-pot reaction, environmental friendliness, economic viability, wide substrate range, ease of operation, quick reaction time, simple workup process, and elevated yields. This chapter concentrates on the many antimicrobial characteristics of isoxazole derivatives, such as their analgesic, antitubercular, anticancer, and antibacterial qualities. We also report the synthesis of isoxazole derivatives using the aqua-mediated reaction of Chalcone hydroxylamine hydrochloride, and p-toluene sulfonic acid as catalyst.

Chapter 13

G. Nagendra, REVA University, Bangalore, India
D. N. Akshitha, REVA University, Bangalore, India
H. S. Lalithamba, Siddaganga Institute of Technology, Tumakuru, India
G. K. Prashanth, Sir M. Visvesvaraya Institute of Technology,
    Bengaluru, India

Heterocycles containing nitrogen, oxygen and sulfur have been under investigation for a long time because of their important medicinal properties. The literature survey has described that thiadiazole moieties serve as analgesic, anti-inflammatory, anti-microbial activities, antihypertensive,anticancer, antituberculosis and vasodilator and this heterocyclic nucleus still possess considerable characteristics to attract the chemists for designing of newer biologically active molecules. Among them 1,2,4-thiadiazole, 1,3,4-thiadiazole and their derivatives are recognized as heterocyclic nuclei of great value in the field of medicinal chemistry. As a heterocyclic unit in a peptidomimetics might add conformational limitations to the structure, influencing the structure-activity-relationship. In this chapter, we are focusing on the synthetic strategies of 1,2,3-thiadiazole, 1,2,5-thiadiazole, 1,2,4-thiadiazole, 1,3,4-thiadiazole and their application in the industrial and biomedical fields.

Preface

In medicinal chemistry, heterocyclic molecules have become fundamental to contemporary drug discovery. Five-membered N and O-heterocycles have attracted considerable interest because of their exceptional bioactive characteristics and adaptability in pharmaceutical applications.

The book, "Five Membered Bioactive N and O-Heterocycles: Models and Medical Applications," seeks to deliver an extensive examination of the synthesis, characteristics, and medical uses of these intriguing molecules. This work unites prominent experts, presenting a distinctive combination of theoretical models, synthetic techniques, and therapeutic applications, so equipping readers with a comprehensive understanding of the contemporary advancements in the field.

CHAPTER OVERVIEW

The book comprises thirteen chapters, each concentrating on a distinct facet of five-membered N and O-heterocycles. This book encompasses a broad spectrum of subjects, from the foundational concepts of heterocyclic chemistry to contemporary advancements in drug discovery and development, including,

Chapter 1: Chemistry and Pharmacological applications of 1,3-Oxazoles: 1,3 Oxazoles

Saurabh Chetia, Lalrem Ruati, Brilliant N. Marak, and Ved Prakash Singh give a thorough rundown of the pharmaceutical uses of oxazoles as well as their chemical characteristics, synthesis techniques, and reactivity. It covers oxazole and its derivatives' traditional and innovative synthesis routes. The special behavior of oxazoles in cycloadditions, reductions, and nucleophilic and electrophilic substitution reactions is examined in the reactivity section. The chapter also discusses the presence of oxazole derivatives in natural goods and their possible uses in medicine.

Chapter 2: Bioactive Pyrazoles- Structure, Function, and Pharmaceutical Potential

This chapter, written by Aditi Boruah, Ariful Islam, and Pranjal K. Baruah, examines the most recent developments in pyrazole chemistry, including novel synthesis techniques and investigations into the relationship between structure and activity. This chapter demonstrates the significance of bioactive pyrazoles in modern medicinal chemistry for creating novel and improved medications by examining their numerous applications.

Chapter 3: Isoxazole A Bioactive Five-Membered Heterocycle with Diverse Applications: Isoxazole A Bioactive Five-Membered Heterocycle

Popat Mohite, Savita Tauro, and Aarati Pawar explore the synthetic methods for creating isoxazole derivatives, encompassing both contemporary methods like microwave-assisted synthesis and more conventional ones like cyclization processes. Each method's benefits and drawbacks are examined, emphasizing how structural differences affect synthesis efficiency. Additionally, the biological activities linked to isoxazole compounds are examined, highlighting their prospective pharmacological roles as anti-inflammatory, antibacterial, antifungal, and anticancer medicines.

Chapter 4: Use of Ionic Liquids and DES for Synthesis of Medicinally Important 1,2,3-Triazole Scaffolds

Diganta sarma, Bishal Bhattacharyya, Ramyata Priyam Borah, Dipankar Nath, and Parishma Gogoi investigate the various environmentally friendly methods that have been devised by various researchers to produce 1,2,3-triazoles using benign methodologies. The synthesis of 1,2,3-triazoles involves the use of volatile organic solvents in azide alkyne cycloaddition processes. This chapter discusses the use of greener solvents such as Deep Eutectic Solvents (DES) and Ionic Liquids (IL) as viable substitutes.

Chapter 5: Pharmacological Significance of 1,2,3-Triazoles: Pharmacological Significance of 1,2,3-Triazoles

Mubarak Hanif Shaikh, Amol A Nagargoje, Dattatraya N Pansare, and Bapurao B Shingate provide a comprehensive overview of the primary synthetic methods currently employed to prepare triazoles, as well as their pharmacological significance, including antifungal, antibacterial, antitubercular, anticancer, anti-inflammatory, and antioxidant properties. Additionally, the literature from 2020 to the present will be included in this book chapter.

Chapter 6: Indazoles Chemistry and Biological Activities: Synthesis, Properties and Biological Activities of Indazole

In this comprehensive review, Navnath Tulshiram Hatvate, Khushbu Bagul, Nandini Anilsingh Gour, and Kalyani S. Sonawane discuss the history, structure, and biological uses of indazole, as well as its role as a nitrogen-containing heterocyclic unit and a useful precursor molecule for the synthesis of various heterocycles. The authors also provide an update on recent patents and ongoing clinical trials pertaining to indazole.

Chapter 7: Bioactive Five-Membered Heterocycles With Two Heteroatoms Fused With A Benzene Ring (a) Benzimidazole

Paran Jyoti Borpatra and Mintu Maan Dutta explain the benzimidazoles that can be manufactured using a variety of synthetic techniques, a wide range of starting materials, reagents, and catalysts, as well as a wide range of solvents and environments that do not include any solvents. Benzimidazole derivatives made synthetically have a variety of therapeutic uses and biological activity. Future studies on the benzimidazole scaffold will yield promising results in the field of medicine to cure life-threatening conditions.

Chapter 8: Benzoxazoles-Diverse Biological Activities and Therapeutic Potential: Benzoxazoles in Medicinal Chemistry

The authors, Prashanth GK, Srilatha Rao, Llalithamba H.S., Mithun Kumar Ghosh, Manoj Gadewar, and Bhagya NP, examine the synthesis of benzoxazoles, their structural properties, and their numerous medical uses, emphasizing their antibacterial, antifungal, antiviral, anticancer, anti-inflammatory, and analgesic properties.

Chapter 9: Synthetic Approaches and Medicinal Attributes of Benzotriazoles

Amol Arjun Nagargoje, Sharad Pandit Panchgalle, Mubarak Hanif Shaikh, and Bapurao B. Shingate give a thorough introduction to benzotriazole in this chapter, emphasizing its synthesis, biological activity, and its uses in medication design. We will emphasize the importance of benzotriazole and its derivatives in medicinal chemistry by examining their historical evolution, contemporary synthesis techniques, and variety of biological activities. Benzotriazole's safety and environmental effects are also covered, highlighting the significance of safe and sustainable methods in its production and use.

Chapter 10: Tetrazoles-Synthetic Strategies and Biomedical Applications: Synthesis of Tetrazole

In their exploration of tetrazoles, Lalithamba H S, Prashanth G K, Aisha Siddekha, Ramya M, and Nagendra G emphasize their synthetic approaches and biomedical uses, which disclose many synthetic techniques for tetrazole manufacture. Aside from the conventional method, the nitrile condensation route to the metal-catalyzed condensation route was also discussed, along with the greener and ionic liquid-mediated approaches. Tetrazoles are thought to be bio isosteres of carboxylic and have a variety of uses in pharmacology and medicine. They also show catalytic activities in organic synthesis.

Chapter 11: Imidazo-Pyridines A Hybrid N-Heterocycles For Their Sustainable Synthetic Approaches And Significant Clinical Diversity: Imidazo-Pyridines: A Hybrid N-Heterocycles

The latest methods used for the synthesis of functionalized imidazo-pyridines throughout the past 20 years, together with their clinical development and applications, are summarized by Kartik Sanghavi, Bonny Patel, Vijay Khedkar, and Khushal M. Kapadiya. The following are the main topics that have been given and discussed: 3-CCR (3-Component Condensation Reaction) based MCRs, classical cyclo-condensation, reaction with nitro olefins, reaction with alkynes, and various elements of imidazo-pyridine along with its biological significance. This chapter will give scientists new ideas for the synthesis of imidazo-pyridines and potential medical uses for the techniques described.

Chapter 12: Aqua Mediated Multicomponent Synthesis and Biological Activity of Fused Isoxazole Derivatives

Shalini Jaiswal reported synthesizing isoxazole derivatives by employing p-toluene sulfonic acid as a catalyst in an aqua-mediated reaction with chalcone hydroxylamine hydrochloride. Additionally, it focuses on the numerous antimicrobial properties of isoxazole derivatives, including their antibacterial, antitubercular, anticancer, and analgesic effects.

Chapter 13: Thiadiazoles- Chemistry and Biological Activities

Examining thiadiazole derivatives, Nagendra G, Akshitha D N, and Lalithamba H S concentrate on the synthesis of 1,2,3-thiadiazole, 1,2,5-thiadiazole, 1,2,4-thiadiazole, and 1,3,4-thiadiazole as well as its uses in the biological and industrial domains.

This book aims to be a significant resource for researchers, students, and practitioners in medicinal chemistry, pharmaceutical sciences, and biochemistry. Through the investigation of five-membered N and O-heterocycles, we seek to stimulate novel discoveries and progress in enhancing human health and well-being.

Shrikaant Kulkarni

Victorian Institute of Technology, Australia

Hemantkumar Akolkar

Rayat Shikshan Sanstha's Abasaheb Marathe Arts and New Commerce Science College, India

Bapurao B. Shingate

Babasaheb Ambedkar Marathwada University, India

Vijay M. Khedkar

Vishwakarma University, India

Chapter 1
Chemistry and Pharmacological Applications of 1,3-Oxazoles

Saurabh Chetia
https://orcid.org/0009-0000-1208-8511
Mizoram University, India

Lalremruati
Mizoram University, India

Brilliant N. Marak
Mizoram University, India

Ved Prakash Singh
Mizoram University, India

ABSTRACT

Oxazoles, a class of five-membered aromatic heterocycles containing nitrogen and oxygen atoms, are of significant importance across various fields, including medicine and agriculture. Renowned for their diverse biological activities, including antibacterial, antiviral, anticancer, and anti-inflammatory properties, oxazoles have garnered significant attention in research. This chapter offers a comprehensive overview of the chemical properties, synthesis methods, and reactivity of oxazoles, alongside their pharmacological applications. It covers both conventional and forefront synthetic pathways of oxazole and its derivatives. The reactivity section explores the unique behavior of oxazoles in cycloadditions, reductions, and nucleophilic and electro-

DOI: 10.4018/979-8-3693-7267-8.ch001

philic substitution reactions. Additionally, the chapter highlights the occurrence of oxazole derivatives in natural products and their potential medicinal applications.

1. INTRODUCTION

5-membered aromatic heterocycles rings have remarkable implications in research related to pharmaceuticals (Rusu et al., 2023) and agriculture (S. Wang et al., 2024). Oxazole is one of the remarkable heterocycles that contain nitrogen and oxygen, which serve as a building block for many highly functional and unique products. The first oxazole was synthesized in 1840 by Zinin, who named the compound azobenzil. This was achieved through the reaction of benzil with alcoholic ammonia (Wiley, 1945). The study of oxazole chemistry commenced in 1876 with the synthesis of 2-methylbenzoxazole (Cornforth & Cornforth, 1947), whereas the synthesis of the parent oxazole was achieved in 1962. Compounds containing oxazoles have been associated with a wide range of biological activities, such as antibacterial(Wales et al., 2015), anti-allergic, anticonvulsant(Song & Deng, 2018), anticancer (Chiacchio et al., 2020), anthelmintic(Laohapaisan et al., 2023), antiviral, depressive, analgesic(Kumar & Singh, 2021), and antioxidant qualities(Putri & Cahyana, 2022). Synthetic Oxazole compounds have good anti-inflammation potential(Kean, 2004), anti-HIV(Deng et al., 2022), antitubercular (Giddens et al., 2005) and TRPV1 antagonist action (Perner et al., 2010a) activities, making them essential in the pharmacological research portfolio. Oxazoles are also utilized in the polymer industries (Grotkopp et al., 2011; Min et al., 2007; Zhao et al., 2009), photography (Turchi & Dewar, 1975), and as fluorescent dyes, agrochemicals, and corrosion inhibitors (El Ibrahimi & Guo, 2021; Tang & Verkade, 1996). This chapter summarizes the recent developments in the synthesis of physiologically active molecules based on oxazoles.

1.2 General Properties of Oxazoles

Oxazoles (Figure 1) are numbered starting from oxygen and numbered around the ring. They are double unsaturated heterocyclic organic molecules featuring an oxygen and nitrogen atom at the 1st and 3rd positions, separated by carbon in between with the chemical formula C_3H_3NO.

Figure 1. Structure of oxazole with numbering

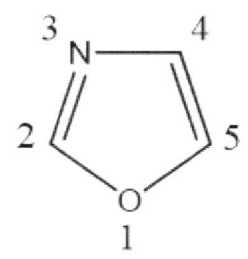

By analysing the structure of oxazoles we can see that the compound has structural similarity to pyridine and furan. According to studies oxazoles also exhibit properties similar to furan and pyridine. Similar to pyridine, they show weak basicity. It has two hydrogen acceptor counts and zero hydrogen bond donor count. They react readily with oxygen or other oxidizing agents. They are more stable in acidic conditions than pyridine (pyridine can be easily protonated by acid), but show some instability similar to that of furans. It is liquid at room temperature and has a boiling point of 69°C. Its density is 1.050g/cm3 with acidity of (pKa) -0.8. (of conjugate acid)

Oxazoles have three carbon, nitrogen, and oxygen on each. All of them are sp^2 and planer in nature. In total, there are 6 non-bonding electrons present in oxazole out of which 2 from nitrogen, 4 from oxygen. Due to the oxygen atom's high electronegativity, delocalization is not very effective. (Figure 2) (Joshi et al., 2023)

Figure 2. The active site of 1,3 oxazole

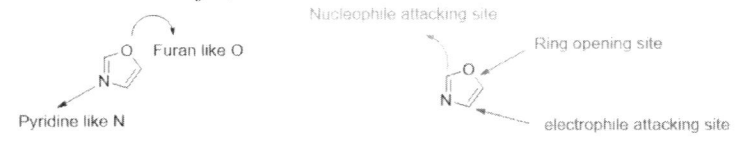

Oxazoles are 5-membered heterocyclic rings with oxygen and nitrogen in their cyclic ring. Hence it shows characteristic ^{1}H NMR and ^{13}C NMR spectra. For ^{1}H NMR, the resonance of hydrogen atoms is to be measured. Oxazoles have distinctive chemical shifts and features like:

a. ^{1}H NMR

- The electronegative oxygen and nitrogen atoms' deshielding effects cause the protons on the oxazole ring to typically appear downfield (at higher ppm values).
- The protons that are immediately bonded next to both heteroatoms (nitrogen and oxygen) typically show distinct chemical shifts.

The resonance (δ) is generally seen in the range between 7.00 to 8.00 ppm in the ^{1}H NMR spectrum, and if there is a substituent present in their moiety, they can change the chemical shift by up to 1 ppm (6-9 ppm).

b. ^{13}C NMR
- The influence of nitrogen and oxygen atoms which imparts electronic effects to carbon in the oxazole ring will give resonances at characteristic chemical sifts. As oxygen and nitrogen are electronegative, the carbon linked to these heteroatoms will generally appear at downfield shifts (higher ppm values).
- The ring's carbon atoms usually exhibit unique chemical shifts that make it possible to identify the oxazole structure.

In the ^{13}C NMR of oxazole, the effect of C(2) substitution on the shielding and deshielding is seen at about <2 ppm on C(4) and C(5) resonances. The oxazole ring's carbon atoms may resonate between 110 and 160 ppm, depending on the substituents and ring structure.

Oxazole has absorbances at 1537, 1498, and 1326 cm^{-1} (ring stretch), 1257 cm^{-1} (C-H in-plane deformation), 1143 and 1080 cm^{-1} (ring breathing), and 1045 cm^{-1} in its infrared spectra. (Informatics.). The substitution pattern has a significant impact on the absorption maximum (λmax) of oxazoles in UV spectroscopy. The parent oxazole ring structure has an absorption maxima in methanol at 205 nm.

Oxazole's aromaticity is due to its presence in a hybrid resonance state of canonical structures. The ring's high reactivity and significant tendency to react with both electrophilic and nucleophilic reagents are associated with its ionic resonance (Figure 3). In practice, during oxidation, the oxazole ring readily cleaves to form acids, amides, and imides. (Zeinali et al., 2020)

Figure 3. Delocalization of π electrons

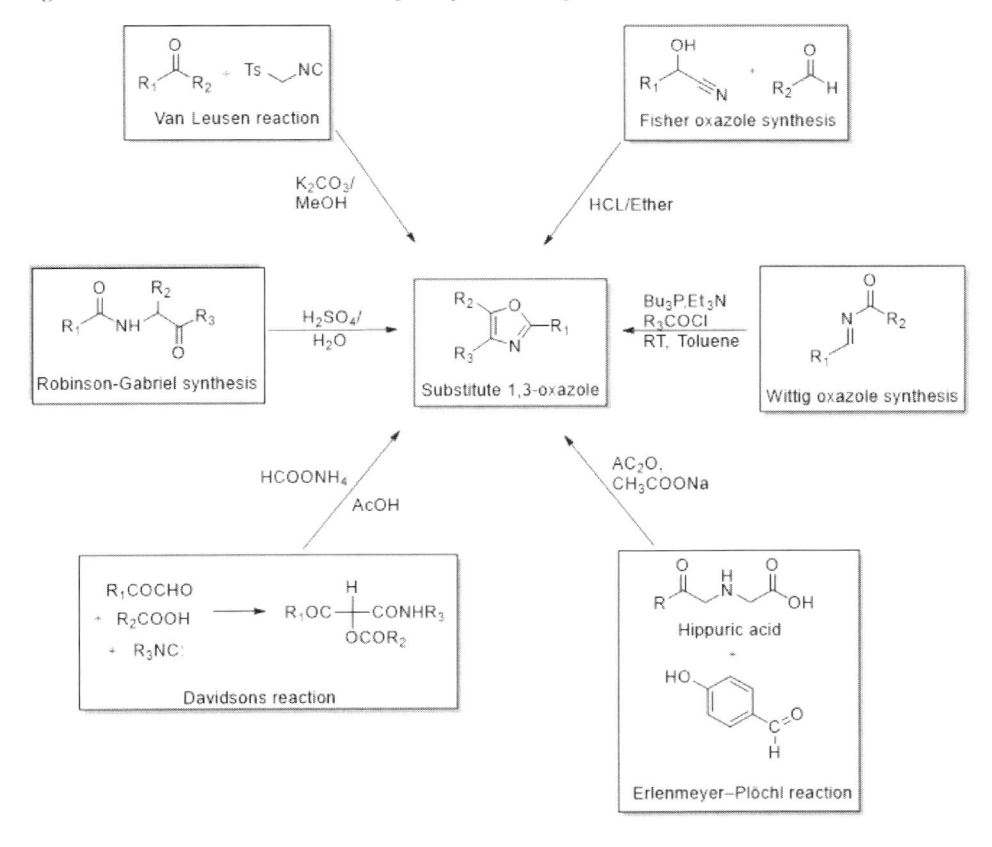

1.3 Synthesis Methods

The conventional and new methods of preparation of oxazoles are mentioned in the schemes given in the following Figure 4 and Figure 5 respectively.

Figure 4. Conventional methods for synthesis of oxazoles

The new method of synthesis uses better reagents, or different reaction methods like microwave or sound, etc., 8(Takeuchi et al., 1989; Thalhammer et al., 2009; Weyrauch et al., 2010; Wipf et al., 2005; Yamada et al., 2017; Ye et al., 2019; Zheng et al., 2012) schemes different types of schemes are mentioned in the synthesis of oxazoles with new method.

Figure 5. Newer methods for the synthesis of 1,3 oxazoles

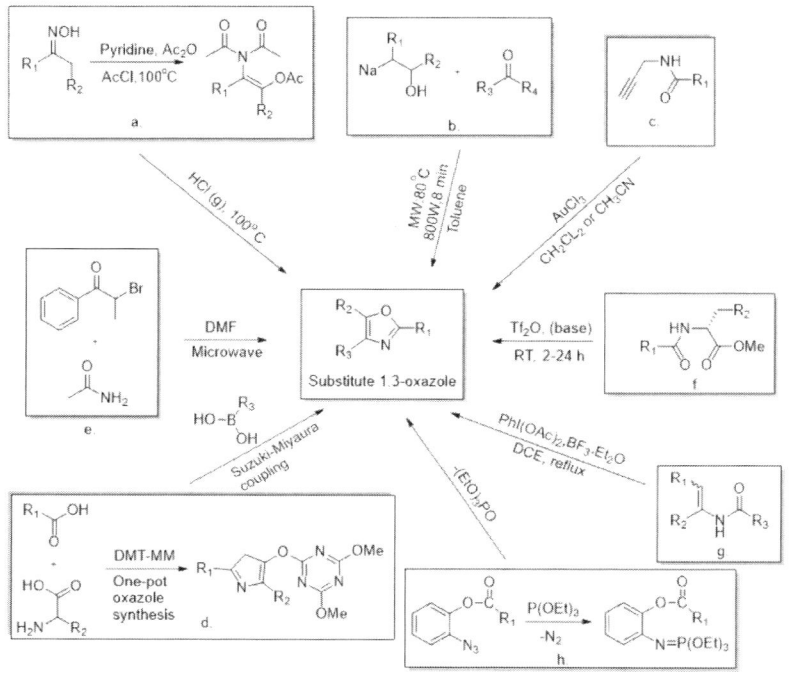

1.3.1 Mechanisms of the Synthesis of Oxazoles

1.3.1.1 Robinson-Gabriel synthesis: 2-acylamino ketones are dehydrated to form oxazole. ("Robinson-Gabriel Oxazole Synthesis," 2010)

Figure 6. Scheme 3

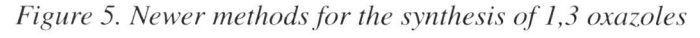

1.3.1.2 The Fisher oxazole synthesis: The cyanohydrin **1**is converted from an iminochloride intermediate **2** by the addition of HCl. After this intermediate interacts with the aldehyde, water is lost, and an intermediate known as chloro-oxazoline **4** is produced. The final product 2,5-diaryloxazole**6** is formed by the isomerization of two protons and the loss of an HCl molecule. (Li & Corey, 2011)

Figure 7. Scheme 4

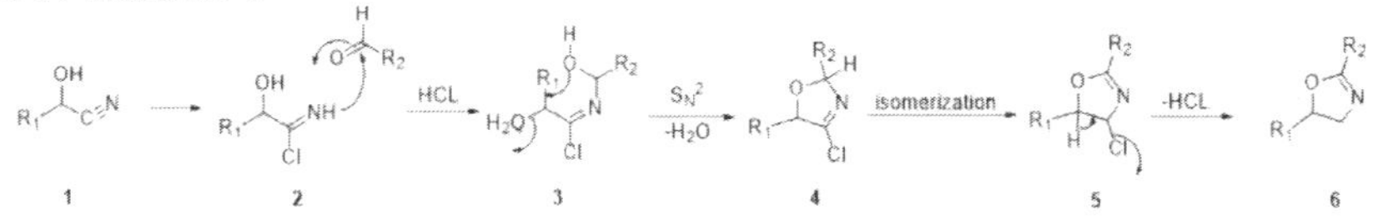

1.3.1.3 Davidson Reaction: The reaction of β-keto ester with ammonium formate. An intermediate's intramolecular reaction yields imine **7**. Now that another intramolecular self-condensation is feasible, the protonated species is produced. **8**. After that, a simple work-up yields oxazole**9**. (Li & Corey, 2011)

Figure 8. Scheme 5

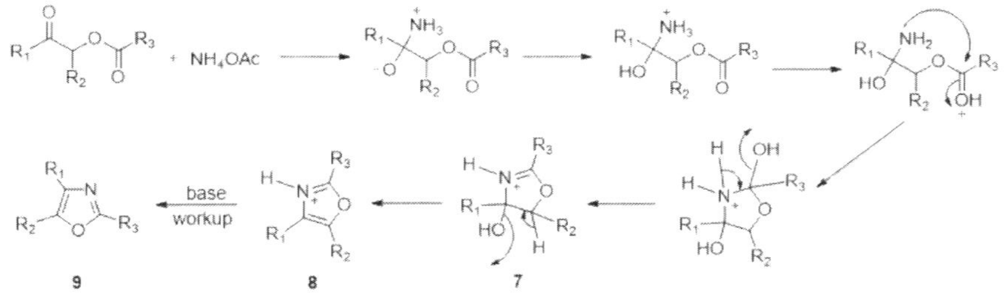

1.3.1.4 Japp Oxazole Synthesis: The process of forming oxazoles from 1,2-aromatic diketones and aromatic aldehydes in the presence of ammonia. (Li & Corey, 2011)

Figure 9. Scheme 6

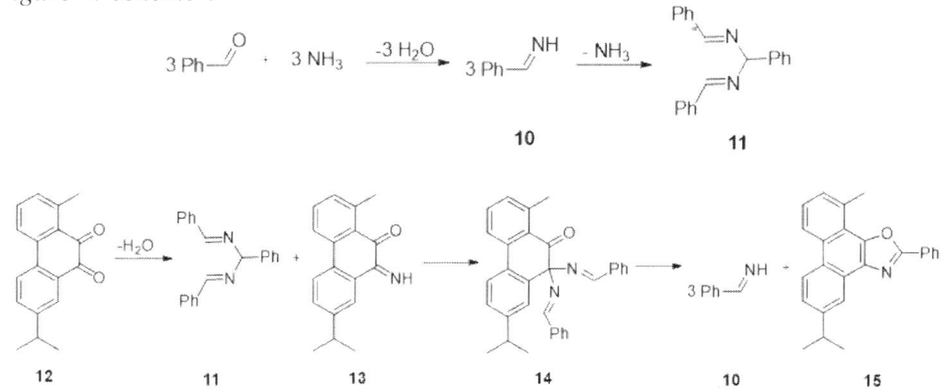

Heterocyclic Ammonia and benzoaldehyde combined to generate imine **10**, which subsequently self-condensed to produce intermediate **11**. Meanwhile, intermediate **11** and imine **(13)**, which were produced by the interaction of diketone **12** and ammonia, interacted to form intermediate **14** and imine **10**. Intermediate **14** was converted to oxazole **15** following the loss of one molecule of imine **10**.

1.3.1.5 Schollkopf Oxazole Synthesis: It is a condensation reaction of an alkyl isocyanide**(16)** and an acylating agent in the presence of a base to give an oxazole with substitution at either (or both) the 4- or 5- position**17**. (Li & Corey, 2011)

Figure 10. Scheme 7

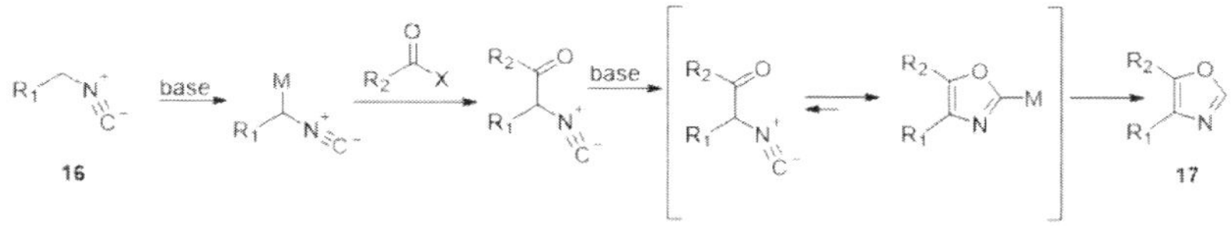

1.3.1.6 Van Leusen reaction: The initial step in the chemical process involves the straightforward deprotonation of TosMIC, facilitated by the electron-withdrawing properties of the isocyanide and sulfone groups. Following this, according to Baldwin's rules, a 5-endo-dig cyclization occurs, resulting in the formation of a 5-membered ring upon attack on the carbonyl group.(Oldenziel et al., 1977)

Elimination of the good tosyl-leaving group might happen easily if the substrate is an aldehyde. An oxazole is the end product of quenching.

Figure 11. Scheme 8

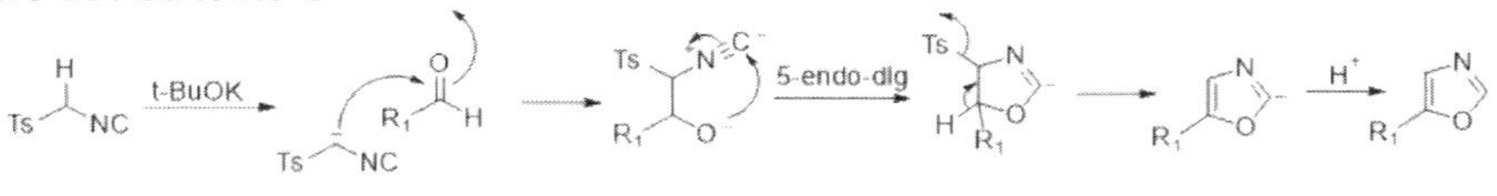

1.3.1.7 Erlenmeyer-Plöchl Azlactone synthesis: Hippuric acid and aromatic aldehydes condensation when in the presence of acetic anhydride is known as Erlenmeyer-Plöchl Azlactone synthesis.("Erlenmeyer-Plöchl Azlactone Synthesis," 2010)

Figure 12. Scheme 9

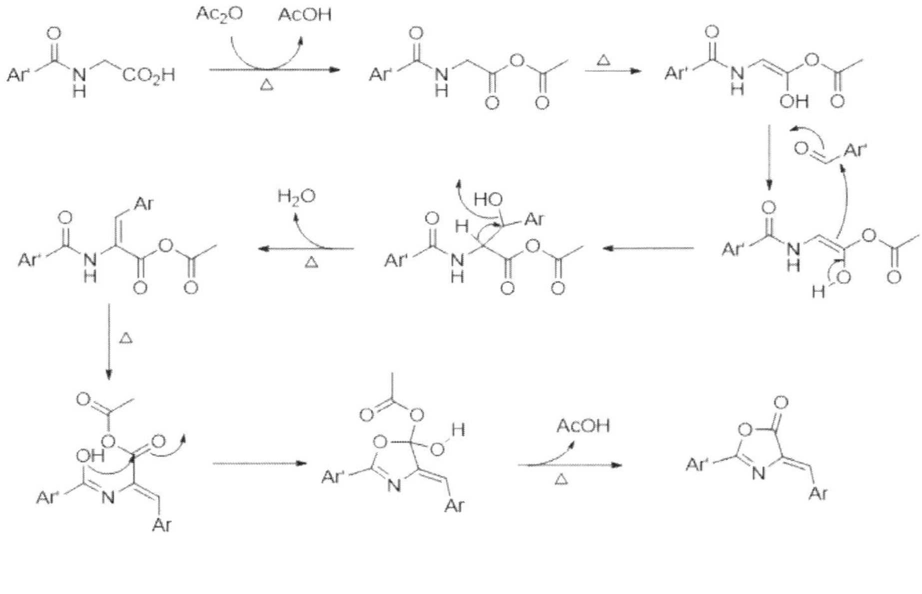

1.3.1.8 The Blumlein-Lewy Oxazole Synthesis: A primary amide **18**, a bromopyruvate ester **19**, and an alcoholic solvent are involved in the reaction, which yields the corresponding oxazoline **20** and then cyclodehydration to produce the oxazole **21(Scheme 10)**. Nevertheless, compared to other methods, this one has not been utilized as much, mostly because of its low conversion rate and undesired side product creation of **22**.

Figure 13. Scheme 10

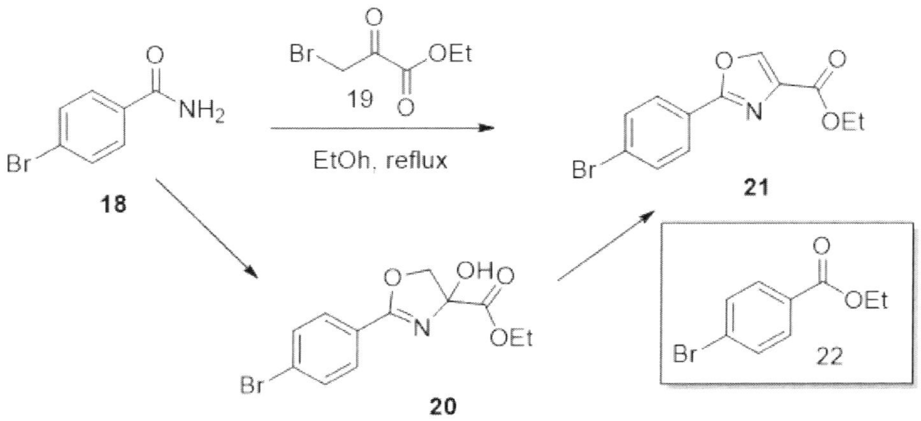

(Ritson et al., 2011) did the modification and did Silver-mediated one-step synthesis of oxazoles. They achieved increased viability of the reaction with modification of the procedure by adding silver salt.
Synthesis:

Figure 14. Scheme 11

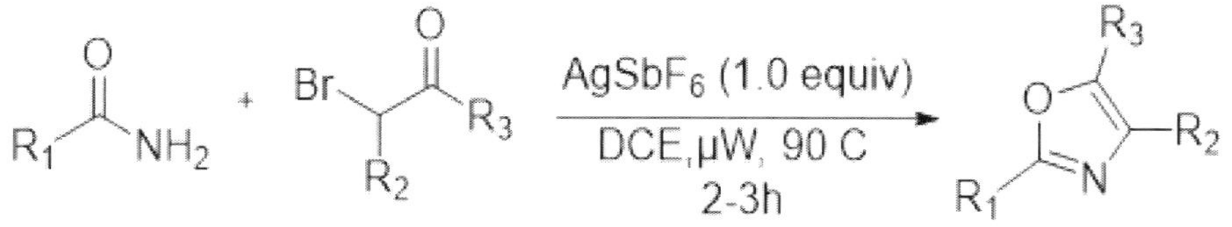

Proposed Reaction Mechanism:(Spiteri et al., 2011)

Figure 15. Reaction mechanism

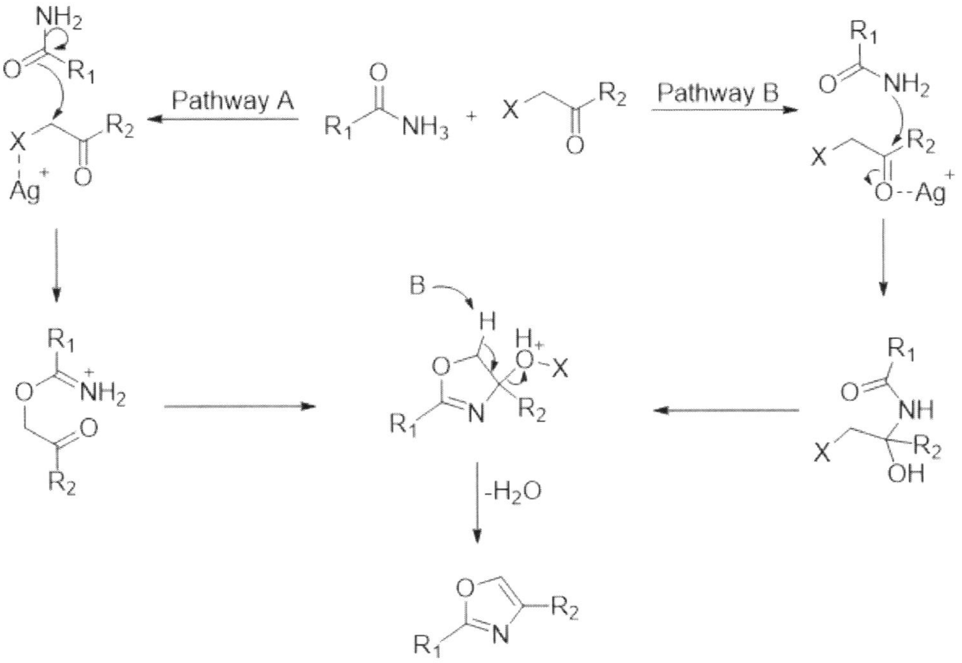

Gao et al., (2013) used molecule iodine as the sole oxidant under basic conditions and shows a broader scope of substrates.

Figure 16. Scheme 12

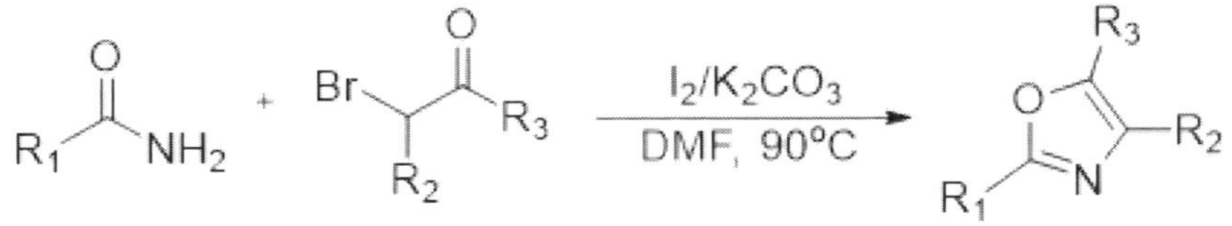

1.3.1.9 From Propargyl Amides: Propargyl amides **23** are helpful intermediates in the production of 5-methyloxazoles **25**, as demonstrated by their cyclization in basic circumstances (sodium hydride) or acidic conditions (with or without additional mercury salts). (Turchi, 1986)

Figure 17. Scheme 13

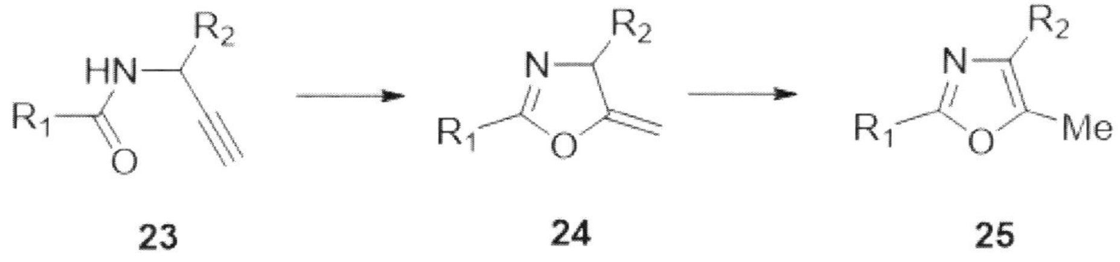

23 **24** **25**

The yields of oxazoles seem to be in the range of 50 to 70%. The 5-methylene-2-oxazolines (**24**) are possible intermediates in this mechanism. From different literature on the reaction obtained the 5-methyleneoxazolinium salts (**26**) (X = Br, I) when N-propargylbenzamide (**23**, R_1 = Ph, R_2 = H) is allowed to react with electrophiles like bromine or iodine, provides evidence for the intermediate of **26**. After treating 4 with aqueous sodium bicarbonate, (X = Br, I) yielded the 5-methyleneoxazolines (**27**), which thermolyze to produce 2-phenyloxazoles (**28**).

Figure 18. Scheme 14

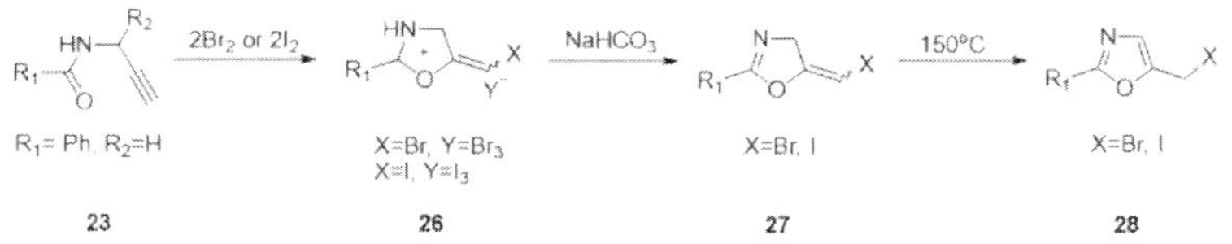

23 **26** **27** **28**

1.4 Reactivity of Oxazole

Oxazole rings provide a wide range of reactions, which permits functionalization at every ring atom except oxygen. Their aromaticity is very weak hence displaying both aromatic substitutions and reactions of double bonds. Additions reaction breaks the aromaticity across the double bond between C (4) and C (5). Oxazoles include two centers that are reactive towards electrophiles: C-4 (or C-5) or the nitrogen atom. On the ring, C-4 is the more susceptible position for electrophilic attack, since the donating resonance effect of the oxygen is most effective in C-4. Due to electron withdrawing effect of oxygen and nitrogen, C-2 will be least reactive towards electrophilic reaction but more susceptible to nucleophilic reaction. These factors are evaluated using MO calculations, which primarily assign a greater electron density at C-4 than C-5 and the lowest electron density at C-2. (Hassner & Fischer, 1993).

1.4.1 Nucleophilic Substitution Reaction

Nucleophilic substitution reactions on oxazole rings are uncommon and typically occur due to the presence of functional groups. When halogens are substituted on the oxazole ring, the preference order is C-2≫C-4>C-5. (Turchi, 1986)

Figure 19. Scheme 15

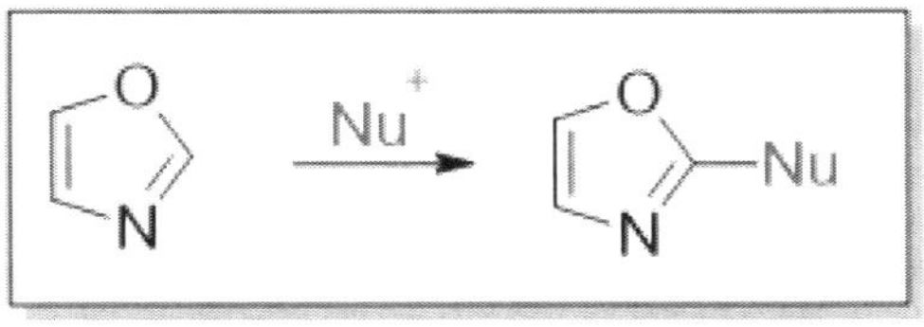

1.4.2 Electrophilic Substitution Reactions

Oxazoles are an electron-deficient ring hence it doesn't undergo electrophilic substitution reaction readily.

a. Since oxazoles have a low pKa of 0.8, powerful acids can protonate them at the N atom. Oxazoles react with alkyl halides to form quaternized compounds.

Figure 20. Scheme 16

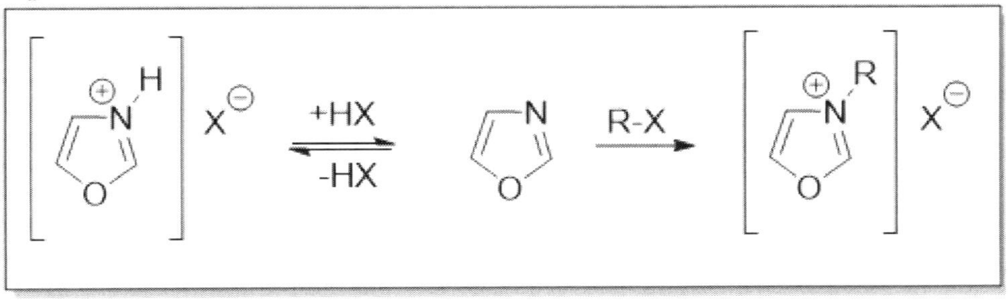

b. Although electrophilic substitution reaction is possible with oxazoles rings are generally activated by an electron-donating substituent (as in furan). For example, using NBS or Br_2 on 4-methyl-2-phenyloxazole undergoes bromination to give bromine in the C-5 position. (Eicher et al., 2003)

Figure 21. Scheme 17

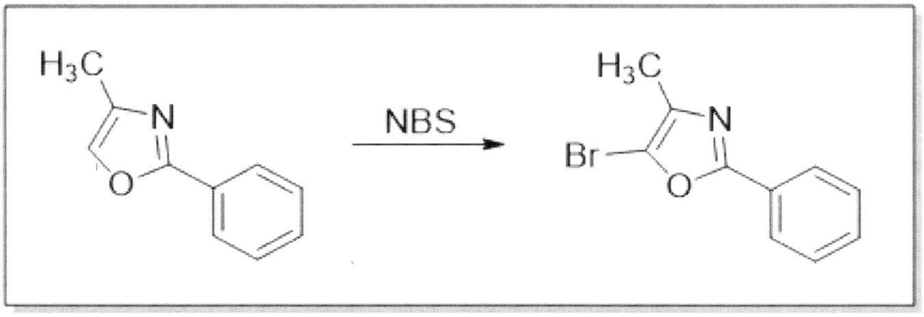

1.4.3 Diels-Alder Reaction with Oxazoles

Reaction of oxazoles with activated alkynes gives furans.

Figure 22. Scheme 18

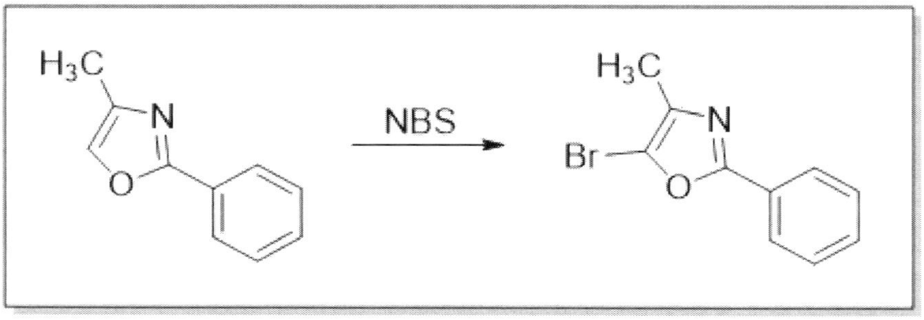

A [4+2] cycloaddition is the first step, while a [4+2] cycloreversion is the second. The creation of furan and thermodynamically very stable acetonitrile is the reason that cycloreversion is not a retroaction of the first step. (Eicher et al., 2012)

1.4.4 N-Acylation of Oxazole

A high degree of reactivity toward the acylation of oxazoles is shown by the reaction of the 3-position acyl group on the oxazole ring. (Joshi et al., 2017)

Figure 23. Scheme 19

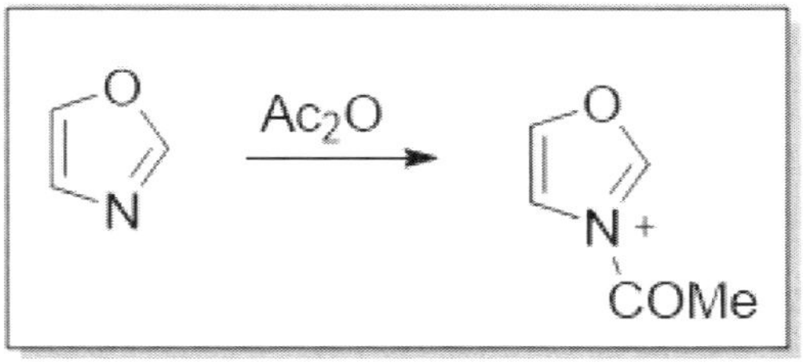

1.4.5 Cycloaddition Reaction (Rymbai et al., 2019)

Figure 24. Scheme 20

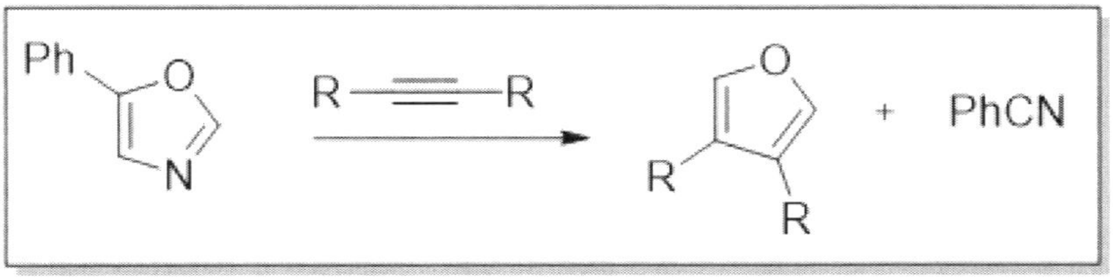

1.4.6 Reduction of Oxazole

In high pressure, high temperature, and hydrogen atmosphere reduction of oxazoles is done using nickel and aluminium. (Lunn, 1987)

Figure 25. Scheme 21

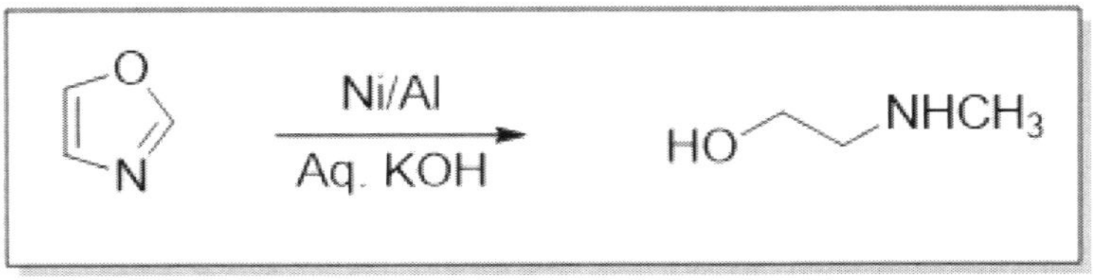

1.4.7 Metallation of Oxazoles

Unsubstituted oxazoles on the 2-position combine with -butyllithium in THF at a temperature of -75°C to form 2-lithioxazoles.

Figure 26. Scheme 22

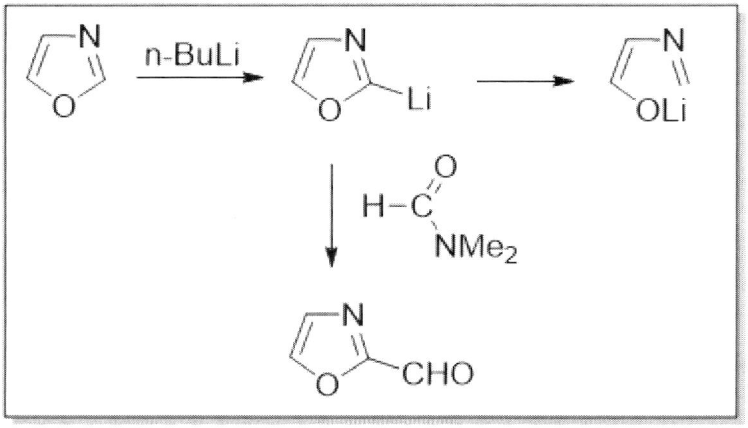

1.4.8 Ozonolysis of Oxazoles

three different classes of products can be observed while ozonolysis of Oxazoles. (Hartner, 1996)

Figure 27. Scheme 23

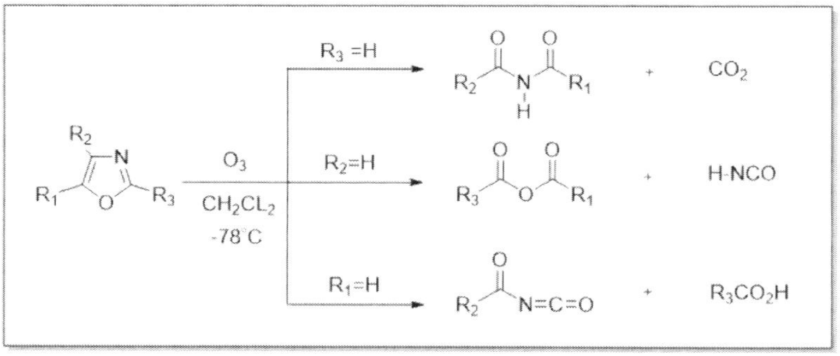

1.5 Pharmaceutical properties of Oxazole Derivatives

Oxazole compoundsas the bioisosteres of thiazoles, tetrazoles, imidazoles, benzimidazoles, and triazoles, have garnered significant interest in medicinal chemistry. Researchers and scientists worldwide have been focusing on oxazole-based compounds, not only as pharmacophores but also as promising candidates for developing novel scaffolds with superior pharmacokinetic properties, a broad spectrum of bioactivity, and low toxicity(Mayer et al., 2017; Sysak & Obmińska-Mrukowicz, 2017). These properties make oxazoles attractive targets for drug design, as they can be tailored to interact with a variety of biological targets. Numerous oxazole-containing natural products have been isolated from a diverse array of microorganisms and marine invertebrates. These natural compounds often serve as core structural motifs in pharmaceuticals, contributing to a wide range of biological activities, including antiproliferative, anticancer, antileukemic, antibacterial, antifungal, antiviral, analgesic, and enzyme inhibitory properties (Antonio & Molinski, 1993; Kobayashi et al., 1997; Li, 2013; Lindquist et al., 1991). The remarkable interactions with biological functions and the widespread presence of oxazoles in both natural products and synthetic drugs have spurred considerable interest in the synthesis, modification, and biological evaluation of oxazole derivatives.

Some oxazole-containing drugs that have been clinically approved for pharmaceutical use, as well as those currently undergoing clinical trials, are listed in Tables 2 and 3, respectively, with their structures presented in Figures 28 and 29.

Table 1. List of oxazole-containing drugs that have been clinically approved for pharmaceutical use.

Drug Name	Therapeutic applications	Mechanism of Action
Rilmenidine (withdrawn)	Treatment of hypertension	Rilmenidine enforces its hypotensive effect by decreasing peripheral sympathetic tone, which is obtained through both central and potentially peripheral mechanisms (Montastruc et al., 1989).
Zolmitriptan	Used to treat acute migraine headaches in adults.	Migraines are headaches and photophobias that are caused by nerve activation in the trigeminocervical complex (TCC). Triptans, such as zolmitriptan, set off this nerve action but do not affect the cerebral vasculature(Napier et al., 1999).
Pemoline	Treat attention-deficit hyperactivity disorder (ADHD) and narcolepsy in children	Pemoline, a central nervous system booster, is used by individuals with ADHD to treat symptoms by altering neurotransmitters, chemicals that brain cells use to communicate (Bacon et al., 2007).

continued on following page

Table 1. Continued

Drug Name	Therapeutic applications	Mechanism of Action
Oxaprozin	Anti-inflammatory drug	Oxiprozin offers anti-inflammatory and fever-lowering benefits by inhibiting cyclooxygenase in platelets, improving blood flow, widening blood vessels, and eliminating heat, with higher selectivity due to its non-specific nature(Greenblatt et al., 1985).
Metaxalone	Central Nervous System Depressants	The exact mechanism of action of metaxalone in humans is not well understood, but it may be attributed to its general depressant effects on the central nervous system(Sahu et al., 2011).
Cabotegravir	Anti-retroviral	Cabotegravir bonds to the active site of HIV integrase and stops the viral genome from being transferred into the host genome. This stops the virus from emulating(Lou et al., 2016).
Tedizolid	Anti- Bacterial	Staphylococcus aureus, resistant to many drugs, poses a threat to antimicrobial trials. Oxazolidinones, a new antibiotic, bind to the A site of the PTC to inhibit protein production(Shorr et al., 2015).
Furazolidone	Anti- Bacterial	Furazolidone, a drug, easily breaks DNA, causing significant changes in bacteria's genes due to its ability to create free radicals that join with DNA(Timperio et al., 2003).
Zoxazolamine	Muscle relaxant	The precise mechanism by which zoxazolamine works is not fully known (McMillen et al., 1992).
Toloxatone	Anti-depressant agent	Toloxatone disrupts MAO-A, causing side effects and drug-food interactions. Daily care is needed for depression treatment, with special MAO-A inhibitors being more effective due to less comorbidity(Berlin et al., 1990).
Linezolid	Anti-Microbial drug	Linezolid, a medication that inhibits bacterial protein translation, can cause resistance due to 23S rRNA mutations, necessitating regular antibiogram checks(Roger et al., 2018).
Trimethadione	Anti-convulsant drug	Dione anticonvulsants reduce T-type calcium currents in thalamic neurons, preventing VDD T-type calcium channels from supplying signals, and causing erratic thalamocortical rhythmicity, a factor in absence seizures(Wildburger et al., 2009).

Table 2. List of oxazole-containing drugs that are currently undergoing clinical trials.

Drug Name	Therapeutic applications	Mechanism of action
Radezolid	Anti-Bacterial Agents	Radezolid is a second-generation antibiotic that blocks protein production by preventing aminoacyl-tRNA binding and locking down the 50S ribosomal subunit on the A, A/P, P–ribosome (Makarov & Reshetnikova, 2021).
Posizolid	Anti-Bacterial Agents	AZD2563 inhibits protein production in bacteria by binding to specific ribosome parts, preventing the formation of a 70S start complex, which is crucial for germ growth and survival(Wookey et al., 2004).
Fenpipalone	Anti-inflammatory activity	Fenpipalone, a medication, blocks D2 dopamine receptors in the brain, potentially aiding mental health issues, but its effectiveness depends on the specific illness and individual responses(Welstead et al., 1973).
Sutezolid	Anti-Tuberculosis	Sutezolid, like Linezolid, inhibits protein production by targeting the 23S rRNA of the 50S ribosomal subunit, rapidly changing from its original form to sulfoxide in plasma (Verma et al., 2022).
Nifuratel	Anti-bacterial, Anti-fungal Agents	The mechanism of action of nitrofurans has not been completely understood(Gruneberg & Leakey, 1976).Top of FormBottom of Form
Haloxazolam	Anxiety	Morphine, a benzodiazepine, manages seizures by allowing chloride to enter nerve cells, reducing neuronal activity, and affecting motor nerves to relax muscles, prevent seizures, and manage seizure duration(Sakai, 1983).
Delpazolid	Anti-bacterial and anti-infective agents	Delpazolid, a drug similar to oxazolidinone, inhibits the production of proteins in bacteria, either killing them or preventing their growth (Cho & Jang, 2020).
Anacetrapib	Anticholesteremic Agents	It binds to CETP and HDL in a way that can be undone, stopping CETP from working (Filippatos et al., 2017).
Inavolisib	Enzyme Inhibitors	Inavolisib, a pill targeting PI3Kα protein in cancer cells, may help advanced HR-positive/HER2-negative and PIK3CA-mutated breast cancer patients maintain better control with less harm(Hanan et al., 2022).
Zoliflodacin	Antineoplastic Agents	The drug, while effective against some Gram-negative bacteria, is currently incapable of destroying germs' DNA, preventing self-copying(Bradford et al., 2020).
Aminorex	Weight loss (anorectic) stimulant drug	Release of catecholamines within the central nervous system (Ramirez et al., 2021).

Figure 28. Chemical structures of oxazole-containing drugs that have been clinically approved for pharmaceutical use.

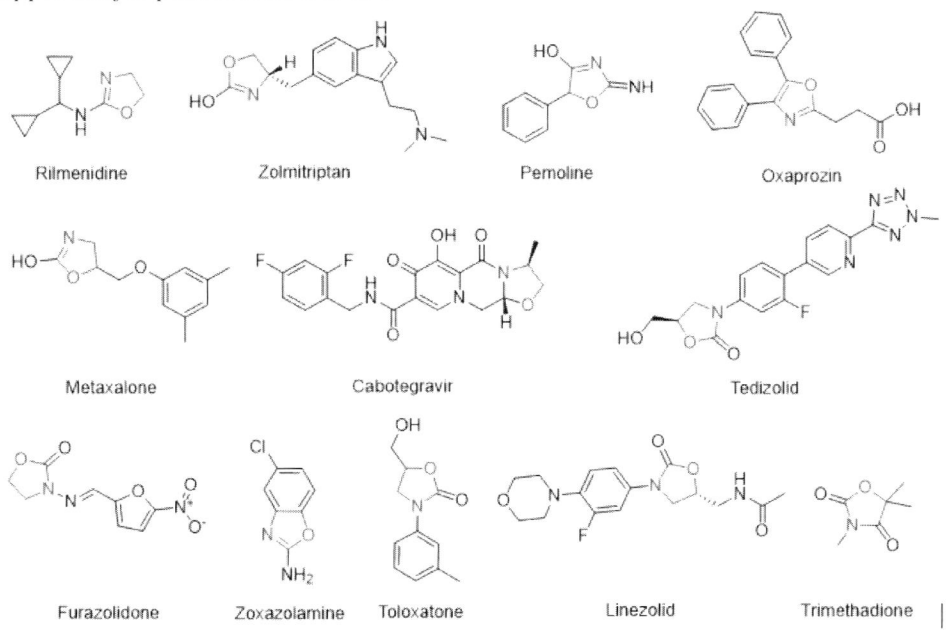

Rilmenidine   Zolmitriptan   Pemoline   Oxaprozin

Metaxalone   Cabotegravir   Tedizolid

Furazolidone   Zoxazolamine   Toloxatone   Linezolid   Trimethadione

Figure 29. Chemical structures of oxazole-containing drugs that are currently undergoing clinical trials or withdrawn from the trial.

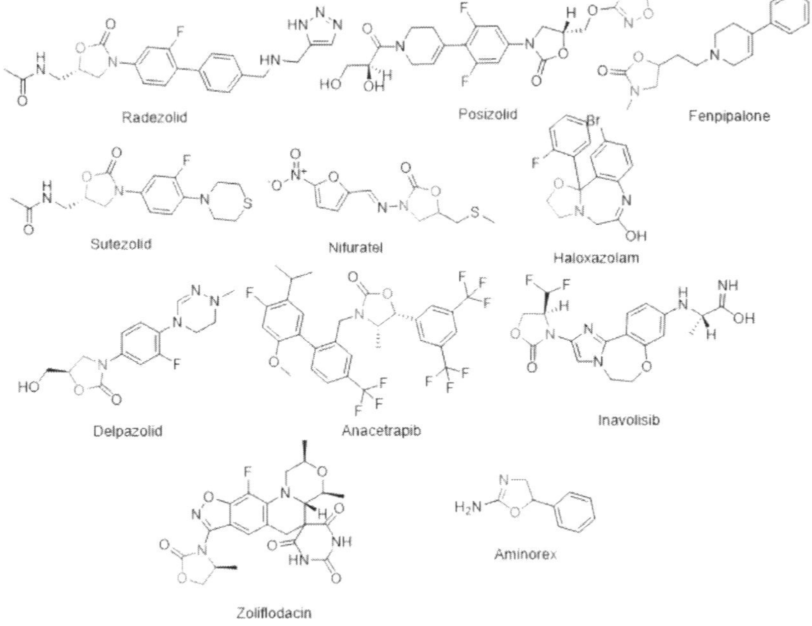

1.5.1 Anticancer Activity of Oxazoles

Despite advances in anticancer treatment, including chemotherapy, many challenges persist. These challenges include severe side effects, drug resistance, and the insufficient sensitivity and efficacy of existing treatments. The search for novel anticancer lead compounds is crucial to address these limitations and improve therapeutic outcomes. Recently, the applications of 1,3-oxazol-4-ylphosphonium salts in cancer treatment have been reported to be promising (Brusnakov et al., 2022). This study focused on how various genetic features influence the ability of the compounds to inhibit cancer cell growth. Among the compounds tested, compounds **29**, **30**, and **31** (Figure 6) were identified as strong anticancer agents, with compound **31** showing particular promise due to its comparable activity with the standard drugs.

Themodification of cis-stilbene combretastatin A-4, a metabolite from the South African bush willow *Combretum caffrum*,by incorporating -chloro groups, along with -methoxy, -ethoxy, and -methyl sulfides (compounds **32**, **33**&**34**) (Figure 6) into the B-ring of the compound, have known to enhance the anticancer and antivascular activity(Mahal et al., 2013). These derivatives showed potent effects against several cancer cell lines, including 518A2 melanoma, HT-29 colon carcinoma, and Ea.hy926 endothelial hybrid cells.In addition, macrocyclic compounds containing oxazole, such as the natural product telomestatin, showedgreat potential as a class of anticancer agents that selectively target G-quadruplex DNA structures(Pilch et al., 2008). Synthetic hexaoxazole-containing macrocyclics, specifically compounds **35** and **36** (Figure 6), exhibited cytotoxic effects against KB3-1 oral carcinoma cells and RPMI 8402 human lymphoblasts, with IC_{50} values ranging from 0.4 to 0.9 µM. Notably, these compounds bind exclusively to the G-quadruplex form of nucleic acids, with no significant interaction with duplex or triplex DNA forms.

The tri-substituted 1,3-oxazoles (**37**, **38**, **39**) (Figure 6) with different groups have been shown to effectively reduce the growth of the Hep-2 cancer cell line(Semenyuta et al., 2016). Compound **37** is the most promising among them, with an IC_{50} value of 60.2 µM, demonstrating strong cytotoxic effects on cancer cells. These compounds showed potential for cancer therapy due to their ability to bind to the oxalate binding site on tubulin, a critical protein in cell division.

Figure 30. Structures of oxazole derivatives as anticancer agents

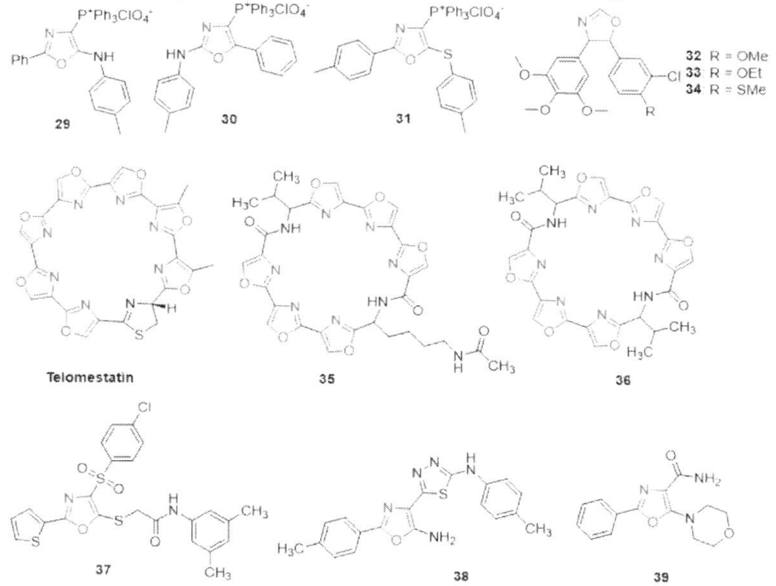

Vascular Endothelial Growth Factor Receptor 2 (VEGFR2) is a crucial receptor for VEGF-mediated angiogenesis signal transduction. Recent studies have indicated that inhibiting VEGFR2 could significantly affect the survival of human glioma cancer stem cells. Compound**40** (Figure 7), which has two pyrid-2-yl moieties, exhibited better enzymatic VEGFR2 tyrosine kinase activity with an IC_{50} value of 15.1 nM, than the standard drugs Sutent and Nexavar (Lintnerová et al., 2014). Additionally, compound **40** was effective in inhibiting endothelial cell growth, tube formation, and migration and it showed high selectivity in inducing apoptosis in endothelial cells.

A novel 2,4-diphenyl oxazole derivative (**41a–c**) (Figure 7) exhibited promising anticancer activity in vitro, where **41a** and **41b** showed good efficacy against the HepG2 cell line, (IC_{50}= 16.89 and 16.65 µM, respectively) (Mathew et al., 2013). In contrast, **41c** demonstrated significant activity against HeLa cells (IC_{50}=56.52 µM). The substituted aryl group at the second position of the oxazole scaffold was crucial for enhancing its anticancer potential.

Indoline sulfonamide oxazole derivatives such as compounds **42a**, **42c**, and **42d**(Figure 31) exhibited strong cytotoxicity, with IC_{50} values ranging from 0.055–0.105 µM and 0.039–0.112 µM, against the cancer cell linesHepG2, HCT116, SK-OV-3 and PC3(Yang et al., 2014). While **42e** and **42f** showed moderate efficacy against the SK-OV-3 and HCT116 cell lines. The SAR analysis suggested that the presence of oxazole, indoline, and cyclopropylgroups is essential for anticancer activity, while ethoxy and methyl substituents at the 5-position greatly enhance the activity.

The tri-substituted oxazoles with methyl, aryl, and piperazinyl groups **43a** and **43b** (Figure 31) exhibited remarkable antiproliferative activity and tumor vascular-disrupting effect in vitro against a range of human cancer cell lines as well as HU-VECs (human umbilical vein endothelial cells) (Choi et al., 2013). With an IC_{50} value of 10.3 nM, compound **43b**showed the most potent activity. Substituting the oxazole ring with either an isoxazole or thiazole moiety resulted in a significant decrease in potency. Modification of the arylpiperazine group by replacing it with other moieties, such as piperidines or piperazines, led to a significant loss of activity. This suggests that the nitrogen atoms within the piperazine ring play a crucial role, likely through potential hydrogen-bond interactions.

Figure 31. Structures of oxazole derivatives as anticancer agents

Figure 31. Structures of oxazole derivatives as anticancer agents

41a, R₁=Br, R₂=H, R₃=H
41b, R₁=H, R₂=H, R₃=OCOCH₃
41c, R₁=NO₂, R₂=H, R₃=H

42a, R₁=H, R₂=H
42c, R₁=H, R₂=5-OCH₃
42d, R₁=H, R₂=5-CH3
42e, R₁=F, R₂=H
42f, R₁=Cl, R₂=H

43a, R=Cl
43b, R=OCH₃

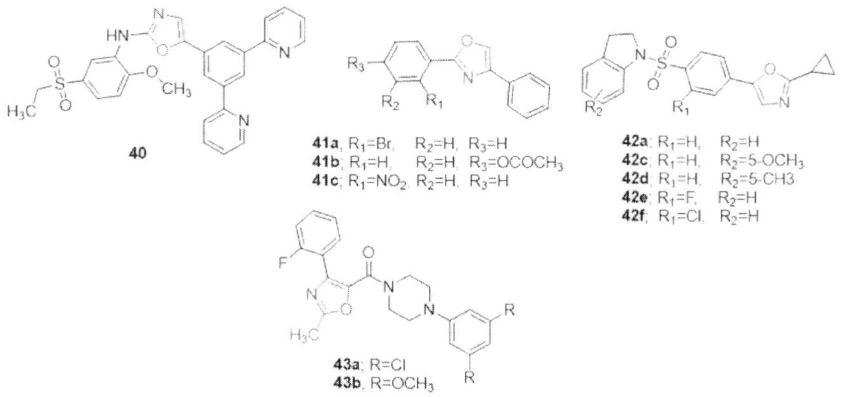

1.5.2 Anti-Microbial Activity of Oxazoles

In recent years, the development of novel antifungal compounds has garnered significant attention due to the increasing prevalence of fungal infections and the limitations of existing treatments. One such promising development is the synthesis and evaluation of **44** (Figure 8), a molecule with rich functionalities and potential biological activities, including antifungal properties (Yin et al., 2022). This compound features two distinct rings: the pyrazole ring, which is substituted with phenyl groups, and the oxazole ring, which includes a tosyl group potentially contributing to its biological efficacy.The antifungal potential of these 1,3-oxazole derivatives was found to be more effective than fluconazole, which is a well-known antifungal drug. These compounds were shown to hinder the creation of ergosterol in Candida species, similar to the mode of action of fluconazole. Ergosterol plays a critical role in maintaining the stability of fungal cell membranes, which is vital for the survival of the fungus. By inhibiting ergosterol synthesis, these derivatives exhibit potent antifungal activity, highlighting their potential as advanced therapies for fungal infections.

Ryu et al. conducted a comprehensive study where they synthesized several benzo[d]oxazole compounds and evaluated their antifungal efficacy against diverse strains,employing 5-fluorocytosine as a reference drug for comparison(Ryu et al., 2009). Remarkably, the antifungal effects of compounds **45** and **46** (Figure 8) were found to be comparable to the standard drug. Similarly, various substituted benzoxazoles synthesized by Anand et. al exhibited potent antimicrobial activity against*E. coli*,*S. aureus, C. glabrata,* and*C. albicans*(Anand et al., 2011). These activities were comparable to those of standard antimicrobial drugs such as trimethoprim and miconazole.Among the investigated compounds, **47** and **48** (Figure 8) demonstrated exceptional antibacterial activity, while compounds **49** and **50** exhibited superior antifungal properties.

Antibiotics, either derived from microorganisms or synthesized chemically, are indispensable in the fight against bacterial infections. Systemic bacterial infections often require treatment with broad-spectrum antibiotics, such as amoxicillin and cephalosporins. Recent research has highlighted the effectiveness of new derivatives, such as methyl 2-(2-arylideneamino) oxazole-4-arylamino benzoxazole-5-carboxylates, in combating bacterial strains like Staphylococcus aureus, Bacillus subtilis, Escherichia coli, and Salmonella typhi (Gadhe et al., 2010). Compounds **51** and **52**(Figure 8)were found to be more effective than ampicillin, while compounds **53** and **54** demonstrated adequate antibacterial activity.

Figure 32. Structures of oxazole derivatives as antimicrobial agents

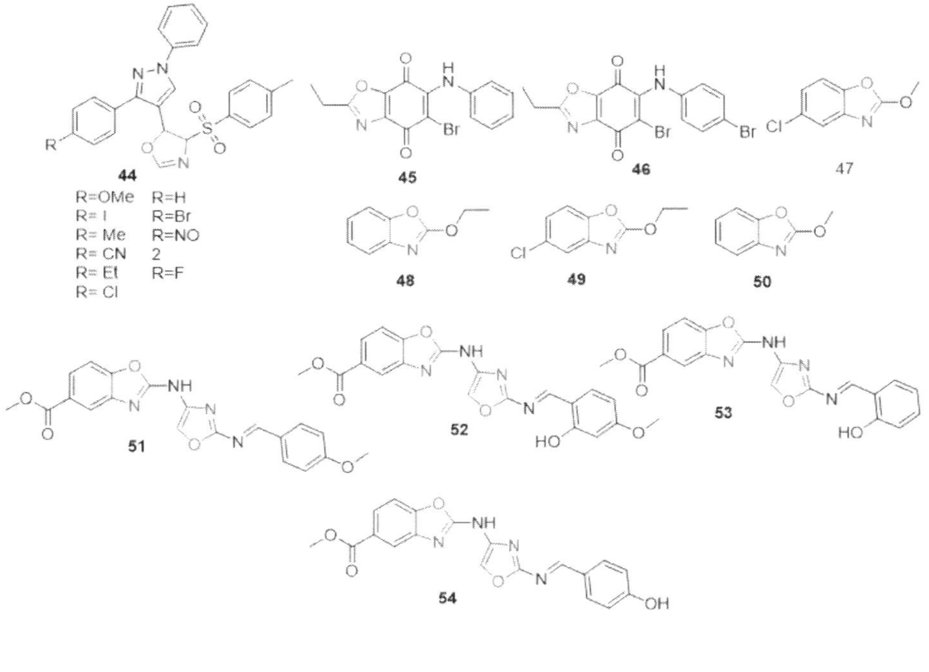

Moreover, a series of synthetic coumarin-based oxazole derivatives and glycosylated forms of oxazole derivatives displayed potent antimicrobial activities. The activity levels of these derivatives were comparable to those of reference drugs such as ciprofloxacin, gatifloxacin, fluconazole, and sulphacetamide(Singh et al., 2010; Taile et al., 2009).Specifically, compounds **55**, **56**, **57**, and **58** (Figure 9) exhibited strong antibacterial activity, while compounds **59**, **60**, and **61** demonstrated highly potent antifungal activity. Further, novel coumarin-based oxazole derivatives**62** and **63**(Figure 9) exhibited notable antibacterial activity against both Gram-negative and Gram-positive bacteria, which iscomparable with the effectiveness of standard drugs like penicillin and streptomycin(Reddy et al., 2010).

Figure 33. Structures of oxazole derivatives as antimicrobial agents

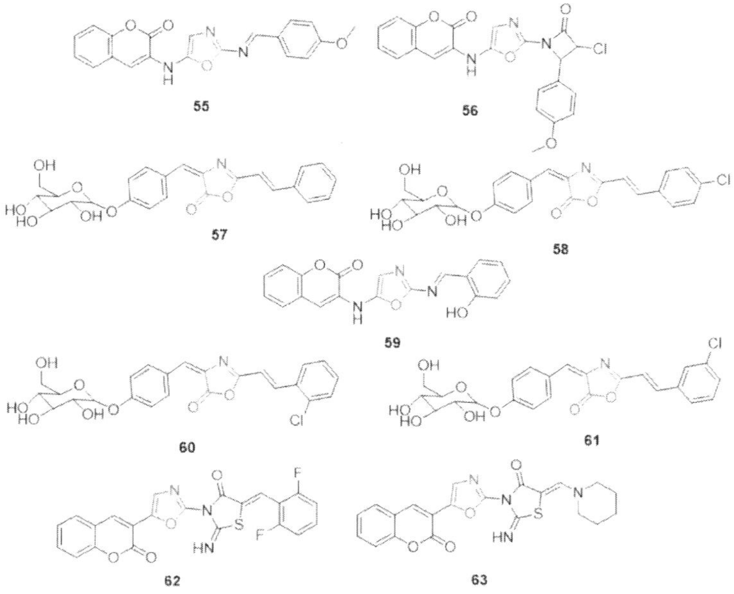

1.5.3 Antitubercular Activity of Oxazoles

Texaline, an oxazole-containing alkaloid, is another notable compound with antitubercular properties. Isolated from Amyris elemifera and Amyris texana, texaline has emerged as a promising candidate for Antitubercular agents. Among its derivatives, compounds **64** and **65**(Figure 10)have demonstrated significant effectiveness as antitubercular agents(Giddens et al., 2005). In addition to that, several synthetic oxazole- and oxazoline-containing compounds have been developed with promising results. Research by Moraski et al. has identified a series of these compounds with excellent inhibitory activity against the Mycobacterium tuberculosis strain MtbH37Rv(Moraski et al., 2010). Notably, compounds **66** and **67** (Figure 10) exhibited the most potent activity against MtbH37Rv when compared to the standard drug rifampicin. Moreover, oxazoline compounds, particularly **68** and **69**, have demonstrated remarkable antitubercular activity, with minimum inhibitory concentration (MIC) values of less than one micromolar.

Figure 34. Structures of oxazole derivatives as antitubercular agents

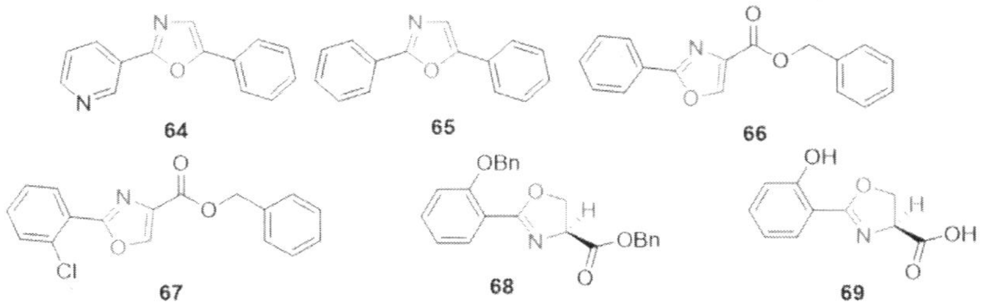

1.5.4 Antimalarial Activity of Oxazoles

Malaria remains one of the most serious and complex health challenges of the 21st century. Caused by four species of protozoan parasites from the Plasmodium genus, malaria is transmitted from person to person by the female Anopheles mosquito. The search for effective treatments is critical, given the global impact of this disease. Among the oxazole-based compounds, 2-(benzhydrylthio) benzo[d]oxazole**70** (Figure 11) has emerged as one of the promising candidates for antimalarial activity (Goyal et al., 2012). Specifically, **70** is of significant interest as it possesses antimalarial activity both in vitro as well in vivo against the multidrug-resistant parasite *Plasmodium yoelii.*

Several studies have demonstrated that compounds containing a quinoline scaffold exhibit antimalarial properties. In particular, quinoline-oxazole hybrid compounds, such as compounds **71** and **72** (Figure 11), have shown moderate activity against *Plasmodium falciparum* in vitro, with potency in the sub-micromolar range(Gordey et al., 2011). Notably, these compounds also exhibited acceptable cytotoxicity levels toward mononuclear leukocytes, indicating a favorable therapeutic index. Furthermore, coupling the oxazole and quinoline scaffolds with different carbon spacers, as seen in compounds **73** and **74** (Figure 11), has been reported to result in significantly higher activity compared to chloroquine against both chloroquine-sensitive 3D7 (CQS) and chloroquine-resistant (CQR) K1 strains(Musonda et al., 2007). This enhanced activity may be attributed to increased basicity from the side chain NH and oxazole N, which could improve compound accumulation in the acidic food vacuole of the parasite through pH trapping. Additionally, the presence of the heterocycle in the side chain may facilitate secondary interactions with targets other than haem, contributing to the observed antiplasmodial effects.

Figure 35. Structures of oxazole derivatives as antimalarial agents

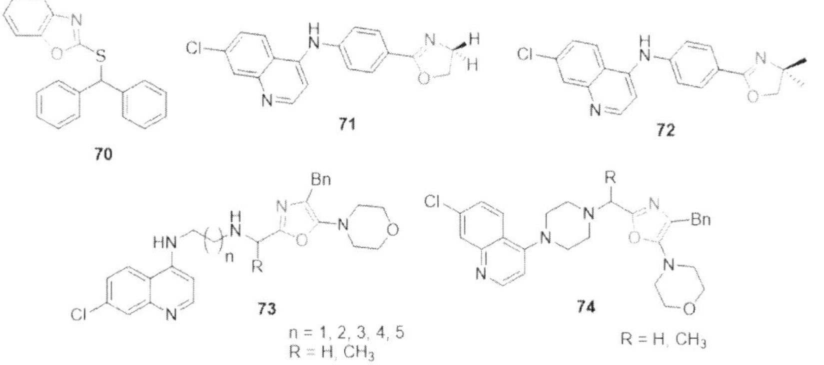

1.5.5 Antidiabetic Activity of Oxazoles

Peroxisome proliferator-activated receptors (PPARs) are pivotal therapeutic targets for the treatment of Type 2 Diabetes Mellitus. A significant number of existing PPAR ligands incorporate a carboxylic acid group or a thiazolidinedione moiety, both of which are essential for their bioactivity. Notably, the 1,2,4-oxadiazole derivative, compound **75** (Figure 12), is hypothesized to bind with PPARα and PPARδ through its acetamide scaffold and an adjacent methyl group(Matsufuji et al., 2013). A 1,3-dioxane carboxylic acid derivative**76**, exhibited PPAR agonistic activityin vitroin vivo glucose and lipid-lowering efficacy in animal models, with an EC_{50} value of 0.0015 μM(Pingali et al., 2008).Similarly, novel thiophene-substituted oxazole derivative**77** containing α-alkoxy-phenylpropanoic acid, functioned as potent dual agonists for PPARα/γ(Raval et al., 2011). Compound **77** can reduce glucose levels by up to 72%.

The G-protein coupled receptor 40 (GPR40) is associated with enhanced insulin secretion in response to elevated glucose levels. Compound **78** (Figure 12), with an Isoxazole core, has demonstrated exceptional efficacy as a GPR40 agonist and it significantly reduced plasma glucose levelsin human GPR40 knock-in mice(J. (Jim) Liu et al., 2014). Moreover, tri-substituted oxazole **79** exhibited notable activity on the GPR40 receptor. Further, an arylsulfonyl 3-(pyridin-2-yloxy) aniline which incorporates a 1,2,4-oxadiazole scaffold derivative**80**, displayed promising activity as a GPR119 agonist(Zhang et al., 2013).

Figure 36. Structures of oxazole derivatives as antidiabetic agents

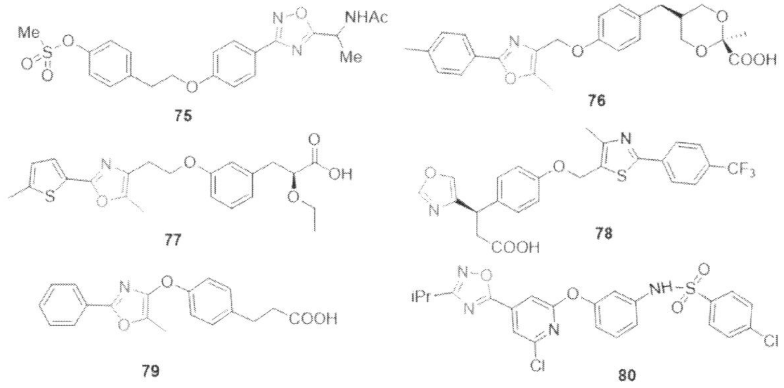

1.5.6 Anti-Inflammatory and Analgesic Activity of Oxazoles

Mofezolac **81** (Figure 37), which features an isoxazole scaffold, has demonstrated effectiveness as a selective COX-1 inhibitor and has been widely used in medication. However, efforts are being made to developanalogs that can reduce their side effects while retaining efficacy as COX-1 inhibitors (Perrone et al., 2015). COX-1, the constitutive enzyme, is crucial for maintaining homeostasis and protecting the gastrointestinal system, while COX-2 is primarily involved in inflammation. Compound **82**is highly selective towards COX-2 inhibition, with a percentage of $70.14\% \pm 1.71$(Dündar et al., 2009). Additionally, substituted quinolyl oxazoles such as compounds **83** and **84** are potent phosphodiesterase 4 (PDE4) inhibitors, which are crucial in the regulation of inflammatory and immune responses via the cAMP pathway(Kuang et al., 2007). Compounds **83** and **84** showed remarkable efficacy against PDE4, with IC_{50} values of 1.4 nM and 1 nM, respectively. Furthermore, oxazole derivatives such as **85** have the potential to inhibit the TRPV1 receptor, with an IC_{50} value of 15 ± 3 nM, which is key in pain signal transmission (Perner et al., 2010b).

Figure 37. Structures of oxazole derivatives as anti-inflammatory and analgesic agents.

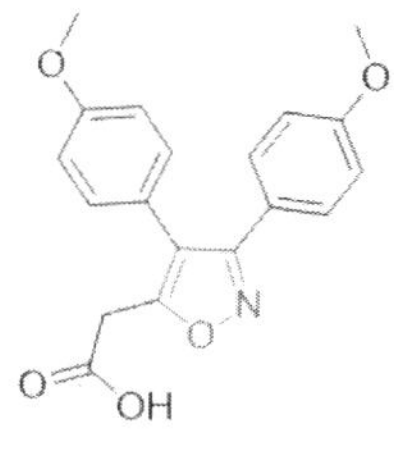

1.5.7 Antiviral Activity of Oxazoles

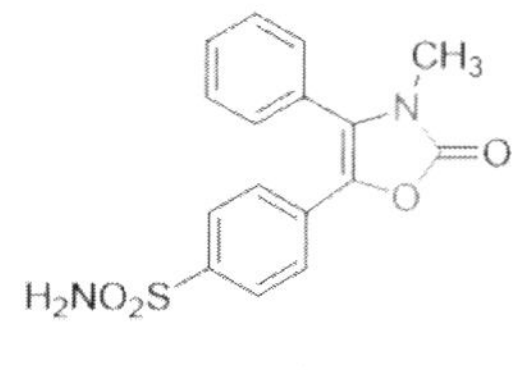

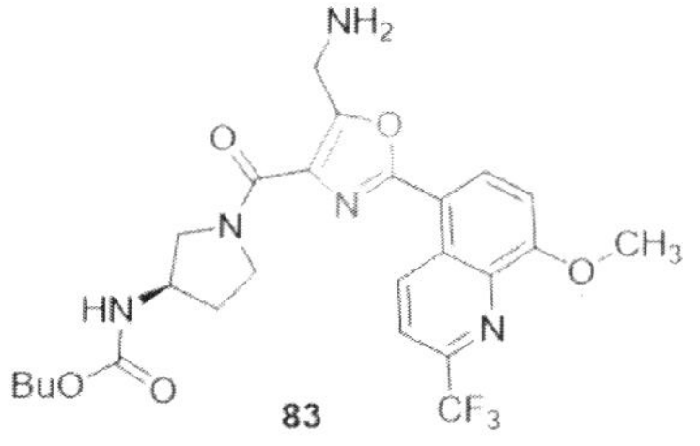

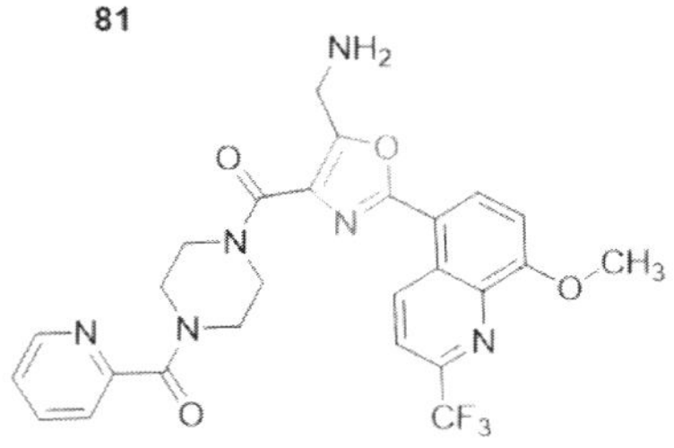

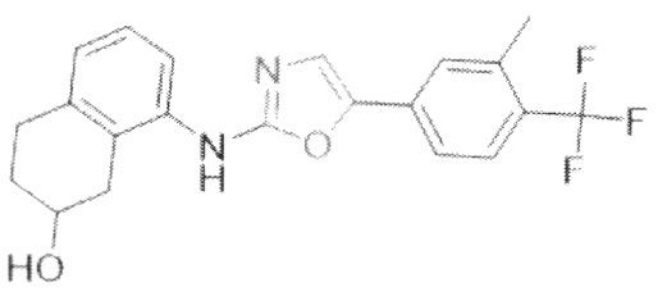

Despite the rising number of Human Immunodeficiency Virus Type 1 (HIV-1) patients, the availability of effective antiviral treatments remains inadequate, underscoring the urgent need for new drugs with novel mechanisms of action. In response to this challenge, extensive research is underway focusing on designing and synthesizing novel antivirals with potent anti-HIV activity. Oxazole derivatives with 4-azaindole moiety **86a-86c**(Figure 14), particularly **86b** exhibited anti-HIV-1 activity with an EC_{50} value of 0.18 μM, with no significant cytotoxicity (T. Wang et al., 2013). Moreover, compound **86c**, with a fluoro substituent, exhibited strong anti-HIV-1 activity and a favorable therapeutic index (Regueiro-Ren et al., 2013). Further, di-substituted oxazole derivatives such as **87a-87d**(Figure 14) showed potency against HIV-1(Kim et al., 2013). Particularly, compound **87d** with the dimethoxy substituent in the phenyl moiety showed exceptionally good activity with an EC_{50} value of 0.42 μM, while compound **87a**also showed good anti-HIV-1 activity with an EC_{50} value of 2 μM. In contrast, compounds **87b** and **87c** with the phenyl substituents of methoxy and 3,4-dichlorophenyl groups showed poor potency with an EC_{50} value of 100 and 35 μM, respectively.

The benzimidazole-based oxazole derivative such as **88**,displayed significant inhibitory activity against HCV genotype-1a (GT-1a) (EC_{50}=3.5 nM) and genotype-1b (GT-1b) (EC_{50}=0.14 nM) (Belema et al., 2014).Moreover, piperazinones with an isoxazole ring **89a and 89b**, also exhibited significant potency against both GT-1a and GT-1b, having EC_{90} values from 0.006 to 0.070 μM(Kakarla et al., 2014). It is hypothesized that the incorporation of a heterocyclic moiety at the C-6 position of the piperazinone ring is crucial for enhancing GT-1 potency.

Figure 38. Structures of oxazole derivatives as antiviral agents

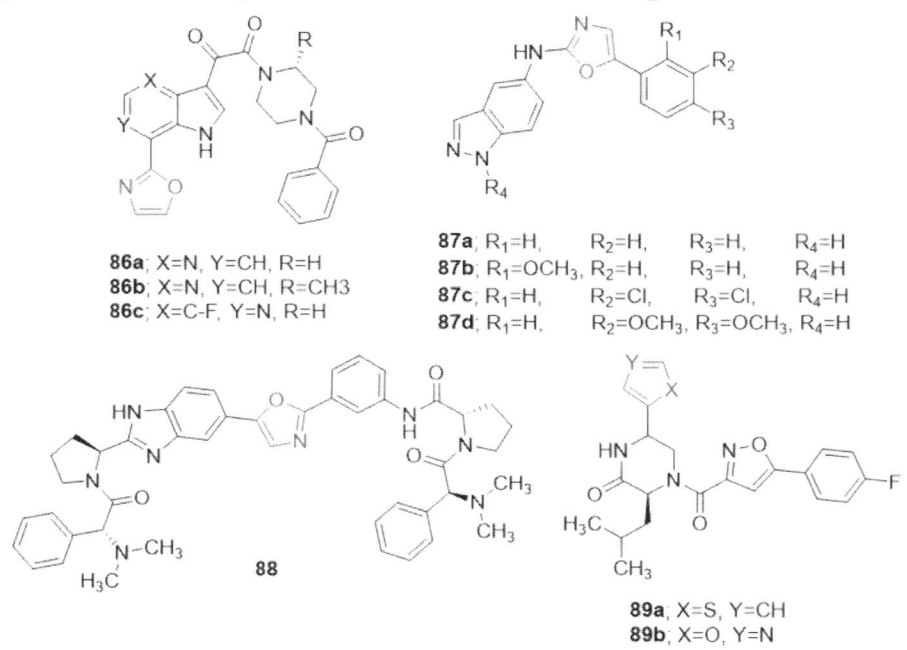

86a; X=N, Y=CH, R=H
86b; X=N, Y=CH, R=CH3
86c; X=C-F, Y=N, R=H

87a, R_1=H,	R_2=H,	R_3=H,	R_4=H
87b; R_1=OCH$_3$,	R_2=H,	R_3=H,	R_4=H
87c, R_1=H,	R_2=Cl,	R_3=Cl,	R_4=H
87d; R_1=H,	R_2=OCH$_3$,	R_3=OCH$_3$,	R_4=H

88

89a; X=S, Y=CH
89b; X=O, Y=N

1.5.8 Antioxidant Activity of Oxazoles

Oxazoles can subdue oxidative stress, a major cause of diseases like cancer, brain issues, heart problems, and aging. They act as antioxidants by eliminating free radicals, holding onto metal ions, stopping unstable enzymes, and altering the body's natural defenses against free radicals. Oxazoles lower reactive oxygen species (ROS) and reactive nitrogen species (RNS) by giving away hydrogen atoms or electrons, preventing lipid breakdown and other oxidation processes. The oxazole ring's redox properties can be influenced by adding hydroxyl or methoxy groups, while nitrogen groups can lower the ring's electrons. 2,5-diphenyloxazole **90** (Figure 15) is a strong antioxidant known for its ability to detect radiation and effectively eliminate free radicals in cells (Kuş et al., 2017). The azlactones **91**(Figure 15) engaged with DPPH by oxidizing it to give 1,1-diphenyl-2-picrylhydrazine and exhibited a strong antioxidant activity (Parveen et al., 2013).

Figure 39. Structures of oxazole derivatives as antioxidant agents

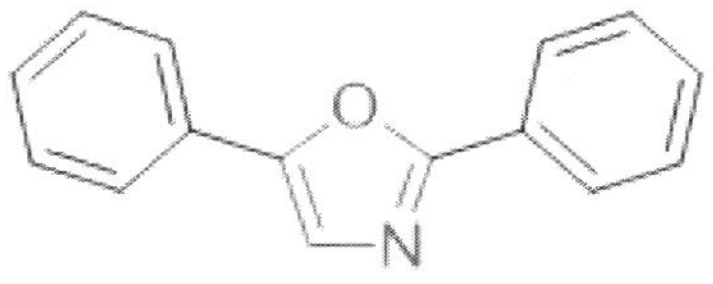

90 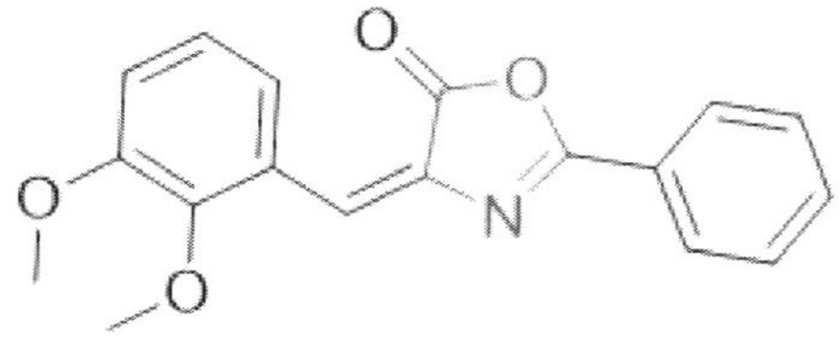   **91**

1.6 Oxazoles in Natural Products

Several natural product compounds have at least one oxazole ring in their molecular structure, especially those that come from marine sources. These substances have a wide range of biological actions that could be used as therapeutics, including ichthyotoxic, antibacterial, peripheral analgesic, HSV-1 inhibitory, anticancer, antileukemic, antifungal, and selective serine-threonine phosphatase inhibitory properties. Some of the selected natural oxazoles are given below in Table 3 and the structures of compounds are given in Figure 40.

Table 3. List of naturally occurring oxazole derivatives

Compounds name	Source	Biological Activity
Phthoxazolin A	Isolated from *Streptomyces* species	Herbicide (Hénaff & Whiting, 1999)
Balsoxin	Isolated from *Amyris* species	Antimycobacterial activity (Ohnmacht et al., 2008)
Texaline	Isolated from *Amyris elemifera*	antimycobacterial activity (Giddens et al., 2005)
Myacalolide	Isolated from marine sponge *Mycale*	Antifungal and anticancer (Panek & Liu, 2000)
venturamide A	Isolated from Oscillatoria	Antimalarial (Linington et al., 2007)
Bengazole A	Isolated from marine sponge Jaspidae Fiji sponge	Antihelminthic activity and antifungal activity (Chandrasekhar & Sudhakar, 2010)

continued on following page

Table 3. Continued

Compounds name	Source	Biological Activity
Leiodolide A	Isolated from *Leiodermatium* species found in deep water sponge	Cytotoxic activity (Bone Relat et al., 2022)
Madumycin II	Isolated from various species of *Streptomyces*	Antibiotics (Tavares et al., 1996)
Thiangazole	Isolated from *Polyangium sp*	HIV-1 inhibitor (Boyce et al., 1995)
Tantazoles	Isolated from blue-green alga *Scytonema Mirabile*	cytotoxic activity (Carmeli et al., 1990)

Figure 40. Structures of some of the naturally occurring oxazole compounds

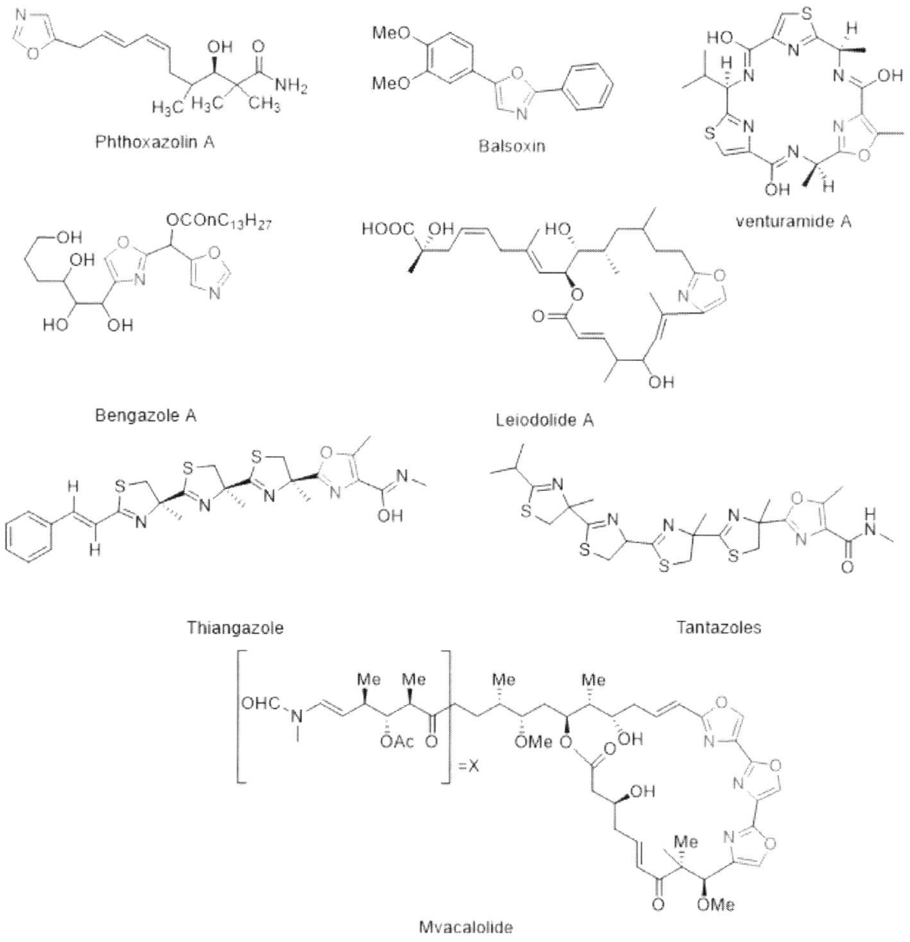

Phthoxazolin A

Balsoxin

venturamide A

Bengazole A

Leiodolide A

Thiangazole

Tantazoles

Mvacalolide

1.6.1 Mono-Oxazoles

Oxazole alkaloids are relatively simple.The oxazole pimprinine was initially identified in 1963 from the filtrates of *Streptomyces pimprina* cultures. Members of this family exhibit a variety of biological actions, including fungicidal, anticancer, antiepileptic, and platelet-aggregation-inhibitory properties. (B. Liu et al., 2019). Studies on their biological activity are not deep enough as these are hard to synthesize because of their low content in nature. But their analogous seems to have biological activity towards the anti-plant virus. (J.-R. Liu et al., 2022) Synthesized pimprinine derivatives gave positive results for antifungal activity. Diaryloxazoles texamine and texaline are oxazoles possessing antimycobacterial properties isolated from *Amyris texana and A. elemifera* (Giddens et al., 2005). Melanoxadin (Hashimoto et al., 1995) and Melanoxazal (Takahashi et al., 1996) are melanin biosynthesis inhibitors that were extracted from the Trichoderma sp. fermentation broth. The pathogenic plant fungus Rhizopus microsporus is the source of the macrolide antibiotic rhizoxin. Additionally, it has antimitotic and antitumor properties. It functions as a metabolite, an antimitotic, and an antineoplastic. It is an epoxide, a macrolide antibiotic, and a member of the 1,3-oxazole class. (Hearn et al., 2007)

Cyclic peptide oxazole containing streptogramin A antibiotic is found in gris-eoviridin(Xie et al., 2012), mitomycin II (Tavares et al., 1996), and Virginiamycin M2("Practical Antimicrobial Therapeutics," 2017). An antibiotic that was isolated from soil bacteria in the *Streptomyces* family

Figure 41. Structures of the naturally occurring oxazole Mono-oxazole.

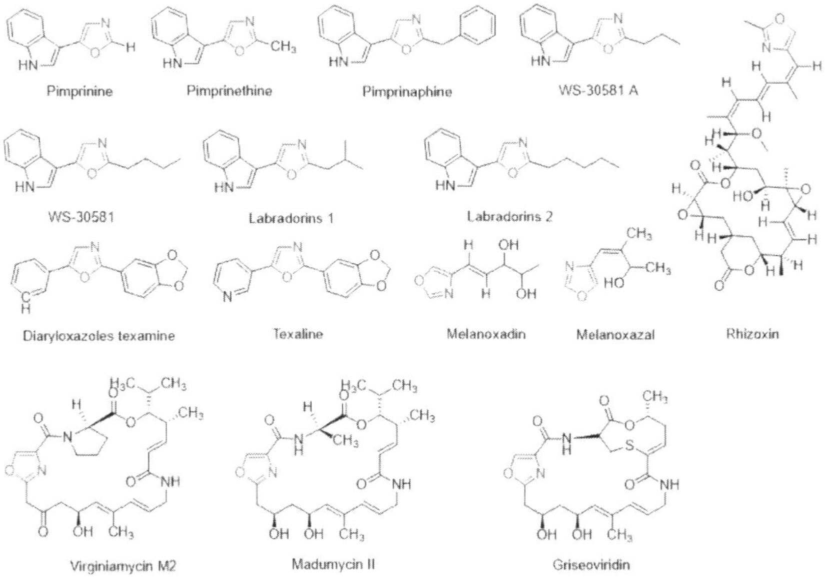

1.6.2 Bis-Oxazoles

Natural goods can have two oxazole rings integrated into their structure in many different kinds of manner. The bengazoles are a class of marine-derived natural compounds distinguished by their distinctive structure with two oxazole rings surrounding a single carbon atom and strong antifungal action(Bull et al., 2007). Bengazoles are composed of a fatty acid ester with a side chain that resembles a polyol. Bengazoles A and Bengazoles B are both strong antihelminthic substances. The source of muscoride A, a novel oxazole alkaloid discovered in 1995, is the freshwater cyanobacterium *Nostocmuscorum*.(Nagatsu et al., 1995), Extremely cytotoxic agents known as phorboxazoles A and B were discovered in 1995 from the marine sponge *Phorbas sp*. While the second 2,4-disubstituted oxazole is an essential part of the side chain, the first 2,4-disubstituted oxazole is integrated into the complicated macrolide ring. Among the most toxic naturally occurring substances, pyrazoles stop cell growth in the S phase of the cell cycle (Searle & Molinski, 1995). Promothiocin A, one of the thiopeptide antibiotics derived from *Streptomyces sp*. SF2741 incorporates both 2,4-disubstituted oxazoles into its macrocycle. (Moody & Bagley, 1998)

Figure 42. Structures of the naturally occurring oxazole Bis-oxazole

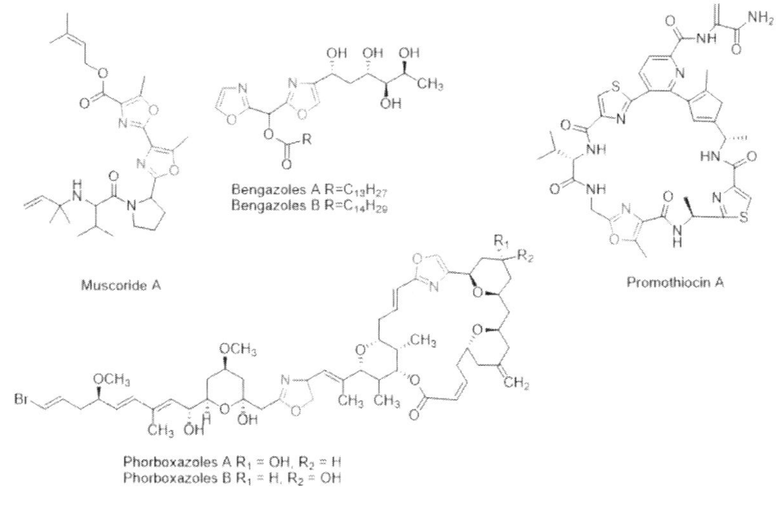

1.6.3 Tris-Oxazoles and Tetra-Oxazoles

The ulapualides such as ulapualide A, are a remarkably unique class of macrolides that contain tris-oxazoles. They were initially identified in the egg masses of the marine nudibranch *Hexabranchus sanguineus*.The structures of ulapualides and their derivatives are based on a side chain with multiple oxy-donor atoms arranged in chelating configurations and a macrocyclic cavity that incorporates oxygen and nitrogen ligands. (Maddock et al., 1993). The antibacterial peptide known as microcin B17 (MccB17) is generated by *Escherichia coli* strains that carry the plasmid-borne mccB17 operon. MccB17 has a lot of noteworthy characteristics. Stabilizing the transitory DNA gyrase–DNA cleavage complex is one of its effective mechanisms of action, which it shares with the incredibly effective fluoroquinolone medications. (Collin & Maxwell, 2019)

Figure 43. Structures of the naturally occurring oxazole Tris-oxazole and tetra-oxazole

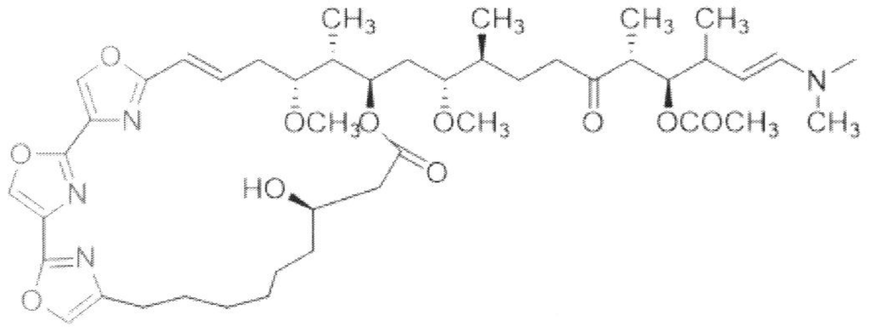

Ulapualides A

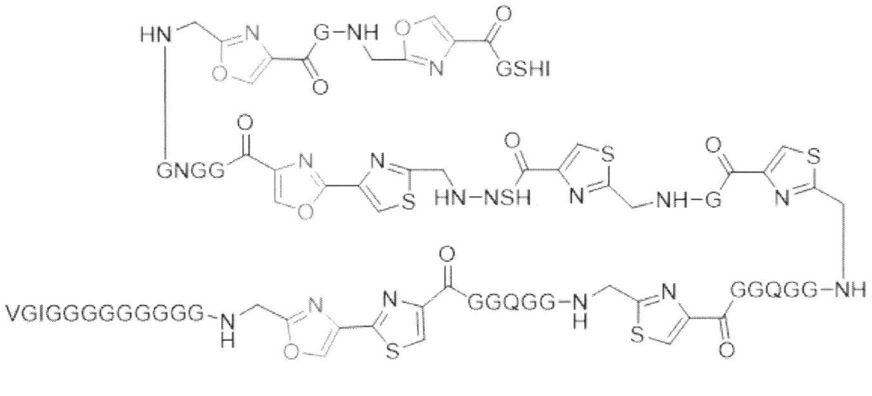

Microcin B17

2. PERSPECTIVE AND CONCLUSION

1,3-Oxazoles are highly versatile compounds with a wide range of applications across diverse fields, including drug development, chemical synthesis, materials science, biology, agrochemicals, and industrial processes. Oxazole-based compounds have demonstrated significant potential in medicinal chemistry, particularly due to their electron-rich heterocyclic structure, which allows for easy modification with various functional groups. This flexibility enables the creation of effective core structures that can combine with bioactive moieties or be incorporated into biomolecules. Oxazole derivatives have become crucial in medicinal research, with substantial efforts leading to numerous noteworthy achievements. Of particular importance is the successful development and commercialization of several oxazole-based drugs, highlighting their impact on the pharmaceutical industry. By altering the oxazole ring and searching for other derivatives of the group, researchers could probably create new medicines to combat cancer and pathogens. Researchers are also investigating its potential application in organic electronic components like OLEDs, OPVs, and chemical and biological monitors.

Although many oxazole-based drugs show great promise, some are associated withseveral toxicities and poor tolerability. To address these challenges, ongoing research should be focused on the design, synthesis, and bioactive evaluation of new oxazole derivatives that offer improved efficacy with reduced side effects.

Thus, the focused exploration of newer oxazole derivatives with higher potency, selectivity, and improved pharmacokinetic profiles is anticipated to be a prime area of research shortly which could bring about breakthroughs transforming the way a variety of diseases are treated. As new oxazole-based compounds with better therapeutic properties are developed, it could lead to more effective and targeted applications that can potentially revolutionize many areas in terms of modern medicine.

REFERENCES

Anand, M., Ranjitha, A., & Himaja, M. (2011). *Silica sulfuric acid catalyzed microwave-assisted synthesis of substituted benzoxazoles and their antimicrobial activity.* https://www.semanticscholar.org/paper/SILICA-SU-LFURIC-ACID-CATALYZED-MICROWAVE-ASSISTED-Anand-Ranjitha/7f84eceddf ec548b61872f59c87f2bf1ea7db2c6

Antonio, J., & Molinski, T. F. (1993). Screening of Marine Invertebrates for the Presence of Ergosterol-Sensitive Antifungal Compounds. *Journal of Natural Products*, 56(1), 54–61. DOI: 10.1021/np50091a008 PMID: 8450321

Bacon, E. R., Chatterjee, S., & Williams, M. (2007). Sleep. In *Comprehensive Medicinal Chemistry II* (pp. 139–167). Elsevier., DOI: 10.1016/B0-08-045044-X/00166-8

Belema, M., Nguyen, V. N., Romine, J. L., St. Laurent, D. R., Lopez, O. D., Goodrich, J. T., Nower, P. T., O'Boyle, D. R., Lemm, J. A., Fridell, R. A., Gao, M., Fang, H., Krause, R. G., Wang, Y.-K., Oliver, A. J., Good, A. C., Knipe, J. O., Meanwell, N. A., & Snyder, L. B. (2014). Hepatitis C Virus NS5A Replication Complex Inhibitors. Part 6: Discovery of a Novel and Highly Potent Biarylimidazole Chemotype with Inhibitory Activity Toward Genotypes 1a and 1b Replicons. *Journal of Medicinal Chemistry*, 57(5), 1995–2012. DOI: 10.1021/jm4016203 PMID: 24437689

Berlin, I., Zimmer, R., Thiede, H., Payan, C., Hergueta, T., Robin, L., & Puech, A. (1990). Comparison of the monoamine oxidase inhibiting properties of two reversible and selective monoamine oxidase-A inhibitors moclobemide and toloxatone, and assessment of their effect on psychometric performance in healthy subjects. *British Journal of Clinical Pharmacology*, 30(6), 805–816. DOI: 10.1111/j.1365-2125.1990.tb05445.x PMID: 1705137

Bone Relat, R. M., Winder, P. L., Bowden, G. D., Guzmán, E. A., Peterson, T. A., Pomponi, S. A., Roberts, J. C., Wright, A. E., & O'Connor, R. M. (2022). High-Throughput Screening of a Marine Compound Library Identifies Anti-Cryptosporidium Activity of Leiodolide A. *Marine Drugs*, 20(4), 240. DOI: 10.3390/md20040240 PMID: 35447913

Boyce, R. J., Mulqueen, G. C., & Pattenden, G. (1995). Total synthesis of thiadiazole, a novel naturally occurring HIV-1 inhibitor from Polyangium sp. *Tetrahedron*, 51(26), 7321–7330. DOI: 10.1016/0040-4020(95)00356-D

Bradford, P. A., Miller, A. A., O'Donnell, J., & Mueller, J. P. (2020). Zoliflodacin: An Oral Spiropyrimidinetrione Antibiotic for the Treatment of Neisseria gonorrhea, Including Multi-Drug-Resistant Isolates. *ACS Infectious Diseases*, 6(6), 1332–1345. DOI: 10.1021/acsinfecdis.0c00021 PMID: 32329999

Brusnakov, M., Golovchenko, O., Velihina, Y., Liavynets, O., Zhirnov, V., & Brovarets, V. (2022). Evaluation of Anticancer Activity of 1,3-Oxazol-4-ylphosphonium Salts *in Vitro. ChemMedChem*, 17(20), e202200319. DOI: 10.1002/cmdc.202200319 PMID: 36037305

Bull, J. A., Balskus, E. P., Horan, R. A. J., Langner, M., & Ley, S. V. (2007). Total Synthesis of Potent Antifungal Marine Bisoxazole Natural Products Bengazoles A and B. *Chemistry (Weinheim an der Bergstrasse, Germany)*, 13(19), 5515–5538. DOI: 10.1002/chem.200700033 PMID: 17440905

Carmeli, S., Moore, R. E., Patterson, G. M. L., Corbett, T. H., & Valeriote, F. A. (1990). Tantazoles, unusual cytotoxic alkaloids from the blue-green alga Scytonema mirabile. *Journal of the American Chemical Society*, 112(22), 8195–8197. DOI: 10.1021/ja00178a070

Chandrasekhar, S., & Sudhakar, A. (2010). Total Synthesis of Bengazole A. *Organic Letters*, 12(2), 236–238. DOI: 10.1021/ol9024138 PMID: 20017536

Chiacchio, M. A., Lanza, G., Chiacchio, U., Giofrè, S. V., Romeo, R., Iannazzo, D., & Legnani, L. (2020). Oxazole-Based Compounds As Anticancer Agents. *Current Medicinal Chemistry*, 26(41), 7337–7371. DOI: 10.2174/0929867326666181203130402 PMID: 30501590

Cho, Y. L., & Jang, J. (2020). Development of Delpazolid for the Treatment of Tuberculosis. *Applied Sciences (Basel, Switzerland)*, 10(7), 2211. DOI: 10.3390/app10072211

Choi, M. J., No, E. S., Thorat, D. A., Jang, J. W., Yang, H., Lee, J., Choo, H., Kim, S. J., Lee, C. S., Ko, S. Y., Lee, J., Nam, G., & Pae, A. N. (2013). Synthesis and Biological Evaluation of Aryloxazole Derivatives as Antimitotic and Vascular-Disrupting Agents for Cancer Therapy. *Journal of Medicinal Chemistry*, 56(22), 9008–9018. DOI: 10.1021/jm400840p PMID: 24160376

Collin, F., & Maxwell, A. (2019). The Microbial Toxin Microcin B17: Prospects for the Development of New Antibacterial Agents. *Journal of Molecular Biology*, 431(18), 3400–3426. DOI: 10.1016/j.jmb.2019.05.050 PMID: 31181289

Cornforth, J. W., & Cornforth, R. H. (1947). 24. A new synthesis of oxazoles and imidazoles including its application to the preparation of oxazole. *Journal of the Chemical Society*, 96, 96. Advance online publication. DOI: 10.1039/jr9470000096

Deng, C., Yan, H., Wang, J., Liu, B., Liu, K., & Shi, Y. (2022). The anti-HIV potential of imidazole, oxazole, and thiazole hybrids: A mini-review. *Arabian Journal of Chemistry*, 15(11), 104242. DOI: 10.1016/j.arabjc.2022.104242

Dündar, Y., Ünlü, S., Banoğlu, E., Entrena, A., Costantino, G., Nunez, M.-T., Ledo, F., Şahin, M. F., & Noyanalpan, N. (2009). Synthesis and biological evaluation of 4,5-diphenyloxazolone derivatives on route towards selective COX-2 inhibitors. *European Journal of Medicinal Chemistry*, 44(5), 1830–1837. DOI: 10.1016/j.ejmech.2008.10.039 PMID: 19084295

Eicher, T., Hauptmann, S., & Speicher, A. (2003). *The Chemistry of Heterocycles: Structure, Reactions, Syntheses, and Applications* (1st ed.). Wiley., DOI: 10.1002/352760183X

Eicher, T., Hauptmann, S., & Speicher, A. (2012). *The chemistry of heterocycles: Structure, reactions, syntheses, and applications* (3rd, completely rev. and updated ed ed.). Wiley-VCH.

El Ibrahimi, B., & Guo, L. (2021). Azole-Based Compounds as Corrosion Inhibitors for Metallic Materials. In Kuznetsov, A. (Ed.), *Azoles—Synthesis, Properties, Applications and Perspectives*. IntechOpen., DOI: 10.5772/intechopen.93040

Erlenmeyer-Plöchl Azlactone Synthesis. (2010). In Z. Wang, *Comprehensive Organic Name Reactions and Reagents* (1st ed., pp. 997–1000). Wiley. DOI: 10.1002/9780470638859.conrr217

Filippatos, T. D., Kei, A., & Elisaf, M. S. (2017). Anacetrapib, a New CETP Inhibitor: The New Tool for the Management of Dyslipidemias? *Diseases (Basel, Switzerland)*, 5(4), 21. DOI: 10.3390/diseases5040021 PMID: 28961179

Gadhe, D., Chilumula, N. R., Gudipati, R., Ampati, S., & Manda, S. (2010). Synthesis of some novel methyl2(2 (arylideneamino) oxazol4 ylamino) benzoxazole5-carboxylate derivatives as antimicrobial agents. *International Journal of Chemistry Research*, 1–6.

Gao, W.-C., Wang, R.-L., & Zhang, C. (2013). Practical oxazole synthesis mediated by iodine from α-bromoketones and benzylamine derivatives. *Organic & Biomolecular Chemistry*, 11(41), 7123. DOI: 10.1039/c3ob41566j PMID: 24057123

Giddens, A. C., Boshoff, H. I. M., Franzblau, S. G., Barry, C. E.III, & Copp, B. R. (2005). Antimycobacterial natural products: Synthesis and preliminary biological evaluation of the oxazole-containing alkaloid texaline. *Tetrahedron Letters*, 46(43), 7355–7357. DOI: 10.1016/j.tetlet.2005.08.119

Gordey, E. E., Yadav, P. N., Merrin, M. P., Davies, J., Ward, S. A., Woodman, G. M. J., Sadowy, A. L., Smith, T. G., & Gossage, R. A. (2011). Synthesis and biological activities of 4-N-(anilinyl-n-[oxazolyl])-7-chloroquinolines (n=3' or 4') against Plasmodium falciparum in in vitro models. *Bioorganic & Medicinal Chemistry Letters*, 21(15), 4512–4515. DOI: 10.1016/j.bmcl.2011.05.131 PMID: 21723121

Goyal, M., Singh, P., Alam, A., Kumar Das, S., Shameel Iqbal, M., Dey, S., Bindu, S., Pal, C., Kumar Das, S., Panda, G., & Bandyopadhyay, U. (2012). Aryl methyl thio arenes prevent multidrug-resistant malaria in mice by promoting oxidative stress in parasites. *Free Radical Biology & Medicine*, 53(1), 129–142. DOI: 10.1016/j. freeradbiomed.2012.04.028 PMID: 22588006

Greenblatt, D., Matlis, R., Scavone, J., Blyden, G., Harmatz, J., & Shader, R. (1985). Oxaprozin pharmacokinetics in the elderly. *British Journal of Clinical Pharmacology*, 19(3), 373–378. DOI: 10.1111/j.1365-2125.1985.tb02656.x PMID: 3986088

Grotkopp, O., Ahmad, A., Frank, W., & Müller, T. J. J. (2011). Blue-luminescent 5-(3-indolyl)oxazoles via microwave-assisted three-component coupling–cycloisomerization–Fischer indole synthesis. *Organic & Biomolecular Chemistry*, 9(23), 8130. DOI: 10.1039/c1ob06153d PMID: 22024934

Gruneberg, R. N., & Leakey, A. (1976). Treatment of candidal urinary tract infection with nifuratel. *British Medical Journal*, 2(6041), 908–910. DOI: 10.1136/bmj.2.6041.908 PMID: 974657

Hanan, E. J., Braun, M.-G., Heald, R. A., MacLeod, C., Chan, C., Clausen, S., Edgar, K. A., Eigenbrot, C., Elliott, R., Endres, N., Friedman, L. S., Gogol, E., Gu, X.-H., Thibodeau, R. H., Jackson, P. S., Kiefer, J. R., Knight, J. D., Nannini, M., Narukulla, R., & Staben, S. T. (2022). Discovery of GDC-0077 (Inavolisib), a Highly Selective Inhibitor and Degrader of Mutant PI3Kα. *Journal of Medicinal Chemistry*, 65(24), 16589–16621. DOI: 10.1021/acs.jmedchem.2c01422 PMID: 36455032

Hartner, F. W. (1996). Oxazoles. In *Comprehensive Heterocyclic Chemistry II* (pp. 261–318). Elsevier., DOI: 10.1016/B978-008096518-5.00062-9

Hashimoto, R., Takahashi, S., Hamano, K., & Nakagawa, A. (1995). A New Melanin Biosynthesis Inhibitor, Melanoxadin from Fungal Metabolite by Using the Larval Haemolymph of the Silkworm, Bombyx mori. *The Journal of Antibiotics*, 48(9), 1052–1054. DOI: 10.7164/antibiotics.48.1052 PMID: 7592054

Hassner, A., & Fischer, B. (1993). New Chemistry of Oxazoles. *Heterocycles*, 35(2), 1441. DOI: 10.3987/REV-92-SR(T)6

Hearn, B. R., Shaw, S. J., & Myles, D. C. (2007). Microtubule Targeting Agents. In *Comprehensive Medicinal Chemistry II* (pp. 81–110). Elsevier., DOI: 10.1016/B0-08-045044-X/00205-4

Hénaff, N., & Whiting, A. (1999). A Convergent Stereoselective Total Synthesis of Racemic Phthoxazolin A. *Organic Letters*, 1(7), 1137–1139. DOI: 10.1021/ol990967b

Informatics, N. O. of D. and. *Oxazole*. National Institute of Standards and Technology. Retrieved August 6, 2024, from https://webbook.nist.gov/cgi/cbook.cgi?ID=C288426&Mask=80,9

Joshi, S., Bisht, A. S., & Juyal, D. (2017). Systematic scientific study of 1, 3-oxazole derivatives as a useful lead for pharmaceuticals. *RE:view*, 6(1).

Joshi, S., Mehra, M., Singh, R., & Kakar, S. (2023). Review on Chemistry of Oxazole Derivatives: Current to Future Therapeutic Prospective. *Egyptian Journal of Basic and Applied Sciences*, 10(1), 218–239. DOI: 10.1080/2314808X.2023.2171578

Kakarla, R., Liu, J., Naduthambi, D., Chang, W., Mosley, R. T., Bao, D., Steuer, H. M. M., Keilman, M., Bansal, S., Lam, A. M., Seibel, W., Neilson, S., Furman, P. A., & Sofia, M. J. (2014). Discovery of a Novel Class of Potent HCV NS4B Inhibitors: SAR Studies on Piperazinone Derivatives. *Journal of Medicinal Chemistry*, 57(5), 2136–2160. DOI: 10.1021/jm4012643 PMID: 24476391

Kean, W. F. (2004). Oxaprozin: Kinetic and dynamic profile in the treatment of pain. *Current Medical Research and Opinion*, 20(8), 1275–1277. DOI: 10.1185/030079904125004420 PMID: 15324530

Kim, S.-H., Markovitz, B., Trovato, R., Murphy, B. R., Austin, H., Willardsen, A. J., Baichwal, V., Morham, S., & Bajji, A. (2013). Discovery of a new HIV-1 inhibitor scaffold and synthesis of potential prodrugs of indazoles. *Bioorganic & Medicinal Chemistry Letters*, 23(10), 2888–2892. DOI: 10.1016/j.bmcl.2013.03.075 PMID: 23566519

Kobayashi, J., Tsuda, M., Fuse, H., Sasaki, T., & Mikami, Y. (1997). Halishigamides A−D, New Cytotoxic Oxazole-Containing Metabolites from Okinawan Sponge *Halichondria* sp. *Journal of Natural Products*, 60(2), 150–154. DOI: 10.1021/np960558d

Kuang, R., Shue, H.-J., Blythin, D. J., Shih, N.-Y., Gu, D., Chen, X., Schwerdt, J., Lin, L., Ting, P. C., Zhu, X., Aslanian, R., Piwinski, J. J., Xiao, L., Prelusky, D., Wu, P., Zhang, J., Zhang, X., Celly, C. S., Minnicozzi, M., & Wang, P. (2007). Discovery of a highly potent series of oxazole-based phosphodiesterase 4 inhibitors. *Bioorganic & Medicinal Chemistry Letters*, 17(18), 5150–5154. DOI: 10.1016/j.bmcl.2007.06.092 PMID: 17683932

Kumar, G., & Singh, N. P. (2021). Synthesis, anti-inflammatory, and analgesic evaluation of thiazole/oxazole substituted benzothiazole derivatives. *Bioorganic Chemistry*, 107, 104608. DOI: 10.1016/j.bioorg.2020.104608 PMID: 33465668

Kuş, C., Uğurlu, E., Özdamar, E. D., & Can-Eke, B. (2017). Synthesis and Antioxidant Properties of New Oxazole-5(4*H*)-one Derivatives. *Turkish Journal of Pharmaceutical Sciences*, 14(2), 174–178. DOI: 10.4274/tjps.70299 PMID: 32454610

Laohapaisan, P., Reamtong, O., Tummatorn, J., Thongsornkleeb, C., Thaenkham, U., Adisakwattana, P., & Ruchirawat, S. (2023). Discovery of N-methylbenzo[d]oxazol-2-amine as a new anthelmintic agent through scalable protocol for the synthesis of N-alkylbenzo[d]oxazol-2-amine and N-alkylbenzo[d]thiazol-2-amine derivatives. *Bioorganic Chemistry*, 131, 106287. DOI: 10.1016/j.bioorg.2022.106287 PMID: 36455482

Li, J. J. (Ed.). (2013). *Heterocyclic chemistry in drug discovery*. Wiley.

Lindquist, N., Fenical, W., Van Duyne, G. D., & Clardy, J. (1991). Isolation and structure determination of diazonamides A and B, unusual cytotoxic metabolites from the marine ascidian Diazona chinensis. *Journal of the American Chemical Society*, 113(6), 2303–2304. DOI: 10.1021/ja00006a060

Linington, R. G., González, J., Ureña, L.-D., Romero, L. I., Ortega-Barría, E., & Gerwick, W. H. (2007). Venturamides A and B: Antimalarial Constituents of the Panamanian Marine Cyanobacterium *Oscillatoria* sp. *Journal of Natural Products*, 70(3), 397–401. DOI: 10.1021/np0605790 PMID: 17328572

Lintnerová, L., García-Caballero, M., Gregáň, F., Meličerčík, M., Quesada, A. R., Dobiaš, J., Lác, J., Sališová, M., & Boháč, A. (2014). Development of chimeric VEGFR2 TK inhibitor based on two ligand conformers from PDB: 1Y6A complex – Medicinal chemistry consequences of a TKs analysis. *European Journal of Medicinal Chemistry*, 72, 146–159. DOI: 10.1016/j.ejmech.2013.11.023 PMID: 24368209

Liu, B., Li, R., Li, Y., Li, S., Yu, J., Zhao, B., Liao, A., Wang, Y., Wang, Z., Lu, A., Liu, Y., & Wang, Q. (2019). Discovery of Pimprinine Alkaloids as Novel Agents against a Plant Virus. *Journal of Agricultural and Food Chemistry*, 67(7), 1795–1806. DOI: 10.1021/acs.jafc.8b06175 PMID: 30681853

Liu, J., Wang, Y., Ma, Z., Schmitt, M., Zhu, L., Brown, S. P., Dransfield, P. J., Sun, Y., Sharma, R., Guo, Q., Zhuang, R., Zhang, J., Luo, J., Tonn, G. R., Wong, S., Swaminath, G., Medina, J. C., Lin, D. C.-H., & Houze, J. B. (2014). Optimization of GPR40 Agonists for Type 2 Diabetes. *ACS Medicinal Chemistry Letters*, 5(5), 517–521. DOI: 10.1021/ml400501x PMID: 24900872

Liu, J.-R., Liu, J.-M., Gao, Y., Shi, Z., Nie, K.-R., Guo, D., Deng, F., Zhang, H.-F., Ali, A. S., Zhang, M.-Z., Zhang, W.-H., & Gu, Y.-C. (2022). Discovery of Novel Pimprinine and Streptochlorin Derivatives as Potential Antifungal Agents. *Marine Drugs*, 20(12), 740. DOI: 10.3390/md20120740 PMID: 36547887

Lou, Y., Buchanan, A. M., Chen, S., Ford, S. L., Gould, E., Margolis, D., Spreen, W. R., & Patel, P. (2016). Effect of Cabotegravir on Cardiac Repolarization in Healthy Subjects. *Clinical Pharmacology in Drug Development*, 5(6), 509–516. DOI: 10.1002/cpdd.272 PMID: 27162089

Lunn, G. (1987). Reduction of heterocycles with nickel-aluminum alloy. *The Journal of Organic Chemistry*, 52(6), 1043–1046. DOI: 10.1021/jo00382a013

Maddock, J., Pattenden, G., & Wight, P. G. (1993). Stereochemistry of ulapualides, a new family of tris-oxazole-containing macrolide ionophores from marine nudibranchs. A molecular mechanics study. *Journal of Computer-Aided Molecular Design*, 7(5), 573–586. DOI: 10.1007/BF00124363 PMID: 8294947

Mahal, K., Biersack, B., & Schobert, R. (2013). New oxazole-bridged combretastatin A-4 analogs as potential vascular-disrupting agents. *International Journal of Clinical Pharmacology and Therapeutics*, 51(01), 41–43. DOI: 10.5414/CPP51041 PMID: 23259996

Makarov, G. I., & Reshetnikova, R. V. (2021). Investigation of radezolid interaction with non-canonical chloramphenicol binding site by molecular dynamics simulations. *Journal of Molecular Graphics & Modelling*, 105, 107902. DOI: 10.1016/j.jmgm.2021.107902 PMID: 33798835

Mathew, J. E., Divya, G., Vachala, S. D., Mathew, J. A., & Jeyaprakash, R. S. (2013). Synthesis and characterization of novel 2,4-diphenyloxazole derivatives and evaluation of their in vitro antioxidant and anticancer activity. *Journal of Pharmacy Research*, 6(1), 210–213. DOI: 10.1016/j.jopr.2012.12.001

Matsufuji, T., Ikeda, M., Naito, A., Hirouchi, M., Kanda, S., Izumi, M., Harada, J., & Shinozuka, T. (2013). Arylpiperazines as fatty acid transport protein 1 (FATP1) inhibitors with improved potency and pharmacokinetic properties. *Bioorganic & Medicinal Chemistry Letters*, 23(9), 2560–2565. DOI: 10.1016/j.bmcl.2013.02.116 PMID: 23528296

Mayer, J. C. P., Sauer, A. C., Iglesias, B. A., Acunha, T. V., Back, D. F., Rodrigues, O. E. D., & Dornelles, L. (2017). Ferrocenylethenyl-substituted 1,3,4-oxadiazolyl-1,2,4-oxadiazoles: Synthesis, characterization and DNA-binding assays. *Journal of Organometallic Chemistry*, 841, 1–11. DOI: 10.1016/j.jorganchem.2017.04.014

McMillen, B. A., Williams, H. L., Lehmann, H., & Shepard, P. D. (1992). On central muscle relaxants, strychnine-insensitive glycine receptors, and two old drugs: Zoxazolamine and HA-966. *Journal of Neural Transmission (Vienna, Austria)*, 89(1–2), 11–25. DOI: 10.1007/BF01245348 PMID: 1329854

Min, J., Lee, J. W., Ahn, Y.-H., & Chang, Y.-T. (2007). Combinatorial Dapoxyl Dye Library and its Application to Site Selective Probe for Human Serum Albumin. *Journal of Combinatorial Chemistry*, 9(6), 1079–1083. DOI: 10.1021/cc0700546 PMID: 17902630

Montastruc, J.-L., Macquin-Mavier, I., Tran, M.-A., Damase-Michel, C., Koenig-Berard, E., & Valet, P. (1989). Recent advances in the pharmacology of rilmenidine. *The American Journal of Medicine*, 87(3), S14–S17. DOI: 10.1016/0002-9343(89)90499-3 PMID: 2571291

Moody, C. J., & Bagley, M. C. (1998). The first synthesis of promothiocin A. *Chemical Communications (Cambridge)*, 18(18), 2049–2050. DOI: 10.1039/a805762a

Moraski, G. C., Chang, M., Villegas-Estrada, A., Franzblau, S. G., Möllmann, U., & Miller, M. J. (2010). Structure-activity relationship of new anti-tuberculosis agents derived from oxazoline and oxazole benzyl esters. *European Journal of Medicinal Chemistry*, 45(5), 1703–1716. DOI: 10.1016/j.ejmech.2009.12.074 PMID: 20116900

Musonda, C. C., Little, S., Yardley, V., & Chibale, K. (2007). Application of multicomponent reactions to antimalarial drug discovery. Part 3: Discovery of aminoxazole 4-aminoquinolines with potent antiplasmodial activity in vitro. *Bioorganic & Medicinal Chemistry Letters*, 17(17), 4733–4736. DOI: 10.1016/j.bmcl.2007.06.070 PMID: 17644333

Nagatsu, A., Kajitani, H., & Sakakibara, J. (1995). Muscoride A: A new oxazole peptide alkaloid from freshwater cyanobacterium Nostoc muscorum. *Tetrahedron Letters*, 36(23), 4097–4100. DOI: 10.1016/0040-4039(95)00724-Q

Napier, C., Stewart, M., Melrose, H., Hopkins, B., McHarg, A., & Wallis, R. (1999). Characterization of the 5-HT receptor binding profile of eletriptan and kinetics of eletriptan binding at human 5-HT1B and 5-HT1D receptors. *European Journal of Pharmacology*, 368(2–3), 259–268. DOI: 10.1016/S0014-2999(99)00026-6 PMID: 10193663

Ohnmacht, S. A., Mamone, P., Culshaw, A. J., & Greaney, M. F. (2008). Direct arylations on water: Synthesis of 2,5-disubstituted oxazoles balsoxin and texaline. *Chemical Communications*, 10(10), 1241. DOI: 10.1039/b719466h PMID: 18309430

Oldenziel, O. H., Van Leusen, D., & Van Leusen, A. M. (1977). Chemistry of sulfonyl methyl isocyanides. 13. A general one-step synthesis of nitriles from ketones using tosylmethyl isocyanide. Introduction of a one-carbon unit. *The Journal of Organic Chemistry*, 42(19), 3114–3118. DOI: 10.1021/jo00439a002

Panek, J. S., & Liu, P. (2000). Total Synthesis of the Actin-Depolymerizing Agent (−)-Mycalolide A: Application of Chiral Silane-Based Bond Construction Methodology. *Journal of the American Chemical Society*, 122(45), 11090–11097. DOI: 10.1021/ja002377a

Parveen, M., Ali, A., Ahmed, S., Malla, A. M., Alam, M., Pereira Silva, P. S., Silva, M. R., & Lee, D.-U. (2013). Synthesis, bioassay, crystal structure, and ab initio studies of Erlenmeyer azlactones. *Spectrochimica Acta. Part A: Molecular and Biomolecular Spectroscopy*, 104, 538–545. DOI: 10.1016/j.saa.2012.11.054 PMID: 23314102

Perner, R. J., Koenig, J. R., DiDomenico, S., Gomtsyan, A., Schmidt, R. G., Lee, C.-H., Hsu, M. C., McDonald, H. A., Gauvin, D. M., Joshi, S., Turner, T. M., Reilly, R. M., Kym, P. R., & Kort, M. E. (2010a). Synthesis and biological evaluation of 5-substituted and 4,5-disubstituted-2-arylamino oxazole TRPV1 antagonists. *Bioorganic & Medicinal Chemistry*, 18(13), 4821–4829. DOI: 10.1016/j.bmc.2010.04.099 PMID: 20570528

Perner, R. J., Koenig, J. R., DiDomenico, S., Gomtsyan, A., Schmidt, R. G., Lee, C.-H., Hsu, M. C., McDonald, H. A., Gauvin, D. M., Joshi, S., Turner, T. M., Reilly, R. M., Kym, P. R., & Kort, M. E. (2010b). Synthesis and biological evaluation of 5-substituted and 4,5-disubstituted-2-arylamino oxazole TRPV1 antagonists. *Bioorganic & Medicinal Chemistry*, 18(13), 4821–4829. DOI: 10.1016/j.bmc.2010.04.099 PMID: 20570528

Perrone, M. G., Vitale, P., Panella, A., Fortuna, C. G., & Scilimati, A. (2015). The general role of the amino and methylsulfonyl groups in selective cyclooxygenase(COX)-1 inhibition by 1,4-diaryl-1,2,3-triazoles and validation of a predictive pharmacometrics PLS model. *European Journal of Medicinal Chemistry*, 94, 252–264. DOI: 10.1016/j.ejmech.2015.02.049 PMID: 25768707

Pilch, D. S., Barbieri, C. M., Rzuczek, S. G., LaVoie, E. J., & Rice, J. E. (2008). Targeting human telomeric G-quadruplex DNA with oxazole-containing macrocyclic compounds. *Biochimie*, 90(8), 1233–1249. DOI: 10.1016/j.biochi.2008.03.011 PMID: 18439430

Pingali, H., Jain, M., Shah, S., Makadia, P., Zaware, P., Goel, A., Patel, M., Giri, S., Patel, H., & Patel, P. (2008). Design and synthesis of novel oxazole containing 1,3-Dioxane-2-carboxylic acid derivatives as PPAR α/γ dual agonists. *Bioorganic & Medicinal Chemistry*, 16(15), 7117–7127. DOI: 10.1016/j.bmc.2008.06.050 PMID: 18625559

Practical Antimicrobial Therapeutics. (2017). *Veterinary Medicine*. Elsevier., DOI: 10.1016/B978-0-7020-5246-0.00006-1

Putri, C. O., & Cahyana, A. H. (2022). *Synthesis and antioxidant activity screening of thiazole and oxazole derivative compounds*. 050004. DOI: 10.1063/5.0072502

Ramirez, R. L.III, Pienkos, S. M., De Jesus Perez, V., & Zamanian, R. T. (2021). Pulmonary Arterial Hypertension Secondary to Drugs and Toxins. *Clinics in Chest Medicine*, 42(1), 19–38. DOI: 10.1016/j.ccm.2020.11.008 PMID: 33541612

Raval, P., Jain, M., Goswami, A., Basu, S., Gite, A., Godha, A., Pingali, H., Raval, S., Giri, S., Suthar, D., Shah, M., & Patel, P. (2011). Revisiting glitazars: Thiophene substituted oxazole containing α-ethoxy phenylpropanoic acid derivatives as highly potent PPARα/γ dual agonists devoid of adverse effects in rodents. *Bioorganic & Medicinal Chemistry Letters*, 21(10), 3103–3109. DOI: 10.1016/j.bmcl.2011.03.020 PMID: 21450468

Reddy, C. S., Rao, L. S., Devi, M. V., Kumar, G. R., & Nagaraj, A. (2010). Synthesis of some new 3-[5-(2-oxo-2H-3-chromenyl)-1,3-oxazol-2-yl]-1,3-thiazolan-4-ones as antimicrobials. *Chinese Chemical Letters*, 21(9), 1045–1048. DOI: 10.1016/j.cclet.2010.03.018

Regueiro-Ren, A., Xue, Q. M., Swidorski, J. J., Gong, Y.-F., Mathew, M., Parker, D. D., Yang, Z., Eggers, B., D'Arienzo, C., Sun, Y., Malinowski, J., Gao, Q., Wu, D., Langley, D. R., Colonno, R. J., Chien, C., Grasela, D. M., Zheng, M., Lin, P.-F., & Kadow, J. F. (2013). Inhibitors of Human Immunodeficiency Virus Type 1 (HIV-1) Attachment. 12. Structure–Activity Relationships Associated with 4-Fluoro-6-azaindole Derivatives Leading to the Identification of 1-(4-Benzoylpiperazin-1-yl)-2-(4-fluoro-7-[1,2,3]triazol-1-yl-1*H*-pyrrolo[2,3-*c*]pyridin-3-yl)ethane-1,2-dione (BMS-585248). *Journal of Medicinal Chemistry*, 56(4), 1656–1669. DOI: 10.1021/jm3016377 PMID: 23360431

Ritson, D. J., Spiteri, C., & Moses, J. E. (2011). A Silver-Mediated One-Step Synthesis of Oxazoles. *The Journal of Organic Chemistry*, 76(9), 3519–3522. DOI: 10.1021/jo1025332 PMID: 21438573

Robinson-Gabriel Oxazole Synthesis. (Robinson-Gabriel Cyclodehydration, Robinson-Gabriel Synthesis). (2010). In Z. Wang, *Comprehensive Organic Name Reactions and Reagents* (1st ed., pp. 2410–2413). Wiley. DOI: 10.1002/9780470638859.conrr543

Roger, C., Roberts, J. A., & Muller, L. (2018). Clinical Pharmacokinetics and Pharmacodynamics of Oxazolidinones. *Clinical Pharmacokinetics*, 57(5), 559–575. DOI: 10.1007/s40262-017-0601-x PMID: 29063519

Rusu, A., Moga, I.-M., Uncu, L., & Hancu, G. (2023). The Role of Five-Membered Heterocycles in the Molecular Structure of Antibacterial Drugs Used in Therapy. *Pharmaceutics*, 15(11), 2554. DOI: 10.3390/pharmaceutics15112554 PMID: 38004534

Rymbai, E. M., Chakraborty, A., Choudhury, R., Verma, N., & De, B. (2019). *Review on Chemistry and Therapeutic Activity of the Derivatives of Furan and Oxazole: The Oxygen Containing Heterocycles. 11*(1). https://www.derpharmachemica.com/pharma-chemica/review-on-chemistry-and-therapeutic-activity-of-the-derivatives-of-furan-and-oxazole-the-oxygen-containing-heterocycles-15521.html

Ryu, C.-K., Lee, R.-Y., Kim, N. Y., Kim, Y. H., & Song, A. L. (2009). Synthesis and antifungal activity of benzo[d]oxazole-4,7-diones. *Bioorganic & Medicinal Chemistry Letters*, 19(20), 5924–5926. DOI: 10.1016/j.bmcl.2009.08.062 PMID: 19733068

Sahu, P. K., Annapurna, M. M., & Kumar, S. D. (2011). Development and Validation of Stability Indicating RP-HPLC Method for the Determination of Metaxalone in Bulk and its Pharmaceutical Formulations. *Journal of Chemistry*, 8(S1). Advance online publication. DOI: 10.1155/2011/645710

Sakai, Y. (1983). Comparative study on the effects of haloxazolam and estazolam, new sleep-inducing drugs, on the α- and γ-motor systems. *Japanese Journal of Pharmacology*, 33(5), 1017–1025. DOI: 10.1016/S0021-5198(19)52447-7 PMID: 6139494

Searle, P. A., & Molinski, T. F. (1995). Phorboxazoles A and B: Potent cytostatic macrolides from marine sponge Phorbas species. *Journal of the American Chemical Society*, 117(31), 8126–8131. DOI: 10.1021/ja00136a009

Semenyuta, I., Kovalishyn, V., Tanchuk, V., Pilyo, S., Zyabrev, V., Blagodatnyy, V., Trokhimenko, O., Brovarets, V., & Metelytsia, L. (2016). 1,3-Oxazole derivatives as potential anticancer agents: Computer modeling and experimental study. *Computational Biology and Chemistry*, 65, 8–15. DOI: 10.1016/j.compbiolchem.2016.09.012 PMID: 27684433

Shorr, A. F., Lodise, T. P., Corey, G. R., De Anda, C., Fang, E., Das, A. F., & Prokocimer, P. (2015). Analysis of Phase 3 ESTABLISH Trials of Tedizolid versus Linezolid in Acute Bacterial Skin and Skin Structure Infections. *Antimicrobial Agents and Chemotherapy*, 59(2), 864–871. DOI: 10.1128/AAC.03688-14 PMID: 25421472

Singh, I., Kaur, H., Kumar, S., Kumar, A., Lata, S., & Kumar, A. (2010). *Synthesis of new coumarin derivatives as antibacterial agents. 2*(3).

Song, M.-X., & Deng, X.-Q. (2018). Recent developments on triazole nucleus in anticonvulsant compounds: A review. *Journal of Enzyme Inhibition and Medicinal Chemistry*, 33(1), 453–478. DOI: 10.1080/14756366.2017.1423068 PMID: 29383949

Spiteri, C., Ritson, D. J., Awaad, A., & Moses, J. E. (2011). Silver mediated one-step synthesis of oxazoles from α-haloketones. *Journal of Saudi Chemical Society*, 15(4), 375–378. DOI: 10.1016/j.jscs.2011.06.021

Sysak, A., & Obmińska-Mrukowicz, B. (2017). The isoxazole ring is a useful scaffold in the search for new therapeutic agents. *European Journal of Medicinal Chemistry*, 137, 292–309. DOI: 10.1016/j.ejmech.2017.06.002 PMID: 28605676

Taile, V., Hatzade, K., Gaidhane, P., & Ingle, V. (2009). Synthesis and Biological Activity of 4-(4-Hydroxybenzylidene)-2- (substituted styryl) oxazole-5-ones and Their o-glucosides. *Turkish Journal of Chemistry*. Advance online publication. DOI: 10.3906/kim-0712-12

Takahashi, S., Hashimoto, R., Hamano, K., Suzuki, T., & Nakagawa, A. (1996). Melanoxazal, a New Melanin Biosynthesis Inhibitor Discovered by Using the Larval Haemolymph of the Silkworm, Bombyx mori. Production, Isolation, Structural Elucidation, and Biological Properties. *The Journal of Antibiotics*, 49(6), 513–518. DOI: 10.7164/antibiotics.49.513 PMID: 8698632

Takeuchi, H., Yanagida, S., Ozaki, T., Hagiwara, S., & Eguchi, S. (1989). Synthesis of novel carbo- and heteropolycycles. 12. A new versatile synthesis of oxazoles by intramolecular aza-Wittig reaction. *The Journal of Organic Chemistry*, 54(2), 431–434. DOI: 10.1021/jo00263a033

Tang, J. S., & Verkade, J. G. (1996). Chiral Auxiliary-Bearing Isocyanides as Synthons: Synthesis of Strongly Fluorescent (+)-5-(3,4-Dimethoxyphenyl)-4-[[*N* - [(4 *S*)-2-oxo-4-(phenylmethyl)-2-oxazolidinyl]]carbonyl]oxazole and Its Enantiomer. *The Journal of Organic Chemistry*, 61(25), 8750–8754. DOI: 10.1021/jo961554g PMID: 11667848

Tavares, F., Lawson, J. P., & Meyers, A. I. (1996). Total Synthesis of Streptogramin Antibiotics. (−)-Madumycin II. *Journal of the American Chemical Society*, 118(13), 3303–3304. DOI: 10.1021/ja954312r

Thalhammer, A., Mecinović, J., & Schofield, C. J. (2009). Triflic anhydride-mediated synthesis of oxazoles. *Tetrahedron Letters*, 50(9), 1045–1047. DOI: 10.1016/j.tetlet.2008.12.080

Timperio, A. M., Kuiper, H. A., & Zolla, L. (2003). Identification of a furazolidone metabolite responsible for the inhibition of amino oxidases. *Xenobiotica*, 33(2), 153–167. DOI: 10.1080/0049825021000038459 PMID: 12623758

Turchi, I. J. (1986). Oxazoles. In Turchi, I. J. (Ed.), *Chemistry of Heterocyclic Compounds: A Series Of Monographs* (1st ed., Vol. 45, pp. 1–341). Wiley., DOI: 10.1002/9780470187289.ch1

Turchi, I. J., & Dewar, M. J. S. (1975). Chemistry of oxazoles. *Chemical Reviews*, 75(4), 389–437. DOI: 10.1021/cr60296a002

Verma, N., Arora, V., Awasthi, R., Chan, Y., Jha, N. K., Thapa, K., Jawaid, T., Kamal, M., Gupta, G., Liu, G., Paudel, K. R., Hansbro, P. M., George Oliver, B. G., Singh, S. K., Chellappan, D. K., Dureja, H., & Dua, K. (2022). Recent developments, challenges, and prospects in advanced drug delivery systems in the management of tuberculosis. *Journal of Drug Delivery Science and Technology*, 75, 103690. DOI: 10.1016/j.jddst.2022.103690

Wales, S. M., Hammer, K. A., Somphol, K., Kemker, I., Schröder, D. C., Tague, A. J., Brkic, Z., King, A. M., Lyras, D., Riley, T. V., Bremner, J. B., Keller, P. A., & Pyne, S. G. (2015). Synthesis and antimicrobial activity of binaphthyl-based, functionalized oxazole and thiazole peptidomimetics. *Organic & Biomolecular Chemistry*, 13(44), 10813–10824. DOI: 10.1039/C5OB01638J PMID: 26349598

Wang, S., Song, H., Cai, Q., & Chen, J. (2024). Research progress of oxazole derivatives in the discovery of agricultural chemicals. *Journal of Heterocyclic Chemistry*, 61(1), 71–85. DOI: 10.1002/jhet.4749

Wang, T., Yang, Z., Zhang, Z., Gong, Y.-F., Riccardi, K. A., Lin, P.-F., Parker, D. D., Rahematpura, S., Mathew, M., Zheng, M., Meanwell, N. A., Kadow, J. F., & Bender, J. A. (2013). Inhibitors of HIV-1 attachment. Part 10. The discovery and structure-activity relationships of 4-azaindole cores. *Bioorganic & Medicinal Chemistry Letters*, 23(1), 213–217. DOI: 10.1016/j.bmcl.2012.10.120 PMID: 23200254

Welstead, W. J.Jr, Helsley, G. C., Taylor, C. R., Turnbull, L. B., Da Vanzo, J. P., Funderburk, W. H., & Alphin, R. S. (1973). 5-(2-Aminoethyl)-2-oxazolidinones with central nervous system depressant and antiinflammatory activity. *Journal of Medicinal Chemistry*, 16(10), 1129–1132. DOI: 10.1021/jm00268a013 PMID: 4795993

Weyrauch, J. P., Hashmi, A. S. K., Schuster, A., Hengst, T., Schetter, S., Littmann, A., Rudolph, M., Hamzic, M., Visus, J., Rominger, F., Frey, W., & Bats, J. W. (2010). Cyclization of Propargylic Amides: Mild Access to Oxazole Derivatives. *Chemistry (Weinheim an der Bergstrasse, Germany)*, 16(3), 956–963. DOI: 10.1002/chem.200902472 PMID: 19938017

Wildburger, N. C., Lin-Ye, A., Baird, M. A., Lei, D., & Bao, J. (2009). Neuroprotective effects of blockers for T-type calcium channels. *Molecular Neurodegeneration*, 4(1), 44. DOI: 10.1186/1750-1326-4-44 PMID: 19863782

Wiley, R. H. (1945). The Chemistry of the Oxazoles. *Chemical Reviews*, 37(3), 401–442. DOI: 10.1021/cr60118a002 PMID: 21013426

Wipf, P., Fletcher, J. M., & Scarone, L. (2005). Microwave-promoted oxazole synthesis: Cyclocondensation cascade of oximes and acyl chlorides. *Tetrahedron Letters*, 46(33), 5463–5466. DOI: 10.1016/j.tetlet.2005.06.063

Wookey, A., Turner, P. J., Greenhalgh, J. M., Eastwood, M., Clarke, J., & Sefton, C. (2004). AZD2563, a novel oxazolidinone: Definition of the antibacterial spectrum, assessment of bactericidal potential and the impact of miscellaneous factors on activity in vitro. *Clinical Microbiology and Infection*, 10(3), 247–254. DOI: 10.1111/j.1198-743X.2004.00770.x PMID: 15008947

Xie, Y., Wang, B., Liu, J., Zhou, J., Ma, J., Huang, H., & Ju, J. (2012). Identification of the Biosynthetic Gene Cluster and Regulatory Cascade for the Synergistic Antibacterial Antibiotics Griseoviridin and Viridogrisein in *Streptomyces griseoviridis*. *ChemBioChem*, 13(18), 2745–2757. DOI: 10.1002/cbic.201200584 PMID: 23161816

Yamada, K., Kamimura, N., & Kunishima, M. (2017). Development of a method for the synthesis of 2,4,5-trisubstituted oxazoles composed of carboxylic acid, amino acid, and boronic acid. *Beilstein Journal of Organic Chemistry*, 13, 1478–1485. DOI: 10.3762/bjoc.13.146 PMID: 28845191

Yang, J., Zhou, S., Ji, L., Zhang, C., Yu, S., Li, Z., & Meng, X. (2014). Synthesis and structure-activity relationship of 4-azaheterocycle benzenesulfonamide derivatives as new microtubule-targeting agents. *Bioorganic & Medicinal Chemistry Letters*, 24(21), 5055–5058. DOI: 10.1016/j.bmcl.2014.09.016 PMID: 25278233

Ye, F., Zhai, Y., Kang, T., Wu, S.-L., Li, J.-J., Gao, S., Zhao, L.-X., & Fu, Y. (2019). Rational design, synthesis, and structure-activity relationship of novel substituted oxazole isoxazole carboxamides as herbicide safener. *Pesticide Biochemistry and Physiology*, 157, 60–68. DOI: 10.1016/j.pestbp.2019.03.003 PMID: 31153478

Yin, W., Cui, H., Jiang, H., Zhang, Y., Liu, L., Wu, T., Sun, Y., Zhao, L., Su, X., Zhao, D., & Cheng, M. (2022). Broadening antifungal spectrum and improving metabolic stability based on a scaffold strategy: Design, synthesis, and evaluation of novel 4-phenyl-4,5-dihydrooxazole derivatives as potent fungistatic and fungicidal reagents. *European Journal of Medicinal Chemistry*, 227, 113955. DOI: 10.1016/j.ejmech.2021.113955 PMID: 34749201

Zeinali, N., Oluwoye, I., Altarawneh, M., & Dlugogorski, B. Z. (2020). Kinetics of Photo-Oxidation of Oxazole and its Substituents by Singlet Oxygen. *Scientific Reports*, 10(1), 3668. DOI: 10.1038/s41598-020-59889-1 PMID: 32111853

Zhang, J., Li, A.-R., Yu, M., Wang, Y., Zhu, J., Kayser, F., Medina, J. C., Siegler, K., Conn, M., Shan, B., Grillo, M. P., Eksterowicz, J., Coward, P., & Liu, J. J. (2013). Discovery and optimization of arylsulfonyl 3-(pyridin-2-yloxy)anilines as novel GPR119 agonists. *Bioorganic & Medicinal Chemistry Letters*, 23(12), 3609–3613. DOI: 10.1016/j.bmcl.2013.04.014 PMID: 23648181

Zhao, Q., Liu, S., Li, Y., & Wang, Q. (2009). Design, Synthesis, and Biological Activities of Novel 2-Cyanoacrylates Containing Oxazole, Oxadiazole, or Quinoline Moieties. *Journal of Agricultural and Food Chemistry*, 57(7), 2849–2855. DOI: 10.1021/jf803632t PMID: 19271709

Zheng, Y., Li, X., Ren, C., Zhang-Negrerie, D., Du, Y., & Zhao, K. (2012). Synthesis of Oxazoles from Enamides via Phenyliodine Diacetate-Mediated Intramolecular Oxidative Cyclization. *The Journal of Organic Chemistry*, 77(22), 10353–10361. DOI: 10.1021/jo302073e PMID: 23106159

Chapter 2
Bioactive Pyrazoles:
Structure, Function, and Pharmaceutical Potential

Aditi Boruah
Gauhati University, India

Ariful Islam
https://orcid.org/0009-0009-2370-1292
Gauhati University, India

Pranjal K. Baruah
https://orcid.org/0000-0002-5829-1440
Gauhati University, India

ABSTRACT

This chapter discusses about bioactive pyrazoles, which have garnered substantial attention in medicinal chemistry and drug development. Pyrazoles are five-membered heterocyclic compounds characterized by a nitrogen-rich structure, imparting them with unique physicochemical and biological and making them useful for treating various diseases. The chapter explores the synthesis, structural diversity, and functional modifications of pyrazoles that enhance their bioactivity. Detailed discussions highlight the mechanisms of action and therapeutic potential of pyrazole derivatives in various pharmacological areas, including anti-inflammatory, anticancer, antimicrobial, antiviral, neuroprotective, and other applications. The chapter also looks at the latest research and advancements in pyrazole chemistry, including new ways to make them and studies on how their structure affects their activity. By exploring the many uses of bioactive pyrazoles, this chapter shows their importance in contemporary medicinal chemistry in developing new and better medicines.

DOI: 10.4018/979-8-3693-7267-8.ch002

1. INTRODUCTION

Heterocyclic compound chemistry is a complex subfield of organic chemistry, encompassing more than half of all known organic compounds. Heterocycles are a diverse and significant class of molecules with a wide range of reactivity, stability, and applications across chemical, biological, and physical domains (Eftekhari-Sis et al., 2013). This diversity makes them intriguing for their synthetic techniques, theoretical implications, and extensive physiological and industrial uses. Heterocycles are abundant in nature, found in natural products like antibiotics, vitamins, and alkaloids, as well as in agrochemicals, medicines, and dyes. They play a crucial role in metabolism and are the fundamental structures of many biologically active molecules, especially nitrogen-containing heterocycles (Ansari et al., 2017; Ju & Varma, 2005).

Pyrazoles, a key class of heteroaromatic ring systems, are widely used in the drug industry. Their fundamental structure consists of a five-membered ring with two neighboring nitrogen atoms and three carbon atoms, giving them an aromatic system (Nisa & Astana, 2019). Due to its electronegativity, the nitrogen atom attracts ring electrons, making the C(3) and C(5) atoms electropositive and prone to nucleophilic reactions. In 1,2-azoles, the pyridine nitrogen and C(4) atoms influence π-electron distribution, altering the charge on C(3) and C(5) depending on the heterocycle (Behr et al., 1967).

Figure 1. Structure of Pyrazole

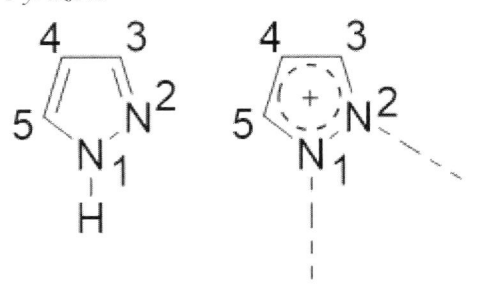

Pyrazole, capable of forming intermolecular hydrogen bonds, has boiling, and melting points of 187–188 °C and 70 °C, respectively. With a tendency towards basic character rather than acidic character, Pyrazole exhibits amphoterism. And hence, in addition to forming sodium and potassium salts, it easily hydrolyzes salts in strong acids (Marinescu & Zalaru, 2021). Pyrazoles, also referred to as azoles, are ligands for several lewis acids (Mukherjee, 2000). Due to the rapid interconversion, pyrazole exists in two identical and inseparable tautomeric forms. In fact, the same anion is formed from both uncharged forms of pyrazole upon protonation and when reacting with bases (Kost & Grandberg, 1996).

The pyrazole nucleus, which contains a five-membered ring with two adjacent nitrogen atoms, has the ability to form cyclic dimers and trimers through intermolecular hydrogen bonding interactions. In these associations, hydrogen atoms from one pyrazole molecule form bonds with the nitrogen atoms of adjacent pyrazole molecules. This leads to the creation of cyclic structures, which are known as dimer and trimer (Koca et al., 2013).

Figure 2. Cyclic dimer and trimer

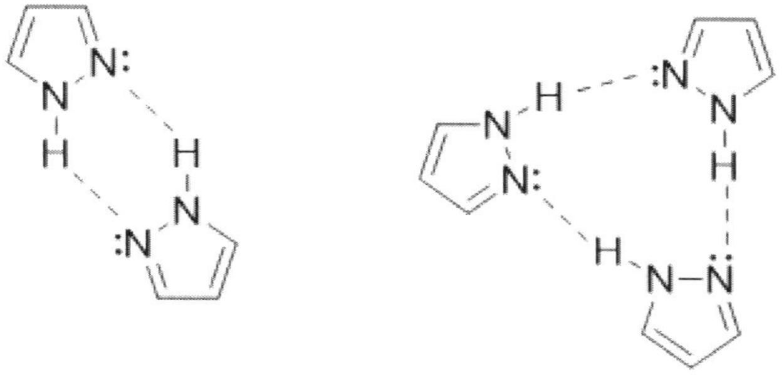

A variety of biological properties are displayed by pyrazole derivatives, including anticancer, (Koca et al., 2013) antimicrobial, analgesic, antipyretic, anti-inflammatory, antidepressant (Gökhan-Kelekçi et al., 2009; Fustero et al., 2011) antidiabetic, angiotensin-converting enzyme inhibition (Alam et al., 2015), sodium channel blocker (Tyagarajan et al., 2010) anti-AIDS (K & Ganguly, 2016). Furthermore, it has been noted that pyrazole prodrugs have the ability to donate nitric oxide (Chowdhury et al., 2010; Abdellatif et al., 2010) and transport bile acid (Bhat et al., 2005). In addition to the pharmaceutical industry, pyrazole also have application in agrochemical industry (Ju & Varma, 2005). Moreover, pyrazole derivatives have interesting photophysical properties, making them suitable for applications in materials science (Deb et al., 2016).

Due to these multifaceted advantages, researchers are increasingly interested in synthesizing new pyrazole derivatives to explore their potential in various fields. By studying the structural modifications and properties of these derivatives, scientists aim to enhance their efficacy and develop new applications, whether in medicine, agriculture, or materials science. The ongoing research into pyrazole derivatives continues to contribute to advancements across multiple disciplines, highlighting their versatility and importance in scientific innovation.

2. VARIOUS METHODS OF SYNTHESIS OF PYRAZOLE AND ITS DERIVATIVES

Various methods for synthesizing pyrazoles and their derivatives

2.1 Cyclocondensation: Reactions for the Synthesis of Pyrazole and its Derivatives

Cyclocondensation, a powerful technique widely utilized by synthetic chemists, plays an essential role as a foundational method for creating the adaptable pyrazole scaffold. This synthetic process is characterized by the intricate condensation reaction between hydrazine derivatives and various 1,3-dicarbonyl compounds, including but not limited to β-diketones, ketoesters, or diesters. The outcome of this chemical interplay yields the distinct five-membered heterocyclic ring structure that is pivotal in the molecular design of diverse compounds. By merging these key building blocks, cyclocondensation serves as a pivotal step in the creation of pyrazole-based molecules with tailored properties and functions, making it a cornerstone in the realm of chemical synthesis. This strategic conjugation of functional groups through the condensation process enables chemists to precisely manipulate the structure and properties of the resulting pyrazole derivatives, unlocking a wide array of potential applications in medicinal chemistry, materials science, and beyond. The ability to selectively construct these pyrazole scaffolds using cyclocondensation provides chemists with a versatile tool for designing novel compounds with specific structural motifs, enhancing the scope of synthetic strategies available in modern chemical research (Nandurkar et al., 2023).

The reaction typically commences with a nucleophilic attack of the hydrazine on the carbonyl carbon of the 1,3-dicarbonyl compound, leading to the formation of a hydrazone intermediate. Subsequent cyclization and elimination of water yield the desired pyrazole product. The electronic properties of the substituents on both the hydrazine and the 1,3-dicarbonyl compound significantly influence the reaction outcome, including regioselectivity and product yield (Castillo et al., 2018).

A cyclocondensation reaction is the main method used to produce substituted pyrazoles. This reaction involves combining a hydrazine, which acts as a bidentate nucleophile, with a carbon unit that includes 1,3-dicarbonyl (I), α,β-unsaturated carbonyl compounds (II, III), or β-enaminones or related compounds (IV).

Figure 3. Examples of α,β-unsaturated carbonyl compounds.

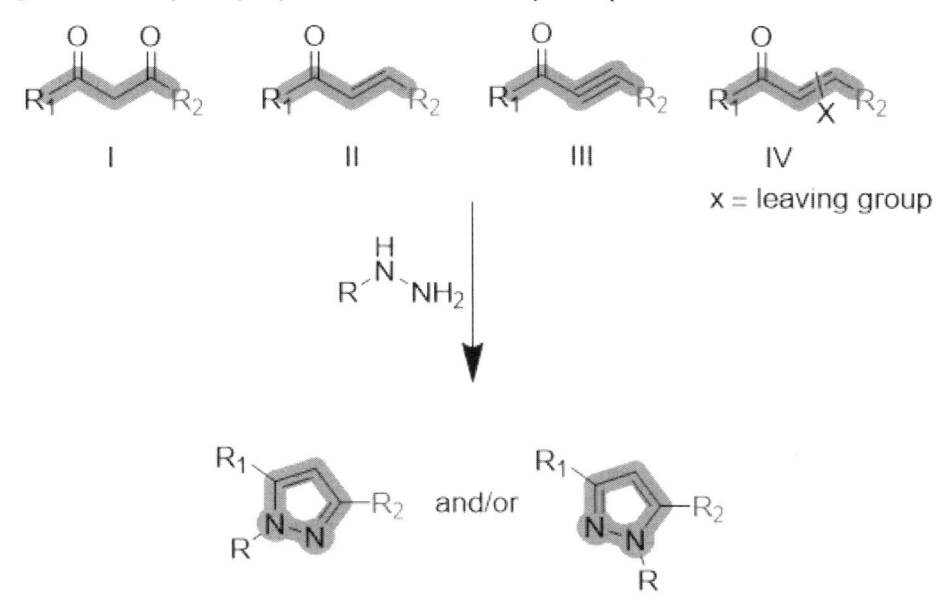

2.1.1 Pyrazoles from 1,3-Diketones

The synthesis of polysubstituted pyrazoles is a widely used method that involves the combination of hydrazine compounds with 1,3 dicarbonyl substances in a cyclo-condensation process. This technique is renowned for its simplicity and efficiency in producing diverse pyrazole derivatives. The credit for the discovery of this process goes to Knorr and his team, who first reported the production of modified pyrazoles back in 1883. Through their experiment, they successfully reacted hydrazine derivatives with β diketones, resulting in the formation of two regioisomers. These regioisomers possess distinct heteroatom placements, specifically adjacent to either the R_1 or R_3 substituents, further enriching the versatility and applicability of this reaction (Hassani et al., 2023).

Figure 4. Synthesis of polysubstituted pyrazoles using hydrazine derivatives and 1,3-dicarbonyl compounds.

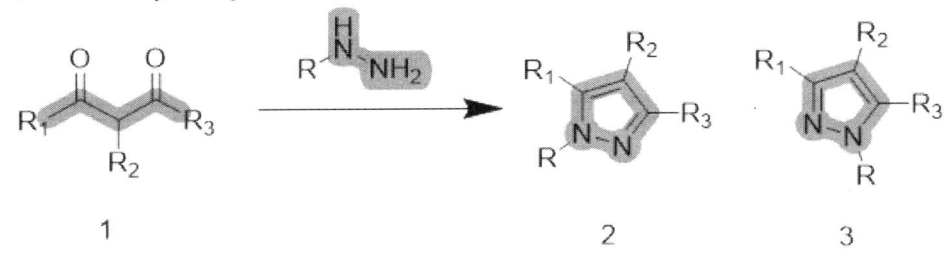

Konwar et. al. (Konwar et al., 2019) has advanced a sustainable method: synthesizing pyrazole derivatives through the use of lithium perchlorate as a Lewis acid catalyst. Firstly, they performed some experiments to the model systems based on acetylacetone and 2,4-dinitriphenylhydrazine that led them to the ideal reaction conditions. Initially, they did the reaction without any catalyst but the solvents were used alone the experiment was not successful hence debugging the necessity of a catalyst. The fabricated route of the reaction between hydrazines and 1,3-diketones using ethylene glycol was also reported. The reaction mechanism led to the production of pyrazoles, yielding 70% to 95% at room temperature.

Figure 5. Synthesis of 1,3,5-substituted pyrazoles from substituted acetylacetone.

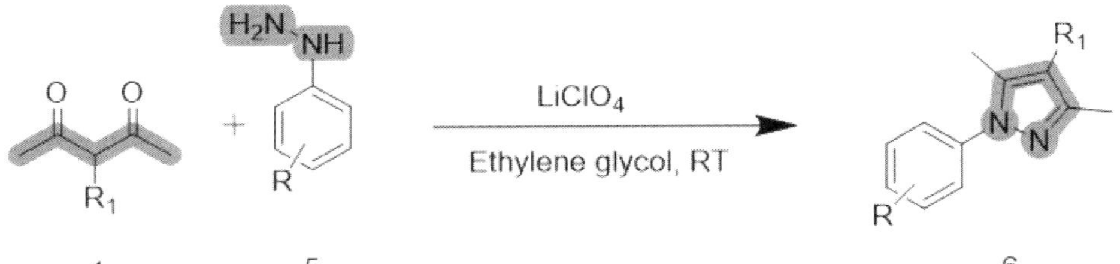

2.2 Cycloaddition Reactions: A Cornerstone for Pyrazole Synthesis

Cycloaddition is a pericyclic process in which "two or more unsaturated molecules (or parts of the same molecule) combine to form a cyclic adduct in which there is a net reduction in bond multiplicity." The subsequent reaction is a cyclization reaction. Many, but not all, cycloadditions are coordinated. Cycloadditions are commonly defined by the size of the participants&; backbones. This would transform the Diels-Alder reaction into a (4 + 2) cycloaddition, the 1,3dipolar cycloaddition into a (3 + 2) cycloaddition, and the cyclopropanation of a carbene with an alkene into a (2+1) cycloaddition. This reaction is a nonpolar addition process (Marichev & Doyle, 2019).

2.2.1 1,3-Dipolar Cycloaddition Reaction

The 1,3-dipolar cycloaddition reaction between alkynes and nitrile imines presents a highly effective and straightforward method for producing functionalized pyrazoles. The key to this reaction lies in the inherent reactivity of nitrile imines as 1,3-dipoles, which can be attributed to their unique electronic structure. Through their reaction with the electron-rich triple bonds found in alkynes, a [3+2] cycloaddition process rapidly forms the five-membered pyrazole ring in one simple step. Such an approach enables the direct incorporation of diverse chemical functionalities onto the pyrazole core, guided by the specific substituents present on both the alkyne and nitrile imine starting materials (Chandanshive, 2013).

Figure 6. 1,3-dipolar cycloaddition reaction between alkynes and nitrile imines

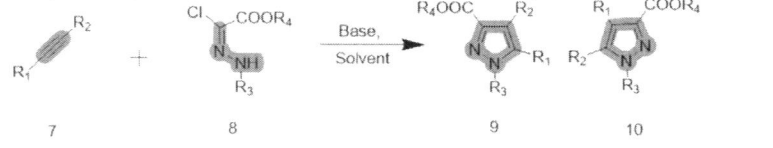

2.2.2 Silver-Mediated [3 + 2] Cycloaddition of Alkynes and N-Isocyanoiminotriphenylphosphorane

Pyrazoles are formed by a silver-mediated [3 + 2] cycloaddition of N-isocyanoiminotriphenylphosphorane, a "CNN" building block, to terminal alkynes. N-isocyanoiminotriphenylphosphorane is a solid isocyanide that is stable, safe, simple to handle, and has no odour. The reaction provides gentle conditions, a broad substrate range, and strong functional group tolerance (Yi et al., 2019).

Figure 7. Silver-Mediated [3 + 2] Cycloaddition of Alkynes and N-Isocyanoimino-triphenylphosphorane

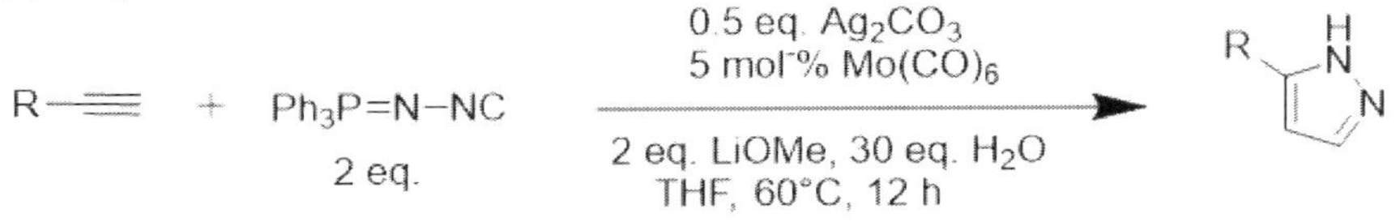

R: Aryl, alkyl

2.3 C-H Functionalization: A Modern Approach for Pyrazole Synthesis

C-H functionalization reactions represent a modern and powerful pyrazole ring construction strategy. Unlike traditional methods that require pre-functionalized substrates, C-H functionalization directly utilizes C-H bonds on starting materials, offering several advantages (Dalton et al., 2021; Rastogi et al., 2019).

2.3.1 Copper-Catalyzed Aerobic C(sp^2)–H Functionalization for Synthesis of Pyrazoles

A novel and efficient Cu-catalyzed direct aerobic oxidative C(sp^2)-H amination strategy has been developed for the expedient construction of pyrazoles. This innovative approach exploits the ubiquity of C-H bonds, circumventing the need for pre-functionalized starting materials and thus streamlining the synthetic process. The reaction proceeds under mild aerobic conditions, rendering it practical for large-scale applications. This methodology showcases a significant advancement in pyrazole synthesis, demonstrating a powerful combination of efficiency, atom economy, and substrate scope (Li et al., 2013).

Figure 8. Synthesis of pyrazoles via copper-catalyzed direct aerobic oxidative $C(sp^2)-H$ amination

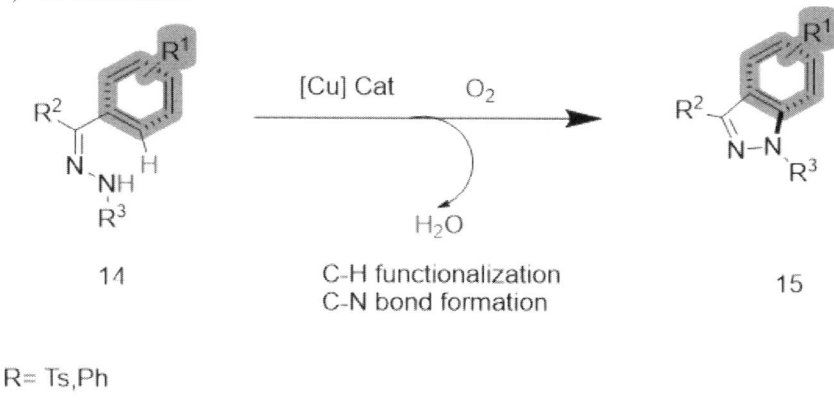

R= Ts,Ph

2.3.2 Ruthenium-Catalyzed Tandem C−H Fluoromethylation

The synthesis of fluorinated heterocycles, notably pyrazoles, has attracted considerable attention in scientific circles due to the intriguing characteristics that fluorine atoms bring to these compounds. The addition of fluorine can lead to significant modifications in their properties, enabling researchers to fine-tune biological activities, enhance lipophilicity, and improve the overall metabolic stability of potential drugs under development. Traditionally, creating fluorinated pyrazole structures has been a complex process, often requiring the use of multi-step procedures that rely on pre-functionalized starting materials. Despite the challenges posed by these conventional methods, recent advancements in C-H functionalization have emerged as promising alternatives, offering more efficient routes for synthesizing these important compounds. By leveraging the power of C-H activation techniques, chemists can now streamline the synthesis of fluorinated pyrazoles, potentially revolutionizing the way these compounds are produced and paving the way for innovative drug discovery strategies that capitalize on the unique properties introduced by fluorine atoms. The ongoing progress in this field suggests exciting opportunities for developing novel pharmaceutical agents with enhanced efficacy and improved pharmacokinetic profiles, ultimately contributing to advancements in the realm of medicinal chemistry and drug design.

This chapter describes a particularly efficient approach for the synthesis of 4-fluoropyrazoles through a double C-H fluoroalkylation reaction. This one-pot methodology utilizes readily available aldehyde-derived N-alkylhydrazones and tribromofluoromethane ($CFBr_3$) as the fluorinating agent. The reaction is catalyzed by $RuCl_2(PPh_3)_3$, a readily accessible ruthenium complex, leading to the desired fluorinated pyrazoles in excellent yields (Prieto et al., 2017).

Figure 9. Ru-Catalyzed Synthesis of 4-Fluoropyrazoles

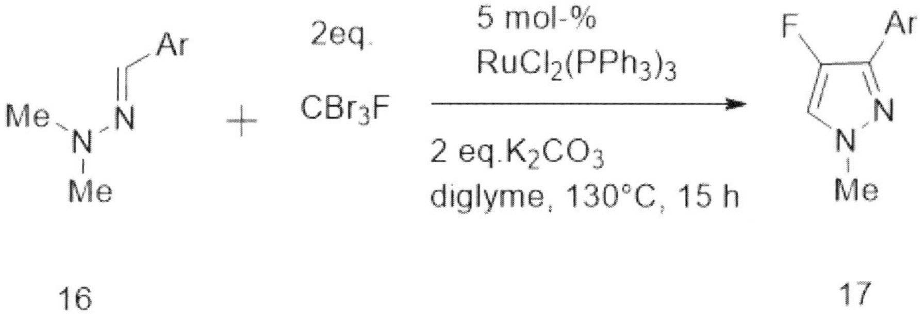

16   17

2.4 Multicomponent Reactions (MCRs)

Multi-component reactions (MCRs) are revolutionizing organic synthesis by offering a powerful and efficient approach to constructing complex molecules like pyrazoles in a single step. This translates to significant advantages. MCRs boast an impressive step economy, minimizing the number of reactions needed and streamlining the entire process. They also excel in atom economy, directly incorporating most of the starting materials into the final product, minimizing waste. Furthermore, MCRs are often simple to execute, requiring less reaction set-up and work-up compared to multi-step processes. This simplicity translates to cost and energy savings, making them an attractive strategy for both research labs and industrial production (Borpatra et al., 2019).

2.4.1 Three-Component Approach to the Synthesis of Polyfunctional Pyrazoles

The three-component reaction (3CR) offers a versatile and efficient strategy for constructing the pyrazole scaffold. This approach involves the condensation of three distinct building blocks in a single synthetic operation, leading to the formation of the desired heterocyclic product. Common components employed in pyrazole synthesis via 3CR include:

Hydrazine derivatives- These serve as the nitrogen source for the pyrazole ring formation.

1,3-Dicarbonyl compounds- β-diketones, ketoesters, and diesters are frequently used as the carbon backbone for the pyrazole ring.

Aldehydes or α,β-unsaturated carbonyl compounds- These components contribute to the diversity of the pyrazole product by introducing additional substituents

Three-component reactions offer significant advantages over traditional multi-step syntheses, including reduced reaction time, improved atom economy, and enhanced product diversity. This approach has gained popularity in recent years due to its potential for rapid library generation and drug discovery efforts (Sabet-Sarvestani et al., 2014).

A very effective easy-to-one-pot method for the preparation of varied pyrazoles is given here through this sentence. The process results from adding vinyl azide, an easy-to-find compound, to the desired aldehyde, along with tosylhydrazine under basic conditions. This method is characterized by the preferential formation of 3,4,5-trisubstituted pyrazoles but in other words, besides that it is a straightforward and high-yielding method. This reaction precisely exhibits a wide spectrum of functional groups as the substrate, which in turn makes the production of a series of pyrazole derivatives broad in scope (Zhang et al., 2013).

Figure 10. Synthesis of polyfunctional pyrazoles

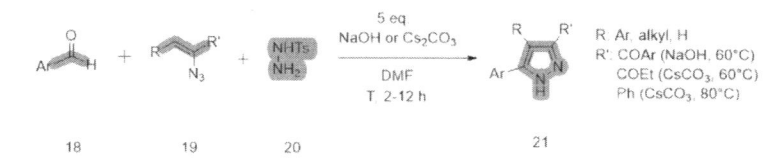

2.4.2 Synthesis of Four-Component Pyrano[2,3-c]pyrazole with Aromatic Aldehydes

Four-component reactions (MCRs) offer a powerful and efficient strategy for the rapid assembly of complex heterocyclic compounds, including pyrazoles. By merging multiple starting materials, such as hydrazines, aldehydes, active methylene compounds, and additional reactive components, in a single synthetic operation, MCRs significantly streamline the synthetic process and enhance product diversity.

This atom-economical approach minimizes waste generation and reduces reaction steps, making it an attractive option for both academic research and industrial applications. The ability to construct complex pyrazole scaffolds with various substitution patterns through MCRs has broadened the scope of potential applications in medicinal chemistry, materials science, and other fields.

A new four-compound response is the main process for making substituted and spiro-conjugated 6-amino-2H,4H-pyrano[2,3-c]pyrazol-5-carbonitriles. The whole operation needs the most easily accessible mild start reagents such as aromatic aldehydes or heterocyclic ketones, malononitriles, β-ketoesters, and hydrazine hydrate. The process becomes a no-brainer way to different 2,4-dihydropyrano[2,3-c]pyrazoles. Further, a modified one-step sequential protocol that can be made out of this method also becomes a possibility for the targeted synthesis of spiro[indoline-3,4'-pyrano[2,3-c]pyrazol]-2-ones. This type of processing multiplies the utility of this method by the construction of complex heterocyclic scaffolds (Litvinov et al., 2009).

Figure 11. Synthesis of Pyrano[2,3-c]pyrazole

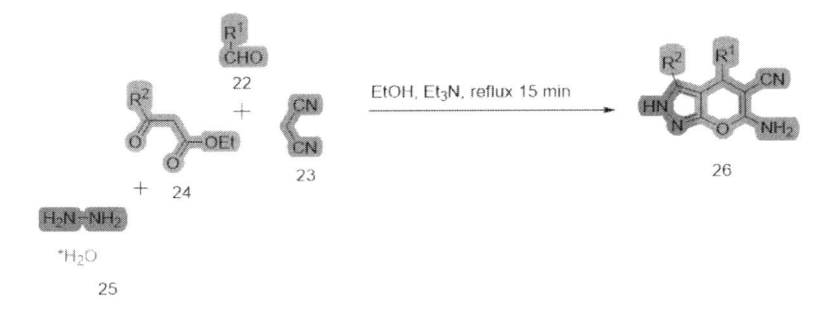

3. BIOACTIVE PYRAZOLE/ PYRAZOLE HAVING MEDICINAL ACTIVITY

3.1 Antiinflammatory

Over the past few decades, inflammation treatment has received a lot of attention. When it comes to treating pain and inflammation, nonsteroidal anti-inflammatory medications (NSAIDs) are the most often utilised therapeutic agents (JH, 1999). NSAIDs are also widely used for respiratory tract infections, soft tissue and oral cavity lesions, rheumatoid arthritis, and fever (Küçükgüzel & Şenkardeş, 2015). By blocking COX enzymes, which catalyse the transformation of arachidonic acid into prostaglandins, prostacyclins, and thromboxanes, NSAIDs achieve their pharmacological effects (Grosser, 2005). The cyclooxygenase enzymes COX-1, COX-2, and COX-3, which are in charge of generating both physiological and pathological prostaglandins, are inhibited by this class of anti-inflammatory drugs (Faria et al., 2017; Kohli et al., 2014). NSAIDs have been associated with a range of side effects, including stroke, gastrointestinal bleeding, and an increased cardiovascular risk in patients with a history of myocardial infarction (Kohli et al., 2014). Because of the connection between COX-2 and induced inflammation, it has been proposed that specifically suppressing COX-2 instead of COX-1 could provide efficient anti-inflammatory benefits with less adverse effects than conventional NSAIDs. As a result, coxibs—including Celecoxib, Rofecoxib, Valdecoxib, and Etoricoxib— selective COX-2 inhibitors with enhanced safety profiles, have been created and marketed (Charlier & Michaux, 2003; Ansari et al., 2017).

For the objective of treating inflammation, Malladi et al. created 3,6-disubstituted-1,2,4-triazolo-[3,4-b]-1,3,4-thiadiazoles containing pyrazole moieties (**30a–30j**). Compound **30h**'s propyl and *p*-chlorophenyl substituents are responsible for its maximum activity (64.7%

Figure 12. Synthesis of 3,6-disubstituted-1,2,4-triazolo-[3,4-b]-1,3,4-thiadiazoles

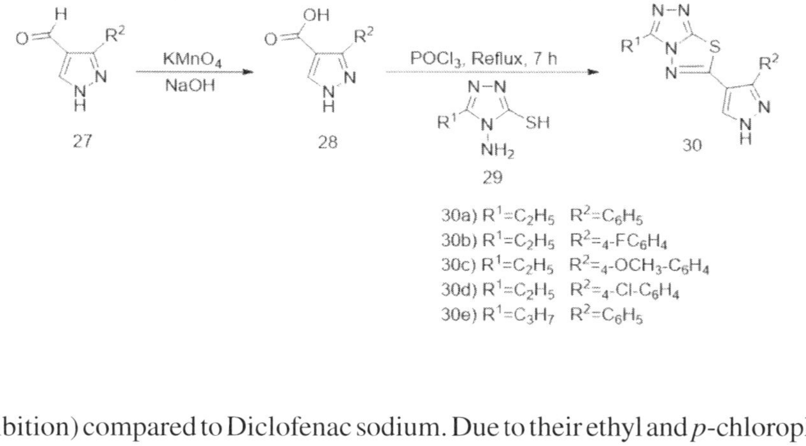

30a) R^1=C_2H_5 R^2=C_6H_5
30b) R^1=C_2H_5 R^2=4-FC_6H_4
30c) R^1=C_2H_5 R^2=4-OCH_3-C_6H_4
30d) R^1=C_2H_5 R^2=4-Cl-C_6H_4
30e) R^1=C_3H_7 R^2=C_6H_5

inhibition) compared to Diclofenac sodium. Due to their ethyl and *p*-chlorophenyl, and propyl and phenyl moieties, respectively, compounds **30d** and **30e** exhibited modest activity (Malladi et al., 2011). Six pyrazole derivatives were synthesised by Macarini et al. after ADME investigations and virtual screening for inhibition of COX-1 and COX-2. With an IC50 value of 0.73 μM, compound **36a** showed the highest activity against COX-2 among them, while Celecoxib, the control, had an IC50 value of 0.88 μM. Compounds with the 3,5-dimethylpyrazole ring form hydrogen bonds with the enzyme due to polar groups. AutoDock Vina docking studies showed that the chalcone moiety's C=O group forms a hydrogen bond with the enzyme, which Celecoxib does not, explaining the higher selectivity (Macarini et al., 2019).

Figure 13. Synthesis of pyrazole chalcone derivatives

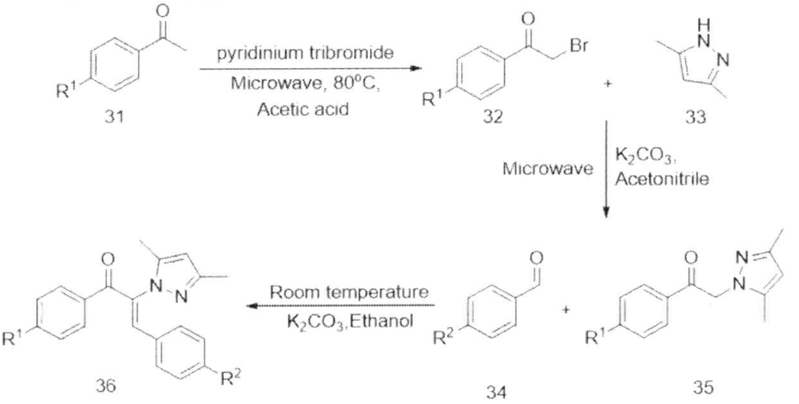

36a) R^1= H R^2= CN
36b) R^1= H R^2= NO_2
36c) R^1= H R^2= $NHCOCH_3$
36d) R^1= H R^2= Cl
36e) R^1= Cl R^2= CN
36f) R^1= OCH_3 R^2= Cl

Somakala et al. conducted a study on pyrazole acetamide derivatives to evaluate their anti-inflammatory potential both in vitro and in vivo. The in vitro assessment used the bovine serum albumin (BSA) anti-denaturation assay to measure the compounds' ability to prevent protein denaturation, a key factor in inflammation. The in vivo evaluation utilized the carrageenan-induced rat paw edema model, a standard for assessing anti-inflammatory activity. Compound **42a** showed the most significant anti-inflammatory effect, achieving 83.1% inhibition, surpassing Diclofenac sodium, a common NSAID, which showed 81.6% inhibition. Notably, compound **42a** was also safe for the gastric mucosa, indicating a lower risk of gastrointestinal side effects compared to typical NSAIDs. Derivatives with chloro, fluoro, and bromo groups demonstrated enhanced anti-inflammatory activity over the standard drug, suggesting that these halogen groups may boost efficacy. Molecular modeling studies using GLIDE version 9.8 revealed strong interactions of these pyrazole derivatives with the target enzyme, indicating their potential as effective anti-inflammatory agents (Somakala et al., 2016).

Figure 14. Synthesis of pyrazole acetamide derivatives

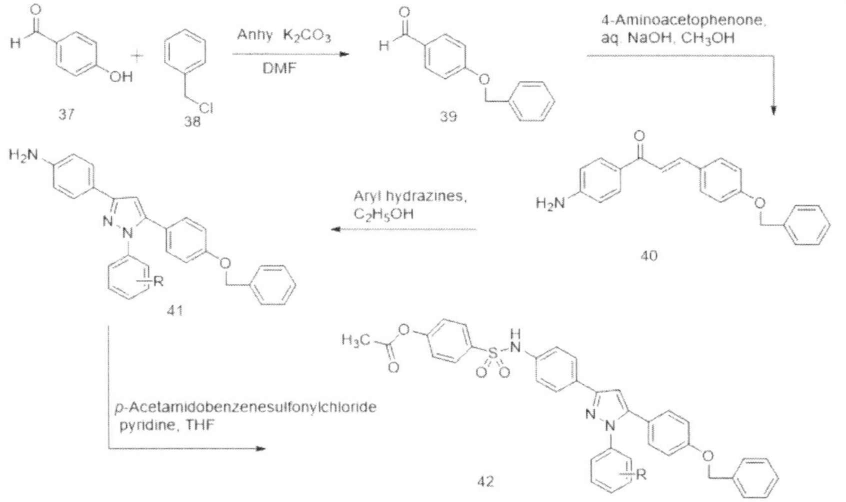

R= 42a) 2-chloro, 42b) 3-chloro, 42c) 4-chloro,42d) 3,4-dichloro
42e) 4-fluoro, 42f) 4-fluoro-3-chloro,42 g) 4-bromo,42 h) 4-nitro,
42 i) 2,4-dinitro,42 j) 2-methyl, 42k) 3-methyl, 42m) 2,6-dimethyl
42n) 2-methoxy, 42o) 4-methoxy

Bansal et al. synthesized pyrazole-containing oxadiazoles and investigated their anti-inflammatory properties, focusing on specific COX-2 inhibition. Among the compounds tested, compound **47g**, which contains a nitro group, exhibited the highest COX-2 inhibitory activity with an IC50 value of 0.31 µM. Molecular docking studies using AutoDock Tools version 1.5.4 demonstrated that compound **47g** had a much stronger binding affinity for COX-2 compared to COX-1. This selectivity was mainly due to the nitro group, which formed hydrogen bonds with COX-2, indicating its crucial role in effectively targeting and inhibiting COX-2 (Bansal et al., 2014).

Figure 15. Synthesis of pyrazole-containing oxadiazoles

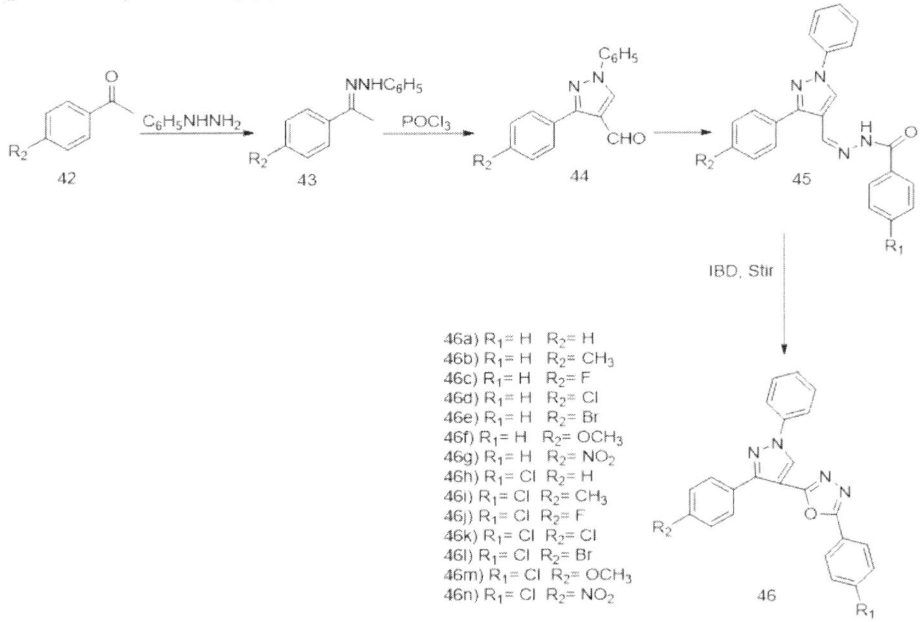

C6H5NHNH2
POCl3
IBD, Stir
NNHC6H5
C6H5
CHO
42
43
44
45
46
46a) R1= H R2= H
46b) R1= H R2= CH3
46c) R1= H R2= F
46d) R1= H R2= Cl
46e) R1= H R2= Br
46f) R1= H R2= OCH3
46g) R1= H R2= NO2
46h) R1= Cl R2= H
46i) R1= Cl R2= CH3
46j) R1= Cl R2= F
46k) R1= Cl R2= Cl
46l) R1= Cl R2= Br
46m) R1= Cl R2= OCH3
46n) R1= Cl R2= NO2

3.2 Anticancer

Owing to alterations in individuals' lifestyles, cancer has emerged as a fatal illness in both developed and developing nations across the globe (Hassan et al., 2014). According to the World Health Organization, cancer causes 8.2 million deaths worldwide annually due to the unchecked proliferation of abnormal cells. The three primary treatments for cancer are radiation therapy, chemotherapy, and surgery. Researchers have synthesized and investigated numerous molecules, including those containing pyrazole rings, for potential use in chemotherapy (Huang et al., 2012; Rai et al., 2015)]. Due to the notable anticancer capabilities of pyrazoles, a prominent class of heterocyclic chemicals, there has been extensive research on developing new and effective anticancer drugs. Because unchecked growth of cancer cells is one of their defining characteristics, blocking proliferative pathways is a potentially effective approach for treating cancer (Ansari et al., 2017).

El-Gamal and his team synthesized a series of novel compounds based on a tri-arylpyrazole scaffold to assess their anticancer potential. These compounds were tested against nearly sixty cancer cell lines, including MDA-MB-435, OVCAR-3, A498, PC-3, RPMI-8226, HOP-92, KM12, SF-295, and MDA-MB-468. All synthesized molecules, **57a-57h** and **58a-58d**, showed notable anticancer activity. Triarylpyrazole **57c** was the most potent, with an IC50 of less than 0.63 µM, indicating high efficacy. These results suggest that triarylpyrazole-based compounds are promising candidates for developing new anticancer therapies (El-Gamal et al., 2013).

Figure 16. Synthesis of triarylpyrazole derivatives

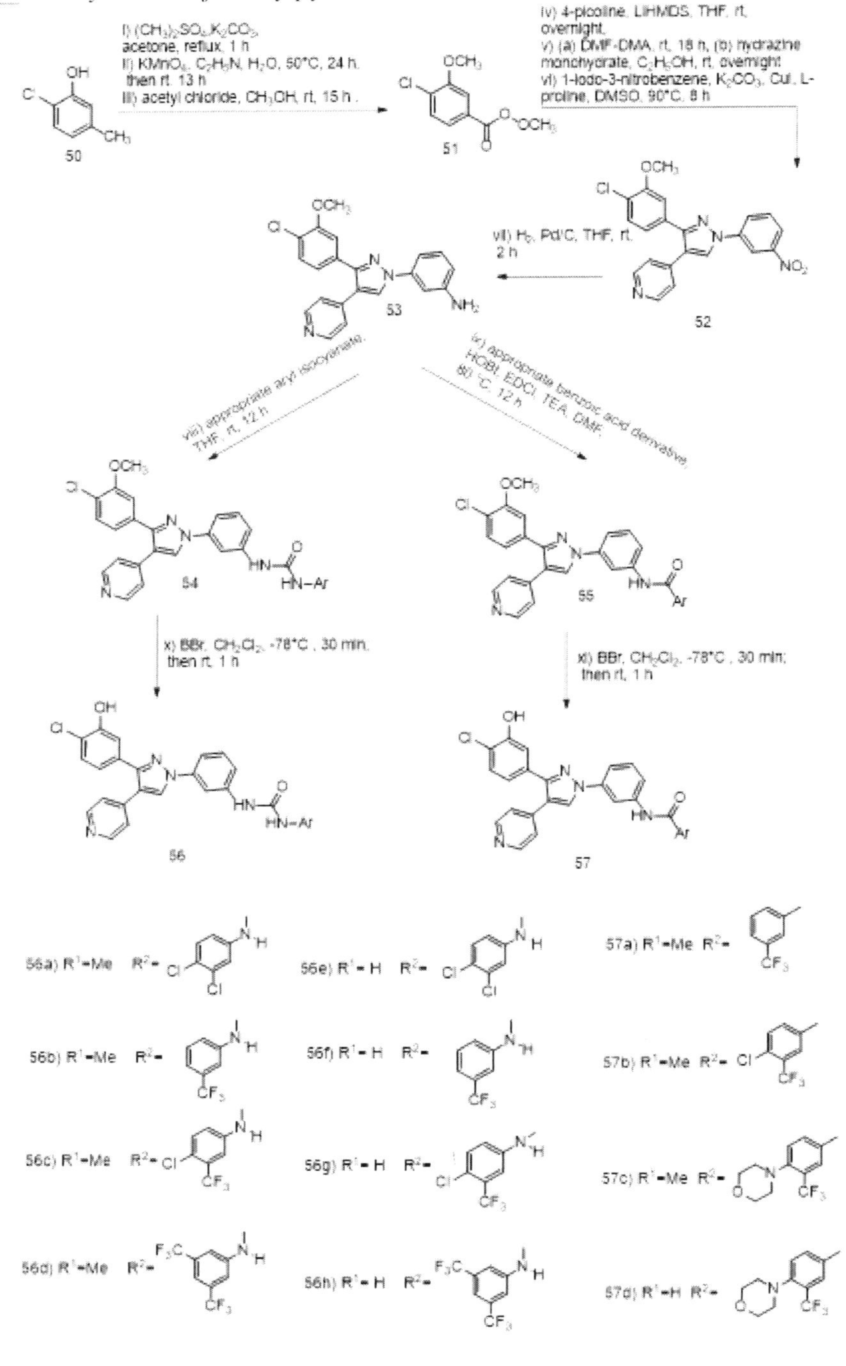

Sankappa Rai U. and colleagues synthesized a series of pyrazole chalcones, **62a-62f**, to evaluate their anticancer potential. The compounds were tested against MCF-7 (breast cancer) and HeLa (cervical cancer) cell lines. Compound **62a** showed the most potent inhibitory effects on both cell lines, attributed to its 4-fluoro-phenyl and 5-fluoro-pyridin groups. These structural features likely enhance its interaction with molecular targets involved in cancer cell proliferation. The study suggests that compound **62a** holds promise as a potent anticancer agent and could be a lead compound for further development (Rai et al., 2015).

Figure 17. Synthesis of pyrazole chalcone derivatives

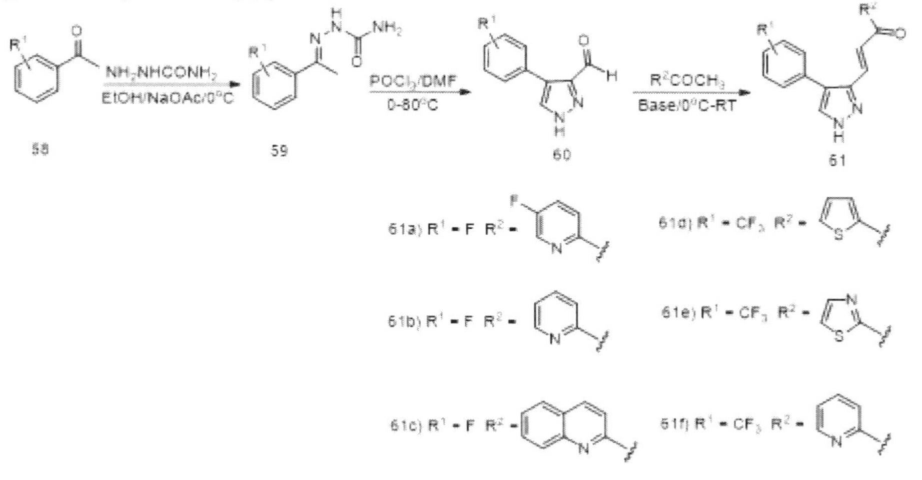

Zhang et al. synthesized a series of 3-(1*H*-indole-3-yl)-1*H*-pyrazole-5-carbohydra-zide derivatives to assess their antiproliferative potential against cancer cell lines, including HepG-2 (liver cancer), BGC823 (gastric cancer), and BT474 (breast cancer). Several compounds demonstrated significant antiproliferative activity compared to the established chemotherapeutic agent 5-fluorouracil. Compounds **68a** and **68b** were particularly effective at arresting the cell cycle in the S phase, indicating their potential to disrupt cancer cell growth and replication, suggesting a promising new approach for cancer treatment (Zhang et al., 2011).

Figure 18. Synthesis of 3-(1H-indol-3-yl)-1H-pyrazole-5-carbohydrazide

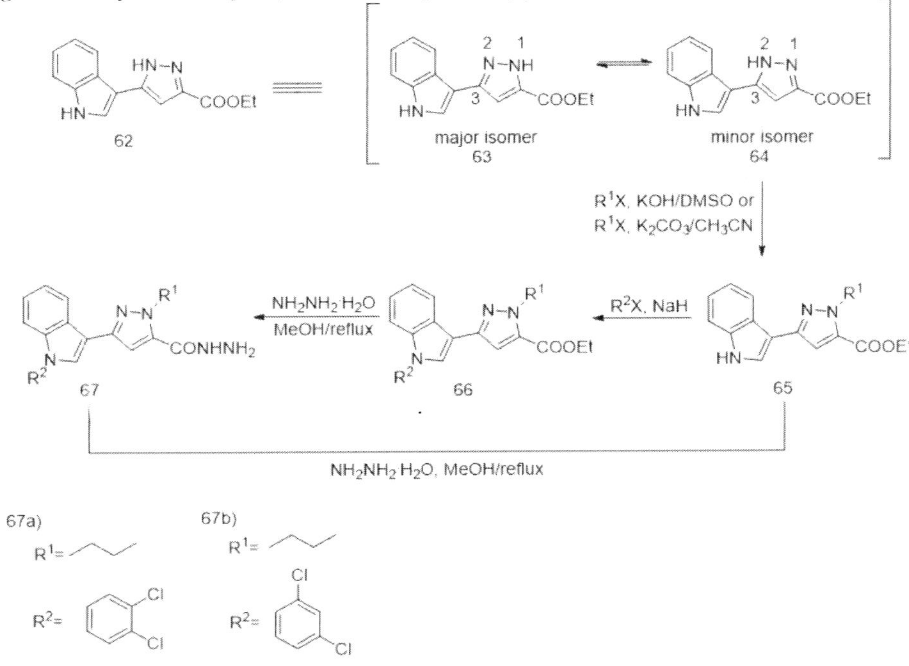

Cankara et al. synthesized novel amide derivatives of 5-(*p*-tolyl)-1-(quinoline-2-yl)pyrazole-3-carboxylic acid to develop new anticancer agents. They tested these compounds' antiproliferative properties against MCF7 (breast cancer), Huh7 (liver cancer), and HCT116 (colon cancer) cell lines using the sulforhodamine B (SRB) assay. Compound **75j**, with a 2-chloro-4-pyridinyl group, showed promising cytotoxic effects with low IC50 values: 1.6 μM for Huh7, 3.3 μM for MCF7, and 1.1 μM for HCT116. These results indicate that compound **75j** is highly effective at inhibiting cancer cell growth at low concentrations, causing cell cycle arrest at the SubG1/G1 phase and inducing apoptosis (Pirol et al., 2014).

Figure 19. Synthesis of amide derivatives of 5-(p-tolyl)-1-(quinoline-2-yl)pyrazole-3-carboxylic acid

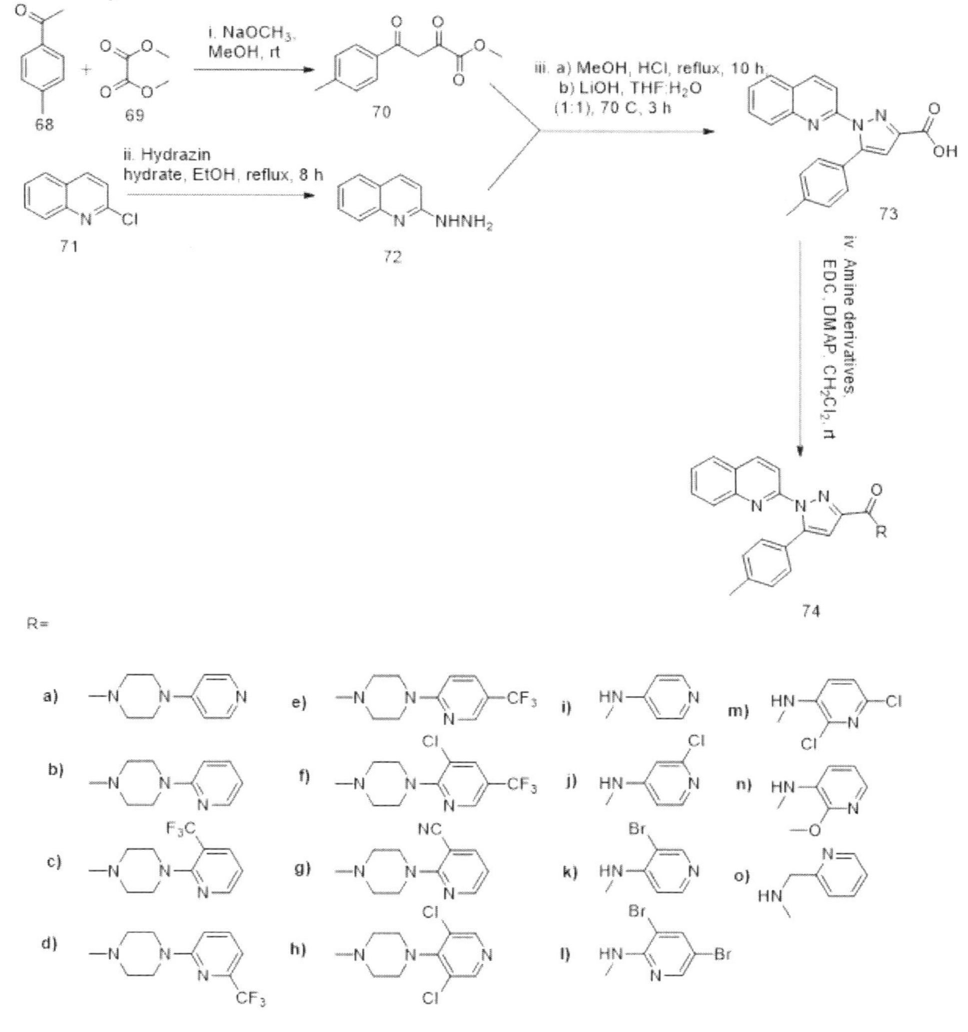

3.3 Antioxidant

Free radicals are highly reactive substances that can damage healthy cells by attacking DNA, proteins, and cell membranes. Antioxidants protect the body by preventing or neutralizing these free radical reactions, thereby slowing down cellular damage (Nimse & Pal 2015; Wright et al., 2001). The body produces endogenous antioxidants, but their effectiveness is affected by age, diet, and health. The body also relies on dietary antioxidants to meet its needs. Free radical damage to DNA can impact the immune system, accelerate aging, and contribute to diseases like diabetes, cancer, Alzheimer's, and cardiovascular disease. It can also cause visual signs of aging, such as wrinkles and age spots (Ansari et al., 2017). Antioxidants are crucial for preventing cellular damage by managing reactive oxygen and nitrogen species, maintaining balance for proper physiological function. Both natural and synthetic dietary antioxidants protect against free radicals, improve quality of life, and help prevent diseases, thus reducing healthcare costs. As a result, developing new, efficient, and low-toxicity antioxidant compounds is a key focus in medicinal chemistry (Silva et al., 2018). Research on small molecules with antioxidant efficacy to prevent harmful effects is an important field. Drug molecules containing a pyrazole core are a preferred choice in medicinal chemistry for developing effective antioxidant agents (Naveen et al., 2021).

The antioxidant activity of pyrazole chalcones was shown by Bandgar et al. . Using the 1,1-diphenyl-2-picrylhydrazyl (DPPH) method, they evaluated the antioxidant properties of the compounds and observed that three compounds **78a-78c** with phenyl, 4-methoxyphenyl, and 3-chlorophenyl groups showed 25–35% DPPH activity (Bandgar et al., 2009).

Figure 20. Synthesis of pyrazole chalcone derivatives

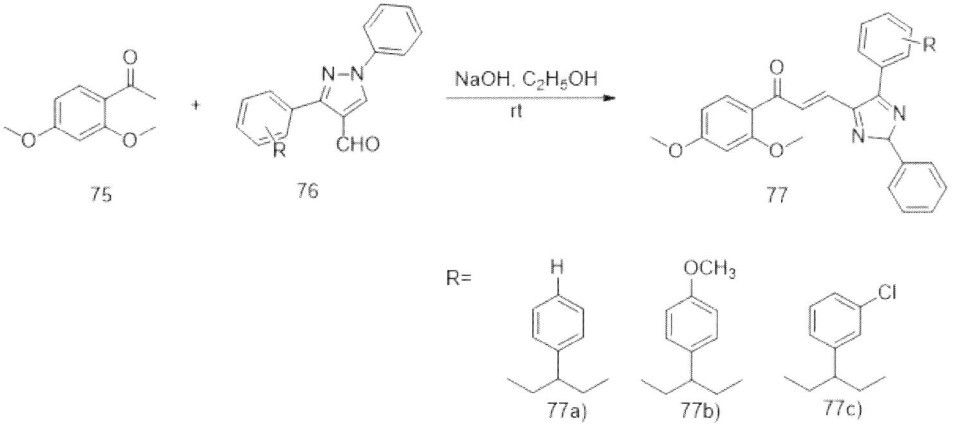

Ambethkar et al. synthesized dihydropyrano[2,3-*c*]pyrazole derivatives (Scheme 17) and evaluated their antioxidant activity using the 2,2-diphenyl-1-picrylhydrazyl (DPPH) radical scavenging assay. All compounds demonstrated radical-scavenging activity. Compounds **84a–84e** and **84g–84i** exhibited moderate scavenging ability, whereas compound **84f** showed superior scavenging capacity. The presence of a hydroxyl group in the para position of **84f** enhances π-conjugation, thereby stabilizing the free radicals produced (Ambethkar et al., 2015).

Figure 21. Synthesis of pyrano pyrazole derivatives

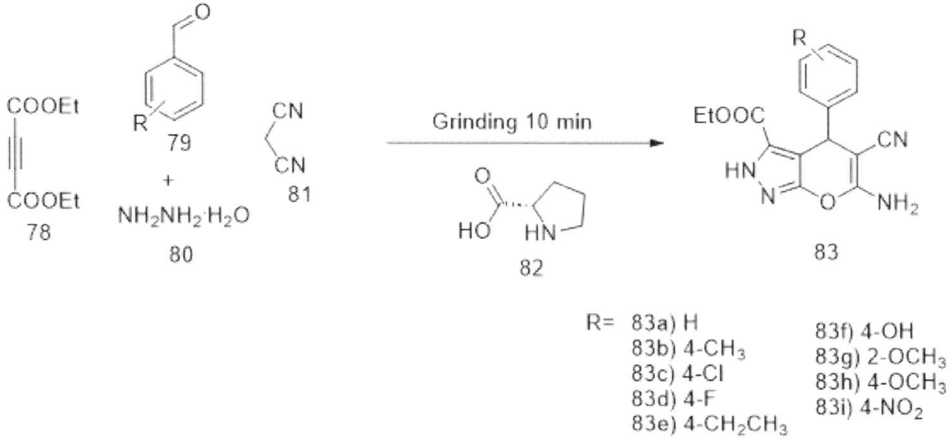
COOEt
COOEt
78
R
79
+
$NH_2NH_2 \cdot H_2O$
80
CN
CN
81
Grinding 10 min
O
HO
HN
82
EtOOC
R
CN
HN
N
O
NH_2
83
R= 83a) H
83b) 4-CH_3
83c) 4-Cl
83d) 4-F
83e) 4-CH_2CH_3
83f) 4-OH
83g) 2-OCH_3
83h) 4-OCH_3
83i) 4-NO_2

Adhikari et al. investigated a series of 3,5-diarylpyrazoline derivatives with a pyrimidine (**88a-88l**)for their free radical scavenging activity using the DPPH method (Chandanshive, 2013). Pyrazolines **88b-88d** and **88i,88j** demonstrated excellent free radical scavenging activity (FRSA) with percentages ranging from 89.79% to 94.11% at 100 mg/mL concentration, surpassing the activity of the standard BHT 7 (93.16%). These compounds also exhibited significant inhibition of hydroxyl radicals, comparable to or higher than BHT 7 (86.98% inhibition). Pyrazolines **88b-88d** and **88i,88j** showed potent superoxide anion scavenging activity, similar to BHT 7. The study highlighted that the presence of halogens such as Cl or Br in one aromatic ring enhances antioxidant activity, whereas strong electron-withdrawing groups like nitro groups are not beneficial (Adhikari et al., 2012).

Figure 22. Synthesis of 3,5-diarylpyrazoline derivatives

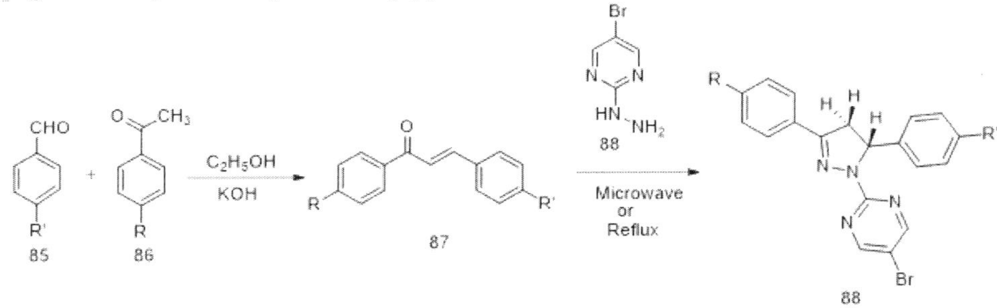

88a) R = H   R^1 = NO$_2$ 88g) R = Br   R^1 = Cl
88b) R = H   R^1 = CH$_3$ 88h) R = Cl   R^1 = CH$_3$
88c) R = H   R^1 = Cl   88i) R = Cl   R^1 = OCH$_3$
88d) R = H   R^1 = Br   88j) R = Cl   R^1 = H
88e) R = Br R^1 = Br   88k) R = Cl   R^1 = Cl
88f) R = Br R^1 = CH$_3$ 88l) R = Cl   R^1 = Br

Ahmad et al. studied the superoxide anion radical scavenging activity (RSA) of compounds **93a-93q**, comparing them with the standard n-propyl gallate. Good to moderate activity was shown by the majority of examined substances. Studies on the structure-activity relationship (SAR) showed that the quantity and strength of oxygen-containing functional groups often boosted activity of the molecule. The most active compound was dihydroxy derivative **93k**, which was followed in order of activity by monomethoxy derivative **93j** and compound having no substituents **93a**. Activity was enhanced by replacing an alkoxy group with a hydroxy group; this was likely due to steric effects. For halogenated derivatives, the free radical scavenging activity was highest in the meta position, followed by ortho, and then para (Ahmad et al., 2010).

Figure 23. Synthesis of N′-arylmethylidene-2-(3,4-dimethyl-5,5-dioxidopyrazolo[4,3-c][1,2]benzothiazine-2(4H)-yl)acetohydrazides

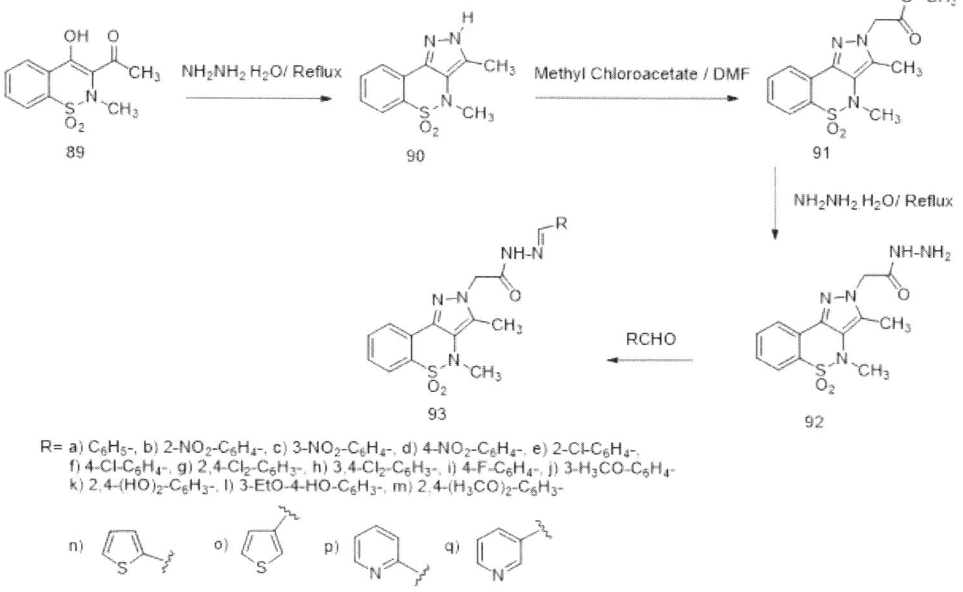
89
NH2NH2 H2O/ Reflux
90
Methyl Chloroacetate / DMF
91
NH2NH2 H2O/ Reflux
92
RCHO
93
R= a) C6H5-, b) 2-NO2-C6H4-, c) 3-NO2-C6H4-, d) 4-NO2-C6H4-, e) 2-Cl-C6H4-,
f) 4-Cl-C6H4-, g) 2,4-Cl2-C6H3-, h) 3,4-Cl2-C6H3-, i) 4-F-C6H4-, j) 3-H3CO-C6H4-
k) 2,4-(HO)2-C6H3-, l) 3-EtO-4-HO-C6H3-, m) 2,4-(H3CO)2-C6H3-
n)
o)
p)
q)

3.4 Antimicrobial

Global microbial infections, including fungal and bacterial, are rising due to antibiotic resistance. Antimicrobials are crucial for both hospitalized and ambulatory patients, with 24% to 40.8% of hospitalized patients using them. Factors like injury, surgery, immunosuppressive medications, aging, and HIV increase susceptibility to infections. The growing resistance to common antibiotics worsens this issue (Danishuddin et al., 2012). Antimicrobial agents are substances that either eliminate microorganisms or hinder their growth. They can be classified based on the specific types of microorganisms they primarily target. Many diseases are caused by these microorganisms, posing a significant threat to humans. The effectiveness of many antimicrobial agents is being compromised by a global rise in resistant microorganisms that were previously susceptible. The rise of both new and old antibiotic-resistant microbial strains in recent decades has generated a considerable demand for new classes of antimicrobial drugs, despite the abundance of antibiotics and chemotherapeutics available for medical use (Ansari et al., 2017). Many nitrogen-containing heterocyclic ring compounds, including pyrazole, pyrrolones, pyridazinones, and pyrazolines, are frequently used to boost vital antibacterial action.

Akbas and coworkers synthesised a number of 1*H*-pyrazole-3-carboxylic acid derivatives and evaluated the antibacterial activities of 1*H*-pyrazole-3-carboxylic acid derivatives against *Pseudomonas putida, Bacillus cereus, Staphylococcus aureus,* and *Escherichia coli*. Based on the results, compound **97** was shown to be the most effective in the series, demonstrating antibacterial activity against both Gram-positive and Gram-negative bacteria (Akbas et al., 2005).

Figure 24. Synthesis of 1H-pyrazole-3-carboxylic acid derivatives

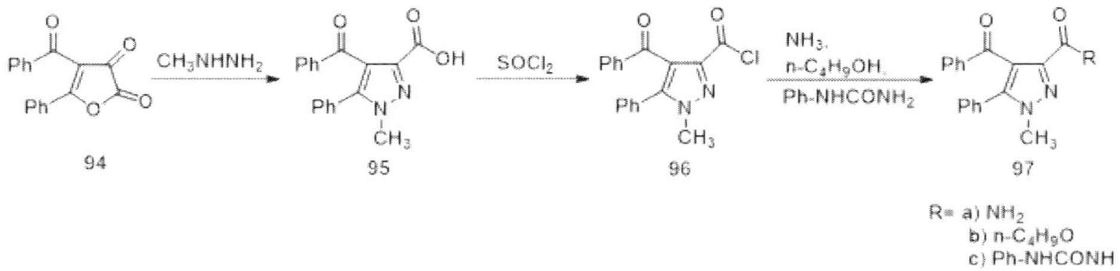

The pyrazole-incorporated imidazole derivatives (**100** and **103**) synthesized by Vijesh et al. show antibacterial activity against a variety of bacteria, including *E. coli, S. aureus,* and *B. subtilis* etc, as well as fungal strains such as *A. niger, C. albicans, M. gypseum* etc. At concentrations of 1 and 0.5 mg/ml, compound **100b**, one of the described compounds, exhibited outstanding antibacterial activity. Among the synthesized compounds, the thioanisyl moiety of compound **100b** contributes to its increased antibacterial activity. The second set of compounds, including **103a–103j**, exhibited moderate antibacterial activity. The biological activity of the compounds is attributed to the imidazole and pyrazole nuclei found in each series, while the presence of additional substituents accounts for the compounds' diverse biological activities (Vijesh et al., 2011).

Figure 25. Synthesis of pyrazole-incorporated imidazole derivatives

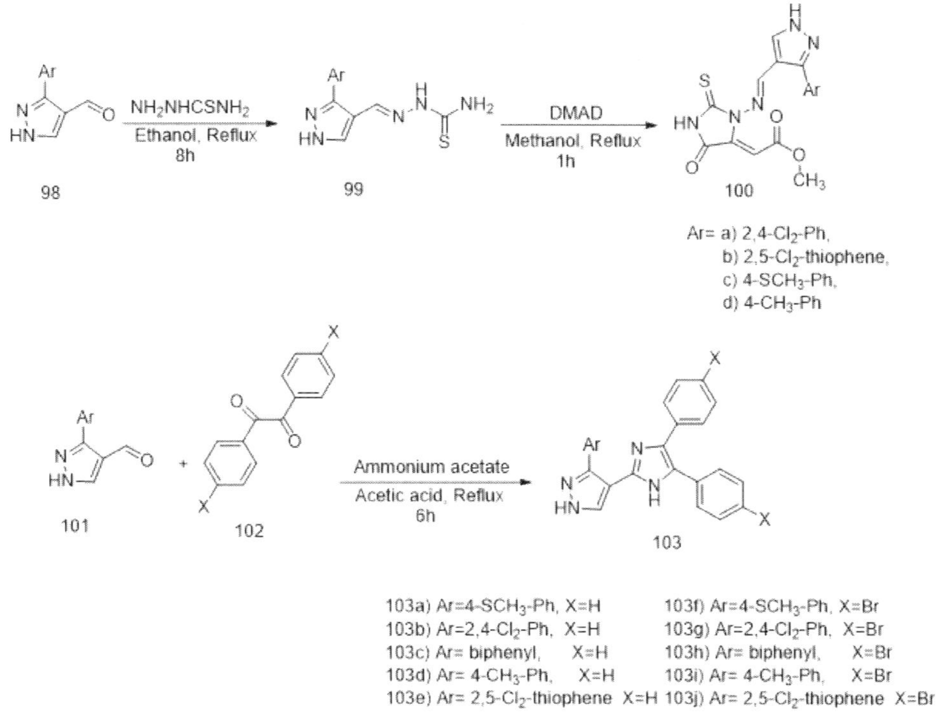

NH2NHCSNH2
Ethanol, Reflux
8h
DMAD
Methanol, Reflux
1h
98
99
100
Ar= a) 2,4-Cl2-Ph,
b) 2,5-Cl2-thiophene,
c) 4-SCH3-Ph,
d) 4-CH3-Ph
Ammonium acetate
Acetic acid, Reflux
6h
101
102
103
103a) Ar=4-SCH3-Ph, X=H
103b) Ar=2,4-Cl2-Ph, X=H
103c) Ar= biphenyl, X=H
103d) Ar= 4-CH3-Ph, X=H
103e) Ar= 2,5-Cl2-thiophene X=H
103f) Ar=4-SCH3-Ph, X=Br
103g) Ar=2,4-Cl2-Ph, X=Br
103h) Ar= biphenyl, X=Br
103i) Ar= 4-CH3-Ph, X=Br
103j) Ar= 2,5-Cl2-thiophene X=Br

The study by Argade et al. explored a novel class of pharmacophores featuring a pyrazole-containing 2,4-disubstituted oxazole ring system, which exhibits broad-spectrum antimicrobial activity. Among the compounds investigated, compound **106** demonstrated significant antibacterial activity. At a concentration of 30 μg/mL, it produced zones of inhibition measuring 7.2 mm, 7.0 mm, and 7.2 mm against three bacterial strains: *Escherichia coli, Pseudomonas aeruginosa*, and *Staphylococcus aureus*, respectively. In addition to its antibacterial activity, Compound **106** also displayed antifungal activity at higher concentrations. At 60 μg/mL, it was effective against *Candida albicans*, a fungal pathogen responsible for various infections, particularly in immunocompromised individuals. This dual antibacterial and antifungal activity highlights the potential of Compound **106** as a broad-spectrum antimicrobial agent (Argade et al., 2008).

Figure 26. Synthesis of (Z)-2-phenyl-4-((1-phenyl-3-(p-tolyl)-1H-pyrazol-4-yl) methylene)oxazol-5(4H)-one

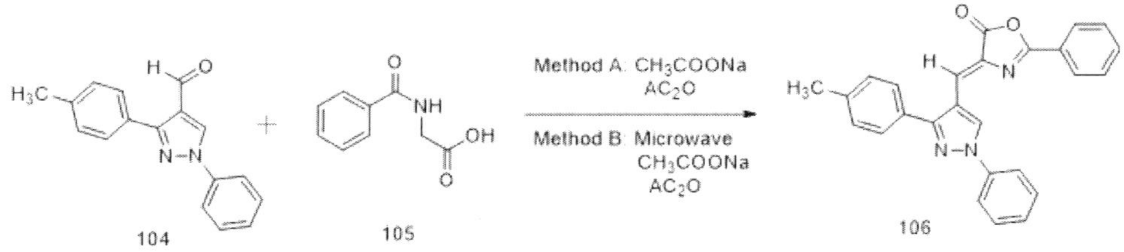

Liu and colleagues developed a series of coumarin-pyrazole carboxamide compounds and tested their antibacterial activity against *Salmonella*, *E. coli*, *S. aureus*, and *L. monocytogenes*. Compounds **109a, 109b, 109c**, and **109d** showed strong antibacterial effects, especially against *Salmonella*, with MIC values of 0.25 mg/L, 0.125 mg/L, 0.05 mg/L, and 0.125 mg/L, respectively. Compound **109c** was the most effective, with an MIC of 0.05 mg/L, outperforming both ciprofloxacin and novobiocin. Substituents like bromine and cyano on compound **109a** improved activity, while a nitro group on **109b** reduced it. The coumarin-pyrazole nucleus also showed significant inhibitory action against Topo-IV and Topo-II (Liu et al., 2018).

Figure 27. Synthesis of coumarin-pyrazole carboxamide derivatives

109a) R^1=Br R^2=H R^3= CN
109b) R^1=NO_2 R^2=H R^3= CN
109c) R^1=H R^2=N(Et)$_2$ R^3=COOEt
109d) R^1=H R^2=N(Et)$_2$ R^3=COOH

3.5 Antiviral

Viruses are significant pathogens that can cause a wide array of serious diseases in humans, other animals, and plants. These microscopic agents invade living cells and hijack their machinery to reproduce, often damaging or killing the host cells in the process. The most common viral infections target the respiratory system, including the nose, upper airways, throat, and lungs. They also target the gastrointestinal tract, causing gastroenteritis, affect the liver, leading to hepatitis, and impact the skin, resulting in warts, rashes, or other blemishes (Vardanyan & Hruby, 2016). Three viruses that produce chronic illnesses with significant rates of sickness and mortality include hepatitis B, hepatitis C, and HIV (Faria et al., 2017). One of the primary causes of chronic liver disease, which can lead to liver cirrhosis and hepatocellular carcinoma (HCC), is an infection with the hepatitis C virus (HCV) (Küçükgüzel & Şenkardeş, 2015). Molecules with a pyrazole core have been emphasized in the quest for new antiviral agents. Pyrazole derivatives have demonstrated activity against HIV, hepatitis C, HSV-1, RSV, and H1N1 (Faria et al., 2017).

Iyer and coworkers conducted research focused on developing inhibitors for the negative regulatory factor (Nef), a multifunctional protein that plays a critical role in the pathogenicity of HIV-1. In their study, the researchers successfully synthesized a compound known as compound **113a**, which exhibited powerful inhibitory effects against Nef. The inhibition of Nef can potentially disrupt the viral replication process and reduce the virulence of HIV-1, thereby contributing to the development of effective treatments against HIV/AIDS. Further investigations into the structure-activity relationship of these compounds revealed that compound **113b** also demonstrated antiviral activity similar to that of compound **113a**. This similarity in antiviral efficacy between the two compounds is attributed to the presence of a pyrazole ring in their chemical structures (Iyer et al., 2014).

Figure 28. Synthesis of diphenylpyrazolodiazene derivatives

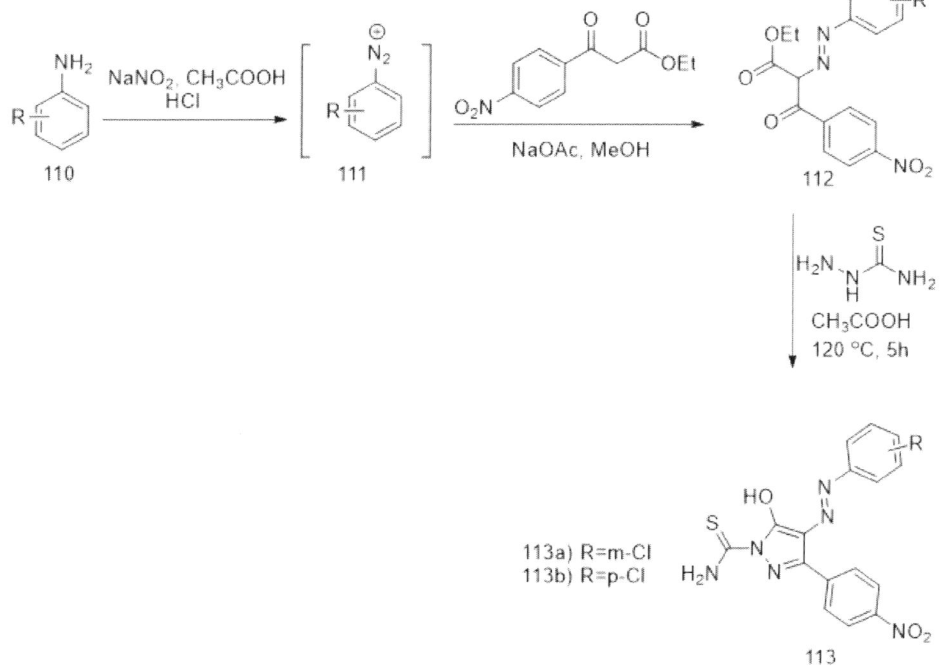

113a) R=m-Cl
113b) R=p-Cl

Using a cell-based complete replication assay, Kim and teammates discovered a novel class of aryl-substituted pyrazole compounds that are potent non-nucleoside reverse transcriptase inhibitors (NNRTIs) with anti-HIV activity. Among all, compound **119** exhibited remarkable efficacy against both viruses containing Y181C and K103N resistance mutations in the reverse transcriptase gene, as well as wild-type HIV-1 (EC50 = 0.2 nM) (Kim et al., 2012).

Figure 29. Synthesis of pyrazole-incorporated phenylaminopyridine derivative

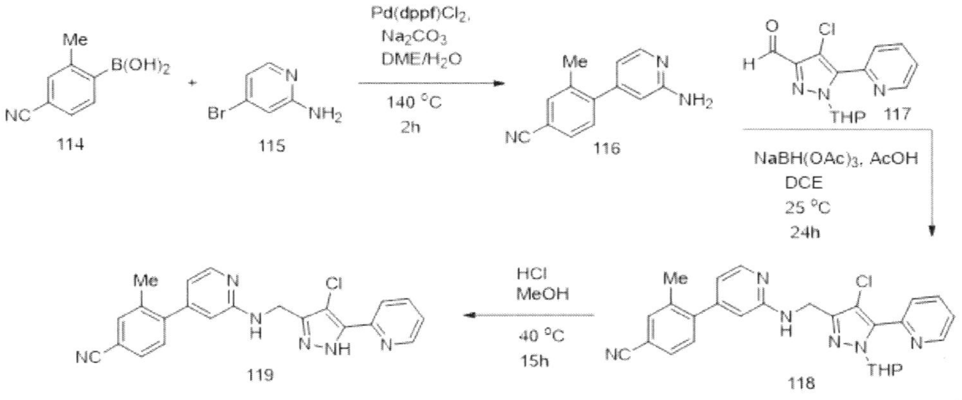

Chuang et al. synthesized a new series of pyridine-pyrazole-sulfonate compounds and subsequently evaluated their anti-HBV activities, establishing the structure-activity relationship in HepG2 2.2.15 cells. Among the synthesized compounds, the one with an ortho nitro group, compound **124d**, exhibited the strongest inhibitory activity with 9.19 μM IC50 value and a high selectivity index of 35.46 (TC50/IC50) (Chuang et al., 2016).

Figure 30. Synthesis of pyridine-pyrazole-sulfonate

R= 124a) p-H 124b) p-CH$_3$ 124c) p-NO$_2$ 124d) o-NO$_2$
124e) p-F 124f) p-Br 124g) p-OMe 124h) OSO$_2$Me

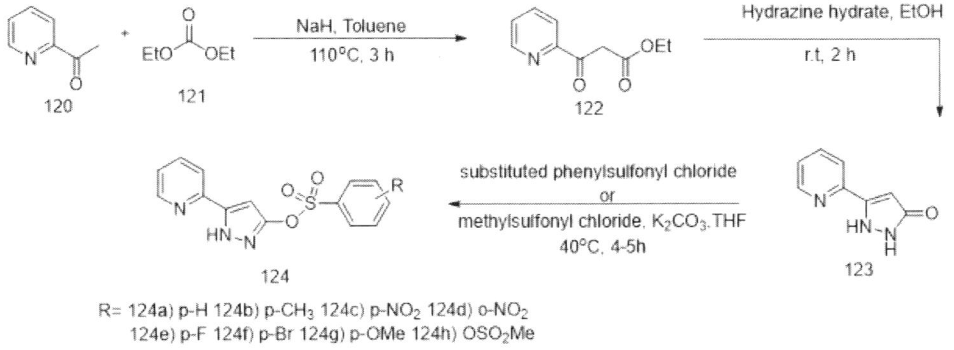

A study was conducted by Han and coworkers to synthesize a series of novel pyrazole-fused heterocyclic derivatives and evaluate their potential biological activities, specifically focusing on their ability to catalytically cleave DNA and their antiviral activities against the bovine viral diarrhoea virus (BVDV). Among the synthesized derivatives, compound 127l demonstrated the most potent antiviral activity, with an EC50 of 0.12 mmol/L. This was a significant finding, as compound 127l's activity was ten times greater than that of ribavirin, a widely used antiviral drug, which had an EC50 of 1.3 mmol/L. This substantial increase in efficacy suggests that compound 127l is a promising candidate for further development as an antiviral agent against BVDV (Han et al., 2015).

Figure 31. Synthesis of pyrazole fused heterocyclic ligand

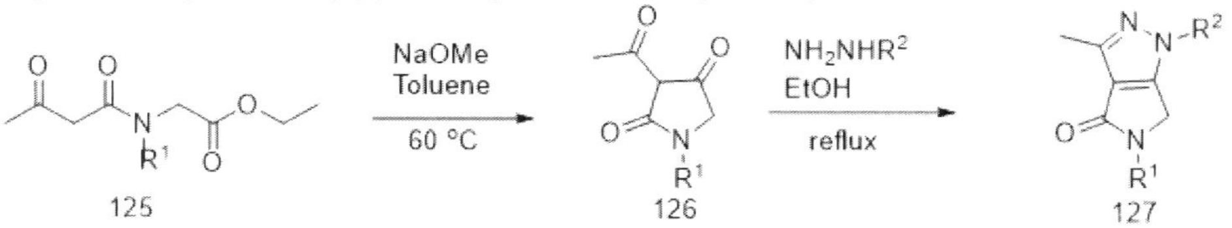

127a) R^1=4-ClC$_6$H$_4$, R^2= H   127e) R^1=Ph, R^2= Ph   127i) R^1=Ph, R^2= 4-FC$_6$H$_4$
127b) R^1=4-MeC$_6$H$_4$, R^2= H   127f) R^1=4-ClC$_6$H$_4$, R^2= Ph   127j) R^1= Ph, R^2= 2-FC$_6$H$_4$
127c) R^1=Ph, R^2= H   127g) R^1=Bn, R^2= Ph   127k) R^1= Ph, R^2= 4-MeC$_6$H$_4$
127d) R^1=Bn, R^2= H   127h) R^1=4-MeC$_6$H$_4$, R^2= Ph   127l) R^1= Ph, R^2= 4-ClC$_6$H$_4$

3.6 Antidiabetic

The chronic illness known as diabetes mellitus (DM) primarily affects how lipids, proteins, and carbohydrates are metabolized. Elevated blood glucose levels are one of its characteristics. Diabetes mellitus has two clinical forms: Type I and Type II diabetes. Type I, or insulin-dependent diabetes, are autoimmune condition arises from the immune system targeting the beta cells in the pancreatic islets of Langerhans that produce insulin and most cases of type I, are identified in children and young people. Because of the Type I diabetes the body produces insufficient insulin, the peptide hormone essential for transporting blood glucose into muscle, liver, and fat tissues, leading to hyperglycemia (Kerru et al., 2018). The more prevalent kind of diabetes, known as type-II diabetes (T2DM) or non-insulin-dependent diabetic mellitus, is mostly caused by dietary and social practices (Mitra, 2008). The primary goal of diabetes mellitus (DM) management is to maintain normal blood glucose levels to reduce the risk of complications from uncontrolled diabetes. Incretin hormones like glucose-dependent insulinotropic polypeptide (GIP) and glucagon-like peptide-1 (GLP-1) play key roles in glucose regulation and are targets for diabetes treatment. Type 1 diabetes is traditionally managed with intravenous insulin and blood glucose monitoring, while GLP-1 analogs are also used. For type 2 diabetes, insulin is crucial, along with strategies to reduce insulin resistance, decrease hepatic glucose output, enhance insulin secretion, and limit glucose reabsorption to improve glucose regulation (Swedberg et al., 2015; Kakinuma et al., 2010). The promising antidiabetic properties of pyrazoles have led to extensive research efforts aimed at synthesizing novel pyrazole derivatives with enhanced efficacy.

Brigance and colleagues conducted a study where they synthesized a series of pyrazolopyrimidines. These compounds were then evaluated for their ability to inhibit the enzyme dipeptidyl peptidase-4 (DPP4), an enzyme that plays a significant role in glucose metabolism and is a target for the treatment of type 2 diabetes. Among the various pyrazolopyrimidines synthesized, two compounds, identified as compounds **134c** and **134f**, demonstrated the highest potency in inhibiting DPP4. The inhibition strength of these compounds was quantified using the Ki value, which indicates the binding affinity of an inhibitor to an enzyme. Compounds **134c** and **134f** had Ki values of 20 nM and 22 nM, respectively, signifying a strong inhibitory effect at very low concentrations. And it also showed and showed excellent selectivity over other dipeptidyl peptidases (Brigance et al., 2010).

Figure 32. Synthesis of pyrazolopyrimidine derivatives

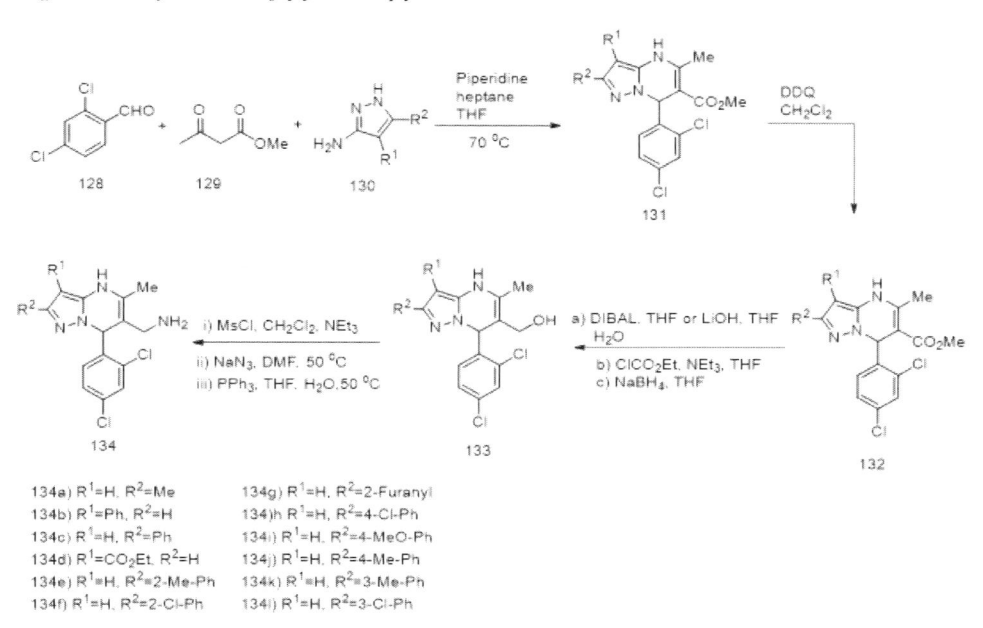

134a) R^1=H, R^2=Me
134b) R^1=Ph, R^2=H
134c) R^1=H, R^2=Ph
134d) R^1=CO$_2$Et, R^2=H
134e) R^1=H, R^2=2-Me-Ph
134f) R^1=H, R^2=2-Cl-Ph
134g) R^1=H, R^2=2-Furanyl
134)h R^1=H, R^2=4-Cl-Ph
134i) R^1=H, R^2=4-MeO-Ph
134j) R^1=H, R^2=4-Me-Ph
134k) R^1=H, R^2=3-Me-Ph
134l) R^1=H, R^2=3-Cl-Ph

Toda and colleagues embarked on research to find new treatments for type 2 diabetes and synthesised a series of novel pyrazole-based compounds that act as insulin secretagogues. Insulin secretagogues are substances that stimulate the pancreas to secrete insulin, helping to regulate blood sugar levels. Among the compounds synthesised, compound **141e**, stood out due to its significant glucose-lowering effects. This compound was tested in oral glucose tolerance tests (OGTT), which are used to assess how effectively the body can manage glucose intake. The tests were conducted in both mice and monkeys to evaluate the compound's efficacy across different species. In these tests, compound **141e** showed a strong ability to lower blood glucose levels, indicating its potential as an effective treatment for type 2 diabetes. This experiment highlights the promise of pyrazole compounds in developing more effective diabetes treatments (Kakinuma et al., 2010).

Figure 33. Synthesis of pyrazole-4-carboxamide derivatives

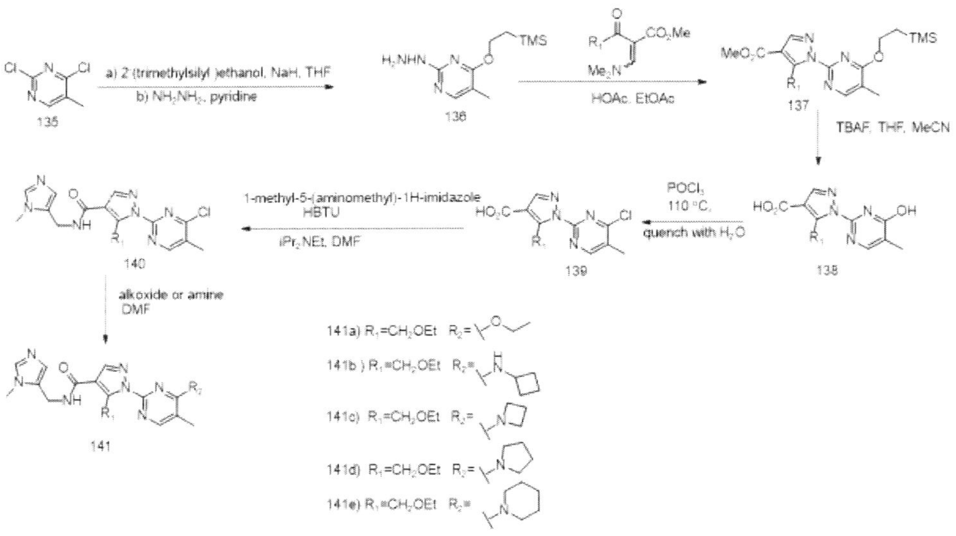

The study conducted by Chaudhry and colleagues involved the synthesis and evaluation of a series of compounds known as imidazolylpyrazoles. These compounds were tested for their ability to inhibit the enzyme α-glucosidase. One specific molecule, compound **145e**, among the synthesised compounds, exhibited strong inhibitory activity against α-glucosidase. The effectiveness of this inhibition was measured using the IC50 value. Compound **145e** had an IC50 value of 23.95 μM, indicating a strong ability to inhibit α-glucosidase compared to the reference drug acarbose, a well-known α-glucosidase inhibitor used in the management of diabetes, according to the in vitro enzyme inhibition (Chaudhry et al., 2017).

Figure 34. Synthesis of imidazolylpyrazole derivatives

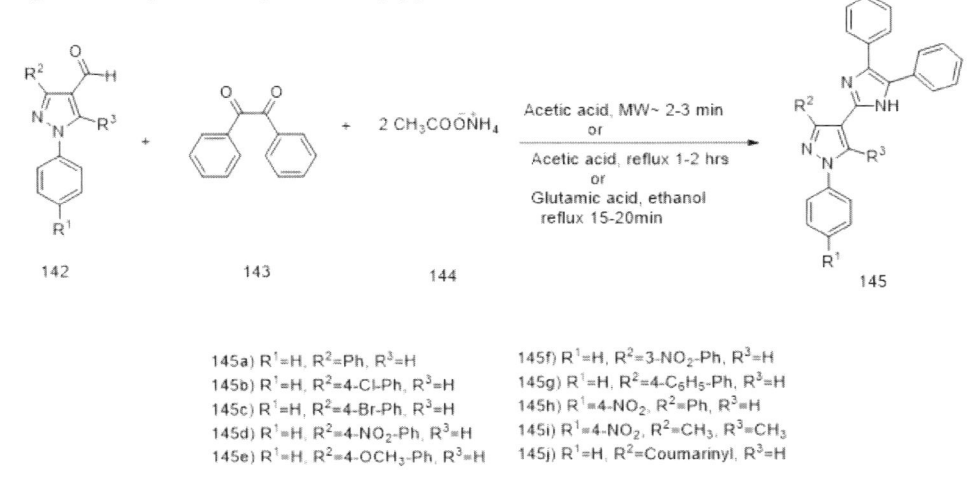

145a) R^1=H, R^2=Ph, R^3=H
145b) R^1=H, R^2=4-Cl-Ph, R^3=H
145c) R^1=H, R^2=4-Br-Ph, R^3=H
145d) R^1=H, R^2=4-NO$_2$-Ph, R^3=H
145e) R^1=H, R^2=4-OCH$_3$-Ph, R^3=H

145f) R^1=H, R^2=3-NO$_2$-Ph, R^3=H
145g) R^1=H, R^2=4-C$_6$H$_5$-Ph, R^3=H
145h) R^1=4-NO$_2$, R^2=Ph, R^3=H
145i) R^1=4-NO$_2$, R^2=CH$_3$, R^3=CH$_3$
145j) R^1=H, R^2=Coumarinyl, R^3=H

Xiong and colleagues conducted research focused on synthesizing new pyrazole compounds with the aim of targeting the glucagon receptor. Inhibiting this receptor can be beneficial in managing conditions such as type 2 diabetes. During their study, the researchers synthesized several novel compounds, among them compound **151m** was found to be a strong and specific glucagon receptor antagonist by means of lead optimisation. Compound **151m** was identified as a reversible and competitive antagonist exhibiting strong functional cAMP activity which has IC50 of 15.7 nM. Furthermore, this compound also exhibited a high binding affinity to the glucagon receptor, with an IC50 value of 6.6 nM. This suggests that the compound binds strongly to the receptor, making it a potent antagonist (Xiong et al., 2012).

Figure 35. Synthesis of pyrazole derivative

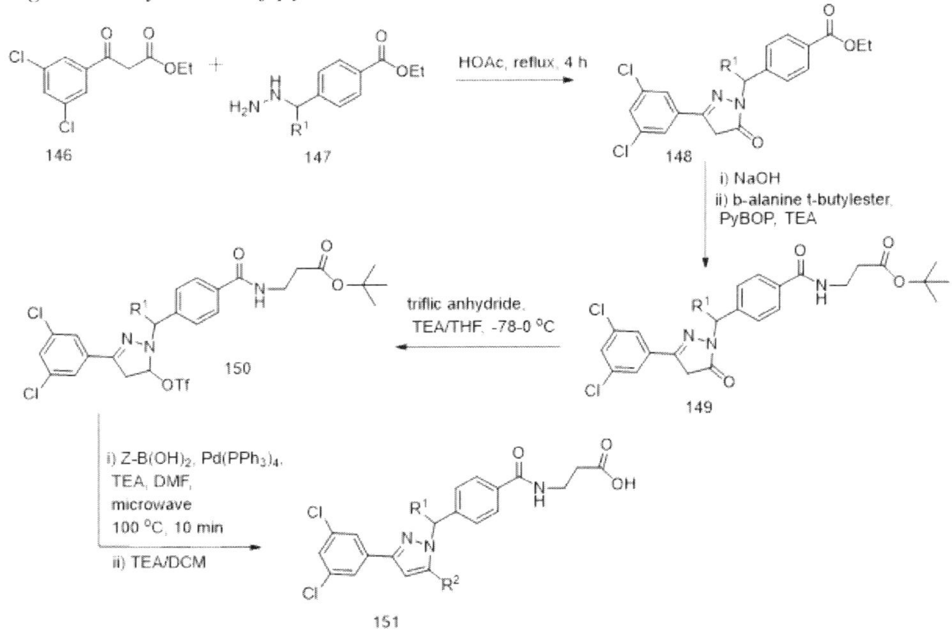

151a) R_1= H	R_2= 4-CF$_3$O-Ph	151l) R_1= Me	R_2= 3-Cl, 4-PrO-Ph
151b) R_1= H	R_2= 2-CF$_3$O-Ph	151m) R_1= (S)-Me	R_2= 6-MeO-naph-2-yl
151c) R_1= H	R_2= 3-CF$_3$O-Ph	151n) R_1= (R)-Me	R_2= 6-MeO-naph-2-yl
151d) R_1= H	R_2= 3-Cl, 4-PrO-Ph	151o) R_1= (S)-Me	R_2= 6-CF$_3$O-naph-2-yl
151e) R_1= H	R_2= naph-2-yl	151p) R_1= (R)-Me	R_2= 6-CF$_3$O-naph-2-yl
151f) R_1= H	R_2= 6-MeO-naph-2-yl	151q) R_1= (S)-Me	R_2= 6-CF$_3$-naph-2-yl
151g) R_1= H	R_2= 5-CF$_3$-O-naph-2yl	151r) R_1= (S)-Me	R_2= 6-Cl-naph-2-yl
151h) R_1= H	R_2= 6-CF$_3$-O-naph-2yl	151s) R_1= (S)-Me	R_2= 6-Me-naph-2-yl
151i) R_1= H	R_2= 7-CF$_3$-O-naph-2yl	151t) R_1= (S)-Me	R_2= 6-iPr-naph-2-yl
151j) R_1= H	R_2= 8-CF$_3$-O-naph-2yl	151u) R_1=(S)-Me	R_2= 6-EtO-naph-2-yl
151k) R_1= Me	R_2= 4-CF$_3$O-Ph	151v) R_1= Me	R_2= 6-cycloexyl-Ph

4. FUTURE SCOPE

In the future, the studies to be adopted on the pyrazole derivatives will be many-sided, giving a constant role to the priority of the synthetic steps optimization. This optimization process will cover three key aspects: efficiency, optimization of synthetic parts for faster reaction times and minimum steps; selectivity, completing the concerned production process with the desirable pyrazole regioisomer; and sustainability, the development of ecologically safe processes that will produce less garbage and recoverable resources and use renewable sources. These upgrades will not only, on the one hand, make the discovery and development of new drugs based on pyrazole incomplete, the topic, but, on the other hand, will also make the drugs be cost-effective and environmentally friendly. This includes investigating innovative catalytic systems and reaction conditions in order to reduce waste and energy consumption. Second, thorough Structure-Activity Relationship (SAR) investigations will be vital for identifying the unique structural aspects contributing to the biological activities of pyrazole derivatives. This will guide the construction of more powerful and selective molecules. Third, knowing the particular biological processes via which pyrazole derivatives exert their effects would increase and accelerate the development of tailored medicines, including examining their interactions with specific enzymes, receptors, and signaling cascades. Pyrazole derivatives have enormous promise as novel therapeutic medicines, considering their various biological functions and ongoing advances in synthetic techniques. Their adaptability enables the development of customised molecules for specific illnesses with great effectiveness and minimum negative effects. For example, in oncology, pyrazole derivatives with focused anticancer properties may lead to more successful therapies with fewer side effects than standard chemotherapy. The discovery of novel pyrazole-based antimicrobials in infectious illnesses might address the rising problem of antibiotic resistance. Furthermore, the neuroprotective effects of pyrazole derivatives provide promise for innovative therapies for neurodegenerative illnesses that now have few therapeutic choices.

5. CONCLUSION

Heterocyclic compounds, particularly those containing nitrogen, have emerged as indispensable building blocks in drug discovery. Pyrazoles, with their versatile scaffold and ability to modulate a wide spectrum of biological targets, exemplify this potential. Their structural diversity and ease of synthesis have made them attractive candidates for developing novel therapeutic agents. This chapter delves into the intricacies of pyrazole chemistry, encompassing their synthesis, structural modifications, and biological evaluation, with a focus on their potential as promising drug candidates. Pyrazole derivatives, renowned for their structural diversity, are pivotal in modern drug discovery. This chapter explores efficient and versatile synthetic routes to access a wide array of functionalized pyrazoles, encompassing both classical and innovative methodologies. Our aim is to unlock the full potential of these compounds for therapeutic applications. Understanding the structure-activity relationships (SARs) of pyrazole derivatives is crucial for optimizing their biological properties. By systematically modifying substituents and exploring diverse scaffolds, researchers can fine-tune the activity and drug-like properties of these compounds. A target-oriented or focused-library approach accelerates this process, enabling the rapid generation and evaluation of pyrazole analog. By employing this approach, researchers can make informed adjustments to optimize the efficacy and suitability of these compounds for therapeutic applications. Recent advancements in technology and the ability to predict drug-like and lead-like properties further enhance the potential of pyrazole derivatives. These advancements may significantly boost the exploration and development of new drugs, underscoring the ongoing relevance and promise of pyrazole-based compounds in drug discovery.

REFERENCES

Abdellatif, K. R., Chowdhury, M. A., Velázquez, C. A., Huang, Z., Dong, Y., Das, D., Yu, G., Suresh, M. R., & Knaus, E. E. (2010). Celecoxib prodrugs possessing a diazen-1-ium-1,2-diolate nitric oxide donor moiety: Synthesis, biological evaluation and nitric oxide release studies. *Bioorganic & Medicinal Chemistry Letters*, 20(15), 4544–4549. DOI: 10.1016/j.bmcl.2010.06.022 PMID: 20576432

Adhikari, A., Kalluraya, B., Sujith, K. V., Gouthamchandra, K., Jairam, R., Mahmood, R., & Sankolli, R. (2012). Synthesis, characterization and pharmacological study of 4,5-dihydropyrazolines carrying pyrimidine moiety. *European Journal of Medicinal Chemistry*, 55, 467–474. DOI: 10.1016/j.ejmech.2012.07.002 PMID: 22877623

Ahmad, M., Siddiqui, H. L., Zia-Ur-Rehman, M., & Parvez, M. (2010). Anti-oxidant and anti-bacterial activities of novel N′-arylmethylidene-2-(3, 4-dimethyl-5, 5-dioxidopyrazolo[4,3- *c*][1,2]benzothiazin-2(4*H*)-yl) acetohydrazides. *European Journal of Medicinal Chemistry*, 45(2), 698–704. DOI: 10.1016/j.ejmech.2009.11.016 PMID: 19962218

Akbas, E., Berber, I., Sener, A., & Hasanov, B. (2005). Synthesis and antibacterial activity of 4-benzoyl-1-methyl-5-phenyl-1*H*-pyrazole-3-carboxylic acid and derivatives. *Il Farmaco*, 60(1), 23–26. DOI: 10.1016/j.farmac.2004.09.003 PMID: 15652364

Alam, M. J., Alam, O., Alam, P., & Naim, M. J. (2015). A review on pyrazole chemical entity and biological activity. *International Journal of Pharmaceutical Sciences and Research*, 6, 1433–1442.

Ambethkar, S., Padmini, V., & Bhuvanesh, N. (2015). A green and efficient protocol for the synthesis of dihydropyrano[2,3-*c*]pyrazole derivatives via a one-pot, four component reaction by grinding method. *Journal of Advanced Research*, 6(6), 975–985. DOI: 10.1016/j.jare.2014.11.011 PMID: 26644936

Ansari, A., Ali, A., Asif, M., & Shamsuzzaman, S. (2017). Review: Biologically active pyrazole derivatives. *New Journal of Chemistry*, 41(1), 16–41. DOI: 10.1039/C6NJ03181A

Argade, N. D., Kalrale, B. K., & Gill, C. H. (2008). Microwave Assisted Improved Method for the Synthesis of Pyrazole Containing 2,4,-Disubstituted Oxazole-5-one and their Antimicrobial Activity. *Journal of Chemistry*, 5(1), 120–129. DOI: 10.1155/2008/265131

Bandgar, B. P., Gawande, S. S., Bodade, R. G., Gawande, N. M., & Khobragade, C. N. (2009). Synthesis and biological evaluation of a novel series of pyrazole chalcones as anti-inflammatory, antioxidant and antimicrobial agents. *Bioorganic & Medicinal Chemistry*, 17(24), 8168–8173. DOI: 10.1016/j.bmc.2009.10.035 PMID: 19896853

Bansal, S., Bala, M., Suthar, S. K., Choudhary, S., Bhattacharya, S., Bhardwaj, V., Singla, S., & Joseph, A. (2014). Design and synthesis of novel 2-phenyl-5-(1,3-diphenyl-1*H*-pyrazol-4-yl)-1,3,4-oxadiazoles as selective COX-2 inhibitors with potent anti-inflammatory activity. *European Journal of Medicinal Chemistry*, 80, 167–174. DOI: 10.1016/j.ejmech.2014.04.045 PMID: 24780593

Behr, L. C., Fusco, R., & Jarboe, C. H. (1967). The chemistry of heterocyclic compounds pyrazoles, pyrazolines, pyrazolidines, indazoles and condensed rings. In R. H. Wiley (Ed.), *Pyrazoles, Pyrazolines, Pyrazolidines, Indazoles and Condensed Rings* (pp. 888-950). Wiley New York: Intersci. Pub.

Bhat, L., Jandeleit, B., Dias, T. M., Moors, T. L., & Gallop, M. A. (2005). Synthesis and biological evaluation of novel steroidal pyrazoles as substrates for bile acid transporters. *Bioorganic & Medicinal Chemistry Letters*, 15(1), 85–87. DOI: 10.1016/j.bmcl.2004.10.027 PMID: 15582416

Borpatra, P. J., Rastogi, G. K., Saikia, B., Deb, M. L., & Baruah, P. K. (2019). Multi-Component Reaction of 6-Aminouracils, Aldehydes and Secondary Amines: Conversion of the Products into Pyrimido[4,5-*d*]pyrimidines through C-H Amination/Cyclization. *ChemistrySelect*, 4(12), 3381–3386. DOI: 10.1002/slct.201900210

Brigance, R. P., Meng, W., Fura, A., Harrity, T., Wang, A., Zahler, R., Kirby, M. S., & Hamann, L. G. (2010). Synthesis and SAR of azolopyrimidines as potent and selective dipeptidyl peptidase-4 (DPP4) inhibitors for type 2 diabetes. *Bioorganic & Medicinal Chemistry Letters*, 20(15), 4395–4398. DOI: 10.1016/j.bmcl.2010.06.063 PMID: 20598534

Castillo, J. C., & Portilla, J. (2018). Recent advances in the synthesis of new pyrazole derivatives. *Targets Heterocycl. Syst, 22*, 194-223. http://dx.medra.org/10.17374/targets.2019.22.194

Chandanshive, J. Z. (2013). Regiocontrolled Synthesis of Pyrazole Derivatives Through 1,3-Dipolar Cycloaddition Reaction and Synthesis of Helicene-Thiourea based and Polymer Supported Soos's Catalyst for Asymmetric Synthesis. DOI: 10.6092/unibo/amsdottorato/5826

Charlier, C., & Michaux, C. (2003). Dual inhibition of cyclooxygenase-2 (COX-2) and 5-lipoxygenase (5-LOX) as a new strategy to provide safer non-steroidal anti-inflammatory drugs. *European Journal of Medicinal Chemistry*, 38(7–8), 645–659. DOI: 10.1016/S0223-5234(03)00115-6 PMID: 12932896

Chaudhry, F., Naureen, S., Huma, R., Shaukat, A., Al-Rashida, M., Asif, N., Ashraf, M., Munawar, M. A., & Khan, M. A. (2017). In search of new α -glucosidase inhibitors: Imidazolylpyrazole derivatives. *Bioorganic Chemistry*, 71, 102–109. DOI: 10.1016/j.bioorg.2017.01.017 PMID: 28160945

Chowdhury, M. A., Abdellatif, K. R., Dong, Y., Yu, G., Huang, Z., Rahman, M., Das, D., Velázquez, C. A., Suresh, M. R., & Knaus, E. E. (2010). Celecoxib analogs possessing a N-(4-nitrooxybutyl)piperidin-4-yl or N-(4-nitrooxybutyl)-1,2,3,6-tetrahydropyridin-4-yl nitric oxide donor moiety: Synthesis, biological evaluation and nitric oxide release studies. *Bioorganic & Medicinal Chemistry Letters*, 20(4), 1324–1329. DOI: 10.1016/j.bmcl.2010.01.014 PMID: 20097072

Chuang, H., Huang, L. S., Kapoor, M., Liao, Y., Yang, C., Chang, C., Wu, C., Hwu, J. R., Huang, T., & Hsu, M. (2016). Design and synthesis of pyridine-pyrazole-sulfonate derivatives as potential anti-HBV agents. *MedChemComm*, 7(5), 832–836. DOI: 10.1039/C6MD00008H

Dalton, T., Faber, T., & Glorius, F. (2021). C–H activation: Toward sustainability and applications. *ACS Central Science*, 7(2), 245–261. DOI: 10.1021/acscentsci.0c01413 PMID: 33655064

Danishuddin, M., Kaushal, L., Baig, M. H., & Khan, A. U. (2012). AMDD: Antimicrobial Drug Database. *Genomics, Proteomics & Bioinformatics*, 10(6), 360–363. DOI: 10.1016/j.gpb.2012.04.002 PMID: 23317704

Deb, M. L., Pegu, C. D., Borpatra, P. J., & Baruah, P. K. (2016). Copper catalyzed oxidative deamination of Betti bases: An efficient approach for benzoylation/formylation of naphthols and phenols. *RSC Advances*, 6(46), 40552–40559. DOI: 10.1039/C6RA04567G

Eftekhari-Sis, B., Zirak, M., & Akbari, A. (2013). Arylglyoxals in synthesis of heterocyclic compounds. *Chemical Reviews*, 113(5), 2958–3043. DOI: 10.1021/cr300176g PMID: 23347156

El-Gamal, M. I., Park, Y. S., Chi, D. Y., Yoo, K. H., & Oh, C. (2013). New tri-arylpyrazoles as broad-spectrum anticancer agents: Design, synthesis, and biological evaluation. *European Journal of Medicinal Chemistry*, 65, 315–322. DOI: 10.1016/j.ejmech.2013.04.067 PMID: 23732996

Faria, J. V., Vegi, P. F., Miguita, A. G. C., Santos, M. S. D., Boechat, N., & Bernardino, A. M. R. (2017). Recently reported biological activities of pyrazole compounds. *Bioorganic & Medicinal Chemistry*, 25(21), 5891–5903. DOI: 10.1016/j.bmc.2017.09.035 PMID: 28988624

Fustero, S., Sánchez-Roselló, M., Barrio, P., & Simón-Fuentes, A. (2011). From 2000 to mid-2010: A fruitful decade for the synthesis of pyrazoles. *Chemical Reviews*, 111(11), 6984–7034. DOI: 10.1021/cr2000459 PMID: 21806021

Ganguly, S. (2016). A battle against AIDS: New pyrazole key to an older lock-reverse transcriptase. *International Journal of Pharmacy and Pharmaceutical Sciences*, 8(11), 75–79.

Gökhan-Kelekçi, N., Koyunoğlu, S., Yabanoğlu, S., Yelekçi, K., Ozgen, O., Uçar, G., Erol, K., Kendi, E., & Yeşilada, A. (2009). New pyrazoline bearing 4(3*H*)-quinazolinone inhibitors of monoamine oxidase: Synthesis, biological evaluation, and structural determinants of MAO-A and MAO-B selectivity. *Bioorganic & Medicinal Chemistry*, 17(2), 675–689. DOI: 10.1016/j.bmc.2008.11.068 PMID: 19091581

Grosser, T. (2005). Biological basis for the cardiovascular consequences of COX-2 inhibition: Therapeutic challenges and opportunities. *The Journal of Clinical Investigation*, 116(1), 4–15. DOI: 10.1172/JCI27291 PMID: 16395396

Han, C., Guo, Y., Wang, D., Dai, X., Wu, F., Liu, H., Dai, G., & Tao, J. (2015). Novel pyrazole fused heterocyclic ligands: Synthesis, characterization, DNA binding/cleavage activity and anti-BVDV activity. *Chinese Chemical Letters*, 26(5), 534–538. DOI: 10.1016/j.cclet.2015.01.006

Hassan, G. S., Abou-Seri, S. M., Kamel, G., & Ali, M. M. (2014). Celecoxib analogs bearing benzofuran moiety as cyclooxygenase-2 inhibitors: Design, synthesis and evaluation as potential anti-inflammatory agents. *European Journal of Medicinal Chemistry*, 76, 482–493. DOI: 10.1016/j.ejmech.2014.02.033 PMID: 24607877

Hassani, I. A. E., Rouzi, K., Assila, H., Karrouchi, K., & Ansar, M. (2023). Recent Advances in the synthesis of pyrazole derivatives: A review. *Reactions*, 4(3), 478–504. DOI: 10.3390/reactions4030029

Huang, X., Lu, X., Zhang, Y., Song, G., He, Q., Li, Q., Yang, X., Wei, Y., & Zhu, H. (2012). Synthesis, biological evaluation, and molecular docking studies of N-((1,3-diphenyl-1*H*-pyrazol-4-yl)methyl)aniline derivatives as novel anticancer agents. *Bioorganic & Medicinal Chemistry*, 20(16), 4895–4900. DOI: 10.1016/j.bmc.2012.06.056 PMID: 22819191

Iyer, P. C., Zhao, J., Emert-Sedlak, L. A., Moore, K. K., Smithgall, T. E., & Day, B. W. (2014). Synthesis and structure–activity analysis of diphenylpyrazolodiazene inhibitors of the HIV-1 Nef virulence factor. *Bioorganic & Medicinal Chemistry Letters*, 24(7), 1702–1706. DOI: 10.1016/j.bmcl.2014.02.045 PMID: 24650642

JH. (1999). Nonsteroidal anti-inflammatory agents. *Drugs of Today (Barcelona, Spain)*, 35(4–5), 225. DOI: 10.1358/dot.1999.35.4-5.552199

Ju, Y., & Varma, R. S. (2005). Aqueous N-Heterocyclization of Primary Amines and Hydrazines with Dihalides: Microwave-Assisted Syntheses of N-Azacycloalkanes, Isoindole, Pyrazole, Pyrazolidine, and Phthalazine Derivatives. *The Journal of Organic Chemistry*, 71(1), 135–141. DOI: 10.1021/jo051878h PMID: 16388628

Kakinuma, H., Oi, T., Hashimoto-Tsuchiya, Y., Arai, M., Kawakita, Y., Fukasawa, Y., Iida, I., Hagima, N., Takeuchi, H., Chino, Y., Asami, J., Okumura-Kitajima, L., Io, F., Yamamoto, D., Miyata, N., Takahashi, T., Uchida, S., & Yamamoto, K. (2010). (1*S*)-1,5-Anhydro-1-[5-(4-ethoxybenzyl)-2-methoxy-4-methylphenyl]-1-thio-d-glucitol (TS-071) is a Potent, Selective Sodium-Dependent Glucose Cotransporter 2 (SGLT2) Inhibitor for Type 2 Diabetes Treatment. *Journal of Medicinal Chemistry*, 53(8), 3247–3261. DOI: 10.1021/jm901893x PMID: 20302302

Kerru, N., Singh-Pillay, A., Awolade, P., & Singh, P. (2018). Current anti-diabetic agents and their molecular targets: A review. *European Journal of Medicinal Chemistry*, 152, 436–488. DOI: 10.1016/j.ejmech.2018.04.061 PMID: 29751237

Kim, J., Lee, D., Park, C., So, W., Jo, M., Ok, T., Kwon, J., Kong, S., Jo, S., Kim, Y., Choi, J., Kim, H. C., Ko, Y., Choi, I., Park, Y., Yoon, J., Ju, M. K., Kim, J., Han, S., & No, Z. (2012). Discovery of phenylaminopyridine derivatives as novel HIV-1 non-nucleoside reverse transcriptase inhibitors. *ACS Medicinal Chemistry Letters*, 3(8), 678–682. DOI: 10.1021/ml300146q PMID: 24900529

Koca, İ., Özgür, A., Coşkun, K. A., & Tutar, Y. (2013). Synthesis and anticancer activity of acyl thioureas bearing pyrazole moiety. *Bioorganic & Medicinal Chemistry*, 21(13), 3859–3865. DOI: 10.1016/j.bmc.2013.04.021 PMID: 23664495

Kohli, P., Steg, P. G., Cannon, C. P., Smith, S. C.Jr, Eagle, K. A., Ohman, E. M., Alberts, M. J., Hoffman, E., Guo, J., Simon, T., Sorbets, E., Goto, S., & Bhatt, D. L. (2014). NSAID Use and Association with Cardiovascular Outcomes in Outpatients with Stable Atherothrombotic Disease. *The American Journal of Medicine*, 127(1), 53–60.e1. DOI: 10.1016/j.amjmed.2013.08.017 PMID: 24280110

Konwar, M., Phukan, P., Chaliha, A. K., Buragohain, A. K., Damarla, K., Gogoi, D., Kumar, A., & Sarma, D. (2019). An Unexplored Lewis Acidic Catalytic System for Synthesis of Pyrazole and its Biaryls Derivatives with Antimicrobial Activities through Cycloaddition-Iodination-Suzuki Reaction. *ChemistrySelect*, 4(35), 10236–10245. DOI: 10.1002/slct.201902266

Kost, A. N. (1996). *Grandberg II. Progress in Pyrazole Chemistry. Adv Het Chem. AR*. Katritzky, Academic Press New York.

Küçükgüzel, Ş., & Şenkardeş, S. (2015). Recent advances in bioactive pyrazoles. *European Journal of Medicinal Chemistry*, 97, 786–815. DOI: 10.1016/j.ejmech.2014.11.059 PMID: 25555743

Li, X., He, L., Chen, H., Wu, W., & Jiang, H. (2013). Copper-Catalyzed Aerobic $C(sp^2)$–H Functionalization for C–N Bond Formation: Synthesis of Pyrazoles and Indazoles. *The Journal of Organic Chemistry*, 78(8), 3636–3646. DOI: 10.1021/jo400162d PMID: 23547954

Litvinov, Y. M., Shestopalov, A. A., Rodinovskaya, L. A., & Shestopalov, A. M. (2009). New Convenient Four-Component Synthesis of 6-Amino-2,4-dihydropyrano[2,3-c]pyrazol-5-carbonitriles and One-Pot Synthesis of 6'-Aminospiro[(3H)-indol-3,4'-pyrano[2,3-c]pyrazol]-(1H)-2-on-5'-carbonitriles. *Journal of Combinatorial Chemistry*, 11(5), 914–919. DOI: 10.1021/cc900076j PMID: 19711896

Liu, H., Ren, Z., Wang, W., Gong, J., Chu, M., Ma, Q., Wang, J., & Lv, X. (2018). Novel coumarin-pyrazole carboxamide derivatives as potential topoisomerase II inhibitors: Design, synthesis and antibacterial activity. *European Journal of Medicinal Chemistry*, 157, 81–87. DOI: 10.1016/j.ejmech.2018.07.059 PMID: 30075404

Macarini, A. F., Sobrinho, T. U. C., Rizzi, G. W., & Corrêa, R. (2019). Pyrazole–chalcone derivatives as selective COX-2 inhibitors: Design, virtual screening, and in vitro analysis. *Medicinal Chemistry Research*, 28(8), 1235–1245. DOI: 10.1007/s00044-019-02368-8

Malladi, S., Isloor, A. M., Shetty, P., Fun, H. K., Telkar, S., Mahmood, R., & Isloor, N. (2011). Synthesis and anti-inflammatory evaluation of some new 3,6-disubstituted-1,2,4-triazolo-[3,4-b]-1,3,4-thiadiazoles bearing pyrazole moiety. *Medicinal Chemistry Research*, 21(10), 3272–3280. DOI: 10.1007/s00044-011-9865-0

Marichev, K. O., & Doyle, M. P. (2019). Catalytic asymmetric cycloaddition reactions of enoldiazo compounds. *Organic & Biomolecular Chemistry*, 17(17), 4183–4195. DOI: 10.1039/C9OB00478E PMID: 30924829

Marinescu, M., & Zalaru, C. M. (2021). *Synthesis, Antibacterial and Anti-Tumor Activity of Pyrazole Derivatives. Recent Trends in Biochemistry.* MedDocs Publishers.

Mitra, A. (2008). Some salient points in Dietary and Life-Style survey of rural Bengal particularly tribal populace in relation to rural diabetes prevalence. *Studies on Ethno-Medicine*, 2(1), 51–56. DOI: 10.1080/09735070.2008.11886315

Mukherjee, R. N. (2000). Coordination chemistry with pyrazole-based chelating ligands: Molecular structural aspects. *Coordination Chemistry Reviews*, 203(1), 151–218. DOI: 10.1016/S0010-8545(99)00144-7

Nandurkar, D., Danao, K., Rokde, V., Shivhare, R., & Mahajan, U. (2023). Pyrazole Scaffold: Strategies toward the Synthesis and Their Applications. In *IntechOpen eBooks*. DOI: 10.5772/intechopen.108764

Naveen, S., Kumara, K., Kumar, A. D., Kumar, K. A., Zarrouk, A., Warad, I., & Lokanath, N. (2021). Synthesis, characterization, crystal structure, Hirshfeld surface analysis, antioxidant properties and DFT calculations of a novel pyrazole derivative: Ethyl 1-(2,4-dimethylphenyl)-3-methyl-5-phenyl-1*H*-pyrazole-4-carboxylate. *Journal of Molecular Structure*, 1226, 129350. DOI: 10.1016/j.molstruc.2020.129350

Nimse, S. B., & Pal, D. (2015). Free radicals, natural antioxidants, and their reaction mechanisms. *RSC Advances*, 5(35), 27986–28006. DOI: 10.1039/C4RA13315C

Nisa, N. U., & Astana, N. P. R. W. (2019). Evaluation of antiurolithic herbal formula for urolithiasis: A randomized open-label clinical study. *Asian Journal of Pharmaceutical and Clinical Research*, ●●●, 88–93. DOI: 10.22159/ajpcr.2019.v12i4.30232

Pirol, Ş. C., Çalışkan, B., Durmaz, I., Atalay, R., & Banoglu, E. (2014). Synthesis and preliminary mechanistic evaluation of 5-(*p*-tolyl)-1-(quinolin-2-yl)pyrazole-3-carboxylic acid amides with potent antiproliferative activity on human cancer cell lines. *European Journal of Medicinal Chemistry*, 87, 140–149. DOI: 10.1016/j.ejmech.2014.09.056 PMID: 25247770

Prieto, A., Bouyssi, D., & Monteiro, N. (2017). Ruthenium-Catalyzed Tandem C–H Fluoromethylation/Cyclization of N-Alkylhydrazones with CBr_3F: Access to 4-Fluoropyrazoles. *The Journal of Organic Chemistry*, 82(6), 3311–3316. DOI: 10.1021/acs.joc.7b00085 PMID: 28263600

Rai, U. S., Isloor, A., Shetty, P., Pai, K., & Fun, H. (2015). Synthesis and in vitro biological evaluation of new pyrazole chalcones and heterocyclic diamides as potential anticancer agents. *Arabian Journal of Chemistry*, 8(3), 317–321. DOI: 10.1016/j.arabjc.2014.01.018

Rastogi, G. K., Saikia, B., Pahari, P., Deb, M. L., & Baruah, P. K. (2019). Cu(I)/Fe(III) promoted dicarbonylation of aminopyrazole via oxidative C H coupling with methyl ketones. *Tetrahedron Letters*, 60(17), 1189–1192. DOI: 10.1016/j.tetlet.2019.03.059

Sabet-Sarvestani, H., Eshghi, H., Bakavoli, M., Izadyar, M., & Rahimizadeh, M. (2014). Theoretical investigation of the chemoselectivity and synchronously pyrazole ring formation mechanism from ethoxymethylenemalononitrile and hydrazine hydrate in the gas and solvent phases: DFT, meta-GGA studies and NBO analysis. *RSC Advances*, 4(82), 43485–43495. DOI: 10.1039/C4RA06316C

Silva, V. L., Elguero, J., & Silva, A. M. (2018). Current progress on antioxidants incorporating the pyrazole core. *European Journal of Medicinal Chemistry*, 156, 394–429. DOI: 10.1016/j.ejmech.2018.07.007 PMID: 30015075

Somakala, K., Amir, M., Sharma, V., & Wakode, S. (2016). Synthesis and pharmacological evaluation of pyrazole derivatives containing sulfonamide moiety. *Monatshefte Für Chemie - Chemical Monthly, 147*(11), 2017–2029. DOI: 10.1007/s00706-016-1694-x

Swedberg, J. E., Schroeder, C. I., Mitchell, J. M., Durek, T., Fairlie, D. P., Edmonds, D. J., Griffith, D. A., Ruggeri, R. B., Derksen, D. R., Loria, P. M., Liras, S., Price, D. A., & Craik, D. J. (2015). Cyclic alpha-conotoxin peptidomimetic chimeras as potent GLP-1R agonists. *European Journal of Medicinal Chemistry*, 103, 175–184. DOI: 10.1016/j.ejmech.2015.08.046 PMID: 26352676

Toda, N., Hao, X., Ogawa, Y., Oda, K., Yu, M., Fu, Z., Chen, Y., Kim, Y., Lizarzaburu, M., Lively, S., Lawlis, S., Murakoshi, M., Nara, F., Watanabe, N., Reagan, J. D., Tian, H., Fu, A., Motani, A., Liu, Q., & Shibuya, S. (2013). Potent and orally bioavailable GPR142 agonists as novel insulin secretagogues for the treatment of type 2 diabetes. *ACS Medicinal Chemistry Letters*, 4(8), 790–794. DOI: 10.1021/ml400186z PMID: 24900747

Tyagarajan, S., Chakravarty, P. K., Zhou, B., Taylor, B., Eid, R., Fisher, M. H., Parsons, W. H., Wyvratt, M. J., Lyons, K. A., Klatt, T., Li, X., Kumar, S., Williams, B., Felix, J., Priest, B. T., Brochu, R. M., Warren, V., Smith, M., Garcia, M., & Duffy, J. L. (2010). Discovery of a novel class of biphenyl pyrazole sodium channel blockers for treatment of neuropathic pain. *Bioorganic & Medicinal Chemistry Letters*, 20(24), 7479–7482. DOI: 10.1016/j.bmcl.2010.10.017 PMID: 21106456

Vardanyan, R., & Hruby, V. (2016). Antiviral drugs, *Synthesis of Best-Seller Drugs* (pp. 687–736). Elsevier eBooks. DOI: 10.1016/B978-0-12-411492-0.00034-1

Vijesh, A. M., Isloor, A. M., Telkar, S., Peethambar, S. K., Rai, S., & Isloor, N. (2011). Synthesis, characterization and antimicrobial studies of some new pyrazole incorporated imidazole derivatives. *European Journal of Medicinal Chemistry*, 46(8), 3531–3536. DOI: 10.1016/j.ejmech.2011.05.005 PMID: 21620535

Wright, J. S., Johnson, E. R., & DiLabio, G. A. (2001). Predicting the activity of phenolic antioxidants: Theoretical method, analysis of substituent effects, and application to major families of antioxidants. *Journal of the American Chemical Society*, 123(6), 1173–1183. DOI: 10.1021/ja002455u PMID: 11456671

Xiong, Y., Guo, J., Candelore, M. R., Liang, R., Miller, C., Dallas-Yang, Q., Jiang, G., McCann, P. E., Qureshi, S. A., Tong, X., Xu, S. S., Shang, J., Vincent, S. H., Tota, L. M., Wright, M. J., Yang, X., Zhang, B. B., Tata, J. R., & Parmee, E. R. (2012). Discovery of a Novel Glucagon Receptor Antagonist *N*-[(4-{(1*S*)-1-[3-(3, 5-Dichlorophenyl)-5-(6-methoxynaphthalen-2-yl)-1*H*-pyrazol-1-yl]ethyl}phenyl) carbonyl]-β-alanine (MK-0893) for the Treatment of Type II Diabetes. *Journal of Medicinal Chemistry*, 55(13), 6137–6148. DOI: 10.1021/jm300579z PMID: 22708876

Yi, F., Zhao, W., Wang, Z., & Bi, X. (2019). Silver-Mediated [3 + 2] cycloaddition of alkynes and N-Isocyanoiminotriphenylphosphorane: Access to monosubstituted pyrazoles. *Organic Letters*, 21(9), 3158–3161. DOI: 10.1021/acs.orglett.9b00860 PMID: 30990050

Zhang, D., Wang, G., Zhao, G., Xu, W., & Huo, L. (2011). Synthesis and cytotoxic activity of novel 3-(1*H*-indol-3-yl)-1*H*-pyrazole-5-carbohydrazide derivatives. *European Journal of Medicinal Chemistry*, 46(12), 5868–5877. DOI: 10.1016/j.ejmech.2011.09.049 PMID: 22000925

Zhang, G., Ni, H., Chen, W., Shao, J., Liu, H., Chen, B., & Yu, Y. (2013). One-Pot Three-Component approach to the synthesis of polyfunctional pyrazoles. *Organic Letters*, 15(23), 5967–5969. DOI: 10.1021/ol402810f PMID: 24255982

Chapter 3
Isoxazole:
A Bioactive Five–Membered Heterocycle With Diverse Applications

Popat Mohite

https://orcid.org/0000-0002-4536-4444

AETs St. John Institute of Pharmacy and Research, Palghar, India

Savita Tauro

AETs St. John Institute of Pharmacy and Research, Palghar, India

Aarati Pawar

AETs St. John Institute of Pharmacy and Research, Palghar, India

ABSTRACT

Isoxazole is a versatile five-membered heterocyclic compound that has attracted significant attention due to its diverse medicinal chemistry and material science applications. This chapter provides a comprehensive overview of synthetic methodologies for preparing isoxazole derivatives, including traditional approaches such as cyclization reactions and modern techniques like microwave-assisted synthesis. The advantages and limitations of each method, highlighting the influence of structural variations on synthesis efficiency, are discussed. Furthermore, the biological activities associated with isoxazole compounds, showcasing their roles as potential pharmacological agents including anti-inflammatory, antibacterial, antifungal, and anticancer properties, are explored. By integrating synthetic strategies with biological evaluation, this chapter aims to illustrate the significance of isoxazole derivatives in drug discovery and development and their potential applications in therapeutic settings.

DOI: 10.4018/979-8-3693-7267-8.ch003

1. INTRODUCTION

Organic compounds containing carbon and a heteroatom are known as Heterocyclic compounds. They have grown in number significantly because to both their synthetic utility and intense synthetic research. Heterocyclic compounds containing Nitrogen and Oxygen play a significant role in medicinal chemistry and due to their wide range of therapeutic activities, they are used for the synthesis of various medicinal agents. Due to the structural diversity of the heterocyclic compounds, these compounds possess a wide spectrum of therapeutic applications. Nitrogen-containing heterocyclic compounds are regarded as an important class of chemicals in medicinal research as it has a wide range of applications (Sysak & Obmińska-Mrukowicz, 2017). Among all, isoxazole plays a vital role in pharmaceutical applications. Isoxazole are unsaturated five-membered heterocyclic compound containing one nitrogen, one oxygen, and carbon in the ring. Since it was the isomer "oxazole" that was initially found, Hantszch proposed the word "isoxazole" to describe the five-membered fully unsaturated heterocycles. The trival name is based on the Hantszch-Widman system of nomenclature, where "iso" represents isomer, the nitrogen atom is represented by "aza" and the oxygen atom by "oxa". The name "isoxazole" is the result of adding the suffix "ole" which indicates that the ring is five-membered(Wiley, 2021). Isoxazole possess various therapeutic applications such as anti-bacterial(Shaik et al., 2020), anti-malarial(Mabasa et al., 2024), anti-viral(Li et al., 2017), anti-fungal((Trefzger et al., 2020), anti-cancer(Eid et al., 2021), anti-convulsant, anti-oxidant, anti-lipidemic, anti-inflammatory and analgesic(Abdelall, 2020).

Figure 1. Graphical representation of therapeutic applications of isoxazole

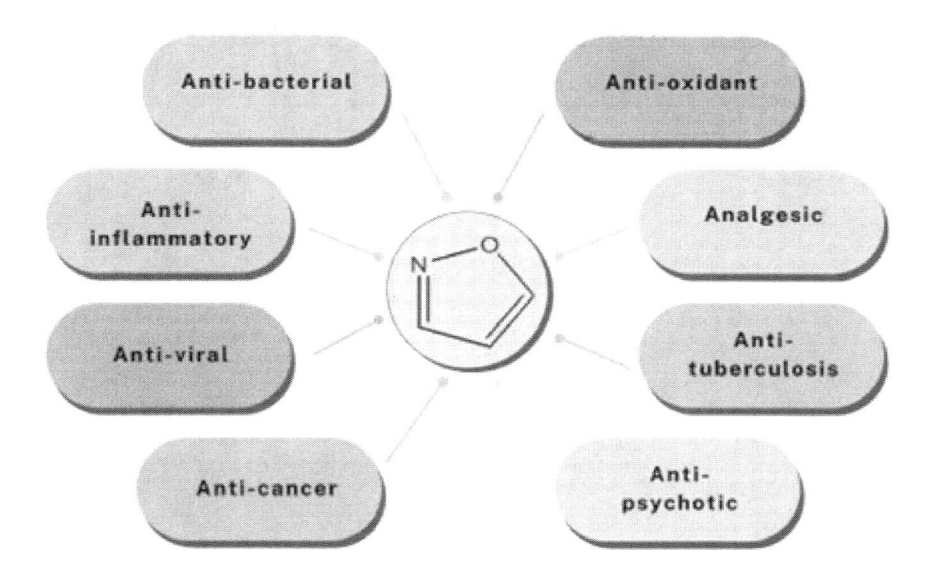

The analogs of isoxazole are isoxazolines which is partially saturated and isoxazolidine which is completely saturated (Figure 2)(• Kumar & Shankar, 2021; Ouzounthanasis et al., 2023). Isoxazole is an azole that has an oxygen atom next to the nitrogen and is a component of several biodynamic agents. It has a broad range of biological action. The versatility of substituted isoxazoles in undergoing chemical transformations to produce valuable synthetic intermediated makes them significant synthons as well. An aromatic and non-aromatic ring fused with an isoxazole is known as a fused isoxazole, a class of heterocyclic compounds(Mallik et al., 2024). It was found that fused isoxazoles had certain unusual structural characteristics. For example, the five-membered ring system in these molecules can cause certain angles to form in small molecules, which enhances the molecule's ability to interact with their target. This is quality that is hardly noticeable in molecules with six members. Medicinal chemists can modify different parts of an isoxazole to increase its potency, selectivity, and pharmacokinetics, as well as to optimize its therapeutic potential. Fused isoxazoles structural variety and adaptability enable the creation of molecules with strong selectivity for certain biological targets (Barmade et al., 2016).

Figure 2. Structure and analogs of isoxazole moiety

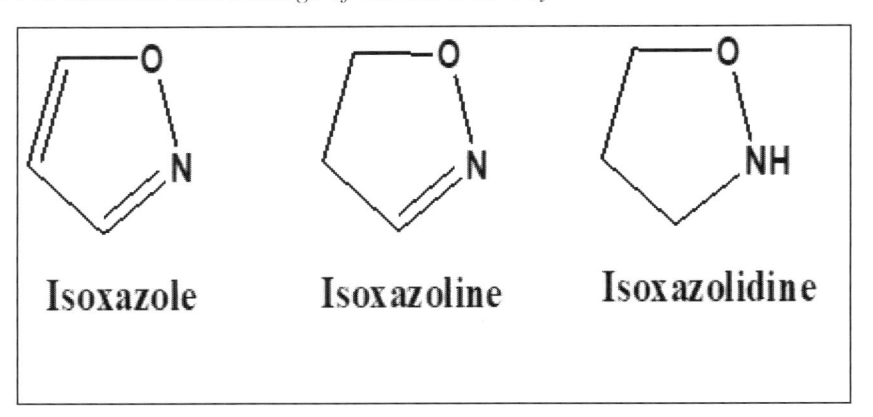

Some marketed drugs containing isoxazole nuclei are given in Table1:

Table 1. Isoxazoles marketed drugs

Drugs	Structure	Pharmacological class
Sulfamethoxazole		Antimicrobial and Anticancer (13)
Sulfisoxazole		Antibacterial (14)
Cycloserine		Antitubercular (15)
Acivicin		Antitumour (16)
Valdecoxib		COX-2 inhibitor (17)
Leflunomide		Antirheumatic (18)

Isoxazole heterocycle in particular has drawn a lot of attention due to its numerous uses in the industrial, agrochemical, pharmaceutical, and catalytic domains. The widespread use of hazardous solvents and additives in the synthetic technique has led to the adoption of green chemistry approaches for the production of isoxazole analogs. These techniques include the use of microwaves, ultrasonic-induced synthesis, and a few more environmentally friendly technologies (Basavanna et al., 2022). While there have been several reviews on isoxazole previously published, they have mostly concentrated on either the chemical interactions of the nucleus or its synthetic methods and their diverse applications. The book chapter provides an overview of the methodical compilation of synthesis techniques and the therapeutic journey that isoxazole and its derivatives have taken us thus far.

2. SYNTHETIC ROUTES

Md Imran Hossain et al., by [3+2] cycloaddition synthesized 3,4,5-trisubstituted isoxazoles. The synthesis was carried out at room temperature under mild basic conditions, where 95% water, 5% methanol is used as solvent mixture and base DIPEA. The reaction between nitrile oxides and 1,3-diketones, β- ketoesters, or β- ketoamides resulted in 3,4,5-trisubstituted isoxazoles (Hossain et al., 2022).

Guolan Dou et al., by reacting 3-(dimethylamino)-1-arylprop-2-en-1-ones with hydroxylamine hydrochloride in aqueous media developed isoxazole derivatives. The reaction has been carried out without using any catalyst (Dou et al., 2013).

Irina A. Mironova et al., by using hypervalent iodine species developed a method for synthesis of fused isoxazoles and isoxazolines via catalytic intramolecular oxidative cycloaddition of aldoximes (Mironova et al., 2022).

Amberlyst 15 or sodium hydrogen sulfate supported on silica gel catalyze the conversion of alpha-nitro ketones into nitrile oxides. Following in situ 1,3-dipolar cycloaddtion to alkynes yields extremly good yields of 3-acylisoxazoles (Itoh et al., 2021).

Figure 3. Synthetic routes of isoxazole

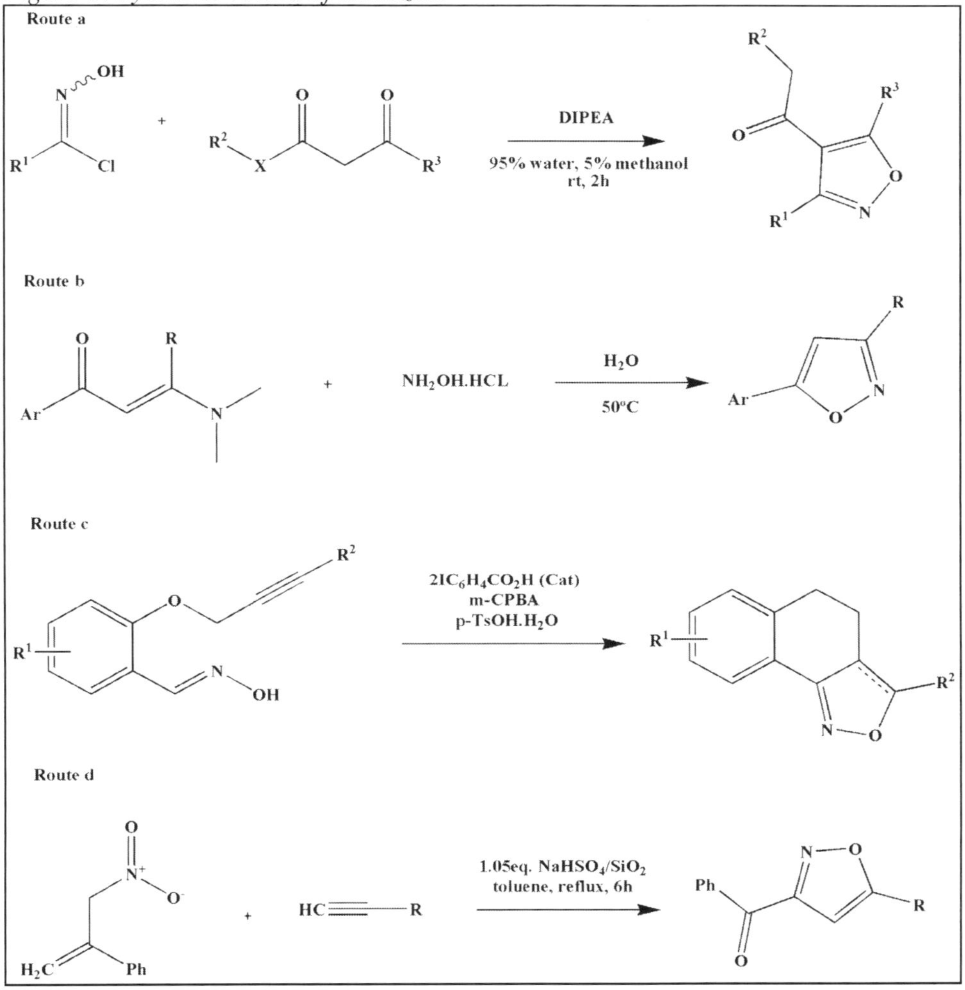

2.1 Synthesis of Fused Isoxazole Derivatives

A cinchona-derived amine catalyzed cyclic 2,4-dienone regio- and diasterodivergent [4+2] cycloadditions were explored by Wei Xiao et.al. Alkylidene isoxazole-5-(4H)-one was created in this reaction by reacting an aromatic aldehyde with a heteroaryl or 2-styryl group in the presence of 2,4-dienone. A [4+2] cycloaddtion process took place. Very basic beginning ingredients were utilized to prepare the Z-4-alkylideneisoxazol-5(4H)-one, a diasteroid Xiao et al., 2018). The oxidative conversion of oximes into 3-aryl isoxazoles by Mansour and co-workers was investigated. The oxidants such as copper chloride, acetonitrile, and manganese dioxide were used with unsaturated oximes under reflux to produce a compound with 50% yield. Lower yields were observed for the procedure that used solvents like THF and DMF (Kerim et al., 2020).

Alexander and colleagues investigated the synthesis of fused isoxazoles that sterically allowed intramolecular nucleophilic NMe2 group substitution. The reaction is an aromatic nucleophilic substitution known as SNAr. At 120ºC, the molecule 1,8-bis(dimethylamino)-2-napthaldehyde oxime was converted, and the NMe2 group was substituted to create the fused isoxazole N,N-dimethylnaptho[2,1-d]isoxazole-9-amine (Antonov et al., 2020).

3. BIOLOGICAL ACTIVITIES

Many researchers worldwide are working to generate pharmacologically active compounds containing the isoxazole ring due to its various and potentially strong pharmacological actions. The talks that follow demonstrate how isoxazole is used pharmacologically to treat different biological characteristics.

3.1 Analgesic and Anti-Inflammatory Isoxazoles

N-(4-methyoxyphenyl) acetamide is the outcome of the ten unique derivatives of isoxazole that Shital S. Sangle et al synthesized from 4-methoxy anilline. This molecules was then hydrolyzed with sodium hydroxide and treated with aromatic aldehydes to yield N-(4-methoxyphenyl)3-phenyl propanamide. Additionally, hydroxylamine hydrochloride in ethanol and N-(4-mehoxyphenyl)-3-substituted propanamides were refluxed on a water bath for six hours. The resulted compounds were derivatives of N-(4-methoxyphenyl)-5-phenyl-4,5-dihydroisoxazol-3-amine. Eddy's hot plate method was used to assess the analgesic efficacy of these synthesized drugs. Among the synthesized derivatives of isoxazole 5-(4-bromophenyl)-N-(4-methoxyphenyl) 4,5-dihydroisoxazole-3-amine i.e **compound 1a** (Fig No. 5) and 5-(4-chloro-3-nitrophenyl)-N-(4-methoxyphenyl) 4,5-dihydroisoxazole-3-amine **compound 1b** (Fig No. 5) showed good analgesic acitivity (Sangale et al., 2023).

A novel family of isoxazole linked quinazoline 4(3H)-one derivatives were synthesized and analyzed by G. Saravanan et.al. In this series of synthesized coumpounds, 2-methyl-3(4-(5-(4-(trifluromethyl)phenylisoxazol-3-yl)quinozolin-4(3H)-one i.e **compound 2** (Figure 5) demonstrated more effective analgesic and anti-inflammatory activity compared to the reference standard Diclofenac (Saravanan et al., 2021).

A novel series of isoxazole was synthsized by perrone et al., who then assessed the compounds selctivity and COX inhibitory action. **Compound 3b** (Figure 5) is a sub-micromolar selective COX-2 inhibitor, and **compound 3a** was discovered to be a selective COX-1 inhibitor (Perrone et al., 2016).

The Claisen Schmit condensation process was utilized to design and manufacture ten distinct isoxazole derivatives. The current study *in vitro* anti-inflammatory activities, or the inhibition of both COX-1 and COX-2 enzymes, revealed that while the tested **compounds 4a, 4b** and **4c** (Fig No. 5) demonstrated a strong effect against the COX-2 enzymes anti inflammatory assay, they were found to be a good candidate for an anti-inflammatory effect towards COX-1 enzymes (Alam et al., 2023).

A new series of isoxazole compounds with anticipated immunosuppressive properties was synthesized by Maczynski M, et al. (2015). Further, the activity in the human cell models *in vitro* and *in vivo* models in mouse of **compound 5** (Figure 5) i.e (ethyl N-{4-[(2,4-dimethoxybenzyl)carbamoyl]-3-methylisoxazol-5-yl} acetamide) was evaluated (Mączyński et al., 2016).

Figure 4. Isoxazole compounds having anti-inflammatory and analgesic activity.

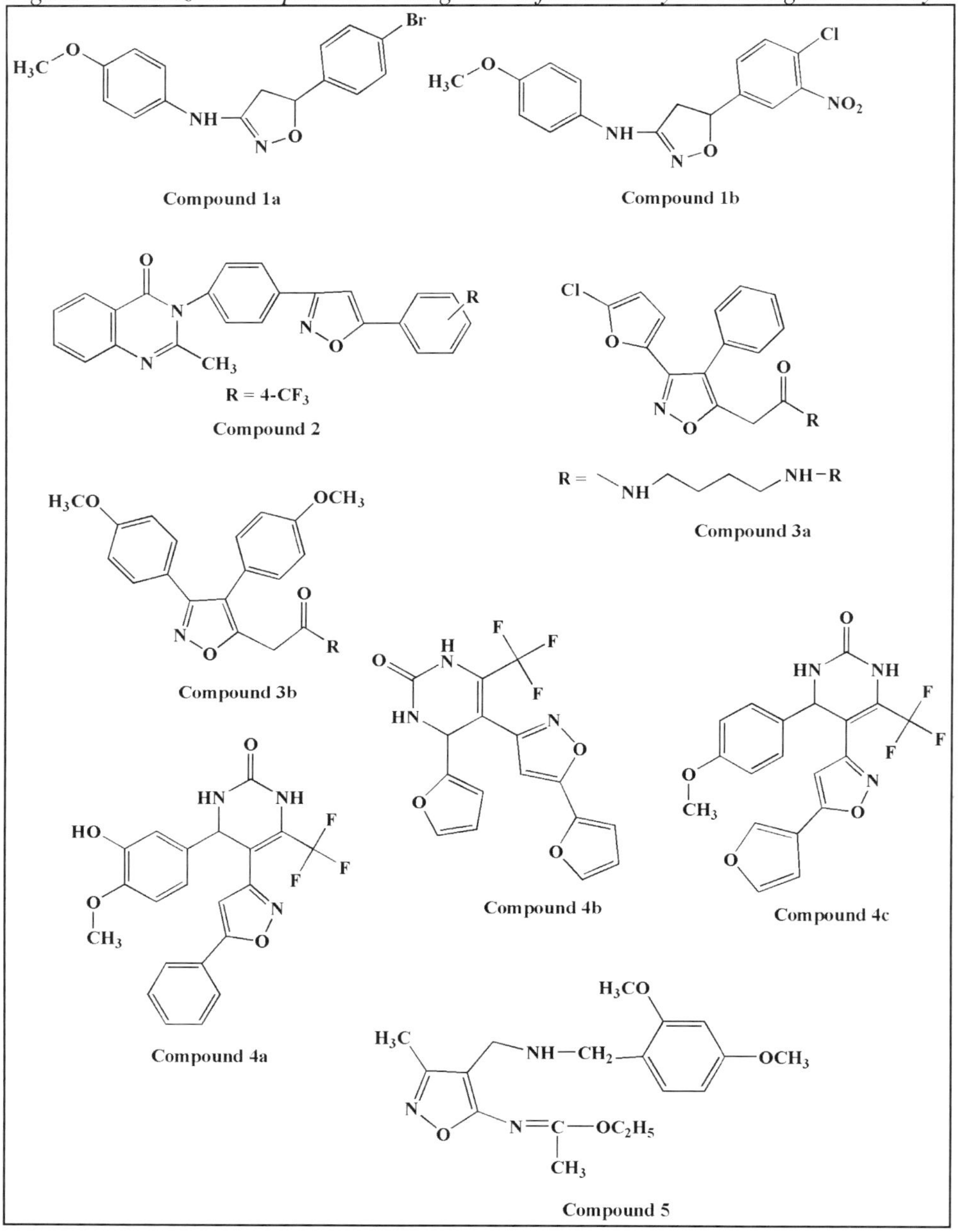

Compound 1a

Compound 1b

Compound 2

R = 4-CF$_3$

Compound 3a

Compound 3b

Compound 4b

Compound 4c

Compound 4a

Compound 5

3.2 Anti-Cancer/Anti-Tumor Activity Isoxazoles

The second greatest cause of death globally is cancer, a group of diseases marked by uncontrolled cell development with the ability to spread to other parts of the body. Ashmad M. Eid. Et.al synthesized several isoxazole-carboxamide derivatives and assessed for cytotoxic activity against breast (MCF-7), cervical (HeLa), and liver (Hep3B) cancer cell lines compared to Doxorubicin used as a standard. **Compounds 6a and 6b** (Figure 5) were active against Hep3B cells and reduced Hep3B secretion of alpha-fetoprotein, both compounds induced a delay in the G2/M phase of 18.07%, which is similar to the doxorubicin positive control. Hence compound indicates potent and promising anticancer activity (Eid et al., 2021).

In this research, Mohammed Hawash et.al, studies synthesized several phenyl-isoxazole-carboxamide derivatives and evaluated in vitro cytotoxic activity against four cancer cell lines: hepatocellular carcinoma (Hep3B and HepG2), cervical adenocarcinoma (HeLa), breast carcinoma (MCF-7) and normal line (Hek293T). The **compound 7** (Fig No. 5) demonstrated strong activity against HeLa and Hep3B cancer cell lines with IC50 values of 0.91 and 8.02 M (Hawash et al., 2021).

Zhang et al. (2017) produced a different series of 4,5-diaryl isoxazoles by substituting varied-length amino acid derivatives to the 3-amino motif. The Hsp90 inhibitory and anti-proliferative properties of these compounds were evaluated in respect to human lung cancer cells H3122 and breast cancer cells BT-474. **Compound 8** (Fig No. 5), which had a linear linker including ethylene glycol and a terminal valine methyl ester, showed strong Hsp90 binding potency (14 nM) and anti-proliferative action against the examined cell lines (Zhang et al., 2017).

Yang et al, developed compounds of pyridinyl-4,5-2H-isoxazole effect against the MCF-7 cell line in vitro. The only compounds that showed strong biological activity were those that have hydrophobic groups on the phenyl. **Compound 9a and 9b** (Fig No. 5) with the strongest antiproliferative activity against the human hepatoma cell line (HepG2) and the cervical cancer cell (Hela) were also shown to have the strongest antitumor efficacy. (Yang et al., 2017).

B. V. Durga Rao et al. synthesized isoxazole-thiadiazole linked carbazole derivatives. Using the MTT assay against the human cancer cell lines MCF-7 (breast), A549 (lung), DU-145 (prostate), and MDA MB-231 (breast), the synthesized **compound 10a and 10b** (Figure 5) were assessed for their anticancer activities. Etoposide was used as a reference drug and this compounds showed higher activity than etoposide. Compounds exhibit very strong activity, among these compound are the most active one (Durga Rao et al., 2019).

Chernysheva et al. CA4 was modified with an isoxazole ring and evaluated using an *in vivo* sea urchin embryo assay. They found that at a minimum effective concentration (MEC) of 0.002 μM, CA4 had anti-microtubule destabilizing effects, and that at 0.005 μM and 0.02 μM, respectively, isoxazole **compound 11** (Figure 5) altered the distribution of sea urchin embryos (Chernysheva et al., 2018).

Figure 5. Isoxazole compounds having anti-cancer/anti-tumor activity

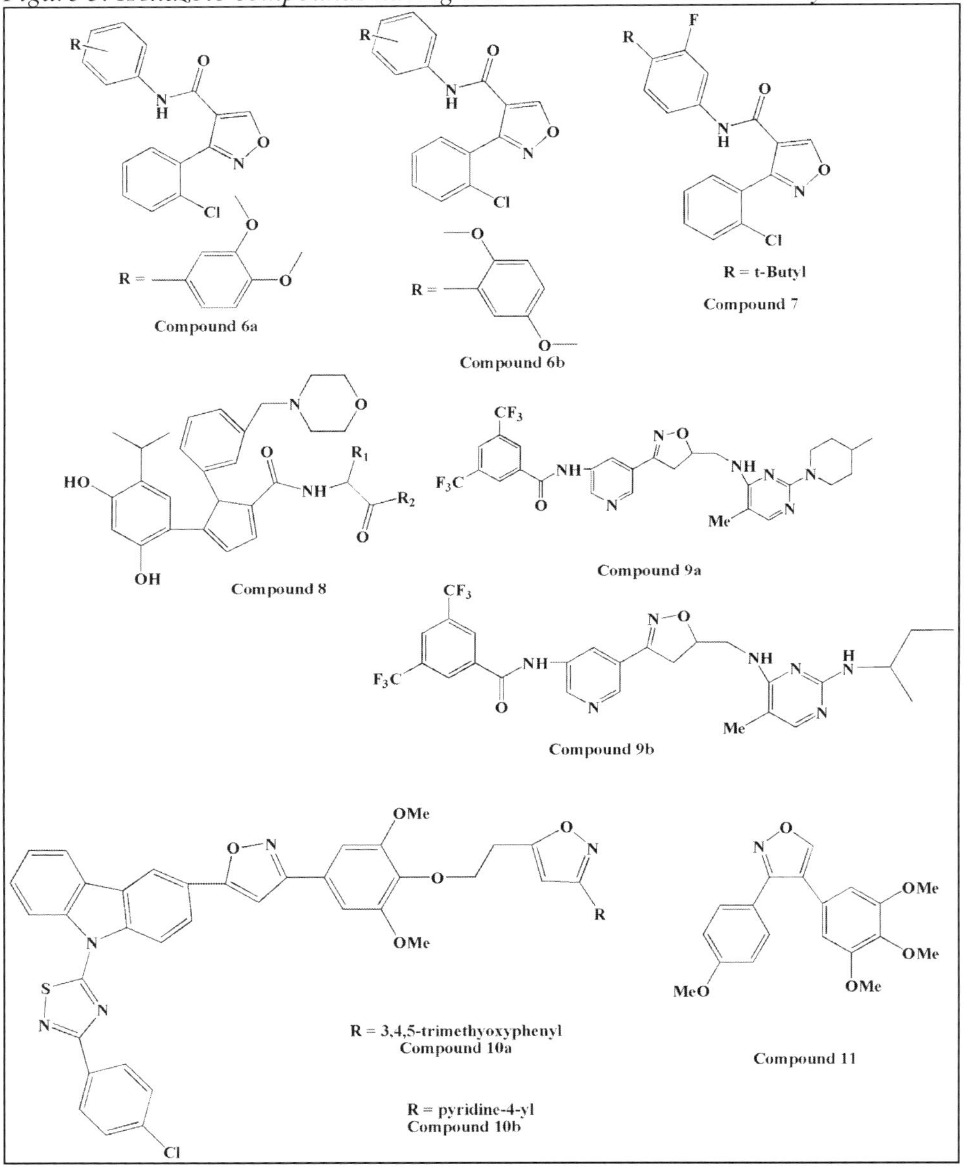

3.3 Anti-Microbial Isoxazoles

Kudryavtseva et al. Isoxazoline-based acridone chemical compounds were synthesized and their antibacterial activity was evaluated. **Compounds 12a** and **12b** (Figure 6) are potent inhibitors of many pathogens and are essentially two times more potent than metronidazole. They exhibit better antibacterial properties than ofloxacin and nitrofurazone and show the best activity against *Candida albicans*. The addition of nitrofuran greatly increased the isoxazoline ring's antibacterial activity (Kudryavtseva et al., 2020).

Li et al. Sampangin derivatives were produced with **compound 13** (Figure 6) showing the best activity. It was superior to voriconazole (MIC80 = 0.12 µg/mL) and fluconazole (MIC80 = 2 µg/mL mL) in antibacterial activity against *Cryptococcus neoformans* H99 (MIC80 = 0.031 µg/mL). Similar to voriconazole, **compound 13** was also effective against *C albicans* (MIC80 = 0.12 µg/mL) (Li et al., 2019).

Imen et al., synthesized several derivatives of coumarin isoxazoline and evaluated the antibacterial activity of these compounds *in vitro* against both Gram-positice and Gram-negative bacteria. Of all the **compounds** tested **14a – 14d** (Figure 6) had the strongest antibacterial activity. The **compound 14a, 14b,** and **14c** (Figure 6) were much more active than the standard drug gentamicin against *Pseudomonas aeruginosa*. Furthermore, compared to other isoxazoline derivatives, 14a exhibited superior antibacterial activity against *Staphylococcus aureus*, *Enterococcus faecalis*, and *Escherichia coli* (Zghab et al., 2017).

Isoxazole ester derivatives are made using naturally occurring octincholic acid as the precursor compound by Sahoo et al. Strong inhibitory effect against Mtb H37Rv is exhibited by derivative 15 (MIC = 0.5 µg/mL). **Compound 15** (Figure 6) is the most powerful clinical isolate of DR-Mtb, with MICs ranging from 1-4 µg/mL. By modifying the structure of compound, **compound 15** was shown to be effective against DR-Mtb (MIC = 0.25-0.5 µg/mL). According to SAR, the molecule has more biological activity after halogen substitution, alkoxy group greatly increases biological activity and quinoline substitution decreases biological activity. Therefore, compounds containing isoxazole ester fragments should be effective anti-inflammatory agents (Kumar Sahoo et al., 2022).

3.4 Antiparkinson Isoxazoles

In order to test for in vitro MAO inhibitory action, a novel series of N-(1-substitutedphenyl)ethylidiene)-5-phenylisoxazole-3-carbohydrazide derivatives was synthesized. The reversible and competitive nature of the inhibitors was demonstrated by the enzyme kinetic tests. Potent inhibitors binding site interactions were further uncovered by docking experiments, which also revealed that certain chemicals fit well in the MAO-B active site close to the FAD cofactor. The produced compounds underwent an in silico ADME assessment. As a result of their psitive ADME profiles, all of the compounds were anticipated to have good oral bioavailability. According to motor behavioral assessment using the footprint and horizontal wire tests demonstrated that compounds could considerably reduce MPTP-induced neurodegeneration. Therefore, the active chemical **compounds 16a and 16b** (Figure 7) found in this derivatives may offer interesting starting points for the creation of strong MAO-B inhibitors based on isoxazoles that might be used to the treatment Parkinson's disease (Agrawal & Mishra, 2019a).

In order to create selective MAO inhibitors, the nonselective MAO-inhibitor isocarboazid was tailored to create isoxazole-based hydrazone derivatives in the current study. No compound was observed to be active against MAO-A, while the majority of the compounds demostrated MAO-B inhibitory activity, indicating the great MAO-B selectivity of the produced compounds. When the compounds were evaluated using in silico ADME, all of them showed a favrouable ADME profile, indicating that the produced compounds would have good oral bioavailability. The most potent molecule, **compound 17** (Figure 7), was found to be competitive and reversible by enzyme kinetic tests. Potent inhibitors binding site interactions were further uncovered by docking experiments, which also revealed that certain chemicals fit well in the MAO-B active site close to the FAD factor. Therefore, it is possible to see active compound as a prospective lead molecule for the creation of powerful MAO-B inhibitors based on isoxazoles that might be utilized to treat Parkinson's disease (Agrawal & Mishra, 2019b).

Figure 7. Isoxazole compounds having antiparkinson's activity

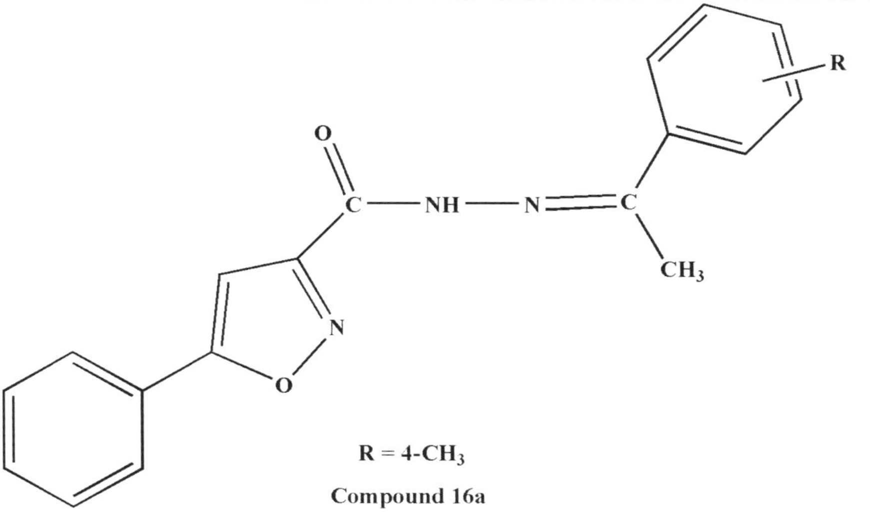

R = 4-CH₃

Compound 16a

R = 3,4-diOCH₃

Compound 16b

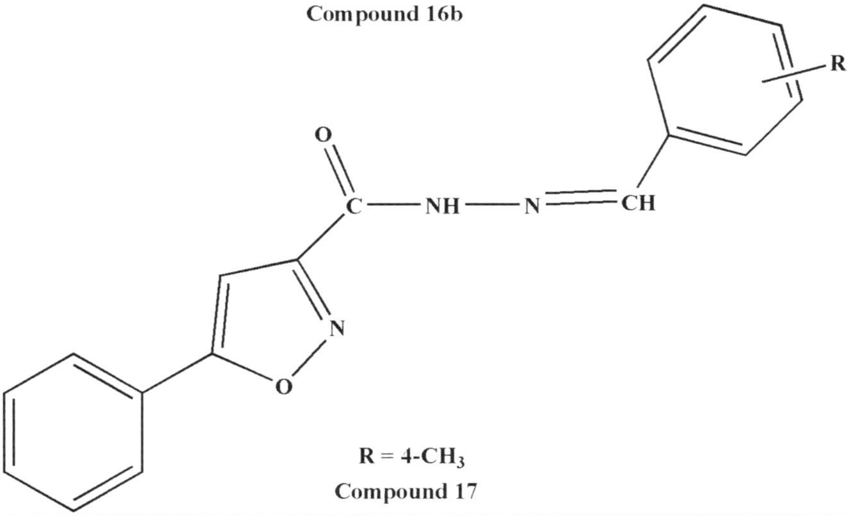

R = 4-CH₃

Compound 17

3.5 Anti-Diabetic Activity

Some flavonoids including isoxazoles were synthesized by Alghtami et al. Compound 18 was detected to have the highest α-amylase inhibitory effect compared to the effective drug acarbose. The inhibitory impact of α-amylase was enhanced by halogen substitution of the phenyl group of the isoxazole ring(F, Cl, Br); **compound 18** (Figure 8), a fluorinated derivative, had the maximum activity. In contrast, methyl or tert-butyl substitutions did not affect α-amylase inhibition; this showed little effect from increased energy consumption (+I) (Algethami et al., 2021).

Rekha et al. Natural drug karanja was successfully engineered to produce isoxazole derivative and its anti-inflammatory properties were tested in xylene-induced rat ear edema model. **Compound 19** (Figure 8) showed 75.45% inhibition, which exceeded the 51.13% inhibition value of Karanja and was comparable to ibuprofen (77.27%). This suggests that isoxazole derivatives can effectively inhibit ear edema while reducing inflammatory mediators (Rekha et al., 2021).

Figure 8. Isoxazole compounds having anti-diabetic activity

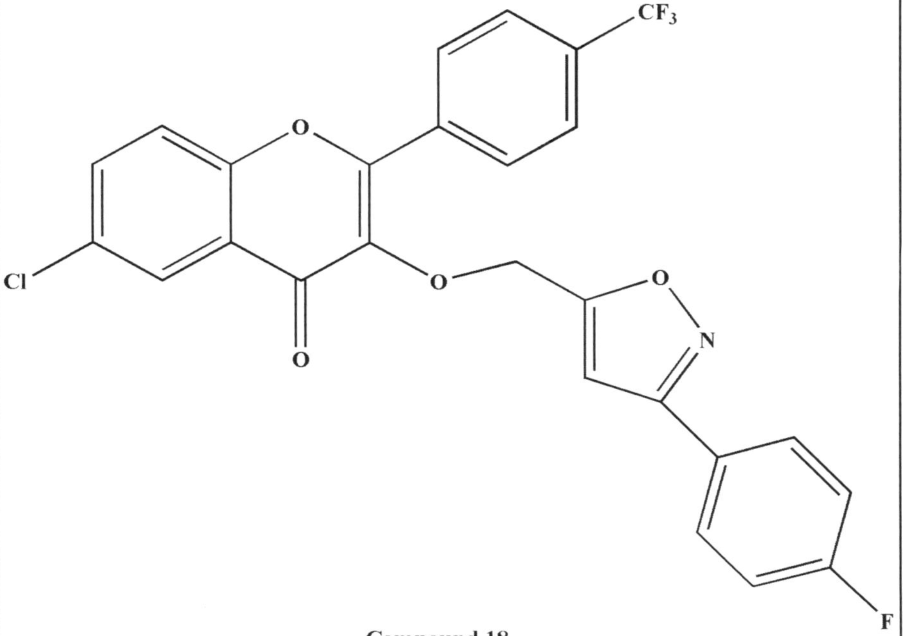

Compound 18

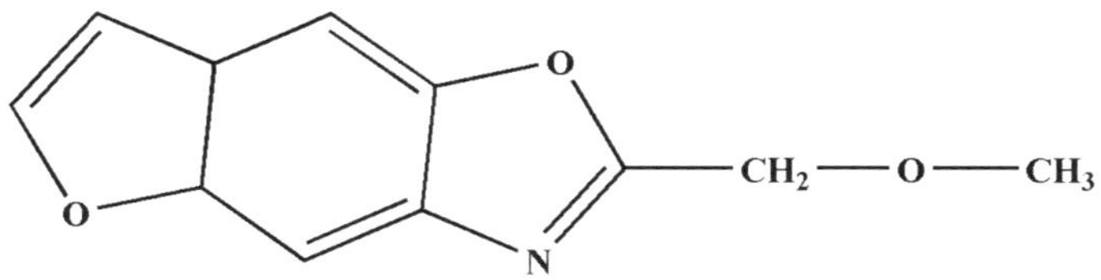

Compound 19

3.6 Anti-Oxiant Activity

Fluorophenylisoxazolecarboxamide was resynthesized and tested for DPPH radical scavenging, lipase and α-amylase inhibition. **Compounds 20** and **21** (Fig No. 9) have higher antioxidant activity than that of Trolox (IC50 = 3.10 ± 0.92 μg/mL) even in the absence of enzyme inhibition in the DPPH assay (Hawash et al., 2022).

Figure 9. Isoxazole compound having anti-oxidant activity.

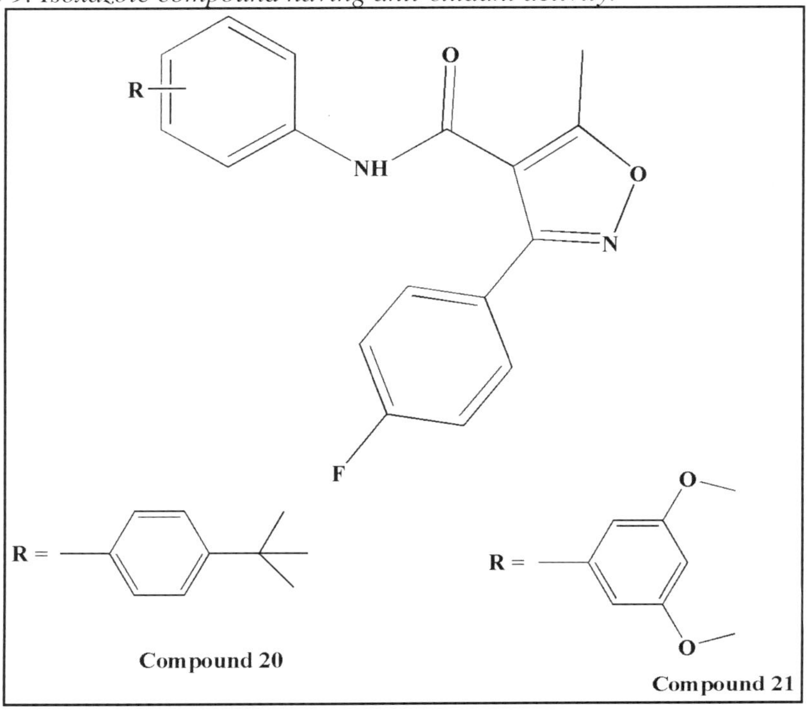

Compound 20

Compound 21

4. FUTURE SCOPES

4.1 Development of Green Synthetic Methods

Green chemistry developments may result in more effective and environmentally friendly process for creating isoxazole derivatives. This covers the application of eco-friendly solvents, renewable resources, and energy-saving procedures. Isoxazole production could become more economical and environmentally benign through green synthesis, expanding its uses across a range of sectors.

Hamid Beyzaei et al., synthesized novel 5-amino-isoxazole-4-carbonitriles derivatives by reacting malononitrile, hydroxylamine hydrochloride and different aryl or heteroaryl aldehydes via green multicomponent reaction. The catalytic reaction media used in this reaction was K2CO3/glycerol. Antimicrobial activity was observed against variety of bacterial strains by compound 22a, 22b, 22c and antioxidant activity was showed by compound 22d (Figure 10) (Beyzaei et al., 2018)

Figure 10. Isoxazole derivatives

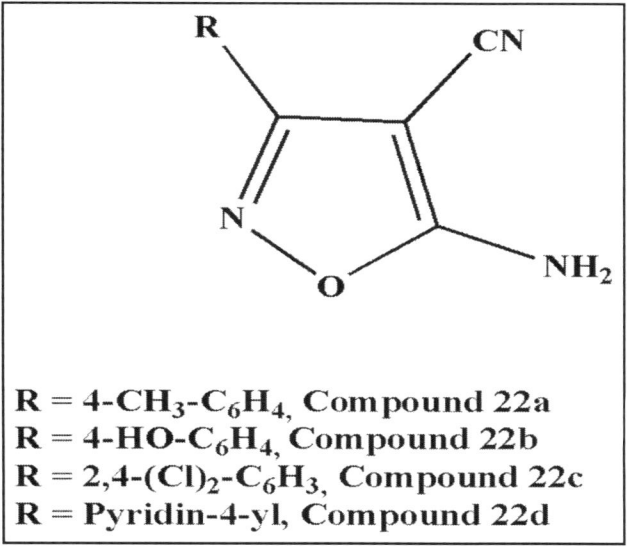

R = 4-CH$_3$-C$_6$H$_4$, Compound 22a
R = 4-HO-C$_6$H$_4$, Compound 22b
R = 2,4-(Cl)$_2$-C$_6$H$_3$, Compound 22c
R = Pyridin-4-yl, Compound 22d

4.2 Agrochemicl Innovations

Derivatives of isoxazoles may be used to create novel, eco-friendly insecticides and herbicides. To minimize environmental damage, research could concentrate on creating chemicals with high specificity for target weeds or pests. Advances in isoxazole-based agrochemicals may result in more environmentally friendly farming methods, which would aid in addressing the problems associated with global food security.

4.3 Materials Science Applications

Materials and polymers containing isoxazole can be created for applications in electrics, nanotechnology, and high-performance coatings. These cutting-edge materials allow for the exploitation of isoxazoles special qualities, including electrical behavior and thermal stability. Further investigation in this field may result in the creation of novel materials that find use in a variety of sectors, including aerospace and electrics.

4.4 Multifunctional Therapeutics

Isoxazole derivative's multifunctionality, which includes their capacity to interact with a variety of biological targets, can be investigated in the creation of polypharmacology medications that treat complicated illnesses including cancer and neurological diseases. This strategy targets several disease pathways at once, which may result in the creation of medicines that are less harmful and more effective.

5. CONCLUSION

The exploration of isoxazole as a bioactive compound has demonstrated its significant potential in various fields of medicinal chemistry. Derivatives of isoxazole show a wide range of biological actions, such as antimicrobial, anti-inflammatory, anticancer, and antiviral properties. These diverse applications are attributed to the unique chemical structure of isoxazole, which allows for versatile interactions with biological targets. The synthetic routes to isoxazole compounds have evolved considerably, offering a range of methodologies that enhance the efficiency and specificity of the desired products. Given the compound's biological relevance and the ongoing advancements in synthetic chemistry, isoxazole remains a promising scaffold for the development of new therapeutic agents. To effectively utilize isoxazole in drug discovery and development, future research should concentrate on refining existing synthetic approaches and investigating novel derivatives.

REFERENCES

A Barmade, M., R Murumkar, P., Kumar Sharma, M., & Ram Yadav, M. (2016). Medicinal chemistry perspective of fused isoxazole derivatives. *Current Topics in Medicinal Chemistry*, 16(26), 2863–2883.

Abdelall, E. K. A. (2020, January). Synthesis and biological evaluations of novel isoxazoles and furoxan derivative as anti-inflammatory agents. *Bioorganic Chemistry*, 94, 103441. DOI: 10.1016/j.bioorg.2019.103441 PMID: 31859011

Agrawal, N., & Mishra, P. (2019, September 1). Novel isoxazole derivatives as potential antiparkinson agents: Synthesis, evaluation of monoamine oxidase inhibitory activity and docking studies. *Medicinal Chemistry Research*, 28(9), 1488–1501. DOI: 10.1007/s00044-019-02388-4

Agrawal, N., & Mishra, P. (2019, April). Synthesis, monoamine oxidase inhibitory activity and computational study of novel isoxazole derivatives as potential antiparkinson agents. *Computational Biology and Chemistry*, 79, 63–72. DOI: 10.1016/j.compbiolchem.2019.01.012 PMID: 30731360

Alam, W., Khan, H., Saeed Jan, M., Rashid, U., Abusharha, A., & Daglia, M. (2023, September 6). Synthesis, in-vitro inhibition of cyclooxygenases and in silico studies of new isoxazole derivatives. *Frontiers in Chemistry*, 11, 11. DOI: 10.3389/fchem.2023.1222047 PMID: 37744065

Algethami, F. K., Saidi, I., Abdelhamid, H. N., Elamin, M. R., Abdulkhair, B. Y., Chrouda, A., & Ben Jannet, H. (2021, August 27). Trifluoromethylated Flavonoid-Based Isoxazoles as Antidiabetic and Anti-Obesity Agents: Synthesis, In Vitro α-Amylase Inhibitory Activity, Molecular Docking and Structure–Activity Relationship Analysis. *Molecules (Basel, Switzerland)*, 26(17), 5214. DOI: 10.3390/molecules26175214 PMID: 34500647

Antonov, A. S., Tupikina, E. Y., Karpov, V. V., Mulloyarova, V. V., & Bardakov, V. G. (2020, December 17). Sterically Facilitated Intramolecular Nucleophilic NMe2 Group Substitution in the Synthesis of Fused Isoxazoles: Theoretical Study. *Molecules (Basel, Switzerland)*, 25(24), 5977. DOI: 10.3390/molecules25245977 PMID: 33348591

Basavanna, V., Doddamani, S., Chandramouli, M., Bhadraiah, U. K., & Ningaiah, S. (2022, August 23). Green approaches for the synthesis of pharmacologically enviable isoxazole analogues: A comprehensive review. *Journal of the Iranian Chemical Society*, 19(8), 3249–3283. DOI: 10.1007/s13738-022-02556-1

Beyzaei, H., Kamali Deljoo, M., Aryan, R., Ghasemi, B., Zahedi, M. M., & Moghaddam-Manesh, M. (2018, December 15). Green multicomponent synthesis, antimicrobial and antioxidant evaluation of novel 5-amino-isoxazole-4-carbonitriles. *Chemistry Central Journal*, 12(1), 114. DOI: 10.1186/s13065-018-0488-0 PMID: 30443685

Chernysheva, N. B., Maksimenko, A. S., Andreyanov, F. A., Kislyi, V. P., Strelenko, Y. A., Khrustalev, V. N., Semenova, M. N., & Semenov, V. V. (2018, February). Regioselective synthesis of 3,4-diaryl-5-unsubstituted isoxazoles, analogues of natural cytostatic combretastatin A4. *European Journal of Medicinal Chemistry*, 146, 511–518. DOI: 10.1016/j.ejmech.2018.01.070 PMID: 29407976

Dadiboyena, S., & Nefzi, A. (2010, November). Recent methodologies toward the synthesis of valdecoxib: A potential 3,4-diarylisoxazolyl COX-II inhibitor. *European Journal of Medicinal Chemistry*, 45(11), 4697–4707. DOI: 10.1016/j.ejmech.2010.07.045 PMID: 20724040

Deng, D., Zhou, J., Li, M., Li, S., Tian, L., Zou, J., Wang, T., Wu, J., Zeng, F., & Yang, J. (2020, July 23). Leflunomide monotherapy versus combination therapy with conventional synthetic disease-modifying antirheumatic drugs for rheumatoid arthritis: A retrospective study. *Scientific Reports*, 10(1), 12339. DOI: 10.1038/s41598-020-69309-z PMID: 32704073

Dou, G., Xu, P., Li, Q., Xi, Y., Huang, Z., & Shi, D. (2013, November 5). Clean and Efficient Synthesis of Isoxazole Derivatives in Aqueous Media. *Molecules (Basel, Switzerland)*, 18(11), 13645–13653. DOI: 10.3390/molecules181113645 PMID: 24196411

Durga Rao, B. V., Sreenivasulu, R., & Basaveswara Rao, M. V. (2019, October 25). Design, Synthesis, and Evaluation of Isoxazole-Thiadiazole Linked Carbazole Hybrids as Anticancer Agents. *Russian Journal of General Chemistry*, 89(10), 2115–2120. DOI: 10.1134/S1070363219100207

Eid, A. M., Hawash, M., Amer, J., Jarrar, A., Qadri, S., Alnimer, I., Sharaf, A., Zalmoot, R., Hammoudie, O., Hameedi, S., & Mousa, A. (2021, March 9). Synthesis and Biological Evaluation of Novel Isoxazole-Amide Analogues as Anticancer and Antioxidant Agents. *BioMed Research International*, 2021(1), 1–9. DOI: 10.1155/2021/6633297 PMID: 33763478

Eid, A. M., Hawash, M., Amer, J., Jarrar, A., Qadri, S., Alnimer, I., Sharaf, A., Zalmoot, R., Hammoudie, O., Hameedi, S., & Mousa, A. (2021, March 9). Synthesis and Biological Evaluation of Novel Isoxazole-Amide Analogues as Anticancer and Antioxidant Agents. *BioMed Research International*, 2021(1), 1–9. DOI: 10.1155/2021/6633297 PMID: 33763478

Hawash, M., Jaradat, N., Abualhasan, M., Thaher, M., Sawalhi, R., Younes, N., Shanaa, A., Nuseirat, M., & Mousa, A. (2022, October 29). In vitro and in vivo assessment of the antioxidant potential of isoxazole derivatives. *Scientific Reports*, 12(1), 18223. DOI: 10.1038/s41598-022-23050-x PMID: 36309576

Hawash, M., Jaradat, N., Bawwab, N., Salem, K., Arafat, H., Hajyousef, Y., Shtayeh, T., & Sobuh, S. (2021, December 27). Design, synthesis, and biological evaluation of phenyl-isoxazole-carboxamide derivatives as anticancer agents. *Heterocyclic Communications*, 27(1), 133–141. DOI: 10.1515/hc-2020-0134

Hossain, M. I., Khan, M. I. H., Kim, S. J., & Le, H. V. (2022, April 22). Synthesis of 3,4,5-trisubstituted isoxazoles in water via a [3 + 2]-cycloaddition of nitrile oxides and 1,3-diketones, β-ketoesters, or β-ketoamides. *Beilstein Journal of Organic Chemistry*, 18, 446–458. DOI: 10.3762/bjoc.18.47 PMID: 35529890

Itoh, K., Hayakawa, M., Abe, R., Takahashi, S., Hasegawa, K., & Aoyama, T. (2021, December 9). Itoh K ichi, Hayakawa M, Abe R, Takahashi S, Hasegawa K, Aoyama T. A Facile Approach to the Synthesis of 3-Acylisoxazole Derivatives with Reusable Solid Acid Catalysts. *Synthesis*, 53(24), 4636–4643. DOI: 10.1055/a-1581-0235

Kerim, M. D., Boyode, P., Garrec, J., & El Kaïm, L. (2020, June 10). Metal-Free Addition of Boronic Acids to Silylnitronates. *Synlett*, 31(09), 856–860. DOI: 10.1055/s-0039-1690845

Kreuzer, J., Bach, N. C., Forler, D., & Sieber, S. A. (2015). Target discovery of acivicin in cancer cells elucidates its mechanism of growth inhibition. *Chemical Science (Cambridge)*, 6(1), 237–245. DOI: 10.1039/C4SC02339K

Kudryavtseva, T. N., Lamanov, A. Yu., Sysoev, P. I., & Klimova, L. G. (2020, January 26). Synthesis and Antibacterial Activity of New Acridone Derivatives Containing an Isoxazoline Fragment. *Russian Journal of General Chemistry*, 90(1), 45–49. DOI: 10.1134/S1070363220010077

Kumar, G., & Shankar, R. (2021, February 4). 2-Isoxazolines: A Synthetic and Medicinal Overview. *ChemMedChem*, 16(3), 430–447. DOI: 10.1002/cmdc.202000575 PMID: 33029886

Kumar Sahoo, S., Naiyaz Ahmad, M., Kaul, G., Nanduri, S., Dasgupta, A., Chopra, S., & Madhavi Yaddanapudi, V. (2022, July 23). Exploration of Isoxazole-Carboxylic Acid Methyl Ester Based 2-Substituted Quinoline Derivatives as Promising Anti-tubercular Agents. *Chemistry & Biodiversity*, 19(7), e202200324. DOI: 10.1002/cbdv.202200324 PMID: 35653161

Li, F., Hu, Y., Wang, Y., Ma, C., & Wang, J. (2017, February 23). Expeditious Lead Optimization of Isoxazole-Containing Influenza A Virus M2-S31N Inhibitors Using the Suzuki–Miyaura Cross-Coupling Reaction. *Journal of Medicinal Chemistry*, 60(4), 1580–1590. DOI: 10.1021/acs.jmedchem.6b01852 PMID: 28182419

Li, Z., Liu, N., Tu, J., Ji, C., Han, G., Wang, Y., & Sheng, C. (2019, March). Discovery of novel simplified isoxazole derivatives of sampangine as potent anti-cryptococcal agents. [Internet]. *Bioorganic & Medicinal Chemistry*, 27(5), 832–840. https://linkinghub.elsevier.com/retrieve/pii/S0968089618321205. DOI: 10.1016/j.bmc.2019.01.029 PMID: 30711309

Mabasa, J. L., Mabasa, T. F., Nyathi, M. L., & Moshapo, P. T. (2024, April). The design, synthesis and antiplasmodial evaluation of novel sulfoximine-isoxazole hybrids as potential antimalarial agents. *European Journal of Medicinal Chemistry Reports*, 10, 100128. DOI: 10.1016/j.ejmcr.2023.100128

Mączyński, M., Artym, J., Kocięba, M., Kochanowska, I., Ryng, S., & Zimecki, M. (2016, October). Anti-inflammatory properties of an isoxazole derivative – MZO-2. *Pharmacological Reports*, 68(5), 894–902. DOI: 10.1016/j.pharep.2016.04.017 PMID: 27351945

Mallik, N. N., Manasa, C., Basavanna, V., Shanthakumar, D. C., Ningaiah, S., & Lingegowda, N. S. (2024). Synthesis of Fused Isoxazoles: A Comprehensive Review. In *RAiSE-2023* (p. 222). MDPI. DOI: 10.3390/engproc2023059222

Mironova, I., Nenajdenko, V., Postnikov, P., Saito, A., Yusubov, M., & Yoshimura, A. (2022, June 16). Efficient Catalytic Synthesis of Condensed Isoxazole Derivatives via Intramolecular Oxidative Cycloaddition of Aldoximes. *Molecules (Basel, Switzerland)*, 27(12), 3860. DOI: 10.3390/molecules27123860 PMID: 35744982

Nasr, T., Bondock, S., & Eid, S. (2016, March 3). Design, synthesis, antimicrobial evaluation and molecular docking studies of some new 2,3-dihydrothiazoles and 4-thiazolidinones containing sulfisoxazole. *Journal of Enzyme Inhibition and Medicinal Chemistry*, 31(2), 236–246. DOI: 10.3109/14756366.2015.1016514 PMID: 25815670

Nomenclature and Application of Heterocyclic Compounds. (2021). *Applied Organic Chemistry*. Wiley.

Ouzounthanasis, K. A., Rizos, S. R., & Koumbis, A. E. (2023, December 22). A Convenient Synthesis of Novel Isoxazolidine and Isoxazole Isoquinolinones Fused Hybrids. *Molecules (Basel, Switzerland)*, 29(1), 91. DOI: 10.3390/molecules29010091 PMID: 38202674

Perrone, M. G., Vitale, P., Panella, A., Ferorelli, S., Contino, M., Lavecchia, A., & Scilimati, A. (2016, June 6). Isoxazole-Based-Scaffold Inhibitors Targeting Cyclooxygenases (COXs). *ChemMedChem*, 11(11), 1172–1187. DOI: 10.1002/cmdc.201500439 PMID: 27136372

Rekha, M. J., Bettadaiah, B. K., Muthukumar, S. P., & Govindaraju, K. (2021, January). Synthesis, characterization and anti-inflammatory properties of karanjin (Pongamia pinnata seed) and its derivatives. *Bioorganic Chemistry*, 106, 104471. DOI: 10.1016/j.bioorg.2020.104471 PMID: 33257003

Robbins, L., Balaram, A., Dejneka, S., McMahon, M., Najibi, Z., Pawlowicz, P., & Conrad, W. H. (2023, May 26). Heterologous production of the D-cycloserine intermediate O-acetyl-L-serine in a human type II pulmonary cell model. *Scientific Reports*, 13(1), 8551. DOI: 10.1038/s41598-023-35632-4 PMID: 37237156

Sangale, S. S., Kale, P. S., Lamkane, R. B., Gore, G. S., Parekar, P. B., & Shivpuje, S. S. (2023, February 23). Synthesis of Novel Isoxazole Derivatives as Analgesic Agents by Using Eddy's Hot Plate Method. *South Asian Research Journal of Pharmaceutical Sciences.*, 5(01), 18–27. DOI: 10.36346/sarjps.2023.v05i01.002

Saravanan, G., Alagarsamy, V., & Dineshkumar, P. (2021, August 24). Synthesis, analgesic, anti-inflammatory and in vitro antimicrobial activities of some novel isoxazole coupled quinazolin-4(3H)-one derivatives. *Archives of Pharmacal Research*, 44(8), 1–11. DOI: 10.1007/s12272-013-0262-8 PMID: 24155019

Shaik, A., Bhandare, R. R., Palleapati, K., Nissankararao, S., Kancharlapalli, V., & Shaik, S. (2020, February 26). Antimicrobial, Antioxidant, and Anticancer Activities of Some Novel Isoxazole Ring Containing Chalcone and Dihydropyrazole Derivatives. *Molecules (Basel, Switzerland)*, 25(5), 1047. DOI: 10.3390/molecules25051047 PMID: 32110945

Sysak, A., & Obmińska-Mrukowicz, B. (2017, September). Isoxazole ring as a useful scaffold in a search for new therapeutic agents. *European Journal of Medicinal Chemistry*, 137, 292–309. DOI: 10.1016/j.ejmech.2017.06.002 PMID: 28605676

Trefzger, O. S., Barbosa, N. V., Scapolatempo, R. L., das Neves, A. R., Ortale, M. L. F. S., Carvalho, D. B., Honorato, A. M., Fragoso, M. R., Shuiguemoto, C. Y. K., Perdomo, R. T., Matos, M. F. C., Chang, M. R., Arruda, C. C. P., & Baroni, A. C. M. (2020, February 16). Design, synthesis, antileishmanial, and antifungal biological evaluation of novel 3,5-disubstituted isoxazole compounds based on 5-nitrofuran scaffolds. *Archiv der Pharmazie*, 353(2), 1900241. DOI: 10.1002/ardp.201900241 PMID: 31840866

Vaickelionienė, R., Petrikaitė, V., Vaškevičienė, I., Pavilonis, A., & Mickevičius, V. (2023, March 23). Synthesis of novel sulphamethoxazole derivatives and exploration of their anticancer and antimicrobial properties. *PLoS One*, 18(3), e0283289. DOI: 10.1371/journal.pone.0283289 PMID: 36952512

Xiao, W., Yang, Q. Q., Chen, Z., Ouyang, Q., Du, W., & Chen, Y. C. (2018, January 5). Regio- and Diastereodivergent [4 + 2] Cycloadditions with Cyclic 2,4-Dienones. *Organic Letters*, 20(1), 236–239. DOI: 10.1021/acs.orglett.7b03598 PMID: 29240444

Yang, H., Xu, G., & Pei, Y. (2017, February 10). Synthesis, preliminary structure-activity relationships and biological evaluation of pyridinyl-4,5-2H-isoxazole derivatives as potent antitumor agents. *Chemical Research in Chinese Universities*, 33(1), 61–69. DOI: 10.1007/s40242-017-6330-8

Zghab, I., Trimeche, B., Ben Mansour, M., Hassine, M., Touboul, D., & Ben Jannet, H. (2017, May). Regiospecific synthesis, antibacterial and anticoagulant activities of novel isoxazoline chromene derivatives. *Arabian Journal of Chemistry*, 10, S2651–S2658. DOI: 10.1016/j.arabjc.2013.10.008

Zhang, C., Wang, X., Liu, H., Zhang, M., Geng, M., Sun, L., Shen, A., & Zhang, A. (2017, January). Design, synthesis and pharmacological evaluation of 4,5-diarylisoxazols bearing amino acid residues within the 3-amido motif as potent heat shock protein 90 (Hsp90) inhibitors. *European Journal of Medicinal Chemistry*, 125, 315–326. DOI: 10.1016/j.ejmech.2016.09.043 PMID: 27688186

Chapter 4
Use of Ionic Liquids and DES for Synthesis of Medicinally Important 1,2,3–Triazole Scaffolds

Bishal Bhattacharyya
https://orcid.org/0009-0009-3112-7227
Dibrugarh University, India

Ramyata Priyam Borah
https://orcid.org/0009-0002-5004-5736
Dibrugarh University, India

Dipankar Nath
https://orcid.org/0009-0005-2322-3914
Dibrugarh University, India

Parishma Gogoi
https://orcid.org/0009-0002-5142-5888
Dibrugarh University, India

Diganta Sarma
Dibrugarh University, India

ABSTRACT

Cu catalysed azide alkyne cycloaddition reactions, known as click reactions, result in the regioselective formation of 1,4 disubstituted 1,2,3-triazoles. 1,2,3-triazoles are five membered nitrogen containing heterocycles that have gained much importance in recent times due to their various applications such as anti-cancer, anti- TB etc.

DOI: 10.4018/979-8-3693-7267-8.ch004

Several greener approaches have been developed by different researchers to prepare 1,2,3-triazoles using benign methodologies. Greener solvents like Deep Eutectic Solvents (DES) and Ionic Liquids (IL) can serve as good alternative for volatile organic solvents used during azide alkyne cycloaddition reactions

INTRODUCTION TO GREEN SOLVENTS

Green solvents are environmentally friendly, sustainable solvents that minimize the hazard of using conventional organic solvents that are harmful to human health and the environment. Conventional solvents are found to be volatile, flammable, carcinogenic and toxic. (Khandelwal, Tailor, & Kumar, 2016) Green solvents are prime alternatives to volatile organic solvents and almost practically reduce chemical waste. Ionic Liquids (ILs) and Deep Eutectic Solvents (DES) are examples of such solvents that are considered 'Green'. Extensive research works have been on going on such green solvents for the past few decades.

ILs are green solvents made up of a discrete (Abbott, Capper, Davies, & Rasheed, 2004) bigger organic cation and a smaller organic or inorganic anion that stay liquid at ambient temperature and cannot form a lattice. They can be designed and tuned to meet the necessity of a reaction, and hence, they are called designer solvents or task-specific solvents. Changing anion-cation combinations of ionic liquids also changes their physicochemical properties. (Clarke, Tu, Levers, Brohl, & Hallett, 2018) ILs are ion conducting, viscous, non-volatile and non-flammable. Hence their application has been seen in various fields such as solar industry, catalysis, separation medium and elements of electrochemical devices. (Abbott, Capper, Davies, & Rasheed, 2004) The low vapour pressure of the ILs infer the electrostatic interaction between the oppositely charged ions. ILs have great chemical stability and low volatility. Hence, they are found to be of very little environmental concern. Most common ILs are made up of quaternary ammonium cations such as of pyridinium, imidazolium, dimethylethylphenylammonium salts and some polyatomic anions such as $[BF_4]^-$, $[PF_6]^-$ etc. (Gabriel & Weiner, 1888) or halide ions such as Cl^-, Br^- etc. These ILs can be recycled and reused after an organic reaction is performed.

The very initial concept of IL was given by Gabriel and Weiner in 1888 with the synthesis of ethanolammonium nitrate. (Gabriel & Weiner, 1888) Succeeding that, the synthesis of ethylammonium nitrate in 1914 by Paul Walden[5] marked the beginning of different approaches to design ILs. It was the first Room Temperature IL (RTIL) prepared by reacting HNO_3 and $C_2H_5NH_2$. As the IL designed by Walden was very unstable and explosive, later scientists discovered new methods to prepare stable, non-volatile, low-vapour pressure, non-toxic ILs. The 1st generation ILs were introduced in the year 1970 by Osteryoung *et al.* which were about $AlCl_3$ and the alkyl pyridinium halide system. (Chum, Koch, Miller, & Osteryoung, 1975) The 2nd generation ILs were introduced in 1992 by Wilkes and Zaworotko to overcome the disadvantages of the 1st generation ILs as they were not water/air stable. They synthesized water and air-stable ILs with 1-ethyl-3-methylimidazolium cations [EtMeim+] and anions viz. CH_3COO^-, NO_3^-, BF_4^-. (Wilkes & Zaworotko, 1992) Since the early years of 2000s, functionallized imidazolium based ILs were used for metal extraction (Visser, Swatloski, Reichert, Davis, Rogers, Mayton, Sheff, & Wierzbicki, 2001) and tunable properties were found in ILs and hence they were termed "Task Specific ILs". These were the 3rd generation ionic liquids.

Ionic liquids can also be classified according to the acidity and basicity of their ion combinations (Figure 1). For example, ILs containing $AlCl_3$ are referred to as Lewis acidic ILs (Wang, Guo, Li, Wang, Wenga, & Wua, 2006) (Hajipour, Azizi, & Ruoho, 2009) and acaetate, formate, etc containing ILs are basic ILs and are amphoteric, having hydrogen sulfate and dihydrogen phosphate anions. The synthesis of ILs depends upon the reaction conditions and is designed accordingly to meet the necessity of the reaction. However, apart from the vast designability and applicability of the ILs, it is seen that the catalytic activity of imidazolium based ionic liquids in presence of base gets disrupted. (Raiguel, Dehaen, & Binnemans, 2020) It is due to the presence of an acidic proton in the C-2 methylene position in the imidazolium moiety, which gets abstracted when a strong base containing anion such as OH- is added, leading to the formation of an N-heterocyclic carbene (NHC). One such example is [BMIM][OH], which loses its basicity due to this reason. The cations are larger and more hindered, making them less mobile compared to the anions. It can be concluded that ionic liquids (ILs) containing basic cationic components are more stable than those containing basic anionic components. (Yoshizawa-Fujita, Johansson, Newman, MacFarlane, & Forsyth, 2006) Acidic Ionic Liquids (Figure 2) (AILs) are of two types- Lewis Acidic ILs (LAILs) and Bronsted Acidic ILs (BAILs) and some are comprised of both acidic functions. LAILs mainly have electron accepting ability in the anionic group, whereas the BAILs show acidity due to ionisable protons.

Figure 1. Some basic anions and cations

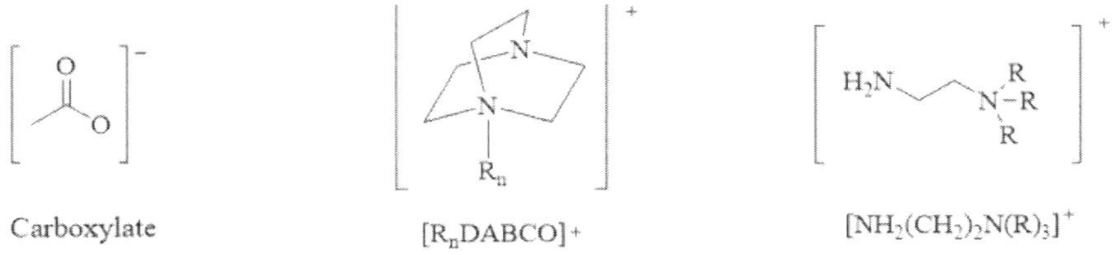

Carboxylate　　　　　　　[R_nDABCO]+　　　　　　[$NH_2(CH_2)_2N(R)_3$]+

Figure 2. Some Acidic Ionic Liquids

1-(3-Propylsulfonic)-3-methylimidazoliumhydrogensulfate

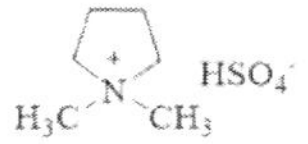

N,N-Dimethylpyrrolidiniumhydrogensulfate

Deep Eutectic Solvents (DES) are another class of green solvents. DESs are analogues to Ionic Liquids (ILs) because of their similar characteristics, such as low melting point, tunability, low pressure, low toxicity etc. but with more advantages related to preparation cost and environmental impact. (Shaibuna & Theresa, 2018) They bear some unique physicochemical properties viz., low preparation cost, tunable physiochemical properties (Smith, Abbott, & Ryder, 2014), wide liquid-range (Smith, Abbott, & Ryder, 2014), non-flammability (Smith, Abbott, & Ryder, 2014), negligible vapour pressure, bio renewability and biodegradability and have maximum synthetic efficiency. They contain large non-symmetric ions that have low lattice energy and hence low melting points. (Khandelwal, Tailor, & Kumar, 2016) For quite a long time, a new kind of green solvent has been synthesized by mixing quaternary ammonium salts with metal salts acting as Hydrogen Bond Donor (HBD) (Khandelwal, Tailor, & Kumar, 2016) having a deep eutectic point with many properties like the ILs but they are found to be different. The decrease in the melting point of the mixture relative to the melting points of the individual components is due to the charge delocalization occurring through hydrogen bonding between the halide ion of the quaternary ammonium salts and the hydrogen bond donor moiety. However, it was found from Nuclear magnetic resonance (NMR) that no reaction occurred between the components. (D'Agostino, Gladden, Mantle, Abbott, Ahmed, Al-Murshedi, & Harris, 2015) (D'Agostino, Harris, Abbott, Gladden, & Mantle, 2011) According to Abbott *et al.* the freezing point depends on the lattice energy of DESs, interaction between couple anion-HBD and entropy changes arising from the formation of the liquid phase. (Khandelwal, Tailor, & Kumar, 2016)

Figure 3. Some HBD and HBA

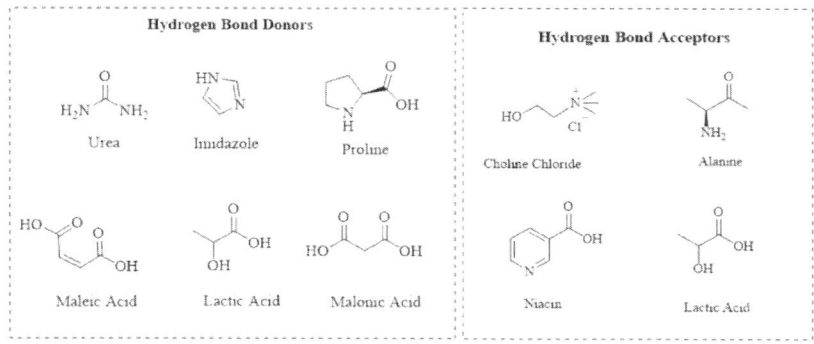

There are traditionally five types of DESs- Type I, Type II, Type III, Type-IV and Type-V DESs. Type I DES are those that are combination of a quaternary ammonium salt and a halide salt, Type II are those that contain a quaternary ammonium salt and a metal halide hydrate, Type III ones contain a quaternary ammonium salt and an HBD to mimic an IL (typically an organic molecular component such as an amide, carboxylic acid, or polyol), Type IV DESs contain metal chloride hydrate and HBD and Type V DESs are obtained from molecular, non-ionic HBD and Hydrogen Bond Acceptor (HBA) (Figure 3). Hydrogen bonding is believed to be especially effective in Type V types of DESs. (Hansen, Spittle, Chen, Poe, Zhang, Klein, Horton, Adhikari, Zelovich, Doherty, Gurkan, Magninn, Ragauskas, & Dadmun, 2021)

All the DESs are very easy to prepare. The most commonly used method is to heat and stir the constituents under an inert atmosphere to form a homogeneous liquid where no additional solvent or purification steps are required. So, 100% atom economy is achieved by simply mixing the HBDs and HBAs as all the materials get incorporated into the final product. They can also be prepared by other methods such as vacuum evaporation, grinding, and freeze-drying. In some cases, DESs are prepared taking water as a solvent and further evaporated under vacuum. (Gutiérrez, Ferrer, Mateo, & Monte, 2009) They appear to be typically viscous, clear liquids with different undertones ranging from white to amber-brown, or as cloudy, opaque solids below the eutectic temperature. (Hansen, Spittle, Chen, Poe, Zhang, Klein, Horton, Adhikari, Zelovich, Doherty, Gurkan, Magninn, Ragauskas, & Dadmun, 2021) These solvents are most commonly used in metal processing and synthetic media. (Khandelwal, Tailor, & Kumar, 2016)

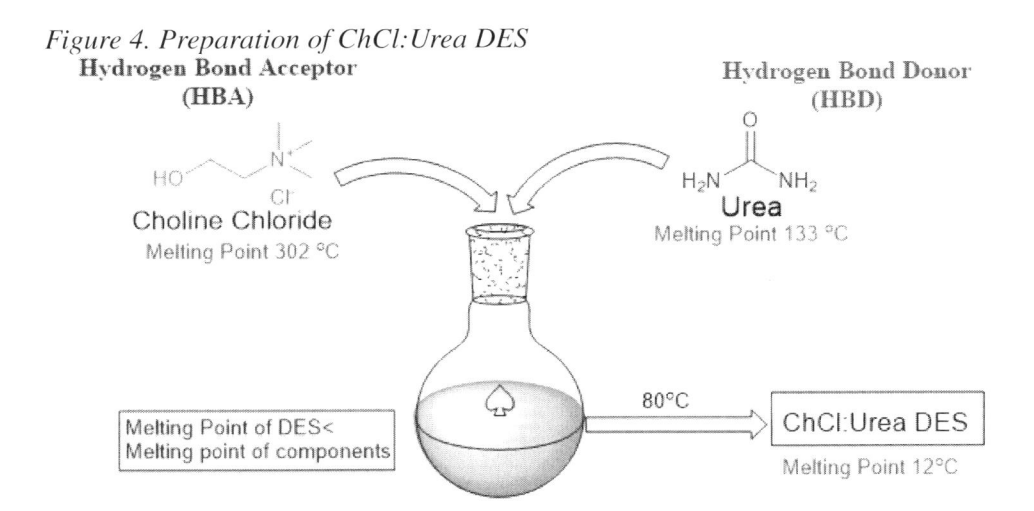

Figure 4. Preparation of ChCl:Urea DES

Hydrogen Bond Acceptor (HBA)

HO—N⁺—Cl⁻
Choline Chloride
Melting Point 302 °C

Hydrogen Bond Donor (HBD)

H_2N—C(=O)—NH_2
Urea
Melting Point 133 °C

Melting Point of DES<
Melting point of components

80°C → ChCl:Urea DES
Melting Point 12°C

The DESs can be tuned like ILs, by changing the temperature and composition of the starting materials to suit the properties for the desired application better. This change in the temperature and the compositions can be monitored by a phase diagram. (Florindo, Branco, & Marrucho, 2019) The first DES was reported by Abbott (Abbott, Capper, Davies, Rasheed, & Tambyrajah, 2003) and et al. using choline chloride as HBA and urea as HBD in 1:2 ratio. Researchers have reported various DESs since then and have been used as a green medium in synthetic chemistry (Figure 4).

Nitrogen containing heterocycles are one of the well-known structural architectures in natural products, possessing interesting physicochemical properties, biological activities and pharmacological value. (Lin, He, Geng, Xiao, Ji, Zheng, Huang, 2021) 1,2,3-triazoles constitute a privileged class of N-heterocyclic compounds which have gained considerable attention in industrial and academic areas. (Bagherzadeh, Amiri, & Sardarian, 2023)These scaffolds have found a wide range of applications in different fields such as medicinal chemistry, (Nehra, Tittal, Ghule, & Lal, 2021) biochemistry, (Agouram, Hadrami, & Bentama, 2021) organic synthesis, (Das, Dey, & Pathak, 2019) materials chemistry, (Kantheti, Narayan, & Raju, 2015) chemical sensing, (Rani, Lal, Shrivastava, & Ghule, 2020) bioconjugation (Zheng, Rouhani-fard, Jalloh, & Wu, 2012) etc. In medicinal chemistry, they are valued for their significant biological activities which include anti-tumor, anti-viral, antitubercular, anti-cancer, anti-inflammatory and anti-microbial properties (Figure 5). (Bangalore, Vagolu, Bollikanda, Veeragoni, Choudante, Misra, Sriram, Sridhar, Kantevari, 2019; Zhang, Xu, Gao, Ren, Chang, Lv, Feng, 2017; Zhang, 2019; Ashwini, Garg, Mohan, Fuchs, Rangappa, Anusha, Swaroop, Rakesh, Kanojia, Madan, & Bender, 2015) Moreover, they exhibit various advantageous properties like excellent chemical stability, hydrogen bonding ability, intense dipole moment, aromatic character and π-stacking interaction etc. (Bagherzadeh, Amiri, & Sardarian, 2023)

Figure 5. Some representative examples of 1,2,3-triazole containing drugs

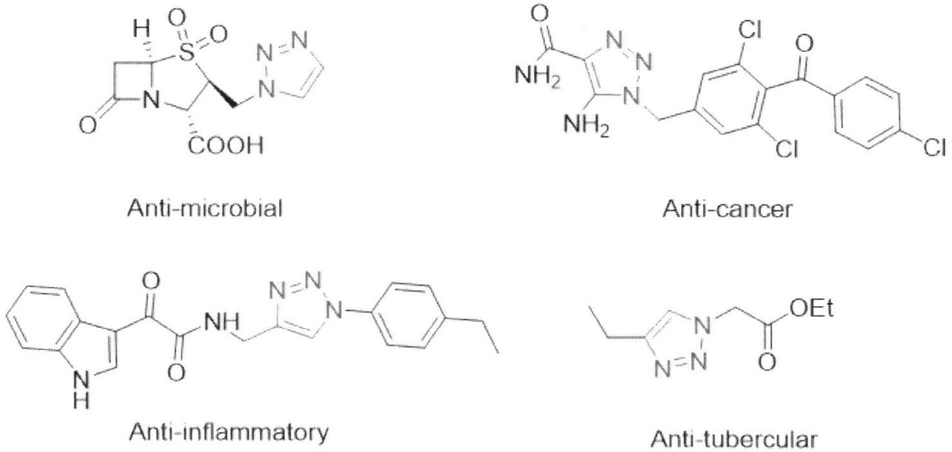

Anti-microbial

Anti-cancer

Anti-inflammatory

Anti-tubercular

These significant characteristics of 1,2,3-triazoles have encouraged researchers to develop efficient methodologies for the synthesis of biologically important 1,2,3-triazoles and the synthesis of 1,2,3-triazole derivatives has come out as an active, evolving and promising area of research. (Vaishnani, Bijani, Rahamathulla, Baldaniya, Jain, Kamal Y. Thajudeen, Ahmed, Farhana, Pasha, 2024)

Various methodologies have been reported for synthesizing 1,2,3-triazoles; however, 1,3-dipolar cycloaddition of azides and alkynes is the most significant and versatile route. In 1960, Huisgen established a great work in the field of triazole synthesis, which provided a straightforward synthetic protocol for the synthesis of 1,2,3-triazoles. The protocol involved 1,3-diploar cycloaddition of organic azides and terminal alkynes under thermal conditions which gave a mixture of corresponding 1,4- and 1,5-disubstituted 1,2,3-triazoles. (Huisgen & Padwa, 1984) But Huisgen's protocol for the synthesis of 1,2,3-triazoles was associated with some demerits such as high temperature requirement, more reaction time, low regioselectivity and low yield of products. (Kumar, Kumar, Singh, Tittal, & Lal, 2024) To overcome these disadvantages, Meldal (Tornøe, Christensen, & Meldal, 2002) and Sharpless (Rostovtsev, Green, Fokin, & Sharpless, 2002) groups in 2002 independently did the modification of the Huisgen 1,3-dipolar cycloaddition process by simply incorporating Cu catalyst. The modified reaction appeared as more straightforward and novel triumphant form, known as Cu catalysed azide-alkyne cycloaddition

(CuAAC), which leads to selective formation of 1,4-disubstituted 1,2,3-triazoles as the only product of the reaction. The CuAAC protocol is advantageous over traditional methods because of having mild reaction condition, high selectivity, high yield, low reaction temperature and broad substrate scope. (Figure 6) (Kumar, Kumar, Singh, Tittal, & Lal, 2024)

Figure 6. Scheme 1

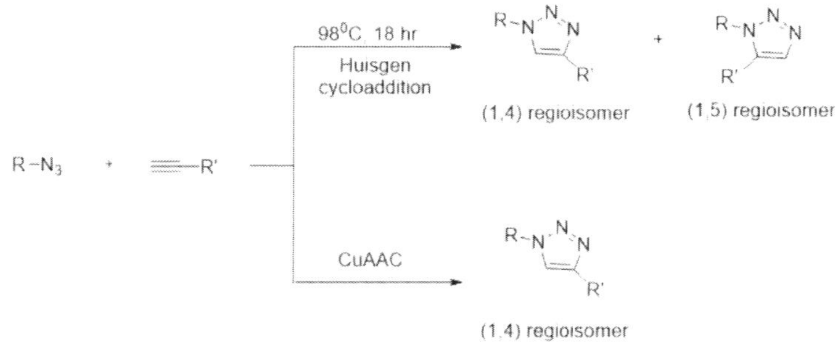

Atom economical CuAAC is one of the most well-known examples of 'click chemistry' concepts due to its specificity, reliability and biocompatibility. (Moses & Moorhouse, 2007) The term 'click chemistry' was first coined by Sharpless (Kolb, Finn, & Sharpless, 2001) in 2001 to define the reactions that- "are modular, high yielding, wide in scope, create only in offensive by-products (that can be removed without column chromatography), stereospecific, simple to perform and require benign or easily removable solvent." The 'click chemistry' concept emerged to design the chemical processes for synthesising highly complex organic structures directly from simple and readily available organic entities. (Nebra & García-Álvarez, 2020) Click chemistry has contributed significantly in revolutionization of various fields of modern chemistry like material science, nanotechnology, organic synthesis, drug discovery and development. The Cu catalysed cycloaddition reactions between azides and terminal alkynes to form 1,2,3-triazoles have become the benchmark of click reactions. (Moses & Moorhouse, 2007) Numerous methodologies have been developed for synthesizing substituted 1,2,3-triazoles adopting CuAAC strategy. (Llgen & König, 2009; Vidal & Garcia-Alvarez, 2014; De La Cerda-Pedro, Rojas-Lima, Santillan, Lopez-Ruiz, 2015) However, most of the reported CuAAC methodologies involve use of volatile conventional hazardous organic solvents and catalysts which is against the requirement (i.e. use of benign solvent) of click reaction. The ascending use of traditional organic solvents and catalysts for the synthesis of 1,2,3-triazoles results adverse impact on human health and environment, which has become a major concern. (Lu, Ma, Liu, Li, Mo, Zhang, 2015; Masood, 2023)

To overcome this harmful situation, researchers are giving their attention towards greener strategy for synthesizing important heterocyclic 1,2,3-triazoles. Moreover, nowadays the scientific community has paid much more attention towards acquiring 'green chemistry principles'; so that the synthetic methodologies cause less harm to the environment and human health. The concept of 'green chemistry' was introduced in 1990s, which focused on fabricating the synthetic processes and the hazardous chemicals to make them safer for people and environment. (Kumar, Kumar, Singh, Tittal, & Lal, 2024) In the research community, initiatives are in progress to substitute the traditional organic solvents and catalysts with environmentally benign or green alternatives. Efforts are given to develop more ecofriendly and newer sustainable CuAAC protocols for the synthesis of 1,4-disubstituted 1,2,3-triazoles employing recyclable catalysts and environment friendly solvents or reaction medium. (Giofre, Tiecco, Ferlazzo, Romeo, Ciancaleoni, Germani, Iannazzo, 2021; Pan, Yan, Osako, & Uozumi, 2017) For solvents to be considered as green, they should be reusable, biodegradable, inexpensive, readily available, non-toxic to both the environment and humans, safe (non-flammable and with low vapor pressure) and can effectively dissolve a wide range of chemicals and prevent waste. With these uncompromising criteria, it is not so surprising that the range of green and sustainable solvents is relatively limited and the most common examples are water, ionic liquids, deep eutectic solvents and solvents derived from biomass such as glycerol, lactic acid (DESs). (Nebra & García-Álvarez, 2020)

CLICK REACTIONS USING IONIC LIQUIDS

Ionic liquids (ILs) are compatible reaction medium or catalysts that uphold the green chemistry principles. Ionic liquids are salts that are liquid at room temperature, consisting entirely of ions (organic cation and inorganic or organic anion) with melting points below 100°C. Researchers and industries have increasingly focused on ILs due to their unique characteristics such as low vapor pressure, excellent electrochemical thermal stability, low toxicity and renewability; are hence accepted as eco-friendly solvents or liquids. (Lei, Chen, Koo, & MacFarlane, 2017; Fabre & Murshed, 2021) The term "designer solvent" is referred to ionic liquids because their structure contains both positively and negatively charged components, enabling them to be specially designed for various reaction conditions. (Sheldon, 2005; Vekariya, 2017) Tremendous attention has been paid towards developing greener methodologies of azide-alkyne cycloaddition reaction (a common example of click reaction) for synthesising 1,2,3-triazoles using ionic liquids which has the potential to improve selectivity, reaction efficiency and environmental sustainability. (Kumar, Kumar, Singh, Tittal, & Lal, 2024)

In the reported literatures regarding synthesis of 1,2,3-triazoles, it has been observed that ionic liquids are employed either as solvents or as catalytic systems. Additionally, in some reports ionic liquids serve both as solvent and catalyst. In 2016 A. Ali and coworkers established a CuAAC reaction employing a basic ionic liquid, 3-butyl-1-methylimidazolium hydroxide [Bmim]OH as solvent without involving any bases, reducing agent and Additives. Under this regioselective protocol, 1,4-disubstituted 1,2,3-triazoles were synthesised in excellent yields with no undesirable by-products. It was observed that there was no improvement in the yield of products on replacement of the anion of ionic liquid [Bmim]OH, indicates that the anion of [Bmim]OH plays a significant role in the reaction. In the reaction process, [Bmim]OH assists rapid deprotonation of the alkyne to form reactive Cu(I)acetylide complex. (Figure 7) (Ali, Konwar, Chetia, & Sarma, 2016)

Figure 7. Scheme 2

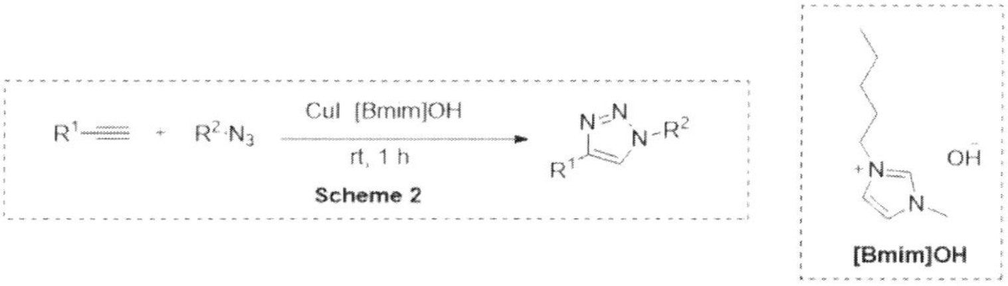

J. D. Patil et al. developed a regioselective and green method for the synthesis of 1,4-disubstituted 1,2,3-trizoles by conducting a reaction between organic azides and terminal alkynes in water in presence of in situ incorporated copper with Merrifield resin (polymer) supported ascorbate functionalized task specific ionic liquid [MR-IMZ-As] at room temperature. This protocol came out as most favourable due to its operational simplicity, high yield of products, room temperature requirement, recyclability of catalyst. The ionic liquid [MR-IMZ-As] acts as efficient catalyst in this protocol giving high yield at room temperature. (Figure 8) (Patil, Patil, & Pore, 2015)

Figure 8. Scheme 3

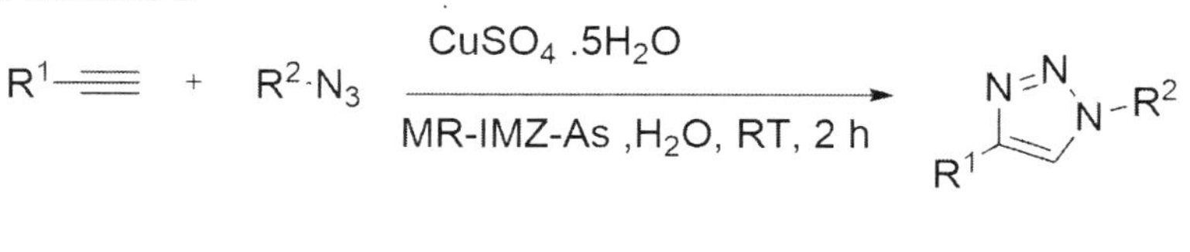

P. Phukan el.al reported an efficient and regioselective protocol for the synthesis of 1,4-disubstituted 1,2,3-triazoles by cycloaddition, using low copper loaded ionic liquid [Bmim][CuCl$_3$] as catalyst. This method demonstrated significant advantages as it minimized the use of additional ligands, bases, and solvents for the cycloaddition of terminal alkynes and organic azides, which resulted in the formation of 1,2,3-triazoles. The catalyst showed its efficacy as being non-cytotoxic due to its very low copper loading and also along with ascorbic acid it exhibited the best catalytic activity by providing the product in 94% yield within a very short period of time under solvent free conditions. (Figure 9) (Phukan, Kulshrestha, Kumar, Chakraborti, Chattopadhyay, Sarma, 2021)

Figure 9. Scheme 4

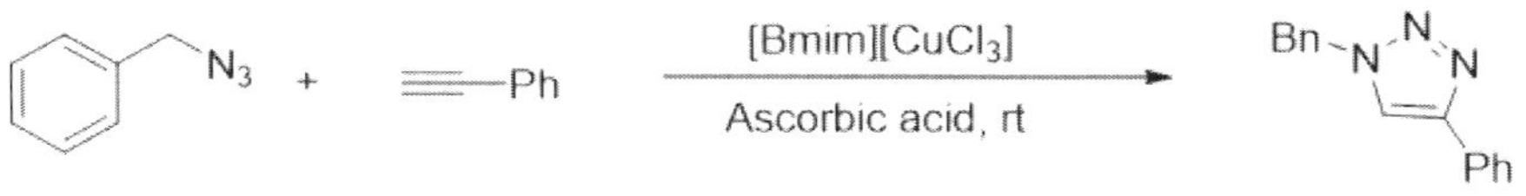

Scheme 4

Raut et al. in 2009 suggested a greener approach to synthesise 1,4-disubstituted 1,2,3-triazoles from terminal alkynes and azides using copper nanoparticles (CuNPs) using an IL: water solution. Initially, the IL-based NPs were obtained from copper acetate by reduction in the presence of hydrazine hydrate in the IL: water solvent system and were then stabilized by the addition of PVA/PVP. [Bmim][BF$_4$] with water gave better yield of the product in less amount of time. (Figure 10) (Raut, Wankhede, Vaidya, Bhilare, Darwatkar, Deorukhkar, Trivedi, Salunkhe, 2009)

Figure 10. Scheme 5

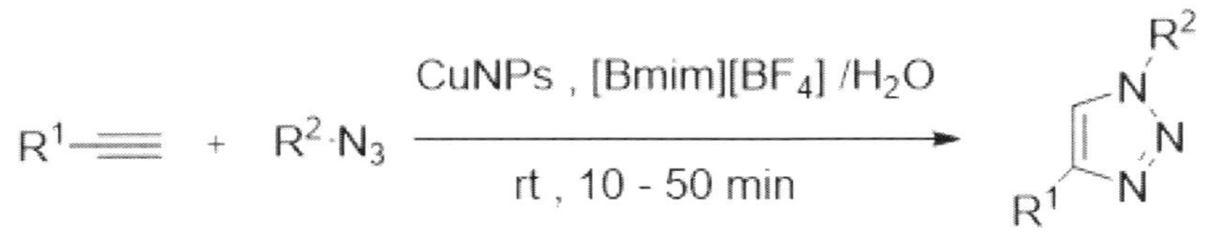

Scheme 5

Javaherian and co-workers in 2014 used IL: Water (1:1) as solvent for Cu catalysed azide alkyne cycloaddition reaction. The reaction was carried out in presence of $CuSO_4$ .5H_2O and sodium ascorbate. Tetra-ethylene-glycol bis-(1-methyl-3-imidazolium) tosylate was used with water for a better yield of the reaction (Figure 11). (Javaherian, Kazemi, & Ghaemi, 2014)

Figure 11. Scheme 6

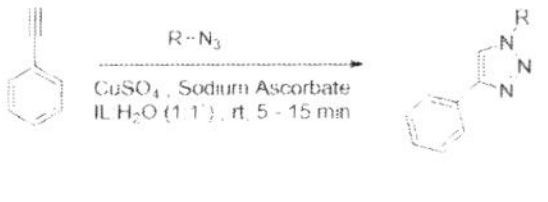

In 2019 Akolkar et al. synthesized 1,4-disubstituted 1,2,3-triazoles under ultrasonic irradiation in the presence of $Cu(OAc)_2 \cdot H_2O$ in $[Et_3NH][OAc]$. This method gives high yield within a short time, reduced the use of toxic solvents, and the solvent $[Et_3NH][OAc]$ is reusable and recyclable. Among all the used catalysts $Cu(OAc)_2 \cdot H_2O$ in $[Et_3NH][OAc]$ provided an excellent yield in a short reaction time (Figure 12). (Akolkar, Nagargoje, Krishna, Sriram, Sangshetti, Damale, Shingate, 2019)

Figure 12. Scheme 7

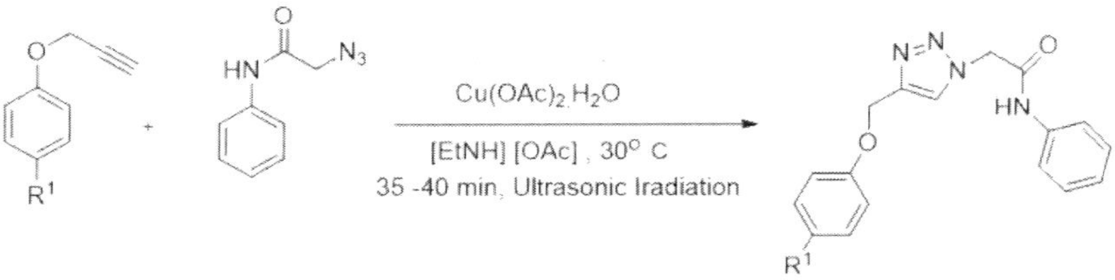

CLICK REACTIONS USING DEEP EUTECTIC SOLVENTS

Deep Eutectic Solvents (DESs) are formed when two or more chemical substances combine to create a new eutectic mixture which has a melting point lower than that of the individual components due to the establishment of a three-dimensional hydrogen-bond network.[38] Ammonium salt choline chloride (ChCl, also known as vitamin B4) which is both biorenewable and biodegradable, is one of the most frequently utilized components in the synthesis of these eutectic mixtures. (Blusztajn, 1998) Choline chloride forms sustainable, liquid eutectic mixtures when combines with various hydrogen bond donors such as: i) naturally occurring polyols (like glycerol, ethylene glycol, or carbohydrates), (Abbott, Harris, Ryder, D'Agostino, Gladden, Mantle, 2011) ii) biorenewable organic acids, (Abbott, Boothby, Capper, Davies, Rasheed, 2004) or iii) urea. (Abbott, Capper, Davies, Rasheed, & Tambyrajah, 2003) DESs have found its application as a novel and green solvents in CuAAC reactions because of their notable advantages such as nontoxicity, non-volatility, non-flammability, ease of recyclability, thermal stability, and most crucially, their ability to dissolve both polar and nonpolar reagents. (Bagherzadeh, Amiri, & Sardarian, 2023) König et al. were the original pathfinders in this field of analyzing CuAAC reaction in DESs; they in 2009 reported the cycloaddition of phenylacetylene and benzyl azide catalysed by CuI, for synthesising 1,4-dibsubstituted 1,2,3-triazoles. In the three-component CuAAC process, employing the ternary eutectic mixture of D-sorbitol/urea/NH_4Cl as a sustainable solvent facilitated the in-situ synthesis of benzyl azide (from $PhCH_2Br$ and NaN_3). Notably, this CuAAC process showed a slight improvement on using the L-carnitine-based eutectic mixture (Figure 13). (Illgen & König, 2009)

Figure 13. Scheme 8

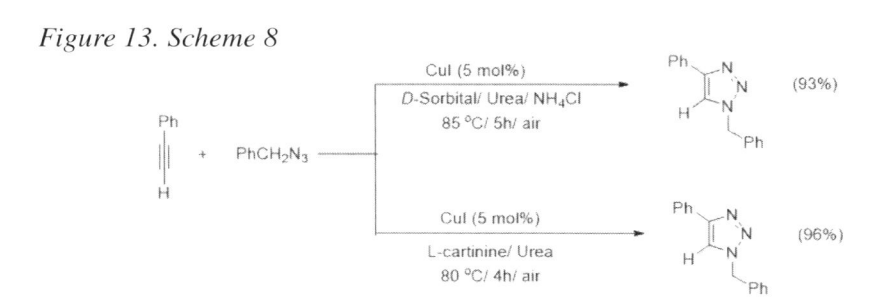

After that, G. Alvarez showed that use of the eutectic mixture of choline chloride and glycerol (1ChCl/2Gly) enabled the CuAAC reaction to proceed under milder conditions, specifically at room temperature and with just 1 mol% of CuI; although this approach required a longer reaction time (Figure 14). (Vidal & García-Álvarez, 2014)

Figure 14. Scheme 9

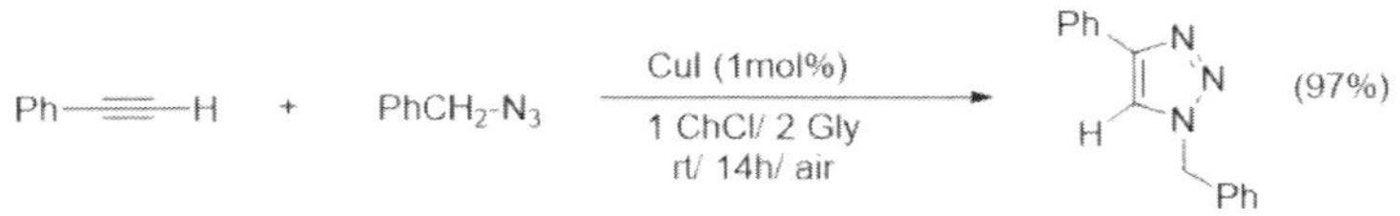

Sadarian and coworkers synthesised a novel and green metal acidic deep eutectic solvent, Cu(II)-ADES having copper salt, gallic acid, and choline chloride. This synthesised M-ADES is recyclable and reusable without loss of its activity even after seven cycles; and demonstrates excellent reactivity as catalyst/solvent system in synthesising a collection of 1,4-disubstituted 1,2,3-triazoles via three-component click reactions under base and reducing agent-free conditions, utilizing a range of benzyl, allyl, ester, and alkyl halides, sodium azides, and terminal alkynes. This triazole synthesis method using M-ADES is notable for its atom efficiency, short reaction times, wide range of substrates scope, and high effectiveness in large-scale production of desired products. (Figure 15) (Bagherzadeh, Amiri, & Sardarian, 2023)

Figure 15. Scheme 10

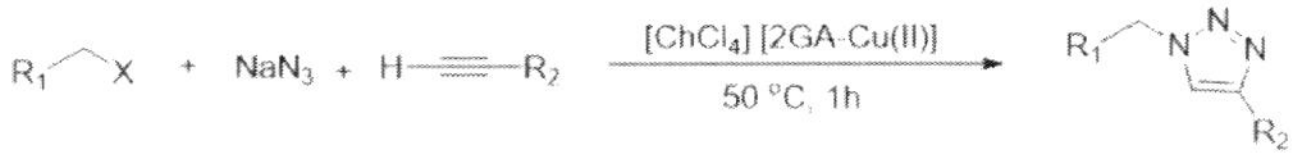

Thus, ILs and DESs play a very significant role as solvent/catalyst system in click reactions for synthesising various medicinally important 1,2,3-triazoles.

Giofre et al. selected a set of differently structured natural deep eutectic solvents (NADES) and investigated the utility of the NADESs as "active" and green reaction media in CuAAC reaction in the absence of bases. During the experiment; the absence of a base, low temperatures, reduced reagent amounts, and solvent recycling indicate that NADESs have great potential for CuAAC. The best results were shown by GA/TMG NADES in base-free conditions as the best potential green solvent for the CuAAC, whereas, the NADES Asc/ChCl proved to be the best green solvent for the non-classical CuAAC base and reducing agent-free reactions. Due to the presence of L-abscorbic acid Asc/ChCl was seen to be a reducing DES. (Figure 16) (Giofre, Tiecco, Ferlazzo, Romeo, Ciancaleoni, Germani, Iannazzo, 2021)

Figure 16. Scheme 11

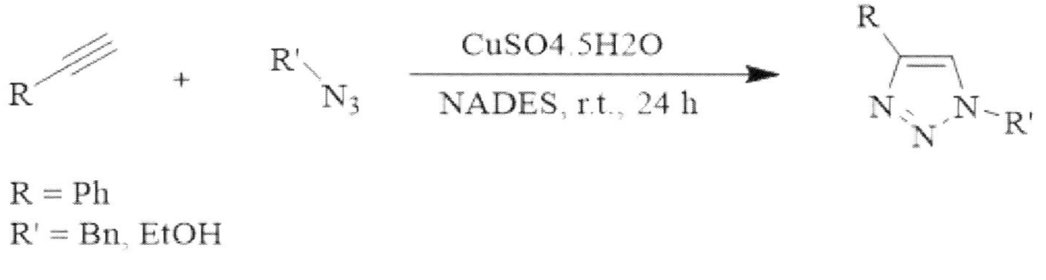

R = Ph
R' = Bn, EtOH

Díez-González and co-workers carried out a regioselective one-pot synthesis of 1,5- and 1,4- disubstituted triazoles using choline chloride: urea (1:2) as DES with a base trimethylguanidine (TMG). TMG was found to be cheaper, relatively non-toxic, and showed better miscibility with the DES than bases such as DBN (1,5-Diazabicyclo[4.3.0]non-5-ene) and DBU (1,8-Diazabycyclo[5.4.0.]undec-7-ene) leading to better removal from the reaction mixture. This metal-free methodology does not involve any purification step and minimizes the usage of volatile organic solvents. (Figure 17) (Sebest, Haselgrove, White, & González, 2020)

Figure 17. Scheme 12

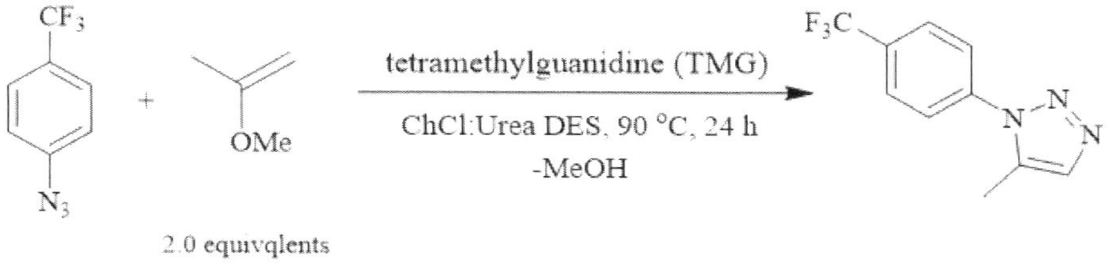

Sebest et al. performed a multigram synthesis of 1,2,3-triazolines via azide-alkyne cycloaddition in DES made up of choline chloride: urea (1:2) where some of the side products were also aziridines at 80 °C for 16 h. When the product triazoline was heated in an organic solvent such as DMSO, decomposition of the product was seen, but in the presence of DES, the products were stable. A range of di- and tri-substituted Δ^2-1,2,3-triazolines have been prepared and fully characterized. (Figure 18) (Sebest, Cassarubios, Rzepa, White, & González, 2018)

Figure 18. Scheme 13

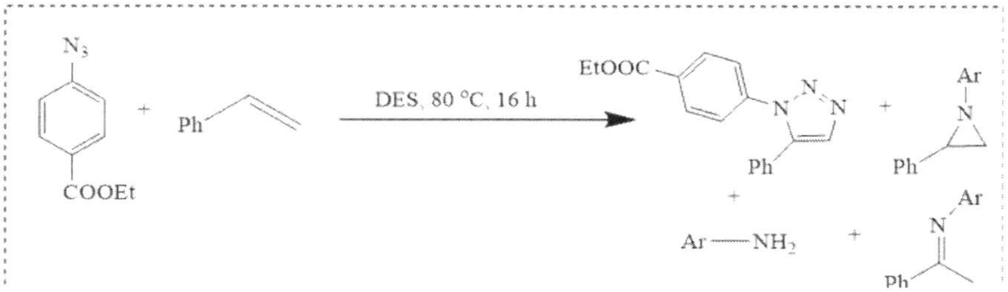

Bidraha Bagh et al. synthesized two air stable copper-(I) halide coordination polymers with NNs and NNO ligand frameworks and were successfully utilized as efficient catalysts in azide-alkyne cycloaddition reaction to prepare 1,2,3-triazoles with good to better yields. After using water and glycerol as reaction media, a deep eutectic solvent composed of choline chloride and glycerol was found to be an excellent alternative. In the DES, complete conversion with excellent isolated yield was achieved in a short period (1 h) with low catalyst loading (1 mol %) at room temperature. Also, complete conversion was achieved within 24 h with ppm-level (50 ppm) catalyst loading at 70 °C. The reaction media was recycled up to 5 times with good product yield. (Figure 19) (Sethi, Jana, Behera, Behera, & Bagh, 2023)

Figure 19. Scheme 14

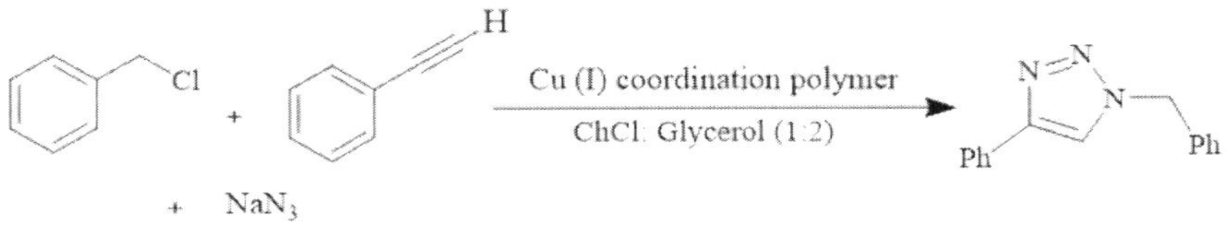

Reaction condition 1: Cu(I) catalyst (1 mol%), DES, r.t., 1 h
Reaction condition 2: Cu(I) catalyst, DES, 70 °C, 24 h

Konig et al. in 2009 developed a methodology for the synthesis of 1,4-disubstituted 1,2,3-triazoles using CuI catalyst in DES. They used D-sorbitol/urea/NH$_4$Cl (7:2:1) melt as a deep eutectic solvent. The reaction was carried out at 85 °C for 5 hours (Figure 20). Interestingly, the above-mentioned CuAAC process proceeded slightly better when the *L*-carnitine-based eutectic mixture L-carnitine /urea is used as a sustainable solvent (Figure 20)**.** (Ilgen & Konig, 2009)

Figure 20. Scheme 15 and Scheme 16

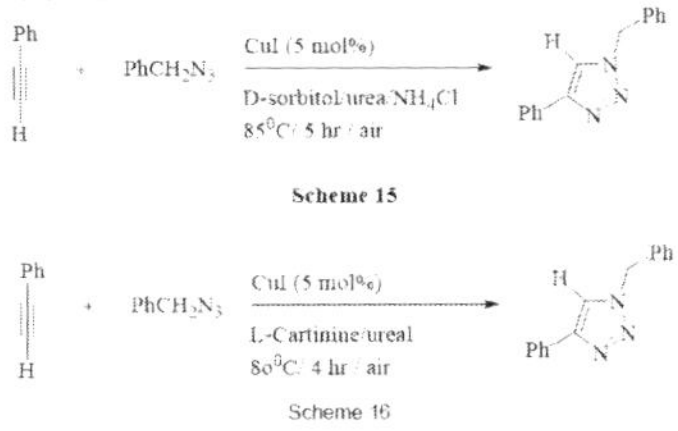

García-Alvarez and co-workers in 2014 demonstrated the use of the eutectic mixture formed by choline chloride and glycerol (1:2). This reaction between azide and alkyne proceeds at room temperature using CuI catalyst (Figure 21). (Vidal & Alvarez, 2014)

Figure 21. Scheme 17

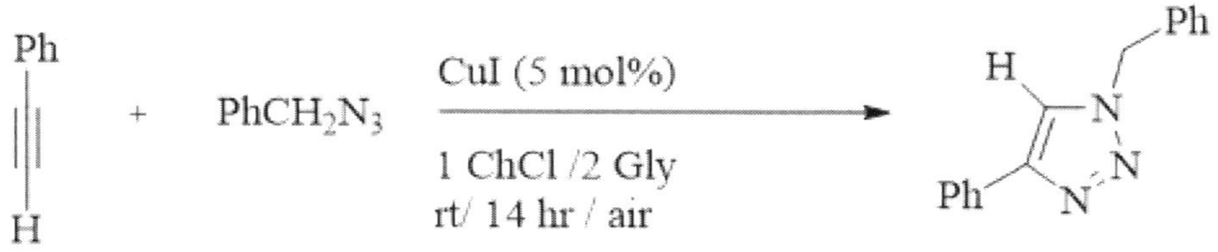

Handy et al. in 2017 reported a one pot synthetic procedure of 1,4-disubstituted 1,2,3-triazoles. The reaction was carried out between alkyl/aryl bromide and sodium azide in presence of an alkyne. The in situ generated azide reacts with the terminal alkyne in the presence of a deep eutectic mixture of choline choride and glycerol, CuI is used as a catalyst and N,N-dimethylethylenediamine (DMEAD) as co-catalyst (Figure 22). (Kafle & Handy, 2017)

Figure 22. Scheme 18

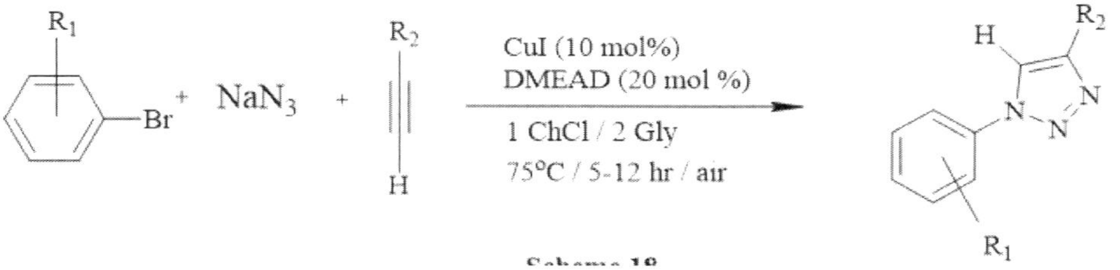

Martins et al. in 2016 introduced a simple approach to obtain 4-acyl-1-substituted-1,2,3- triazoles in excellent yields by employing a DES composed of choline chloride and ethylene glycol in 1: 2 ratio as green reaction media (Figure 23). (Martins, Paveglio, Rodrigues, Frizzo, Zanatta, Bonacorso, 2016)

Figure 23. Scheme 19

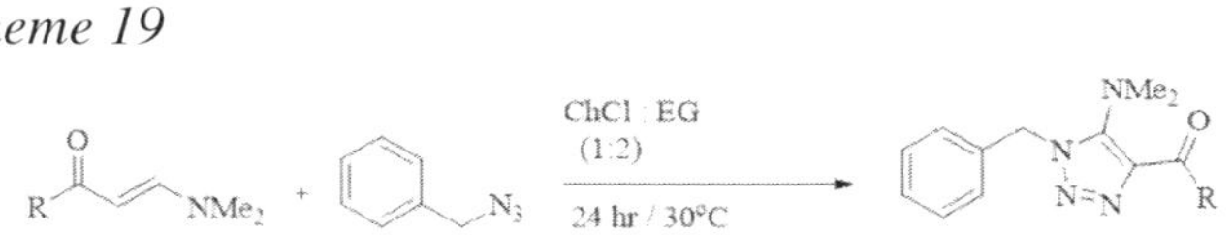

Recently in 2022, an eco-friendly metal-free protocol was developed by P. Vitale and co-workers for the regioselective synthesis of densely functionalized 1,2,3-triazoles through a 1,3-dipolar cycloaddition reaction of alkanone enolates with azides. The starting material 1-phenylpropan-2-one was treated with ^{t}BuOK for 1 hour at room temperature in a choline chloride:urea (1:2) eutectic mixture. The enolate formed is then treated with phenyl azide in choline acetate:urea (1:2) eutectic mixture. The maximum yield was found to be 98%. (Figure 24). (Cicco, Perna, Falcicchio, Altomare, Messa, Salomone, Capriati, Vitale, 2022)

Figure 24. Scheme 20

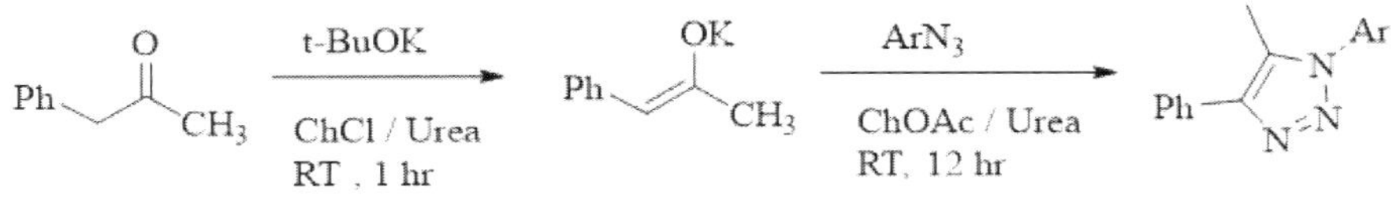

CONCLUSION

Over the past few years, 1,2,3-triazoles have gained tremendous attention because of their wide application in pharmaceuticals, supramolecular chemistry, chemical biology, and material sciences. 1,2,3-triazole bearing compounds are used as anticancer, antimicrobial, anti-infective, antioxidant drugs etc. One of the simplest methods for production of 1,2,3- triazoles is a "click reaction", featuring an azide-alkyne [3+2] cycloaddition catalysed by Cu(II). The CuAAC reaction can be considered as greener and more efficient if the reaction can be done using sustainable solvents like DES or Ionic Liquids. Low cost of preparation, renewability, biodegradability etc. are the convincing characteristics that make these solvents a very greener alternative to conventional organic solvents.

REFERENCES

Abbott, A. P., Boothby, D., Capper, G., Davies, D., & Rasheed, R. K. (2004). Deep Eutectic Solvents Formed between Choline Chloride and Carboxylic Acids: Versatile Alternatives to Ionic Liquids. *Journal of the American Chemical Society*, 126(29), 9142–9147. DOI: 10.1021/ja048266j PMID: 15264850

Abbott, A. P., Capper, G., Davies, D. L., & Rasheed, R. (2004). Ionic liquids based upon metal halide/substituted quaternary ammonium salt mixtures. *Inorganic Chemistry*, 43(11), 3447–3452. DOI: 10.1021/ic049931s PMID: 15154807

Abbott, A. P., Capper, G., Davies, D. L., Rasheed, R. K., & Tambyrajah, V. (2003). Novel solvent properties of choline chloride/urea mixturesElectronic supplementary information (ESI) available: spectroscopic data. See http://www.rsc.org/suppdata/cc/b2/b210714g/. *Chemical Communications (Cambridge)*, (1), 70–71. DOI: 10.1039/b210714g

Abbott, A. P., Capper, G., Davies, D. L., Rasheed, R. K., & Tambyrajah, V. (2003). Novel solvent properties of choline chloride/urea mixturesElectronic supplementary information (ESI) available: spectroscopic data. See http://www.rsc.org/suppdata/cc/b2/b210714g/. *Chemical Communications (Cambridge)*, (1), 70–71. DOI: 10.1039/b210714g

Abbott, A. P., Harris, R. C., Ryder, K. S., D'Agostino, C., Gladden, L. F., & Mantle, M. D. (2011). Glycerol eutectics as sustainable solvent systems. *Green Chemistry*, 13(1), 82–90. DOI: 10.1039/C0GC00395F

Agouram, N., Hadrami, E. M., & Bentama, A. (2021). 1,2,3-Triazoles as Biomimetics in Peptide Science. *Molecules (Basel, Switzerland)*, 26(10), 2937. DOI: 10.3390/molecules26102937 PMID: 34069302

Akolkar, S., Nagargoje, A., Krishna, V., Sriram, D., Sangshetti, J., Damale, M., & Shingate, B. (2019). New *N*-phenylacetamide-incorporated 1,2,3-triazoles: [Et$_3$NH][OAc]-mediated efficient synthesis and biological evaluation. *RSC Advances*, 9(38), 22080–22091. DOI: 10.1039/C9RA03425K PMID: 35518861

Ali, A. A., Konwar, M., Chetia, M., & Sarma, D. (2016). [Bmim]OH mediated Cu-catalyzed azide–alkyne cycloaddition reaction: A potential green route to 1,4-disubstituted 1,2,3-triazoles. *Tetrahedron Letters*, 57(50), 5661–5665. DOI: 10.1016/j.tetlet.2016.11.014

Ashwini, N., Garg, M., Mohan, C. D., Fuchs, J. F., Rangappa, S., Anusha, S., Swaroop, T. R., Rakesh, K. S., Kanojia, D., Madan, V., Bender, A., & Koeffler, H. P., Basappa., & Rangappa, K. S. (2015). Synthesis of 1,2-benzisoxazole tethered 1,2,3-triazoles that exhibit anticancer activity in acute myeloid leukemia cell lines by inhibiting histone deacetylases, and inducing p21 and tubulin acetylation. [PubMed]. *Bioorganic & Medicinal Chemistry*, 23(18), 6157–6165. DOI: 10.1016/j.bmc.2015.07.069

Bagherzadeh, N., Amiri, M., & Sardarian, A. R. (2023). Novel Cu(ii) acidic deep eutectic solvent as an efficient and green multifunctional catalytic solvent system in base-free conditions to synthesize 1,4-disubstituted 1,2,3-triazoles. *RSC Advances*, 13(51), 36403–36415. DOI: 10.1039/D3RA06570G PMID: 38099257

Bangalore, P. K., Vagolu, S. K., Bollikanda, R. K., Veeragoni, D. K., Choudante, P. C., Misra, S., Sriram, D., Sridhar, B., & Kantevari, S. (2019). Usnic Acid Enaminone-Coupled 1,2,3-Triazoles as Antibacterial and Antitubercular Agents. [PubMed]. *Journal of Natural Products*, 83(1), 26–35. DOI: 10.1021/acs.jnatprod.9b00475

Blusztajn, J. K. (1998). Choline, a Vital Amine. *Science*, 284(5378), 794–795. DOI: 10.1126/science.281.5378.794 PMID: 9714685

Chum, H. (1975). L.; Koch, V, R.; Miller, L, L.; Osteryoung, R, A. *Journal of the American Chemical Society*, 97, 3264–3265. DOI: 10.1021/ja00844a081

Cicco, L., Perna, F. M., Falcicchio, A., Altomare, A., Messa, F., Salomone, A., Capriati, V., & Vitale, P. (2022). 1,3-Dipolar Cycloaddition of Alkanone Enolates with Azides in Deep Eutectic Solvents for the Metal-Free Regioselective Synthesis of Densely Functionalized 1,2,3-Triazoles. *European Journal of Organic Chemistry*, 2022(36), e202200843. DOI: 10.1002/ejoc.202200843

Clarke, C. J., Tu, W. C., Levers, O., Brohl, A., & Hallett, J. P. (2018). Green and sustainable solvents in chemical processes. *Chemical Reviews*, 118(2), 747–800. DOI: 10.1021/acs.chemrev.7b00571 PMID: 29300087

D'Agostino, C., & Gladden, L. (2015). F.; Mantle, M, D.; Abbott, A, P.; Ahmed, E, I.; Al-Murshedi, A, Y, M.; Harris, R, C. *Physical Chemistry Chemical Physics*, 17, 15297–15304. PMID: 25994171

D'Agostino, C., & Harris, R. (2011). C.; Abbott, A, P.; Gladden, L, F.; Mantle, M, D. *Physical Chemistry Chemical Physics*, 13, 21383–21391. DOI: 10.1039/c1cp22554e PMID: 22033601

Das, J., Dey, S., & Pathak, T. (2019). Metal-Free Route to Carboxylated 1,4-Disubstituted 1,2,3-Triazoles from Methoxycarbonyl-Modified Vinyl Sulfone. *The Journal of Organic Chemistry*, 84(23), 15437–15447. DOI: 10.1021/acs.joc.9b02443 PMID: 31657567

De La Cerda-Pedro, J. E., Rojas-Lima, S., Santillan, R., & Lopez-Ruiz, H. (2015).... *Journal of the Mexican Chemical Society*, 59, 130–136.

Fabre, E., & Murshed, S. S. (2021). A review of the thermophysical properties and potential of ionic liquids for thermal applications. *Journal of Materials Chemistry. A, Materials for Energy and Sustainability*, 9(29), 15861–15879. DOI: 10.1039/D1TA03656D

Florindo, C., Branco, L. C., & Marrucho, I. (2019). M. *ChemSusChem*, 12, 1549–1559. DOI: 10.1002/cssc.201900147 PMID: 30811105

Gabriel, S., & Weiner, J. (1888)... *Chemische Berichte*, 21, 2669–2679.

Giofre, S. V., Tiecco, M., Ferlazzo, A., Romeo, R., Ciancaleoni, G., Germani, R., & Iannazzo, D. (2021). Base-Free Copper-Catalyzed Azide-Alkyne Click Cycloadditions (CuAAc) in Natural Deep Eutectic Solvents as Green and Catalytic Reaction Media**. *European Journal of Organic Chemistry*, 2021(34), 4777–4789. DOI: 10.1002/ejoc.202100698

Giofre, S. V., Tiecco, M., Ferlazzo, A., Romeo, R., Ciancaleoni, G., Germani, R., & Iannazzo, D. (2021). Base-Free Copper-Catalyzed Azide-Alkyne Click Cycloadditions (CuAAc) in Natural Deep Eutectic Solvents as Green and Catalytic Reaction Media**. *European Journal of Organic Chemistry*, 2021(34), 4777–4789. DOI: 10.1002/ejoc.202100698

Guti rrez, M, C.; Ferrer, M, L.; Mateo, C, R.; Monte, F. 2009*Langmuir*-2555095515DOI: 10.1021/la900552b PMID: 19432491

Hajipour, A. (2009). R.; Azizi, G.; Ruoho, A, E. *Synthetic Communications*, 39, 242–250. DOI: 10.1080/00397910802120916

Hansen, B. B., Spittle, S., Chen, B., Poe, D., Zhang, Y., Klein, J. M., Horton, A., Adhikari, L., Zelovich, T., Doherty, B. W., Gurkan, B., Magninn, E. J., Ragauskas, A., Dadmun, M., Zawodzinski, T. A., Baker, G. A., Tuckerman, M. E., Savinell, R. F., & Sangoro, J. R. (2021). Deep Eutectic Solvents: A Review of Fundamentals and Applications. *Chemical Reviews*, 121(3), 1232–1285. DOI: 10.1021/acs.chemrev.0c00385 PMID: 33315380

Huisgen, R. (1984). 1, 3-Dipolar Cycloaddition Chemistry.

Ilgen, F., & Konig, B. (2009). Organic reactions in low melting mixtures based on carbohydrates and l-carnitine—A comparison. *Green Chemistry*, 11(6), 848–854. DOI: 10.1039/b816551c

Illgen, F., & König, B. (2009).. . *Green Chemistry*, 11, 848–854. DOI: 10.1039/b816551c

Javaherian, M., Kazemi, F., & Ghaemi, M. (2014). A dicationic, podand-like, ionic liquid water system accelerated copper-catalyzed azide-alkyne click reaction. *Chinese Chemical Letters*, 25(12), 1643–1647. DOI: 10.1016/j.cclet.2014.09.005

Kafle, A., & Handy, T. (2017). A one-pot, copper-catalyzed azidation/click reaction of aryl and heteroaryl bromides in an environmentally friendly deep eutectic solvent. *Tetrahedron*, 73(50), 7024–7029. DOI: 10.1016/j.tet.2017.10.050

Kantheti, S., Narayan, R., & Raju, K. V. S. N. (2015). The impact of 1,2,3-triazoles in the design of functional coatings. *RSC Advances*, 5(5), 3687–3708. DOI: 10.1039/C4RA12739K

Khandelwal, S., Tailor, Y. K., & Kumar, M. (2016). Deep eutectic solvents (DESs) as eco-friendly and sustainable solvent/catalyst systems in organic transformations. *Journal of Molecular Liquids*, 215, 345–386. DOI: 10.1016/j.molliq.2015.12.015

Khandelwal, S., Tailor, Y. K., & Kumar, M. (2016). Deep eutectic solvents (DESs) as eco-friendly and sustainable solvent/catalyst systems in organic transformations. *Journal of Molecular Liquids*, 215, 345–386. DOI: 10.1016/j.molliq.2015.12.015

Kolb, H. C., Finn, M. G., & Sharpless, K. B. (2001). Click Chemistry: Diverse Chemical Function from a Few Good Reactions. *Angewandte Chemie International Edition*, 40(11), 2004–2021. DOI: 10.1002/1521-3773(20010601)40:11<2004::AID-ANIE2004>3.0.CO;2-5 PMID: 11433435

Kumar, A., Kumar, V., Singh, P., Tittal, R. M., & Lal, K. (2024). Ionic liquids for the green synthesis of 1,2,3-triazoles: A systematic review. *Green Chemistry*, 26(7), 3565–3594. DOI: 10.1039/D3GC04898E

Lei, Z., Chen, B., Koo, Y. M., & MacFarlane, D. R. (2017). Introduction: Ionic Liquids. [PubMed]. *Chemical Reviews*, 117(10), 6633–6635. DOI: 10.1021/acs.chemrev.7b00246

iLin, Y., He, S. F., Geng, H., Xiao, Y. C., Ji, K. L., Zheng, J. F., & Huang, P. Q. (2021). Chemoselective Reactions of Isocyanates with Secondary Amides: One-Pot Construction of 2,3-Dialkyl-Substituted Quinazolinones. *The Journal of Organic Chemistry*, 86(7), 5345–5353. DOI: 10.1021/acs.joc.0c02929 PMID: 33710879

Llgen, F., & König, B. (2009).. . *Green Chemistry*, 11, 848–854. DOI: 10.1039/b816551c

Lu, J., Ma, E. Q., Liu, Y. H., Li, Y. M., Mo, L. P., & Zhang, Z. H. (2015). One-pot three-component synthesis of 1,2,3-triazoles using magnetic $NiFe_2O_4$–glutamate–Cu as an efficient heterogeneous catalyst in water. *RSC Advances*, 5(73), 59167–59185. DOI: 10.1039/C5RA09517D

Martins, A. P., Paveglio, G. C., Rodrigues, L. V., Frizzo, C. P., Zanatta, N., & Bonacorso, H. G. (2016). Promotion of 1,3-dipolar cycloaddition between azides and β-enaminones by deep eutectic solvents. *New Journal of Chemistry*, 40(7), 5989–5992. DOI: 10.1039/C5NJ03654B

Moses, J. E., & Moorhouse, A. D. (2007). The growing applications of click chemistry. *Chemical Society Reviews*, 36(8), 1249–1262. DOI: 10.1039/B613014N PMID: 17619685

Nebra, N., & García-Álvarez, J. (2020). Recent Progress of Cu-Catalyzed Azide-Alkyne Cycloaddition Reactions (CuAAC) in Sustainable Solvents: Glycerol, Deep Eutectic Solvents, and Aqueous Media. *Molecules (Basel, Switzerland)*, 25(9), 2015. DOI: 10.3390/molecules25092015 PMID: 32357387

Nehra, N., Tittal, R. K., Ghule, V. D., & Lal, K. (2021). Synthesis, antifungal studies, molecular docking, ADME and DNA interaction studies of 4-hydroxyphenyl benzothiazole linked 1,2,3-triazoles. *Journal of Molecular Structure*, 1245, 131013. DOI: 10.1016/j.molstruc.2021.131013

Pan, S., Yan, S., Osako, T., & Uozumi, Y. (2017). Batch and Continuous-Flow Huisgen 1,3-Dipolar Cycloadditions with an Amphiphilic Resin-Supported Triazine-Based Polyethyleneamine Dendrimer Copper Catalyst. *ACS Sustainable Chemistry & Engineering*, 5(11), 10722–10734. DOI: 10.1021/acssuschemeng.7b02646

Patil, J. D., Patil, S. A., & Pore, D. M. (2015). A polymer supported ascorbate functionalized task specific ionic liquid: An efficient reusable catalyst for 1,3-dipolar cycloaddition. *RSC Advances*, 5(27), 21396–21404. DOI: 10.1039/C4RA16481D

Phukan, P., Kulshrestha, A., Kumar, A., Chakraborti, S., Chattopadhyay, P., & Sarma, D. (2021). Cu(II) ionic liquid promoted Simple and Economical Synthesis of 1,4-disubstituted-1,2,3-triazoles with Low Catalyst Loading. *Journal of Chemical Sciences*, 133(4), 131. DOI: 10.1007/s12039-021-01980-9

Raiguel, S., Dehaen, W., & Binnemans, K. (2020). Stability of ionic liquids in Brønsted-basic media. *Green Chemistry*, 22(16), 5225–5252. DOI: 10.1039/D0GC01832E

Rani, P., Lal, K., Shrivastava, R., & Ghule, V. D. (2020). Synthesis and characterization of 1,2,3-triazoles-linked urea hybrid sensor for selective sensing of fluoride ion. *Journal of Molecular Structure*, 1203, 127437. DOI: 10.1016/j.molstruc.2019.127437

Raut, D., Wankhede, K., Vaidya, V., Bhilare, S., Darwatkar, N., Deorukhkar, A., Trivedi, G., & Salunkhe, M. (2009). Copper nanoparticles in ionic liquids: Recyclable and efficient catalytic system for 1,3-dipolar cycloaddition reaction. *Catalysis Communications*, 10(8), 1240–1243. DOI: 10.1016/j.catcom.2009.01.027

Rostovtsev, V. V., Green, L. G., Fokin, V. V., & Sharpless, K. B. (2002)...*Angewandte Chemie*, 114(14), 2708–2711. DOI: 10.1002/1521-3757(20020715)114:14<2708::AID-ANGE2708>3.0.CO;2-0

Sebest, F., Cassarubios, L., & Rzepa, H. (2018). S.; White, A. J. P.; González, S. D. *Green Chemistry*, 20, 4023. DOI: 10.1039/C8GC01797B

Sebest, F., Haselgrove, S., White, A. J. P., & González, S. D. (2020). Metal-Free 1,2,3-Triazole Synthesis in Deep Eutectic Solvents. *Synlett*, 31(6), 605–609. DOI: 10.1055/s-0039-1690736

Sethi, S., Jana, N. C., Behera, S., Behera, R. R., & Bagh, B. (2023). Azide–Alkyne Cycloaddition Catalyzed by Copper(I) Coordination Polymers in PPM Levels Using Deep Eutectic Solvents as Reusable Reaction Media: A Waste-Minimized Sustainable Approach. *ACS Omega*, 8(1), 868–878. DOI: 10.1021/acsomega.2c06231 PMID: 36643452

Shaibuna, M., Theresa, L. C., & Sreekumar, K. (2018). A New Green and Efficient Brønsted: Lewis Acidic DES for Pyrrole Synthesis. *Catalysis Letters*, 148(8), 2359–2372. DOI: 10.1007/s10562-018-2414-4

Sheldon, A. (2005). Green solvents for sustainable organic synthesis: State of the art. *Green Chemistry*, 7(5), 267–278. DOI: 10.1039/b418069k

Smith, E., Abbott, A. P., & Ryder, K. S. (2014). Deep Eutectic Solvents (DESs) and Their Applications. *Chemical Reviews*, 114(21), 11060–11082. DOI: 10.1021/cr300162p PMID: 25300631

Tornøe, C. W., Christensen, C., & Meldal, M. (2002). Peptidotriazoles on Solid Phase: [1,2,3]-Triazoles by Regiospecific Copper(I)-Catalyzed 1,3-Dipolar Cycloadditions of Terminal Alkynes to Azides. *The Journal of Organic Chemistry*, 67(9), 3057–3306. DOI: 10.1021/jo011148j PMID: 11975567

Vaishnani, M. J., Bijani, S., Rahamathulla, M., Baldaniya, L., Jain, V., & Kamal, Y. (2024). Thajudeen, K.Y.; Ahmed, M.M.; Farhana, S; Pasha, I. *Green Chemistry Letters and Reviews*, 17, 2307989. DOI: 10.1080/17518253.2024.2307989

Vekariya, R. L. (2017). A review of ionic liquids: Applications towards catalytic organic transformations. *Journal of Molecular Liquids*, 227, 44–60. DOI: 10.1016/j.molliq.2016.11.123

Vidal, C., & Alvarez, J. C. (2014). Glycerol: A biorenewable solvent for base-free Cu(i)-catalyzed 1,3-dipolar cycloaddition of azides with terminal and 1-iodoalkynes. Highly efficient transformations and catalyst recycling. *Green Chemistry*, 16(7), 3515–3521. DOI: 10.1039/c4gc00451e

Vidal, C., & Garcia-Alvarez, J. (2014). Glycerol: A biorenewable solvent for base-free Cu(i)-catalyzed 1,3-dipolar cycloaddition of azides with terminal and 1-iodoalkynes. Highly efficient transformations and catalyst recycling. *Green Chemistry*, 16(7), 3515–3521. DOI: 10.1039/c4gc00451e

Vidal, C., & García-Álvarez, J. (2014). Glycerol: A biorenewable solvent for base-free Cu(i)-catalyzed 1,3-dipolar cycloaddition of azides with terminal and 1-iodoalkynes. Highly efficient transformations and catalyst recycling. *Green Chemistry*, 16(7), 3515–3521. DOI: 10.1039/c4gc00451e

Visser, A. (2001). E.; Swatloski, R, P.; Reichert, W, M.; Davis, J. H.; Rogers, R, D.; Mayton, R.; Sheff, S.; Wierzbicki, A. *Chemical Communications (Cambridge)*, 135–136. DOI: 10.1039/b0080411

Walden, P. (1914).. . *Bull. Acad. Sci. St. Petersburg.*, 1800, 405–422.

Wang, C., Guo, L., Li, H., Wang, Y., Wenga, J., & Wua, L. (2006). Preparation of simple ammonium ionic liquids and their application in the cracking of dialkoxy-propanes. *Green Chemistry*, 8(7), 603–607. DOI: 10.1039/b600041j

Wilkes, J. (1992). S.; Zaworotko, M, J. *Journal of the Chemical Society. Chemical Communications*, 13, 965–967. DOI: 10.1039/c39920000965

Yoshizawa-Fujita, M., Johansson, K., Newman, P., & MacFarlane, D. (2006). R.; and Forsyth, M. *Tetrahedron Letters*, 47, 2755–2758. DOI: 10.1016/j.tetlet.2006.02.073

Zhang, B. (2019). Comprehensive review on the anti-bacterial activity of 1,2,3-triazole hybrids. [PubMed]. *European Journal of Medicinal Chemistry*, 168, 357–372. DOI: 10.1016/j.ejmech.2019.02.055

Zhang, S., Xu, Z., Gao, C., Ren, Q. C., Chang, L., Lv, J. S., & Feng, L. S. (2017). Triazole derivatives and their anti-tubercular activity. [PubMed]. *European Journal of Medicinal Chemistry*, 138, 501–513. DOI: 10.1016/j.ejmech.2017.06.051

Zheng, T., Rouhanifard, S. H., Jalloh, A. S., & Wu, P. (2012). Click triazoles for bioconjugation. Click triazoles, 163-183.

Chapter 5
Pharmacological Significance of 1,2,3–Triazoles

Mubarak Hanif Shaikh
https://orcid.org/0000-0002-1190-2371
Radhabai Kale Mahila Mahavidyalaya, Ahmednagar, India

Amol A. Nagargoje
https://orcid.org/0000-0001-6689-3436
K.M.C. College of Arts, Science, and Commerce, Khopoli, India

Dattatraya N. Pansare
Deogiri College, India

Bapurao B. Shingate
Dr. Babasaheb Ambedkar Marathwada University, India

ABSTRACT

Click chemistry is not a single specific reaction, but was meant to mimic nature, which also generates substances by joining small modular units. The 1,3-dipolar azide, alkyne cycloaddition (CuAAC) reaction catalyzed by copper, as nearly quantitative and easy to execute has emerged as the leading example of "click chemistry". Given the importance of the triazole scaffold in medicinal chemistry, its synthesis has attracted the attention of the drug discovery and development community. This book chapter will summarizes the major synthetic methods currently used for the preparation of triazole and pharmacological significance such as antifungal, antibacterial, antitubercular, anticancer, anti-inflammatory, antioxidant and many more properties will discussed. Furthermore, this book chapter will comprise the

DOI: 10.4018/979-8-3693-7267-8.ch005

literature from 2020 to till date

1. INTRODUCTION

Triazoles and the heterocyclic derivatives they are connected to have attracted a lot of attention recently because of their significance in bioactivity and synthetic processes. Due to their effective use in medicinal chemistry, azolic derivatives, such as thiazole, triazole, oxadiazole, and thiadiazole, are pharmacologically active compounds that have been the subject of substantial investigation for a range of biological purposes.[1]

Triazole is a heterocyclic ring containing two carbon atoms and three nitrogen atoms, having the chemical formula $C_2H_3N_3$. Compounds **1** and **2** are its two isomeric forms; it is sometimes referred to as pyrrodiazole (**Figure 1**).

Figure 1. Isomeric forms of triazoles

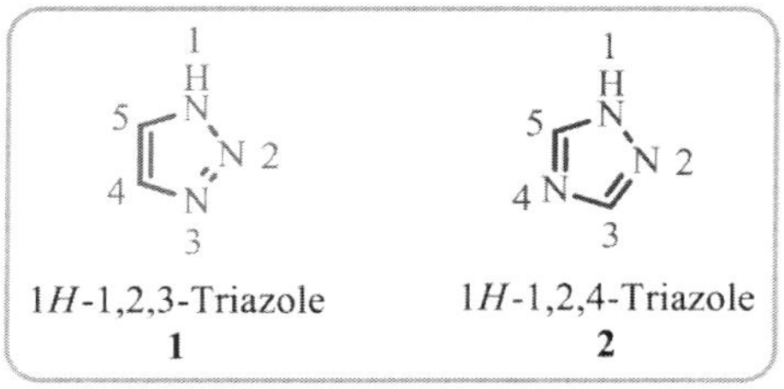

Figure 1. Isomeric forms of triazoles

1.1 Physical Properties

The parent 1*H*-1,2,3-triazole is a colorless liquid, highly soluble in water, with a density of 1.192, an mp of 23-25 °C, and a bp of 203 °C. In a solid state it exists as a 1:1 mixture of 1*H*- and 2*H*-tautomers. The dipole moment of a tautomeric mixture in benzene at 25 °C is 1.85 D. A pK_a of 1.17 of protonated 1*H*-1,2,3-triazole indicates that it is a weak base, while a conjugated base of 9.4 indicates that it is a weak acid.

1.2 Chemical Reactivity

The 1,2,3-triazole ring is very stable and normally not cleaved either by hydrolysis or oxidation, but reductive cleavages do occur. Due to the presence of two pyridine-type nitrogen atoms in the ring, ease of quaternization is decreased and requires vigorous conditions.

1,2,3-Triazoles are unique in their structure and bonding patterns, which significantly influence their interactions with other molecules, including biological entities. The 1,4-disubstituted-1,2,3-triazole moiety, in particular, exhibits a notable similarity to a Z-amide bond. The lone pair on the N_3 nitrogen atom mimics the carbonyl oxygen of the amide bond. Additionally, the C(5)-H bond in triazoles can engage in intermolecular interactions as a hydrogen bond donor group,[2] similar to the N-H bond of amides. The electrophilic nature of the C(5) carbon in triazoles is also electronically akin to the carbonyl carbon of an amide (**Figure 2**).[3]

Furthermore, the overall dipole moment of a triazole system is much stronger compared to an amide bond. This stronger dipole moment can enhance the hydrogen bond donor and acceptor abilities of the triazole, thereby improving its potential to mimic peptide bonds (**Figure 2**).[4] There are some structural differences between triazoles and amide bonds. One significant difference is that the additional atom in the triazole backbone increases the R_1-R_2 distance by 1.1 Å over the classical amide bond. The 1,5-substitution pattern of triazoles resembles the E-amide bond, with the substituent links and relative positions of the hydrogen bonding acceptor and donor sites being identical in terms of the atoms involved.[5] However, due to minor variations in atom polarization, the electrophilic carbonyl carbon in amides is replaced by a negatively polarized nitrogen atom in triazoles.

Figure 2. Topological and electronic similarities of 1,2,3-triazoles and amides[6]

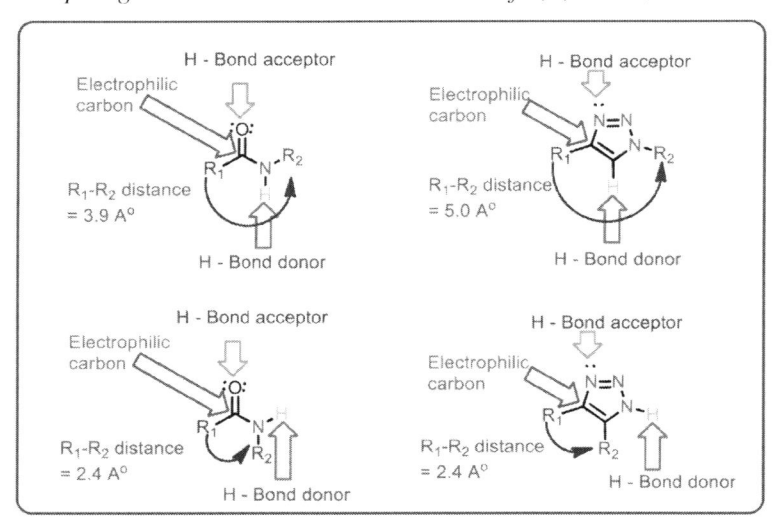

Figure 2. Topological and electronic similarities of 1,2,3-triazoles and amides

The ability of 1,2,3-triazoles to mimic specific features of peptide bonds allows them to interact productively with various molecules, including biological targets.[7] They are expected to act as bio-isosteres for acyl-phosphate and trans-olefinic moieties. For instance, in the preparation of siderophores **4** as inhibitors of enzymes endogenous to *Mycobacterium tuberculosis*,[8] an acyl-phosphate group, which is a good leaving group, was replaced with a stable triazole to mimic the intermediate **3** (**Figure 3**).

Furthermore, the replacement of a trans-olefinic group by a triazole moiety[9] in resveratrol **5**, known for its anticancer and anti-aging properties, resulted in a compound that was even more biologically active. Molecular modeling studies revealed that the spatial arrangements of the phenolic hydroxyls, which regulate biological activity, remained unaltered despite the substitution.

Figure 3. Triazole as a bioisostere of acylphosphate and trans-olefinic moieties

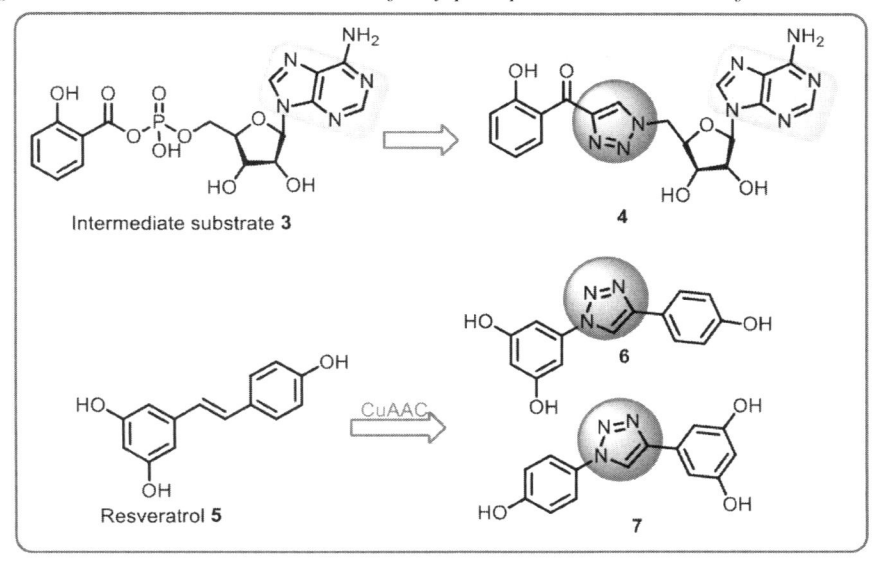

Figure 3. Triazole as a bioisostere of acylphosphate and *trans*-olefinic moieties

1,2,3-Triazoles can easily form hydrogen bonds and π-π stacking interactions. They are also stable to metabolic degradation and can be used as a bioisostere for amide bond, disulphide bond, ester bond, carboxylic acid, aromatic rings and olefins rigid analogs.[10] 1,2,3-Triazole is one of the key structural units found in a large variety of bioactive molecules including anticancer,[11-12] antimicrobial,[13] antifungal, antiviral,[14] antileshmanial,[15] antitubercular,[16] anti-inflammatory,[17] anticonvulsant,[18] antidepressant,[19] antioxidant,[20] and anti-HIV activities.[21] They are also used as neuroprotective agents[22] and in the treatment of Alzheimer's disease.[23]

2. SYNTHETIC METHODS FOR THE SYNTHESIS OF 1,2,3-TRIAZOLES

2.1 Earlier Synthesis of 1,2,3-Triazoles

Although Huisgen[24] did not completely realise the generality, scope, and mechanism of cycloadditions until the 1960s, Dimroth was the first to report on the synthesis of triazoles by the cycloaddition of azides and acetylenes in the early 1900s. The synthesis of 1*H*-1,2,3-triazole *via* cycloaddition of acetylene and hydrazoic acid was reported by Wiley and colleagues,[25] however, the mechanism of this reaction remained unclear until Huisgen's comprehensive investigation.[26] These reactions are not regio-selective and proceed slowly in the absence of a transition-metal catalyst. They also need high temperatures to achieve satisfactory yields. An 80-95% yield of 1,4 and 1,5-disubstituted triazoles is produced by the process (**Scheme 1**).

Figure 4. Huisgen 1,3-dipolar azide-alkyne cycloaddition reaction

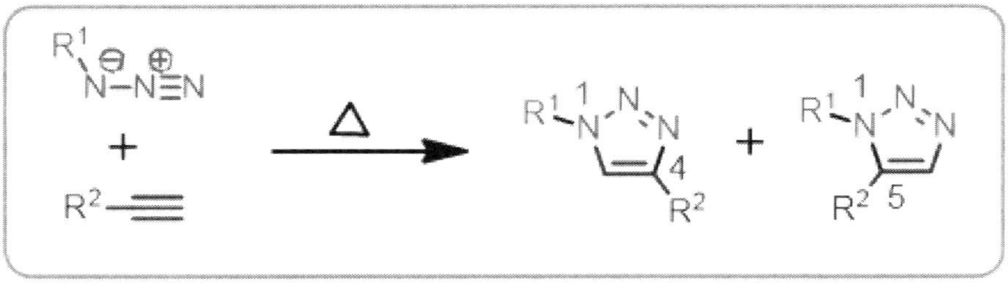

There have been several documented attempts to regulate regioselectivity. Until Sharpless and Meldal discovered the copper (I)-catalyzed reaction[27-28] in 2002, these efforts were mainly unsuccessful. With an 81-97% yield, this process only produces the 1,4-disubstituted 1,2,3-triazole. In this context, Sharpless[29] introduced the term "click chemistry" to refer to the ready production of bonds in reactions that have a high degree of selectivity, a high chemical yield, a broader scope, and the ability to generate molecules by attaching small units together in a fast and reliable manner. The most prominent example of "click chemistry" is the 1,3-dipolar azide/alkyne cycloaddition reaction catalysed by copper (CuAAC, **Scheme 2**, example given in **Scheme 3**),[30-32] which is almost quantitative and simple to carry out.

Figure 5. Regioselectivity of the 1,3-dipolar cycloaddition between an azide and an alkyne

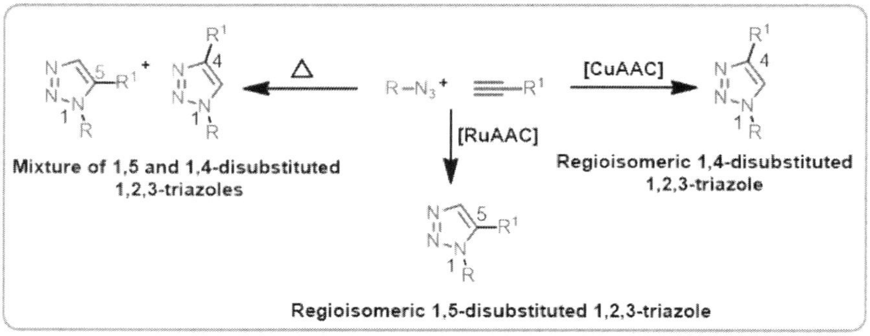

For the synthesis of bio-conjugates, the 1,3-dipolar cycloaddition to create a 1,4-disubstituted triazole bridge has gained popularity. CuAAC has been successfully applied in the production of various molecular compounds with a wide range of uses and enzyme inhibitors.[33] Following Sharpless and colleagues' successful implementation of (CuAAC), further researchers developed the ruthenium-catalyzed azide-alkyne cycloaddition (RuAAC), which yields only 1,5-disubstituted 1,2,3-triazoles (**Scheme 2**, example shown in **Scheme 3**).[30-32,34]

Figure 6. Copper- and ruthenium-catalyzed 1,3-dipolar cycloaddition reactions

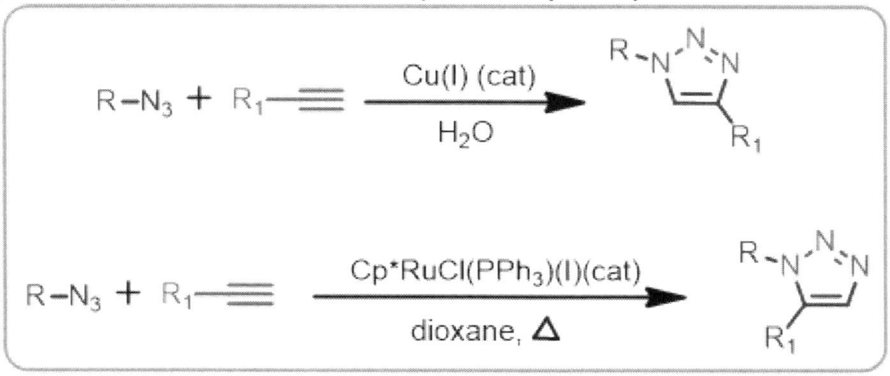

2.2 Synthesis of 1,2,3-Triazoles After 2020

Azido genipin reactions with various alkynes were investigated by Silalai *et. al.*[35] Genepin equivalents, which carry aliphatic, phthalimide, benzylamine, phenyl, and benzyl ether substituted triazoles with varying carbon chain lengths, were successfully synthesized in good to outstanding yields (**Scheme 4**).

Figure 7. Synthesis of Genepin derivatives

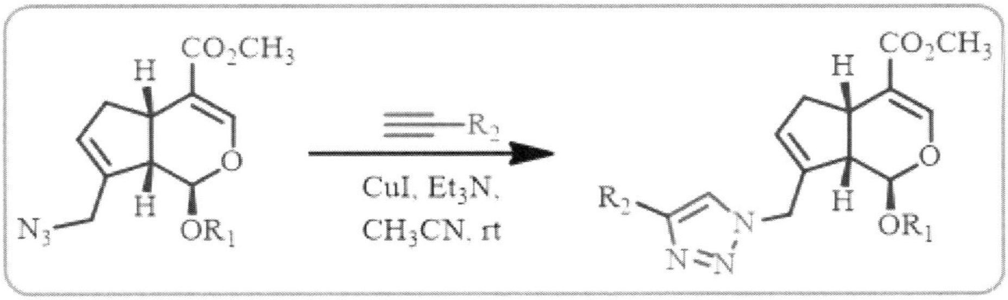

Click chemistry was used to create a new series of 4′-((4-substituted-4,5-dihydro-1H-1,2,3-triazol-1-yl)methyl)-[1,1′-biphenyl]-2-carbonitrile (OTBN-1,2,3-triazole) derivatives[36] (**Scheme 5**). A range of alkynes and N3-OTBN were used to design and synthesise a number of bioactive compounds utilising copper (II) acetate mono-hydrate in aqueous dimethylformamide at room temperature. In addition to being extremely economical and minimising the amount of synthesis required, the reaction produced 91-28% of the desired products without the need for chromatographic procedures or extra stages.

Figure 8. Synthesis of OTBN-1,2,3-triazole derivatives

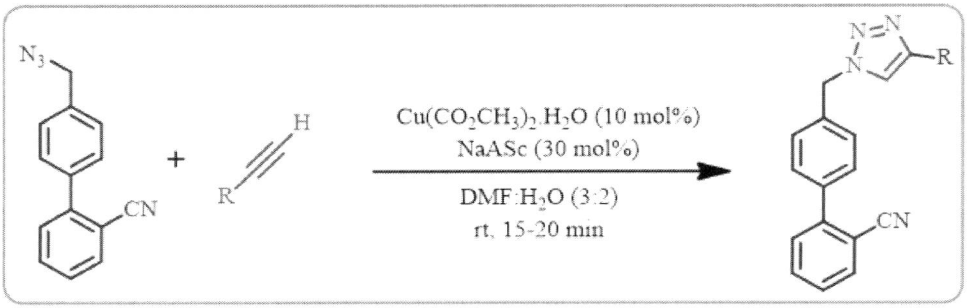

Guo[37] and coworkers synthesized a series of novel compounds by introducing 1,2,3-triazole moieties to cabotegravir (**Scheme 6**). The compounds synthesized by the reaction of substituted alkyne and azide in presence of TBA, chalcanthite, sodium ascorbate in 1:1 water:THF solvent system at 70°C.

Figure 9. Synthesis of 1,2,3-triazole moieties to cabotegravir

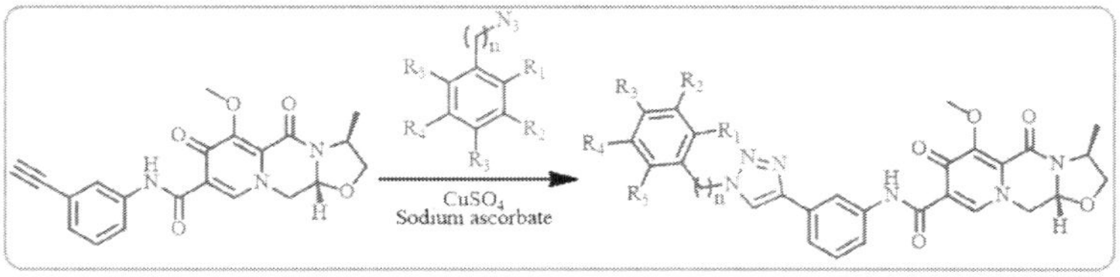

Using K_2CO_3 as a catalyst, methyl 1-cyclopropyl-6-fluoro-4-oxo-7-(4-(3-oxobutanoyl) piperazin-1-yl)-1,4-dihydroquinoline-3-carboxylate was combined with a variety of organic azides to create a library of 1,2,3-triazoles that included the ciprofloxacin core in good to exceptional yield [38] (**Scheme 7**).

Figure 10. Synthesis of of 1,2,3-triazoles included ciprofloxacin core

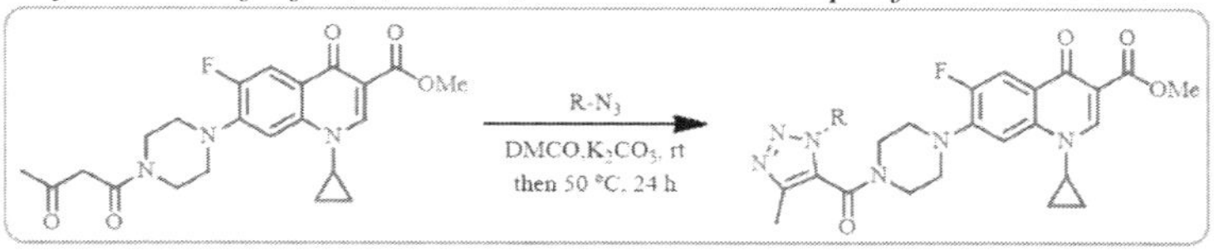

The reaction of substituted alkyne with 2-azido-1-phenyl ethanone precursors, in the presence of CuI/sodium *L*-ascorbate/$CuSO_4 \cdot 5H_2O$ in $DMF:H_2O$ (3:2 v/v) as the solvent afforded after chromatographic purification (eluent: 5% EtOAc: n-hexane) compounds in 39-83% yield, results in the formation of novel nitroimidazole-piperazine-1,2,3-triazole hybrids[39] (**Scheme 8**).

Figure 11. Synthesis of of novel nitroimidazole-piperazine-1,2,3-triazole hybrids

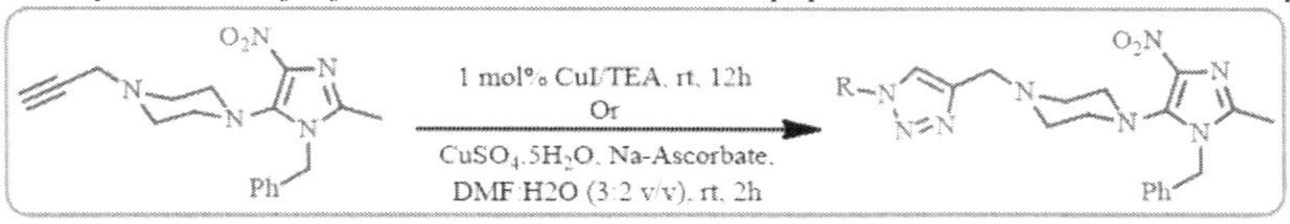

The 2-azido-*N*-(4-(5-(4,6-dimorpholino-1,3,5-triazin-2-yl)-1,3,4-oxadiazol-2-yl) phenyl) acetamide was subjected to a "Click reaction," which produced the final target compounds by reacting with different terminal alkynes using Na-ascorbic acid and $CuSO_4 \cdot H_2O$ in BuOH/water over the course of 12 hours at room temperature[40] (**Scheme 9**).

Figure 12. Synthesis of 1,2,3-triazole-incorporated 1,3,4-oxadiazole-triazine derivatives

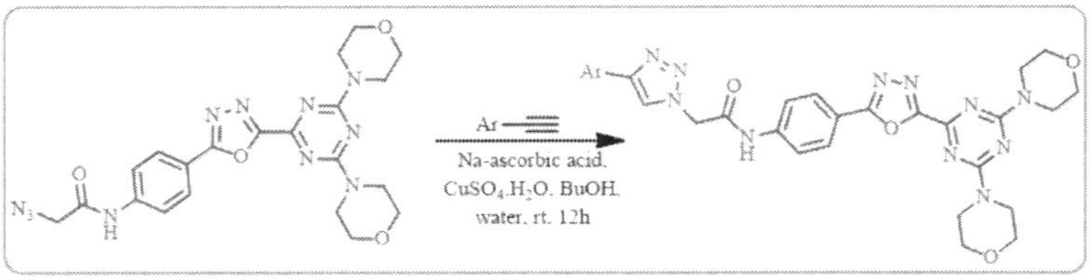

A one-pot synthetic method that does not require catalyst has been presented for the synthesis of substituted 1,2,3-triazole. Two additional C-N bonds are formed in addition to the basic click chemistry that drives the reaction forward. Maleimides and 1,4-naphthoquinone were both used as coupling partners to produce a large range of substituted 1,2,3-triazoles with good yield[41] (**Scheme 10**).

Figure 13. Synthesis of maleimides and 1,4-naphthoquinone substituted 1,2,3-triazoles

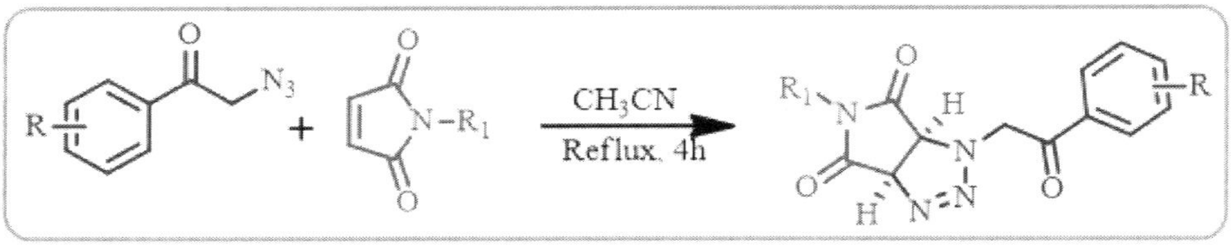

Beginning with the suitable *S*-propargylated 1,2,4-triazoles, new conjugates of substituted 1,2,3-triazoles connected to 1,2,4-triazoles were synthesised (**Scheme 11**).[42] Cu(I)-Catalyzed cycloaddition of 1,2,4-triazole-based alkyne side chain with several un/functionalized alkyl- and/or aryl-substituted azides allowed for the ligation of 1,2,4-triazoles to the 1,2,3-triazole core, resulting in the desired 1,4-disubstituted 1,2,3-triazoles using both conventional and microwave techniques.

Figure 14. Synthesis of 1,4-disubstituted 1,2,3-triazoles-1,2,4-triazoles conjugates

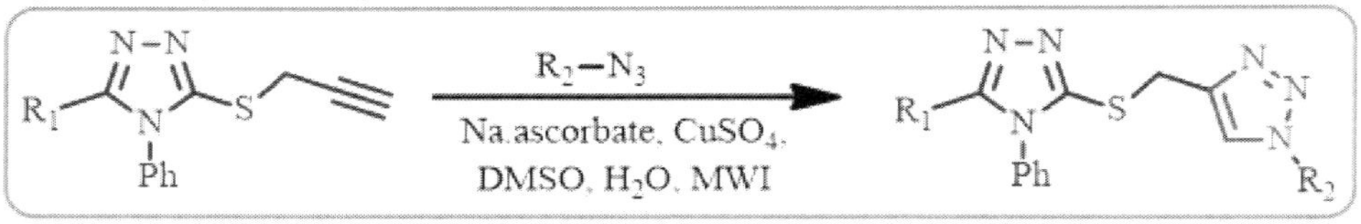

A novel 1,2,3-triazole derivative, 4-((1-(3,4-dichlorophenyl)-1*H*-1,2,3-triazol-4-yl)-methoxy)-2-hydroxybenzaldehyde, was produced *via* azide-alkyne cycloaddition (CuAAC) catalysed by copper(I) (**Scheme 12**).[43]

Figure 15. Synthesis of novel 1,2,3-triazole-2-hydroxybenzaldehyde derivatives

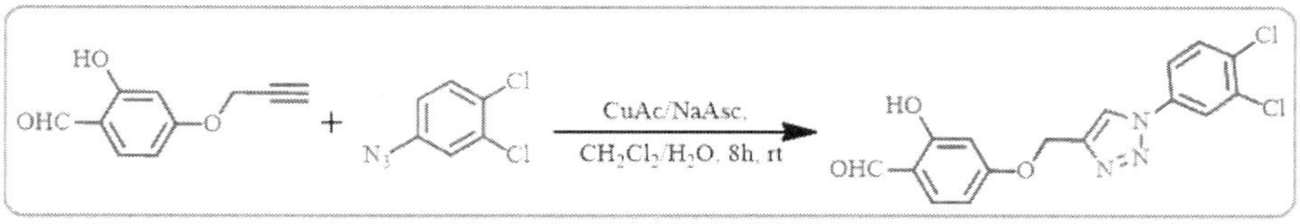

1,3-Dipolar cycloaddition between an aromatic azide and terminal alkyne was catalysed by copper to generate the 1,2,3-triazole ring in 65-88% yield (**Scheme 13**).[44]

Figure 16. Synthesis of novel 1,2,3-triazole derivatives

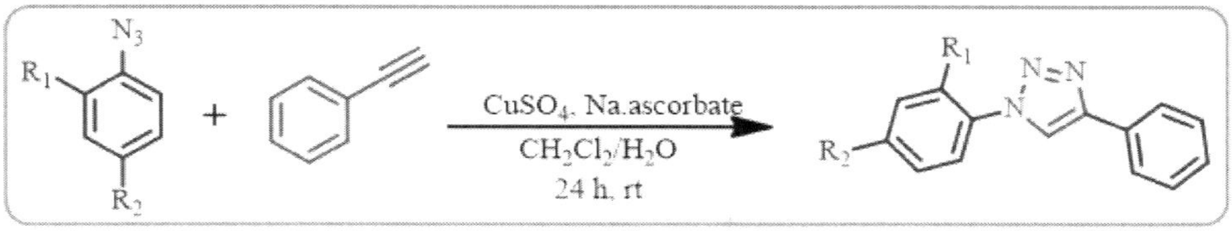

Using capsaicin's one- and two-point alteration, a number of new 1,2,3-triazole derivatives and its structural isomer (a new natural product hybrid capsaicinoid) were synthesized without changing the amide bond (**Scheme 14**).[45]

Figure 17. Synthesis of new 1,2,3-triazole-capsaicinoid derivatives

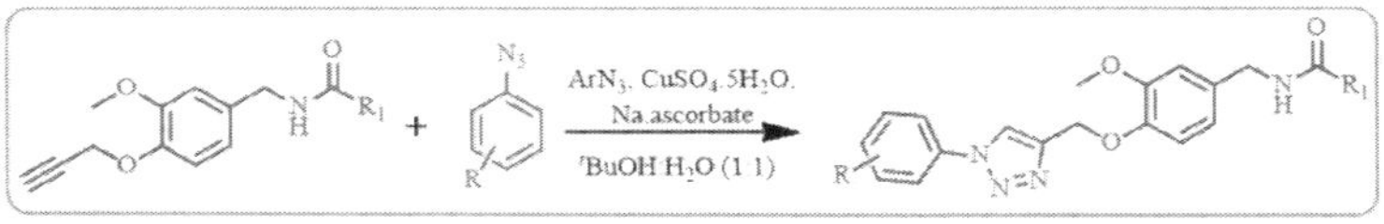

Acetophenones were treated with DMA-DMF in toluene for four hours in order to produce the appropriate enaminones. The compounds next underwent a one-hour reaction with various aryl azides at 150 °C, yielding pure triazoles with yields as high as 88%. This work also described the use of $NaBH_4$ in the acyl reduction process to yield hydroxy-1,2,3-triazoles (**Scheme 15**).[46]

Figure 18. Synthesis of hydroxy-1,2,3-triazoles

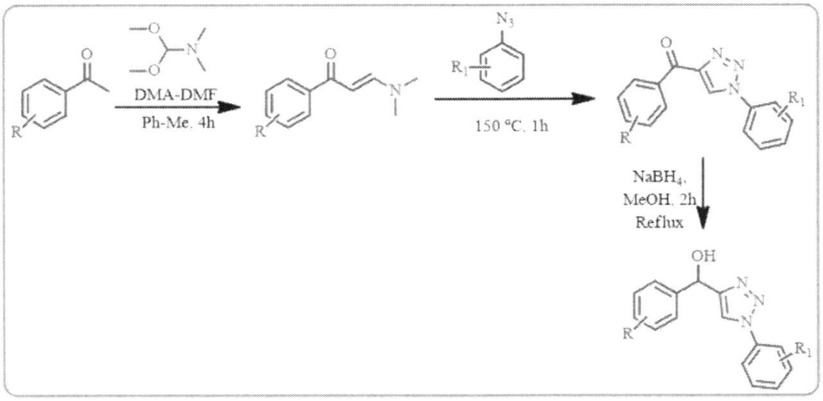

CuNPs/C served as the catalyst to produce triazole **9** in a high yield (80%) from propargyl alcohol and 1-(chloromethyl)naphthalene **8** *via* the multicomponent CuAAC in water.[47] In order to produce bromide **C** (80%), compound **9** was next treated with CBr_4 and PPh_3 in dichloromethane (DCM). Using two distinct *H*-phosphonates, the Michaelis-Becker reaction was carried out on compound **10**, yielding phosphonates **11** in good yields (70-76%). Trimethylsilyl bromide (TMSBr) was used to hydrolyse dimethyl phosphonate **11** in DCM, yielding the equivalent phosphonic acid **12** (85%) (**Scheme 16**).

Figure 19. Synthesis of phosphonic acid-1,2,3-triazole derivatives

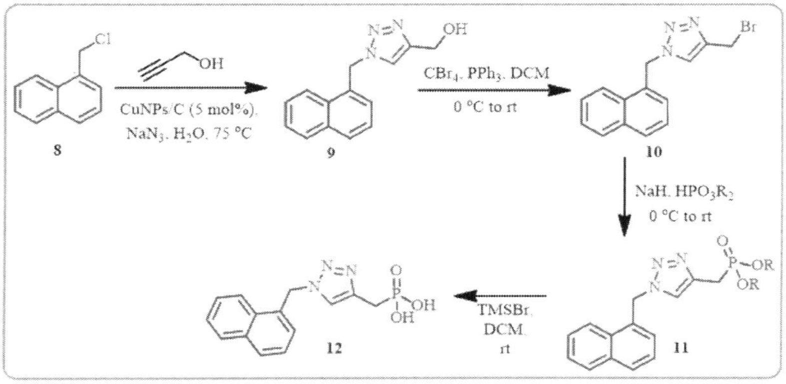

By logically integrating a pharmacophoric active heterocyclic ring containing indole and triazole moieties in one molecular framework using both conventional and microwave irradiation methods, a series of *N*-substituted 1,2,3-triazolylmethyl indole derivatives with 72-95% yield were synthesized[48] (**Scheme 17**).

Figure 20. Synthesis of of N-substituted 1,2,3-triazolylmethyl indole derivatives

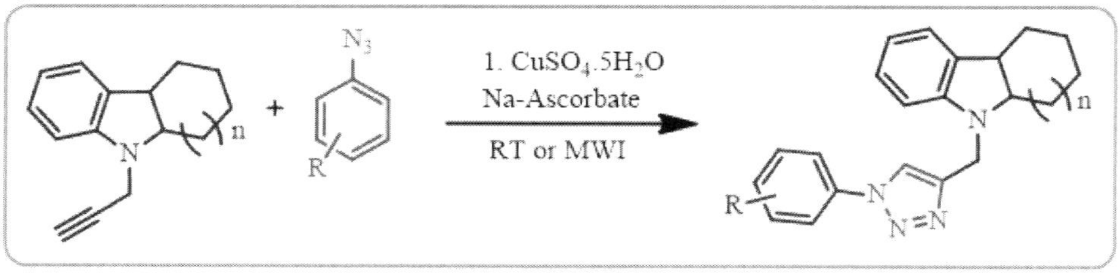

Through a three-component click reaction involving an alkyl halide, sodium azide, and terminal alkyne in the presence of a monophosphine Cu(I) complex containing bis(pyrazolyl)methane (L_1) $(CuIL_1PPh_3)$, 1,4-disubstituted 1,2,3-triazoles were produced in up to 93% yield (**Scheme 18**).[49] Oxygen and water compatibility: The catalyst is highly stable, performs well in ultrasonic conditions.

Figure 21. Monophosphine Cu(I) complex catalysed synthesis of 1,2,3-triazoles

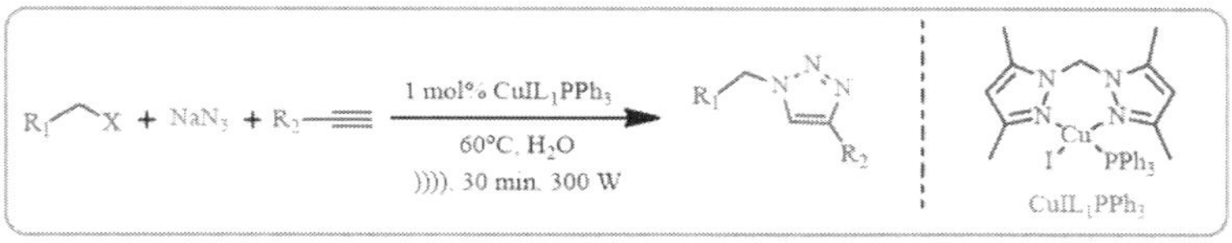

For the synthesis of elaborated kojic acid derivatives in a CuI-catalyzed azide/alkyne cycloaddition reaction under ultrasonic conditions, magnetic nanoparticles coated with carbon quantum dot and copper(I) iodide $(Fe_3O_4@CQD@CuI)$[50] were employed as an environmentally friendly heterogeneous Lewis/Bronsted acid sites in 88-93% yields (**Scheme 20**).

Figure 22. Copper(I) iodide $(Fe_3O_4@CQD@CuI)$ catalyzed synthesis of 1,2,3-triazoles

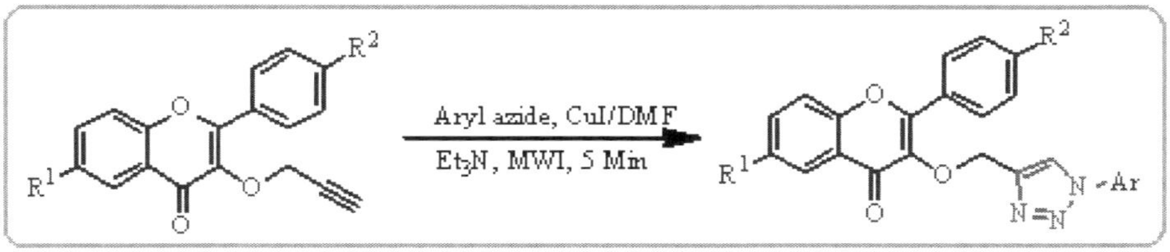

By using Cu(I) to catalyse the azide-alkyne cycloaddition (CuAAC) under both conventional and microwave irradiation conditions, Znati et al.[51] identified a new class of deadly 1,2,3-triazole linked flavonol hybrids with 78-93% yields (Figure 23. **Scheme 21**).

Figure 23. Synthesis of 1,2,3-triazole linked flavonols hybrids

Figure 23. Synthesis of 1,2,3-triazole linked flavonols hybrids

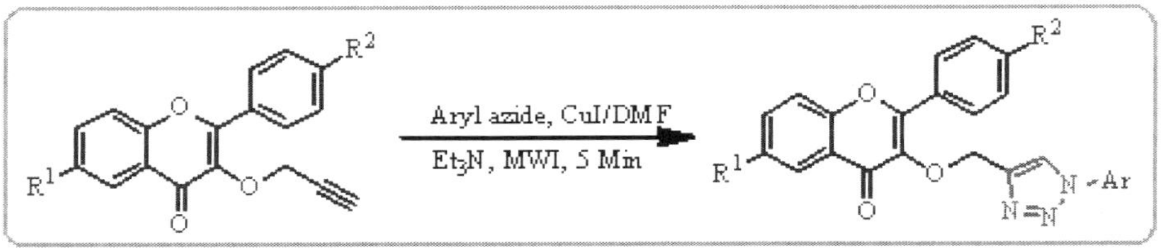

Using a cycloaddition-elimination sequence, it was reported that metal-free regio-selective synthesis of 1,5- and 1,4-disubstituted triazoles could be achieved in up to 93% yield (**Scheme 22**).[52] A deep eutectic solvent (DES) that is safe for the environment was used to conduct the reactions.

Figure 24. Metal-free region-selective formation of 1,5- and 1,4-disubstituted triazoles

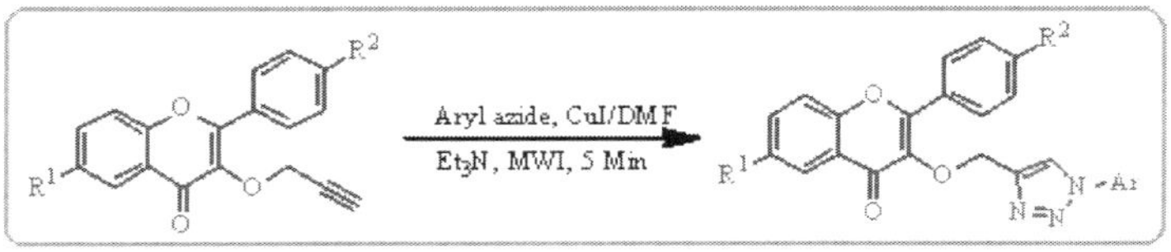

The RuAAC aided synthesis of 1,2,3-triazoles for the cholecystokinin-2 receptor (CCK2R) was described by Grob *et. al.*[53] (**Scheme 23**) Up to 90% yield of an amino-propargyl derivative protected by Fmoc and an α-azido Bn-protected ester combine to create the triazolyl moiety.

Figure 25. RuAAC assisted synthesis of 1,2,3-triazoles

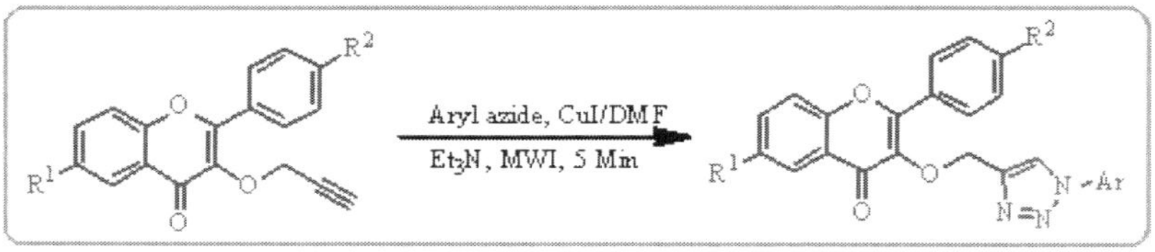

Through the organocatalytic enolate-mediated azide-carbonyl [3+2] cycloaddition, Shingate et al.[17b] synthesised a library of novel 1,2,3-triazole derivatives and produced a highly functionalised triazole core structure in 85-95% yield (**Scheme 24**).

Figure 26. Synthesis of 1,2,3-triazole derivatives via the organocatalytic enolate-mediated azide-carbonyl [3 + 2] cycloaddition

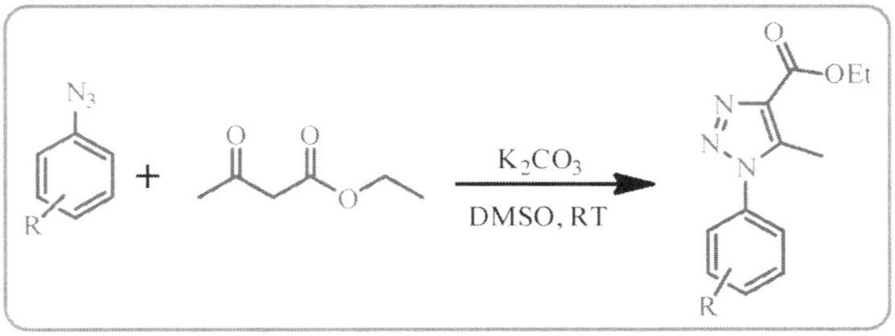

The synthesis of 1-mono- and 1,4-disubstituted 1H-1,2,3-triazoles was investigated using copper-on-charcoal as a heterogeneous catalyst in a continuous flow environment (**Scheme 25**).[54] This method proved to be a stable and versatile approach. A range of substituted 1,2,3-triazoles with good yields and functional group tolerance could be produced using this method. The procedure described was used to generate rufinamide, an antiepileptic medication, with a 96% isolated yield.

Figure 27. Synthesis of 1,4-disubstituted 1,2,3-triazoles under continuous flow conditions

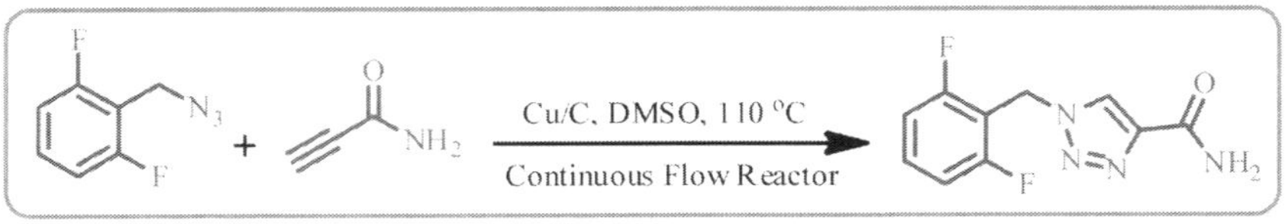

3. BIOLOGICAL ACTIVITY

The synthesis and evaluation of a novel series of 1,2,3-triazole-based hybrids and conjugates has demonstrated the potential of 1,2,3-triazoles as "linkers." These novel chemicals are being investigated as potential lead candidates for a range of biological uses. Their effectiveness is wide-ranging and includes functions as drugs against leishmaniasis, diabetes, bacteria, viruses, TB, cancer, malaria, and neuroprotection.[55]

1,2,3-Triazoles have gained special attention in the drug discovery because several drug molecules contain 1,2,3-triazole group such as Tazobactam, Carboxyamido-triazole (CAI), Cefatrizine, Radezolid and Mubritinib (**Figure 28**).

Figure 28. Commercially available 1,2,3-Triazole heterocyclic drugs

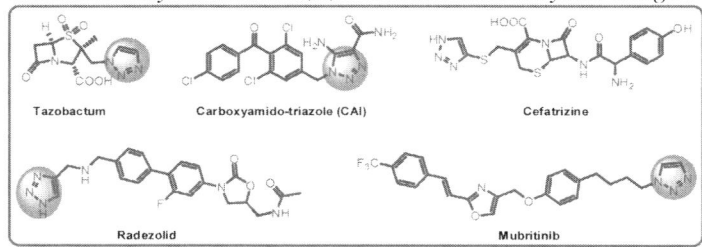

3.1. Anti-Cancer Activity

Derivatives of 1,2,3-triazoles have been studied for a long time, largely because of **Figure 29**'s illustration of their potential use in cancer treatment.

Figure 29. 1,2,3-Triazole containing FDA approved drugs

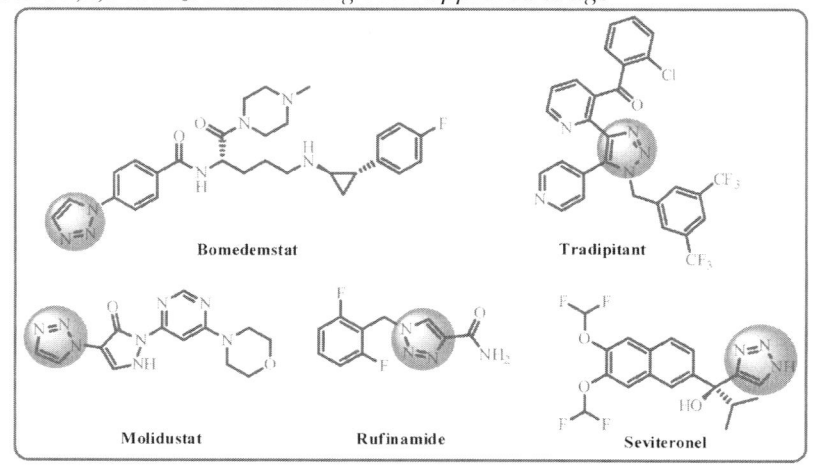

Numerous institutions have carried out focused research in the development of the anticancer drugs. Antineoplastic properties and DNA interactions of kinetically inert dicobalt(III) $[Co_2(Lpytrz)_3](OTf)_6$, have explored six types of cylinders with ligands made up of 1,2,3-triazoles bonded to either methylene (**13**) or a 1,4-xylyl (**14**) (**Figure 30**).[56] The bigger cylinder, a racemic (rac-2) mixture of the ($\Delta\Delta$ and $\Lambda\Lambda$) helicates, and the smaller cylinder, a meso-1 ($\Delta\Lambda$) isomer, were selected for this investigation. Rac-2 and meso-1 both have strong and targeted anti-cancer effects on cells. Greater activity is exhibited by the larger and more lipophilic rac-2, which has IC_{50} values in the low micromolar range, similar to cisplatin, a drug utilised in clinical settings. Remarkably, for both cylinders, the concentration of cobalt in the nuclei matches the concentration of cobalt in the entire cell, and a substantial portion of the total cobalt in the cell builds up in the nucleus.

The OTBN-1,2,3-triazole analogues can be employed as bridges to build molecular libraries of 1,2,3-triazole molecules with bioactive functionalisation.[36] Two interesting candidates were found during the initial screening process: **15**, a pan-cancer medication, and **16** (**Figure 30**), a lung cancer-specific treatment. Both **15** and **16** did not show any cytotoxicity to RPE cells that were not malignant. It has been confirmed that **16** is a lung-cancer-specific cytostatic medication with good potential for usage in combination with current cytotoxic medicines as a lung-cancer treatment in the future. However, it was shown that **15** had a strong specificity against STK33. It has been demonstrated that the advancement of several cancer types is correlated with high STK33 expression.

Twelve bis-1,2,3-triazole-based chalcones (78-90% yield) synthesized in five steps. In the MTT cell viability assay, seven hybrid compounds **17** & **18** (**Figure 30**) exhibited good cytotoxicity against A-549 lung cancer cells ($IC_{50} = 44.72\pm0.66$ to 86.22 ± 1.06 μM), which is equivalent to the IC_{50} of the anticancer medication doxorubicin (39.86 ± 1.15 μM).[57]

Using a naphtho[2,3-d][1,2,3]triazole-4,9-dione scaffold, a series of new hD-HODH inhibitors with the ability to induce ROS were developed and synthesised.[58] Compounds **19** and **20** (**Figure 30**) had favourable cellular and enzymatic activities (Raji: $IC_{50} = 0.48$ and 0.16 μM, respectively), with hDHODH having an enzymatic activity of 9.0 and 4.5 nM. The R136 residue and ligand-protein cocrystal structures show a unique H-bonding relationship. Compounds **19** and **20** caused ROS generation, mitochondrial malfunction, apoptosis, and cell cycle Sphase arrest, according to a mechanism investigation. Synthesized derivatives, astonishingly, demonstrated strong growth suppression and a good safety profile *in vivo*.

A set of twenty-eight new cabotegravir derivatives containing the 1,2,3-triazole moiety were synthesized.[37] Four distinct human cancer cell lines were used to assess the potential anticancer activity of the synthesised compounds: HuH-7 (hepatocellular), MCF-7 (breast), SKOV3 (ovarian), and HCT-116 (colon). Several substances shown strong anti-proliferative action against various cancer cell lines, according to preliminary biological evaluations. With IC_{50} values of 6.59 and 7.83 μM in HuH-7 cells, 27.24 and 8.59 μM in MCF-7 cells, 4.46 and 6.30 μM in SKOV3 cells, and 23.90 and 17.00 μM in HCT-116 cells, respectively, compounds **21** and **22** (**Figure 30**) were shown to have the most noticeable impacts. Subsequent research revealed that compounds **21** and **22** caused cell death by inducing apoptosis in cells and raising reactive oxygen species (ROS) levels. Furthermore, after treatment with compounds **21** and **22**, changes in the expression of proteins related to autophagy and DNA damage were detected by western blot analysis.

Using a click reaction, six unique ciprofloxacin-1,2,3-triazole hybrids were synthesized.[38] All the synthesized compounds were tested against three cell lines, glioblastoma (U-87 MG), non-small cell lung cancer (A549), and breast cancer (MCF7), to determine their *in vitro* anticancer activities. The anti-proliferative effect of hybrids **23** and **24** (**Figure 30**) against all three cell lines was impressive. All cancer cell lines had IC_{50} values of **24** that were far lower than the IC_{50} of the common reference substance. With a selectivity index of 142.3 for the U87 cell line, the IC_{50} of **24** for the normal cell (HDF) line was much higher than the published IC_{50} for cisplatin [IC_{50} = 170.7 ± 8.1 μM/mL (HDF), (p < 0.001)]. This indicates that the drug is less toxic to normal cells and has a greater therapeutic index.

Click Chemistry is used to develop and synthesise 4-((4-(1-benzyl-2-methyl-4-nitro-1*H*-imidazole-5-yl)piperazine-1-yl)methyl)-1-substituted-1*H*-1,2,3-triazole motifs.[39] The antiproliferative potency of the newly created chromophores is evaluated against three human cancer cell lines (MCF-7, HepG2, and PC3) as well as one normal cell line (Dermal/Fibroblast). Strong effects against the MCF-7 cell line have been demonstrated by compounds **25** and **26** (**Figure 30**), whose IC_{50} values are (2.00 ± 0.03 μM and 5.00 ± 0.01 μM, respectively). In order to study the ligand-protein interactions and free binding energies at the atomic level, ADMET studies and molecular docking investigations are carried out on the most active hybrid nitroimidazole derivatives, **25** and **26**, with 4-hydroxytamoxifen (4-OHT) at the human oestrogen receptor alpha (hER) during binding active sites. At a distance of 3.2 Å, the triazole ring in the **25** derivative establishes a hydrogen bond with Asp58. Furthermore, polar interaction with the amino acid residue His231 is discovered.

Using the MTT assay and etoposide as the control medication, a new library of 1,2,3-triazole-incorporated 1,3,4-oxadiazole-triazine derivatives were synthesised, and evaluated *in vitro* for anticancer activity against the cancer cell lines PC3 and DU-145 (prostate cancer), A549 (lung cancer), and MCF-7 (breast cancer).[40] With IC_{50} values ranging from 0.16 ± 0.083 μM to 11.8 ± 7.46 μM, the compounds demonstrated impressive anticancer activity; in contrast, the positive control showed values between 1.97 0.45 μM and 3.08 0.135 μM. Against PC3, A549, MCF-7, and DU-145 cell lines, compound **27** (**Figure 30**) with a 4-pyridyl moiety had remarkable anticancer activity; its IC_{50} values were 0.17 ± 0.063 μM, 0.19 ± 0.075 μM, 0.51 ± 0.083 μM, and 0.16 ± 0.083 μM, in that order.

Synthetic 1,2,4-triazole conjugates of substituted 1,2,3-triazoles were synthesized using both classical and microwave methods.[42] Human cervical carcinoma (HeLa), human breast adenocarcinoma (MCF-7), and human colon cancer (Caco-2 and HCT116) were among the cancer cell lines that were subjected to an anticancer screening. Based on the standard reference medication, doxorubicin, compounds **28**, **29**, and **30** (**Figure 30**) had substantial anticancer activity against MCF-7 and Caco-2 cancer cell lines, with IC_{50} values of 0.31 and 4.98 μM, respectively.

4-((1-(3,4-Dichlorophenyl)-1H-1,2,3-triazol-4-yl)-methoxy)-2-hydroxy benzaldehyde derivatives were synthesized by Gokturk *et. al.*[43] Research on the compound's DNA/bovine serum albumin (BSA) binding activity both *in vitro* and *in silico* revealed that the compound's CT-DNA binding activity was mediated by intercalation, while its BSA binding activity was mediated by both polar and hydrophobic contacts. The 3-(4,5-dimethylthiazol-2-yl)-2,5-diphenyltetrazolium bromide (MTT) assay was also used to assess the compound's anticancer potential utilising human cell lines, including MDA-MB-231, LNCaP, Caco-2, and HEK-293. After 48 hours, the compound's IC_{50} value of 16.63 ± 0.27 μM showed more cytotoxic action on Caco-2 cancer cell lines than both cisplatin and etoposide. In Caco-2 cells, compound **31** (**Figure 30**) markedly increased the loss of mitochondrial membrane potential (MMP) levels. The reactive oxygen species (ROS) experiment demonstrated that compound **31** might cause apoptosis through the formation of ROS.

The structural isomer of capsaicin and its unique 1,2,3-triazole derivatives (new natural product hybrid capsaicinoid) were synthesised.[45] At a single dose of 10 μM, the newly synthesised compounds were tested for their antiproliferative efficacy against an NCI panel consisting of 60 cancer cell lines. Lung cancer cell lines (A549, NCI-H460) were shown to be more sensitive to the majority of the synthesised compounds among the cell lines evaluated. In A549 cell lines, compound **32** (**Figure 30**) showed possible antiproliferative action with IC_{50} value 2.91 μM. Compound **32** was shown to exhibit the highest level of activity among the compounds, with an IC_{50} value of 2.91 μM against A549. In addition, **32** causes cell cycle arrest at the *S*-phase, alters the potential of the mitochondrial membrane, reduces the ability of cell migration by causing cellular apoptosis and increased generation of reactive oxygen species, and alters the surface and nuclear morphology of A549 non-small cell lung cancer cell lines by reducing the number and shrinking of cells and exhibiting nuclear blabbing, which is a sign of apoptosis.

Cu(I)-Catalyzed "click" reaction was used to design and synthesise a series of 1,4-disubstituted 1,2,3-triazoles with a 10-demethoxy-10-N-methylaminocolchicine core.[59] The compounds were then tested for their in vitro cytotoxicity against four cancer cell lines (A549, MCF-7, LoVo, and LoVo/DX) and one noncancerous cell line (BALB/3T3). In order to evaluate the analogues' ability to overcome the LoVo/DX cells' drug resistance and confirm their selectivity for killing cancer cells rather than normal cells, selectivity (SI) and resistance (RI) indices were also calculated. In comparison to unmodified colchicine or doxorubicin and cisplatin, the compounds containing an ester or amide moiety in the fourth position of the 1,2,3-triazole of 10-*N*-methylaminocolchicine **33** (**Figure 30**) were found to have the best therapeutic potential (low IC$_{50}$ values and favourable SI values).

Figure 30. 1,2,3-Triazole conjugates exhibiting anticancer activity

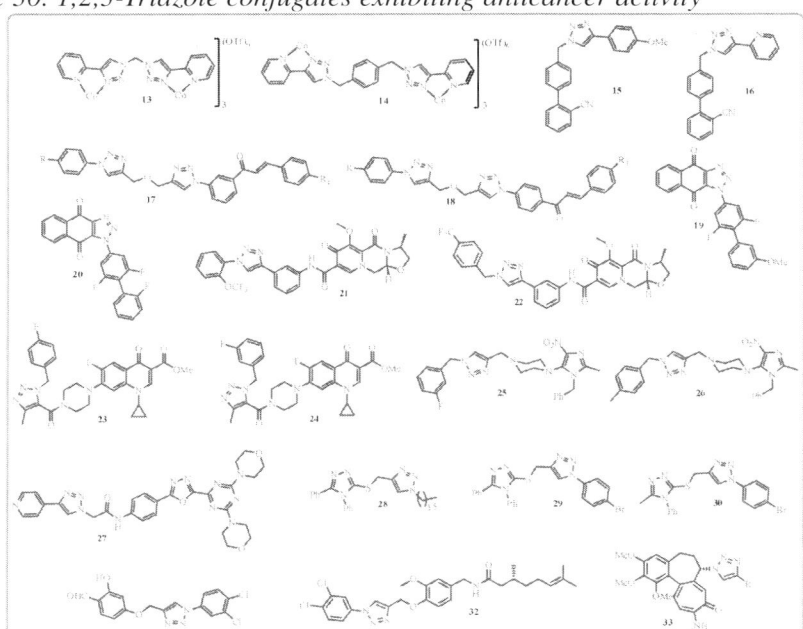

3.2. Anti-Tubercular Activity

With varying substitutions at the C-2 and/or C6 locations of the benzimidazole ring, new benzimidazole-1,2,3-triazole-quinoline hybrids and their intermediates were effectively synthesised in yields ranging from 55 to 80%.[60] All the synthesised compounds were evaluated for their *in vitro* anti-mycobacterial activity against the H37Rv strain of Mycobacterial TB. A subset of these compounds was then examined for cytotoxicity on TZM-bl cell lines. All hybrid compounds were more effective than ethambutol (MIC$_{90}$ = 9.54 μM), the first-line reference medication, and displayed good MIC$_{90}$ activities ranging from 1.07 to 8.66 μM. The hybrid compounds that showed the most promise were **34** (MIC_{90} = 1.54 μM, CC_{50} = 58.89 μM and % cell viability = 14.07) and **35** (MIC_{90} = 1.49 μM, CC_{50} = 4.62 μM and % cell viability = 44.03). Notably, **34** and **35** (**Figure 31**) showed cytotoxicity towards TZM-bl cell-lines but were nearly six times more effective than ethambutol.

1,8-Diazabicyclo[5.4.0]undec-7-ene (DBU) acetate ionic liquid has been used in an effective, environmentally friendly method that produces 1,4-disubstituted-1,2,3-triazole in a solvent- and external base-free environment.[61] This procedure is also used to synthesise unique amino acid compounds that contain 1,2,3-triazole molecules. Using the agar-based proportion test, all of the synthesised compounds **36** (**Figure 31**) were assessed for their ability to suppress the growth of M. tuberculosis H37Ra (ATCC 25177 strain).

Figure 31. 1,2,3-Triazole conjugates exhibiting antitubercular activity

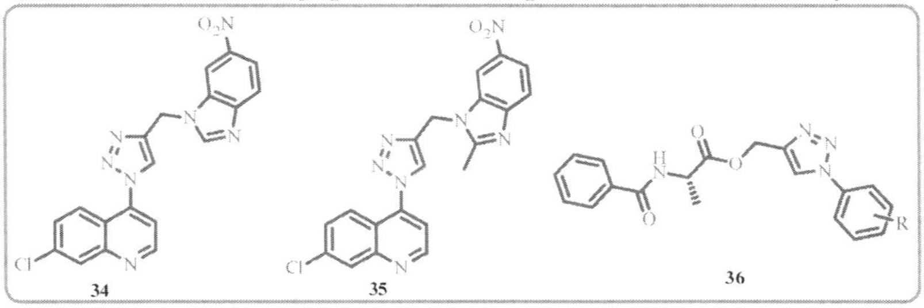

3.3. Anti-Alzheimer Activity

A unique set of 1,2,3-triazole-genipin analogues was created, synthesised, and tested for their ability to inhibit butyrylcholinesterase (BuChE) and acetylcholinesterase (AChE) as well as for their neuroprotective properties.[35] When compared to galantamine (IC_{50} = 34.05 μM), the genipin analogues containing bromoethyl- and diphenylhydroxy-triazole **37** (**Figure 32**) shown strong inhibitory effect on BuChE in addition to *in vitro* neuroprotective qualities against H_2O_2 toxicity.

Figure 32. 1,2,3-Triazole conjugates exhibiting antialzheimer activity

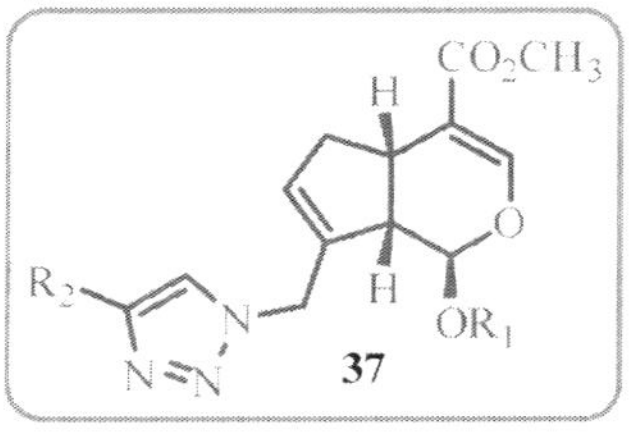

3.4 Antidiabetic Activity

Ebrahimi *et. al.*,[62] synthesized novel ten derivatives of indole-acrylamide-1,2,3-triazole. The ability of these substances to inhibit α-glucosidase was assessed. Every single one of the investigated compounds was more effective than the conventional α-glucosidase inhibitor acarbose. Comparing them to acarbose ($IC_{50} = 750.0 \pm 1.5$ µM). The most effective molecule, **38** (**Figure 33**), has a K_i value of 63 µM and exhibits competitive inhibitory behaviour, according to kinetic studies. The most active compounds were found to interact with the target enzyme's active site as well, according to a docking research conducted in the α-glucosidase active site.

A variety of novel 1*H*-1,2,3-triazole derivatives of Meldrum's acid were synthesised with great efficiency using the "click" method, and they were evaluated for *in vitro* α-glucosidase inhibitory activity.[63] Compound **39** (**Figure 33**) was shown to be several times more potent than acarbose and to have exceptionally potent inhibitory activity among all the investigated drugs. Moreover, molecular docking analysis of the active molecules' binding pattern demonstrated that the triazole-substituted group modifies the position and induces conformational changes that impact the binding of the Meldrum moiety, thereby altering the compounds' inhibitory potency. The Meldrum moiety is the primary constituent and crucial component for interaction with active site residues.

Figure 33. 1,2,3-Triazole conjugates exhibiting antidiabetic activity

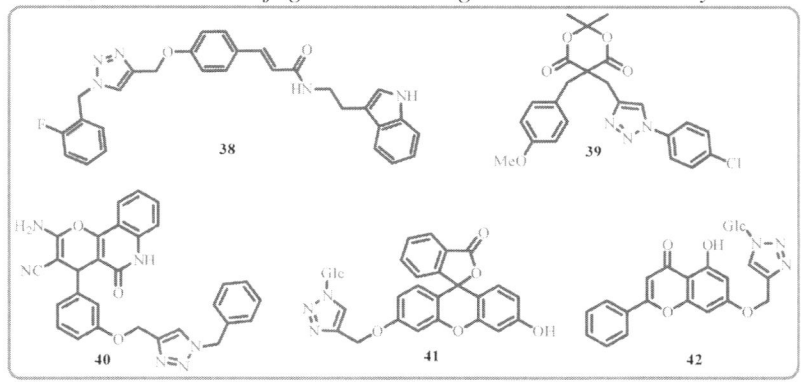

The synthesis of pyrano[3,2-*c*]quinoline-1,2,3-triazole hybrids greatly reduced *S. cerevisiae*'s α-glucosidase activity.[64] With an IC_{50} value of 1.19 ± 0.05, compound **40** (**Figure 33**) was determined to be the most active drug against this enzyme in a competitive manner. Compound **40**'s two R and S enantiomers interacted with important amino acids in the enzyme's active region, according to an *in silico* research.

A series of newly discovered 1,4-substituted 1,2,3-triazole-α-D-glucoside derivatives that were produced quickly *via* protecting-group-free synthesis.[65] The produced derivatives showed that all of the derivatives inhibited lysosomal α-glucosidase, and that the IC_{50} of **41** and **42** (**Figure 33**) on human lysosomal α-glucosidase were 60 times lower than those of acarbose.

3.5. Anti-Inflammatory Activity

16 Unique UA-GA derivatives that were altered at the UA C-28 site were created and produced.[66] The synthesized derivatives assessed *in vivo* and *in vitro* anti-inflammatory and *in vitro* cytotoxicity efficacy. Through modification of the free carboxyl group in hybrids, the toxicity of UA can be removed. Both the ether connection between the 1,2,3-triazole and gallate fragment and the C-28 ester or amide bond between the triterpenoid and linker group that resists hydrolysis guarantee the stability of new derivatives. Furthermore, the addition of gallate and 1,2,3-triazole substituents gave amphiphilic hybrids superior oxidation and inflammation resistance. Compound **43** (**Figure 34**) significantly reduced the expression of iNOS and COX-2, effectively blocked LPS-stimulated oxidation, and further prevented the activation of the PI3K-Akt signal pathway by downregulating ROS levels. As a result, the creation of UA-GA hybrids might offer a chemical foundation and an

anti-inflammatory mechanism for the creation and improvement of more potent medicines.

Novel thiazolidinone compounds based on 1,2,3-triazoles were synthesized using a click chemistry technique, and their anti-inflammatory and antioxidant properties were assessed.[17a] Comparing compounds **44** and **45** (**Figure 34**) to the common medication diclofenac sodium, the former has more anti-inflammatory efficacy. Compounds **44** and **45** shows potential antioxidant activity (IC_{50} = 12.55-16.30 μg/mL) when compared with standard drug BHT. The molecular docking analysis was also conducted against the active site of the inflammatory enzyme PPARγ in order to explain the reported biological activity data. The results showed a substantial association between the binding score and biological activity for these drugs. The triazole-incorporated 2,4-thiazolidinedione derivatives may have the perfect structural prerequisites for the future development of novel therapeutic medicines, according to the findings of the *in vitro* and *in silico* investigation.

Figure 34. 1,2,3-Triazole conjugates exhibiting anti-inflammatory activity

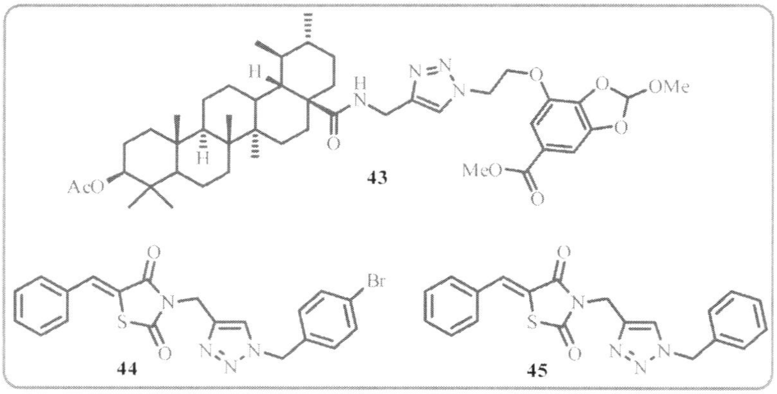

3.6. Antioxidant Activity

Reactive oxygen species are produced and accumulated in cells and tissue in an unbalanced manner by oxidative stress, which can result in a number of chronic and degenerative diseases, including ageing, cataracts, cancer, rheumatoid arthritis, autoimmune disorders, cardiovascular diseases, and neurological diseases. Using the 1,1-diphenyl-2-picrylhydrazyl radical scavenging assay, Shingate *et. al.*[67] have screened a library of bis-1,2,3-triazolyl-*N*-phenylacetamides **46** (**Figure 35**) for their *in vitro* antioxidant activity. After conducting a thorough structural activity relationship analysis to determine the primary substitution influencing the antioxidant activity, a molecular docking investigation was conducted against the myeloperoxidase (MPO) enzyme.

A straightforward and effective procedure has been created to synthesise novel 1,2,3-triazole-2,3-dihydroquinazolin-4[1H]-one (DHQ) conjugates 47 (**Figure 35**) using a solvent-free, ultrasound-assisted ionic liquid [HDBU][HSO$_4$] catalysed in excellent yields.[68] Using the 1,1-diphenyl-2-picryl hydrazyl (DPPH) assay, the newly synthesised derivatives were evaluated for antioxidant activity and were discovered to be strong scavengers. Significant antioxidant activity was demonstrated by compounds 47.

From basic aromatic aldehydes, half- and full-structured monocarbonyl curcumin analogues linked 1,2,3-triazole moiety were successfully synthesised.[69] According to the antioxidant evaluation, the finished goods had a modest level of DPPH radical inhibition. At a 100 ppm sample concentration, Compound 48 (**Figure 35**), which has a methoxy group and a structural motif similar to a half-curcumin analogue, was found to be the most active drug, with an inhibition of 32.04 ± 0.30%.

The novel ethyl 2-methyl-4-phenyl-7-((1-phenyl-1H-1,2,3-triazol-4-yl)oxy)-4H-chromene-3-carboxylate analogues were synthesized and screened for antioxidant activity by DPPH and hydrogen peroxide radical scavenging methods.[70] Compared to ascorbic acid, which had an IC$_{50}$ value of 1.46 ± 0.52 μM, the novel compounds 49 and 50 (**Figure 35**) demonstrated a higher radical quenching percentage, with IC$_{50}$ values of 1.29 ± 0.35 and 1.23 ± 0.34 μM, respectively.

Using the click chemistry technique, a series of novel 1,2,3-triazole-tethered coumarin conjugates connected by N-phenylacetamide were synthesised efficiently and in excellent yields.[71] The conjugates that were created were assessed for their antioxidant and antifungal properties in vitro. When compound 51 (**Figure 35**) was compared to the conventional antioxidant butylated hydroxytoluene, it showed potential for radical scavenging. The compound's antifungal activity was determined against strains of Candida albicans, Fusarium oxysporum, Aspergillus flavus, Aspergillus niger, and Cryptococcus neoformans. Some of the synthesized compounds exhibited lower minimum inhibitory concentration values and greater potency than the conventional medication miconazole.

Figure 35. 1,2,3-Triazole conjugates exhibiting antioxidant activity

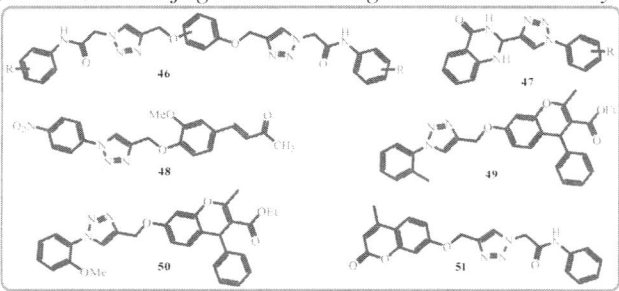

3.7. Antimicrobial Activity

To search for potent antibiotic substitutes for untreatable plant bacterial infections, a series of unique 1,2,3-triazole-tailored carbazoles was created.[72] According to the bioassay results, compound **52** (**Figure 36**) significantly inhibited the development of pathogens *Xoo* and *Xac*, with EC_{50} values of 3.36 and 2.87 µg/mL, respectively. Compound **52** shown efficaciousness in treating rice bacterial blight *in vivo* testing, with 50.78% and 53.23% curative and protective efficiency at 200 µg/mL, respectively. It's interesting to note that adding 0.1% auxiliaries, like organic silicon and orange oil, may boost the effectiveness of the control by increasing compound **52**'s surface wettability towards rice leaves and providing 61.38% and 65.50% of curative and protective effects, respectively.

The 1-aryl-5-substituted-1*H*-1,2,3-triazole-4 carboxyamides were designed, synthesised, and assessed for their antimicrobial potential.[73] The following pathogens were chosen for the antimicrobial evaluation: *Escherichia coli*, *Klebsiella pneumonia*, *Acinetobacter baumannii*, *Pseudomonas aeruginosa*, *Staphylococcus aureus*, *Cryptococcus neoformans var. grubii*, and *Candida albicans*. Against *S. aureus*, a number of 5-methyl-1*H*-1,2,3-triazole-4-carboxamides shown strong antibacterial activity. Conversely, pathogenic yeast *C. albicans* was effectively inhibited by 5-amino-1*H*-1,2,3-triazole-4-carboxamide **53** and [1,2,3]triazolo[1,5-*a*]quinazoline-3-carboxamide **54** (**Figure 36**). Therefore, compound **55** (**Figure 36**) showed 50% growth inhibition against *S. aureus* at 1 µM. Compound **54** killed about 40% of *C. albicans* cells at the same dosage. These substances generally showed selective activity and had no appreciable effect on the viability of human keratinocytes from the HaCaT line.

Benzooxapine-based triazolyl chalcones **56** (**Figure 36**) have been synthesised and evaluated for their *in vitro* biological activities, including antibacterial activity against Gram-(+ve) and Gram-(-ve) microorganisms, antifungal activity, and antitubercular activity against the H37Rv strain of all 1,2,3-triazoles.[74] *In vitro* biological tests demonstrated that the compounds have several active ingredients and shown efficacy as antibacterial, antifungal, and anti-tubercular agents, among other variants.

A range of 1,2,3-triazole derivatives of quinoxalinone were synthesised using the active substructure splicing approach and their antifungal activities were evaluated against 6 different fungal stains *in vitro* and *vivo*.[75] In a biological test, compound **57** (**Figure 36**) showed outstanding antifungal efficacy. Using the mycelial growth method, **57**'s EC_{50} value against *C. gloeosporioides* was 1.17 µg/mL, and at 1.91 µg/mL, it demonstrated excellent spore germination inhibition. In an *in vivo* test, **57** showed efficacious protection against pepper anthrax disease.

Using a click reaction, a unique sequence of hybrid derivatives of 1,2,3-triazole and benzimidazolidinone was created. Every synthetic derivative has had its antibacterial and anti-inflammatory properties assessed.[76] Tests for biological activity revealed that compounds **58**, **59**, and **60** (**Figure 36**), particularly derivative **59**, have strong antibacterial activity against target structures. Furthermore, compound **59** demonstrated the highest inhibition against *A. brasiliensis*, *A. fumigatus*, and *C. albicans* when compared to other products, according to the *in vitro* antifungal results. An assessment of the synthetic compounds' biological activities revealed that they have modest anti-inflammatory properties. In terms of potency and effectiveness, compound **58** was the most successful.

By using chemical techniques, ten triazole compounds with thymol moieties have been created.[77] The results of the antimicrobial activity demonstrate that the addition of the triazole moiety to the thymol nucleus **61** (**Figure 36**), along with the substitution of fluoro (-F), nitro (-NO_2), and chloro (-Cl) groups on the aromatic nucleus of some of the synthesised compounds, significantly increased the broad-spectrum antimicrobial activity when compared to the parent compound, thymol.

A novel ciprofloxacin series comprising 1,2,3-triazole conjugates of cipro-floxacin was synthesised,[78] and tested *in vitro* for antimicrobial activity against a range of strains, including *Staphylococcus aureus* (ATCC25923), *Enterococcus faecalis* (clinical isolate), *Staphylococcus epidermidis* (ATCC3594), *Escherichia coli* (ATCC25922), *Pseudomonas aeruginosa* (ATCC27853), *Salmonella typhi* (clinical isolate), *Salmonella typhimurium* (clinical isolate, *Acinetobacter bauman-nii* (ATCC19606), *Aeromonas hydrophila* (ATCC7966), *Plesiomonas shigelloides* (ATCC14029), and *Sphingo biiumpaucimobilis* (MTCC6362). It's interesting to note that some of the conjugates exhibited more antibacterial activity than the antibiotic ciprofloxacin, while compound **62** (**Figure 36**) demonstrated MIC 1.56 µM against *S. typhi* (clinical).

A library of new 1,2,3-triazole-appended bis-pyrazoles **63** (**Figure 36**) were synthesized and evaluated for their antifungal activity against different fungal strains, namely, *Candida albicans*, *Cryptococcus neoformans*, *Candida glabrata*, *Candida* *tropicalis*, *Aspergillus niger*, and *Aspergillus fumigatus*.[79] All of the compounds demonstrated good minimum inhibitory concentration values and broad-spectrum action against the tested fungal strains. The molecular docking study conducted against sterol 14α-demethylase (CYP51) may offer significant understanding into the binding affinities and mechanisms of these chemicals. Additionally, these compounds were assessed for their antioxidant properties, yielding encouraging results as well.

Using the techniques of disc diffusion and minimal inhibition concentration (MIC), a new series of 1,2,3-triazole-8-quinolinol hybrids was synthesised[80] in good yields and assessed for *in vitro* antibacterial activity against *Escherichia coli* (E. coli), *Xanthomonasfragariae* (X. fragariae), *Staphylococcus aureus* (S. aureus), *and Bacillus subtilis* (B. subtilis). With MIC values of 10 µg/mL against *S. aureus* and 20 µg/mL against *B. subtilis*, *E. coli*, and *X. fragariae*, Hybrid **64** (**Figure 36**) demonstrated outstanding antibacterial ability. These values were comparable to those of the standard antibiotic, nitroxoline.

A novel family of 1,2,3-triazole hybrids were synthesized, comprising either naphthalen-1-ol or 8-hydroxyquinoline (8-HQ) and 2- or 4-hydroxyphenyl benzo-thiazole (2- or 4-HBT),[81] and evaluated for their antimicrobial activity. Out of all the synthesized derivatives, compound **65** (**Figure 36**) demonstrated significant antibacterial activity (zone of inhibition 15.5-17.6 mm) against ciprofloxacin and superior antifungal activity (zone of inhibition 33.7 and 30.8 mm) against fluco-nazole. Quinoline- and 2-HBT-linked 1,2,3-triazoles of shorter alkyl linkers, like **65**, have a higher binding affinity (3.90×10^5 L mol^{-1}) with hs-DNA than naphthol- and 4-HBT-linked 1,2,3-triazoles bound to longer alkyl linkers, according to *in vitro* DNA binding studies using herring fish sperm DNA (hs-DNA).

2-Amino-3,5-dicyano-6-phenylthiopyridine was synthesised using a one-pot multicomponent method, incorporating 1,2,3-triazole derivatives from triazolyl aldehydes, malononitrile, and thiophenol in high yields.[82] All of the synthetic compounds' antitubercular, antifungal, and antioxidant properties were assessed. The findings of the antifungal activity demonstrate that, with a MIC value of less than 25 µg/mL, compounds **66**, **67**, and **68** (**Figure 36**) were shown to be more potent than micanazole. In contrast, the antitubercular screening against the MTB H37Rv strain yielded no encouraging findings. Additionally, all of the compounds' in silico ADME features were examined and showed promise for development as oral medication candidates.

Figure 36. 1,2,3-Triazole conjugates exhibiting antimicrobial activity

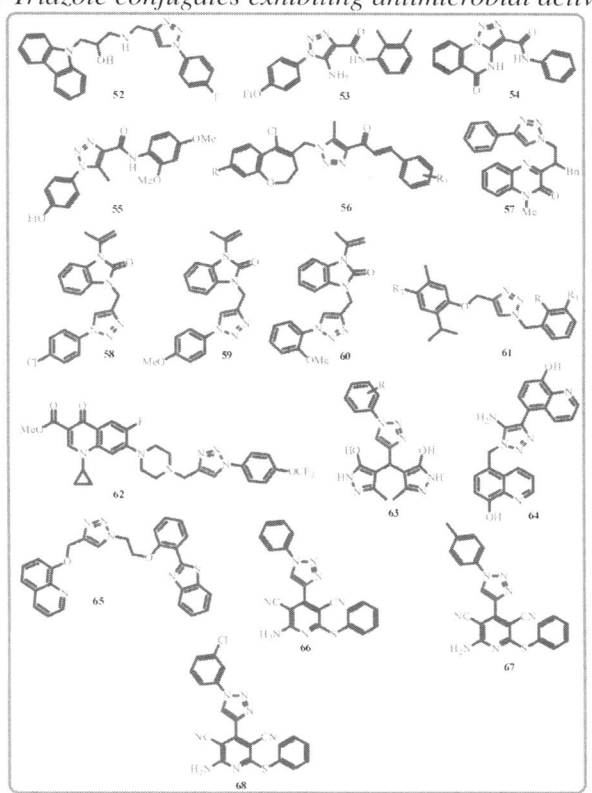

3.8. Miscellaneous Biological Activity of 1,2,3-Triazole Conjugates

Disrupting the homodimeric interface of Leishmania infantum trypanothione disulphide reductase (LiTryR) has been shown to be an alternative and underutilised tactic in the hunt for new antileishmanial drugs. Revuelto *et. al.*,[83] synthesized a new twenty six 1,2,3-triazole-based chemotype compounds which are noncompetitive, slow-binding inhibitors of *Li*TryR. Their overall K_i^* values of 0.5 μM (**69, Figure 37**) indicate that they function as slow-binding, noncompetitive inhibitors, much like prototype peptide. Additionally, these molecules have better selectivity indices than earlier in-house imidazole-based compounds and strong antileishmanial activity against both external and intracellular parasites.

Hydroxy-1,2,3-triazoles were synthesized significantly and in a dose-dependent way, these hydroxy-triazoles inhibited ZIKV and CHIKV replication at low concentrations.[46] Particularly, substances like **70** and **71** (**Figure 37**) revealed considerable inhibitory capability against ZIKV adsorption and CHIKV reproduction, respectively.

A series of phosphonate-functionalized 1,4-disubstituted 1,2,3-triazoles using supported copper nanoparticles were synthesized and evaluated their activity on $\alpha7$ receptors by single-channel and whole-cell recordings.[47] There were other triazole compounds that showed positive allosteric modulator (PAM) activity; the most effective one was the one functionalized with the methyl phosphonate group. The actions of type I PAMs were recapitulated by $\alpha7$ potentiation, which was observed at the macroscopic level as an increase in the maximum currents induced by acetylcholine with negligible effects on desensitization. The active chemicals considerably lengthened channel openings and activation episodes, but they had no effect on channel amplitude at the single-channel level. Determined the transmembrane amino acids necessary for potentiation, categorized the most active molecule as a type I PAM, deciphered the chemical mechanism of action, and used multiple SAR techniques based on the structure. Nielsen *et. al.*,[47] suggest using phosphonate-functionalized 1,2,3-triazoles **72** (**Figure 37**) as a scaffold to create novel therapeutic medicines for neurological illnesses, in addition to being a novel pharmacophore with intrinsic $\alpha7$-PAM action.

Rahman *et. al.*,[84] have shown that 1,2,3-triazoles, with a favourable balance between potency and lipophilicity, can bioisosterically substitute the amide functionality as the main pharmacophore in the 2-AMPP scaffold. With an EC_{50} of 14 nM in the cell-based functional cAMP test and GPR88-specific agonist activity in the mouse striatal tissues, compound **73** (**Figure 37**) emerged as the most potent one in this investigation. Furthermore, **73** showed a favourable in vivo PK profile and good permeability. All things considered, the compound **73** as a viable agonist probe for understanding the roles of GPR88 and assessing GPR88's therapeutic potential for treating striatal-associated illnesses such drug addiction, schizophrenia, and Parkinson's disease.

Leucine-Zipper and Sterile-α Motif Kinase (ZAK) is a novel and intriguing target for the development of medications that have antihypertrophic cardiomyopathy (HCM)-preventing properties. As specific ZAK inhibitors, a group of 1,2,3-triazole benzenesulfonamides were synthesised.[85] Among these, compound **74** (**Figure 37**) has a strong binding to the ZAK protein (K_d = 8.0 nM) and a robust suppression of ZAK's kinase activity with single-digit nM (IC_{50} = 4.0 nM). It also shows good selectivity against a panel of 403 wildtype kinases on a KINOMEscan screening platform. Oral treatment of this drug in a spontaneous hypertensive rat (SHR) paradigm exhibits promise *in vivo* anti-HCM efficacy, as it dose-dependently blocks both p38/GATA-4 and JNK/c-Jun signalling. Compound **74** might be the starting point for the development of novel anti-HCM drugs.

One of the main receptors linked to the immunosuppressive effects of the inflammatory mediator PGE2 in the tumour microenvironment is the prostanoid EP4 receptor. In order to improve immunity-mediated tumour eradication, blocking EP4 signalling has recently become a viable cancer immunotherapy tactic. Yang *et. al.*,[86] synthesised a family of 1*H*-1,2,3-triazole-based ligands with low nanomolar antagonist activity towards the human EP4 receptor and good subtype selectivity in an attempt to find new subtype-selective EP4 antagonists. The most promising molecule, compound **75** (**Figure 37**), efficiently reduces the expression of several immunosuppression-related genes in macrophage cells and demonstrates single-digit nanomolar efficacy in the EP4 calcium flux and cAMP-response element reporter tests. For additional *in vivo* biological testing, compound **75** was selected due to its advantageous ADMET characteristics. In the mouse CT26 colon cancer model, compound **75** administered orally dramatically reduced the growth of the tumour along with increased cytotoxic T lymphocyte infiltration in the tumour tissue.

Even with the most current advancements in cancer treatment, one of the biggest problems in the profession is still therapy resistance. Within this framework, signalling molecules-like cytokines-have surfaced as potentially fruitful avenues for drug discovery research. Macrophage migration inhibitory factor (MIF) and its closely related counterpart D-dopachrome tautomerase (D-DT) are examples of cytokines. Osipyan *et. al.*,[87] designed a novel class of D-DT binders which produces a dual-targeted inhibitor that may be able to use the Proteolysis Targeting Chimaera (PROTAC) technology to initiate D-DT breakdown. Osipyan *et. al.*,[87] synthesized a brand-new 1,2,3-triazole library that targets D-DT. A cereblon (CRBN) ligand was coupled to the most powerful derivative, **76** (IC_{50} of 0.5 ± 0.04 µM with strong selectivity towards D-DT), *via* aliphatic amides, which were produced by an incredibly practical and efficient solvent-free process. Through enzyme inhibition tests, the compound **77** was found to have a moderate inhibitory potency (IC_{50} of 5.9 ± 0.7 µM). Regretfully, D-DT degradation assays revealed no activity from **77** (**Figure 37**).

Haidar *et. al.*,[88] were able to synthesise several compounds which are derived from the naturally occurring biologically active diterpene totarol. They also found that these compounds have substantial sub-micromolar inhibitory action on human PIP5K1α. The study findings indicate that compound **78** (**Figure 37**), which was the most active in cells and significantly suppressed PIP5K1α activity in the prepared series, has promise for further development. According to docking tests, the active ingredient fitted snugly into the ATP binding site of the enzyme's zebrafish crystal structure.

Agriculture utilising nitrogen fertiliser has detrimental effects on the environment, such as nitrous oxide (N_2O) production, nitrate (NO_3^-) contamination of groundwater, and eutrophication of rivers. The conversion of ammonia to NO_3^- can be slowed down by adding inhibitors to fertilisers to improve the efficiency of nitrogen utilisation. Regrettably, for unclear reasons, commercial inhibitors have not shown consistent results in a range of agro ecosystems. Yildirim *et. al.*,[89] showed that 4-methyl-1-(prop-2-yn-1-yl)-1*H*-1,2,3-triazole (**79**) has superior nitrification inhibitory capabilities through a combination of bacterial research and soil incubations. Compound **79** (**Figure 37**) functions as a mechanistic, irreversible inhibitor of ammonia monooxygenase, the key enzyme, in contrast to commercial reversible inhibitors. This allows for the effective retention of ammonium (NH_4^+) and the suppression of NO_3^- and N_2O production over a 21-day period in a variety of agricultural soils with pH values ranging from 4.7 to 7.5. A battery of tests for fresh water and terrestrial ecotoxicity, along with a bacterial viability stain, revealed no evidence of acute or long-term toxicity. Compound **79** had a stronger inhibitory effect on both ammonia-oxidizing bacteria and archaea, according to real-time quantitative polymerase chain reaction (qPCR) study. As a result, **79** works better than nitrification inhibitors that are currently on the market and has a lot of potential for use in a variety of agricultural contexts.

Two sets of triazoles were synthesised by Silva *et. al.*,[90] one was produced by ethyl cyanoacetate and various phenyl azides undergoing a 1,3-dipolar cycloaddition process, resulting in 1*H*-1,2,3-triazoles, and the other *via* rearrangement of Dimroth, resulting in 2*H*-1,2,3-triazoles. It was demonstrated that both series were effective against Trypanosoma cruzi's epimastigote form. The most potent substances were 1,2,3-triazoles **80** (S.I. between 100 and 200), **81**, and **82** (S.I. >200) (**Figure 37**), which could rupture the trypomastigotes' plasma membranes, inhibit CYP51, and stop the synthesis of ergosterol. When interacting with CYP51, compound **80** displayed the best and most favourable profile.

By using sodium azide and triazolyl nitriles in a [2+3]-cycloaddition reaction with [DBU][OAc] under ultrasonic irradiation, 1,2,3-triazole-based tetrazole derivatives **83** (**Figure 37**) have been synthesized.[91] As compared to the common medication albendazole, the majority of the compounds have demonstrated superior anthelmintic activity. Additionally, a molecular docking analysis demonstrated that each of these drugs bound to the β-tubulin receptor's active site, offering a logical explanation for their anthelmintic effect.

Figure 37. 1,2,3-Triazole conjugates exhibiting miscellaneous biological activity

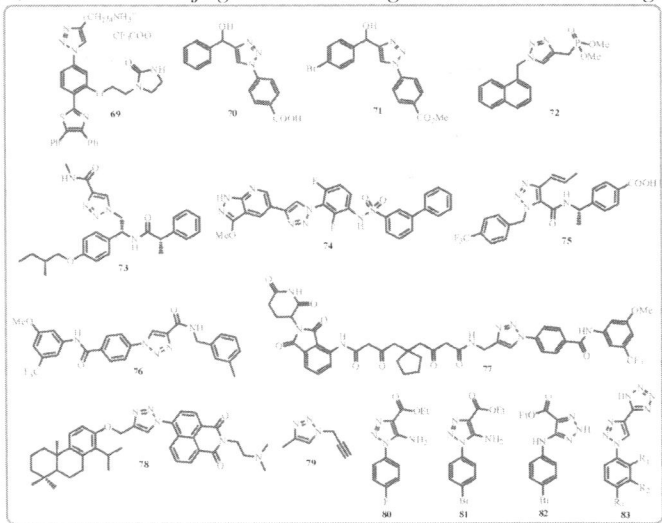

4. CONCLUSION

In this book chapter, we have summarized different synthetic methodologies to prepare 1,2,3-triazoles. Click chemistry was designed to resemble nature, which similarly creates chemicals by connecting small modular components; it is not a single unique reaction. The most well-known example of "click chemistry" is the 1,3-dipolar azide, alkyne cycloaddition (CuAAC) reaction, which is catalysed by copper and is almost quantitative and simple to carry out. The drug research and discovery community has become interested in the synthesis of the triazole scaffold due to its significance in medicinal chemistry. The main synthetic techniques now employed to synthesise triazoles are summarised in this book chapter, along with their pharmacological significance-including their anticancer, antitubercular, antialzheimer, antidiabetic, anti-inflammatory, antioxidant, antifungal, antibacterial, and many other qualities.

4.1 Abbreviations

Table 1.

AChE	Acetylcholinesterase
ATCC	American Type Culture Collection
BHT	Butylated Hydroxy Toluene
BSA	Bovine Serum Albumin
BuChE	Butyrylcholinesterase
CAI	Carboxyamido-triazole
CCK2R	Cholecystokinin-2 Receptor
CHIKV	Chikungunya Virus
COX	Cyclooxygenase
CuAAC	Copper-Catalyzed Azide-Alkyne Cycloaddition
CYP	Cytochrome P
DBU	1,8-Diazabicyclo[5.4.0]undec-7-ene
DCM	Dichloromethane
D-DT	D-dopachrome Tautomerase
DES	Deep Eutectic Solvent
DHODH	Dihydroorotate Dehydrogenase
DHQ	2,3-Dihydroquinazolin-4[1H]-one
DMA	Dimethylacetamide
DMF	Dimethylformamide
DNA	Deoxyribonucleic acid
DPPH	2,2-diphenyl-1-picrylhydrazyl
HBT	Hydroxyphenyl Benzothiazole
HCM	Hypertrophic Cardiomyopathy
HDF	Human Dermal Fibroblasts
hER	Human Oestrogen Receptor Alpha
HIV	Human Immunodeficiency Virus
8-HQ	8-Hydroxyquinoline
iNOS	Inducible nitric oxide synthase
LiTryR	Leishmania infantum Trypanothione Disulphide Reductase
MIC	Minimum Inhibitory Concentration
MIF	Macrophage Migration Inhibitory Factor
MMP	Mitochondrial Membrane Potential

continued on following page

Table 1. Continued

AChE	Acetylcholinesterase
MPO	Myeloperoxidase Enzyme
MTT	3-(4,5-Dimethylthiazol-2-yl)-2,5-Diphenyltetrazolium Bromide
4-OHT	4-Hydroxytamoxifen
PAM	Positive Allosteric Modulator
PROTAC	Proteolysis Targeting Chimaera
qPCR	Quantitative Polymerase Chain Reactor
RI	Resistance Indices
ROS	Reactive Oxygen Species
RPE	Retinal Pigment Epithelium
RuAAC	Ruthenium-Catalyzed Azide-Alkyne cycloaddition
SHR	Spontaneous Hypertensive Rat
SI	Selectivity Indices
STK33	Serine/Threonine Kinase 33
TB	Tuberculosis
TBA	Tertiary Butyl Alcohol
TEA	Triethylamine
TMSBr	Bromotrimethylsilane
ZAK	Zipper and Sterile-α Motif Kinase
ZIKV	Zika Virus

REFERENCES

Addo, J. K., Ansah, E. O., Dayie, N. T. K. D., Cheseto, X., & Torto, B. (2022). Synthesis of 1,2,3-triazole-thymol derivatives as potential antimicrobial agents. *Heliyon*, 8(10), e10836. DOI: 10.1016/j.heliyon.2022.e10836 PMID: 36217474

Agarwal, A., Singh, P., Maurya, A., Patel, U. K., Singh, A., & Nath, G. (2022). Ciprofloxacin-tethered 1,2,3-triazole conjugates: New quinolone family compounds to upgrade our antiquated approach against bacterial infections. *ACS Omega*, 7(3), 2725–2736. DOI: 10.1021/acsomega.1c05303 PMID: 35097270

Aitha, S., Thumma, V., Matta, R., Ambala, S., Jyothi, K., Manda, S., & Pochampally, J. (2023). Antioxidant activity of novel 4*H*-chromene tethered 1,2,3-triazole analogues: Synthesis and molecular docking studies. *Results in Chemistry*, 5, 100987. DOI: 10.1016/j.rechem.2023.100987

Akolkar, S. V., Nagargoje, A. A., Shaikh, M. H., Warshagha, M. Z. A., Sangshetti, J. N., Damale, M. G., & Shingate, B. B. (2020). New *N*-phenylacetamide-linked 1,2,3-triazole-tethered coumarin conjugates: Synthesis, bioevaluation, and molecular docking study. *Archiv der Pharmazie*, 2020(11), e2000164. DOI: 10.1002/ardp.202000164 PMID: 32776355

Al-Ghulikah, H., Ghabi, A., Haouas, A., Mtiraoui, H., Jeanneau, E., & Msaddek, M. (2023). Synthesis of new 1,2,3-triazole linked benzimidazolidinone: Single crystal X-ray structure, biological activities evaluation and molecular docking studies. *Arabian Journal of Chemistry*, 16(3), 104566. DOI: 10.1016/j.arabjc.2023.104566

Al-Taweel, S., Al-Saraireh, Y., Al-Trawneh, S., Alshahateet, S., Al-Tarawneh, R., Ayed, N., Alkhojah, M., Al-Khaboori, W., Zereini, W., & Al-Qaralleh, O. (2023). Synthesis and biological evaluation of ciprofloxacin-1,2,3-triazole hybrids as antitumor, antibacterial, and antioxidant agents. *Heliyon*, 9(12), e22592. DOI: 10.1016/j.heliyon.2023.e22592 PMID: 38125538

Ali, A. L. S., Khan, D., Naqvi, A., Al-blewi, F. F., Rezki, N., Aouad, M. R., & Hagar, M. (2021). Design, synthesis, molecular modeling, anticancer studies, and density functional theory calculations of 4-(1,2,4-triazol-3-ylsulfanylmethyl)-1,2,3-triazole derivatives. *ACS Omega*, 6(1), 301–316. DOI: 10.1021/acsomega.0c04595 PMID: 33458482

Ardiansah, B., Hardhani, M. R., Putera, D. D. S. R., Wukirsari, T., Cahyana, A. H., Jia, J. W., & Khan, M. M. (2023). Design, synthesis, and antioxidant evaluation of monocarbonyl curcumin analogues tethered 1,2,3-triazole scaffold. *Case Studies in Chemical and Environmental Engineering*, 8, 100425. DOI: 10.1016/j.cscee.2023.100425

Ashok, D., Thara, G., Kumar, B. K., Srinivas, G., Ravinder, D., Vishnu, T., Sarasijae, M., & Sushmitha, B. (2023). Microwave-assisted synthesis, molecular docking studies of 1,2,3-triazole-based carbazole derivatives as antimicrobial, antioxidant and anticancer agents. *RSC Advances*, 13(1), 25–40. DOI: 10.1039/D2RA05960F PMID: 36545291

Avula, S. K., Ullah, S., Halim, S. A., Khan, A., Anwar, M. U., Csuk, R., Al-Harrasi, A., & Rostami, A. (2023). Meldrum-based-1*H*-1,2,3-triazoles as antidiabetic agents: Synthesis, *in vitro* α-glucosidase inhibition activity, molecular docking studies, and *in silico* approach. *ACS Omega*, 8(28), 24901–24911. DOI: 10.1021/acsomega.3c01291 PMID: 37483205

Baddam, S. R., Avula, M. K., Akula, R., Battula, V. R., Kalagara, S., Buchikonda, R., Ganta, S., Venkatesan, S., & Allaka, T. R. (2024). Design, synthesis and *in silico* molecular docking evaluation of novel 1,2,3-triazole derivatives as potent antimicrobial agents. *Heliyon*, 10(7), e27773. DOI: 10.1016/j.heliyon.2024.e27773 PMID: 38590856

Banoji, V., Angajala, K. K., Vianala, S., Manne, S., Ravulapelly, K. R., & Vannada, J. (2024). Synthesis, characterization, cytotoxic evaluation, and molecular docking studies of novel 1,2,3-triazole-based chalcones for potential anticancer applications. *Results in Chemistry*, 7, 101294. DOI: 10.1016/j.rechem.2023.101294

Binder, W. H., & Kluger, C. (2006). Azide/alkyne-"click" reactions: Applications in material science and organic synthesis. *Current Organic Chemistry*, 10(14), 1791–1815. DOI: 10.2174/138527206778249838

Boruah, D. J., Kathirvelan, D., Bora, K., Maurya, R. A., & Yuvaraj, P. (2023). Efficient and environmentally friendly synthesis of 1,2,3-triazole derivatives via [3+2] cycloaddition and their potential as lung cancer inhibitors: An in silico study. *Results in Chemistry*, 5, 100903. DOI: 10.1016/j.rechem.2023.100903

Bozorov, K., Zhao, J., & Aisa, H. A. (2019). 1,2,3-Triazole-containing hybrids as leads in medicinal chemistry: A recent overview. *Bioorganic & Medicinal Chemistry*, 27(16), 3511–3531. DOI: 10.1016/j.bmc.2019.07.005 PMID: 31300317

Castillo, J. C., Bravo, N. F., Tamayo, L. V., Mestizo, P. D., Hurtado, J., Macias, M., & Portilla, J. (2020). Water-compatible synthesis of 1,2,3-triazoles under ultrasonic conditions by a Cu(I) complex-mediated click reaction. *ACS Omega*, 5(46), 30148–30159. DOI: 10.1021/acsomega.0c04592 PMID: 33251449

Crlikova, H., Malina, J., Novohradsky, V., Kostrhunova, H., Vasdev, R. A. S., Crowley, J. D., Kasparkova, J., & Brabec, V. (2020). Antiproliferative activity and associated dna interactions of $[Co_2L_3]^{6+}$ cylinders derived from bis(bidentate) 2-pyridyl-1,2,3-triazole ligands. *Organometallics*, 39(8), 1448–1455. DOI: 10.1021/acs.organomet.0c00146

Danne, A. B., Deshpande, M. V., Sangshetti, J. N., Khedkar, V. M., & Shingate, B. B. (2021). New 1,2,3-triazole-appended bis-pyrazoles: Synthesis, bioevaluation, and molecular docking. *ACS Omega*, 6(38), 24879–24890. DOI: 10.1021/acsomega.1c03734 PMID: 34604669

Deshmukh, T. R., Khedkar, V. M., Sangshetti, J. N., & Shingate, B. B. (2023). Exploring the antioxidant potential of bis-1,2,3-triazolyl-N-phenylacetamides. *Research on Chemical Intermediates*, 49(2), 635–653. DOI: 10.1007/s11164-022-04915-2

Dheer, D., Singh, V., & Shankar, R. (2017). Medicinal attributes of 1,2,3-triazoles: Current developments. *Bioorganic Chemistry*, 71, 30–54. DOI: 10.1016/j.bioorg.2017.01.010 PMID: 28126288

Drelinkiewicz, D., & Whitby, R. J. (2022). A practical flow synthesis of 1,2,3-triazoles. *RSC Advances*, 12(45), 28910–28915. DOI: 10.1039/D2RA04727F PMID: 36320728

Ebrahimi, S. E. S., Babania, H., Khanaposhtani, M. M., Asgari, M. S., Mojtabavi, S., Faramarzi, M. A., Meymandi, A. Y., Zareie, S., Larijani, B., Biglar, M., Rastgar, H., Foroumadi, A., & Mahdavi, M. (2022). Design, synthesis, and biological evaluation of new indole-acrylamide-1,2,3-triazole derivatives as potential α-glucosidase inhibitors. [REMOVED HYPERLINK FIELD]. *Polycyclic Aromatic Compounds*, 42(6), 3157–3165. DOI: 10.1080/10406638.2020.1854323

Esmaili, S., Ebadi, A., Khazaei, A., Ghorbani, H., Faramarzi, M. A., Mojtabavi, S., Mahdavi, M., & Najafi, Z. (2023). Novel pyrano[3,2-*c*]quinoline-1,2,3-triazole hybrids as potential anti-diabetic agents: In vitro α-glucosidase inhibition, kinetic, and molecular dynamics simulation. *ACS Omega*, 8(26), 23412–23424. DOI: 10.1021/acsomega.3c00133 PMID: 37426262

Faydy, M. E., Lakhrissi, L., Dahaieh, N., Ounine, K., Tuzun, B., Chahboun, N., Boshaala, A., Alobaid, A., Warad, I., Lakhrissi, B., & Zarrouk, A. (2024). Synthesis, biological properties, and molecular docking study of novel 1,2,3-triazole-8-quinolinol hybrids. *ACS Omega*, 9(23), 25395–25409. DOI: 10.1021/acsomega.4c03906 PMID: 38882066

Garg, A., Borah, D., Trivedi, P., Gogoi, D., Chaliha, A. K., Ali, A. Z., Chetia, D., Chaturvedi, V., & Sarma, D. (2020). A simple work-up-free, solvent-free approach to novel amino acid linked 1,4-disubstituted 1,2,3-triazoles as potent antitubercu-losis agents. *ACS Omega*, 5(46), 29830–29837. DOI: 10.1021/acsomega.0c03862 PMID: 33251417

Gokturk, T., Cetin, E. S., Hokelek, T., Pekel, H., Sensoy, O., Aksu, E. N., & Gup, R. (2023). Synthesis, structural investigations, DNA/BSA interactions, molecular docking studies, and anticancer activity of a new 1,4-disubstituted 1,2,3-triazole derivative. *ACS Omega*, 8(35), 31839–31856. DOI: 10.1021/acsomega.3c03355 PMID: 37692230

Gorantla, J. N., Maniganda, S., Pengthaisong, S., Ngiwsara, L., Sawangareetrakul, P., Chokchaisiri, S., Kittakoop, P., Svasti, J., & Cairns, J. R. K. (2021). Chemoenzymatic and protecting-group-free synthesis of 1,4-substituted 1,2,3-triazole-α-D-glucosides with potent inhibitory activity toward lysosomal α-glucosidase. *ACS Omega*, 6(39), 25710–25719. DOI: 10.1021/acsomega.1c03928 PMID: 34632227

Grob, N. M., Schibli, R., Behe, M., Valverde, I. E., & Mindt, T. L. (2021). 1,5-Disubstituted 1,2,3-triazoles as amide bond isosteres yield novel tumor-targeting minigastrin analogs. *ACS Medicinal Chemistry Letters*, 12(4), 585–592. DOI: 10.1021/acsmedchemlett.0c00636 PMID: 33859799

Guimaraes, T. T., Pinto, M. D. C. F. R., Lanza, J. S., Melo, M. N., Neto, R. L. M., Melo, I. M. M., Diogo, E. B. T., Ferreira, V. F., Camara, C. A., Valenca, W. O., Oliveira, R. N., Frezard, F., & Silva, E. N.Jr. (2013). Potent naphthoquinones against antimony-sensitive and -resistant Leishmania parasites: Synthesis of novel α- and nor-α-lapachone-based 1,2,3-triazoles by copper-catalyzed azide-alkyne cycload-dition. *European Journal of Medicinal Chemistry*, 63, 523–530. PMID: 23535320

Guo, Y., Hou, J., Wu, H., Chen, Y., Liu, G., Wang, D., Wang, H., Mao, L., Li, S., & Wang, T. (2024). Synthesis and discovery of novel 1,2,3-triazole based cabotegravir derivatives with potent anticancer activity. *Journal of Molecular Structure*, 1298, 137042. DOI: 10.1016/j.molstruc.2023.137042

Haidar, S., Amesty, A., Royo, S. O., Gotz, C., El-Awaad, E., Kaiser, J., Bodecker, S., Arnold, A., Aichele, D., Amaro-Luis, J. M., Braun, A. E., & Jose, J. (2024). 1,2,3-Triazole-totarol conjugates as potent PIP5K1α lipid kinase inhibitors. *Bioorganic & Medicinal Chemistry*, 105, 117727. DOI: 10.1016/j.bmc.2024.117727 PMID: 38669736

Haldon, E., Nicasio, M. C., & Perez, P. J. (2015). Copper-catalysed azide-alkyne cycloadditions (CuAAC): An update. *Organic & Biomolecular Chemistry*, 13(37), 9528–9550. DOI: 10.1039/C5OB01457C PMID: 26284434

Hein, J. E., & Fokin, V. V. (2010). Copper-catalyzed azide–alkynecycloaddition (CuAAC) and beyond: New reactivity of copper(i) acetylides. *Chemical Society Reviews*, 39(4), 1302–1315. DOI: 10.1039/b904091a PMID: 20309487

Heravi, M. M., & Vavsari, V. F. (2015). Recent advances in application of amino acids: Key building blocks in design and syntheses of heterocyclic compounds. *Advances in Heterocyclic Chemistry*, 114, 77–145.

Horne, W. S., Yadav, M. K., Stout, C. D., & Ghadiri, M. R. (2004). Heterocyclic peptide backbone modifications in an alpha-helical coiled coil. *Journal of the American Chemical Society*, 126, 15366. PMID: 15563148

Huang, X., Liu, H. W., Long, Z. Q., Li, Z. X., Zhu, J. J., Wang, P. Y., Qi, P. Y., Liu, L. W., & Yang, S. (2021). Rational optimization of 1,2,3-triazole-tailored carbazoles as prospective antibacterial alternatives with significant *in vivo* control efficiency and unique mode of action. *Journal of Agricultural and Food Chemistry*, 69(16), 4615–4627. DOI: 10.1021/acs.jafc.1c00707 PMID: 33855856

Huisgen, R. (1961). 1,3-Dipolar cycloaddition. *Proceedings of the Chemical Society*, 357-396.

Huisgen, R. (1963). Kinetics and mechanism of 1,3-dipolar cycloadditions. *Angewandte Chemie International Edition in English*, 2(11), 633–645. DOI: 10.1002/anie.196306331

Jiang, Y., Zhu, C. H., Xia, Z. H., & Zhao, H. Q. (2024). Novel quinoxalinone-1,2,3-triazole derivatives as potential antifungal agents for plant anthrax disease: Design, synthesis, antifungal activity and SAR study. *Advanced Agrochem*, 3(3), 222–228. DOI: 10.1016/j.aac.2023.11.003

Kadaba, P. K. (1986). 1,2,3,-Triazole Anticonvulsant Drugs. 9, https://uknowledge.uky.edu/ps_patents

Kanishchev, O. S., Gudz, G. P., Shermolovich, Y. G., Nesterova, N. V., Zagorodnya, S. D., & Golovan, A. V. (2011). Synthesis and biological activity of the nucleoside analogs based on polyfluoroalkyl-substituted 1,2,3-triazoles. *Nucleosides, Nucleotides & Nucleic Acids*, 30, 768–783. PMID: 21967288

Khan, A., Naaz, F., Basit, R., Das, D., Bisht, P., Shaikh, M., Lone, B. A., Pokharel, Y. R., Ahmed, Q. N., Parveen, S., Ali, I., Singh, S. K., Chashoo, G., & Shafi, S. (2024). Correction to 1,2,3-triazole tethered hybrid capsaicinoids as antiproliferative agents active against lung cancer cells (A549). *ACS Omega*, 9(9), 11026–11026. DOI: 10.1021/acsomega.4c00155 PMID: 38463328

Khan, I., Tantray, M. A., Hamid, H., Alam, M. S., Kalam, A., Hussain, F., & Dhulap, A. (2016). Synthesis of pyrimidin-4-one-1,2,3-triazole conjugates as glycogen synthase kinase-3β inhibitors with anti-depressant activity. *Bioorganic Chemistry*, 68, 41–55. PMID: 27454617

Khare, S. P., Deshmukh, T. R., More, D. D., Kute, A. M., Sangshetti, J. N., Khedkar, V. M., & Shingate, B. B. (2024). Synthesis and bioevaluation of 1,2,3-triazole linked highly substituted pyridine derivatives. *Chemico-Biological Interactions*, 14, 1–12.

Kolb, H. C., Finn, M. G., & Sharpless, K. B. (2001). Click chemistry: Diverse chemical function from a few good reactions. *Angewandte Chemie International Edition*, 40(11), 2004–2021. DOI: 10.1002/1521-3773(20010601)40:11<2004::AID-ANIE2004>3.0.CO;2-5 PMID: 11433435

Krzywik, J., Goldeman, A. N., Mozga, W., Wietrzyk, J., & Huczynski, A. (2021). Novel double-modified colchicine derivatives bearing 1,2,3-triazole: Design, synthesis, and biological activity evaluation. *ACS Omega*, 6(40), 26583–26600. DOI: 10.1021/acsomega.1c03948 PMID: 34661013

Lopez, H. H., Ramos, S. L., Duran, C. F. A. G., Alvarez, A. P., Gutierrez, I. R. R., Peralta, M. A. L., & Hernandez, R. S. R. (2020). Synthesis of 1,4-Biphenyl-triazole Derivatives as Possible 17β-HSD1 Inhibitors: An *in Silico* Study. *ACS Omega*, 5(23), 14061–14068. DOI: 10.1021/acsomega.0c01519 PMID: 32566872

Mamidyala, S. K., & Finn, M. G. (2010). In situ click chemistry: Probing the binding landscapes of biological molecules. *Chemical Society Reviews*, 39(4), 1252. DOI: 10.1039/b901969n PMID: 20309485

Meldal, M., & Tornoe, C. W. (2008). Cu-Catalyzed azide-alkyne cycloaddition. *Chemical Reviews*, 108(8), 2952–3015. DOI: 10.1021/cr0783479 PMID: 18698735

Moses, J. E., & Moorhouse, A. D. (2007). The growing applications of click chemistry. *Chemical Society Reviews*, 36, 1249. PMID: 17619685

Moses, J. E., & Moorhouse, A. D. (2007). The growing applications of click chemistry. *Chemical Society Reviews*, 36(8), 1249–1262. DOI: 10.1039/B613014N PMID: 17619685

Najaf, Z., Esmaili, S., Khaleseh, B., Babaee, S., Khoshneviszadeh, M., Chehardoli, G., & Akbarzadeh, T. (2022). Ultrasound-assisted synthesis of kojic acid-1,2,3-triazole baseddihydropyrano[3,2-*b*]pyran derivatives using Fe_3O_4@CQD@CuI as a novelnanomagnetic catalyst. *Scientific Reports*, 12(1), 19917. DOI: 10.1038/s41598-022-24089-6 PMID: 36402826

Najafi, Z., Mahdavi, M., Saeedi, M., Karimpour-Razkenari, E., Edraki, N., Sharifza-deh, M., Khanavi, M., & Akbarzadeh, T. (2019). Novel tacrine-coumarin hybrids linked to 1,2,3-triazole as anti-Alzheimer's compounds: *In vitro* and *in vivo* biological evaluation and docking study. *Bioorganic Chemistry*, 83, 303–316. DOI: 10.1016/j.bioorg.2018.10.056 PMID: 30396115

Nehra, N., Tittal, R. K., & Ghule, V. D. (2021). 1,2,3-Triazoles of 8-hydroxyquin-oline and HBT: Synthesis and studies (DNA binding, antimicrobial, molecular docking, ADME, and DFT). *ACS Omega*, 6(41), 27089–27100. DOI: 10.1021/acsomega.1c03668 PMID: 34693129

Nielsen, B. E., Stabile, S., Vitale, C., & Bouzat, C. (2020). Design, synthesis, and functional evaluation of a novel series of phosphonate-functionalized 1,2,3-triazoles as positive allosteric modulators of α7 nicotinic acetylcholine receptors. *ACS Chemical Neuroscience*, 11(17), 2688–2704. DOI: 10.1021/acschemneuro.0c00348 PMID: 32786318

Nyoni, N. T. P., Ncube, N. B., Kubheka, M. X., Mkhwanazi, N. P., Senzani, S., Singh, T., & Tukulula, M. (2023). Synthesis, characterization, in vitro antimycobacterial and cytotoxicity evaluation, DFT calculations, molecular docking and ADME studies of new isomeric benzimidazole-1,2,3-triazole-quinoline hybrid mixtures. *Bioorganic Chemistry*, 141, 106904. DOI: 10.1016/j.bioorg.2023.106904 PMID: 37832224

Oggu, S., Akshinthala, P., Katari, N. K., Nagarapu, L. K., Malempati, S., Gundla, R., & Jonnalagadda, S. B. (2023). Design, synthesis, anticancer evaluation and molecular docking studies of 1,2,3-triazole incorporated 1,3,4-oxadiazole-Triazine derivatives. *Heliyon*, 9(5), e15935. DOI: 10.1016/j.heliyon.2023.e15935 PMID: 37206039

Osipyan, A., Bulai, R. G., Wu, Z., de Witte, J., van der Velde, J. J. H., Kader, M., van der Wouden, P. E., Poelarends, G. J., & Dekker, F. J. (2024). The synthesis of 1,2,3-triazoles as binders of D-dopachrome tautomerase (D-DT) for the development of dual-targeting inhibitors. *European Journal of Medicinal Chemistry*, 276, 116665. DOI: 10.1016/j.ejmech.2024.116665 PMID: 39013358

Pagliai, F., Pirali, T., Grosso, E. D., Brisco, R. D., Tron, G. C., Sorba, G., & Genazzani, A. A. (2006). Rapid synthesis of triazole-modified resveratrol analogues via click chemistry. *Journal of Medicinal Chemistry*, 49, 467. PMID: 16420033

Palmer, M. H., Findlay, R. H., & Gaskell, A. J. (1974). Electronic charge distribution and moments of five- and six-membered heterocycles. *Journal of the Chemical Society, Perkin Transactions 2: Physical Organic Chemistry*, 4(4), 420. DOI: 10.1039/p29740000420

Pokhodylo, N., Manko, N., Finiuk, N., Klyuchivska, O., Matiychuk, V., Obushak, M., & Stoika, R. (2021). Primary discovery of 1-aryl-5-substituted-1*H*-1,2,3-triazole-4-carboxamides as promising antimicrobial agents. *Journal of Molecular Structure*, 1246, 131146. DOI: 10.1016/j.molstruc.2021.131146

Purcell, W. P., & Singer, J. A. (1967). Electronic and molecular structure of selected unsubstituted and dimethyl amides from measurements of electric moments and nuclear magnetic resonance. *Journal of Physical Chemistry*, 71, 4316.

Qi, Z., Xie, P., Wang, Z., Zhou, H., Tao, R., Popov, S. A., Yang, G., Shults, E. E., & Wang, C. (2024). Synthesis of novel ursolic acid-gallate hybrids via 1,2,3-triazole linkage and its anti-oxidant and anti-inflammatory activity study. *Arabian Journal of Chemistry*, 17(5), 105762. DOI: 10.1016/j.arabjc.2024.105762

Rahman, M. T., Decker, A. M., Laudermilk, L., Maitra, R., Ma, W., Hamida, S. B., Darcq, E., Kieffer, B. L., & Jin, C. (2021). Evaluation of amide bioisosteres leading to 1,2,3-triazole containing compounds as GPR88 agonists: Design, synthesis, and structure-activity relationship studies. *Journal of Medicinal Chemistry*, 64(16), 12397–12413. DOI: 10.1021/acs.jmedchem.1c01075 PMID: 34387471

Revuelto, A., de Lucio, H., Soriano, J. C. G., Murcia, P. A. S., Gago, F., Ruiz, A. J., Camarasa, M. J., & Velazquez, S. (2021). Efficient dimerization disruption of leishmania infantum trypanothione reductase by triazole-phenyl-thiazoles. *Journal of Medicinal Chemistry*, 64(9), 6137–6160. DOI: 10.1021/acs.jmedchem.1c00206 PMID: 33945281

Rostovtsev, V. V., Green, L. G., Fokin, V. V., & Sharpless, K. B. (2002). A stepwise huisgen cycloaddition process: Copper(I)-catalyzed regioselective "ligation" of azides and terminal alkynes. *Angewandte Chemie International Edition*, 41(14), 2596–2599. DOI: 10.1002/1521-3773(20020715)41:14<2596::AID-ANIE2596>3.0.CO;2-4 PMID: 12203546

Saber, S. O. W., Al-Qawasmeh, R. A., Abu-Qatouseh, L., Shtaiwi, A., Khanfar, M. A., & Al-Soud, Y. A. (2023). Novel hybrid motifs of 4-nitroimidazole-piperazinyl tagged 1,2,3-triazoles: Synthesis, crystal structure, anticancer evaluations, and molecular docking study. *Heliyon*, 9(9), e19327. DOI: 10.1016/j.heliyon.2023.e19327 PMID: 37681149

Santos, C. C., Batista, R. R. S., Barros, C. S., Azevedo, M. F. M. F., Ronconi, C. M., Buarque, C. D., & Paixao, I. C. N. P. (2024). *In vitro* study of the inhibitory potential of hydroxy-1,2,3-triazoles on the replication of ZIKA and chikungunya arboviruses. *Results in Chemistry*, 8, 101589. DOI: 10.1016/j.rechem.2024.101589

Sebest, F., Haselgrove, S., White, A. J. P., & Diez-Gonzales, S. (2020). Metal-free 1,2,3-triazole synthesis in deep eutectic solvents. *Synlett*, 31(6), 605–609. DOI: 10.1055/s-0039-1690736

Shaikh, M. H., Subhedar, D. D., Akolkar, S. V., Nagargoje, A. A., Asrondkar, A., Khedkar, V. M., & Shingate, B. B. (2023). New 1,2,3-Triazole-Tethered Thiazolidinedione Derivatives: Synthesis, Bioevaluation and Molecular Docking Study. *Polycyclic Aromatic Compounds*, 43, 3353–3379.

Shaikh, M. H., Subhedar, D. D., Akolkar, S. V., Nagargoje, A. A., Khedkar, V. M., Sarkar, D., & Shingate, B. B. (2022). Tetrazoloquinoline-1, 2, 3-triazole derivatives as antimicrobial agents: Synthesis, biological evaluation and molecular docking study. *Polycyclic Aromatic Compounds*, 42, 1920–1941.

Shaikh, M. H., Subhedar, D. D., Khan, F. A. K., Sangshetti, J. N., & Shingate, B. B. (2016). 1,2,3-Triazole incorporated coumarin derivatives as potential antifungal and antioxidant agents. *Chinese Chemical Letters*, 27, 295–301.

Shaikh, M. H., Subhedar, D. D., Nawale, L., Sarkar, D., Khan, F. A., Sangshetti, J. N., & Shingate, B. B. (2019). Novel Benzylidenehydrazide-1,2,3-Triazole Conjugates as Antitubercular Agents: Synthesis and Molecular Docking. *Mini-Reviews in Medicinal Chemistry*, 19, 1178–1194. PMID: 30019644

Siddiqui, M. A., Nagargoje, A. A., Shaikh, M. H., Siddiqui, R. A., Pund, A. A., Khedkar, V. M., Asrondkar, A., Deshpande, P. P., & Shingate, B. B. (2023). Design, Synthesis and bioevaluation of highly functionalized 1,2,3-triazole-guanidine conjugates as anti-inflammatory and antioxidant agents. *Polycyclic Aromatic Compounds*, 43, 5567–5581.

Siddiqui, M. A., Shaikh, M. H., Nagargoje, A. A., Shaikh, T. T., Khedkar, V. M., Deshpande, P. P., & Shingate, B. B. (2022). [DBU][OAc]-mediated synthesis and anthelmintic activity of triazole–tetrazole conjugates. *Research on Chemical Intermediates*, 48(12), 5187–5208.

Siddiqui, M. M., Nagargoje, A. A., Akolkar, S. V., Sangshetti, J. N., Khedkar, V. M., Pisal, P. M., & Shingate, B. B. (2022). [HDBU][HSO$_4$]-catalyzed facile synthesis of new 1,2,3-triazole-tethered 2,3-dihydroquinazolin-4[1*H*]-one derivatives and their DPPH radical scavenging activity. *Research on Chemical Intermediates*, 48(3), 1199–1225. DOI: 10.1007/s11164-021-04639-9

Silalai, P., Jaipea, S., Tocharus, J., Athipornchai, A., Suksamrarn, A., & Saeeng, R. (2022). New 1,2,3-Triazole-genipin analogues and their anti-alzheimer's activity. *ACS Omega*, 7(28), 24302–24316. DOI: 10.1021/acsomega.2c01593 PMID: 35874205

Silva, T. B., Ji, K. N. K., Pauli, F. P., & Galvao, R. M. S. (2021). Synthesis and *in vitro* and *in silico* studies of 1*H*- and 2*H*-1,2,3-triazoles as antichagasic agents. *Bioorganic Chemistry*, 116, 105250. DOI: 10.1016/j.bioorg.2021.105250 PMID: 34469833

Somu, R. V., Boshoff, H., Qiao, C., Bennett, E. M., Barry, C. E., & Aldrich, C. C. J. (2005). Rationally designed nucleoside antibiotics that inhibit siderophore biosynthesis of Mycobacterium tuberculosis. *Journal of Medicinal Chemistry*, 49, 31. PMID: 16392788

Stefely, J. A., Palchaudhuri, R., Miller, P. A., Peterson, R. J., Moraski, G. C., Hergenrother, P. J., & Miller, M. J. (2010). *N*-((1-Benzyl-1*H*-1,2,3-triazol-4-yl)methyl) arylamide as a new scaffold that provides rapid access to antimicrotubule agents: Synthesis and evaluation of antiproliferative activity against select cancer cell lines. *Journal of Medicinal Chemistry*, 53(8), 3389–3395. DOI: 10.1021/jm1000979 PMID: 20334421

Sweeney, J. B., Rattray, M., Pugh, V., & Powell, L. A. (2018). Riluzole-triazole hybrids as novel chemical probes for neuroprotection in amyotrophic lateral sclerosis. *ACS Medicinal Chemistry Letters*, 9(6), 552–556. DOI: 10.1021/acsmedchemlett.8b00103 PMID: 29937981

Syam, S., Abdelwahab, S. I., Al-Mamary, M. A., & Mohan, S. (2012). Synthesis of chalcones with anticancer activities. *Molecules (Basel, Switzerland)*, 17(6), 6179–6195. DOI: 10.3390/molecules17066179 PMID: 22634834

Tornoe, C. W., Christensen, C., & Meldal, M. (2002). Peptidotriazoles on solid phase: [1,2,3]-triazoles by regiospecific copper(i)-catalyzed 1,3-dipolar cycloadditions of terminal alkynes to azides. *The Journal of Organic Chemistry*, 67(9), 3057–3064. DOI: 10.1021/jo011148j PMID: 11975567

Tron, G. C., Pirali, T., Billigton, R. A., Canonico, P. L., Sorba, G., & Genazzani, A. A. (2008). Click chemistry reactions in medicinal chemistry: Applications of the 1,3-dipolar cycloaddition between azides and alkynes. *Medicinal Research Reviews*, 28, 278. PMID: 17763363

Usachev, B. I. (2018). Chemistry of fluoroalkyl-substituted 1,2,3-triazoles. *Journal of Fluorine Chemistry*, 210, 6–45.

Vala, D. P., Miller, A. D., Atmasidha, A., Parmar, M. P., Patel, C. D., Upadhyay, D. B., Bhalodiya, S. S., Gonzalez-Bakker, A., Khan, A. N., Nogales, J., Padron, J. M., Banerjee, S., & Patel, H. M. (2024). Click-chemistry mediated synthesis of OTBN-1,2,3-Triazole derivatives exhibiting STK33 inhibition with diverse anti-cancer activities. *Bioorganic Chemistry*, 149, 107485. DOI: 10.1016/j.bioorg.2024.107485 PMID: 38824700

Wiley, R. H., Hussung, K. F., & Moffat, J. (1956). The preparation of 1,2,3-triazole. *The Journal of Organic Chemistry*, 21(2), 190–192. DOI: 10.1021/jo01108a010

Yang, J., Shibu, M. A., Kong, L., & Luo, J., BadrealamKhan, F., Huang, Y., ... & Lu, X. (2019). Design, synthesis, and structure–activity relationships of 1, 2, 3-triazole benzenesulfonamides as new selective leucine-zipper and sterile-α motif Kinase (ZAK) Inhibitors. *Journal of Medicinal Chemistry*, 63(5), 2114–2130.

Yang, J. J., Yu, W. W., Hu, L. L., Liu, W. J., Lin, X. H., Wang, W., Zhang, Q., Wang, P. L., Tang, S. W., Wang, X., Liu, M., Lu, W., & Zhang, H. K. (2020). Discovery and characterization of 1*H*-1,2,3-triazole derivatives as novel prostanoid ep4 receptor antagonists for cancer immunotherapy. *Journal of Medicinal Chemistry*, 63(2), 569–590. DOI: 10.1021/acs.jmedchem.9b01269 PMID: 31855426

Yildirim, S. C., Nathanael, J. G., Frindte, K., Leal, O. A., Walker, R. M., Roessner, U., Knief, C., Bruggemann, N., & Wille, U. (2024). 4-Methyl-1-(prop-2-yn-1-yl)-1*H*-1,2,3-triazole (MPT): A novel, highly efficient nitrification inhibitor for agricultural applications. *ACS Agricultural Science & Technology*, 4(2), 255–265. DOI: 10.1021/acsagscitech.3c00506

Znati, M., Horchani, M., Latapie, L., Jannet, H. B., & Bouajila, J. (2021). New 1,2,3-triazole linked flavonoid conjugates: Microwave-assisted synthesis, cytotoxic activity and molecular docking studies. *Journal of Molecular Structure*, 1246, 131216. DOI: 10.1016/j.molstruc.2021.131216

Zuo, Z., Liu, X., Qian, X., Zeng, T., Sang, N., Liu, H., Zhou, Y., Tao, L., Zhou, X., Su, N., Yu, Y., Chen, Q., Luo, Y., & Zhao, Y. L. (2020). Bifunctional Naphtho[2,3-*d*][1,2,3]triazole-4,9-dione compounds exhibit antitumor effects *in vitro* and *in vivo* by inhibiting dihydroorotate dehydrogenase and inducing reactive oxygen species production. *Journal of Medicinal Chemistry*, 63(14), 7633–7652. DOI: 10.1021/acs.jmedchem.0c00512 PMID: 32496056

Chapter 6
Indazoles Chemistry and Biological Activities:
Synthesis, Properties, and Biological Activities of Indazole

Navnath Tulshiram Hatvate
https://orcid.org/0000-0002-3388-3247
Institute of Chemical Technology, Mumbai, India

Khushbu Bagul
https://orcid.org/0009-0009-0139-8165
Institute of Chemical Technology, Marathwada, India

Nandini Anilsingh Gour
Institute of Chemical Technology, Marathwada, India

Kalyani S. Sonawane
https://orcid.org/0009-0002-7170-0234
Institute of Chemical Technology, Marathwada, India

ABSTRACT

In organic and medicinal chemistry research, indazole is an essential nitrogen-containing heterocyclic unit that is also a helpful precursor molecule for the synthesis of several kinds of heterocycles. The diverse tautomeric forms and unique chemical properties make it a versatile scaffold in medicinal chemistry. Indazole's relevance in the pharmaceutical industry is underscored by its presence in currently marketed drugs and investigational compounds, highlighting its therapeutic potential. In addition, the present ring structure has already been explored for diverse biological activity, including anti-microbial, anti-viral, anti-protozoal, anti-cancer, anti-inflammatory, analgesic, antipyretic, anti-oxidant, anti-convulsant, anti-depressant, anti-emetic,

DOI: 10.4018/979-8-3693-7267-8.ch006

anti-diabetic, neuroprotective, antihypertensive, and anti-arrhythmic properties. This chapter comprehensively reviews indazole's synthesis, properties, and biological applications, along with an update on recent patents and ongoing clinical trials.

1. INTRODUCTION

The world's best-selling drugs are nitrogen-containing heterocyclic moieties due to their adaptability to various biological scaffolds as well as medicinal products. Various nitrogen-containing heterocycles like indole, indazole, quinolone, quinazoline, and carbazole are essential components for numerous synthetic and semi-synthetic pharmaceuticals. Indazoles are aromatic heterocyclic organic compounds with the chemical formula $C_7H_6N_2$ (Tan et al., 2022). This moiety has $10\text{-}\pi$ electrons in its bicyclic aromatic framework. Like the pyrazole molecule, it resembles pyridine and pyrrole, and its reactivity reflects its dual behaviour(Teixeira et al., 2006). This moiety is a member of the azoles family, first introduced by Emil Fisher in 1889. They consist of a bicyclic structure formed by fusing a benzene ring and a pyrazole ring. It is also called isoindazole and benzopyrazole (Kumar et al., 2022). Indazole is a naturally occurring alkaloid, so only a few naturally isolated indazole alkaloids from Nigella have been reported. Nigellicine and Nigellidine are the main indazole phytonutrients present in Nigella sativa seeds (Niu et al., 2020). The diverse bioactivities exhibited by indazole cores, attributed to the interaction of their two successive nitrogen atoms with enzyme sites, make them valuable for drug development. These scaffolds are desirable target compounds in chemical synthesis because of their intriguing biological characteristics. Numerous indazole derivatives with biological and medicinal qualities are possible due to the remarkable selectivity with which the indazole ring can be functionalized at various positions. Compounds with indazole rings have been commonly reported to display anti-microbial (J. R. Saketi et al., 2023), anti-viral (Yin et al., 2021), anti-protozoal (Rodríguez-Villar et al., 2021), anti-cancer (Cao et al., 2022), anti-inflammatory (Cheekavolu, 2016), anti-oxidant (Sapnakumari et al., 2014), anti-convulsant (Matsumura et al., 2013), anti-depressant (Degnan et al., 2016a), anti-emetic (Basak et al., 2020), anti-diabetic (Bushra et al., 2021), neuroprotective (Jismy et al., 2021), antihypertensive (Sączewski et al., 2016), anti-arrhythmic (Uppulapu et al., 2022), analgesic (Abbady et al., 2014) and antipyretic activity (Badawey et al., 1998).

1.1. Structure and Isomers of Indazole

The IUPAC name for indazoles is 1H-indazoles, and the numbering of the structure begins in the same manner as the numbering of the pyrazole ring, with the number 1 specified to the nitrogen, which includes the hydrogen atom and the number 2 to the remaining nitrogen atom (Figure 1). The aromatic character of indazole gives it stability and affects how it behaves in chemical interactions. Its distinct structural characteristics make it an intriguing topic for research in contemporary organic chemistry (Schmidt et al., 2012).

Figure 1. Tautomer's of indazole

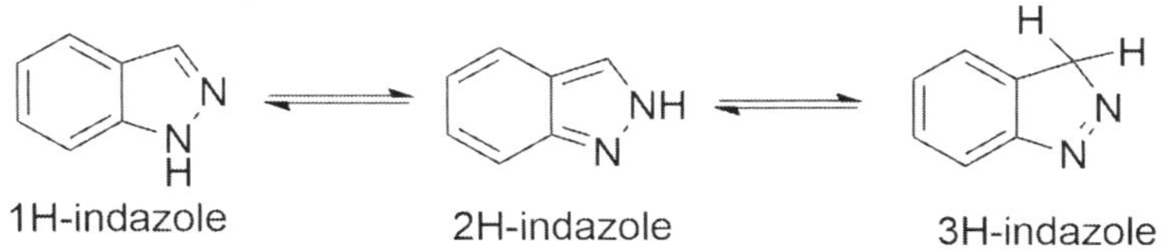

1H-indazole   2H-indazole   3H-indazole

Indazole exhibits three isomer forms, namely 1H-, 2H- and 3H-indazoles (Figure 2). The 1H-tautomer (benzenoid form) is more stable than the 2H-form (quinonoid form) and 3H-form. Three tautomeric forms of indazoles are generated *via* proto-tropic annular tautomerism (Dong et al., 2018).

Figure 2. numbering of Indazoles

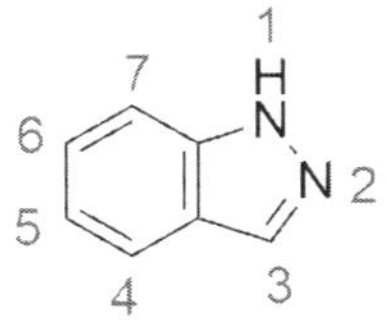

2. SYNTHESIS OF INDAZOLE

2.1. Conventional Methods for Synthesis of Indazoles

2.1.1. From Nitroarene

The novel approach for synthesizing 1*H*-indazoles **3** from nitrobenzene **1** and *N*-tosylhydrazone **2** by using cesium carbonate (Cs_2CO_3), DMF, with heating while the reaction occurred without heating in the presence of sodium hydride (NaH), and DMF, with the help of transition metals (Figure 3). Substrates employed included *m*-trifluoromethyl, *m*-ethoxycarbonyl, and *m*-cyano substituted nitrobenzene **1,** with *p/o*-substituted nitrobenzene **1,** which requires a strong base sodium hydride. Reactions with nitrobenzene **1** proceeded effectively; however, the ones involving powerful electron-withdrawing substituents decreased the yields of its desired derivatives. The heteroaromatic aldehydes had negligible effects on the yield (Zhenxing Liu et al., 2014).

Figure 3. Scheme 1: Synthesis of Indazole from nitroarene

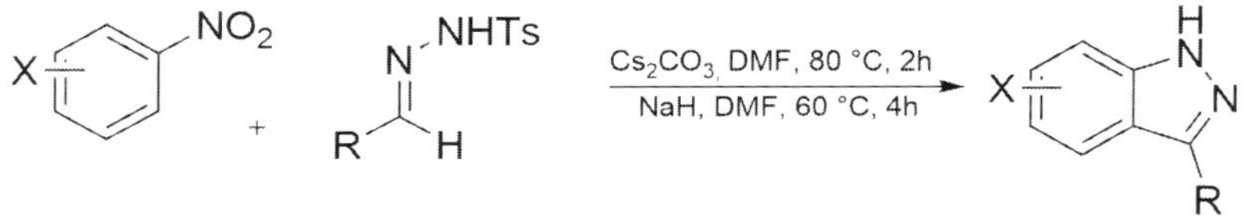

2.1.2. From Diazo Compounds

The reaction between aryl diazoesters **4** and diazonium salt **5** proceeds *via* cyclization under reaction conditions of 0.1 M DMF under heating, yielding 1*H*-indazole **6** (Figure 4). Aryl diazoesters **4** with electron-withdrawing substituents generally afforded favorable and efficient yields, whereas those with electron-donating substituents did not yield the desired products. *p*-substituted aryl diazoesters **4** reacted smoothly, while *m*-substituted aryl diazoesters **4** yielded two regioisomers. (Li et al., 2020).

Figure 4. Scheme 2: Synthesis of Indazole from Diazo Compounds

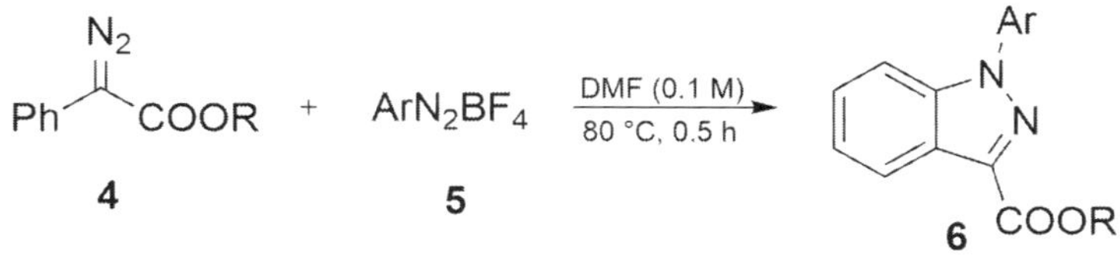

2.1.3. From *o*-Halo-(het)Aryldehydes or -Phenones

In this cross-coupling reaction, *o*-halo aryldehydes **7** react with imine **8** under conditions involving benzophenone hydrazone, facilitating *N*-arylation and cyclization to produce various substituted $1H$ indazoles **9** (Figure 5). It can be applied to substituted benzaldehydes, phenones, and heterocyclic analogues, yielding novel indazole derivatives. To selectively obtain *N*-1-substituted indazoles, an additional alkylation step can be incorporated into this procedure (Dubost et al., 2014).

Figure 5. Scheme 3: Synthesis of Indazole from o-halo(het)aryldehydes or -phenones

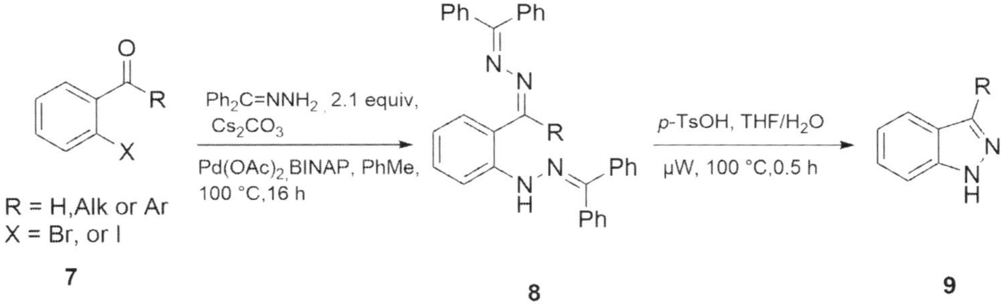

2.1.4. From 2-Halobenzonitriles

The researchers describe a regioselective approach to producing 3-unsubstituted 1-alkyl-$1H$-indazoles **12** from *o*-halobenzonitrile **10** under reaction conditions. The two-step reaction used 1-methyl-3-amino-$1H$-indazoles **11** as intermediates, after which comes reductive deamination. When *o*-halobenzonitrile **10** with electron-donating or -withdrawing groups, R are combined by methylhydrazine (MeNH-NH$_2$) within ethanol (EtOH), they successfully yield 3-amino-1-methyl-1H-indazoles. Concise deamination of 1-methyl-3-amino-1H-indazoles **11** with *tert*-butyl nitrite (*t*-BuONO) in chloroform (CHCl$_3$), DMF, or THF resulted in a large quantity of 1-alkyl-1H-indazoles **12** (Figure 6) (H.-J. Liu et al., 2013).

Figure 6. Scheme 4: Synthesis of Indazole from 2-halobenzonitriles

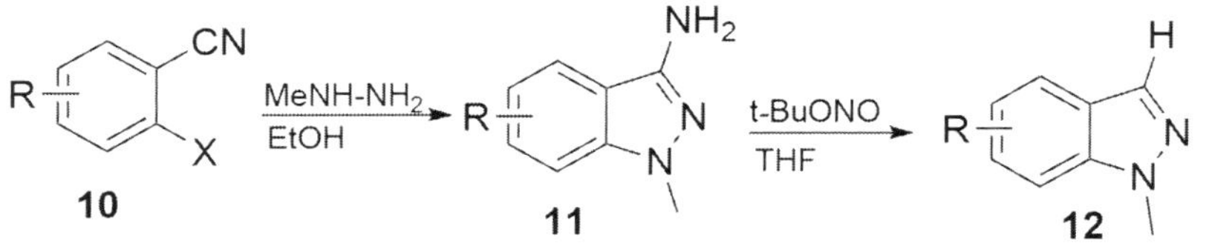

2.1.5. From Pyrazole

Miura, Satoh et al. demonstrated that benzannulation of indoles produces carbazoles using a Pd-H benzannulation approach that requires directing groups. They showed that the benzannulation of 4-bromopyrazoles **13** with alkynes **14** results in functionalized indazoles **15**. Compound 4-bromopyrazoles 13 was modified to improve regioselectivity to promote oxidative addition at the C-4 position. The catalytic reaction involving Pd(OAc)$_2$, P(*t*Bu)$_3$H·BF$_4$, K$_2$CO$_3$, PivOH and toluene produced high yields of the desired indazoles **15** illustrated in Figure 7 (Kim et al., 2017).

Figure 7. Scheme 5: Synthesis of Indazole from Pyrazole

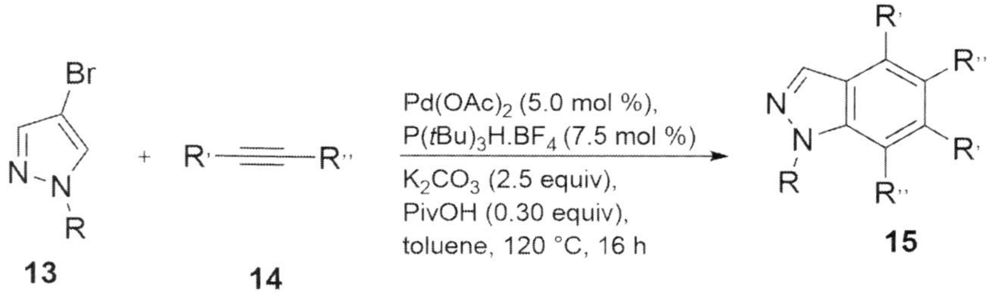

R = Me,CH$_2$Ph,CH$_2$CH$_2$Ph
R' = Me, Et
R'' = Ph

2.1.6. From Substituted Formylboronic Acids

The synthesis strategy involved using the Chan-Evans-Lam reaction to form C-N bonds. To extend this approach for the synthesis of indazoles **19** and **20**, the copper-catalyzed addition of phenylboronic acids to azodicarboxylates **17** was investigated, following the method reported by Uemura and Chatani. By employing 2-formylphenylboronic acids **16** as substrates, the addition to the N=N bond in azodicarboxylates **17** would yield *N*-arylhydrazine intermediates **18**, which could be further transformed into indazoles **19** and **20** (Figure 8).

Figure 8. Scheme 6: Synthesis of Indazole from substituted formylboronic acids

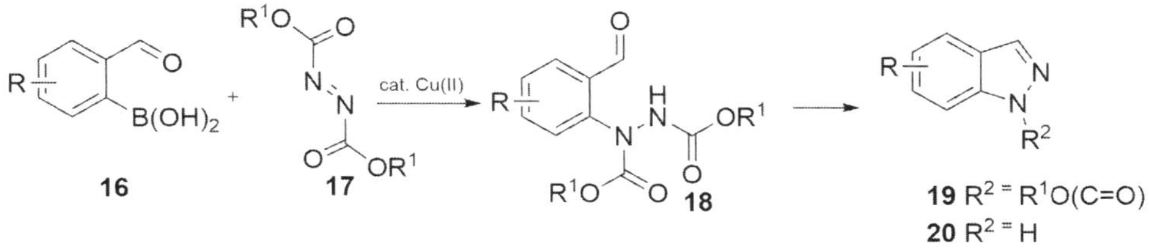

2.1.7. From Aldehydes or Ketones

Hari, Y. et al. developed a novel synthetic approach using trimethylsilyldiazomethane (Me_3SiCHN_2) and discovered that the corresponding magnesium bromide salt [$Me_3SiC(MgBr)N_2$] successfully reacted with simple carbonyl compounds to produce 2-diazo-(2-trimethylsilyl)ethanol. They used $Me_3SiC(MgBr)N_2$ to activate a broader range of electrophilic carbonyl compounds. In Figure 9, aldehydes or ketones **21** were initially transformed into 2-diazo-(2-trimethylsilyl)ethyl alcohol derivatives **23** using the Grignard reagent, diazo(trimethylsilyl)methyl magnesium bromide **22**. Subsequently, substituted 2-(trimethylsilyl)phenyl trifluoromethanesulfonate (OTf) was reacted with potassium fluoride (KF) and Crown-6 to generate 1*H*-indazoles **24** (Hari et al., 2009).

Figure 9. Scheme 7: Synthesis of Indazole from aldehydes or ketones

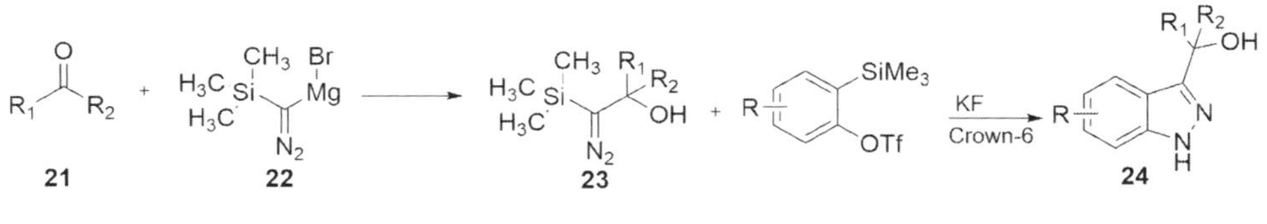

2.1.8. From Arynes

In Larock's study of the 1,3-cycloaddition reaction between arynes and dicarbonyl-containing diazo compounds, 3H-indazoles were initially generated as intermediates. These intermediates underwent acyl migration to produce 1-acyl-1H-indazoles. The practical synthesis of 3-alkyl/aryl-1H-indazoles **27** was achieved through a 1,3-dipolar cycloaddition reaction between α-substituted α-diazomethyl-phosphonates **25** and arynes **26** under simple conditions: CsF (4.0 equiv) and CH_3CN at 40 °C (Figure 10) (Chen et al., 2018).

Figure 10. Scheme 8: Synthesis of Indazole from arynes

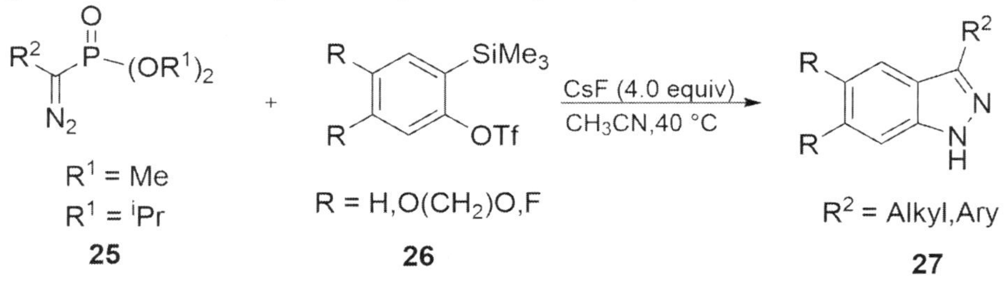

2.1.9. From Arylhydrazones

A novel synthesis method for substituted 1H-indazoles from arylhydrazones **28** has been developed, utilizing $FeBr_3$ and O_2 to facilitate C-H activation and C-N bond formation. The reactions were conducted under mild conditions: phenylhydrazone (0.25 mmol), $FeBr_3$ (0.025 mmol), and p-xylene (2.0 mL) were used in a dry oxygen environment at 140 °C. This process produced the corresponding 1,3-diaryl-substituted indazoles **29** in moderate to good yields (Figure 11) (Zhang et al., 2013).

Figure 11. Scheme 9: Synthesis of Indazole from arylhydrazones

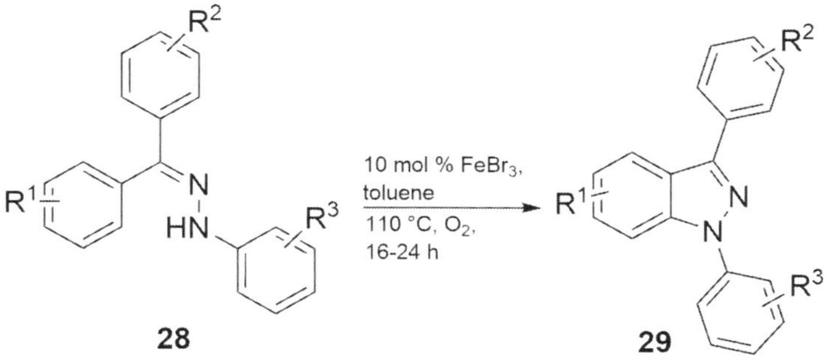

2.2. Green Methods for the Synthesis of Indazole

Green indazole synthesis techniques aim to reduce waste and energy consumption by using eco-friendly reagents and procedures. This involves combining sustainable reaction conditions with bio-based or renewable starting materials, such as waste products or plant-derived chemicals. Greener indazole synthesis can also be achieved through catalysts or processes that produce fewer byproducts or require less hazardous solvents (Karmakar et al., 2023).

2.2.1. Microwave-Assisted Synthesis

Non-traditional approaches, such as grinding, water-based biphasic reactions, and microwave chemistry, have prepared Schiff bases **32** from amines **31** and 2-nitrobenzaldehydes **30**. Under microwave irradiation, these nitro compounds reacted with triethyl phosphite, forming nitrenes and indazoles **33** through insertion reactions. This process uses less energy and is more environmentally friendly than the Cadogan reaction for creating nitrogen heterocycles from nitrenes (Figure 12) (Kaur et al., 2018; Varughese et al., 2006).

Figure 12. Scheme 10: Synthesis Indazole by microwave using nitrenes

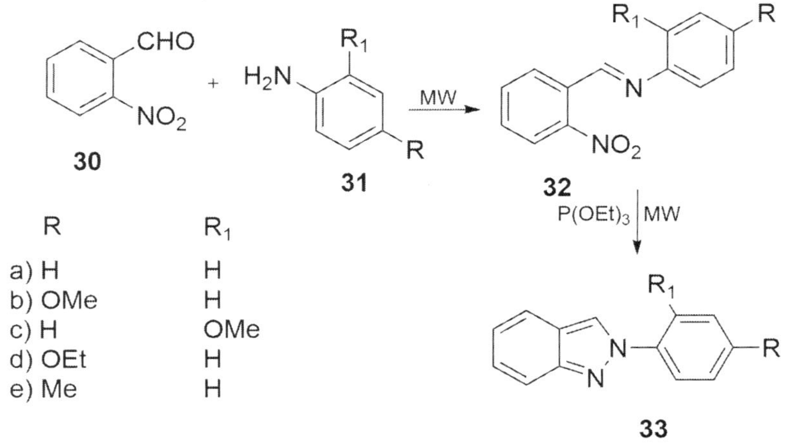

R	R$_1$
a) H	H
b) OMe	H
c) H	OMe
d) OEt	H
e) Me	H

2.2.2. One-Pot Synthesis

Dar B. A. *et al.* developed a straightforward and effective one-pot, three-component synthesis of 2H-indazoles **37** utilizing sodium azide **36**, 2-bromobenzaldehyde **34**, and 2-primary amines/aniline **35**. This synthesis is facilitated by a heterogeneous Cu(II)-Clay catalyst, which promotes the formation of C–N and N–N bonds during the domino condensation process (Figure 13) (Dar et al., 2018).

Figure 13. Scheme 11: Synthesis of indazoles using Cu(II)-Clay catalyst

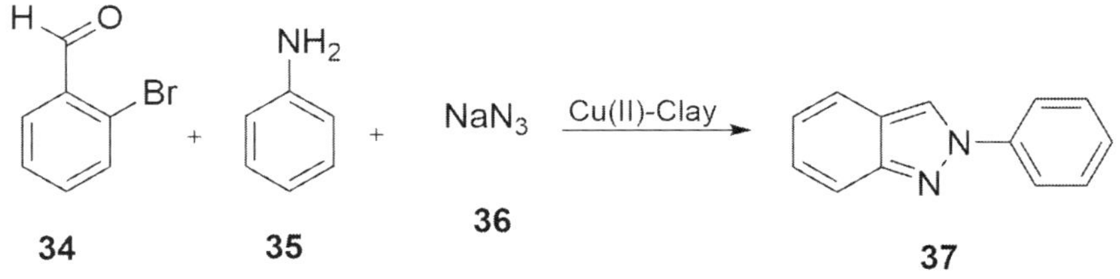

Reddy G, T *et al.* citric acid used a one-pot synthesis technique as a biodegradable catalyst that produces *N*-alkylated indazoles **39** with good to outstanding isolated yields (~78-96%) from easily accessible raw materials such as ethyl chloroacetate, 2-methylanilines, and citric acid, NaNO$_2$ from substituted aniline **38** (Figure 14) (Reddy G et al., 2018).

Figure 14. Scheme 12: Synthesis of N-alkylated indazoles by using biodegradable citric acid catalyst

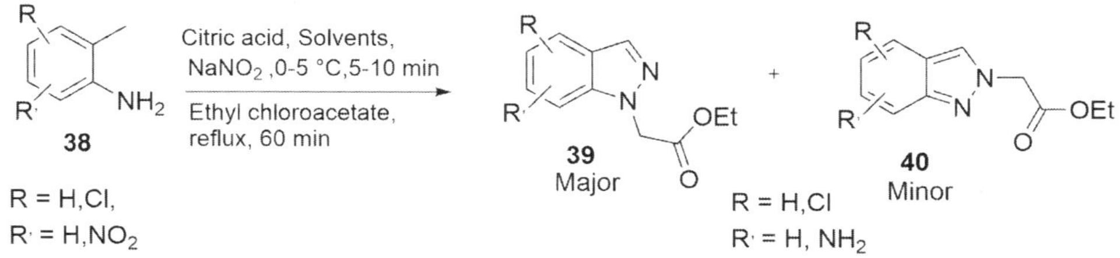

2.2.3. Three-Component Synthesis by Using Green Solvent

The work describes a three-component one-pot reaction between sodium azide **43**, primary amines **42**, and 2-halobenzaldehydes **41** to synthesize 2*H*-indazole 44 derivatives (Figure 15) efficiently. Under ligand-free circumstances, copper(I) oxide nanoparticles (Cu_2O-NP) in polyethylene glycol (PEG 300) accelerate the chemical reaction. (Sharghi et al., 2014).

Figure 15. Scheme 13: Synthesis of indazoles by using PEG as a green solvent

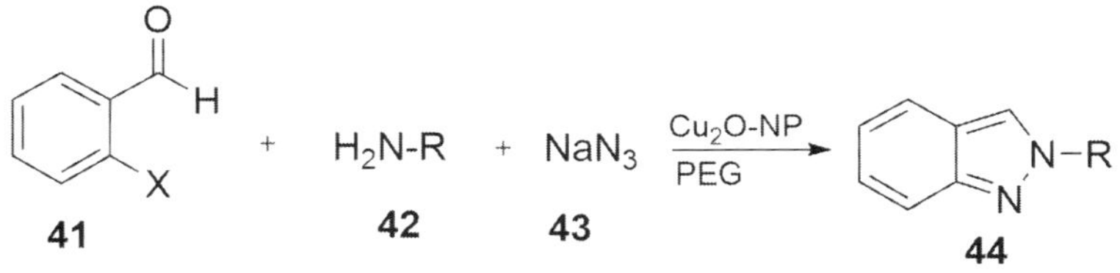

3. CURRENT MARKETED DRUGS/INVESTIGATIONAL APPLICATIONS OF INDAZOLE

Numerous research efforts have led to the development of compounds incorporating the indazole core, either as standalone drugs or as components in hybrid molecules, showcasing its potential in addressing a spectrum of diseases.

Anti-cancer drugs such as **Lonidamine (45),** a hexokinase inhibitor, show potential in cancer treatment by disrupting glycolysis in cancer cells, thereby reducing tumor growth and metastasis. It may enhance the effectiveness of other treatments, such as chemotherapy and radiation therapy, while exhibiting low toxicity in normal cells (Pal et al., 2022; Tang et al., 2016). **Axitinib (46)** is one of the tyrosine kinase inhibitors (TKIs) that has been utilized for the treatment of renal cell carcinoma (RCC) due to its anti-angiogenic qualities. Nevertheless, there is no proof that axitinib has a direct cytotoxic anti-tumor effect in RCC (Morelli et al., 2015; Tang et al., 2016). Currently, at least forty-three different types of indazole-based medicines are being employed in clinical trials or applications.

Figure 16. currently marketed drugs containing indazole

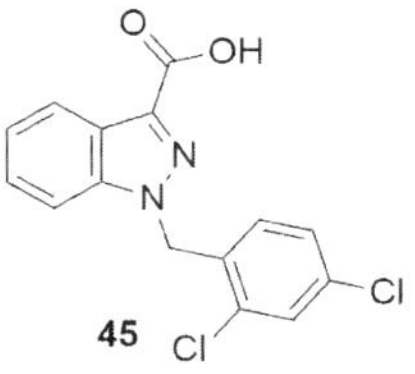

45

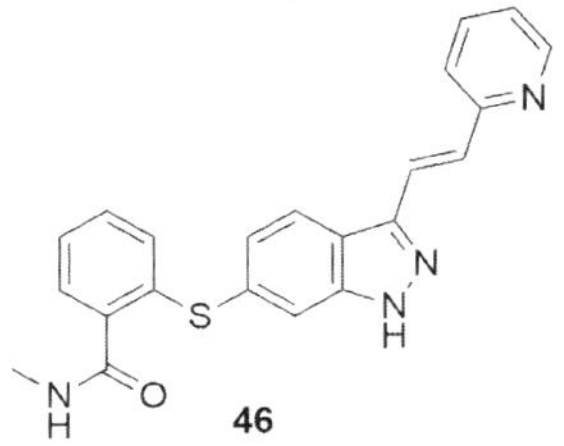

46

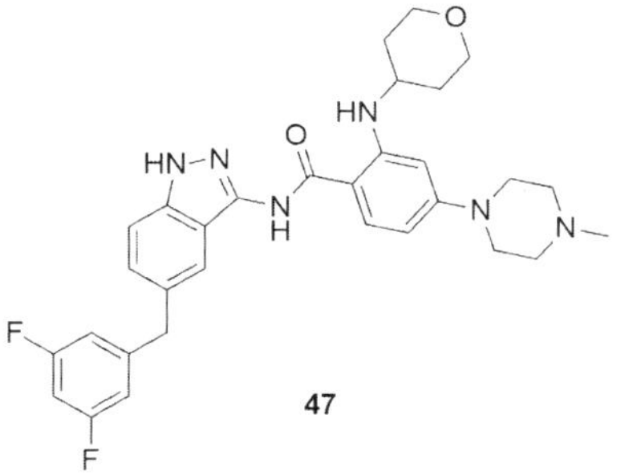

47

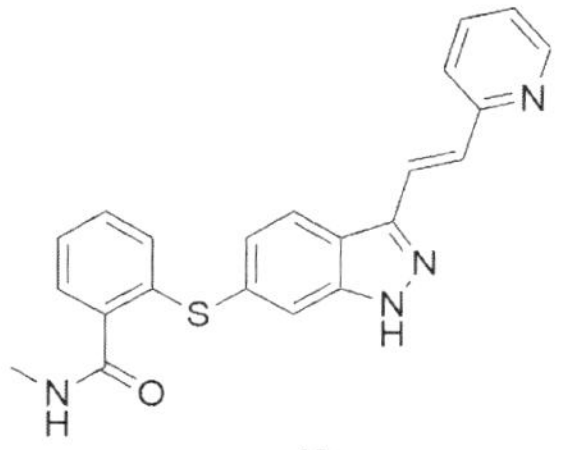

48

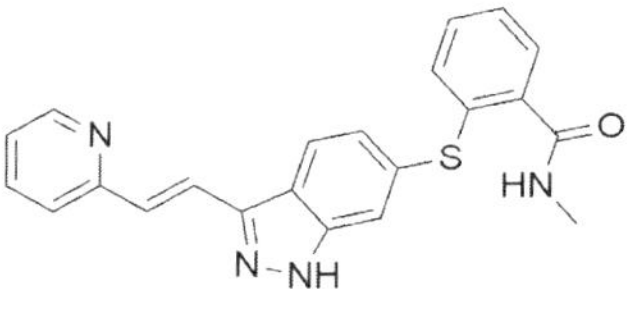

49

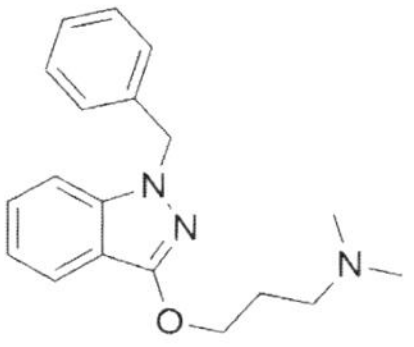

50

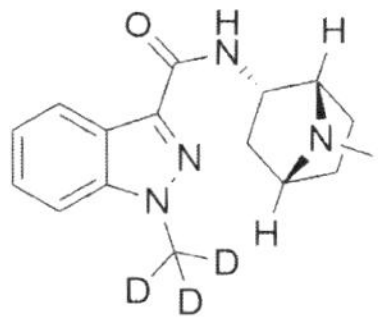

51

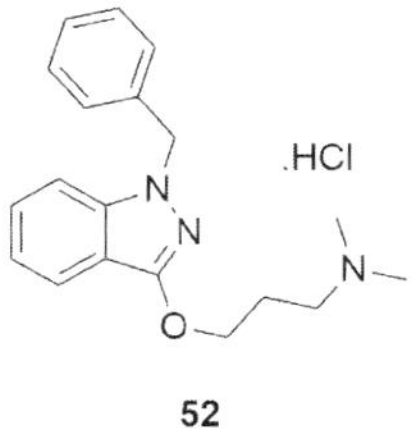

52

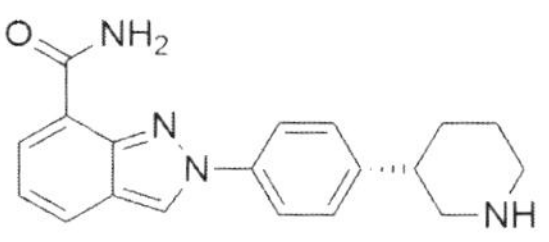

53

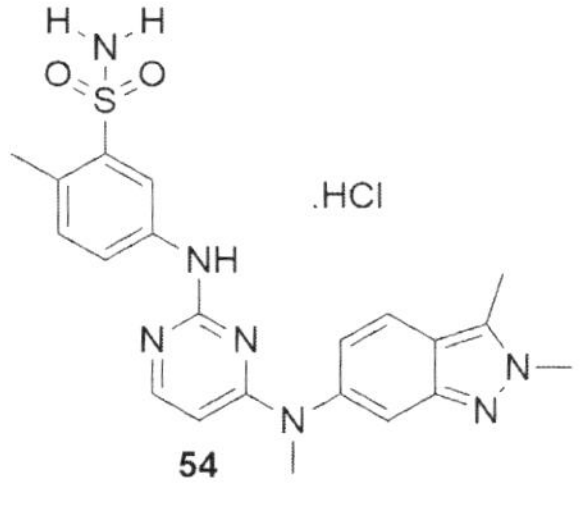

54

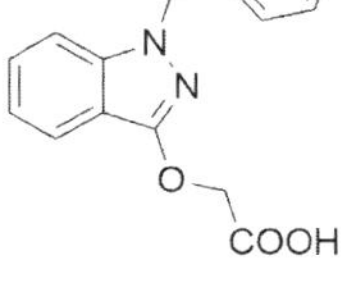

55

For instance, **Entrectinib (47)** is an inhibitor of anaplastic lymphoma kinase (ALK) in the typical compound depicted in Figure 16 (Puri et al., 2023). **Asitinib (48)** is an intravascular epidermal growth factor receptor (VEGFR) inhibitor, and **Linifanib (49)** is a multi-targeted ATP-competitive tyrosine kinase inhibitor. **Benzydamine (50)**, a non-steroidal anti-inflammatory medication, and the serotonin receptor antagonist **Granisetron (51)** are utilized as well for the relief of chemotherapy-induced vomiting in cancer patients (C. Wang et al., 2023). **Benzydamine Hydrochloride (52)**, a selective inhibitor of cyclooxygenase, has been used as a nonsteroidal anti-inflammatory drug (NSAID) since 1966 for pain relief (Cao et al., 2021). Recurrent epithelial ovarian, fallopian tube, primary peritoneal, breast, and prostate cancers are all treated with the anti-cancer medication **Niraparib (53)**.

The FDA has approved **Pazopanib hydrochloride (54)**, a second-generation tyrosine kinase inhibitor, for treating soft tissue sarcoma and advanced renal cell carcinoma. The compound **54** is available as a tablet (Votrient®) from Novartis (Jadhav et al., 2024). **Bendazac (55)** is a commercially marketed anti-inflammatory medication with a 1*H*-indazole scaffold. It is an oxyacetic acid with choleretic, anti-inflammatory, anti-necrotic, and anti-lipidemic activities. Bendazac's primary function is to prevent protein denaturation, helping to control and delay the development of ocular cataracts. However, due to potential hepatotoxicity, it has been removed or discontinued in several regions. Bendazac is available for topical, oral, and eye drop use. It has been shown to lower cholesterol, enhance the blood-retinal barrier, and prevent phytohemagglutinin-induced lymphocyte conversion (Kumar et al., 2024).

4. BIOLOGICAL ACTIVITIES OF INDAZOLE

4.1. Anti-Microbial

Ananda Kumar Dunga *et al.* effectively developed and synthesized a series of dual 1,2,3-triazole and 1,3,4-oxadiazole heterocyclic compounds connected to 3-bromo-1*H*-indazole as prospective antibacterial drugs. Most of the compounds exhibited notable inhibitory activity, the compound **56** demonstrating exceptional antibacterial activity against *S. pneumoniae*, a Gram-positive bacterium (Dunga et al., 2022).

Figure 17. indazoles derivitives as anti-microbial agents

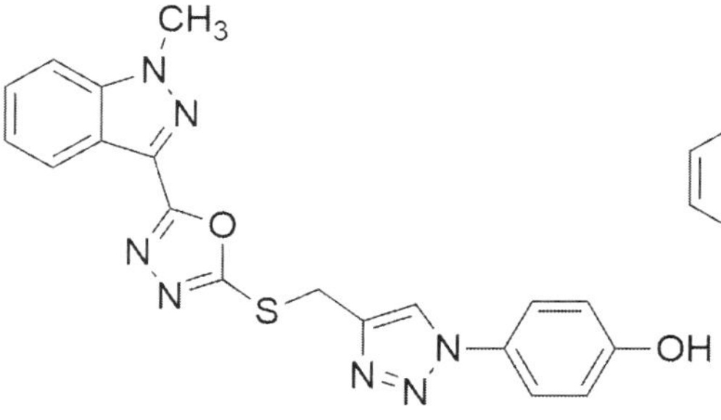

56

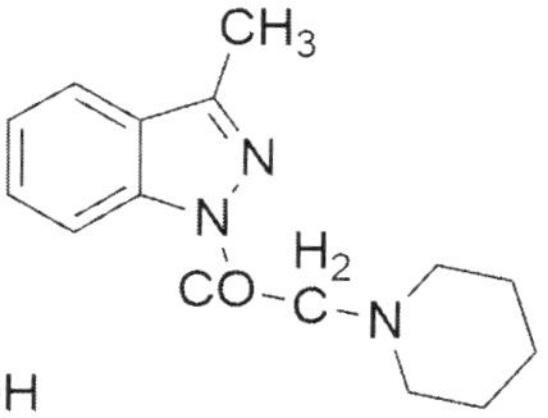

57

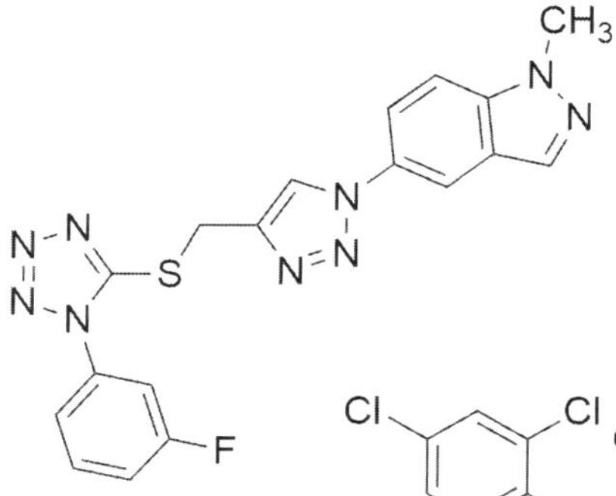

58

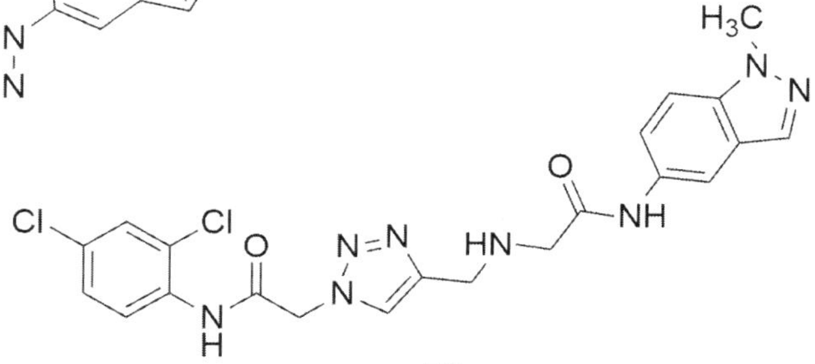

59

Farrukh Shaikh *et al.* produced novel heterocyclic compounds, including derivatives of 3-methyl-1*H*-indazole **57** for antibacterial action. Their antibacterial activity on *Bacillus subtilis* and *Escherichia coli*. Compound **57** demonstrated the highest antibacterial effectiveness compared to ciprofloxacin (Shaikh et al., 2024).

Venkanna Gujja *et al.* developed novel 1,2,3-triazolyl tetrazoles containing indazoles as prospective antibacterial agents. These pyrrolidine-containing 1,2,3-triazole and tetrazole derivatives exhibit significant biological activity and have applications in medicinal chemistry. Compounds featuring 1,2,3-triazolyl tetrazole scaffolds **58** demonstrated higher antibacterial activity against Gram-positive and Gram-negative microbes, with MICs ranging from 5.03 to 18.02 µM. Among the synthesized hybrids, compound **58** exhibited the most potent antibacterial activity, with MIC values of 5.03 µM, 10.01 µM, 14.02 µM, 12.03 µM, and 13.01 µM for the bacterial strains *S. aureus*, *B. subtilis*, *M. luteus*, *E. coli*, and *P. aeruginosa*, respectively. This indicates a higher activity level than Gemifloxacin, which had MIC values of 6.02 µM, 9.01 µM, and 7.01 µM, respectively (Gujja et al., 2023).

Siva Kumar Gandham *et al.* investigated the synthesis of novel derivatives of 1,2,3-triazolyl–indazoles, employing docking techniques and biological evaluation to identify active molecules. Initial antibacterial investigations revealed that derivatives **59** exhibited the most potent activity against *S. epidermidis* and *P. aeruginosa* microorganisms. Among all the derivatives studied, compound **59** demonstrated superior antibacterial activity (Gandham et al., 2024).

4.2 Anti-Viral

Anna Egorova *et al.* researched pyrrolo[2,3-*e*]indazoles, a novel class of small molecule inhibitors that target the neuraminidase of the *influenza A virus*. Research on structure-activity reveals that the R_1-phenyl ring is critical in suppressing *H3N2* and *H1N1* subtypes. They found that compound **60** have shown promising antiviral activity. These derivatives are promising for further development since they show dual activity against bacterial and viral neuraminidases, and some even suppress bacterial growth (Egorova et al., 2023).

Figure 18. Indazole derivatives as anti-viral agents

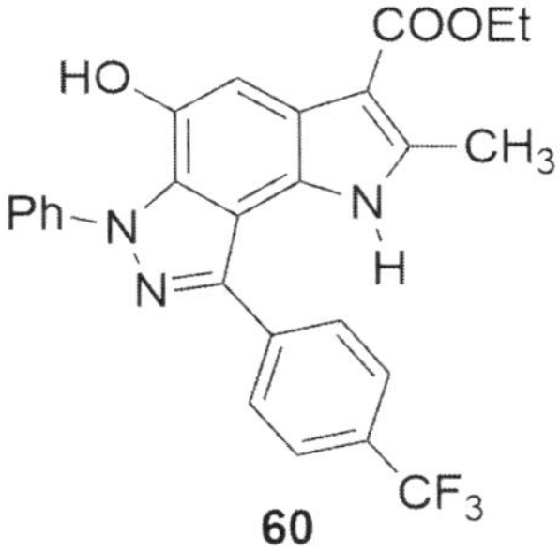

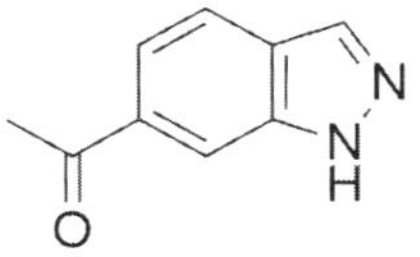

Aïssatou Aïcha Sow *et al.* investigated compound **61** as a potent new inhibitor of DENV, a common arthropod-borne virus. The compound exhibited high specificity to DENV and reduced the formation of infectious DENV particles by 1,000-fold without causing cytotoxicity. The half maximal effective concentration (EC_{50}) of compound **61** against DENV was 6.5 µM, highlighting its effectiveness in combating the viruses (Sow et al., 2023).

4.3 Anti-Protozoal

Karen Rodríguez-Villar *et al.* recently reported on the anti-protozoal activity of indazole derivatives. Compound **62** was evaluated *in vitro* against *E. histolytica*, *G. intestinalis*, and *T. vaginalis* with IC$_{50}$ values of 0.074 ± 0.003 µM against *E. histolytica*, 0.023 ± 0.009 µM against *G. intestinalis*, and 0.067 ± 0.006 µM against *T. vaginalis*, indicating potent anti-protozoal activity (Rodríguez-Villar et al., 2021).

Figure 19. Indazole derivatives as anti-protozal agents

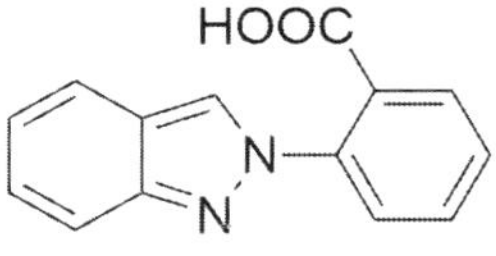

62

4.4 Anti-Inflammatory Activity

Chakrapani Cheekavolu *et al.* studied indazole's *in vivo* and *in vitro* anti-inflammatory activity and its derivatives. They examined indazoles, which have anti-inflammatory effects through COX-2 interaction. At 50 µM, the maximum suppressive effect of compound **64** was 78%. Celecoxib's IC$_{50}$ value was determined to be 5.10 µM using the same assay procedure (Cheekavolu, 2016).

Kantlam Chamakuri *et al.* synthesized and assessed the anti-inflammatory activities of 7-azaindazole-chalcone derivatives from 5-bromo-1*H*-pyrazolo[3,4-b] pyridine-3-carbaldehyde. Derivatives **65** exhibited the most excellent anti-inflammatory effect, with oedema inhibition of 70% among all synthesized compounds (Chamakuri et al., 2016). Vishal Kumar *et al.* synthesized the 1,3-substituted 1*H*-indazole and assessed it for anti-inflammatory activity in Sprague Dawley rats. Compound **66** had shown an excellent anti-inflammatory effect amongst all the synthesized derivatives (Kumar et al., 2024).

Zhiguo Liu *et al.* developed and synthesized a novel class of 3-(indol-5-yl)-indazoles and examined their anti-inflammatory effects in macrophages. Among all the derivatives, **67** was found to be an excellent inhibitor of lipopolysaccharide (LPS)-induced expression of tumour necrosis factor-α as well as interleukin-6 (IL-6) in macrophages with IC$_{50}$ values of 0.89 and 0.53µM, respectively (Zhiguo Liu et al., 2019).

Figure 20. Indazole derivatives as anti-infammatory agents

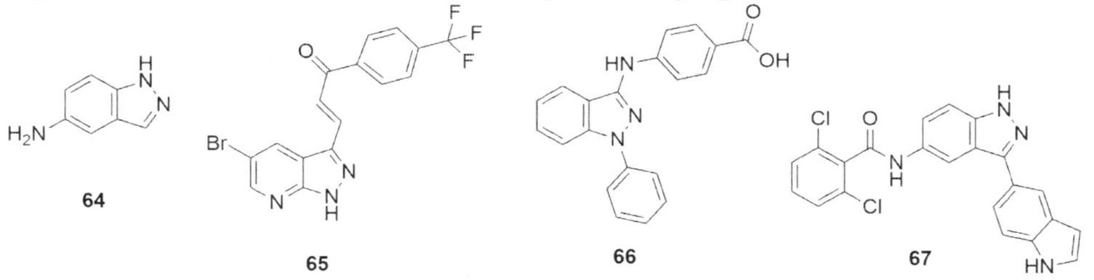

64

65

66

67

4.5 Anti-Cancer Activity

Using pharmacophore models and molecular docking, Qian *et al.* announced the discovery of a novel class of 1*H*-indazole derivatives as IDO1 suppressors. Based on biological research, compound **68** exhibited the strongest anti-IDO1 activity (IC_{50} is 5.3 µM) (Qian et al., 2016).

To identify a new class of highly effective Pim kinase blockers, the researchers refined the piperidine along with 2,6-difluorophenyl portions of compound **69** with an IC_{50} value of 1.4 µM in KMS-12 BM cell assays, compound **70**, which demonstrated sub-nanomolar to nanomolar inhibitory activities against Pim kinases, also inhibited BAD phosphorylation (Wu et al., 2015).

Song *et al.* identified a new class of 3-(pyrrolopyridin-2-yl)indazole derivatives as blockers of *Aurora A*. Compound **71** demonstrated nanomolar IC_{50} values against HL60, HCT116, and HCT116 cells at the G2/M phase (Song et al., 2015).

Chang *et al.* used sub-structure screening to search their internal library and found indazole compound **72** with moderate blocking action against Aurora A (IC_{50} is 13.56 µM) (Chang et al., 2016). Shan *et al.* discovered a novel class of *N,N'*-dibenzoylpiperazine derivatives that function as Bcr-Abl blockers and contain 1H-indazol-3-amine.

Figure 21. Indazole derivatives as anti-cancer agents

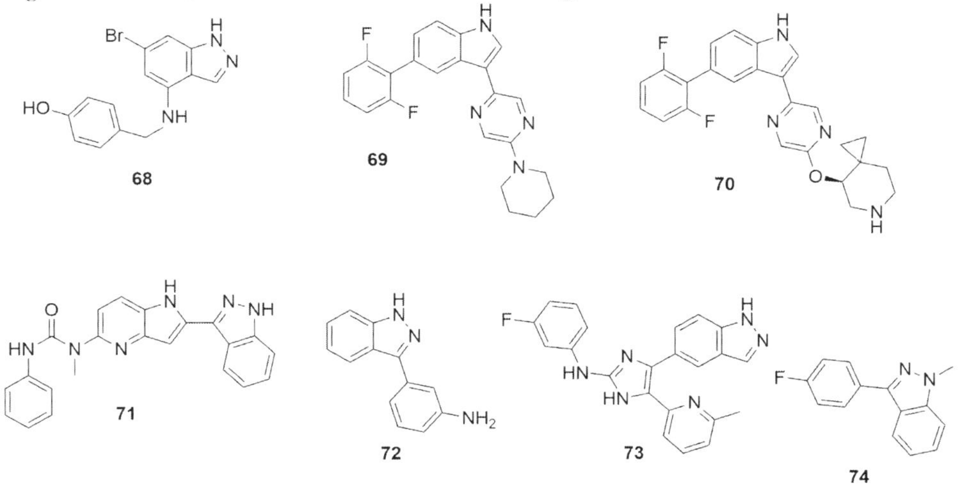

A sequence of 4-(1H-indazol-5-yl)-5-(6 methylpyridin-2-yl)-1H-imidazoles and 4-(1-methyl-1H-indazol-5-yl-5-(6-methylpyridin-2-yl)-1*H*-imidazoles were synthesized and assessed by Yue Ying Liu *et al.* for their ability to impede ALK5 as well as p38a mitogen-activated protein kinases. The most potent derivative was found to be **73** suppressed ALK5- as well as p38a-mediated phosphorylation with IC$_{50}$ values of 0.004 mM and 0.004 mM, accordingly, in the enzymatic analysis (Y. Y. Liu et al., 2021).

Jagan Mohana Rao Saketi *et al.* synthesized a series of 3-aryl-1*H*-indazolesas and *N*-methyl-3*H*-indazoles derivatives. All the synthesized indazoles were evaluated for their *in vitro* anti-cancer effects using the cell lines HCT-116 and MDA-MB-231. The anti-cancer studies of the tested compounds showed that compound **74** exhibited significant suppressive activity on the two tested cancer cell lines amongst all the synthesized derivatives (J. M. R. Saketi et al., 2020).

4.6 Anti-Oxidant Actions

ChakraPani Cheekavolu *et al.* studied the *in vivo* and *in vitro* anti-inflammatory activity of indazoles. Compound **75** inhibited DPPH free radical generation at concentrations ranging from 39.35% to 51.26%, with 6-nitroindazole showing the highest inhibition. However, nitric oxide scavenging activity was observed only at low concentrations (Cheekavolu, 2016).

Figure 22. Indazole derivatives as anti-oxidant agents

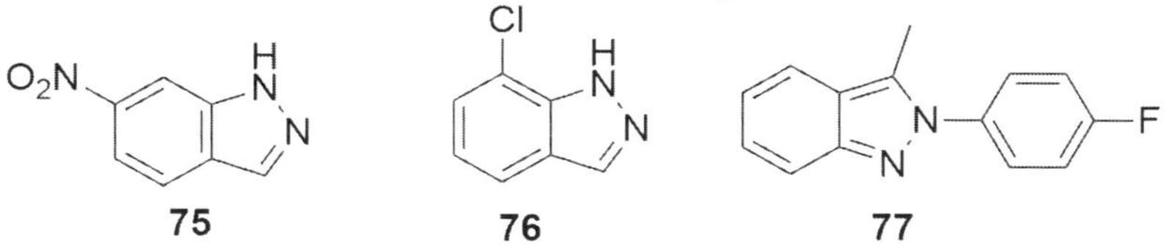

75   **76**   **77**

Zuhal Alim *et al.* studied some indazoles that lower the impact of human serum paraoxonase, an anti-oxidant enzyme. In this study, the authors aimed to investigate the effects of some indazole molecules on hPON1 activity. According to the results, compound **76** has an IC_{50} 72.9 µM blocks the catalytic action of hPON1 more than other molecules and has a greater affinity for hPON1 (Alım et al., 2019).

In addition, Efrain Polo *et al.* demonstrated anti-oxidant characteristics of indazole derivatives. The DPPH and ABTS methods were used to assess the impact of the *in vitro* anti-oxidant. Compound **77** demonstrated the most impact in the ABTS assay among these assays (Polo et al., 2016).

4.7 Anti-Diabetic Activity

Rafaila Rafique *et al.* synthesized the new 19 *N*-sulfonohydrazide substituted indazoles and evaluated the *in vitro* α-amylase blocking potentials. Compounds **78** were discovered to be the most influential radical scavenger and α-amylase enzyme blocker (DPPH and BTS) (Taha et al., 2024). Bushra *et al.* synthesized 24 Schiff bases of indazole and screened for *in vitro* α-glucosidase enzyme suppressive impact. Compound **79** (IC_{50} is 9.43 ± 0.1 µM) was found to be the most potent molecule of this series as compared to the standard acarbose (IC_{50} is 750 ± 10 µM) (Bushra et al., 2021).

Muhammad Taha *et al.* synthesized the 27 analogous schiff bases based on indazoles by using (1-methyl-1*H*-indazole-3-carboxylic acid) as a starting material and by proceeding through a three-step reaction pathway. Compound **80** is the most effective in the series because it contains three OH groups. For α-glucosidase as well as α-amylase, the IC_{50} values are 0.40 ±0.01 µM and 0.70 ±0.01 µM respectively (Taha et al., 2024).

Figure 23. Indazole derivatives as anti-diabetic activity

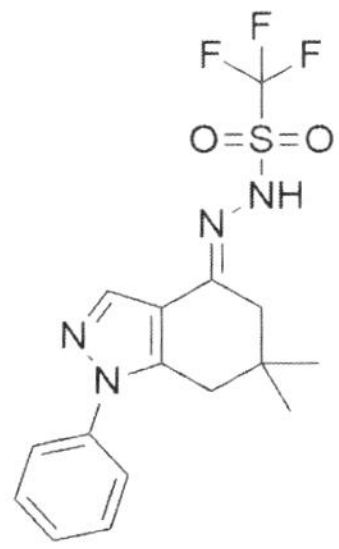

78 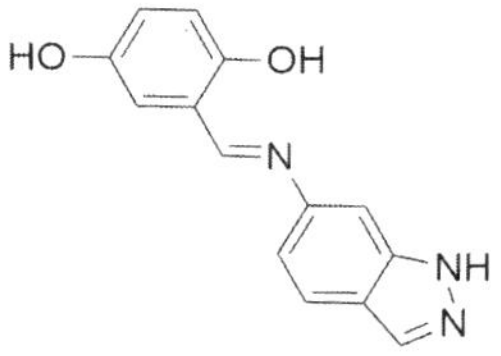 79 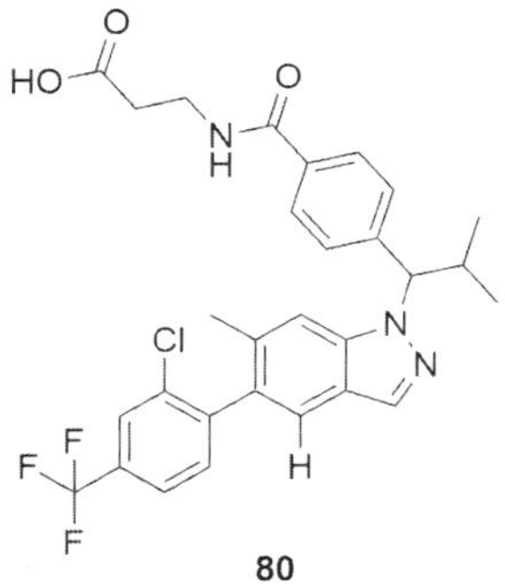 80

4.8 Neuroprotective Effect

Mariagrazia Rullo *et al.* designed a series of 5-substituted-1H-indazoles derivatives. They assessed them *in vitro* as blockers of human monoamine oxidase (hMAO) A and B. Among all derivatives, compound **81** was found to be the most effective hMAO B blocker with IC_{50} of 52 nM as well as excellent selectivity over MAO-A (Rullo et al., 2023).

Pedro Gonzalez-Naranjo *et al.* synthesized the novel family of 5-substituted indazoles and assessed for the *in vitro* inhibitory assays on AChE/BuChE. Pharmacological test findings have demonstrated that 5-substituted indazole-based derivatives **82** (IC_{50} value >10 38%) behave as the most potent AChE/BuChE blockers (González-Naranjo et al., 2022).

Figure 24. Indazole derivatives as neuroprotective action

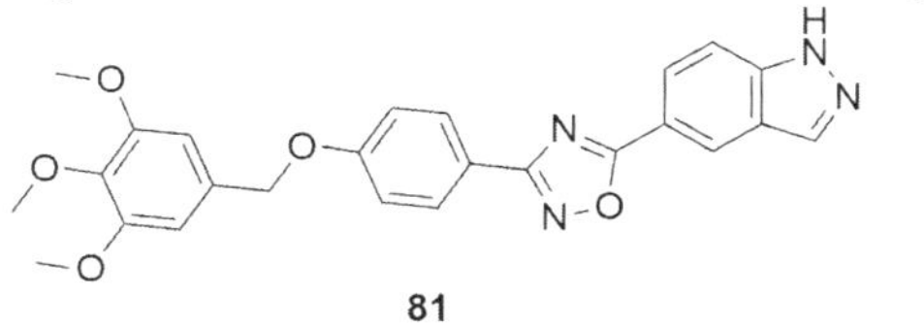

Wen Shuai *et al.* pioneered the new indazole derivatives as selective JNK inhibitors for the treatment of PD. By using SAR analysis and docking-based digital screening, they identified compound **83** demonstrated extraordinary kinase selectivity, exceeding 100-fold isoform selectivity for JNK3 over JNK1/2 as well as excellent inhibitory action against JNK3 (IC_{50} is 85.21 nM) (Shuai et al., 2023).

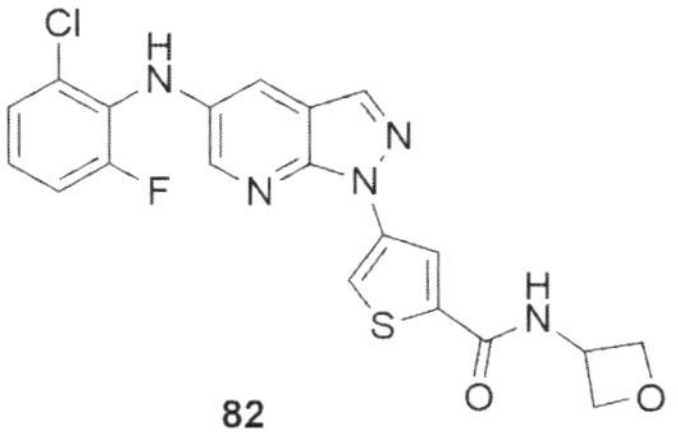

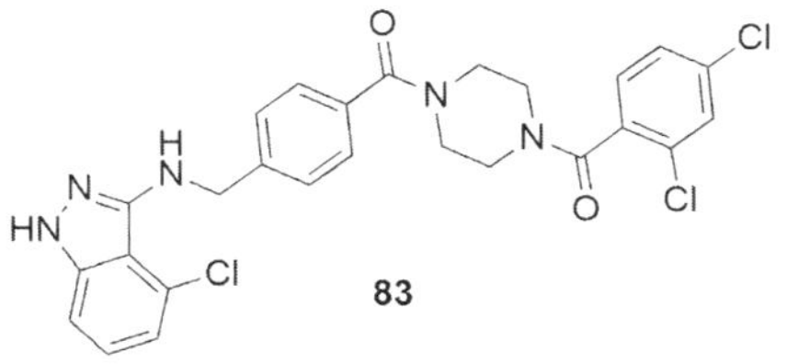

4.9 Analgesic Activity

Guanglin Luo *et al.* identified the Indazole-acylsulfonamides as powerful and Selective NaV1.7 blockers for treating pain. Compound **84** demonstrated effectiveness in the mouse formalin test and decreased neuropathic pain (Luo et al., 2019).

Kantlam Chamakuri *et al.* synthesized and assessed the analgesic effect of 7-azaindazole-chalcone derivatives from 5-bromo-1*H*-pyrazolo[3,4-b] pyridine-3-carbaldehyde. Compound **85** exhibited fantastic analgesic activity compared with the activity of standard drug diclofenac sodium (Chamakuri et al., 2016). V. L. Gein *et al.* synthesized a novel class of 4,5,6,7-tetrahydro-2H-indazole chemotypes via the reaction of N^1, N^3, 2-triaryl-6- hydroxy-6-methyl-4-oxocyclohexane-1,3-dicarboxamides with hydrazine hydrate. The synthesized derivatives were assessed for analgesic activity. Compound **86** was found to have the most excellent analgesic activity (Gein et al., 2019).

Figure 25. Indazole derivatives as anagesic activity

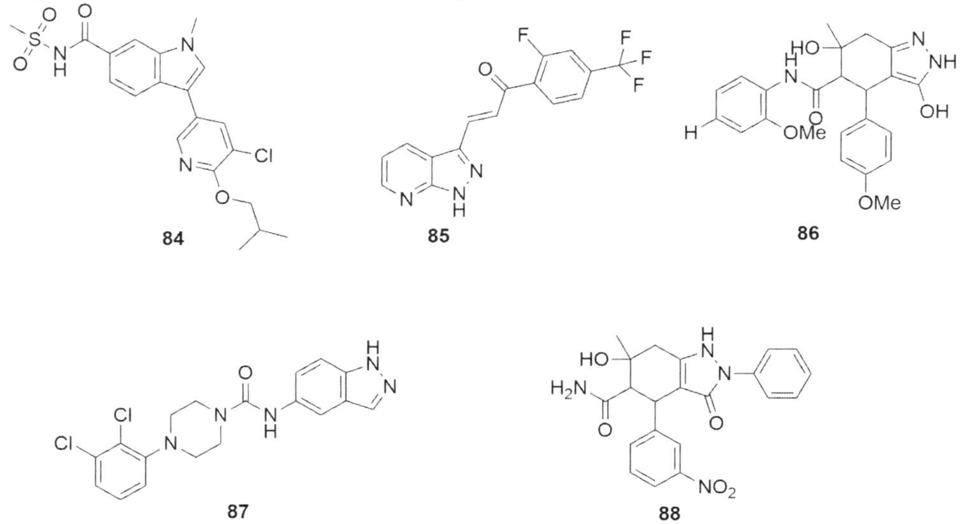

Qianqian Liang *et al.* discovered that the potent and selective transient receptor potential vanilloid 1 (TRPV1) agonist had analgesic effects. According to the whole-cell clamp patch assay, compound **87** was found to be a potent and selective TRPV1 agonist, and it relieves inflammatory and thermal pain by desensitizing the native TRPV1 current in the dorsal root ganglion (DRG) in mice (Liang et al., 2022).

V. L. Gein *et al.* synthesized the new series of hydrazones as well as hexahydro-1*H*-indazole-5-carboxamides by the reaction of 2-aryl-4-hydroxy-4-methyl-6-oxocyclohexane-1,3-dicarboxamides with tosylhydrazide, 2,4-DNP and phenylhydrazine. The synthesized derivatives were assessed for analgesic activity. The results show that compound **88** has excellent antinociceptive activity with defensive reflex time of 24.60 ± 0.98 (Gein et al., 2022).

4.10 Anti-Depressant Activity

Agnieszka A. Kaczor *et al.* identified a compound **89** with affinity to dopamine D2 receptor of 115 nM. Additionally, D2AAK3 shows affinity for D_1, D_3, 5-HT_{1A}, 5-HT_{2A} as well as 5-HT_7 receptors ranging from nanomolar or low micromolar (Kaczor et al., 2021). Piotr Sterpnick *et al.* designed and synthesized indazole and piperazine scaffolds and further screened them for anti-convulsant activity. The compound **90** was found to be the have greatest anti-convulsant effect (Stępnicki et al., 2023).

Figure 26. Indazole derivatives as anti-depressant activity

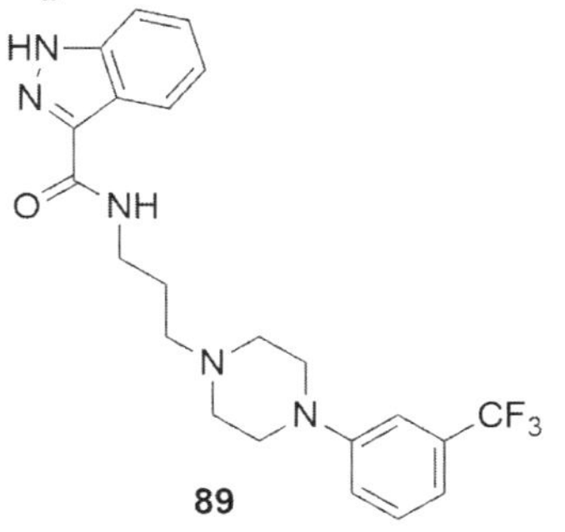

89

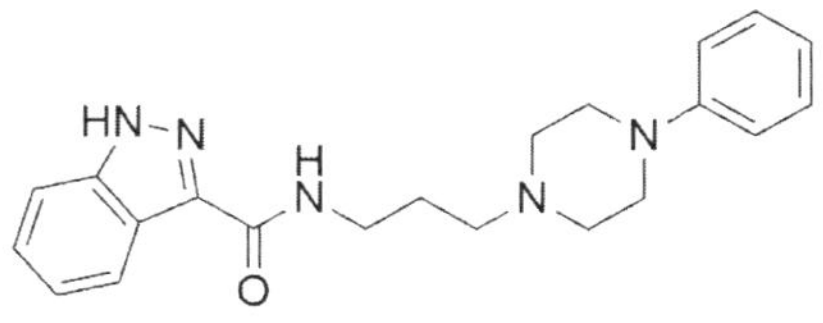

90

4.11 Antihypertensive Agents

Guido Furlotti *et al.* designed and synthesized a series of novel bicyclic and tricyclic 2*N*-alkyl-indazole-amide hybrids. They identified compound **91** potent 5-HT_{2A} antagonists with high affinity for the α-1 receptor and greater than 100-fold selectivity over other serotonin receptor subtypes (Furlotti et al., 2018).

Yangyang Yao *et al.* discovered novel *N*-substituted procainamide-indazoles derivatives as a potent inhibitor of Rho kinase enzyme. The synthesized derivatives were screened for vasorelaxant activity. Among the synthesized derivatives, compound **92** (IC_{50} 0.27 μM) was found to be the most active. *Ex-vivo* tests revealed that compound **92** has significant vasorelaxant activity in the rat (Yao et al., 2017).

Figure 27. Indazole derivatives as Antihypertensive agents

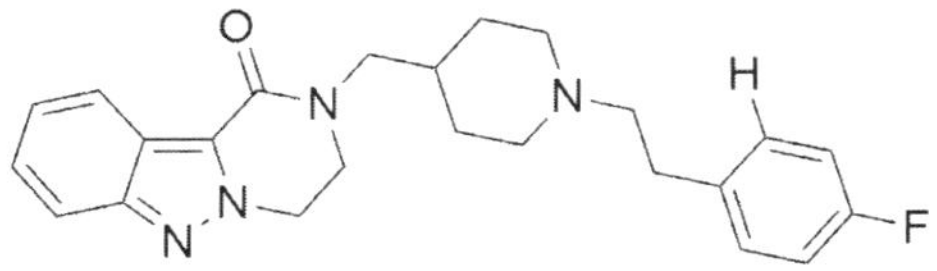

91

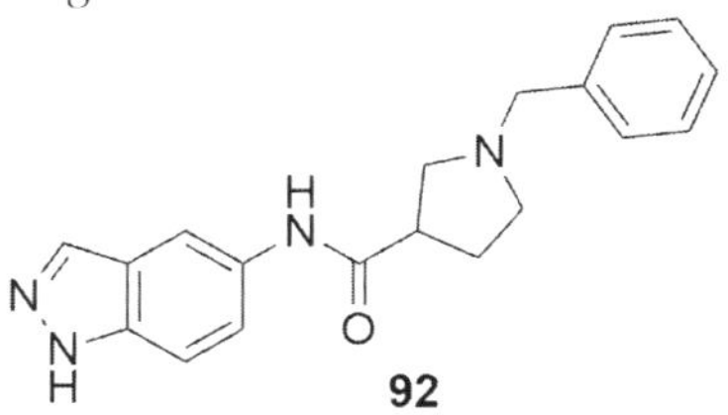

92

4.12 Anticonvulsant Activity

A new series of indazole substituted-1,3,4-thiadiazole derivatives was synthesized and evaluated for anti-convulsant activity by Kikkeri P. Harish *et al.* The compounds **93** and **94** being the most potent, displaying no neurotoxicity at a maximum dose of 100 mg/kg (Harish et al., 2013).

Figure 28. Indazole derivatives as anti-convusant agents

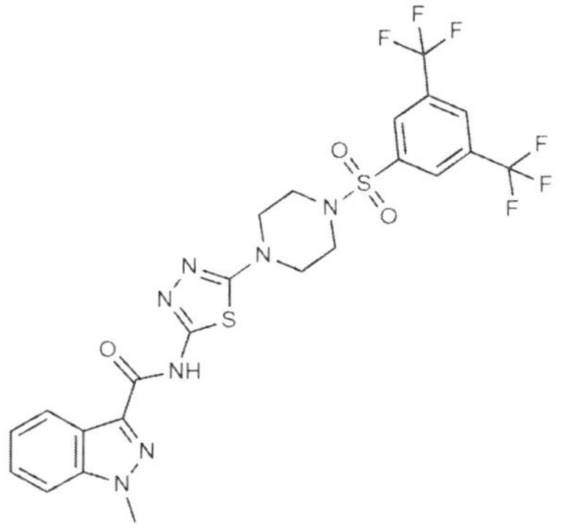

93

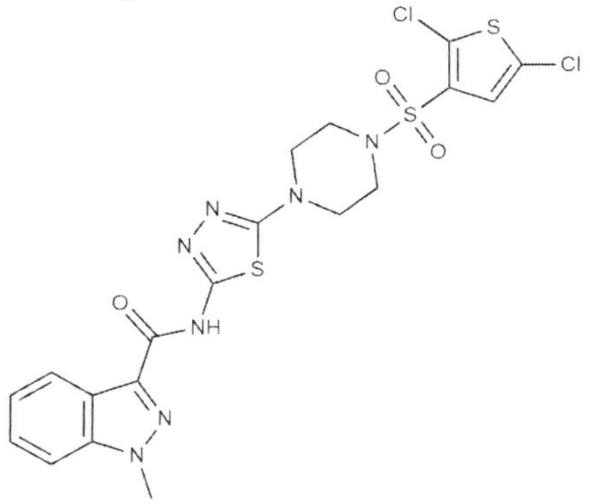

94

5. PATENT UPDATES ON INDAZOLE

There are multiple patents filed on Indazole derivatives. The most recent updates on these patents are summarized in Table 1 (*Google Patents*, n.d.).

Table 1. Patent updates on indazole

Patent Id	Title	Assignee	Inventor/Author	Grant date Y/M/Dte
KR-102083857-B1	New substituted indazoles, methods for the production thereof, pharmaceutical preparations that contain said new substituted indazoles, and use of said new substituted indazoles to produce drugs	Bayer Pharma Actien Gesellschaft	Ulrich Bothe, Holger Siebeneicher, Nikol Schmidt, Reinhard Neuvemeyer, Wolf Böhmer, Judith Günther, Holger Steuber, Martin Lange, Christian Stegmann, Andreas Sutter, Alexandra Rausch, Christian Friedrich, Peter Auff.	2020-03-03
CN-101952256-B	1-benzyl-3-hydroxymethyl indazole derivatives and the purposes in the disease for the treatment of based on MCP-1, CX3CR1 and p40 expression thereof	Francis Angelique Chemical Associates, Inc.	A. Guglielmonti, G. Folotti, G. Mangano, N. Cazzurla, B. Garofalo	2016-03-16
ES-2526323-T3	New derivative of indazole or a salt thereof, intermediate product in its production, anti-oxidant that uses it, and use in an indazole derivative or salt thereof	Ube Industries, Ltd.	Masahiko Hagihara, Ken-Ichi Komori, Hidetoshi Sunamoto, Hiroshi Nishida, Yasunori Tsuzaki, Akira Takama, Kazutaka Kido, Tomokazu Fujimoto, Takeshi Matsugi	2015-01-09
ES-2600318-T3	New indazoles for the treatment and prophylaxis of respiratory syncytial virus infection	F. Hoffmann-La Roche Ag	Song Feng, Lu Gao, Di Hong, Lisha Wang, Hongying Yun, Shu-Hai Zhao	2017-02-08
KR-101551187-B1	Indazole derivatives selectively inhibiting the activity of Janus kinase 1	Konkuk University Industry-Academic Cooperation Foundation	Jeong Yu-hoon, Kim Mi-kyung	2015-09-08
CN-108947970-B	Indazole derivative and preparation method and application thereof	Sichuan University	Yu Luoting, Wei Yuquan	2022-04-05

continued on following page

Table 1. Continued

Patent Id	Title	Assignee	Inventor/Author	Grant date Y/M/Dte
CN-102439005-B	Novel substituted indazole and aza-indazole derivatives as gamma secretase modulators	Janssen Pharmaceuticals Inc., Selzom Ltd.	F.P. Bischoff, H.J.M. Gissen, S.M.A. Peters, G.B. Meaney	2015-07-22
JP-5250029-B2	Transition metal catalyzed synthesis of 2H-indazole	Sanofi	Nis Haarland, Marc Nazare, Andreas Lindenschmidt, Jorge Alonso, Omar Lukaik, Matthias Ullmann	2013-07-31
EP-3423446-B1	New 2-substituted indazoles, methods for producing same, pharmaceutical preparations that contain same, and use of same to produce drugs	Bayer Pharma Aktiengesellschaft	Ulrich Bothe, Holger Siebeneicher, Nicole Schmidt, Judith GÜNTHER, Holger STEUBER, Ulf Börner, Martin Lange, Reinhard Nubbemeyer, Nicholas Charles Ray, Pascal Savy	2020-09-16
CN-109152771-B	Use of 2-substituted indazoles for the treatment and prevention of autoimmune diseases	Bayer Pharma AG	A. Rausch, S. J. Yoder, J. Kretschmar, U. Bote, N. Schmidt	2022-07-19
JP-6530406-B2	Novel indazole carboxamides, processes for their preparation, pharmaceutical preparations containing them and their use for the preparation of medicaments	Bayer Pharma Aktiengesellschaft	Ulrich Bothe, Holger Siebenaicher, Nicole Schmidt, Andrea Rothgeri, Ulf Boehmer, Zven Ring, Horst Irlbacher, Judit Günther, Holger Steuber, Martin Lange, Martina Schaefer	2019-06-12

continued on following page

Table 1. Continued

Patent Id	Title	Assignee	Inventor/Author	Grant date Y/M/Dte
EP-2613781-B1	Indazole derivatives for use in the treatment of influenza virus infection	GlaxoSmithKline Intellectual Property Development Limited	Ian Robert Baldwin, Kenneth David Down, Paul Faulder, Simon Gaines, Julie Nicole Hamblin, Zoe Alicia Harrison, Katherine Louise Jones, Paul Spencer Jones, Suzanne Elaine Keeling, Joelle Le, Christopher James Lunniss, Charlotte Jane Mitchell, Nigel James Parr, Timothy John Ritchie, John Edward Robinson, Juliet Kay Simpson, Christian Alan Paul Smethurst, Yoshiaki Washio	2016-08-24
EP-2613782-B1	Indazole derivatives useful as erk inhibitors	Merck Sharp & Dohme Corp.	Yongqi Deng, Gerald W. Shipps, Jr., Sie-Mun Lo, Liang Zhu, Alan B. Cooper, Kiran Muppalla	2016-11-02
JP-5855253-B2	Indazole compounds, compositions and methods of use	F. Hoffmann-La Roche AG F. Hoffmann-La Roche Aktiengesellschaft, F. Hoffmann-La Roche AG	Birch, Jason, Goldsmith, Richard Ay, Ortwine, Daniel Fred, Pastor, Richard, Pei, Zhonghua	2016-02-09
CN-105037355-B	The indazole inhibitors and its therapeutical uses of Wnt signal transduction paths	saummed ltd.	J. Hood, D. M. Wallace, S. K. Kersey	2017-06-06

continued on following page

Table 1. Continued

Patent Id	Title	Assignee	Inventor/Author	Grant date Y/M/Dte
US-10266548-B2	Substituted benzylindazoles for use as Bub1 kinase inhibitors in the treatment of hyperproliferative diseases	Bayer Intellectual Property Gmbh, Bayer Pharma Aktiengesellschaft	Marion Hitchcock, Anne Mengel, Vera Pütter, Gerhard Siemeister, Antje Margret Wengner, Hans Briem, Knut Eis, Volker Schulze, Amaury Ernesto Fernandez-Montalvan, Stefan Prechtl, Simon Holton, Jörg Fanghänel, Philip Lienau, Cornelia Preusse, Mark Jean Gnoth	2019-04-23
EP-2766352-B1	Indazole compounds as kinase inhibitors and method of treating cancer with same	University Health Network (UHN)	Heinz W. Pauls, Radoslaw Laufer, Yong Liu, Sze-Wan Li, Bryan T. Forrest, Yunhui Lang, Narendra Kumar B. Patel, Louise G. Edwards, Grace Ng, Peter Brent Sampson, Miklos Feher, Donald E. Awrey	2018-06-06
CN-104395308-B	The preparation method of N-[5-(3,5-difluoro-benzyl)-1H-indazole-3-base]-4-(4-thyl-piperazin-1-base)-2-(ttetrahydro-pyran-4-base amino)-Benzoylamide	Nerviano Medical Sciences Ltd.	N.A. Balbugan, R. Folino, T. Formagali, P. Orsini	2016-08-24
JP-6411345-B2	N-alkylated indole and indazole compounds as RORγT inhibitors and their use	Merck Sharp & Dohme Corp., Merck Sharp & Dohme Corporation	Barr, Kenneth, Jay, Bienstock, Corey, MacLean, Jiyoung, Jiang, Hornjuiin, Beresis, Richard, Thomas, Anthony, Neville, Lapointe, Blair, Siametsuta, Nunzio.	2018-10-24

continued on following page

Table 1. Continued

Patent Id	Title	Assignee	Inventor/Author	Grant date Y/M/Dte
TW-I810182-B	Fused indazole pyridone compounds as anti-virals	Novartis	Jiping Fu, Yuxi Han, Shabrunian Karul, Lu Peichao, Keith Bruce Pfister, Joseph Michael Yang	2023-08-01
ES-2882495-T7	1H-indazole-3-carboxamide derivatives and related compounds as inhibitors of factor D for the treatment of diseases characterized by abnormal activity of the complement system, such as immune disorders	Biocryst Pharm Inc	Pravin L Kotian, Yarlagadda S Babu, Weihe Zhang, Lakshminarayana Vogeti, Minwan Wu, Venkat R Chintareddy, Krishnan Raman	2022-04-26
RU-2665462-C2	5-azaindazole compounds and methods for use thereof	F. Hoffmann-La Roche Ag	Stephen DEW, Huiyong HU, Alexander KOLESNIKOV, Vicki H. TsUI, Xiaojing Wang	2018-08-30
ES-2637238-T3	Indole and indazole compounds that activate AMPK	Pfizer Inc.	Samit Kumar Bhattacharya, Kimberly Okeefe Cameron, Matthew Scott Dowling, David Christopher Ebner, Dilinie Prasadhini Fernando, Kevin James Filipski, Daniel Wei-Shung Kung, Esther Cheng Yin LEE, Aaron Christopher Smith, Meihua Mike Tu	2017-10-11
US-8569511-B2	Substituted 3-(1H-benzo[d]imidazol-2-yl)-1H-indazole analogs as inhibitors of the PDK1 kinase	University Of Utah Research Foundation	David J. Bearss, Hariprasad Vankayalapati, Venkataswamy Sorna, Steven L. Warner, Sunil Sharma	2013-10-29

continued on following page

Table 1. Continued

Patent Id	Title	Assignee	Inventor/Author	Grant date Y/M/Dte
CA-3037728-C	4,6-indazole compounds and methods for ido and tdo modulation, and indications therefor	Plexxikon Inc., Jiazhong Zhang, Hannah POWERS, Aaron ALBERS, Phuongly Pham, Guoxian Wu, John BUELL, Wayne Spevak, Zuojun GUO, Jack Walleshauser, Ying Zhang	Jiazhong Zhang, Hannah POWERS, Aaron ALBERS, Phuongly Pham, Guoxian Wu, John BUELL, Wayne Spevak, Zuojun GUO, Jack Walleshauser, Ying Zhang	2023-10-24
EP-2566328-B1	Indazoles	GlaxoSmithKline LLC	Celine Duquenne, Neil Johnson, Steven D. Knight, Louis Lafrance, William H. Miller, Kenneth Newlander, Stuart Romeril, Meagan B. Rouse, Xinrong Tian, Sharad Kumar Verma	2015-03-04
US-10745388-B2	Indazole compounds and uses thereof	Incyte Corporation	Oleg Vechorkin, Alexander Sokolsky, Qinda Ye, Wenqing Yao	2020-08-18
US-11352341-B2	2H-indazole derivatives as CDK4 and CDK6 inhibitors and therapeutic uses thereof	Beta Pharma, Inc.	Michael Nicholas Greco, Michael John Costanzo, Jirong Peng, Don Zhang	2022-06-07
EP-3280419-B1	Indazole and azaindazole btk inhibitors	Merck Sharp & Dohme Corp.	Tony Siu, Michael D. Altman, Brian M. Andresen, Jian Liu, Joseph Kozlowski, Sobhana Babu Boga, Younong Yu, Rajan Anand, Jiaqiang Cai, Dahai Wang, Shilan Liu	2019-10-30

continued on following page

Table 1. Continued

Patent Id	Title	Assignee	Inventor/Author	Grant date Y/M/Dte
JP-6982748-B2	2- (1H-Indazole-3-yl) -3H-imidazole [4,5-c] pyridine and their anti-inflammatory use	Biosplice Therapeutics Inc.	John Hood, David Mark Wallace, Sunil Kumar Casey, Yusuf Yazju, Christopher Swearingen, Lewis A. Delamarie	2021-12-17
ES-2904544-T3	Indazole compounds as FGFR kinase inhibitors, preparation and use thereof	Shanghai Haihe Pharmaceutical Co Ltd, Shanghai Inst Materia Medica Cas	Meiyu Geng, Lei Liu, Lei Jiang, Min Huang, Chuantao Zha, Jing Ai, Lei Wang, Jianhua Cao, Jian Ding	2022-04-05
AU-2010280827-B8	Process for the preparation of 1-benzyl-3-hydroxymethyl-1H-indazole and its derivatives and required magnesium intermediates	Aziende Chimiche Riunite Angelini Francesco A.C.R.A.F. S.P.A.	Giuliano Caracciolo Torchiarolo, Guido Furlotti, Tommaso Iacoangeli	2016-04-07

6. ONGOING CLINICAL TRIALS

The clinical trial updates for indazole analogs are summarized in Table 2. These trials have primarily focused on treating various diseases, highlighting the compound's versatility in therapeutic research (*ClinicalTrials.gov*, n.d.).

Table 2. Ongoing Clinical trials

Sr.no	Study title	Study type	Study phase	Study design	Conditions	Status	Clinical trials.gov identifier
1	A Study Comparing Two Different Capsules, APL-101 and PLB-1001 Capsules, in Healthy Chinese and Caucasian Participants	Interventional	Phase 1	Randomized	Bioequivalence	Recruiting	NCT05367388
2	A Study of PLB1001 in Non-small Cell Lung Cancer With c-Met Dysregulation (KUNPENG)	Interventional	Phase 2	N/A	Non-small Cell Lung Cancer	Recruiting	NCT04258033
3	Vebreltinib Plus PLB1004 in EGFR-mutated, Advanced NSCLC With MET Amplification or MET Overexpression Following EGFR-TKI	Interventional	Phase 1 Phase 2	Non-Randomized	Non-Small-Cell Lung Cancer	Recruiting	NCT06343064
5	A Study of Ramucirumab (LY3009806) or Merestinib (LY2801653) in Advanced or Metastatic Biliary Tract Cancer	Interventional	Phase 2	Randomized	Biliary Tract Cancer, Metastatic Cancer, Advanced Cancer	Active, not recruiting	NCT02711553
6	A Study of Anti-PD-L1 Checkpoint Antibody (LY3300054) Alone and in Combination in Participants With Advanced Refractory Solid Tumors	Interventional	Phase 1	Non-Randomized	Solid Tumor, Microsatellite Instability-High (MSI-H) Solid Tumors, Cutaneous Melanoma,Pancreatic Cancer, Breast Cancer (HR+HER2-)	Active, not recruiting	NCT02791334

continued on following page

Table 2. Continued

Sr.no	Study title	Study type	Study phase	Study design	Conditions	Status	Clinical trials.gov identifier
7	Study of ARRY-614 Plus Either Nivolumab or Nivolumab+Ipilimumab	Interventional	Phase 1 Phase 2	Non-Randomized	Renal Cell Carcinoma, Melanoma, Solid Tumor, Non-small Cell Lung Cancer, Head and Neck Squamous Cell Carcinoma	Active, not recruiting	NCT04074967
8	NP-G2-044 as Monotherapy and Combination Therapy in Patients With Advanced or Metastatic Solid Tumor Malignancies	Interventional	Phase 1 Phase 2	Non-Randomized	Advanced or Metastatic Solid Tumor Malignancies	Recruiting	NCT05023486
9	Study of H3B-6545 in Japanese Women With Estrogen Receptor (ER)-Positive, Human Epidermal Growth Factor Receptor 2 (HER2)-Negative Breast Cancer	Interventional	Phase 1	Randomized	Breast Neoplasms	Active, not recruiting	NCT04568902
10	A Study of H3B-6545 in Combination With Palbociclib in Women With Advanced or Metastatic Estrogen Receptor-Positive Human Epidermal Growth Factor Receptor-2 (HER2)-Negative Breast Cancer	Interventional	Phase 1	N/A	Receptors, Estrogen, Genes, Erbb-2, Breast Neoplasms	Active, not recruiting	NCT04288089

continued on following page

Table 2. Continued

Sr.no	Study title	Study type	Study phase	Study design	Conditions	Status	Clinical trials.gov identifier
11	Trial of H3B-6545, in Women With Locally Advanced or Metastatic Estrogen Receptor-positive, HER2 Negative Breast Cancer	Interventional	Phase 1 Phase 2	Non-Randomized	Breast Neoplasms, Breast Cancer, Estrogen-receptor Positive Breast Cancer Cancer, Breast, Breast Cancer Female, Breast Adenocarcinoma, Estrogen Receptor Positive Tumor, ER Positive	Completed	NCT03250676
12	Trial of H3B-6545, in Women With Locally Advanced or Metastatic Estrogen Receptor-positive, HER2 Negative Breast Cancer	Interventional	Phase 1 Phase 2	Non-Randomized	Breast Neoplasms, Breast Cancer, Estrogenreceptor Positive Breast Cancer Breast Cancer Female, Breast Adenocarcinoma, Estrogen Receptor Positive Tumor, ER Positive	Completed	NCT03250676
13	A Study Evaluating the Safety, Tolerability, and Pharmacokinetics of Multiple Ascending Doses of SM04755 Following Topical Administration to Healthy Subjects	Interventional	Phase 1	Randomized	Tendinopathy	Completed	NCT03229291
14	Safety and Pharmacokinetics of SM04755 in Subjects With Advanced Colorectal, Gastric, Hepatic, or Pancreatic Cancer	Interventional	Phase 1	N/A	Colorectal Cancer, Gastric Cancer, Hepatic Cancer, Pancreatic Cancer	Completed	NCT02191761

continued on following page

Table 2. Continued

Sr.no	Study title	Study type	Study phase	Study design	Conditions	Status	Clinical trials.gov identifier
15	A Repeat Insult Patch Test (RIPT) Study Evaluating the Sensitization Potential of Topical SM04755 Solution in Healthy Volunteers	Interventional	Phase 1	Randomized	Tendinopathy	Completed	NCT03502434
16	Clinical Trial of Metastasis Inhibitor NP-G2-044 in Patients With Advanced or Metastatic Treatment-Refractory Solid Tumor Malignancies	Interventional	Phase 1	N/A	Breast Cancer, Pancreas Cancer Prostate Cancer, Lung Cancer, Colon Cancer, Esophagus Cancer, Liver Cancer, Ovary Cancer, Advanced or Metastatic Treatment-refractory Solid Tumor Malignancies	Completed	NCT03199586
17	Phase 2 Neoadjuvant Doxorubicin and Cyclophosphamide -> Docetaxel With Lapatinib in Stage II/III Her2Neu+ Breast Cancer	Interventional	Phase 2	N/A	Breast Cancer, Metastatic Breast Cancer	Completed	NCT00404066
18	Phase 3 Study on the Efficacy and Safety of Tanezumab in Patients With Cancer Pain Due to Bone Metastasis Who Are Taking Background Opioid Therapy	Interventional	Phase 3	Randomized	Bone Metastasis, Cancer Pain	Completed	NCT02609828

continued on following page

Table 2. Continued

Sr.no	Study title	Study type	Study phase	Study design	Conditions	Status	Clinical trials.gov identifier
19	Long Term Safety and Efficacy Study of Tanezumab in Subjects With Osteoarthritis of the Hip or Knee	Interventional	Phase 3	Randomized	Chronic Pain, Osteoarthritis, Hip, Osteoarthritis, Knee	Completed	NCT02528188
20	Study of a c-Met Inhibitor PLB1001 in Patients With *PTPRZ1-MET* Fusion Gene Positive Recurrent High-grade Glioma	interventional	Phase 1	N/A	Glioma	Completed	NCT02978261
21	First-in-human Phase I Study of a Selective c-Met Inhibitor PLB1001	Interventional	Phase 1	N/A	Non-Small Cell Lung Cancer	Completed	NCT02896231
22	APL-501 or Nivolumab in Combination With APL-101 in Locally Advanced or Metastatic HCC and RCC	Interventional	Phase 1 Phase 2	Non-Randomized	Hepatocellular Carcinoma, Renal Cell Carcinoma	Terminated	NCT03655613
23	A Study of Merestinib (LY2801653) in Healthy Participants	Interventional	Phase 1	Randomized	Healthy	Completed	NCT02779738
24	Combination Merestinib and LY2874455 for Patients With Relapsed or Refractory Acute Myeloid Leukemia	Interventional	Phase 1	N/A	Relapsed Adult Acute Myeloid Leukemia, Refractory Adult Acute Myeloid Leukemia	Completed	NCT03125239

continued on following page

Table 2. Continued

Sr.no	Study title	Study type	Study phase	Study design	Conditions	Status	Clinical trials.gov identifier
25	A Study of Merestinib (LY2801653) in Japanese Participants With Advanced or Metastatic Cancer	Interventional	Phase 1	Non-Randomized	Advanced Cancer, Metastatic Cancer, Biliary Tract Carcinoma, Cholangiocarcinoma, Gall Bladder Carcinoma, Solid Tumor, Non-Hodgkin's Lymphoma	Completed	NCT03027284
26	A Study in Advanced Cancers Using Ramucirumab (LY3009806) and Other Targeted Agents	Interventional	Phase 1	Non-Randomized	Advanced Cancer, Colorectal Cancer, Mantle Cell Lymphoma	Completed	NCT02745769
27	Merestinib on Bone Metastases in Subjects With Breast Cancer	Interventional	Phase 1	N/A	Bone Metastases, Breast Cancer	Terminated	NCT03292536
28	Merestinib In Non-Small Cell Lung Cancer And Solid Tumors	Interventional	Phase 2	Non-Randomized	Carcinoma, Non-Small-Cell Lung, Solid Tumor	Terminated	NCT02920996
29	A Study of ARRY-614 in Patients With Low or Intermediate-1 Risk Myelodysplastic Syndromes	Interventional	Phase 1	N/A	Myelodysplastic Syndromes	Completed	NCT01496495
30	A Study of ARRY-614 in Patients With Low or Intermediate-1 Risk Myelodysplastic Syndrome	Interventional	Phase 1	N/A	Myelodysplastic Syndromes	Completed	NCT00916227

7. CONCLUSION

Indazole stands out as a remarkably versatile and potent scaffold in medicinal chemistry, demonstrating a broad spectrum of biological activities and therapeutic potential. Advances in both conventional and green synthesis methods have facilitated its efficient synthesis, while ongoing research continues to uncover novel applications in drug development. The compound's presence in marketed drugs and clinical trials underscores its significance in modern therapeutics. As future studies delve deeper into its structure-activity relationships and explore innovative derivatives. Indazole is poised to be increasingly crucial in addressing diverse medical challenges and advancing healthcare.

REFERENCES

Abbady, M. S., & Youssef, M. S. K. (2014). Synthesis and biological activity of some new pyridines, pyrans, and indazoles containing pyrazolone moiety. *Medicinal Chemistry Research*, 23(7), 3558–3568. DOI: 10.1007/s00044-014-0935-y

Alım, Z., Kılıç, D., & Demir, Y. (2019). Some indazoles reduced the activity of human serum paraoxonase 1, an antioxidant enzyme: In vitro inhibition and molecular modeling studies. *Archives of Physiology and Biochemistry*, 125(5), 387–395. DOI: 10.1080/13813455.2018.1470646 PMID: 29741961

Angapelly, S., Sri Ramya, P. V., Angeli, A., Supuran, C. T., & Arifuddin, M. (2017). Sulfocoumarin-, Coumarin-, 4-Sulfamoylphenyl-Bearing Indazole-3-carboxamide Hybrids: Synthesis and Selective Inhibition of Tumor-Associated Carbonic Anhydrase Isozymes IX and XII. *ChemMedChem*, 12(19), 1578–1584. DOI: 10.1002/cmdc.201700446 PMID: 28940980

Badawey, E. S. A. M., & El-Ashmawey, I. M. (1998). Nonsteroidal anti-inflammatory agents - Part 1: Anti-inflammatory, analgesic and antipyretic activity of some new 1-(pyrimidin-2-yl)-3- pyrazolin-5-ones and 2-(pyrimidin-2-yl)-1,2,4,5,6,7-hexahydro-3H-indazol-3- ones. *European Journal of Medicinal Chemistry*, 33(5), 349–361. DOI: 10.1016/S0223-5234(98)80002-0

Basak, S., Kumar, A., Ramsey, S., Gibbs, E., Kapoor, A., Filizola, M., & Chakrapani, S. (2020). High-resolution structures of multiple 5-HT3AR-setron complexes reveal a novel mechanism of competitive inhibition. *eLife*, 9, e57870. Advance online publication. DOI: 10.7554/eLife.57870 PMID: 33063666

Burroughs, S. K., Kaluz, S., Wang, D., Wang, K., Van Meir, E. G., & Wang, B. (2013). Hypoxia inducible factor pathway inhibitors as anti-cancer therapeutics. *Future Medicinal Chemistry*, 5(5), 553–572. DOI: 10.4155/fmc.13.17 PMID: 23573973

Bushra, S., Shamim, S., Khan, K. M., Ullah, N., Mahdavi, M., Faramarzi, M. A., Larijani, B., Salar, U., Rafique, R., Taha, M., & Perveen, S. (2021). Synthesis, in vitro, and in silico evaluation of Indazole Schiff bases as potential α-glucosidase inhibitors. *Journal of Molecular Structure*, 1242, 130826. DOI: 10.1016/j.molstruc.2021.130826

Cao, Y., Luo, C., Yang, P., Li, P., & Wu, C. (2021). Indazole scaffold: A generalist for marketed and clinical drugs. *Medicinal Chemistry Research*, 30(3), 501–518. DOI: 10.1007/s00044-020-02665-7

Cao, Y., Yang, Y., Ampomah-Wireko, M., Obaid Arhema Frejat, F., Zhai, H., Zhang, S., Wang, H., Yang, P., Yuan, Q., Wu, G., & Wu, C. (2022). Novel indazole skeleton derivatives containing 1,2,3-triazole as potential anti-prostate cancer drugs. *Bioorganic & Medicinal Chemistry Letters*, 64, 128654. DOI: 10.1016/j.bmcl.2022.128654 PMID: 35259487

Chamakuri, K., Muppavarapu, S. M., & Yellu, N. R. (2016). Synthesis and analgesic and anti-inflammatory activities of 7-azaindazole chalcone derivatives. *Medicinal Chemistry Research*, 25(10), 2392–2398. DOI: 10.1007/s00044-016-1671-2

Chang, C.-F., Lin, W.-H., Ke, Y.-Y., Lin, Y.-S., Wang, W.-C., Chen, C.-H., Kuo, P.-C., Hsu, J. T. A., Uang, B.-J., & Hsieh, H.-P. (2016). Discovery of novel inhibitors of Aurora kinases with indazole scaffold: In silico fragment-based and knowledge-based drug design. *European Journal of Medicinal Chemistry*, 124, 186–199. DOI: 10.1016/j.ejmech.2016.08.026 PMID: 27573544

Cheekavolu, C. (2016). In vivo and In vitro Anti-Inflammatory Activity of Indazole and Its Derivatives. *Journal of Clinical and Diagnostic Research : JCDR*, 10(9), FF01–FF06. DOI: 10.7860/JCDR/2016/19338.8465 PMID: 27790461

Chen, G., Hu, M., & Peng, Y. (2018). Switchable Synthesis of 3-Substituted 1 H -Indazoles and 3,3-Disubstituted 3 H -Indazole-3-phosphonates Tuned by Phosphoryl Groups. *The Journal of Organic Chemistry*, 83(3), 1591–1597. DOI: 10.1021/acs.joc.7b02857 PMID: 29283256

Dar, B. A., Safvi, S. W., & Rizvi, M. A. (2018). Microwave Assisted Expeditious and Green Cu(II)-Clay Catalyzed Domino One-Pot Three Component Synthesis of 2H-indazoles. *Bulletin of Chemical Reaction Engineering & Catalysis*, 13(1), 82–88. DOI: 10.9767/bcrec.13.1.963.82-88

Degnan, A. P., Tora, G. O., Huang, H., Conlon, D. A., Davis, C. D., Hanumegowda, U. M., Hou, X., Hsiao, Y., Hu, J., Krause, R., Li, Y.-W., Newton, A. E., Pieschl, R. L., Raybon, J., Rosner, T., Sun, J.-H., Taber, M. T., Taylor, S. J., Wong, M. K., & Gillman, K. W. (2016a). Discovery of Indazoles as Potent, Orally Active Dual Neurokinin 1 Receptor Antagonists and Serotonin Transporter Inhibitors for the Treatment of Depression. *ACS Chemical Neuroscience*, 7(12), 1635–1640. DOI: 10.1021/acschemneuro.6b00337 PMID: 27744678

Degnan, A. P., Tora, G. O., Huang, H., Conlon, D. A., Davis, C. D., Hanumegowda, U. M., Hou, X., Hsiao, Y., Hu, J., Krause, R., Li, Y.-W., Newton, A. E., Pieschl, R. L., Raybon, J., Rosner, T., Sun, J.-H., Taber, M. T., Taylor, S. J., Wong, M. K., & Gillman, K. W. (2016b). Discovery of Indazoles as Potent, Orally Active Dual Neurokinin 1 Receptor Antagonists and Serotonin Transporter Inhibitors for the Treatment of Depression. *ACS Chemical Neuroscience*, 7(12), 1635–1640. DOI: 10.1021/acschemneuro.6b00337 PMID: 27744678

Dong, J., Zhang, Q., Wang, Z., Huang, G., & Li, S. (2018). Recent Advances in the Development of Indazole-based Anti-cancer Agents. *ChemMedChem*, 13(15), 1490–1507. DOI: 10.1002/cmdc.201800253 PMID: 29863292

Dubost, E., Stiebing, S., Ferrary, T., Cailly, T., Fabis, F., & Collot, V. (2014). A general synthesis of diversely substituted indazoles and hetero-aromatic derivatives from o-halo-(het)arylaldehydes or -phenones. *Tetrahedron*, 70(44), 8413–8418. DOI: 10.1016/j.tet.2014.07.092

Dunga, A. K., Allaka, T. R., Kethavarapu, Y., Nechipadappu, S. K., Pothana, P., Ravada, K., Kashanna, J., & Kishore, P. V. V. N. (2022). Design, synthesis and biological evaluation of novel substituted indazole-1,2,3-triazolyl-1,3,4-oxadiazoles: Anti-microbial activity evaluation and docking study. *Results in Chemistry*, 4(October), 100605. DOI: 10.1016/j.rechem.2022.100605

Egorova, A., Richter, M., Khrenova, M., Dietrich, E., Tsedilin, A., Kazakova, E., Lepioshkin, A., Jahn, B., Chernyshev, V., Schmidtke, M., & Makarov, V. (2023). Pyrrolo[2,3- e]indazole as a novel chemotype for both influenza A virus and pneumococcal neuraminidase inhibitors. *RSC Advances*, 13(27), 18253–18261. DOI: 10.1039/D3RA02895J PMID: 37350858

Furlotti, G., Alisi, M. A., Cazzolla, N., Ceccacci, F., Garrone, B., Gasperi, T., La Bella, A., Leonelli, F., Loreto, M. A., Magarò, G., Mangano, G., Bettolo, R. M., Masini, E., Miceli, M., Migneco, L. M., & Vitiello, M. (2018). Targeting Serotonin 2A and Adrenergic α1 Receptors for Ocular Antihypertensive Agents: Discovery of 3,4-Dihydropyrazino[1,2-b]indazol-1(2H)-one Derivatives. *ChemMedChem*, 13(15), 1597–1607. DOI: 10.1002/cmdc.201800199 PMID: 29873449

Gandham, S. K., Kudale, A. A., Allaka, T. R., Chepuri, K., & Jha, A. (2024). New indazole–1,2,3–triazoles as potent anti-microbial agents: Design, synthesis, molecular modeling and in silico ADME profiles. *Journal of Molecular Structure*, 1295(P1), 136714. DOI: 10.1016/j.molstruc.2023.136714

Gavara, L., Suchaud, V., Nauton, L., Théry, V., Anizon, F., & Moreau, P. (2013). Identification of pyrrolo[2,3-g]indazoles as new Pim kinase inhibitors. *Bioorganic & Medicinal Chemistry Letters*, 23(8), 2298–2301. DOI: 10.1016/j.bmcl.2013.02.074 PMID: 23499503

Gein, V. L., Lezhnina, D. D., Nosova, N. V., Makhmudov, R. R., & Dmitriev, M. V. (2022). Reaction of 6-Oxocyclohexane-1, 3-dicarboxamides with Binucleophilic Reagents. *Antinociceptive Activity.*, 58(11), 1610–1616.

Gein, V. L., Yankin, A. N., Nosova, N. V., Levandovskaya, E. B., Novikova, V. V., & Rudakova, I. P. (2019). Synthesis and Biological Activity of 4,5,6,7-Tetrahydro-2H-indazole Derivatives. *Russian Journal of General Chemistry*, 89(6), 1169–1176. DOI: 10.1134/S1070363219060112

González-Naranjo, P., Pérez, C., González-Sánchez, M., Gironda-Martínez, A., Ulzurrun, E., Bartolomé, F., Rubio-Fernández, M., Martin-Requero, A., Campillo, N. E., & Páez, J. A. (2022). Multitarget drugs as potential therapeutic agents for alzheimer's disease. A new family of 5-substituted indazole derivatives as cholinergic and BACE1 inhibitors. *Journal of Enzyme Inhibition and Medicinal Chemistry*, 37(1), 2348–2356. DOI: 10.1080/14756366.2022.2117315 PMID: 36050834

Gujja, V., Sadineni, K., Epuru, M. R., Rao Allaka, T., Banothu, V., Gunda, S. K., & Koppula, S. K. (2023). Synthesis and in Silico Studies of Some New 1,2,3-Triazolyltetrazole Bearing Indazole Derivatives as Potent Anti-microbial Agents. *Chemistry & Biodiversity*, 20(12), e202301232. Advance online publication. DOI: 10.1002/cbdv.202301232 PMID: 37988365

Hari, Y., Sone, R., & Aoyama, T. (2009). Facile two-step synthesis of 3-substituted indazoles using diazo(trimethylsilyl)methylmagnesium bromide. *Organic & Biomolecular Chemistry*, 7(13), 2804. DOI: 10.1039/b907796k PMID: 19532998

Harish, K. P., Mohana, K. N., & Mallesha, L. (2013). Synthesis of indazole substituted-1,3,4-thiadiazoles and their anticonvulsant activity. *Drug Invention Today*, 5(2), 92–99. DOI: 10.1016/j.dit.2013.06.002

Jadhav, K., Sirvi, A., Janjal, A., Kashyap, M. C., & Sangamwar, A. T. (2024). Utilization of Lipophilic Salt and Phospholipid Complex in Lipid-Based Formulations to Modulate Drug Loading and Oral Bioavailability of Pazopanib. *AAPS PharmSciTech*, 25(3), 59. DOI: 10.1208/s12249-024-02780-3 PMID: 38472682

Jismy, B., El Qami, A., Pišlar, A., Frlan, R., Kos, J., Gobec, S., Knez, D., & Abarbri, M. (2021). Pyrimido[1,2-b]indazole derivatives: Selective inhibitors of human monoamine oxidase B with neuroprotective activity. *European Journal of Medicinal Chemistry*, 209, 112911. DOI: 10.1016/j.ejmech.2020.112911 PMID: 33071056

Kaczor, A. A., Targowska-Duda, K. M., Stępnicki, P., Silva, A. G., Koszła, O., Kędzierska, E., Grudzińska, A., Kruk-Słomka, M., Biała, G., & Castro, M. (2021). N-(3-{4-[3-(trifluoromethyl)phenyl]piperazin-1-yl}propyl)-1H-indazole-3-carboxamide (D2AAK3) as a potential antipsychotic: In vitro, in silico and in vivo evaluation of a multi-target ligand. *Neurochemistry International*, 146, 105016. DOI: 10.1016/j.neuint.2021.105016 PMID: 33722679

Karmakar, R., & Mukhopadhyay, C. (2023). Green Synthetic Approach: A Well-organized Eco-friendly Tool for Synthesis of Bio-active Fused Heterocyclic Compounds. *Current Green Chemistry*, 10(1), 5–24. DOI: 10.2174/2213346110666230120154516

Kaur, M., & Kumar, R. (2018). C-N and N-N bond formation via Reductive Cyclization: Progress in Cadogan /Cadogan-Sundberg Reaction‡. *ChemistrySelect*, 3(19), 5330–5340. DOI: 10.1002/slct.201800779

Khan, Y., Khan, S., Hussain, R., Rehman, W., Maalik, A., Gulshan, U., Attwa, M. W., Darwish, H. W., Ghabbour, H. A., & Ali, N. (2023). Identification of Indazole-Based Thiadiazole-Bearing Thiazolidinone Hybrid Derivatives: Theoretical and Computational Approaches to Develop Promising Anti-Alzheimer's Candidates. *Pharmaceuticals (Basel, Switzerland)*, 16(12), 1667. Advance online publication. DOI: 10.3390/ph16121667 PMID: 38139795

Kim, O. S., Jang, J. H., Kim, H. T., Han, S. J., Tsui, G. C., & Joo, J. M. (2017). Synthesis of Fluorescent Indazoles by Palladium-Catalyzed Benzannulation of Pyrazoles with Alkynes. *Organic Letters*, 19(6), 1450–1453. DOI: 10.1021/acs.orglett.7b00410 PMID: 28271896

Kumar, V., Gupta, K., Sirbaiya, A. K., & Rahman, A. (2022). *A Review of Indazole derivatives in Pharmacotherapy of inflammation A REVIEW OF INDAZOLE DERIVATIVES IN PHARMACOTHERAPY OF INFLAMMATION. April 2023*, 1–7.

Kumar, V., Sirbaiya, A. K., Nematullah, M., Haider, M. F., & Rahman, M. A. (2024). Synthesis of 1,3-substituted 1H-indazole derivatives and evaluation of anti-inflammatory activity in Sprague Dawley rats. *Intelligent Pharmacy*, 2(1), 40–44. DOI: 10.1016/j.ipha.2023.09.009

Li, X., Ye, X., Wei, C., Shan, C., Wojtas, L., Wang, Q., & Shi, X. (2020). Diazo Activation with Diazonium Salts: Synthesis of Indazole and 1,2,4-Triazole. *Organic Letters*, 22(11), 4151–4155. DOI: 10.1021/acs.orglett.0c01232 PMID: 32463244

Liang, Q., Qiao, Z., Zhou, Q., Xue, D., Wang, K., & Shao, L. (2022). Discovery of Potent and Selective Transient Receptor Potential Vanilloid 1 (TRPV1) Agonists with Analgesic Effects In Vivo Based on the Functional Conversion Induced by Altering the Orientation of the Indazole Core. *Journal of Medicinal Chemistry*, 65(17), 11658–11678. DOI: 10.1021/acs.jmedchem.2c00469 PMID: 36008373

Liu, H.-J., Hung, S.-F., Chen, C.-L., & Lin, M.-H. (2013). A method for the regioselective synthesis of 1-alkyl-1H-indazoles. *Tetrahedron*, 69(19), 3907–3912. DOI: 10.1016/j.tet.2013.03.042

Liu, J., Peng, X., Dai, Y., Zhang, W., Ren, S., Ai, J., Geng, M., & Li, Y. (2015). Design, synthesis and biological evaluation of novel FGFR inhibitors bearing an indazole scaffold. *Organic & Biomolecular Chemistry*, 13(28), 7643–7654. DOI: 10.1039/C5OB00778J PMID: 26080733

Liu, Y. Y., Guo, Z., Wang, J. Y., Wang, H. M., Da Qi, J., Ma, J., Piao, H.-R., Jin, C. H., & Jin, X. (2021). Synthesis and evaluation of the epithelial-to- mesenchymal inhibitory activity of indazole-derived imidazoles as dual ALK5/p38α MAP inhibitors. *European Journal of Medicinal Chemistry*, 216, 113311. DOI: 10.1016/j.ejmech.2021.113311 PMID: 33677350

Liu, Z., Chen, L., Yu, P., Zhang, Y., Fang, B., Wu, C., Luo, W., Chen, X., Li, C., & Liang, G. (2019). Discovery of 3-(Indol-5-yl)-indazole Derivatives as Novel Myeloid Differentiation Protein 2/Toll-like Receptor 4 Antagonists for Treatment of Acute Lung Injury. *Journal of Medicinal Chemistry*, 62(11), 5453–5469. DOI: 10.1021/acs.jmedchem.9b00316 PMID: 30998353

Liu, Z., Wang, L., Tan, H., Zhou, S., Fu, T., Xia, Y., Zhang, Y., & Wang, J. (2014). Synthesis of 1H-indazoles from N-tosylhydrazones and nitroaromatic compounds. *Chemical Communications (Cambridge)*, 50(39), 5061–5063. DOI: 10.1039/C4CC00962B PMID: 24714999

Luo, G., Chen, L., Easton, A., Newton, A., Bourin, C., Shields, E., Mosure, K., Soars, M. G., Knox, R. J., Matchett, M., Pieschl, R. L., Post-Munson, D. J., Wang, S., Herrington, J., Graef, J., Newberry, K., Sivarao, D. V., Senapati, A., Bristow, L. J., & Dzierba, C. (2019). Discovery of Indole- and Indazole-acylsulfonamides as Potent and Selective NaV1.7 Inhibitors for the Treatment of Pain. *Journal of Medicinal Chemistry*, 62(2), 831–856. DOI: 10.1021/acs.jmedchem.8b01550 PMID: 30576602

Matsumura, N., Kikuchi-Utsumi, K., Sakamaki, K., Watabe, M., Aoyama, K., & Nakaki, T. (2013). Anticonvulsant action of indazole. *Epilepsy Research*, 104(3), 203–216. DOI: 10.1016/j.eplepsyres.2012.11.001 PMID: 23219048

Morelli, M. B., Amantini, C., Santoni, M., Soriani, A., Nabissi, M., Cardinali, C., Santoni, A., & Santoni, G. (2015). Axitinib induces DNA damage response leading to senescence, mitotic catastrophe, and increased NK cell recognition in human renal carcinoma cells. *Oncotarget*, 6(34), 36245–36259. DOI: 10.18632/oncotarget.5768 PMID: 26474283

Niu, Y., Zhou, L., Meng, L., Chen, S., Ma, C., Liu, Z., & Kang, W. (2020). Recent Progress on Chemical Constituents and Pharmacological Effects of the Genus Nigella. *Evidence-Based Complementary and Alternative Medicine, 2020*(Figure 1), 1–15. DOI: 10.1155/2020/6756835

Pal, D., Song, I., Dashrath Warkad, S., Song, K., Seong Yeom, G., Saha, S., Shinde, P. B., & Balasaheb Nimse, S. (2022). Indazole-based microtubule-targeting agents as potential candidates for anti-cancer drugs discovery. *Bioorganic Chemistry*, 122(January), 105735. DOI: 10.1016/j.bioorg.2022.105735 PMID: 35298962

Patel, M. R., Pandya, K. G., Lau-Cam, C. A., Singh, S., Pino, M. A., Billack, B., Degenhardt, K., & Talele, T. T. (2012). Design and synthesis of N-substituted indazole-3-carboxamides as poly(ADP-ribose)polymerase-1 (PARP-1) inhibitors. *Chemical Biology & Drug Design*, 79(4), 488–496. DOI: 10.1111/j.1747-0285.2011.01302.x PMID: 22177599

Pérez-Villanueva, J., Yépez-Mulia, L., González-Sánchez, I., Palacios-Espinosa, J., Soria-Arteche, O., Sainz-Espuñes, T., Cerbón, M., Rodríguez-Villar, K., Rodríguez-Vicente, A., Cortés-Gines, M., Custodio-Galván, Z., & Estrada-Castro, D. (2017). Synthesis and Biological Evaluation of 2H-Indazole Derivatives: Towards Antimicrobial and Anti-Inflammatory Dual Agents. *Molecules (Basel, Switzerland)*, 22(11), 1864. DOI: 10.3390/molecules22111864 PMID: 29088121

Polo, E., Trilleras, J., Ramos, J., Galdámez, A., Quiroga, J., & Gutierrez, M. (2016). Efficient MW-Assisted Synthesis, Spectroscopic Characterization, X-ray and Antioxidant Properties of Indazole Derivatives. *Molecules (Basel, Switzerland)*, 21(7), 903. DOI: 10.3390/molecules21070903 PMID: 27409599

Puri, S., Sawant, S., & Juvale, K. (2023). A comprehensive review on the indazole based derivatives as targeted anti-cancer agents. *Journal of Molecular Structure*, 1284, 135327. DOI: 10.1016/j.molstruc.2023.135327

Qian, S., He, T., Wang, W., He, Y., Zhang, M., Yang, L., Li, G., & Wang, Z. (2016). Discovery and preliminary structure–activity relationship of 1H-indazoles with promising indoleamine-2,3-dioxygenase 1 (IDO1) inhibition properties. *Bioorganic & Medicinal Chemistry*, 24(23), 6194–6205. DOI: 10.1016/j.bmc.2016.10.003 PMID: 27769672

Reddy, A. V., Gogireddy, S., Dubey, P. K., B, M. R., & B, V.REDDY. (2015). Design, synthesis and characterization of 1 H-pyridin-4-yl-3, 5-disubstituted indazoles and their anti-inflammatory and analgesic activity. *Journal of Chemical Sciences*, 127(3), 433–438. DOI: 10.1007/s12039-015-0792-3

Reddy, G. (2018). Citric acid Mediated One-pot Regioselective Synthesis of N-Alkylated Indazoles: An Efficient Green Strategy. *Trends in Green Chemistry*, 04(01). Advance online publication. DOI: 10.21767/2471-9889.100021

Rodríguez-Villar, K., Yépez-Mulia, L., Cortés-Gines, M., Aguilera-Perdomo, J. D., Quintana-Salazar, E. A., Olascoaga Del Angel, K. S., Cortés-Benítez, F., Palacios-Espinosa, J. F., Soria-Arteche, O., & Pérez-Villanueva, J. (2021). Synthesis, Antiprotozoal Activity, and Cheminformatic Analysis of 2-Phenyl-2H-Indazole Derivatives. *Molecules (Basel, Switzerland)*, 26(8), 2145. DOI: 10.3390/molecules26082145 PMID: 33917871

Rooney, L., Vidal, A., D'Souza, A. M., Devereux, N., Masick, B., Boissel, V., West, R., Head, V., Stringer, R., Lao, J., Petrus, M. J., Patapoutian, A., Nash, M., Stoakley, N., Panesar, M., Verkuyl, J. M., Schumacher, A. M., Petrassi, H. M., & Tully, D. C. (2014). Discovery, optimization, and biological evaluation of 5-(2- (trifluoromethyl)phenyl)indazoles as a novel class of transient receptor potential A1 (TRPA1) antagonists. *Journal of Medicinal Chemistry*, 57(12), 5129–5140. DOI: 10.1021/jm401986p PMID: 24884675

Rullo, M., La Spada, G., Miniero, D. V., Gottinger, A., Catto, M., Delre, P., Mastromarino, M., Latronico, T., Marchese, S., Mangiatordi, G. F., Binda, C., Linusson, A., Liuzzi, G. M., & Pisani, L. (2023). Bioisosteric replacement based on 1,2,4-oxadiazoles in the discovery of 1H-indazole-bearing neuroprotective MAO B inhibitors. *European Journal of Medicinal Chemistry*, 255(April), 115352. DOI: 10.1016/j.ejmech.2023.115352 PMID: 37178666

Sączewski, J., Hudson, A., Scheinin, M., Wasilewska, A., Sączewski, F., Rybczyńska, A., Ferdousi, M., Laurila, J. M., Boblewski, K., Lehmann, A., Watts, H., & Ma, D. (2016). Transfer of SAR information from hypotensive indazole to indole derivatives acting at α-adrenergic receptors: In vitro and in vivo studies. *European Journal of Medicinal Chemistry*, 115, 406–415. DOI: 10.1016/j.ejmech.2016.03.026 PMID: 27031216

Saketi, J. M. R., Boddapati, S. N. M., M, R., Adil, S. F., Shaik, M. R., Alduhaish, O., Siddiqui, M. R. H., & Bollikolla, H. B. (2020). Pd(PPh3)4 Catalyzed Synthesis of Indazole Derivatives as Potent Anti-cancer Drug. *Applied Sciences (Basel, Switzerland)*, 10(11), 3792. DOI: 10.3390/app10113792

Saketi, J. R., Boddapati, S. N. M., Raghuram, M., Koduri, G. B., & Bollikolla, H. (2023). A Recent Study on Anti-microbial Properties of Some Novel Substituted Indazoles. In Novel Aspects on Chemistry and Biochemistry Vol. 3 (pp. 1–12). B P International (a part of SCIENCEDOMAIN International). DOI: 10.9734/bpi/nacb/v3/5724B

Sapnakumari, M., Narayana, B., Sarojini, B. K., & Madhu, L. N. (2014). Synthesis of new indazole derivatives as potential antioxidant agents. *Medicinal Chemistry Research*, 23(5), 2368–2376. DOI: 10.1007/s00044-013-0835-6

Schmidt, A., & Guan, Z. (2012). Mesomeric Betaines and N-Heterocyclic Carbenes of Pyrazole and Indazole. *Synthesis*, 44(21), 3251–3268. DOI: 10.1055/s-0032-1316787

Shaikh, F., Arif, M., Khushtar, M., Nematullah, M., & Rahman, M. A. (2024). Synthesis and evaluation of antibacterial activity of novel 3-methyl-1H-indazole derivatives. *Intelligent Pharmacy*, 2(1), 12–16. DOI: 10.1016/j.ipha.2023.09.003

Shan, Y., Dong, J., Pan, X., Zhang, L., Zhang, J., Dong, Y., & Wang, M. (2015). Expanding the structural diversity of Bcr-Abl inhibitors: Dibenzoylpiperazin incorporated with 1H-indazol-3-amine. *European Journal of Medicinal Chemistry*, 104, 139–147. DOI: 10.1016/j.ejmech.2015.09.034 PMID: 26451772

Sharghi, H., & Aberi, M. (2014). Ligand-Free Copper(I) Oxide Nanoparticle Catalyzed Three-Component Synthesis of 2H-Indazole Derivatives from 2-Halobenzaldehydes, Amines and Sodium Azide in Polyethylene Glycol as a Green Solvent. *Synlett*, 25(08), 1111–1115. DOI: 10.1055/s-0033-1340979

Sheng, R., Li, S., Lin, G., Shangguan, S., Gu, Y., Qiu, N., Cao, J., He, Q., Yang, B., & Hu, Y. (2015). Novel potent HIF-1 inhibitors for the prevention of tumor metastasis: Discovery and optimization of 3-aryl-5-indazole-1,2,4-oxadiazole derivatives. *RSC Advances*, 5(100), 81817–81830. DOI: 10.1039/C5RA15191K

Shuai, W., Bu, F., Zhu, Y., Wu, Y., Xiao, H., Pan, X., Zhang, J., Sun, Q., Wang, G., & Ouyang, L. (2023). Discovery of Novel Indazole Chemotypes as Isoform-Selective JNK3 Inhibitors for the Treatment of Parkinson's Disease. *Journal of Medicinal Chemistry*, 66(2), 1273–1300. DOI: 10.1021/acs.jmedchem.2c01410 PMID: 36649216

Sisay, M., & Edessa, D. (2017). PARP inhibitors as potential therapeutic agents for various cancers: Focus on niraparib and its first global approval for maintenance therapy of gynecologic cancers. *Gynecologic Oncology Research and Practice*, 4(1), 18. DOI: 10.1186/s40661-017-0055-8 PMID: 29214031

Song, P., Chen, M., Ma, X., Xu, L., Liu, T., Zhou, Y., & Hu, Y. (2015). Identification of novel inhibitors of Aurora A with a 3-(pyrrolopyridin-2-yl)indazole scaffold. *Bioorganic & Medicinal Chemistry*, 23(8), 1858–1868. DOI: 10.1016/j.bmc.2015.02.004 PMID: 25771484

Sow, A. A., Pahmeier, F., Ayotte, Y., Anton, A., Mazeaud, C., Charpentier, T., Angelo, L., Woo, S., Cerikan, B., Falzarano, D., Abrahamyan, L., Lamarre, A., Labonté, P., Cortese, M., Bartenschlager, R., LaPlante, S. R., & Chatel-Chaix, L. (2023). N-Phenylpyridine-3-Carboxamide and 6-Acetyl-1H-Indazole Inhibit the RNA Replication Step of the Dengue Virus Life Cycle. *Antimicrobial Agents and Chemotherapy*, 67(2), e01331-22. Advance online publication. DOI: 10.1128/aac.01331-22 PMID: 36700643

Stępnicki, P., Wronikowska-Denysiuk, O., Zięba, A., Targowska-Duda, K. M., Bartyzel, A., Wróbel, M. Z., Wróbel, T. M., Szałaj, K., Chodkowski, A., Mirecka, K., Budzyńska, B., Fornal, E., Turło, J., Castro, M., & Kaczor, A. A. (2023). Novel multi-target ligands of dopamine and serotonin receptors for the treatment of schizophrenia based on indazole and piperazine scaffolds-synthesis, biological activity, and structural evaluation. *Journal of Enzyme Inhibition and Medicinal Chemistry*, 38(1), 2209828. DOI: 10.1080/14756366.2023.2209828 PMID: 37184096

Taha, M., Gilani, S. J., Kazmi, I., Rahim, F., Adalat, B., Ullah, H., Nawaz, F., Wadood, A., Ali, Z., Shah, S. A. A., & Khan, K. M. (2024). Synthesis, biological evaluation and molecular docking study of indazole based schiff base analogues as new anti-diabetic inhibitors. *Journal of Molecular Structure, 1300*(September 2023), 137189. DOI: 10.1016/j.molstruc.2023.137189

Tan, C., Yang, S.-J., Zhao, D.-H., Li, J., & Yin, L.-Q. (2022). Antihypertensive activity of indole and indazole analogues: A review. *Arabian Journal of Chemistry*, 15(5), 103756. DOI: 10.1016/j.arabjc.2022.103756

Tang, M., Kong, Y., Chu, B., & Feng, D. (2016). Copper(I) Oxide-Mediated Cyclization of o -Haloaryl N -Tosylhydrazones: Efficient Synthesis of Indazoles. *Advanced Synthesis & Catalysis*, 358(6), 926–939. DOI: 10.1002/adsc.201500953

Teixeira, F. C., Ramos, H., Antunes, I. F., Curto, M. J. M., Duarte, M. T., & Bento, I. (2006). Synthesis and structural characterization of 1- and 2-substituted indazoles: Ester and carboxylic acid derivatives. *Molecules (Basel, Switzerland)*, 11(11), 867–889. DOI: 10.3390/11110867 PMID: 18007393

Uppulapu, S. K., Alam, M., Kumar, S., & Banerjee, S. K. (2022). Indazole and its Derivatives in Cardiovascular Diseases: Overview, Current Scenario, and Future Perspectives. *Current Topics in Medicinal Chemistry*, 22(14), 1177–1188. DOI: 10.2174/1568026621666211214151534 PMID: 34906057

Varughese, D. J., Manhas, M. S., & Bose, A. K. (2006). Microwave enhanced greener synthesis of indazoles via nitrenes. *Tetrahedron Letters*, 47(38), 6795–6797. DOI: 10.1016/j.tetlet.2006.07.062

Wang, C., Zhu, M., Long, X., Wang, Q., Wang, Z., & Ouyang, G. (2023). Design, Synthesis and Antitumor Activity of 1H-indazole-3-amine Derivatives. *International Journal of Molecular Sciences*, 24(10), 8686. DOI: 10.3390/ijms24108686 PMID: 37240028

Wang, Y., Yan, M., Ma, R., & Ma, S. (2015). Synthesis and Antibacterial Activity of Novel 4-Bromo-1 H -Indazole Derivatives as FtsZ Inhibitors. *Archiv der Pharmazie*, 348(4), 266–274. DOI: 10.1002/ardp.201400412 PMID: 25773717

Wu, B., Wang, H.-L., Cee, V. J., Lanman, B. A., Nixey, T., Pettus, L., Reed, A. B., Wurz, R. P., Guerrero, N., Sastri, C., Winston, J., Lipford, J. R., Lee, M. R., Mohr, C., Andrews, K. L., & Tasker, A. S. (2015). Discovery of 5-(1H-indol-5-yl)-1,3,4-thiadiazol-2-amines as potent PIM inhibitors. *Bioorganic & Medicinal Chemistry Letters*, 25(4), 775–780. DOI: 10.1016/j.bmcl.2014.12.091 PMID: 25616902

Xiao, T., Tang, J.-F., Meng, G., Pannecouque, C., Zhu, Y.-Y., Liu, G.-Y., Xu, Z.-Q., Wu, F.-S., Gu, S.-X., & Chen, F.-E. (2020). Indazolyl-substituted piperidin-4-yl-aminopyrimidines as HIV-1 NNRTIs: Design, synthesis and biological activities. *European Journal of Medicinal Chemistry*, 186, 111864. DOI: 10.1016/j.ejmech.2019.111864 PMID: 31767136

Xiao-Feng, L., Wen-Ting, Z., Yuan-Yuan, X., Chong-Fa, L., Lu, Z., Jin-Jun, R., & Wen-Ya, W. (2016). Protective role of 6-Hydroxy-1-H-Indazole in an MPTP-induced mouse model of Parkinson's disease. *European Journal of Pharmacology*, 791, 348–354. DOI: 10.1016/j.ejphar.2016.08.011 PMID: 27614126

Xu, G., Gaul, M. D., Song, F., Du, F., Liang, Y., DesJarlais, R. L., DiLoreto, K., Shook, B., Rentzeperis, D., Santulli, R., Eckardt, A., & Demarest, K. (2019). Discovery of potent and orally bioavailable indazole-based glucagon receptor antagonists for the treatment of type 2 diabetes. *Bioorganic & Medicinal Chemistry Letters*, 29(20), 126668. DOI: 10.1016/j.bmcl.2019.126668 PMID: 31519374

Yao, Y., Li, R., Liu, X., Yang, F., Yang, Y., Li, X., Shi, X., Yuan, T., Fang, L., Du, G., Jiao, X., & Xie, P. (2017). Discovery of Novel N-Substituted Prolinamido Indazoles as Potent Rho Kinase Inhibitors and Vasorelaxation Agents. *Molecules (Basel, Switzerland)*, 22(10), 1766. DOI: 10.3390/molecules22101766 PMID: 29048389

Yin, C., Zhong, T., Zheng, X., Li, L., Zhou, J., & Yu, C. (2021). Direct synthesis of indazole derivatives via Rh(iii)-catalyzed C–H activation of phthalazinones and allenes. *Organic & Biomolecular Chemistry*, 19(35), 7701–7705. DOI: 10.1039/D1OB01458G PMID: 34524333

Zhang, T., & Bao, W. (2013). Synthesis of 1 H -Indazoles and 1 H -Pyrazoles via FeBr 3 /O 2 Mediated Intramolecular C–H Amination. *The Journal of Organic Chemistry*, 78(3), 1317–1322. DOI: 10.1021/jo3026862 PMID: 23297649

Chapter 7
Bioactive Five–Membered Heterocycles With Two Heteroatoms Fused With a Benzene Ring (a) Benzimidazole

Paran Jyoti Borpatra

Institute of Science, Banaras Hindu University, India

Mintu Maan Dutta

Arya Vidyapeeth College, India

ABSTRACT

Heterocyclic compounds play a crucial role in medicinal chemistry, serving as key components in the development of pharmacologically active molecules. The therapeutic promise of many synthesized drugs can be attributed to their heterocyclic scaffolds, wherein even minor modifications in the heterocyclic structure can significantly impact the drug's efficacy. Among these, benzimidazoles are particularly significant. These class of compounds comprises a combination of the aromatic benzene ring and an imidazole ring. A significant natural form of benzimidazole found in nature is N-ribosyl-dimethyl benzimidazole, which plays a crucial role in coordinating to the cobalt metal in vitamin B12. Extensive biochemical and pharmacological research has demonstrated that benzimidazoles are highly effective against various strains of microorganisms. Furthermore, they have exhibit a broad spectrum of biological activities, including anti-inflammatory, anticancer, antihistamine, antimicrobial, antifungal, antioxidant, antidiabetic and antiviral activities.

DOI: 10.4018/979-8-3693-7267-8.ch007

INTRODUCTION

Benzimidazole is an aromatic heterocyclic organic compound with the molecular formula $C_7H_6N_2$. Benzimidazoles are also referred to as benzoglyoxalines or benziminazoles.

Particularly in the early literature, they have also been identified as *o*-phenylene-diamine derivatives (Wright, 1951).

The compound benzimidazole is a six-membered bicyclic hetero aromatic molecule in which the imidazole ring's 4- and 5-positions are fused to the benzene ring (Gaba and Mohan, 2016). Particularly the heterocycles containing nitrogen exhibits a wide variety of biological activities, partly because of their resemblance to numerous naturally occurring and artificially produced compounds with known biological activity (DeSimone *et al.*, 2004).

Figure 1. Imidazole ring's 4- and 5-positions are fused to the benzene ring

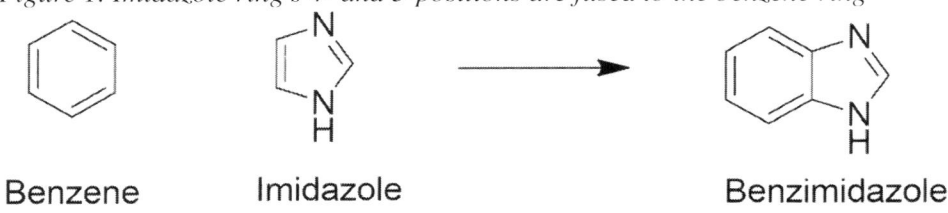

Benzene   Imidazole   Benzimidazole

Hoebrecker synthesized first benzimidazole in 1872 by reducing 2-nitro-4-methylacetanilide to 2,5(or 2,6) dimethylbenzimidazole (Wright 1951). In the 1950s, increased interest in benzimidazole-based chemistry emerged due to the discovery that 5,6-dimethyl-1-(α-D-ribofuranosyl) benzimidazole, which was a crucial component of vitamin B12 structure (Wright 1951; Barker *et al* 1960). The nitrogen atoms in the structure of imidazole and benzimidazole rings cause them to have basic and acidic properties. The hydrogen atom in these rings can be found on either of the two nitrogen atoms in one of two equivalent tautomeric configurations (Gaba and Mohan, 2016). It is advantageous for imidazole and benzimidazole derivatives to easily bind with a range of therapeutic targets and exhibit broad pharmacological actions because of the unique electron-rich property in the structure of these rings (Wright, 1951; Gaba *et al.*, 2010; Bhatnagar *et al.*, 2011; Ingle and Magar, 2011; Gaba and Mohan, 2016).

Derivatives of Benzimidazole have been linked to a wide range of biological activities with positive responses towards anti-viral, anti-oxidant, anti-cancer, anti-inflammatory, anti-urease, anti-fungal, anti-bacterial and proton pump inhibitor properties (Swami *et al.* 2017). Derivatives of benzomidazole have been shown to possess significant antiviral and antitumor properties against a variety of viruses, including the human immune deficiency virus (HIV) and herpes simplex virus type 1 (HSV-1) (Porcari *et al.* 1998; Migawa *et al.* 1998; Denny *et al.* 1990; Forseca *et al.* 2001; Gaba *et al.* 2016).

Synthesis of Benzimidazole using various catalysts:

Owing to the vast range of applications for which these heterocycles are useful, numerous synthesis techniques have been showcased by the various researchers. The most widely employed method for the synthesis of Benzimidazole involves reaction of 1,2-phenylenediamine with aldehydes (Azizian *et al.*, 2016; Dandia *et al.*, 2012; Shelkar *et al.*, 2013). Some common approaches involve the condensation of *o*-phenylenediamine with aldehyde, (Landge *et al.*, 2008; Yu *et al.*, 2009; Azarifar *et al.*, 2010; Malakooti *et al.*, 2014) carboxylic acids, (Preston 1974; Bahrami *et al.*, 2007; Gorepatil *et al.*, 2013; Rambabu *et al.*, 2013; Niknam *et al.*, 2007) alcohols, (Sluiter *et al.*, 2009) nitriles, (Hein *et al.*, 1957; Trivedi *et al.*, 2006) and acyl chlorides (Heravi *et al.*, 2008; Patil *et al.*, 2016) in presence of catalyst.

Literature survey reveals that a variety of catalysts and reagents have been utilized for the synthesis of heterocycles. Few catalysts includes Zn-proline, (Ravi *et al.*, 2007) $FeCl_3.6H_2O$, (Singh *et al.*, 2000) $Na_2S_2O_4$, (Kamal *et al.*, 2014) $ZrCl_4$, (Zhang *et al.*, 2007) $PhI(OAc)_2$, (Du *et al.*, 2007) $CoCl_2.6H_2O$, (Khan *et al.*, 2009) $MgCl_2.6H_2O$, (Ghosh *et al.*, 2015) Perborate sodium, (Yuan *et al.*, 2013) sulfamic acid, (Chakrabarty *et al.*, 2006) nano In_2O_3, (Santra *et al.*, 2012) $Sc(OTf)_3$, (Itoh *et al.*, 2004) $In(OTf)_3$, (Trivedi *et al.*, 2005) $Cu(OTf)_2$, (Chari *et al.*, 2010) $TiCl_3OTf$, (Torabi *et al.*, 2016) SDS, (Bahrami *et al.*, 2010) iron(III)sulfate-silica, (Paul *et al.*, 2012) $MnFe_2O_4$, (Brahmachari *et al.*, 2013) trichloroisocyanuric acid, (Xiao *et al.*, 2009) cetyltrimethyl ammonium bromide, (Yang *et al.*, 2010) Bakers yeast, (Pratap *et al.*, 2009) $HClO_4$/PANI, (Abdollahi-Alibeik *et al.*, 2009) glucose oxidase–peroxidase, (Kumar *et al.* 2010) Dowex 50W, (Mukhopadhyay *et al.*, 2009) silica sulfuric acid, (Chen *et al.*, 2013) Montmorillonite K-10, (Chen *et al.*, 2015) $VOSO_4$, (Digwal *et al.*, 2016) sulfonic-acid-functionalized activated carbon (MTLAC–SA), (Goswami *et al.*, 2018) $CoFe_2O_4@SiO_2$–SO_3H (CF-SA), (Dutta *et al.*, 2019) CuI nanoparticles (Reddy *et al.*, 2016) and so on.

Synthetic Approaches of Benzimidazole Derivatives

Numerous methods have been devised and validated for synthesizing benzimidazoles. One such strategy involves condensing *o*-phenylenediamine directly with formic acid, providing a straightforward route for benzimidazole preparation. Additionally, various strategies exist for synthesizing benzimidazoles. However, some of these approaches encounter disadvantages, including the need for excessive oxidative catalysts, the presence of metals, prolonged reaction times, laborious workup procedures, low yields, some cases usage of high temperatures, handling difficulties, side product formation, stringent reaction conditions, and the utilisation of hazardous solvents. The pharmaceutical industry seeks metal-free, eco-friendly, stable, and cost-effective catalysts and reagents.

1. CONDENSATION OF *O*-PHENYLENEDIAMINE WITH ALDEHYDE

Benzimidazole scaffolds can be synthesized through condensing various aldehydes with *o*-phenylenediamines under several conditions. This reaction proceeds *via* an oxidation step, which can occur in either the presence of atmospheric oxygen or more effectively with various oxidizing agents as demonstrated in Figure 2 (Hashem *et al.*, 2021).

Figure 2. A general synthetic approach to benzimidazoles: condensing o-phenylenediamine with aldehydes

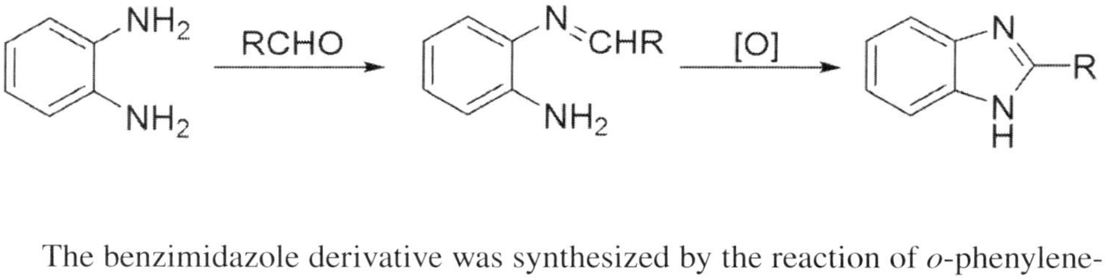

The benzimidazole derivative was synthesized by the reaction of *o*-phenylenediamine and 2,4-dichlorobenzaldehyde in the presence of sodium metabisulfite, as depicted in Figure 3 (Suheyla *et al.*, 2002).

Figure 3. Sodium metabisulfite-catalyzed reaction for benzimidazoles

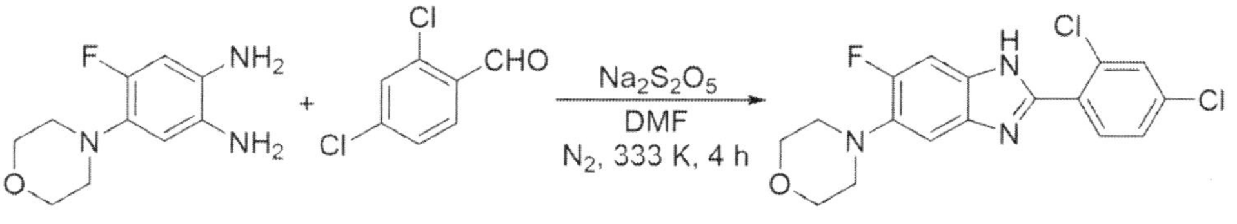

In 2006, De *et al.* devised a methodology for synthesizing benzimidazole employing a recoverable catalyst, indium triflate i.e., $In(OTf)_3$. This procedure is notable for its ease of work-up and good product yields under solvent-free conditions. They also experimented with various catalysts including $Cu(OTf)_2$, $La(OTf)_3$, $Ce(OTf)_3$, $Lu(OTf)_3$ and $Nd(OTf)_3$. Among these, indium triflate proved to be the most effective, delivering excellent yields of benzimidazole scaffolds (Scheme 3a) (Trivedi *et al.*, 2006). Building on these advancements, Kidwai and co-workers (2010) developed an innovative strategy utilizing ceric ammonium nitrate (CAN) as a catalyst for synthesizing benzimidazole scaffolds. In this method, *o*-phenylenediamine reacts with several aldehydes in a polyethylene glycol (PEG) medium, catalyzed by CAN, to produce benzimidazole derivatives with excellent yields (Scheme 3b). This novel method highlights the efficiency and effectiveness of the PEG/CAN system in promoting high-yielding benzimidazole formation (Kidwai *et al.*, 2010). In 2011, Karimi-Jaberi *et al.* proposed a straightforward, eco-friendly one-pot synthetic strategy for the synthesis of benzimidazole derivatives. This method involves reacting *o*-phenylenediamine with various aromatic aldehydes in water at ambient temperature, using boric acid as a promoter for the reaction. This approach furnished good yields of the anticipated product as demonstrated in Scheme 3c. (Karimi-Jaberi *et al.*, 2012) Similarly, in the same year, Rekha *et al.* investigated a catalytic system using alumina, zirconia, manganese oxide/alumina, and manganese oxide/zirconia for the intermolecular cyclization of *o*-phenylenediamine with various aldehydes in ethanol. Among these, $MnZrO_2$ and $MnAl_2O_3$ proved to be the most effective catalysts, yielding excellent yields of benzimidazole derivatives (Scheme 3d) (Rekha *et al.*, 2012). Further advancing the field, in 2012, Shaikh and co-worker reported an approach for the formation of benzimidazole derivatives. The reaction entails the oxidation of the carbon-nitrogen bond between *o*-phenylenediamine and aldehydes utilizing $K_4[Fe(CN)_6]$ catalyst as shown in Figure 4 (Shaikh *et al.*, 2012).

Figure 4. Condensing o-phenylenediamine with aldehydes under various conditions.

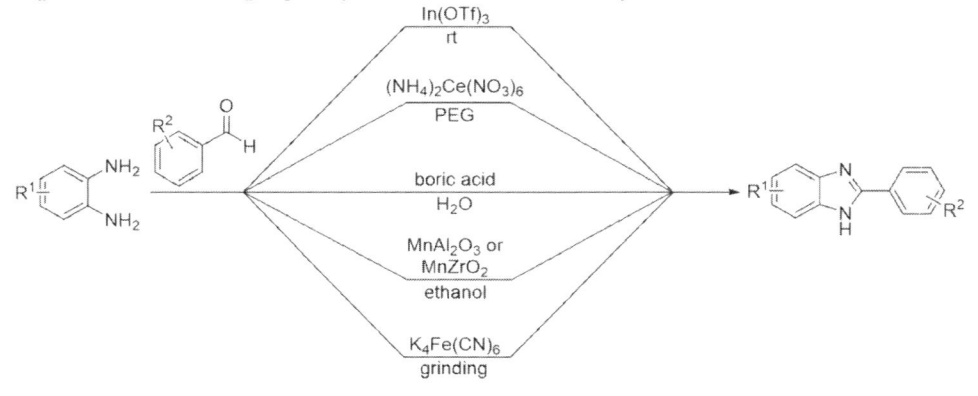

Venkateswarlu and co-workers devised a synthetic strategy for benzimidazole scaffolds using *o*-phenylenediamine with various aldehydes in the presence of a lanthanum chloride catalyst as shown in Figure 5. A variety of aldehydes with different functional groups were condensed effectively with *o*-phenylenediamine, furnishing excellent yields of benzimidazole derivatives (Venkateswarlu *et al.*, 2013).

Figure 5. Lanthanum chloride catalysed reaction for benzimidazoles

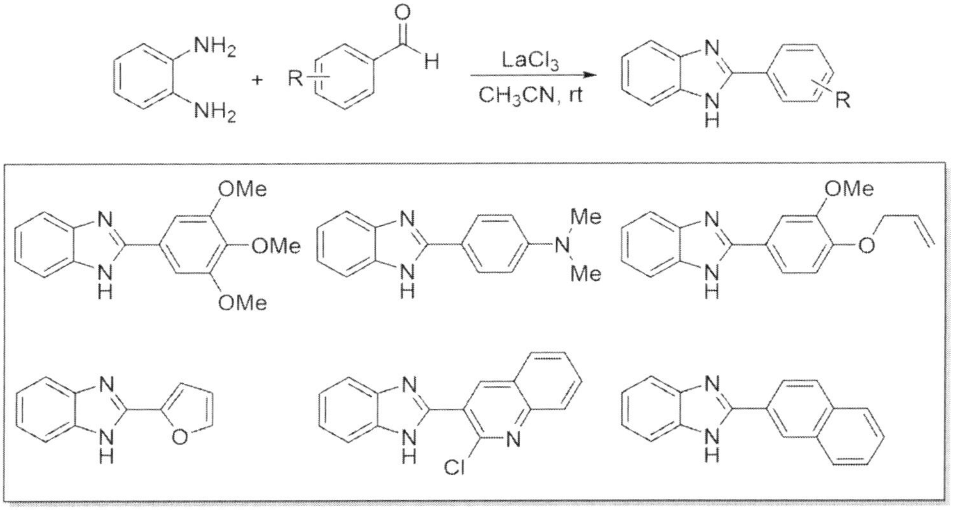

Following this, Kathirvelan and co-workers reported a one-pot synthetic strategy for benzimidazole scaffolds, achieving good yields (Figure 6). This approach entails the condensation of *o*-phenylenediamine with various aldehydes using ammonium chloride as a cost-effective catalyst at 80–90°C, resulting in good yields (Kathirvelan *et al.*, 2013).

Figure 6. Ammonium chloride catalysed reaction for benzimidazoles

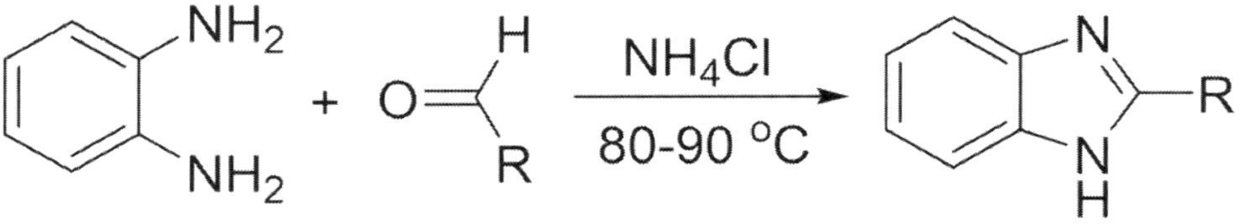

Expanding on these methods, Khunt *et al.* developed a strategy for synthesizing benzimidazole scaffolds by reacting *o*-phenylenediamine with several aldehydes using environment friendly solvents such as polyethylene glycol (PEG)-200, 400, 600, and 800. The optimal yield was achieved using PEG-400 at 80–85 °C (Figure 7). This novel approach employs PEG and utilizes a glycerol/water reaction method to synthesize the target molecules (Khunt *et al.*, 2014).

Figure 7. Benzimidazole synthesis in the presence of glycerol-water and PEG

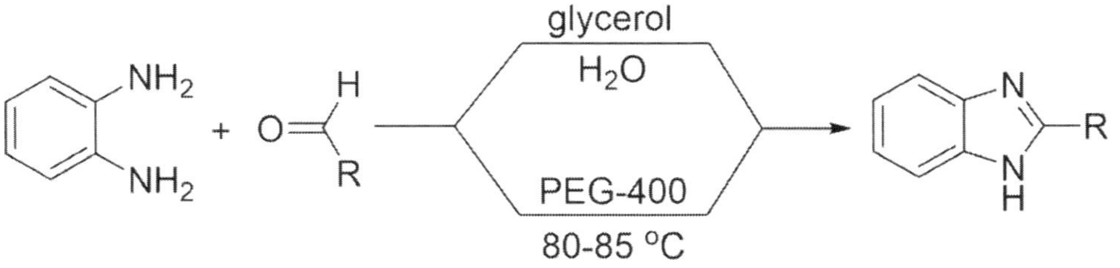

Hao and co-workers devised a highly effective methodology for synthesizing benzimidazoles using gold nanocomposites, including Au/ZnO, Au/TiO$_2$, and Au/Al$_2$O$_3$ as shown in Figure 8 (Hao *et al.*, 2014).

Figure 8. Au/TiO$_2$ catalysed reaction for benzimidazoles

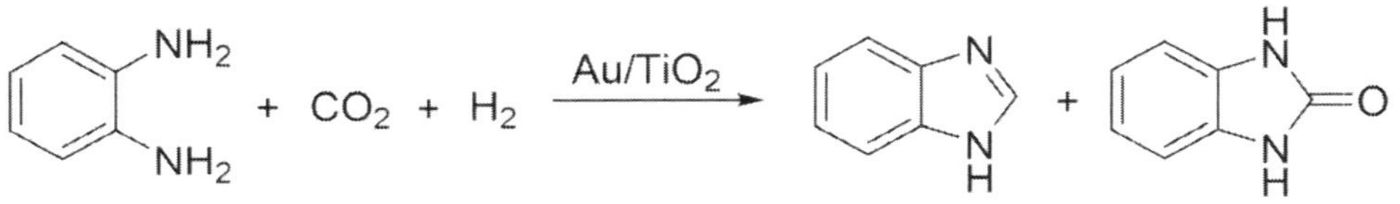

In 2014, Srinivasulu *et al.* devised a one-pot intermolecular cyclization method for synthesizing benzimidazoles. In this process, *o*-phenylenediamine reacts with several aldehydes in the presence of zinc triflate to form benzimidazole. Despite a variety of solvents were tested, the highest yields were achieved using ethyl alcohol with 10-15% zinc triflate over 8 hours (Figure 9). Substrates with both electron-donating and electron-withdrawing substituents offered their corresponding derivatives with excellent yield (Srinivasulu *et al.*, 2014).

Figure 9. Zinc catalysed reaction for benzimidazoles

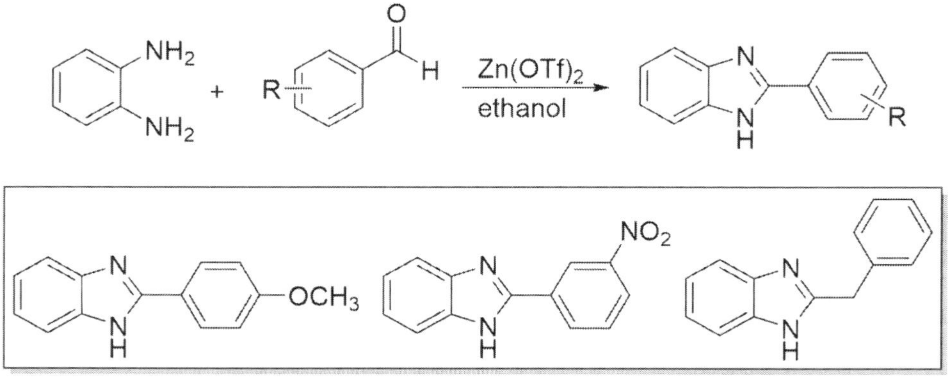

Recent studies have concentrated on developing eco-friendly conditions for these reactions by employing transition metal catalysts. In a different study, benzimidazoles were successfully synthesised through the reaction of *o*-phenylenediamine with numerous aldehydes, employing different catalysts based on transition metal nanoparticles. In the presence of a nanoparticle (CuOnp-SiO$_2$) catalyst, Inamdar and co-workers synthesised novel benzimidazole derivatives with good yields (Scheme 9a). Although, Nasr-Esfahani *et al.* developed a synthetic approach with high yields of benzimidazoles employing a nano catalyst composed Cu(II)-TD@nSiO$_2$ (Nasr-Esfahani *et al.*, 2013). Similarly, Bardajee and co-workers also conducted the identical reaction with Fe(III)-Schiff base/SBA-15 as a catalyst in water, obtaining high yields. (Bardajee *et al.*, 2015). Kommula *et al.* synthesized 2-arylbenzimidazoles with good yields by employing a nano-Fe$_2$O$_3$ catalyst (Kommula *et al.*, 2017). Furthering these advancements, Kumara and co-workers developed benzimidazole derivatives employing the cross-coupling approach of *o*-phenylenediamine and various aldehydes, with cobalt ferrite nanocatalyst and silica-coated cobalt ferrite nanocatalyst (Figure 10) (Kumara *et al.*, 2017).

Figure 10. Various nanoparticle catalysed reaction for benzimidazole

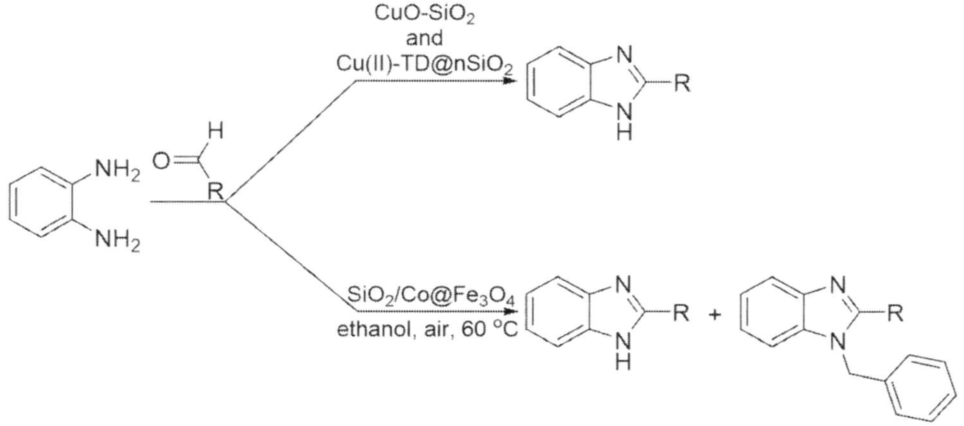

On the contrary, brønsted acidic ionic liquids have proven to be useful, non-toxic, and eco-friendly reagents for synthesizing benzimidazoles. This strategy involves the straightforward condensation between *o*-phenylenediamine with an aldehyde in ethanol, with an acidic ionic liquid acting as the catalyst. The ionic liquid catalyst activates the carbonyl compound by coordinating with its oxygen atom, facilitating ring closure through dehydration to form the target molecule, as illustrated in Figure 11 (Senapak *et al.*, 2019).

Figure 11. Bronsted acidic ionic liquid catalysed reaction for benzimidazoles

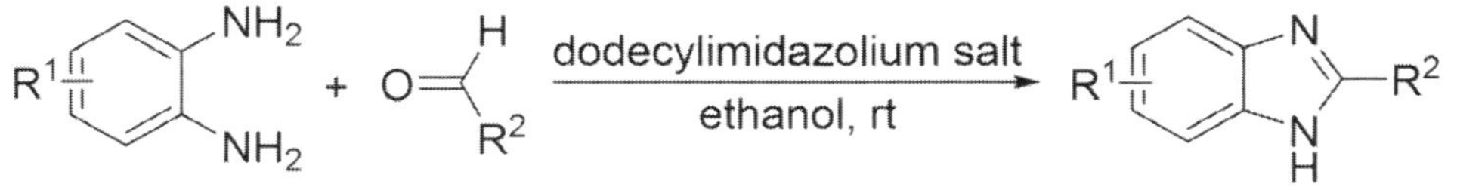

2. CONDENSATION OF *O*-PHENYLENEDIAMINE WITH KETONES

An extensive array of benzimidazole scaffolds have been produced *via* condensing *o*-phenylenediamine with various ketones (Figure 12). This reaction initially produces 2-disubstituted benzimidazolines, which are highly unstable and subsequently decompose into a mixture of benzimidazole derivatives. The specific benzimidazole products formed depend on the alkyl group is eliminated during the decomposition process. In a similar manner, Ladenburg and Rugheimer also described the production of methylbenzimidazole, which entailed heating 3,4-diaminotoluene with acetophenone at 180 °C. The approach furnished benzimidazole derivatives by eliminating the methyl group. However, this method has several drawbacks, including prolonged reaction times, challenging work-up processes, and the generation of by-products. To address these issues, alternative reaction conditions including solvent-free environments, and various catalysts have been explored. Unfortunately, these alternative approaches yielded other heterocyclic compounds instead of the anticipated benzimidazole derivatives as demonstrated in Figure 13 (Alaqeel, 2017).

Figure 12. Condensation of o-phenylenediamine and ketones: general synthetic strategy for benzimidazoles

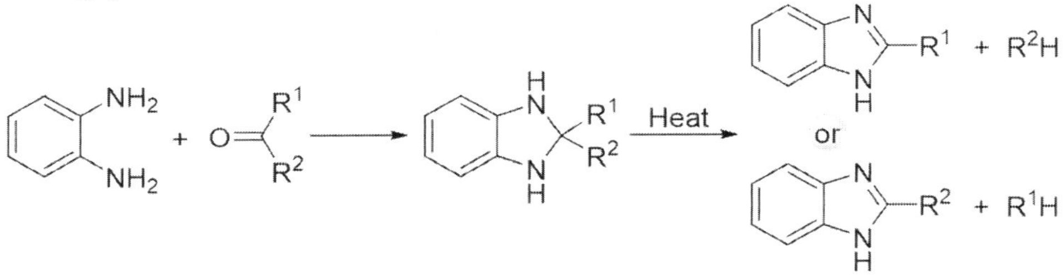

Figure 13. Benzimidazoles synthesis via coupling of o-phenylenediamine and acetophenone

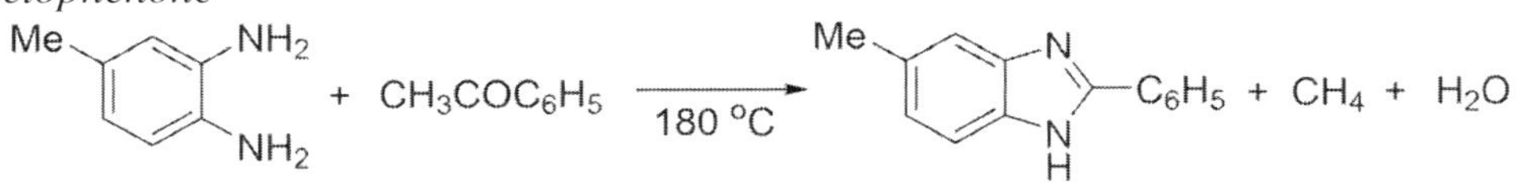

Moreover, Dhanalakshmi *et al.* successfully synthesized benzimidazoles by reacting *o*-phenylenediamine with α, β-unsaturated ketones, achieving good yields through both thermal and microwave irradiation methods (Figure 14) (Dhanalakshmi *et al.*, 2014).

Figure 14. Microwave-assisted reaction for benzimidazoles

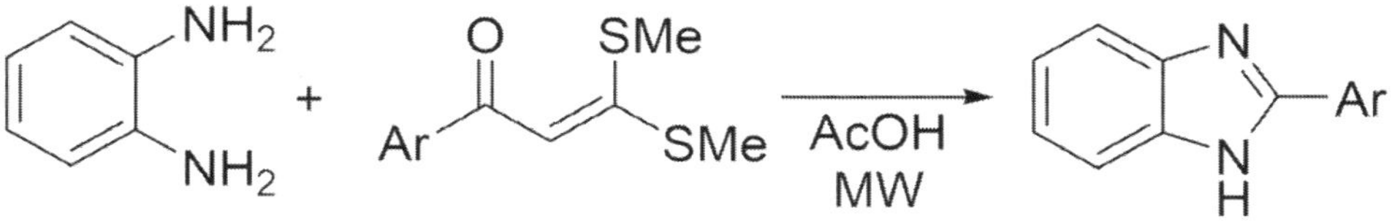

However, heating of *o*-phenylenediamine and 1,3-dicarbonyl derivatives in alcohol induces cyclisation, which yields corresponding derivatives of benzimidazole scaffolds (Figure 15). (Abdallah *et al.*, 2015; Mohareb *et al.*, 2020)

Figure 15. Benzimidazoles synthesis via coupling of o-phenylenediamine and 1,3-dicarbonyl compounds.

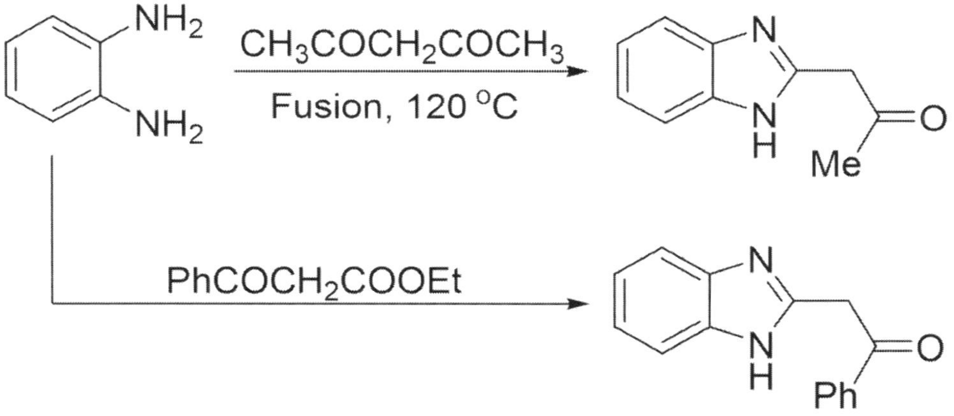

3. CONDENSATION OF *O*-PHENYLENEDIAMINE WITH CARBOXYLIC ACIDS

Numerous reports have shown that *o*-phenylenediamine derivatives can be condensed with carboxylic acids to produce benzimidazole and its derivatives, using hydrochloric acid as a catalyst (Alaqeel, 2017). Rithe *et al.* further demonstrated a novel, one-pot, cost-effective and eco-friendly strategy for synthesizing benzimidazole scaffolds utilizing ammonium chloride as a catalyst by condensing *o*-phenylenediamine and aromatic acid at 80-90 °C as displayed in Figure 16 (Rithe *et al.*, 2015).

Figure 16. Benzimidazoles synthesis via coupling of o-phenylenediamine and car-boxylic acid

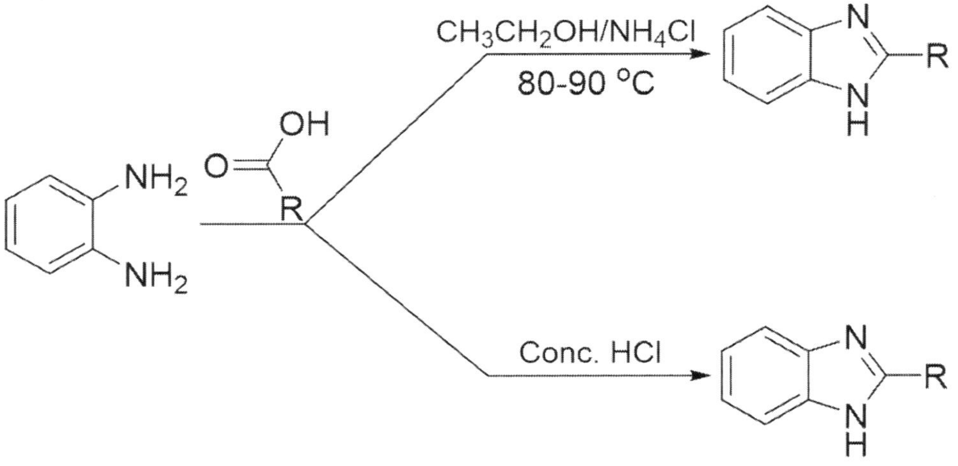

Benzimidazole scaffolds can be synthesized by reacting *o*-phenylenediamine and ibuprofen in a sodium hydroxide solution, as illustrated in Figure 17. This research also highlights the potent antimicrobial properties of benzimidazole (Ibrahim *et al.*, 2011).

Figure 17. Base catalysed reaction for benzimidazoles

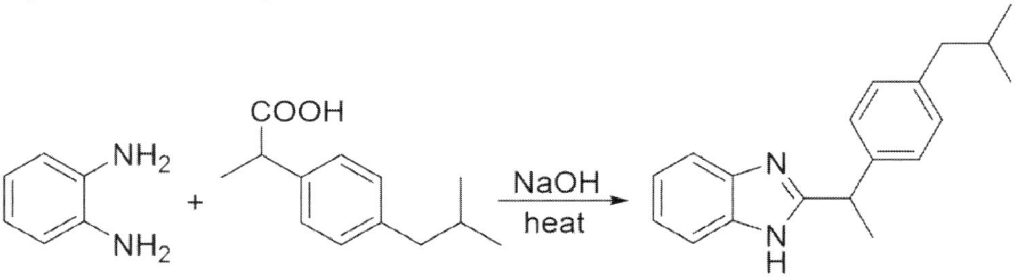

In 2015, Saberi reported a method for synthesizing benzimidazoles utilizing microwave irradiation and catalysts like alumina, silica gel, or zeolite, which eliminates the need for solvents. The method involved mixing *o*-phenylenediamine with various carboxylic acids and 50 g of catalyst, grinding the crude mixture, and then irradiating it in a microwave for 5–9 minutes (Figure 18) (Saberi *et al.*, 2015).

Figure 18. Al,O,, SiO,, or zeolite catalysed reaction for benzimidazoles

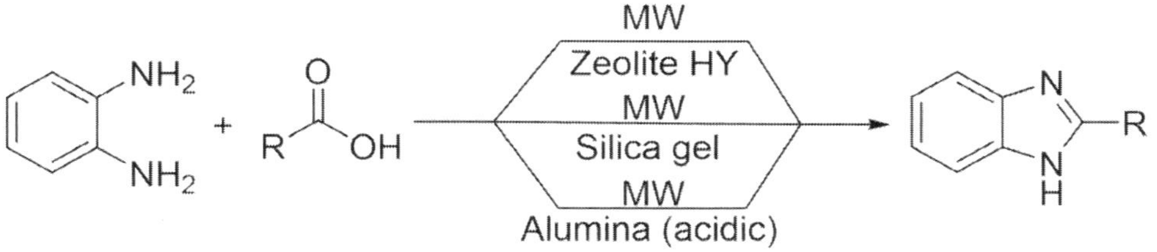

Similarly, Ahmed and co-workers synthesized high yields of benzimidazole derivatives by refluxing *o*-phenylenediamine and *p*-aminobenzoic acid in presence of polyphosphoric acid under xylene with for 6 hours (Figure 19) (Alam *et al.*, 2017).

Figure 19. Benzimidazoles synthesis via coupling of o-phenylenediamines and carboxylic acids

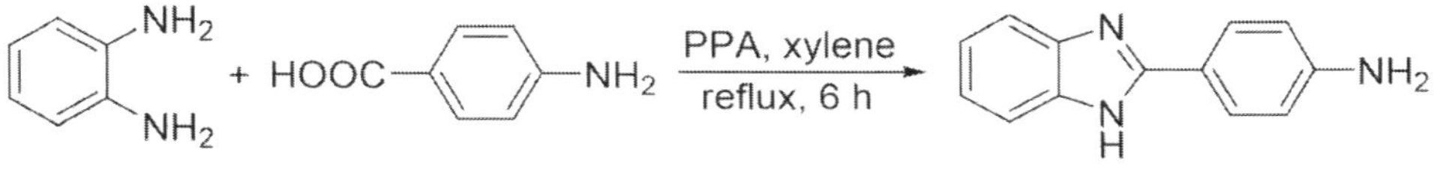

Utilising microwave irradiation to condense *o*-phenylenediamine with monocarboxylic acids, Huynh *et al.* have developed a novel, catalyst-free strategy for synthesizing benzimidazole derivatives (Figure 20) (Huynh *et al.*, 2020).

Figure 20. Benzimidazoles synthesis using microwave irradiation

$R = -CH_2Cl, -CH_3, -CH_2CH_3, -CH_2CH_2CH_3$

4. SYNTHESIS OF BENZIMIDAZOLE *VIA* REARRANGEMENT

Mamedov *et al.* described a unique approach for synthesizing benzimidazole derivatives *via* rearrangement (Figure 21). The approach begins with hydrazine hydrate reacting with 3-arylacylidene-3,4-dihydroquinoxalin-2(1*H*)-ones to produce

3'-aryl-1,2,3,4,4',5'-hexahydrospiro[quinoxalin-2,5'-pyrazol]-3-ones, which were subsequently rearranged to produce the anticipated product (Mamedov *et al.*, 2009).

Figure 21. Acid catalyzed quinoxaline–benzimidazole rearrangements for benzimidazoles

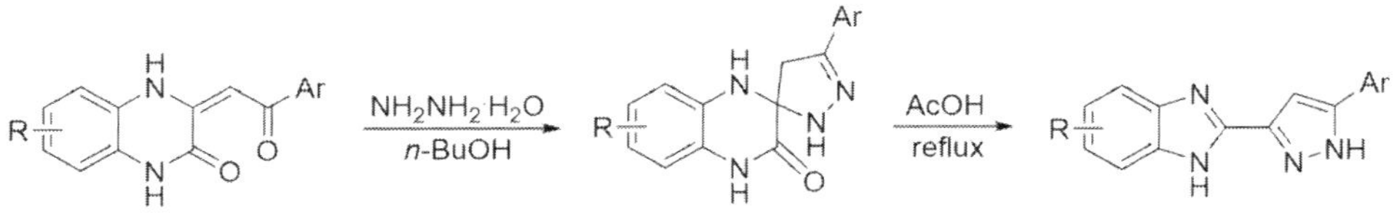

Furthermore, in a similar transformation, *o*-phenylenediamine and 3-aroylquinoxalin-2(1*H*)-ones reacted to form benzimidazole when refluxed in acetic acid (Figure 22) (Mamedov *et al.*, 2010).

Figure 22. Acid catalysed quinoxalinone-o-phenylenediamine rearrangements for benzimidazoles

R = Ph, alkyl

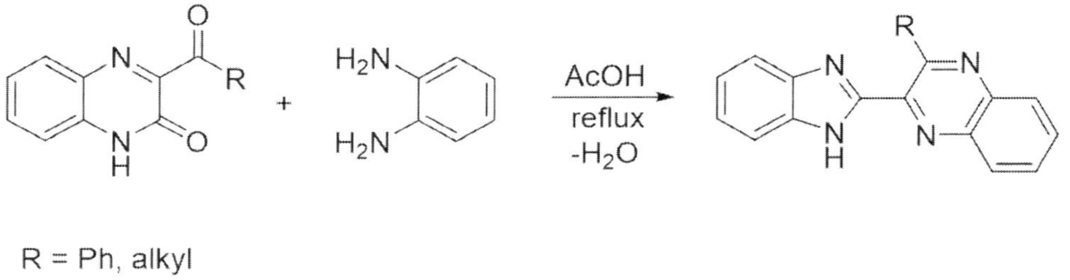

Dicarbonyl compounds are also suitable for this reaction, enhancing its applicability. Eleftheriadis *et al.* reported that dimethylacetone dicarboxylate reacts with *o*-phenylenediamine to form a diazepine derivative, which, upon treatment with mercaptoacetic acid, yields a benzimidazolone as demonstrated in Figure 23 (Eleftheriadis *et al.*, 2013).

Figure 23. Rearrangement of benzodiazepinones for benzimidazolones

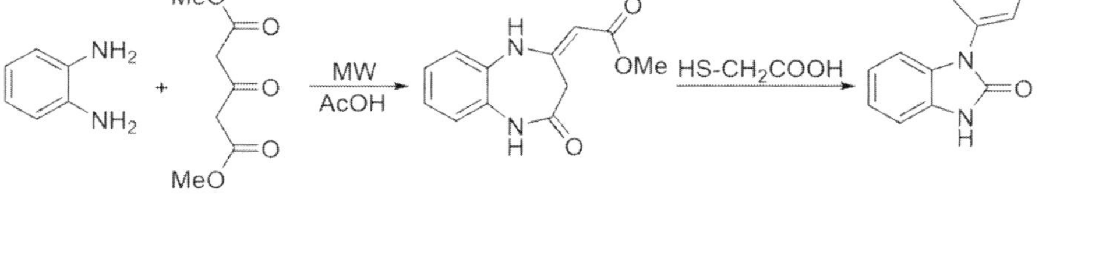

By reacting 3-(2-aminophenyl)-quinoxalin-2(1*H*)-one and ethyl acetoacetate in an acidic medium, Mamedov and co-workers were able to successfully synthesized benzimidazole derivatives, including the polynuclear fused benzimidazoles, which are challenging to synthesis using conventional approaches (Figure 24) (Mamedov *et al.*, 2014).

Figure 24. Acid-catalyzed quinoxalinone–benzimidazole rearrangements

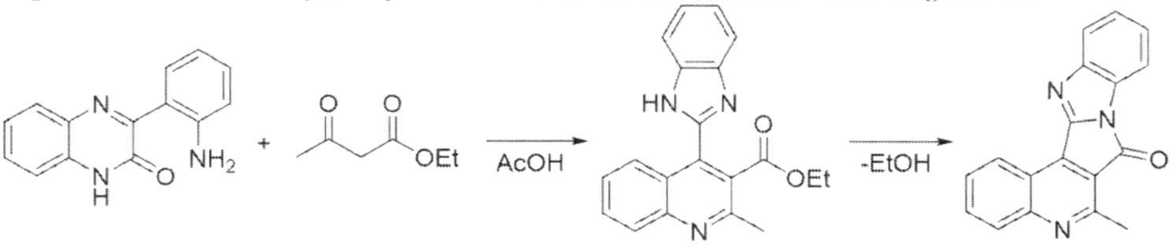

Based on previous research, recent studies have utilized quinoxalin-2(1*H*)-ones for synthesizing various benzimidazoles *via* an acid-catalyzed rearrangement (Figure 25). It was found that benzimidazole derivative forms only with at least one hydrogen atom in the spiro intermediate, while the absence of this hydrogen, the spiro moiety migrates to the 1-position, forming benzimidazolone. The strategy offers significant benefits, such as high reaction rates, good yields, mild reaction conditions and compatibility with an array of substrates (Hashem *et al.*, 2021).

Figure 25. Benzimidazoles and benzimidazolones synthesis via rearrangement

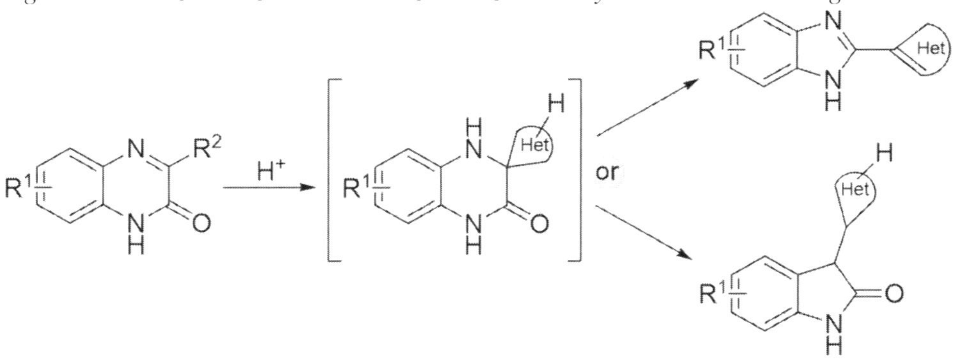

In order to synthesise benzimidazolone, 5-amino-4-carbamoyl-1-(2-nitrophenyl)-1*H*-1,2,3-triazole was also employed as a rearrangement precursor. When triazole derivative was refluxed under DMF, the anticipated benzimidazole scaffold was produced as depicted in Figure 26 (Hashem *et al.*, 2021).

Figure 26. Rearrangement of triazole for benzimidazoles

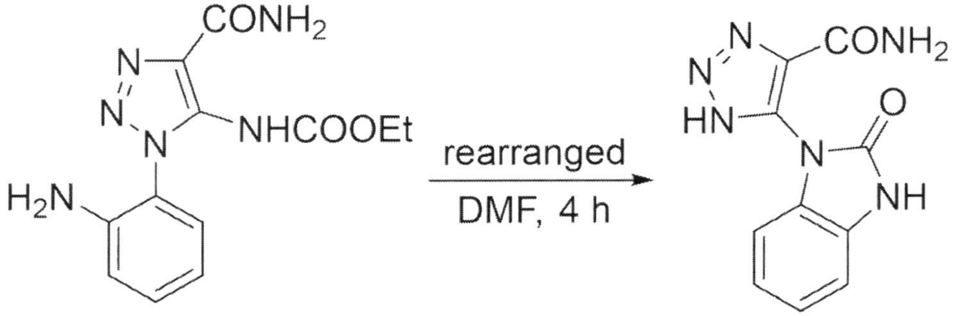

5. GREEN SYNTHESIS OF BENZIMIDAZOLE

Presently, many chemical and pharmaceutical industries are currently grappling with significant environmental issues such as excessive solvent waste, solvent/reagent losses, and catalyst contamination etc. To address these challenges, eco-friendly synthetic methods have emerged as effective solutions, which include microwave-assisted reactions, utilize of eco-friendly catalysts, or solvent-free conditions. In 2013, Nguyen *et al.* devised a green strategy for synthesizing benzimidazoles that eliminates the use of both solvents and metals (Figure 27). They initiated their reaction with *o*-phenylenediamine and benzylamine at 110 °C temperature in the presence of O_2 and acetic acid to achieve best yields (Nguyen *et al.*, 2013).

Figure 27. Benzimidazoles synthesis in the presence of O_2 and acetic acid

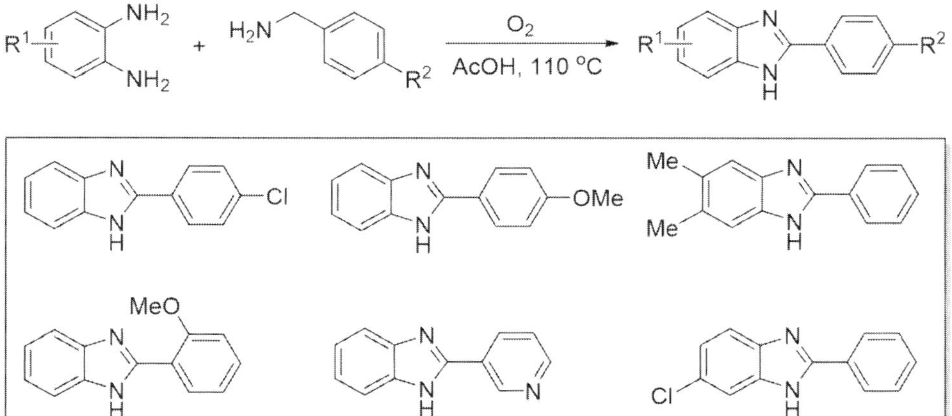

Again, Yu *et al.* described a one-pot synthetic strategy for synthesizing benzimidazole scaffolds *via* cyclo-condensation of *o*-phenylenediamine and various aldehydes in a biocompatible solvent such as ethyl lactate, lactic acid or ethanol in 4-6 hours at room temperature, and the excellent yields were achieved using lactic acid as shown in Scheme 27a (Yu *et al.*, 2016). Furthermore, Azarifar and co-workers utilized microwave irradiation to developed benzimidazoles *via* condensation of *o*-phenylenediamine and aldehydes under acidic medium (Figure 28). The benefits of this strategy include excellent product yields, the utilize of non-toxic solvents and short reaction times (Azarifar *et al.*, 2010).

Figure 28. Different strategies for benzimidazole synthesis

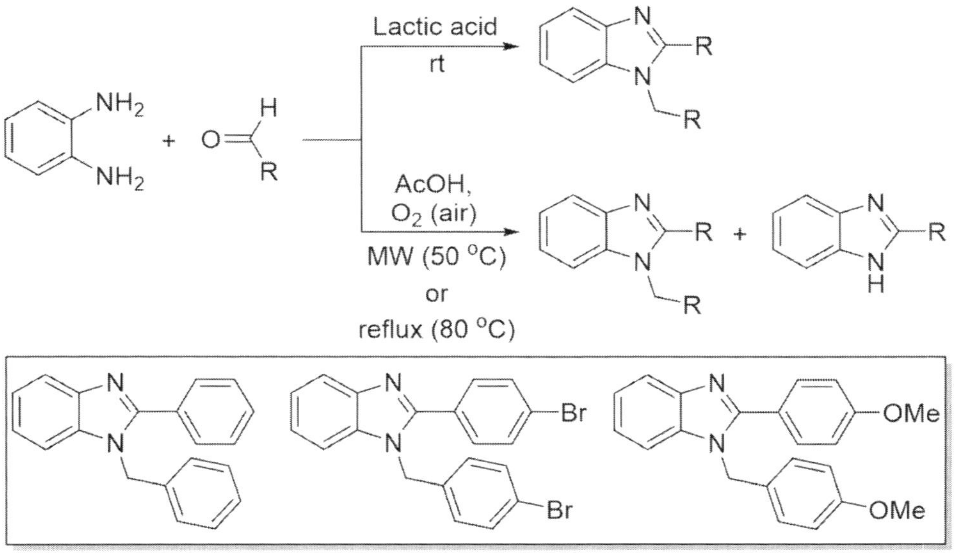

Benzimidazole scaffolds were synthesized by the reaction of *o*-phenylenediamine and formic acid using zinc oxide nanoparticle catalyst at 70°C (Figure 29). This method offers excellent reaction yields, an inexpensive catalyst, mild and solvent-free reaction conditions. (Alinezhad *et al.*, 2012)

Figure 29. ZnO-NP catalysed reaction for benzimidazoles

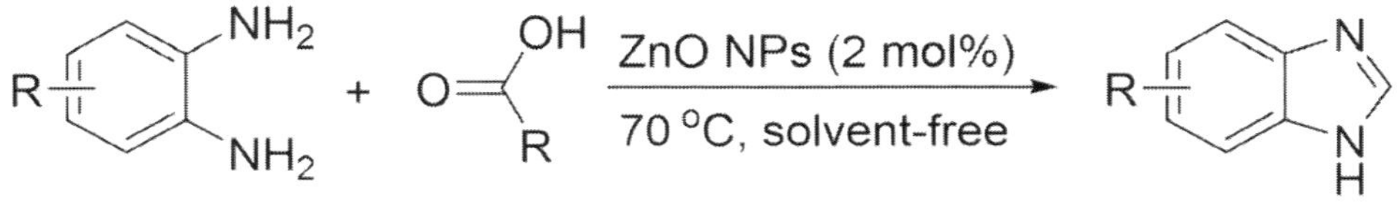

In 2012, Peng *et al.* developed a strategy for synthesizing benzimidazoles *via* intermolecular cyclization of N-(2-iodoaryl) benzamidine with potassium carbonate and cesium carbonate as bases at 100 °C, achieving the best yields (Figure 30). (Chen *et al.*, 2012)

Figure 30. Base catalysed reaction for benzimidazoles

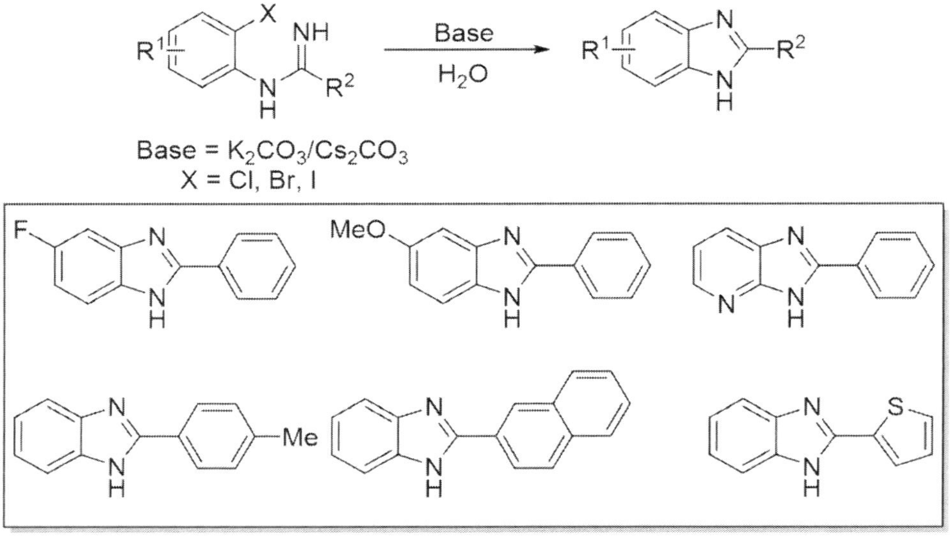

In 2014, Kumar *et al.* prepared 2-aryl-1-arylmethyl-1*H*-benzimidazole derivatives by sonication of *o*-phenylenediamine and various aldehydes empolying silica-gel-supported trichloroacetic acid under heating conditions (Figure 31) (Kumar *et al.*, 2014).

Figure 31. Benzimidazoles synthesis under sonication

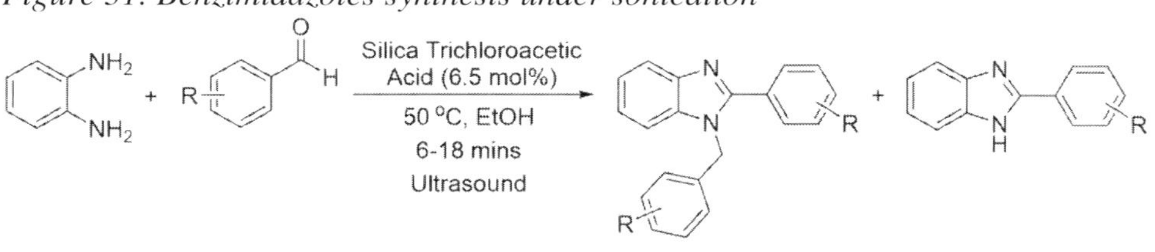

6. PHOTO-CATALYZED REACTION FOR BENZIMIDAZOLE SYNTHESIS

Currently, solar energy is a highly valuable alternative due to its abundance and cost-free availability. With growing energy concerns, researchers are focusing on developing synthetic methods using sustainable energy sources. Since the solar light comprises 43% visible light, various visible light-absorbing chromophores can convert solar energy into chemical energy for use in numerous organic reactions. Recent advancements in photoredox catalysis have garnered significant attention from researchers. Numerous innovative visible light photochemical processes have been reported, utilizing sensitizers such as organic dyes, mpg-C_3N_4, polypyridyl metal complexes, and other photocatalysts. In 2011, a one-pot synthesis of benzimidazoles was devised using a photo-catalyzed aerobic coupling of o-phenylenediamine with benzylamine in presence a photocatalyst mpg-C_3N_4, visible light, and of O_2 (Figure 32) (Su $et\ al.$, 2011)

Figure 32. Photo-catalysed aerobic reaction for benzimidazoles

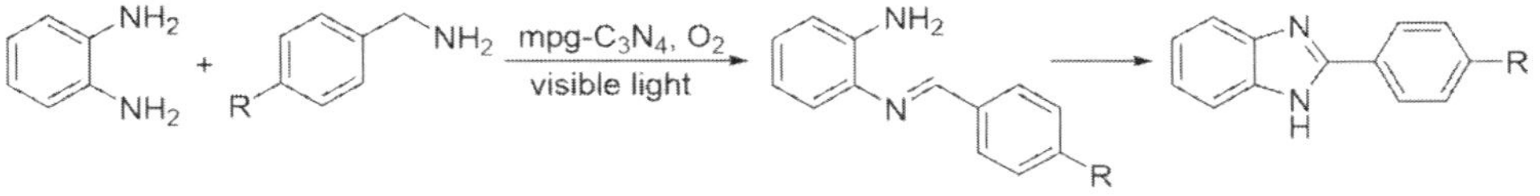

Samanta and co-workers developed a synthetic strategy for synthesizing benzimidazole using 3,6-di(pyridine-2-yl)-1,2,4,5-tetrazine (pytz) as a reducing agent under visible light (Figure 33). In this approach, several aldehydes were reacted with o-phenylenediamine in ethanol, when pytz and sunlight were present, to achieve excellent yields within 2-3 h (Samanta $et\ al.$, 2013).

Figure 33. Benzimidazole synthesis under pytz and sunlight

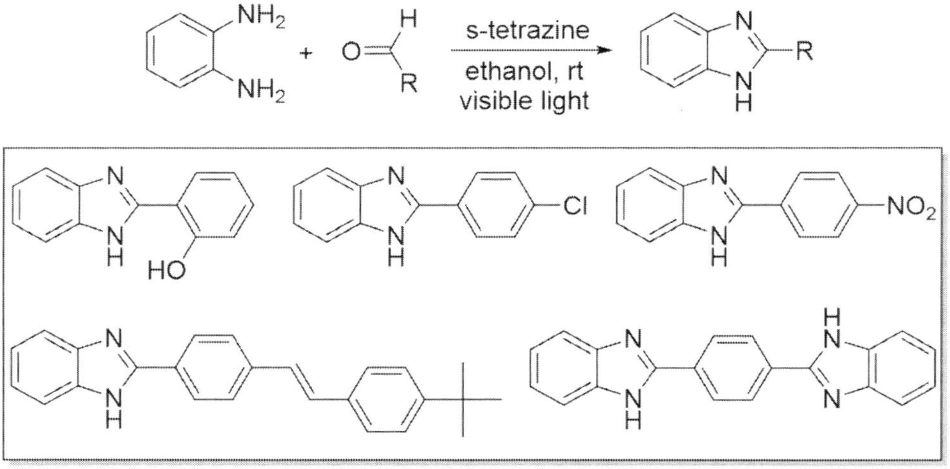

o-Phenylenediamines react with aldehydes with a rose bengal photocatalyst under visible light irradiation, yielding benzimidazoles in excellent yields as depicted in Figure 34 (Kovvuri *et al*., 2016).

Figure 34. Rose bengal photo-catalysed for benzimidazoles

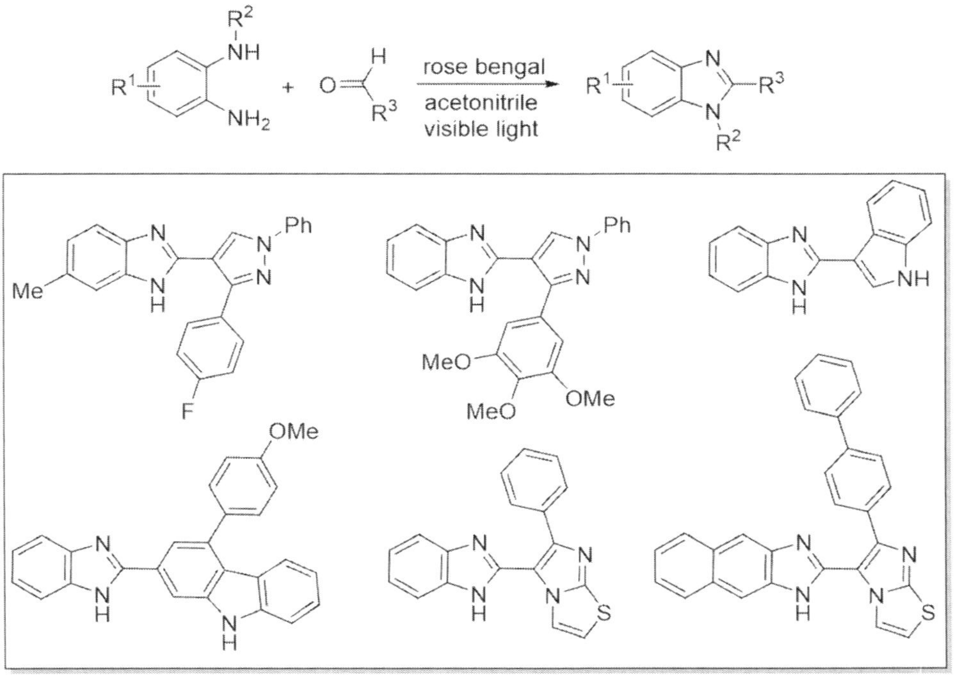

Next, Feizpour *et al.* devised a straightforward method employing a cobalt ascorbic acid complex coated on TiO_2 nanoparticles to produce benzimidazoles through aerobic photooxidative cyclization (Figure 35). After the reaction is completed, the photo-catalyst can be reused. (Feizpour *et al.*, 2018)

Figure 35. TiO2/AA/Co nanohybrid catalysed reaction for benzimidazoles

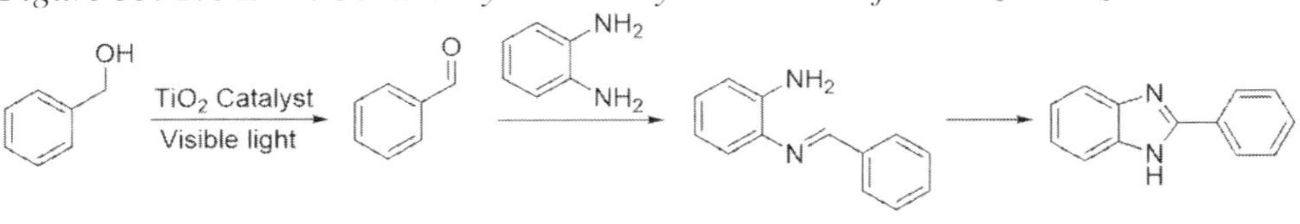

Recently, Skolia *et al.* described a photochemical strategy for synthesizing benzimidazole derivatives. Utilizing photoinitiator like 2,2-dimethoxy-2-phenylacetophenone under CFL lamps, they were able to condense *o*-phenylenediamine with various aldehydes at ambient temperature (Figure 36). This green and economical approach yielded benzimidazole derivatives efficiently, without the production of environmental pollutants, under visible light in a short time (Mohammed *et al.*, 2023).

Figure 36. Visible light promoted reaction for benzimidazoles

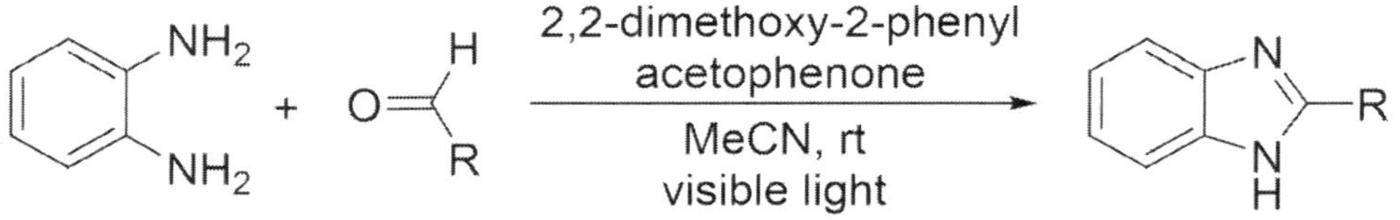

7. MISCELLANEOUS REACTIONS FOR BENZIMIDAZOLE SYNTHESIS

Fu and co-workers produced benzimidazoles via an intramolecular cyclization of *o*-haloacetoanilides, employing Cs_2CO_3 and CuBr in DMSO solvents (Scheme 36a) (Yang *et al.*, 2008). Similarly, Peng *et al.* devised an efficient stretagy for synthesizing benzimidazoles using copper oxide in presence of K_2CO_3 in water solvent. This strategy furnished excellent yields through the intramolecular cyclisation of *o*-halophenylamidine and the cross-coupling reaction of carbon-nitrogen bonds (Figure 37). In this instance, the authors also used N,N'-dimethylethane-1,2-diamine as a promoting ligand (Peng *et al.*, 2010).

Figure 37. Copper catalysed reaction for benzimidazoles

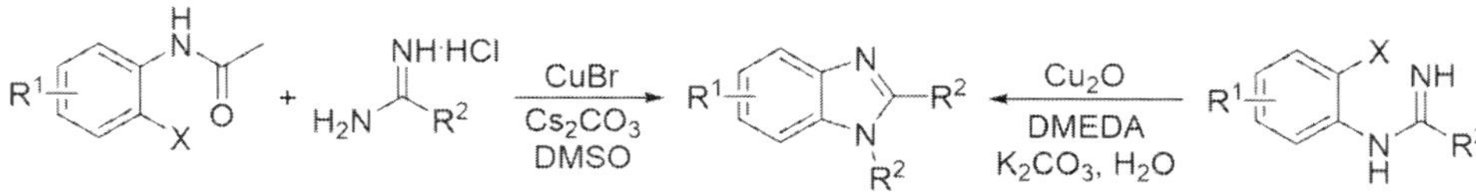

In the same year, Wrobel *et al.* presented a method for synthesizing benzimidazoles *via* base-promoted condensation of nitroarenes. Initially, N-aryl-2-nitrosoanilines were formed from aniline and nitrobenzene, then coupled with benzyl sulfone under basic conditions (Figure 38). The best yields were achieved with bases such as DMU and K_2CO_3 in DMF and MeCN solvents, respectively (Wrobel *et al.*, 2011).

Figure 38. Ehrlich-Sachs condensation reaction

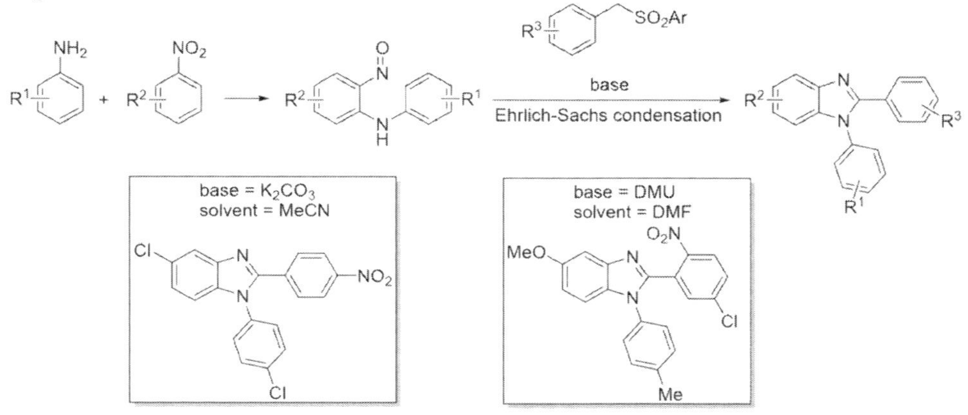

Nguyen *et al.* described a one-pot, metal- and solvent-free strategy for synthesizing benzimidazoles with excellent yields, employing a cross-coupling reaction between *o*-phenylmethanamines and aliphatic amines mediated by elemental sulfur as shown in Figure 39. (Nguyen *et al.*, 2012)

Figure 39. Elemental sulfur mediated synthesis benzimidazoles

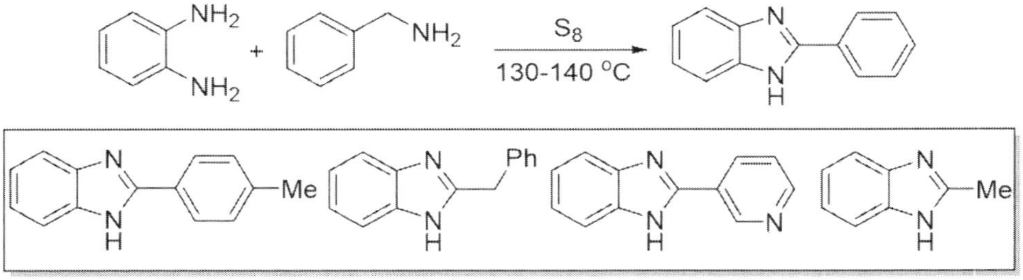

The same group devised a solvent-free method for constructing benzimidazoles using an Iron-Sulfur catalyst (Figure 40). The approach involved *o*-nitroaniline derivatives and substituted 4-picoline at 150 °C, yielding benzimidazole derivatives (Nguyen *et al.*, 2013).

Figure 40. Iron-sulfur catalysed reaction for benzimidazoles

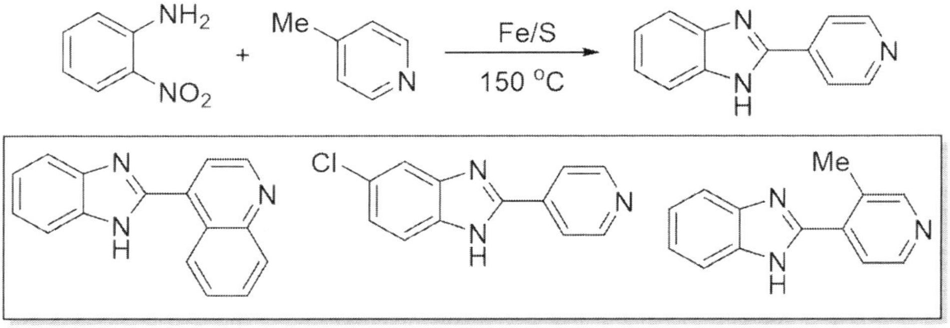

Nguyen *et al.* again developed an approach involving condensation of *o*-nitroaniline with benzylamines in the same year (Scheme 41). They used several metal catalysts and found that $FeCl_3 \cdot 6H_2O$ and $CoBr_2 \cdot xH_2O$ produced the highest yields (Nguyen *et al.*, 2013).

Figure 41. Fe and Co catalysed reaction for benzimidazoles

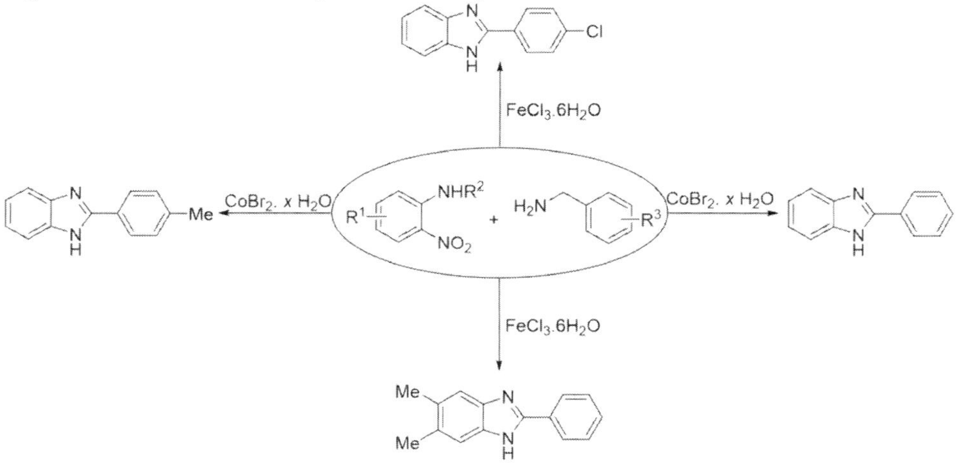

In 2014, the same group condensed phenyl acetic acid with *o*-nitroaniline using an iron-sulfur catalyst ($FeCl_2 \cdot 4H_2O$) and N-methylpiperidine at 130 °C to produce benzimidazole scaffolds as demonstrated in Figure 42 (Nguyen *et al.*, 2014).

Figure 42. Iron catalysed synthesis of benzimidazoles

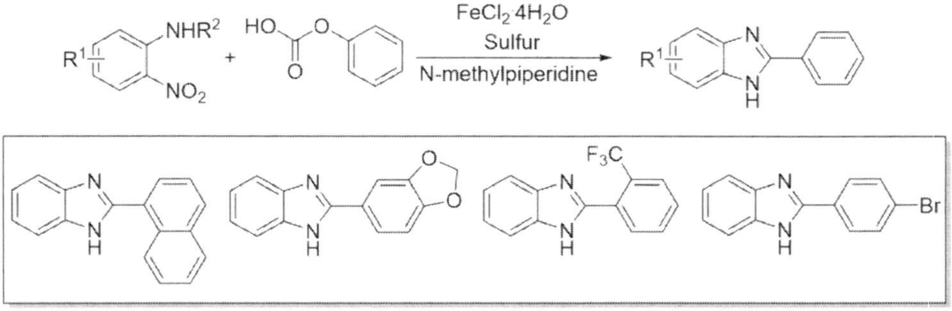

Later that year, Tran *et al.* reported a strategy for synthesizing benzimidazole derivatives with excellent yields using quinones and guanidine at 0 °C. The objective of this study was to synthesize benzosceptrin analogues via a multi-step reaction carried out in dichloromethane (Figure 43) (Tran *et al.*, 2014).

Figure 43. Benzimidazoles synthesis via multistep reaction

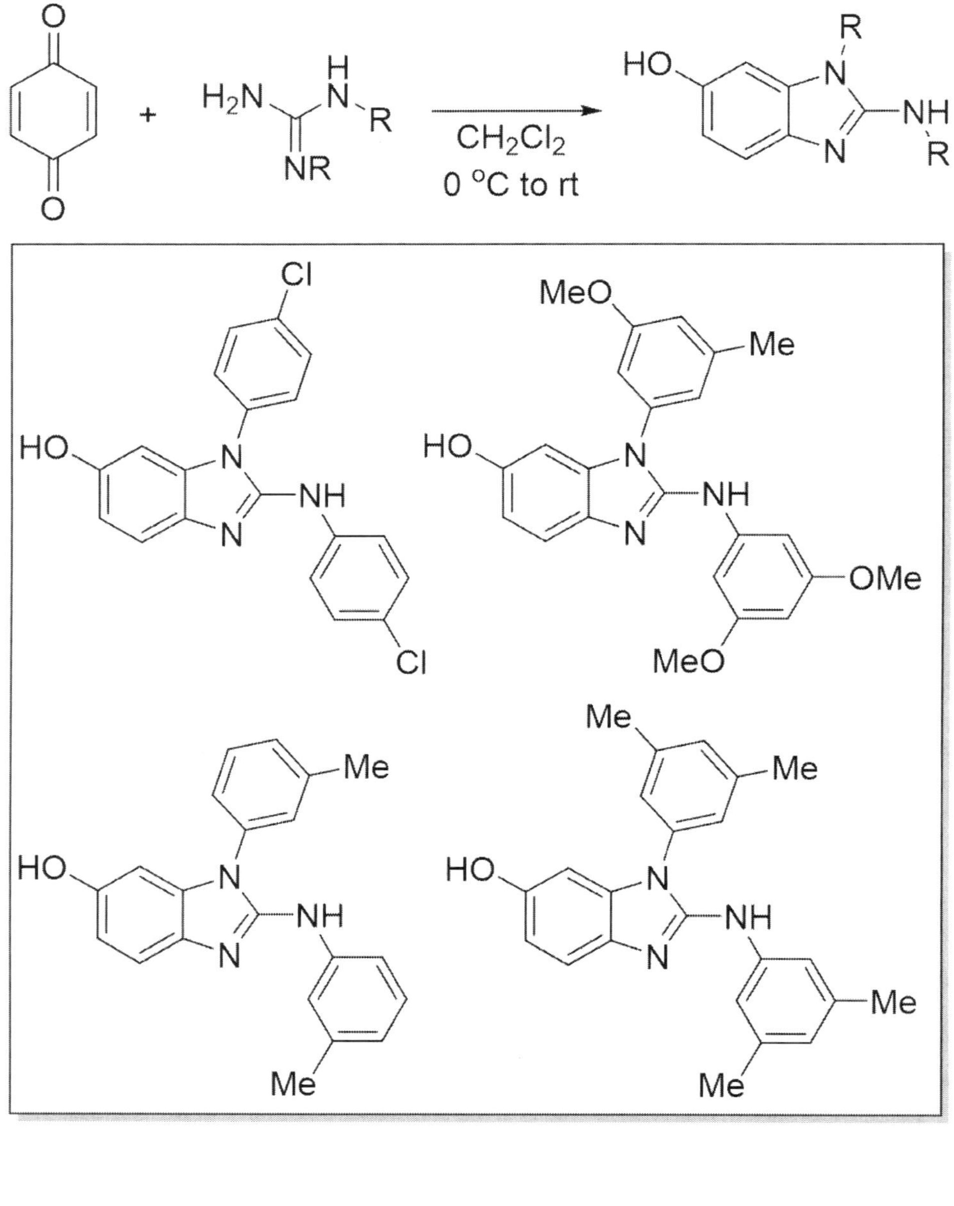

Nguyen *et al.* further investigated a synthetic procedure in the next year, utilizing sodium sulfide and iron chloride as catalysts (Figure 44). They condensed *o*-nitroaniline with phenyl alcohol, producing benzimidazole scaffolds *via* a redox condensation reaction with high yields (Nguyen *et al.*, 2015).

Figure 44. Iron-Sulfur catalysed reaction for benzimidazoles

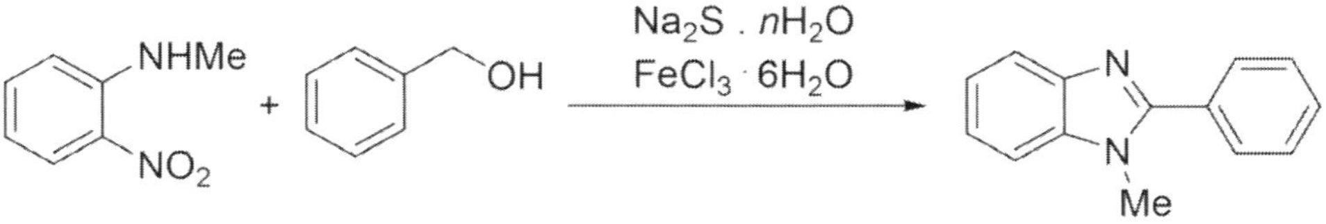

Again, Largeron and co-worker introduced a one-pot, multistep synthetic approach for synthesizing benzimidazoles *via* aerobic oxidation (Figure 45). They used copper salts as electron transfer mediators to convert amines to imines, achieving good yields (Nguyen *et al.*, 2015). In 2016, Liu *et al.* employed copper iodide as a catalyst for producing benzimidazole derivatives. The process involved the aerobic oxidation of *o*-substituted aniline to produce imine in presence of Cu-catalyst and air, which then hydrolyzed to aldehyde and reacted with *o*-aminoaniline, resulting in benzimidazoles as demonstrated in Figure 45(Liu *et al.*, 2016).

Figure 45. Copper catalysed reaction for benzimidazoles

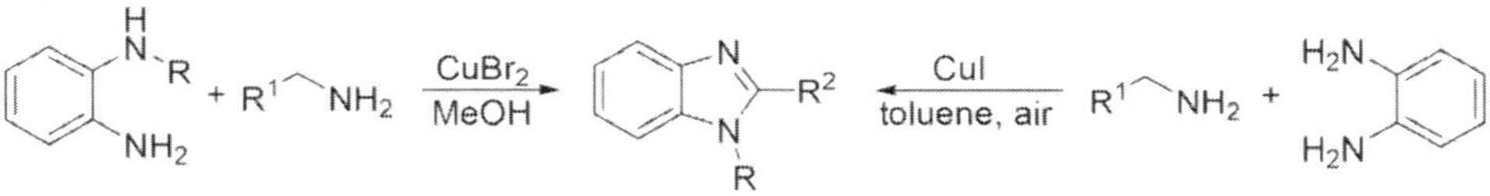

Nguyen and co-workers reported an iodine catalysed approach for constructing benzimidazole derivatives using *o*-(substitutedamino) nitrobenzene and formic acid. Here, formic acid served as the carbon building block, and in the presence of I_2, it underwent cyclisation with derivatives of *o*-nitro aniline as shown in Figure 46 (Nguyen *et al.*, 2016).

Figure 46. Iodine catalysed reaction for benzimidazoles

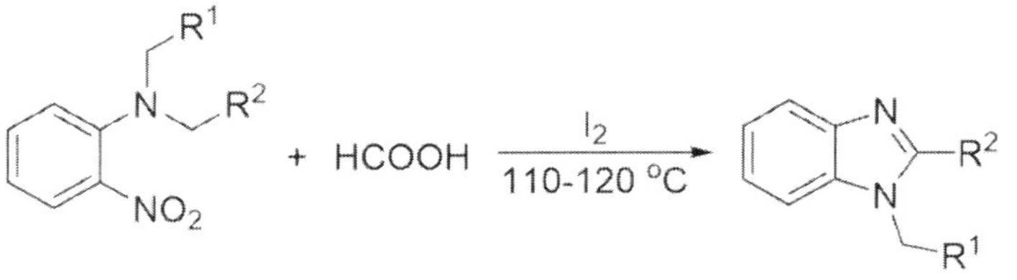

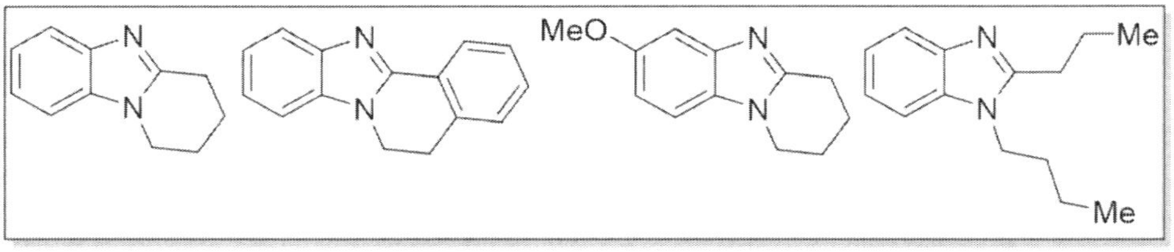

Sun and co-workers devised a strategy that involved the reaction of *o*-phenylene-diamine with dimethylformamide (DMF) utilizing phenylhydrosilane as an initiator and activates the carbonyl group of DMF, forming a Si-O bond (Figure 47). The complex formed from activated DMF and hydrosilane then interacts with the amine group of *o*-phenylenediamine, leading to the cyclization and production of benzim-idazole with good yield. Based on these findings, a tentative reaction mechanism was proposed which depicted in Scheme 46. Initially, DMF is activated by $PhSiH_3$, forming intermediate **I** with the formation of Si-O bond. Intermediate **I** then reacts with *o*-phenylmethanamine to produce intermediate **II**, which releases silanol with the formation intermediate **III**. Subsequently, in the presence of $PhSiH_3$, compound **III** undergoes cyclization, yielding the formation of benzimidazole with the removal of $HNMe_2$. (Zhu *et al.*, 2017).

Figure 47. Synthesis and tentative mechanism for benzimidazole formation using a hydrosilicon initiator

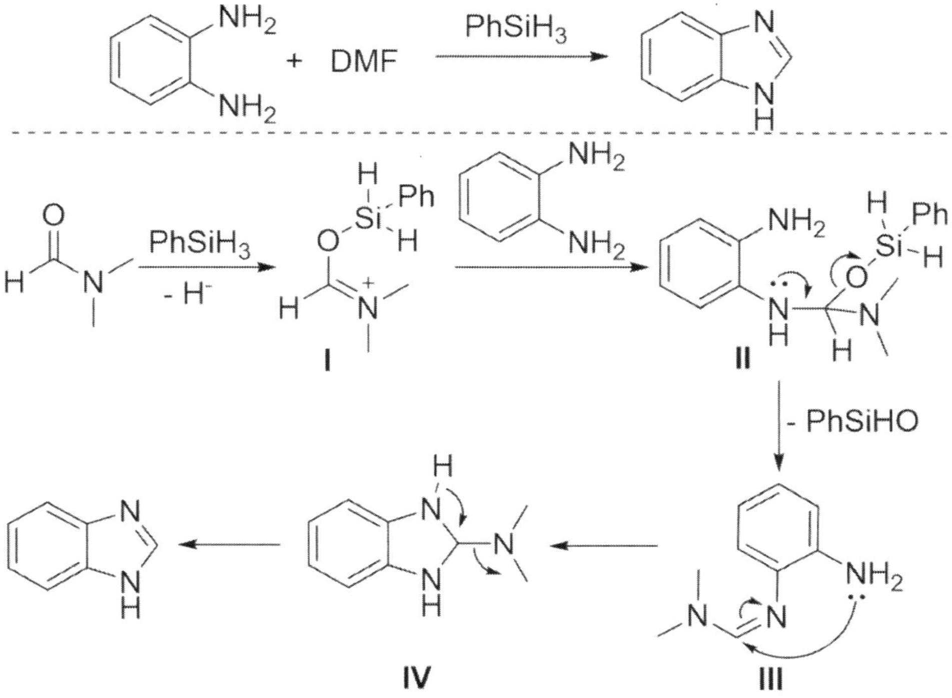

Mostafavi *et al.* also described a strategy for synthesizing benzimidazole. In this strategy, hexamethyldisilazane served as a reagent to convert DMF and *o*-phenylenediamine into benzimidazole, achieving excellent yields without the need for transition metals, acid or solvents (Mostafavi *et al.*, 2018). Similarly, Yuan and co-workers devised a synthetic strategy for synthesizing benzimidazoles by reacting *o*-phenylenediamine with dimethylformamide derivatives, employing imidazolium hydrochloride as an initiator to activate the formamide as demonstrated in Figure 48 (Gan *et al.*, 2018).

Figure 48. Benzimidazoles synthesis in presence of HMDS and Imidazolium chloride

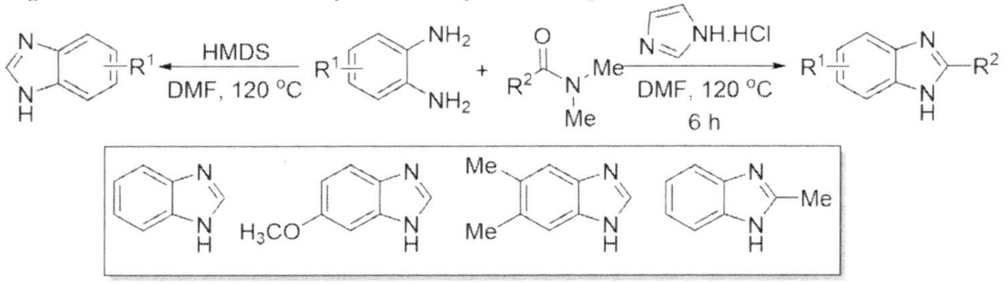

In 2018, Das *et al.* devised a selective strategy for synthesizing benzimidazoles *via* coupling of *o*-phenylenediamine with primary alcohols (Figure 49). This strategy utilised inexpensive bases like KOH, K_2CO_3 and *t*-BuOK, along with phosphine-free NNS-manganese(I) complex catalyst, to achieve excellent yields (Das *et al.*, 2018).

Figure 49. Manganese catalysed reaction for benzimidazoles

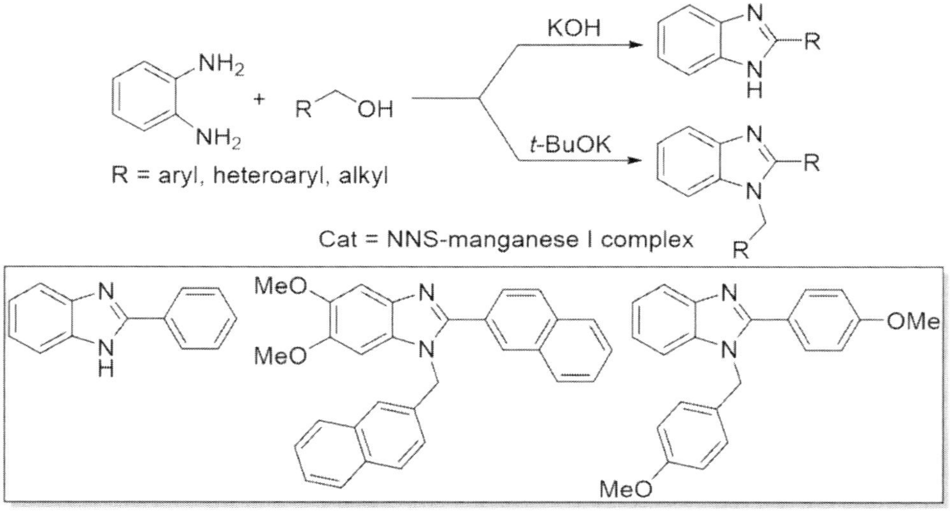

In the meantime, Mokhtari and co-worker developed a solvent-free dehydrogenative coupling method for synthesizing benzimidazole scaffolds utilizing *o*-phenylenediamine and benzyl alcohol (Figure 50). The reaction employs palladium nanoparticles or a Cu-based metalorganic framework of copper ($Cu_2(BDC)_2(DAB$-CO)- MOF) as catalysts (Mokhtari *et al.*, 2018).

Figure 50. Copper catalysed reaction for benzimidazoles

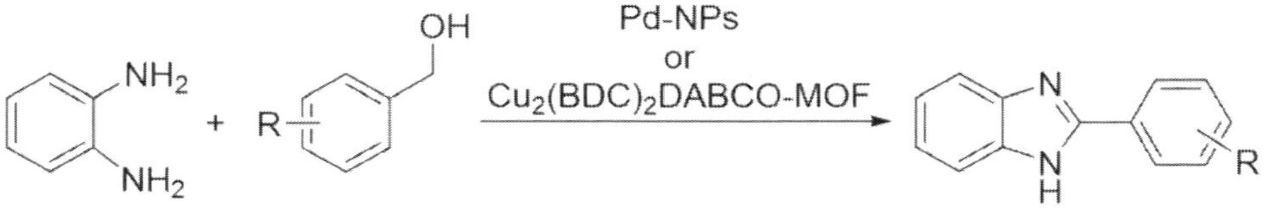

Later that year, Liu *et al.* demonstrated that benzimidazoles could be synthesized without the use of metal catalysts by employing only microwave radiation in DMSO solvent (Figure 51). The reaction involves a S_NAr strategy between primary amine derivatives and fluoro-aryl formamidines, resulting in good yields (Liu *et al.*, 2019).

Figure 51. Benzimidazoles synthesis under metal-free condition using microwave radiation

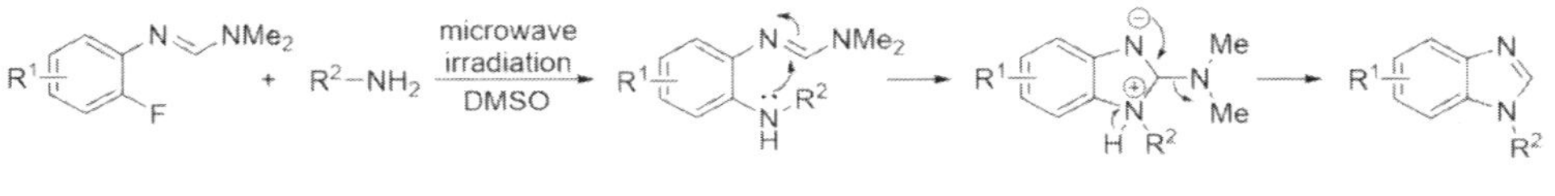

In the same year, Nguyen *et al.* reported a sulfur catalysed oxidative coupling strategy for synthesizing benzimidazole scaffolds using *o*-phenylenediamine and 1,2-diphenyldisulfane in DMSO or NMP at 100 °C as shown in Figure 52 (Nguyen *et al.*, 2019).

Figure 52. Sulfur catalyst reaction for benzimidazoles

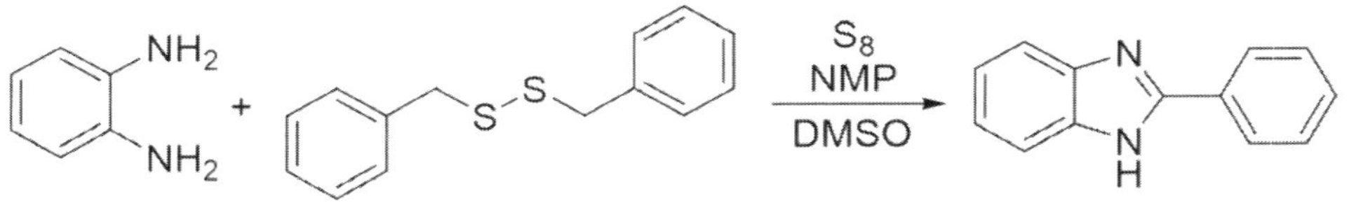

Again, Nguyen *et al.* described a multi-component strategy promoted by elemental sulfur under DMSO to synthesize benzimidazoles (Figure 53). This method utilized redox condensation reactions between *o*-phenylenediamine, maleic anhydride, and phenylmethanamine, involving ring opening and decarboxylative oxidative thioamidation steps to form the benzimidazole scaffolds (Nguyen *et al.*, 2019).

Figure 53. Sulfur catalysed multi-component reaction for benzimidazoles

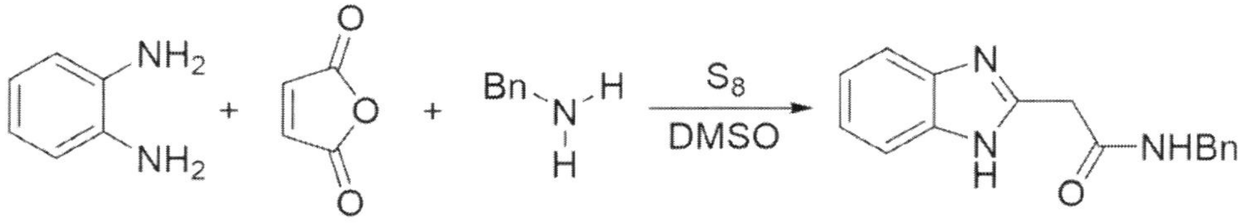

Kammela and co-workers. synthesized benzimidazoles using N-(2-iodophenyl)-N'-benzoyl guanidine with copper acetate and Cs_2CO_3 in DMSO, achieving best yields (Figure 54). As the reaction progresses *via* a C-N cross-coupling mechanism, Cu(II) is converted into Cu(I) and then regenerated to Cu(II) (Shaik *et al.*, 2019).

Figure 54. Copper catalysed reaction for benzimidazoles in presence of base

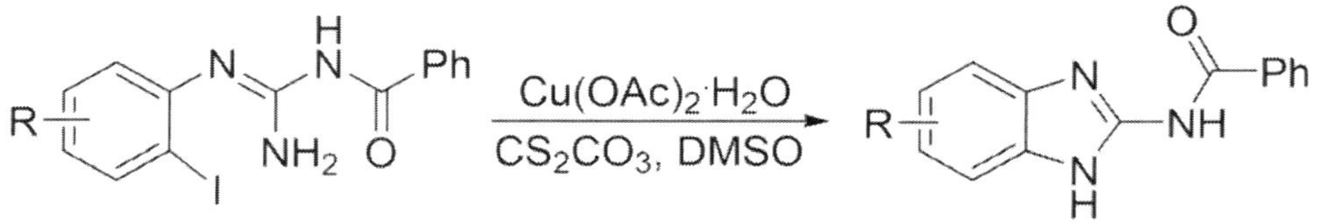

Biological Activity of Benzimidazole Scaffolds

Benzimidazole Scaffolds have been significant in medicine since their initial synthesis, particularly due to their resemblance to purine structures. Notably, 5,6-dimethylbenzimidazole, which originates from the breakdown of Vitamin B12, underscores their biological relevance. These compounds exhibit diverse pharmacological properties, including antitumor, antimicrobial, anti-inflammatory, and anticonvulsant activities. On the other hand, recent research has shown that benzimidazole carbamates, notably Albendazole, Fenbendazole, Flubendazole, and Mebendazole reduce the growth of Trichomonas vaginalis and G. lamblia and are effective in treatment of giardiasis. The most significant biological applications of benzimidazole derivatives as core unit are listed below.

1. Anticancer Activities of Benzimidazole Scaffolds

Cancer is the most prevalent cause of death globally, prompting researchers to develop highly selective anticancer drugs that target cancer cells while sparing normal cells. Many existing drugs, though effective against cancer, but also harm normal cells, leading to significant side effects and often resulting in the discontinuation of chemotherapy. Heterocyclic compounds, particularly benzimidazole derivatives, have emerged as promising anticancer agents due to their outstanding selectivity and low side effects. Researchers are now concentrating significantly on this issue to address the growing demand for effective cancer treatment. Morais *et al.* developed an array of benzimidazole scaffolds featuring hydroxylated or fluorinated alkyl groups to evaluate anticancer activity. Among these compounds, compound **2** exhibited notable anticancer effects (Morais *et al.*, 2017). Again, Wang and co-workers also investigated chrysin benzimidazole derivatives for their anticancer properties. Compound **1** exhibited the most potent antiproliferative effect on MFC tumor cells. The study revealed that the dosing regimen significantly increased MFC cell death. Additionally, the compound has also been studied in mice, where compound **1** effectively prevented tumor growth (Wang *et al.*, 2018). Recently, benzimidazole, structural analogue of the purine base, interacts with both DNA and RNA, offering a wide range of biological activities. Shinde and co-workers developed sugar-based modified benzimidazole nucleosides with various substituents at 2-position, which showed promising anticancer properties (Figure 55). (Shinde *et al.*, 2020)

Indole and benzimidazole rings are important components in many current cancer drugs. Phenylindole scaffolds have been effective against breast cancer, while various 2-benzimidazole scaffolds core unit show potent activity against several cancer cell. Karadayi and co-worker found that indole benzimidazole derivatives having small alkyl groups or *p*-fluorobenzyl at R^1, and electron-withdrawing groups at R^2, exhibit strong anticancer properties. It was found that Compound **4**, featuring a *p*-fluorobenzyl group at position R^1 and Br atom at position R^2, was effective against MCF-7 cells (Karadayi *et al.*, 2020). Additionally, Shaldam and co-workers developed pyrimido[1,2-*a*]benzimidazole scaffolds and were screened against tumor growth *in vitro* (Figure 55). Derivative **6** exhibited strong activity against several cancer cell, especially leukemia, while derivatives **5** and **6** demonstrated promising effects on acute leukemia cell lines (Shaldam *et al.*, 2023).

Figure 55. Anticancer agent

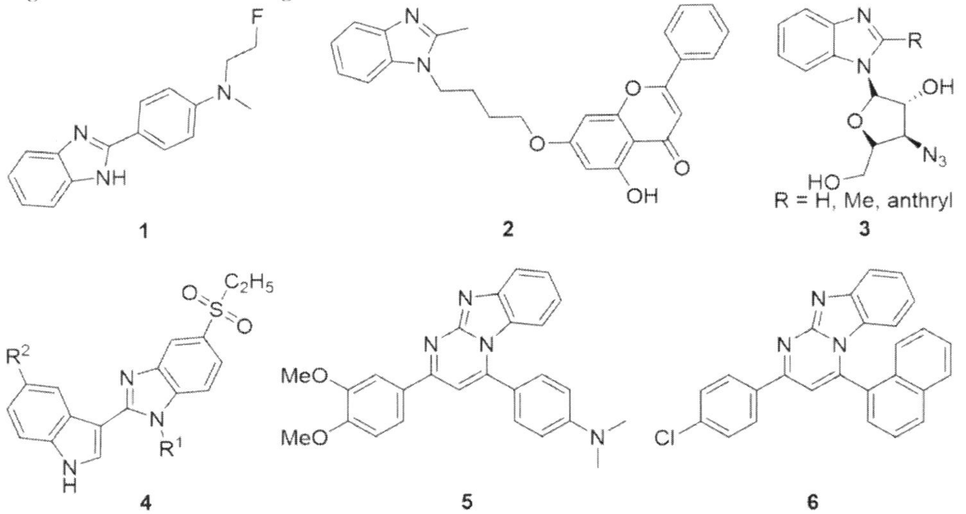

In 2023, Othman and co-workers reported various benzimidazole-triazole hybrids targeting inhibitors of VEGFR-2, EGFR and Topo II. Compounds **7** and **8** exhibited the most effective inhibition against MCF-7, HCT-116, HepG-2, and HeLa cancer cell. Compound **7** showed significant inhibition against both EGFR and Topo II, as well as moderate inhibition of VEGFR-2, while compound **8** was a moderate inhibitor of VEGFR-2 and EGFR, with weaker Topo II inhibition (Othman *et al.*, 2023). Again, Vaddiraju *et al.* synthesized many 3-(1*H*-benzo[*d*]imidazol-2-yl)-3,4-dihydro-2*H*-benzo[*e*][1,3]oxazines and tested their anticancer activity against breast cancer cell lines, including MCF-7 and MDA-MB-231. Scaffold **9** exhibited the excellent efficacy against both MDA-MB-231 and MCF-7. Compound **10** demonstrated strong activity against MDA-MB-231, whereas compound **11** was effective against MCF-7 (Figure 56). Docking studies indicated strong EGFR binding affinities, suggesting these compounds have promising drug-like properties for cancer therapy (Gali *et al.*, 2024).

Figure 56. Potential anticancer compounds

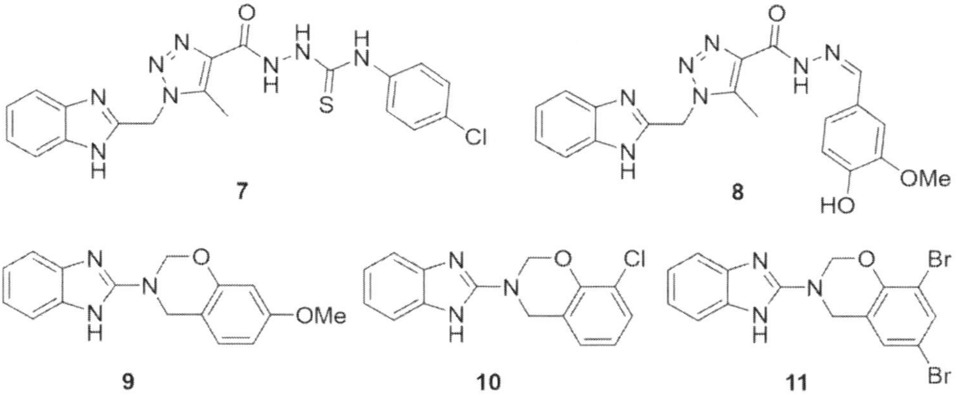

2. Antimicrobial Activity of Benzimidazole Scaffolds

Antimicrobial materials destroy or inhibit the growth of microorganisms, including harmful bacteria and fungi. They are classified by their target such as antibiotics for bacteria and antifungals for fungi. Based on function they are further subdivided like microbicides kill microbes, while biostatic agents prevent their growth. Benzimidazole scaffolds inhibit microbial protein synthesis, with 2-substituted benzimidazole derivatives typically exhibiting excellent pharmacological activity. El-Gohary and co-worker evaluated the synthesized benzimidazole derivatives for antimicrobial activity toward *A. fumigatus* 293, *B. cereus*, *C. albicans*, *E. coli* and *S. aureus*. Among the studied scaffolds, **12** and **13** demonstrated good effectiveness against *S. aureus*, while scaffold **14** showed favourable activity toward *B. cereus*. Additionally, compound **13** displayed positive results against *A. fumigatus* 293 (Figure 57) (El-Gohary *et al.*, **2017**). Again, Bistrovic´ and co-workers synthesized benzimidazole-triazole compounds and tested their antibacterial efficacy *in vitro*. These compounds demonstrated efficacy against Gram-positive bacteria, including *S. aureus*, *E. faecalis*, *VREF*, *MSSA* and *MRSA*, as well as Gram-negative bacteria such as *A. baumannii*, *E. coli*, *P. aeruginosa* and *K. pneumonia* as depicted in Figure 57. Studies on the structure-activity relationship indicated that nonsubstituted amidino benzimidazole scaffolds exhibited the best activity (Bistrović *et al.*, 2018).

Figure 57. Antimicrobial compounds

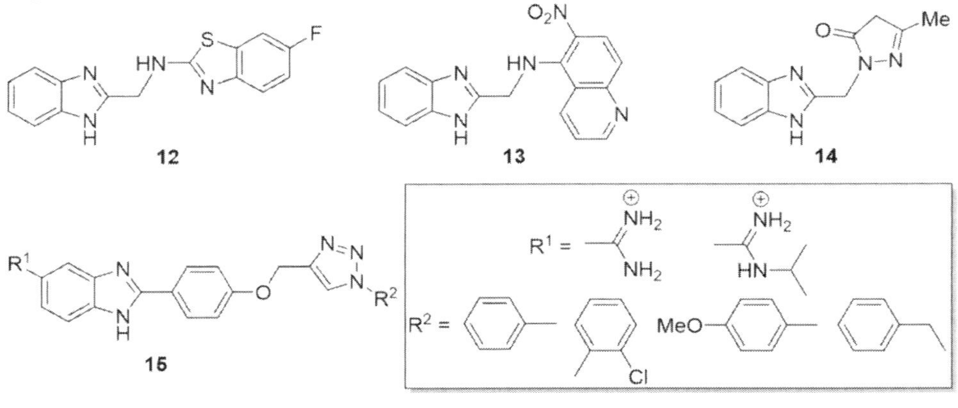

In order to evaluated the antibacterial and anticancer properties, Ersan *et al.* synthesised 2-phenyl benzimidazole scaffolds. These compounds exhibited moderate level of antimicrobial properties towards both gram-positive and gram-negative bacteria, as well as fungi. Notably, phenoxy methyl derivatives **16** and **17** demonstrated the highest activity against *Candida* as shown in Figure 58 (Ersan *et al.*, 2022). In 2023, Jasim and co-workers studied the antibacterial activity of benzimidazole compounds featuring a fluoro-benzene moiety (Figure 58). As compared to the unsubstituted compounds, the fluoro-substituted ones showed superior antibacterial and antifungal properties. Compound **19**, which containing fluorine atom in *meta*-position, has shown significant efficacy against gram-negative bacteria. Additionally, compounds **18** and **19** showed good activity against *B. subtilis*. Structure-activity analysis revealed that the presence of a methyl group at the 5-position enhances antifungal behaviour against *C. parapsilosis*. The efficiency of these scaffolds highlights its considerable possibilities for developing new antimicrobial agents (Jasim *et al.*, 2023). In similar year, Rajagopal co-workers developed bis-benzimidazole derivatives with the goal to assess the antimicrobial activity. They observed that compound **20** exhibited the most potent activity, demonstrating the lowest minimum inhibitory concentration against fungal infections as demonstrated in Figure 58 (Rajagopal *et al.*, 2023).

Figure 58. Antimicrobial agents

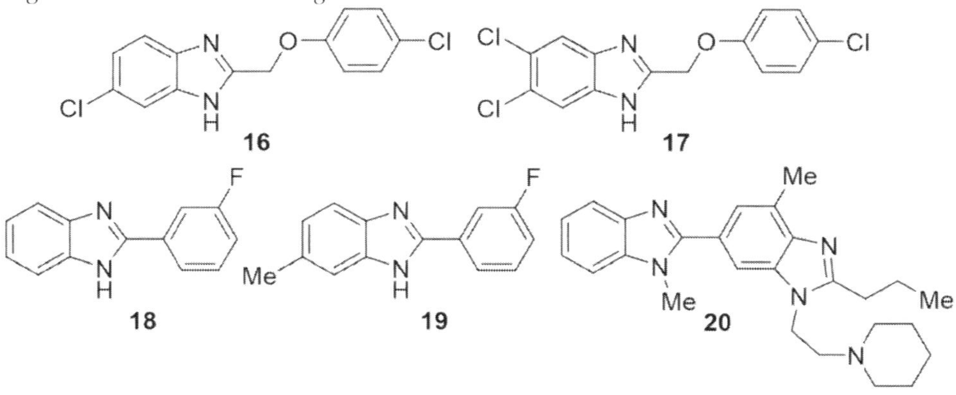

Currently, numerous studies focus on substituted benzimidazole derivatives, highlighting their significant antimicrobial properties. Many of these synthesized derivatives demonstrate effectiveness that rivals or even surpasses that of conventional drugs used for similar purposes.

3. Antidiabetic activity of Benzimidazole Scaffolds

Diabetes is a disorder where the body either fails to generate insulin or cannot utilize it effectively. Diabetes is also recognized as a common cause of affects and death a significant percentage of the worldwide population. To deal with diabetes, numerous synthetic and natural drug molecules have been manufactured. Shingalapur and co-workers devised 2-mercapto benzimidazole scaffolds and assessed their antidiabetic properties. On day nine, the reduction in blood glucose levels was superior for scaffolds **21** and **22** compared to glibenclamide as shown in Figure 5 (Shingalapur *et al.*, 2010). In a similar manner, Nair *et al.* developed a novel N-[(2-amino-5-methylene)-1,3,4-thiadiazole]-2-methyl benzimidazole scaffold. Scaffold **23** was picked among all synthesised derivatives for *in vitro* antidiabetic testing based on its Libdock score. At the same dosage, compound **23** displayed 49.25% inhibition while the standard drug exhibited 68.61% inhibition (Pathare *et al.*, 2021).

Figure 59. Antidiabetic compounds

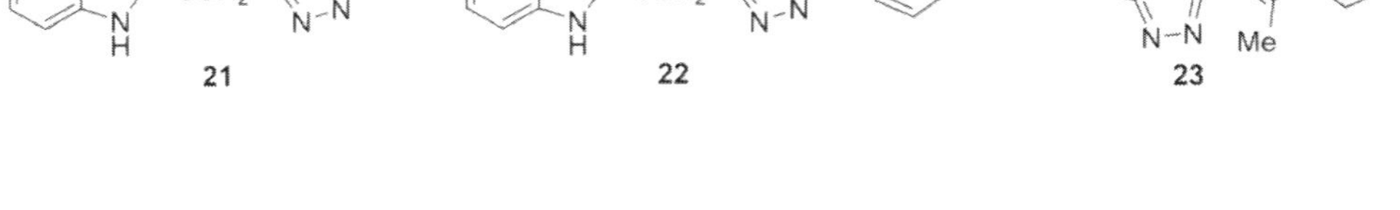

4. Antioxidant Activity of Benzimidazole Scaffolds

Antioxidants are substances that have the ability to inhibit or prevent the oxidation caused by free radicals, potentially protecting an organism's cells from damage. Since air-breathing organisms depend on aerobic oxidation for survival, the production of free radicals is inevitable. These free radicals are extremely toxic and can lead to cell death, cancerous transformations, or cellular malfunctions. Oxidative stress damages proteins, lipids, and DNA, potentially causing genetic mutations that result in cancer, diabetes, neurodegenerative diseases, and cardiovascular disorders. To prevent such damage, it is important to eliminate free radicals through antioxidants, which are derived from organic molecules and found in fruits as vitamins A, C, and E. Benzimidazole scaffolds are effective antioxidants because their structure, which includes benzene and imidazole rings, allows free radicals to resonate with the ring electrons, reducing or neutralizing their harmful effects. Abd *et al.* screened the antioxidant properties of synthesized benzimidazole scaffolds using DPPH method. They observed that scaffolds **24** and **25** exhibited antioxidant activities comparable to the reference used (Figure 60). Additionally, they observed that the pyrimidine thione and epoxide ring are the active components that for the antioxidant activity rather than the pyrazoline ring (Abd *et al.*, **2015**,).

Figure 60. Antioxidant compounds

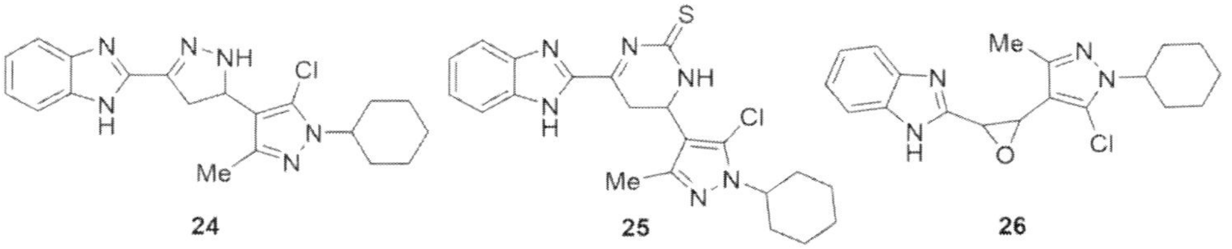

Archie and co-workers developed numerous benzimidazole scaffolds and screened their antioxidant activity. All of the evaluated scaffolds were found to be efficacious compared to the standard butylated hydroxytoluene, with compound **27** yielding the most excellent results as demonstrated in Figure 61 (Archie *et al.*, 2017).

Figure 61. Anti-oxidant agent

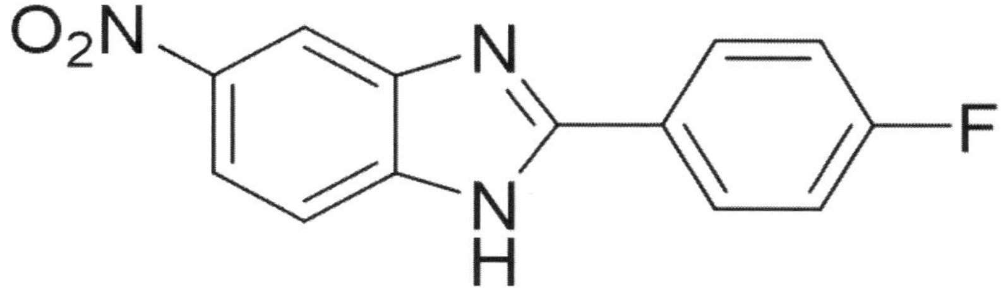

27

To investigate antioxidant properties, Bellam and co-workers devised various benzimidazole scaffolds as core units and screened their capacity to scavenge free radicals in comparison with H_2O_2 and DPPH. Derivatives **28**, **29**, and **30**, which feature a benzyl group on the N-atom of the benzimidazole, showed significant activity with both approaches (Bellam *et al.*, 2017). Again, Küçükoğlu and colleagues developed benzimidazole-1,3,4-oxadiazole scaffolds to examine their antioxidant properties and inhibitory effects on the human carbonic anhydrase isoforms hCA I and hCA II. Scaffolds **31a**, **31d**, and **31g** were particularly effective against hCA II, with scaffold **31a**, containing a 4-bromophenyl group, demonstrating significant inhibition for both isoforms (Figure 62) (Küçükoğlu *et al.*, **2022**). Recently, Ghate and co-workers developed benzimidazolecoumarin-3-carboxamide analogues and evaluated their antioxidant properties employing DPPH and ABTS approaches. They also utilised the DNSA approach to assess α-amylase inhibition. Derivative **32**, with a picoline substitution, exhibited excellent DPPH radical scavenging ability, whereas derivative **33**, featuring a 3,5-dimethoxyphenyl substitution, had the strongest scavenging ability against ABTS radicals as shown in Figure 62 (Patagar *et al.*, **2023**).

Figure 62. Antioxidant compounds

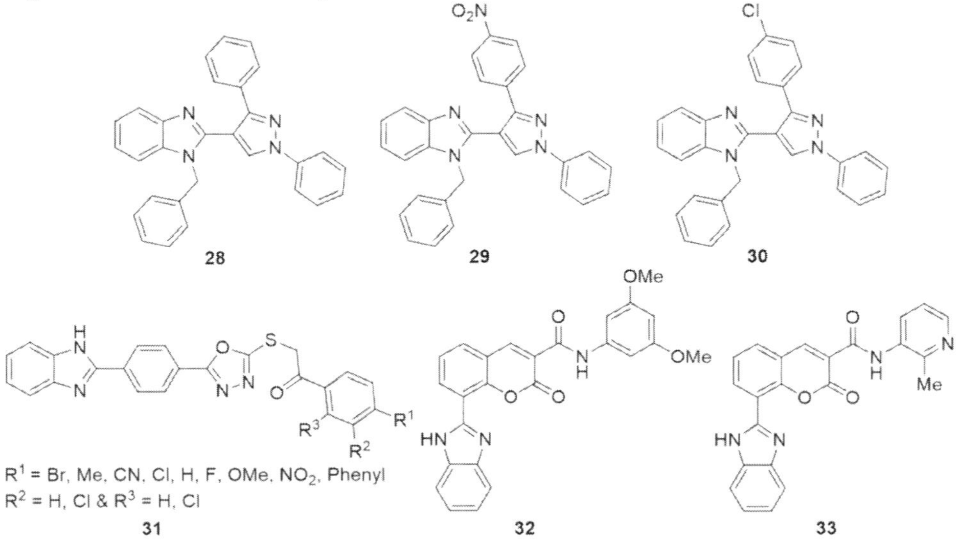

R^1 = Br, Me, CN, Cl, H, F, OMe, NO$_2$, Phenyl
R^2 = H, Cl & R^3 = H, Cl

5. Anti-inflammatory Activity of Benzimidazole Scaffolds

Anti-inflammatories are drugs that reduce the impact of several types of inflammations in the body. Numerous analgesic substances also exhibit anti-inflammatory activities, which serve to reduce pain by decreasing inflammation. Benzimidazole derivatives, which feature with unique aromatic ring structure, have become prominent in medicinal chemistry for developing anti-inflammatory and analgesic agents used in various therapeutically approved drugs to treat different types of pain and inflammation. Gaba and co-workers investigated substituted benzimidazole derivatives, such as compound **34**, for their ability to inhibit certain enzymes in the kidneys and stomach (Figure 63). The majority of the synthesized scaffolds demonstrated significant activity as compared to the reference drug (Gaba *et al.*, **2014**).

Figure 63. Anti-inflammatory compound

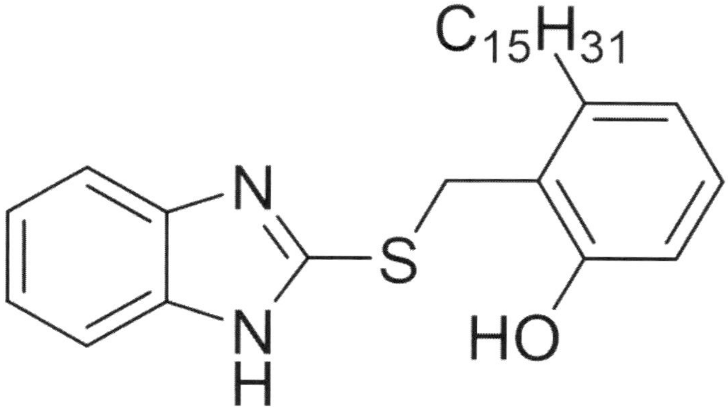

34

Saha *et al.* evaluated substituted benzimidazole derivatives against anti-inflammatory properties using a carrageenan-induced rat paw edema model. Compounds **35**, **36**, and **37** displayed significant activity (Saha *et al.*, 2021). Again, Moharana and co-workers developed non-toxic benzimidazole derivatives with notable anti-inflammatory activities. In silico, derivatives **38** and **39** possessed significant effects and good oral bioavailability. Their safety was validated by toxicity tests conducted *in vitro* and *in vivo* (Moharana *et al.*, 2022). Again, to enhance efficacy and reduce adverse effects, Nagesh *et al.* synthesized non-steroidal anti-inflammatory drugs, targeting on dual COX/5-LOX inhibition. These scaffolds demonstrated notable analgesic and anti-inflammatory activities (Figure 64). (Nagesh *et al.*, 2022).

Figure 64. Anti-inflammatory compounds

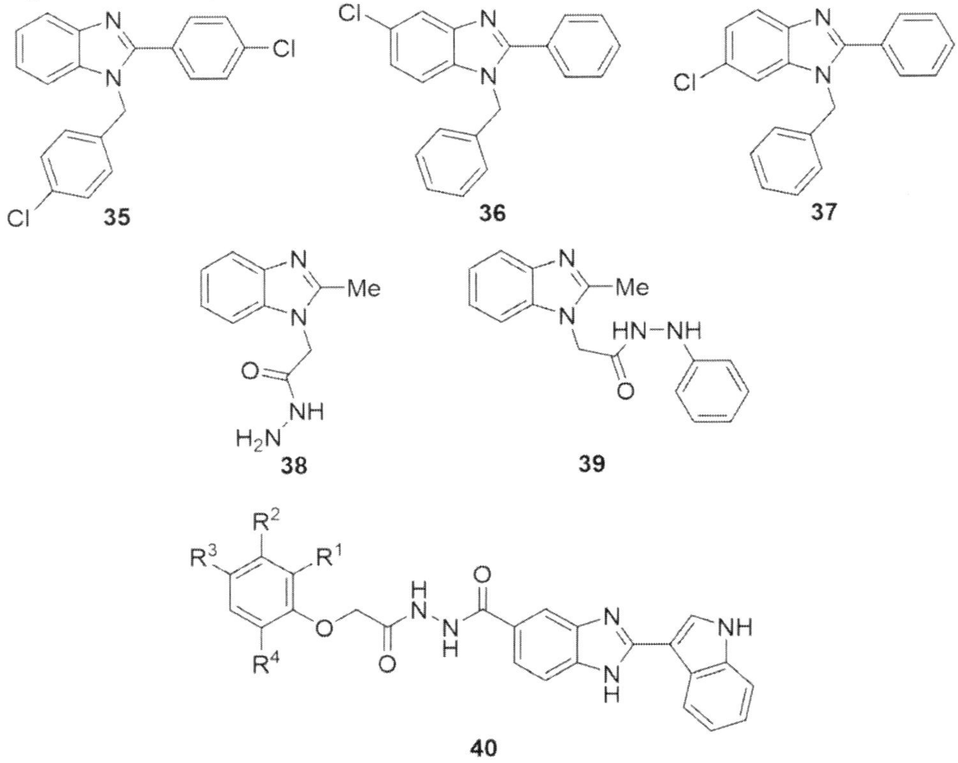

6. Anticonvulsant Activity of Benzimidazole Scaffolds

Anticonvulsants are prescribed to alleviate neuropathic pain, treat epileptic seizures, and act as mood stabilizers. They are also helpful in preventing the brain from spreading seizures. Benzimidazole scaffolds have proven effective as anti-convulsants in several studies. Shingalapur and co-workers explored the anticon-vulsant effects of benzimidazole scaffolds having 2-mercapto benzimidazole and 4-thiazolidinones employing the MES model. Scaffolds **41**, **42**, **43**, and **44** exhibits excellent anticonvulsant activity, with pharmacophore analysis indicating that an OH group was a crucial component in all active molecules as displayed in Figure 65 (Shingalapur *et al.*, 2010).

Figure 65. Anticonvulsants compounds

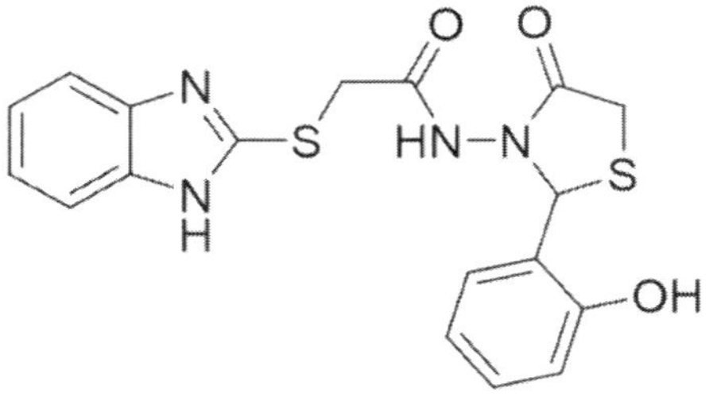

41

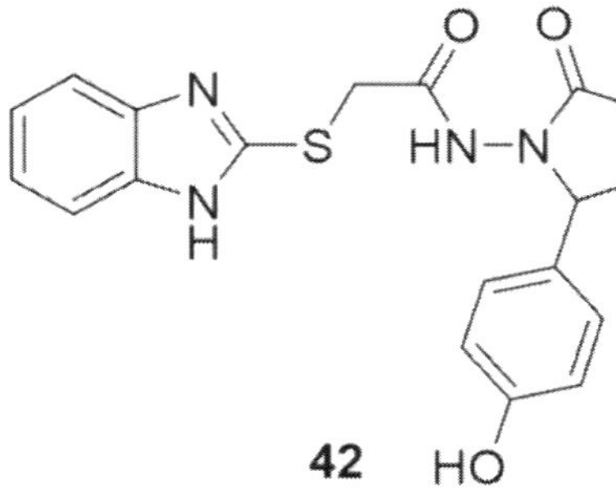

42

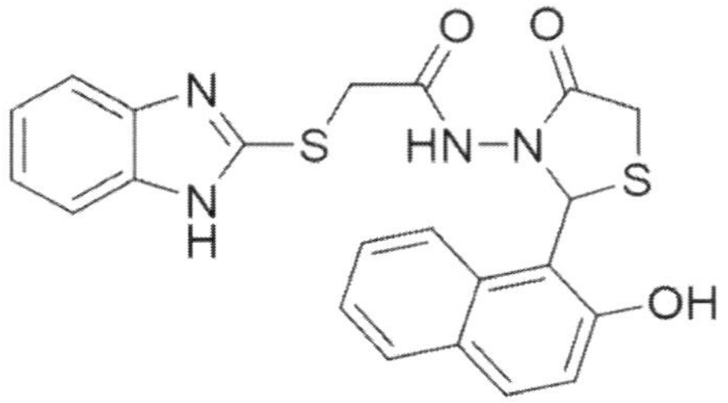

43

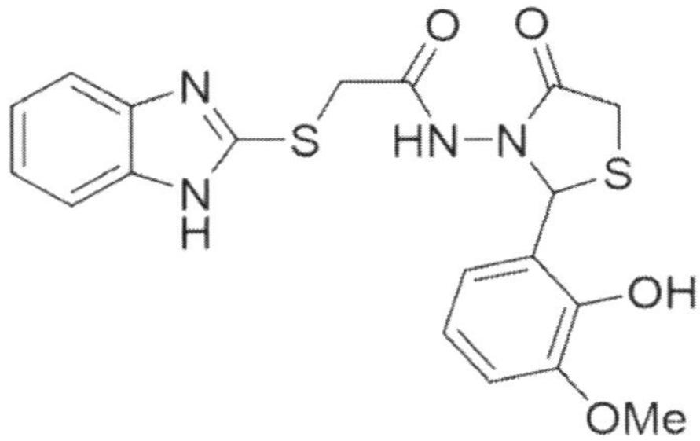

44

Shaharyar and co-workers examined the anticonvulsant properties of synthesized 2-[2-(phenoxymethyl)-1*H*-benzimidazol-1-yl]acetohydrazide scaffolds. Among these, compounds **45** and **46** demonstrated superior anticonvulsant effects compared to standard drugs, with protection indices of 40.5 and 24.7, respectively, indicating lower neurotoxicity. The study highlighted the significance of electron-withdrawing groups, especially fluorine and nitro, in enhancing activity, whereas electron-donating groups had no impact on the biological activities studied (Shaharyar *et al.*, **2016**). Again, Siddiqui and co-workers assessed the anticonvulsant ability of newly synthesized benzimidazole scaffolds, specifically 2-[(1-(2-substituted-benzyl)-1H-benzo[*d*]imidazole-2-yl)methyl]-N-substituted-phenylhydrazinecarb-othioamides. In tests on albino rats employing MES and scPTZ models, scaffold **47** exhibited excellent anticonvulsant effects (Figure 66) (Siddiqui *et al.*, 2016).

Figure 66. Anticonvulsants

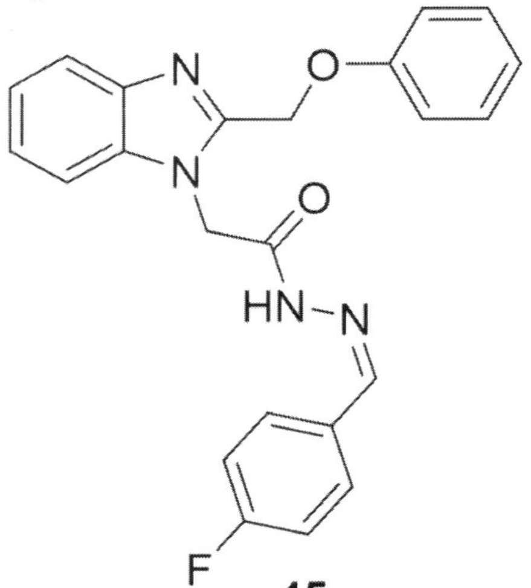

45

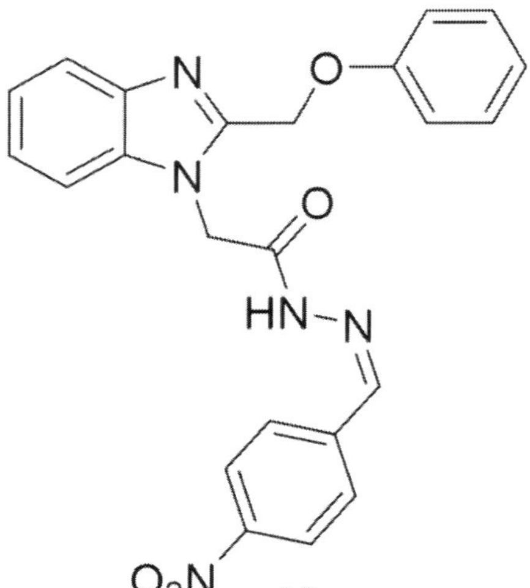

46

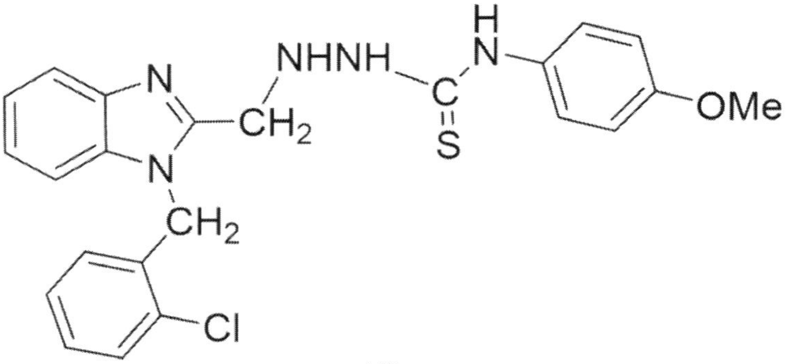

47

7. Anticoagulant Activity of Benzimidazole Scaffolds

Anticoagulants, or blood thinners, are chemicals that prevent blood clotting. These substances naturally occur in blood-feeding animals such as leeches and mosquitoes, where they help temporarily prevent formation of blood clot at the bite site so that blood can be absorbed. Medically, anticoagulants are prescribed to treat various conditions, including heart attacks, pulmonary embolism, deep vein thrombosis and atrial fibrillation. Additionally, they are used in sample tubes, blood transfusion bags, heart-lung machines, and dialysis equipment. Wang and co-worker synthesized fluorinated derivatives of 1-ethyl-1H-benzimidazole and evaluated the antithrombin property. All of the evaluated compounds exhibited strong anticoagulant properties, especially compound **48** being the most potent thrombin inhibitor (Wang *et al.*, **2016**). Again, Yang *et al.* devised and tested the anticoagulant activities of 1,2,5-trisubstituted benzimidazole fluorinated scaffolds. Scaffolds **49**, **50**, and **51** demonstrated high thrombin inhibitory activity (Figure 67). Further, the study also found that an *ortho*-position methyl group on the benzene ring enhances anticoagulant effectiveness (Yang *et al.*, **2016**).

Figure 67. Anticoagulants compounds

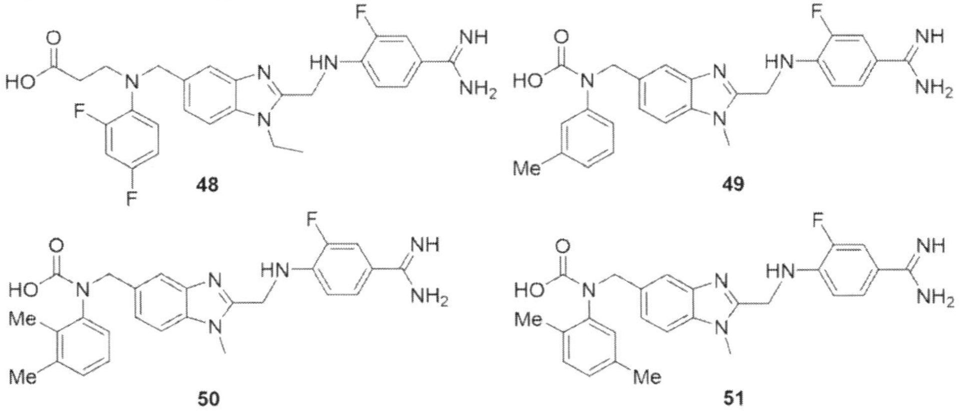

CONCLUSION

Benzimidazole is a highly significant heterocyclic scaffold that contains nitrogen and has a favorable role in biological processes. As a result, it has numerous pharmacological applications. Benzimidazoles can be synthesized employing a variety of synthetic approaches, diverse starting materials, reagents, and catalysts, as well as a variety of solvents and solvent-free environments. Synthetically produced derivatives of Benzimidazole exhibit biological activity and a range of medicinal applications. Positive findings in the realm of medicine to treat life-threatening illnesses will come from the futuristic research on Benzimidazole scaffold.

REFERENCES

Abd, S. N., & Soliman, F. M. A. (2015). Synthesis, some reactions, cytotoxic evaluation and antioxidant study of novel benzimidazole derivatives. *Der Pharma Chemica*, 7, 71–84.

Abdallah, A. E. M., Helal, M. H. E., & Elakabawy, N. I. I. (2015). Heterocyclization, dyeing applications and anticancer evaluations of benzimidazole derivatives: Novel synthesis of thiophene, triazole and pyrimidine derivatives. *Egyptian Journal of Chemistry*, 58(6), 699–719. DOI: 10.21608/ejchem.2015.1015

Abdollahi-Alibeik, M., & Poorirani, S. (2009). Perchloric acid–doped polyaniline as an efficient and reusable catalyst for the synthesis of 2-substituted benzothiazoles. *Phosphorus, Sulfur, and Silicon and the Related Elements*, 184(12), 3182–3190. DOI: 10.1080/10426500802705453

Alam, S. A. M. F., Ahmad, T., Nazmuzzaman, M., Ray, S. K., Sharifuzzaman, M., Karim, M. R., Alam, M. G., Ajam, M. M., Maitra, P., Mandol, D., Uddin, M. E., & Ahammed, T. (2017). Synthesis of benzimidazole derivatives containing schiff base exhibiting antimicrobial activities. *International Journal of Research Studies in Biosciences*, 5, 18–24.

Alaqeel, S. I. (2017). Synthetic approaches to benzimidazoles from *o*-phenylene-diamine: A literature review. *Journal of Saudi Chemical Society*, 21(2), 229–237. DOI: 10.1016/j.jscs.2016.08.001

Alinezhad, H., Salehian, F., & Biparva, P. (2012). Synthesis of benzimidazole derivatives using heterogeneous ZnO nanoparticles. *Synthetic Communications*, 42(1), 102–108. DOI: 10.1080/00397911.2010.522294

Archie, S. R., Das, B. K., Hossain, M. S., Kumar, U., & Rouf, A. S. S. (2017). Synthesis AND antioxidant activity of 2-substituted-5-nitro benzimidazole derivatives. *International Journal of Pharmacy and Pharmaceutical Sciences*, 9(1), 308–310. DOI: 10.22159/ijpps.2017v9i1.14972

Azarifar, D., Pirhayati, M., Maleki, B., Sanginabadi, M., & Yami, N. R. (2010). Acetic acid-promoted condensation of *o*-phenylenediamine with aldehydes into 2-aryl-1-(arylmethyl)-1*H*-benzimidazoles under microwave irradiation. *Journal of the Serbian Chemical Society*, 75(9), 1181–1189. DOI: 10.2298/JSC090901096A

Azizian, J., Torabi, P., & Noei, J. (2016). Synthesis of benzimidazoles and benzox-azoles using $TiCl_3OTf$ in ethanol at room temperature. *Tetrahedron Letters*, 57(2), 185–188. DOI: 10.1016/j.tetlet.2015.11.092

Bahrami, K., Khodaei, M. M., & Kavianinia, I. (2007). A Simple and efficient one-pot synthesis of 2-substituted benzimidazoles. *Synthesis*, 4(4), 547–550. DOI: 10.1055/s-2007-965878

Bahrami, K., Khodaei, M. M., & Nejatia, A. (2010). Synthesis of 1,2-disubstituted benzimidazoles, 2-substituted benzimidazoles and 2-substituted benzothiazoles in SDS micelles. *Green Chemistry*, 12(7), 1237–1241. DOI: 10.1039/c000047g

Bardajee, G. R., Mohammadi, M., Yari, H., & Ghaedi, A. (2015). Simple and efficient protocol for the synthesis of benzoxazole, benzoimidazole and benzothiazole heterocycles using Fe(III)-Schiff base/SBA-15 as a nanocatalyst. *Chinese Chemical Letters*, 27(2), 265–270. DOI: 10.1016/j.cclet.2015.10.011

Barker, H. A., Smyth, R. D., Weissbach, H., Toohey, J. I., Ladd, J. N., & Volcani, B. E. (1960). Isolation and properties of crystalline cobamide coenzymes containing benzimidazole or 5,6 dimethylbenzimidazole. *The Journal of Biological Chemistry*, 235(2), 480–488. DOI: 10.1016/S0021-9258(18)69550-X PMID: 13796809

Bellam, M., Gundluru, M., Sarva, S., Chadive, S., Netala, V. R., Tartte, V., & Cirandur, S. R. (2017). Synthesis and antioxidant activity of some new *N*-alkylated pyrazole-containing benzimidazoles. *Chemistry of Heterocyclic Compounds*, 53(2), 173–178. DOI: 10.1007/s10593-017-2036-6

Bhatnagar, A., Sharma, P. K., & Kumar, N. (2011). A review on "Imidazoles": Their chemistry and pharmacological potentials. *International Journal of Pharm Tech Research*, 3, 268–282.

Bistrović, A., Krstulović, L., Stolić, I., Drenjančević, D., Talapko, J., Taylor, M. C., Kelly, J. M., Bajić, M., & Raić-Malić, S. (2018). Synthesis, anti-bacterial and anti-protozoal activities of amidinobenzimidazole derivatives and their interactions with DNA and RNA. *Journal of Enzyme Inhibition and Medicinal Chemistry*, 33(1), 1323–1334. DOI: 10.1080/14756366.2018.1484733 PMID: 30165753

Brahmachari, G., Laskar, S., & Barik, P. (2013). Magnetically separable $MnFe_2O_4$ nano-material: An efficient and reusable heterogeneous catalyst for the synthesis of 2-substituted benzimidazoles and the extended synthesis of quinoxalines at room temperature under aerobic conditions. *RSC Advances*, 3(34), 14245–14253. DOI: 10.1039/c3ra41457d

Chakrabarty, M., Karmakar, S., Mukherji, A., Arima, S., & Harigaya, Y. (2006). Application of sulfamic acid as an eco-friendly catalyst in an expedient synthesis of benzimidazoles. *Heterocycles*, 68(5), 967–974. DOI: 10.3987/COM-06-10692

Chari, M. A., Sadanandam, P., Shobha, D., & Mukkanti, K. (2010). A simple, mild, and efficient procedure for high-yield synthesis of benzimidazoles using copper-triflate as catalyst. *Journal of Heterocyclic Chemistry*, 47(1), 153–155. DOI: 10.1002/jhet.287

Chen, C., Chen, C., Li, B., Tao, J., & Peng, J. (2012). Aqueous Synthesis of 1-*H*-2-substituted benzimidazoles via transition-metal-free intramolecular amination of aryl iodides. *Molecules (Basel, Switzerland)*, 17(11), 12506–12520. DOI: 10.3390/molecules171112506 PMID: 23095894

Chen, G. F., Xiao, N., Yang, J. S., Li, H. Y., Chen, B. H., & Han, L. F. (2015). A simple and eco-friendly process catalyzed by montmorillonite K-10, with air as oxidant, for synthesis of 2- substituted benzothiazoles. *Research on Chemical Intermediates*, 41(8), 5159–5166. DOI: 10.1007/s11164-014-1619-4

Chen, G. F., Zhang, L. Y., Jia, H. M., Chen, B. H., Li, J. T., Wang, S. X., & Bai, G. Y. (2013). Eco-friendly synthesis of 2-substituted benzothiazoles catalyzed by silica sulfuric acid. *Research on Chemical Intermediates*, 39(5), 2077–2086. DOI: 10.1007/s11164-012-0739-y

Dandia, A., Parewa, V., & Rathore, K. S. (2012). Synthesis and characterization of CdS and Mn doped CdS nanoparticles and their catalytic application for chemoselective synthesis of benzimidazoles and benzothiazoles in aqueous medium. *Catalysis Communications*, 28, 90–94. DOI: 10.1016/j.catcom.2012.08.020

Das, K., Mondal, A., & Srimani, D. (2018). Selective synthesis of 2-substituted and 1,2-disubstituted benzimidazoles directly from aromatic diamines and alcohols catalyzed by molecularly defined nonphosphine manganese(I) Complex. *The Journal of Organic Chemistry*, 83(16), 9553–9560. DOI: 10.1021/acs.joc.8b01316 PMID: 29993244

Denny, W. A., Rewcastle, G. W., & Bagley, B. C. (1990). Potential antitumor agents. Structure-activity relationships for 2-phenylbenzimidazole-4-carboxamides, a new class of minimal DNA-intercalating agents, which may not act via topoisomerase II. *Journal of Medicinal Chemistry*, 33(2), 814–819. DOI: 10.1021/jm00164a054 PMID: 2153829

DeSimone, R. W., Currie, K. S., Mitchell, S. A., Darrow, J. W., & Pippin, D. A. (2004). Privileged structures: Applications in drug discovery. *Combinatorial Chemistry & High Throughput Screening*, 7(5), 473–494. DOI: 10.2174/1386207043328544 PMID: 15320713

Dhanalakshmi, P., Thimmarayaperumal, S., & Shanmugam, S. (2014). Metal catalyst free one-pot synthesis of 2-arylbenzimidazoles from α-aroylketene dithioacetals. *RSC Advances*, 4(23), 12028–12036. DOI: 10.1039/C3RA47761D

Digwal, C. S., Yadav, U., Sakla, A. P., Ramya, P. V. S., Aaghaz, S., & Kamal, A. (2016). $VOSO_4$ catalyzed highly efficient synthesis of benzimidazoles, benzothiazoles, and quinoxalines. *Tetrahedron Letters*, 57(36), 4012–4016. DOI: 10.1016/j.tetlet.2016.06.074

Du, L. H., & Wang, Y. G. (2007). A Rapid and efficient synthesis of benzimidazoles using hypervalent iodine as oxidant. *Synthesis*, 5, 675–678.

Dutta, M. M., Goswami, M., & Phukan, P. (2019). Sulfonic acid functionalized $CoFe_2O_4$ magnetic nanocatalyst for the synthesis of benzimidazoles and benzothiazoles. *Indian Journal of Chemistry*, 58B, 811–819.

El-Gohary, N. S., & Shaaban, M. I. (2017). Synthesis and biological evaluation of a new series of benzimidazole derivatives as antimicrobial, antiquorum-sensing and antitumor agents. *European Journal of Medicinal Chemistry*, 131, 255–262. DOI: 10.1016/j.ejmech.2017.03.018 PMID: 28334654

Eleftheriadis, N., Neochoritis, C. G., Tsoleridis, C. A., Stephanidou-Stephanatou, J., & Iakovidou-Kritsi, Z. (2013). One-pot microwave assisted synthesis of new 2-alkoxycarbonylmethylene-4-oxo-1,5-benzo-, naphtho-, and pyridodiazepines and assessment of their cytogenetic activity. *European Journal of Medicinal Chemistry*, 67, 302–309. DOI: 10.1016/j.ejmech.2013.06.028 PMID: 23871910

Ersan, R. H., Kuzu, B., Yetkin, D., Alagoz, M. A., Dogen, A., Burmaoglu, S., & Algul, O. (2022). 2-Phenyl substituted benzimidazole derivatives: Design, synthesis, and evaluation of their antiproliferative and antimicrobial activities. *Medicinal Chemistry Research*, 31(7), 1192–1208. DOI: 10.1007/s00044-022-02900-3

Feizpour, F., Jafarpour, M., & Rezaeifard, A. (2018). A tandem aerobic photocatalytic synthesis of benzimidazoles by cobalt ascorbic acid complex coated on TiO_2 nanoparticles under visible light. *Catalysis Letters*, 148(1), 30–40. DOI: 10.1007/s10562-017-2232-0

Forseca, T., Gigante, B., & Gilchrist, T. L. (2001). A short synthesis of phenanthro[2,3-d]imidazoles from dehydroabietic acid. Application of the methodology as a convenient route to benzimidazoles. *Tetrahedron*, 57(9), 1793–1799. DOI: 10.1016/S0040-4020(00)01158-3

Gaba, M., & Mohan, C. (2016). Development of drugs based on imidazole and benzimidazole bioactive heterocycles: Recent advances and future directions. *Medicinal Chemistry Research*, 25(2), 173–210. DOI: 10.1007/s00044-015-1495-5

Gaba, M., Singh, D., Singh, S., Sharma, V., & Gaba, P. (2010). Synthesis and pharmacological evaluation of novel 5-substituted-1-(phenylsulphonyl)-2-methyl benzimidazole derivatives as anti- inflammatory and analgesic Agents. *European Journal of Medicinal Chemistry*, 45(6), 2245–2249. DOI: 10.1016/j.ejmech.2010.01.067 PMID: 20172630

Gaba, M., Singh, S., & Mohan, C. (2014). Benzimidazole: An emerging scaffold for analgesic and anti-inflammatory agents. *European Journal of Medicinal Chemistry*, 76, 494–505. DOI: 10.1016/j.ejmech.2014.01.030 PMID: 24602792

Gali, S., Raghu, D., Mallikanti, V., Thumma, V., & Vaddiraju, N. (2024). Design, synthesis of benzimidazole tethered 3,4-dihydro-2*H*-benzo[*e*] [1,3]oxazines as anticancer agents. *Molecular Diversity*, 28(3), 1347–1361. DOI: 10.1007/s11030-023-10661-3 PMID: 37233952

Gan, Z., Tian, Q., Shang, S., Luo, W., Dai, Z., Wang, H., Li, D., Wang, X., & Yuan, J. (2018). Imidazolium chloride-catalyzed synthesis of benzimidazoles and 2-substituted benzimidazoles from *o*-phenylenediamines and DMF derivatives. *Tetrahedron*, 74(52), 7450–7456. DOI: 10.1016/j.tet.2018.11.014

Ghosh, P., & Subba, R. (2015). $MgCl_2 \cdot 6H_2O$ catalyzed highly efficient synthesis of 2-substituted-1*H*-benzimidazoles. *Tetrahedron Letters*, 56(21), 2691–2694. DOI: 10.1016/j.tetlet.2015.04.001

Gorepatil, P. B., Mane, Y. D., & Ingle, V. S. (2013). Samarium(III) triflate as an efficient and reusable catalyst for facile synthesis of benzoxazoles and benzothiazoles in aqueous medium. *Journal of Chemistry*, 24, 2241–2244.

Goswami, M., Dutta, M. M., & Phukan, P. (2018). Sulfonic-acid-functionalized activated carbon made from tea leaves as green catalyst for synthesis of 2-substituted benzimidazole and benzothiazole. *Research on Chemical Intermediates*, 44(3), 1597–1615. DOI: 10.1007/s11164-017-3187-x

Hao, L., Zhao, Y., Yu, B., Zhang, H., Xu, H., & Liu, Z. (2014). Au catalyzed synthesis of benzimidazoles from 2-nitroanilines and CO_2/H_2. *Green Chemistry*, 16(6), 3039–3044. DOI: 10.1039/c4gc00153b

Hashem, H. E., & El Bakri, Y. (2021). An overview on novel synthetic approaches and medicinal applications of benzimidazole compounds. *Arabian Journal of Chemistry*, 14(11), 103418. DOI: 10.1016/j.arabjc.2021.103418

Hein, D. W., Alheim, R. J., & Leavitt, J. J. (1957). The use of polyphosphoric acid in the synthesis of 2-aryl- and 2-alkyl-substituted benzimidazoles, benzoxazoles and benzothiazoles. *Journal of the American Chemical Society*, 79(2), 427–429. DOI: 10.1021/ja01559a053

Heravi, M. M., Sadjadi, S., Oskooie, H. A., Shoar, R. H., & Bamoharram, F. F. (2008). Heteropolyacids as heterogeneous and recyclable catalysts for the synthesis of benzimidazoles. *Catalysis Communications*, 9(4), 504–507. DOI: 10.1016/j. catcom.2007.03.011

Huynh, T.-K.-C., Nguyen, T.-H.-A., Tran, N.-H.-S., Nguyen, T.-D., & Hoang, T.-K.-D. (2020). A facile and efficient synthesis of benzimidazole as potential anticancer agents. *Journal of Chemical Sciences*, 132(1), 84. DOI: 10.1007/s12039-020-01783-4

Ibrahim, K. S., & Begum, J. (2011). Synthesis and antimicrobial activity of some benzimidazole derivative with ibuprofen. *International Journal of Pharma Sciences*, 2, 298–302.

Ingle, R. G., & Magar, D. D. (2011). Heterocyclic chemistry of benzimidazoles and potential activities of derivatives. *International Journals of Drug Research and Technology*, 1, 26–32.

Itoh T, Nagata K, Ishikawa H & Ohsawa A. (2024) Synthesis of 2-arylbenzothiazoles and imidazoles using scandium triflate as a catalyst for both a ring closing and an oxidation steps. *Heterocycles*, 63, 2004, 2769–2783.

Jasim, K. H., Ersan, R. H., Sadeeq, R., Salim, S., Mahmood, S., & Fadhil, Z. (2023). Fluorinated benzimidazole derivatives: In vitro antimicrobial activity. *Bioorganic & Medicinal Chemistry Reports*, 6, 1–8.

Kamal, A., Rao, A. V. S., Nayak, V. L., Reddy, N. V. S., Swapna, K., Ramakrishna, G., & Alvala, M. (2014). Synthesis and biological evaluation of imidazo[1,5-*a*] pyridine-benzimidazole hybrids as inhibitors of both tubulin polymerization and PI3K/Akt pathway. *Organic & Biomolecular Chemistry*, 12(48), 9864–9880. DOI: 10.1039/C4OB01930J PMID: 25354805

Karadayi, F. Z., Yaman, M., Kisla, M. M., Keskus, A. G., Konu, O., & Ates-Alagoz, Z. (2020). Design, synthesis and anticancer/antiestrogenic activities of novel indole-benzimidazoles. *Bioorganic Chemistry*, 100, 103929. DOI: 10.1016/j. bioorg.2020.103929 PMID: 32464404

Karimi-Jaberi, Z., & Amiri, M. (2012). An efficient and inexpensive synthesis of 2-substituted benzimidazoles in water using boric acid at room temperature. *Journal of Chemistry*, 9(1), 167–170. DOI: 10.1155/2012/793978

Kathirvelan, D., Yuvaraj, P., Babu, K. A., Nagarajan, S., & Reddy, B. S. (2013). A green synthesis of benzimidazoles. *Indian Journal of Chemistry*, 52B, 1152–1156.

Keri, R. S., Hosamani, K. M., Reddy, H. S., & Shingalapur, R. V. (2009). Wells–Dawson Heteropolyacid: An Efficient Recyclable Catalyst for the Synthesis of Benzimidazoles Under Microwave condition. *Catalysis Letters*, 131(3-4), 552–559. DOI: 10.1007/s10562-009-9966-2

Khan, A. T., Parvin, T., & Choudhury, L. H. (2009). A Simple and Convenient One-Pot Synthesis of Benzimidazole Derivatives Using Cobalt (II) Chloride Hexahydrate as Catalyst. *Synthetic Communications*, 39(13), 2339–2346. DOI: 10.1080/00397910802654815

Khunt, M. D., Kotadiya, V. C., Viradiya, D. J., Baria, B. H., & Bhoya, U. C. (2014). International Letters of Chemistry. *Physics and Astronomy*, 6, 61–68.

Kidwai, M., Jahan, A., & Bhatnagar, D. (2010). Polyethylene glycol: A recyclable solvent system for the synthesis of benzimidazole derivatives using CAN as catalyst. *Journal of Chemical Sciences*, 122(4), 607–612. DOI: 10.1007/s12039-010-0095-7

Kommula, D., Rama, S., & Madugula, M. (2017). Synthesis of benzimidazoles/benzothiazoles by using recyclable, magnetically separable nano-Fe_2O_3 in aqueous medium. *Journal of the Iranian Chemical Society*, 14(8), 1665–1671. DOI: 10.1007/s13738-017-1107-z

Kovvuri, J., Nagaraju, B., Kamal, A., & Srivastava, A. K. (2016). An efficient synthesis of 2-substituted benzimidazoles *via* photocatalytic condensation of *o*-phenylenediamines and aldehydes. *ACS Combinatorial Science*, 18(10), 644–650. DOI: 10.1021/acscombsci.6b00107 PMID: 27631587

Küçükoğlu, K., Çevik, U. A., Nadaroglu, H., Celik, I., Işık, A., Bostancı, H. E., Özkay, Y., & Kaplancıklı, Z. A. (2022). Design, synthesis and molecular docking studies of novel benzimidazole-1,3,4-oxadiazole hybrids for their carbonic anhydrase inhibitory and antioxidant effects. *Medicinal Chemistry Research*, 31(10), 1771–1782. DOI: 10.1007/s00044-022-02943-6

Kumar, A., Sharma, S., & Maurya, R. A. (2010). Bienzymatic synthesis of benzothia/(oxa)zoles in aqueous medium. *Tetrahedron Letters*, 51(48), 6224–6226. DOI: 10.1016/j.tetlet.2010.06.012

Kumar, B., Smita, K., Kumar, B., & Cumbal, L. (2014). Ultrasound promoted and SiO_2/CCl_3COOH mediated synthesis of 2-aryl-1-arylmethyl-1*H*-benzimidazole derivatives in aqueous media: An eco-friendly approach. *Journal of Chemical Sciences*, 126(6), 1831–1840. DOI: 10.1007/s12039-014-0662-4

Kumara, K. S. J., Krishnamurthy, G., Kumar, N. S., Naik, N., & Praveen, T. M. (2017). Sustainable synthesis of magnetically separable $SiO_2/Co@Fe_2O_4$ nanocomposite and its catalytic applications for the benzimidazole synthesis. *Journal of Magnetism and Magnetic Materials*, 451, 808–821. DOI: 10.1016/j.jmmm.2017.10.125

Landge, S. M., & Török, B. (2008). Synthesis of Condensed Benzo[*N,N*]-Heterocycles by Microwave-Assisted Solid Acid Catalysis. *Catalysis Letters*, 122(3-4), 338–343. DOI: 10.1007/s10562-007-9385-1

Liu, J., Wang, C., Ma, X., Shi, X., Wang, X., Li, H., & Xu, Q. (2016). Simple synthesis of benzazoles by substrate-promoted CuI-catalyzed aerobic oxidative cyclocondensation of *o*-thio/amino/hydroxyanilines and amines under air. *Catalysis Letters*, 146(10), 2139–2148. DOI: 10.1007/s10562-016-1818-2

Liu, X., Cao, H., Bie, F., Yan, P., & Han, Y. (2019). C-N bond formation and cyclization: A straightforward and metal-free synthesis of *N*-1-alkyl-2-unsubstituted benzimidazoles. *Tetrahedron Letters*, 60(15), 1057–1059. DOI: 10.1016/j.tetlet.2019.03.028

Malakooti, R., Rostami-Nasab, M., Mahmoudi, H., Oskooie, H. A., Heravi, M. M., Karimi, N., Amouchi, A., & Kohansal, G. (2014). Synthesis of 2-substituted benzimidazoles and 2-aryl-1*H*- benzimidazoles using $[Zn(bpdo)_2 \cdot 2H_2O]^{2+}$/MCM-41 catalyst under solvent-free conditions. *Reaction Kinetics, Mechanisms and Catalysis*, 111(2), 663–677. DOI: 10.1007/s11144-013-0672-0

Mamedov, V. A., Galimullina, V. R., Zhukova, N. A., Kadyrova, S. F., Mironova, E. V., & Latypov, S. K. (2014). Quinoxalinone–benzimidazole rearrangement: An efficient strategy for the synthesis of structurally diverse quinoline derivatives with benzimidazole moieties. *Tetrahedron Letters*, 55(31), 4319–4324. DOI: 10.1016/j.tetlet.2014.06.023

Mamedov, V. A., Murtazina, A. M., Gubaidullin, A. T., Hafizova, E. A., & Rizvanov, I. K. (2009). Efficient synthesis of 2-(pyrazol-3-yl)benzimidazoles from 3-arylacylidene-3,4-dihydroquinoxalin-2(1*H*)-ones and hydrazine hydrate *via* a novel rearrangement. *Tetrahedron Letters*, 50(37), 5186–5189. DOI: 10.1016/j.tetlet.2009.05.116

Mamedov, V. A., Zhukova, N. A., Beschastnova, T. N., Gubaidullin, A. T., Balandina, A. A., & Latypov, S. K. (2010). A reaction for the synthesis of benzimidazoles and 1*H*-imidazo[4,5-*b*] pyridines via a novel rearrangement of quinoxalinones and their aza-analogues when exposed to 1,2-arylenediamines. *Tetrahedron*, 66(51), 9745–9753. DOI: 10.1016/j.tet.2010.10.026

Migawa, M. T., Giradet, J. L., Walker, J. A., Koszalka, G. W., Chamber-Lain, S. D., Drach, J. C., & Townsend, L. B. (1998). Design, Synthesis, and Antiviral Activity of α-Nucleosides: D- and l-Isomers of Lyxofuranosyl- and (5-Deoxylyxofuranosyl) benzimidazoles. *Journal of Medicinal Chemistry*, 41(8), 1242–1251. DOI: 10.1021/jm970545c PMID: 9575044

Mohammed, L. A., Farhan, M. A., Dadoosh, S. A., Alheety, M. A., Majeed, A. H., Mahmood, A. S., & Mahmoud, Z. H. (2023). A review on benzimidazole heterocyclic compounds: Synthesis and their medicinal activity applications. *SynOpen*, 7(4), 652–673. DOI: 10.1055/a-2155-9125

Moharana, A. K., Dash, R. N., Mahanandia, N. C., & Subudhi, B. B. (2022). Synthesis and anti-inflammatory activity evaluation of some benzimidazole derivatives. *Pharmaceutical Chemistry Journal*, 56(8), 1070–1074. DOI: 10.1007/s11094-022-02755-3 PMID: 36405379

Mohareb, R. M., Kamel, M. M., & Milad, Y. R. (2020). Uses of β-diketones for the synthesis of novel heterocyclic compounds and their antitumor evaluations. *Bulletin of the Chemical Society of Ethiopia*, 34, 385–405. DOI: 10.4314/bcse.v34i2.15

Mokhtari, J., & Bozcheloei, A. H. (2018). One-pot synthesis of benzoazoles *via* dehydrogenative coupling of aromatic 1,2-diamines/2-aminothiophenol and alcohols using Pd/Cu-MOF as a recyclable heterogeneous catalyst. *Inorganica Chimica Acta*, 482, 726–731. DOI: 10.1016/j.ica.2018.07.017

Morais, G. R., Palma, E., Marques, F., Gano, L., Oliveira, M. C., Abrunhosa, A., Miranda, H. V., Outeiro, T. F., Santos, I., & Paulo, A. (2017). Synthesis and biological evaluation of novel 2-aryl benzimidazoles as chemotherapeutic agents. *Journal of Heterocyclic Chemistry*, 54(1), 255–267. DOI: 10.1002/jhet.2575

Mostafavi, H., Islami, M. R., Ghonchepour, E., & Tikdari, A. M. (2018). Synthesis of 1*H*-1,3-benzimidazoles, benzothiazoles and 3*H*-imidazo[4,5-*c*]pyridine using DMF in the presence of HMDS as a reagent under the transition-metal-free condition. *Chemical Papers*, 72(12), 2973–2978. DOI: 10.1007/s11696-018-0540-5

Mukhopadhyay, C., & Datta, A. (2009). Water-promoted Dowex 50W catalyzed highly efficient green protocol for 2-arylbenzothiazole formation. *Journal of Heterocyclic Chemistry*, 46(1), 91–95. DOI: 10.1002/jhet.9

Nagesh, K. M. J., Prashanth, T., Khamees, H. A., & Khanum, S. A. (2022). Synthesis, analgesic, anti-inflammatory, COX/5-LOX inhibition, ulcerogenic evaluation, and docking study of benzimidazole bearing indole and benzophenone analogs. *Journal of Molecular Structure*, 1259, 132741. DOI: 10.1016/j.molstruc.2022.132741

Nasr-Esfahani, M., Mohammadpoor-Baltork, I., Khosropour, A. R., Moghadam, M., Mirkhani, V., & Tangestaninejad, S. (2013). Synthesis and characterization of Cu(II) containing nanosilica triazine dendrimer: A recyclable nanocomposite material for the synthesis of benzimidazoles, benzothiazoles, bis-benzimidazoles and bisbenzothiazoles. *Journal of Molecular Catalysis A Chemical*, 379, 243–254. DOI: 10.1016/j.molcata.2013.08.009

Nguyen, K. M. H., & Largeron, M. (2015). A bioinspired catalytic aerobic oxidative C-H functionalization of primary aliphatic amines: Synthesis of 1,2-disubstituted benzimidazoles. *Chemistry (Weinheim an der Bergstrasse, Germany)*, 21(36), 12606–12610. DOI: 10.1002/chem.201502487 PMID: 26206475

Nguyen, L. A., Retailleau, P., & Nguyen, T. B. (2019). Elemental sulfur/DMSO-promoted multicomponent one-pot synthesis of malonic acid derivatives from maleic anhydride and amines. *Advanced Synthesis & Catalysis*, 361(12), 2864–2869. DOI: 10.1002/adsc.201900160

Nguyen, T. B., Bescont, J. L., Ermolenko, L., & Al-Mourabit, A. (2013). Cobalt- and iron-catalyzed redox condensation of *o*-substituted nitrobenzenes with alkylamines: A step- and redox-economical synthesis of diazaheterocycles. *Organic Letters*, 15(24), 6218–6221. DOI: 10.1021/ol403064z PMID: 24228936

Nguyen, T. B., Ermolenko, L., & Al-Mourabit, A. (2013). Selective autoxidation of benzylamines: Application to the synthesis of some nitrogen heterocycles. *Green Chemistry*, 2013(15), 2713–2717. DOI: 10.1039/c3gc41186a

Nguyen, T. B., Ermolenko, L., & Al-Mourabit, A. (2013). Iron sulfide catalyzed redox/condensation cascade reaction between 2-amino/hydroxy nitrobenzenes and activated methyl groups: A straightforward atom economical approach to 2-hetaryl-benzimidazoles and -benzoxazoles. *Journal of the American Chemical Society*, 135(1), 118–121. DOI: 10.1021/ja311780a PMID: 23249371

Nguyen, T. B., Ermolenko, L., & Al-Mourabit, A. (2015). Sodium sulfide: A sustainable solution for unbalanced redox condensation reaction between *o*-nitroanilines and alcohols catalyzed by an iron-sulfur system. *Synthesis*, 47(12), 1741–1748. DOI: 10.1055/s-0034-1380134

Nguyen, T. B., Ermolenko, L., & Al-Mourabit, A. (2016). Formic acid as a sustainable and complementary reductant: An approach to fused benzimidazoles by molecular iodine-catalyzed reductive redox cyclization of *o*-nitro-*t*-anilines. *Green Chemistry*, 18(10), 2966–2970. DOI: 10.1039/C6GC00902F

Nguyen, T. B., Ermolenko, L., Corbin, M., & Al-Mourabit, A. (2014). Fe/S-catalyzed decarboxylative redox condensation of arylacetic acids with nitroarenes. *Organic Chemistry Frontiers: An International Journal of Organic Chemistry / Royal Society of Chemistry*, 1(10), 1157–1160. DOI: 10.1039/C4QO00221K

Nguyen, T. B., Ermolenko, L., Dean, W. A., & Al-Mourabit, A. (2012). Benzazoles from aliphatic amines and *o*-amino/mercaptan/hydroxyanilines: Elemental sulfur as a highly efficient and traceless oxidizing agent. *Organic Letters*, 14(23), 5948–5951. DOI: 10.1021/ol302856w PMID: 23171411

Nguyen, T. B., Nguyen, L. P., & Nguyen, T. T. (2019). Sulfur-catalyzed oxidative coupling of dibenzyl disulfides with amines: Access to thioamides and aza heterocycles. *Advanced Synthesis & Catalysis*, 361(8), 1787–1791. DOI: 10.1002/adsc.201801695

Niknam, K., & Fatehi-Raviz, A. (2007). Synthesis of 2-Substituted Benzimidazoles and bis-Benzimidazoles by Microwave in the Presence of Alumina-Methanesulfonic Acid. *Journal of the Iranian Chemical Society*, 4(4), 438–443. DOI: 10.1007/BF03247230

Othman, D. I. A., Hamdi, A., Tawfik, S. S., Elgazar, A. A., & Mostafa, A. S. (2023). Identification of new benzimidazole-triazole hybrids as anticancer agents: Multitarget recognition, *in vitro* and *in silico* studies. *Journal of Enzyme Inhibition and Medicinal Chemistry*, 38(1), 2166037. DOI: 10.1080/14756366.2023.2166037 PMID: 36651111

Patagar, D. N., Batakurki, S. R., Kusanur, R., Patra, S. M., Saravanakumar, S., & Ghate, M. (2023). Synthesis, antioxidant and anti-diabetic potential of novel benzimidazole substituted coumarin-3-carboxamides. *Journal of Molecular Structure*, 1274, 134589. DOI: 10.1016/j.molstruc.2022.134589

Pathare, B., & Bansode, T. (2021). Review-biological active benzimidazole derivatives. *Results in Chemistry*, 3, 100200. DOI: 10.1016/j.rechem.2021.100200

Patil, S. V., Patil, S. S. V. D., & Bobade, V. D. (2016). A simple and efficient approach to the synthesis of 2-substituted benzimidazole via sp^3 C–H functionalization. *Arabian Journal of Chemistry*, 9, S515–S521. DOI: 10.1016/j.arabjc.2011.06.017

Paul, S., & Basu, B. (2012). Highly selective synthesis of libraries of 1,2-disubstituted benzimidazoles using silica gel soaked with ferric sulfate. *Tetrahedron Letters*, 53(32), 4130–4133. DOI: 10.1016/j.tetlet.2012.05.129

Peng, J., Ye, M., Zong, C., Hu, F., Feng, L., Wang, X., Wang, Y., & Chen, C. (2010). Copper-catalyzed intramolecular C−N bond formation: A straightforward synthesis of benzimidazole derivatives in water. *The Journal of Organic Chemistry*, 76(2), 716–719. DOI: 10.1021/jo1021426 PMID: 21175149

Porcari, A. R., Devivar, R. V., Kucera, L. S., Drach, J. C., & Townsend, L. B. (1998). Design, Synthesis, and Antiviral Evaluations of 1-(Substituted benzyl)-2-substituted-5,6-dichlorobenzimidazoles as Nonnucleoside Analogues of 2,5,6-Trichloro-1-(β-d-ribofuranosyl)benzimidazole. *Journal of Medicinal Chemistry*, 41(8), 1252–1262. DOI: 10.1021/jm970559i PMID: 9548815

Pratap, U. R., Mali, J. R., Jawale, D. V., & Mane, R. A. (2009). Bakers' yeast catalyzed synthesis of benzothiazoles in an organic medium. *Tetrahedron Letters*, 50(12), 1352–1354. DOI: 10.1016/j.tetlet.2009.01.032

Preston, P. N. (1974). Synthesis, reactions, and spectroscopic properties of benzimidazoles. *Chemical Reviews*, 74(3), 279–314. DOI: 10.1021/cr60289a001

Rajagopal, K., Dhandayutham, S., Nandhagopal, M., & Narayanasamy, M. (2023). Study on new series of bis-benzimidazole derivatives synthesis, characterization, single crystal XRD, biological activity and molecular docking. *Journal of Molecular Structure*, 1283, 135253. DOI: 10.1016/j.molstruc.2023.135253

Rambabu, D., Murthi, P. R. K., Dulla, B., Basaveswara Rao, M. V., & Pal, M. (2013). Amberlyst-15–Catalyzed Synthesis of 2-Substituted 1,3-Benzazoles in Water under Ultrasound. *Synthetic Communications*, 43(22), 3083–3092. DOI: 10.1080/00397911.2013.769605

Ravi, V., Ramu, E., Vijay, K., & Rao, A. S. (2007). Zn-Proline Catalyzed Selective Synthesis of 1,2-Disubstituted Benzimidazoles in Water. *Chemical & Pharmaceutical Bulletin*, 55(8), 1254–1257. DOI: 10.1248/cpb.55.1254 PMID: 17666854

Reddy, P. L., Arundhathi, R., Tripathia, M., & Rawata, D. S. (2016). CuI nanoparticles mediated expeditious synthesis of 2-substituted benzimidazoles using molecular oxygen as the oxidant. *RSC Advances*, 6(58), 53596–53601. DOI: 10.1039/C6RA11678G

Rekha, M., Hamza, A., Venugopal, B., & Nagaraju, N. (2012). Synthesis of 2-substituted benzimidazoles and 1,5-disubstituted benzodiazepines on alumina and zirconia catalysts. *Chinese Journal of Catalysis*, 33(2-3), 439–446. DOI: 10.1016/S1872-2067(11)60338-0

Rithe, S. R., Jagtap, R. S., & Ubarhande, S. S. (2015). One pot synthesis of substituted benzimidazole derivatives and their characterization. *Rasayan Journal of Chemistry*, 8, 213–217.

Saberi, A. (2015). Efficient synthesis of Benzimidazoles using zeolite, alumina and silica gel under microwave irradiation. *Indian Journal of Science and Technology*, 39, 7–10.

Saha, P., Brishty, S. R., & Rahman, S. M. A. (2021). Pharmacological screening of substituted benzimidazole derivatives. *Dhaka University Journal of Pharmaceutical Sciences*, 20(1), 95–102. DOI: 10.3329/dujps.v20i1.54037

Samanta, S., Das, S., & Biswas, P. (2013). Photocatalysis by 3,6-disubstituted-*s*-tetrazine: Visible-light driven metal-free green synthesis of 2-substituted benzimidazole and benzothiazole. *The Journal of Organic Chemistry*, 78(22), 11184–11193. DOI: 10.1021/jo401445j PMID: 24134516

Santra, S., Majee, A., & Hajra, A. (2012). Nano indium oxide: An efficient catalyst for the synthesis of 1,2-disubstituted benzimidazoles in aqueous media. *Tetrahedron Letters*, 53(15), 1974–1977. DOI: 10.1016/j.tetlet.2012.02.021

Senapak, W., Saeeng, R., Jaratjaroonphong, J., Promarak, V., & Sirion, U. (2019). Metal-free selective synthesis of 2-substituted benzimidazoles catalyzed by Brönsted acidic ionic liquid: Convenient access to one-pot synthesis of *N*-alkylated 1,2-disubstituted benzimidazoles. *Tetrahedron*, 75(26), 3543–3552. DOI: 10.1016/j.tet.2019.05.014

Shaharyar, M., Mazumder, A., Garg, R., & Pandey, R. D. (2016). Synthesis, characterization and pharmacological screening of novel benzimidazole derivatives. *Arabian Journal of Chemistry*, 9, S342–S347. DOI: 10.1016/j.arabjc.2011.04.013

Shaik, vali B., Seelam, M., Tamminana, R., & Kammela, P. R. (2019) Copper promoted *C-S* and *C-N* cross-coupling reactions: The synthesis of 2-(*N*-aryolamino) benzothiazoles and 2-(*N*-aryolamino)benzimidazoles. *Tetrahedron, 75*, 3865–3874.

Shaikh, K. A., & Patil, V. A. (2012). An efficient solvent-free synthesis of imidazolines and benzimidazoles using $K_4[Fe(CN)_6]$ catalysis. *Organic Communications*, 5, 12–17.

Shaldam, M. A., Hendrychová, D., El-Haggar, R., Vojáčková, V., Majrashi, T. A., Elkaeed, E. B., Masurier, N., Kryštof, V., Tawfik, H. O., & Eldehna, W. M. (2023). 2,4-Diaryl-pyrimido[1,2-a]benzimidazole derivatives as novel anticancer agents endowed with potent anti-leukemia activity: Synthesis, biological evaluation and kinase profiling. *European Journal of Medicinal Chemistry*, 258, 115610. DOI: 10.1016/j.ejmech.2023.115610 PMID: 37437350

Shelkar, R., Sarode, S., & Nagarkar, J. (2013). Nano ceria catalyzed synthesis of substituted benzimidazole, benzothiazole, and benzoxazole in aqueous media. *Tetrahedron Letters*, 54(51), 6986–6990. DOI: 10.1016/j.tetlet.2013.09.092

Shinde, V. S., Lawande, P. P., Sontakke, V. A., & Khan, A. (2020). Synthesis of benzimidazole nucleosides and their anticancer activity. *Carbohydrate Research*, 498, 108178. DOI: 10.1016/j.carres.2020.108178 PMID: 33045644

Shingalapur, R. V., Hosamani, K. M., Keri, R. S., & Hugar, M. H. (2010). Derivatives of benzimidazole pharmacophore: Synthesis, anticonvulsant, antidiabetic and DNA cleavage studies. *European Journal of Medicinal Chemistry*, 45(5), 1753–1759. DOI: 10.1016/j.ejmech.2010.01.007 PMID: 20122763

Shingalapur, R. V., Hosamani, K. M., Keri, R. S., & Hugar, M. H. (2010). Derivatives of benzimidazole pharmacophore: Synthesis, anticonvulsant, antidiabetic and DNA cleavage studies. *European Journal of Medicinal Chemistry*, 45(5), 1753–1759. DOI: 10.1016/j.ejmech.2010.01.007 PMID: 20122763

Siddiqui, N., Alam, M., Ali, R., Yar, M. S., & Alam, O. (2016). Synthesis of new benzimidazole and phenylhydrazinecarbothiomide hybrids and their anticonvulsant activity. *Medicinal Chemistry Research*, 25(7), 1390–1402. DOI: 10.1007/s00044-016-1570-6

Singh, M. P., Sasmal, S., Lu, W., & Chatterjee, M. N. (2000). Synthetic utility of catalytic Fe(iii)/Fe(ii) redox cycling towards fused heterocycles: A facile access to substituted benzimidazole, bisbenzimidazole and imidazopyridine derivatives. *Synthesis*, 10(10), 1380–1390. DOI: 10.1055/s-2000-7111

Sluiter, J., & Christoffers, J. (2009). Synthesis of 1-methylbenzimidazoles from carbonitriles. *Synlett*, 1, 63–66.

Srinivasulu, R., Kumar, K. R., & Satyanarayana, P. V. V. (2014). Facile and efficient method for synthesis of benzimidazole derivatives catalyzed by zinc triflate. *Green and Sustainable Chemistry*, 4(1), 33–37. DOI: 10.4236/gsc.2014.41006

Su, F., Mathew, S. C., Mohlmann, L., Antonietti, M., Wang, X., & Blechert, S. (2011). Aerobic Oxidative Coupling of Amines by Carbon Nitride Photocatalysis with Visible Light. *Angewandte Chemie International Edition*, 50(3), 657–660. DOI: 10.1002/anie.201004365 PMID: 21226146

Suheyla, O., Kaynak, F. B., Kus, C., & G ker, H. (2002). 2-(2,4-Dichlorophenyl)-5-fluoro-6-morpholin-4-yl-1H-benzimidazole monohydrate. *Acta Crystallographica. Section E, Structure Reports Online*, 58(10), 1062–1064. DOI: 10.1107/S1600536802015842

Swami, M. B., Jadhav, A. H., Mathpati, S. R., Ghuge, H. G., & Patil, S. G. (2017). Eco-friendly highly efficient solvent free synthesis of benzimidazole derivatives over sulfonic acid functionalized graphene oxide in ambient condition. *Research on Chemical Intermediates*, 43(4), 2033–2053. DOI: 10.1007/s11164-016-2745-y

Torabi, P., Azizian, J., & Noei, J. (2016). Synthesis of benzimidazoles and benzoxazoles using $TiCl_3OTf$ in ethanol at room temperature. *Tetrahedron Letters*, 57(11), 185–188. DOI: 10.1016/j.tetlet.2016.02.038

Tran, M. Q., Ermolenko, L., Retailleau, P., Nguyen, T. B., & Al-Mourabit, A. (2014). Reaction of quinones and guanidine derivatives: Simple access to bis-2-aminobenzimidazole moiety of benzosceptrin and other benzazole motifs. *Organic Letters*, 16(3), 920–923. DOI: 10.1021/ol403672p PMID: 24479902

Trivedi, R., De, S. K., & Gibbs, R. A. (2006). A convenient one-pot synthesis of 2-substituted benzimidazoles. *Journal of Molecular Catalysis A Chemical*, 245(1-2), 8–11. DOI: 10.1016/j.molcata.2005.09.025

Venkateswarlu, Y., Kumar, S. R., & Leelavathi, P. (2013). Facile and efficient one-pot synthesis of benzimidazoles using lanthanum chloride. *Bioorganic & Medicinal Chemistry Letters*, 3, 1–8. PMID: 23919542

Wang, F., & Ren, Y.-J. (2016). Design, synthesis, biological evaluation and molecular docking of novel substituted 1-ethyl-1*H*-benzimidazole fluorinated derivatives as thrombin inhibitors. *Journal of the Iranian Chemical Society*, 13(6), 1155–1166. DOI: 10.1007/s13738-016-0830-1

Wang, Z., Deng, X., Xiong, S., Xiong, R., Liu, J., Zou, L., Lei, X., Cao, X., Xie, Z., Chen, Y., Liu, Y., Zheng, X., & Tang, G. (2018). Design, synthesis and biological evaluation of chrysin benzimidazole derivatives as potential anticancer agents. *Natural Product Research*, 32(24), 2900–2909. DOI: 10.1080/14786419.2017.1389940 PMID: 29063798

Wright, J. B. (1951). The chemistry of the benzimidazoles. *Chemical Reviews*, 48(3), 397–541. DOI: 10.1021/cr60151a002 PMID: 24541208

Wrobel, Z., Stachowska, K., Grudzień, K., & Kwast, A. (2011). *N*-aryl-2-nitrosoanilines as intermediates in the two-step synthesis of substituted 1,2-diarylbenzimidazoles from simple nitroarenes. *Synlett*, 2011(10), 1439–1443. DOI: 10.1055/s-0030-1260764

Xiao, H. L., Chen, J. X., Liu, M. C., Zhu, D. J., Ding, J. C., & Wu, H. Y. (2009). Trichloroisocyanuric Acid (TCCA) as a Mild and Efficient Catalyst for the Synthesis of 2-Arylbenzothiazoles. *Chemistry Letters*, 38(2), 170–171. DOI: 10.1246/cl.2009.170

Yang, D., Fu, H., Hu, L., Jiang, Y., & Zhao, Y. J. (2008). Copper-catalyzed synthesis of benzimidazoles via cascade reactions of *o*-haloacetanilide derivatives with amidine hydrochlorides. *The Journal of Organic Chemistry*, 73(19), 7841–7844. DOI: 10.1021/jo8014984 PMID: 18754576

Yang, H., Ren, Y., Gao, X., & Gao, Y. (2016). Synthesis and anticoagulant bioactivity evaluation of 1,2,5-trisubstituted benzimidazole fluorinated derivatives. *Chemical Research in Chinese Universities*, 32(6), 973–978. DOI: 10.1007/s40242-016-6205-4

Yang, X. L., Xu, C. M., Lin, S. M., Chen, J. X., Ding, J. C., Wu, H. Y., & Su, W. K. (2010). Eco-friendly synthesis of 2-substituted benzothiazoles catalyzed by cetyltrimethyl ammonium bromide (CTAB) in water. *Journal of the Brazilian Chemical Society*, 21(1), 37–42. DOI: 10.1590/S0103-50532010000100007

Yu, C., Guo, P., Jin, C., & Su, W. (2009). The synthesis of benzimidazole derivatives in the absence of solvent and catalyst. *Journal of Chemical Research*, 5(5), 333–336. DOI: 10.3184/030823409X447763

Yu, Z. Y., Zhou, J., Fang, Q. S., Chen, L., & Song, Z. B. (2016). Chemoselective synthesis of 1,2-disubstituted benzimidazoles in lactic acid without additive. *Chemical Papers*, 70(9), 1293–1298. DOI: 10.1515/chempap-2016-0056

Yuan, J., Zhao, Z., Zhu, W., Li, H., Qian, X., & Xu, Y. (2013). New strategy for the synthesis of 2-phenylbenzimidazole derivatives with sodium perborate (SPB) as oxidant. *Tetrahedron*, 69(34), 7026–7030. DOI: 10.1016/j.tet.2013.06.045

Zhang, Z. H., Yin, L., & Wang, Y. M. (2007). An expeditious synthesis of benzimidazole derivatives catalyzed by Lewis acids. *Catalysis Communications*, 8(7), 1126–1131. DOI: 10.1016/j.catcom.2006.10.022

Zhu, J., Zhang, Z., Miao, C., Liu, W., & Sun, W. (2017). Synthesis of benzimidazoles from *o*-phenylenediamines and DMF derivatives in the presence of PhSiH$_3$. *Tetrahedron*, 73(25), 3458–3462. DOI: 10.1016/j.tet.2017.05.018

Chapter 8
Benzoxazoles:
Diverse Biological Activities and Therapeutic Potential

G. K. Prashanth

https://orcid.org/0000-0001-6691-4030

Sir M. Visvesvaraya Institute of Technology, India

Srilatha Rao

Nitte Meenakshi Institte of Technology, India

H. S. Lalithamba

Siddaganga Institute of Technology, India

K. V. Rashmi

Sir M. Visvesvaraya Institute of Technology, India

N. P. Bhagya

Sai Vidya Institute of Technology, India

Mithun Kumar Ghosh

Medi-Caps University, India

Manoj Gadewar

KR Mangalam University, India

Nirmala R. Darekar

Department of Chemistry, Radhabai Kale Mahila Mahavidyalaya, India

ABSTRACT

Benzoxazoles are heterocyclic compounds featuring a benzene ring fused to an oxazole ring. They have attracted significant attention in medicinal chemistry due to their diverse biological activities. Many pharmacological compounds have a core structure called a heterocyclic scaffold, which is essential to their therapeutic actions. Modest alterations to this fundamental structure may result in notable variations in the medication's mechanism of action. Benzoxazole and its derivatives have shown substantial and noteworthy therapeutic effects. This chapter explores the synthesis, structural characteristics, and various medical applications of ben-zoxazoles, highlighting their roles as antibacterial, antifungal, antiviral, anticancer,

DOI: 10.4018/979-8-3693-7267-8.ch008

anti-inflammatory, and analgesic agents.

INTRODUCTION

Organic heterocyclic compounds with sulfur, oxygen, and nitrogen are crucial due to their diverse applications in various fields. These compounds are integral to pharmaceuticals, where they play roles in drug design and development, providing therapeutic benefits. In industrial applications, they contribute to the production of dyes, plastics, and agrochemicals. In medicine, they serve as the backbone for many antibiotics and antiviral drugs. Additionally, they are used in sensing technologies for detecting environmental pollutants and in food packaging to enhance preservation and safety. Their versatile nature makes them invaluable across multiple disciplines (Al-Saidi & Khan, 2024)(Al-Saidi & Khan, 2022)(Aljaar et al., 2019)(Alrooqi et al., 2022)(Gemili et al., 2019)(Gul et al., 2024)(Kabi et al., 2022)(Kabi et al., 2022)(Khan et al., 2021)(Khan et al., 2022)(Khan et al., 2022)(Mohammad Abu-Taweel et al., 2024)(Muhammad et al., 2022).

Benzoxazole is a heterocyclic compound featuring a benzene ring fused with an oxazole ring. The oxazole component consists of a five-membered ring with an oxygen atom and a nitrogen atom in the pyridine position. This compound was first synthesized by Hantzsch in 1887. Benzoxazole has the molecular formula C_7H_5NO, with a molar mass of 119.12 g/mol. It has a melting point range of 27–30°C and a boiling point of 182°C. This structure is significant due to its diverse applications in pharmaceuticals and materials science (Dinakaran et al., 2012)

(Parvathi et al., 2013)(Katritzky & Weeds, 1967)(Kaur et al., 2015)(Singh et al., 2010)(Jyothi & Merugu, 2017)(Cornforth & Cornforth, 1947). The ability to modify the benzoxazole structure at various positions has led to the development of numerous derivatives with specific and enhanced pharmacological profiles. These compounds exhibit remarkable biological activities, making them valuable in various medical applications. They possess anti-ulcer, anticancer, and antihypertensive properties, providing crucial benefits in treating ulcers, cancer, and high blood pressure. Their antifungal and anti-inflammatory capabilities help combat fungal infections and reduce inflammation. They are effective against tuberculosis and parasites, showcasing antitubercular and antiparasitic effects. Additionally, they offer anti-obesity and antimalarial benefits, aiding in weight management and malaria prevention. Their antiglycation and antiviral properties contribute to diabetes management and viral infection control. Moreover, they have analgesic, antioxidant, antihistaminic, and antibacterial potential, making them useful for pain relief, fighting oxidative stress, managing allergies, and treating bacterial infections (Aljaar et al., 2023)(Gujjarappa et al., 2022)(Aljaar et al., 2015)(Gujjarappa et al., 2020)(Aljaar et al.,

2013)(Kaldhi et al., 2019)(Shafiei et al., 2020) (Vodnala et al., 2018). Benzoxazole derivatives are increasingly used as fluorescent and colorimetric sensors to detect various pollutants. This emerging field is gaining interest in chemical biology and analytical chemistry for its potential to enhance environmental monitoring (Xu et al., 2014) (Yang et al., 2017). Additionally, the metal complexes formed with these derivatives exhibit impressive biological and catalytic activities, making them valuable in developing new pharmaceuticals and catalysts. These properties highlight their significance in both scientific research and practical applications (Balaghi et al., 2013) (Sun et al., 2022).

The mechanisms underlying the therapeutic effects of benzoxazoles are diverse and complex (Tian et al., 2024)(Abdullahi et al., 2024)(Faydalı & Arpacı, 2024) (Jarrar et al., 2024). They often involve interactions with key biological targets, such as enzymes, receptors, and DNA, leading to the inhibition of disease-related pathways (Yin et al., 2024). The ability to fine-tune the chemical structure of benzoxazoles has allowed researchers to optimize their interactions with these targets, enhancing their efficacy and reducing potential side effects (Yin et al., 2024)(Kumar et al., 2022)(Liu et al., 2024)(Fu et al., 2023).

In recent years, significant advancements have been made in the synthesis of novel benzoxazole derivatives, as well as in drug delivery systems that improve their bioavailability and therapeutic efficacy (Sharma et al., 2020)(Omar et al., 2020)(Muhammed et al., 2022). These developments have paved the way for the translation of benzoxazole-based compounds from the laboratory to the clinic, with several derivatives currently undergoing clinical trials for various indications (IMSA, 2024)(Han et al., 2023).

This chapter aims to provide a comprehensive overview of the medical applications of benzoxazoles. It will cover their pharmacological activities, mechanisms of action, and the latest advancements in benzoxazole-based drug development. By highlighting the therapeutic potential of these compounds, this review seeks to underscore the importance of benzoxazoles in modern medicinal chemistry and their promising future in drug discovery and development.

CHEMICAL STRUCTURE AND PROPERTIES

Benzoxazoles consist of a benzene ring fused to an oxazole ring. This fusion imparts significant stability and biological activity to the molecule. The basic structure can be modified at various positions, allowing for the development of a wide range of derivatives with specific pharmacological properties.

Pharmacological Activities

Antimicrobial Activity

Benzoxazoles have demonstrated potent antimicrobial activity against a variety of bacterial and fungal pathogens. Several derivatives have been developed as potential antibiotics, showing efficacy against resistant strains. The mechanism of action typically involves the inhibition of key bacterial enzymes or disruption of cell wall synthesis.

In a study by Kakkar et al., a series of benzoxazole compounds were synthesized and evaluated for their antimicrobial efficacy. The researchers used the tube dilution technique to screen these compounds against four Gram-negative bacteria (*Pseudomonas aeruginosa*, *Escherichia coli*, *Salmonella typhi* and *Klebsiella pneumoniae*) and one Gram-positive bacterium (*Bacillus subtilis*). The compounds were tested against two fungal strains, *Aspergillus niger* and *Candida albicans*. The results of this study demonstrated that the synthesized compounds, specifically **1** and **2**, exhibited significant antimicrobial activity. For comparison, ofloxacin was used as a positive control for antibacterial activity, and fluconazole served as the positive control for antifungal activity. The study's findings suggest that these benzoxazole derivatives hold promise as potential antimicrobial agents (Kakkar et al., 2018).

Yernale et al. (2022) synthesized novel indole-based complexes of Cu^{+2}, Co^{+2}, Ni^{+2}, and Zn^{+2}, designated as **3-6** (**Figure 5**). Various analytical techniques were employed to elucidate the structural characteristics of these metal complexes. The structural analysis revealed that the Zn^{+2} complex (**6**) adopts a tetrahedral geometry, whereas the Cu^{+2}, Co^{+2}, and Ni^{+2} complexes (**3-5**) exhibit octahedral geometries. The biological activities of these complexes were thoroughly investigated, including their antituberculosis, antifungal, antioxidant, and antibacterial properties. Among the complexes, **3** and **4** demonstrated notable antimicrobial activity against Escherichia coli and Bacillus subtilis, with minimum inhibitory concentrations (MICs) of 12.50 µg/mL. The **3-5** complexes showed significant activity against Mycobacterium tuberculosis, with MIC values of 3.125 µg/mL, comparable to the standard drug Ciprofloxacin. These results highlight the potential of these metal complexes as promising therapeutic agents. The study's findings could be instrumental in the design and development of new antimicrobial and antituberculosis drugs (Yernale et al., 2022).

Srivastava et al. synthesized a benzoxazole derivative, specifically 2-[1-benzyl-4(4-methoxyphenyl)-1H-1,2,3-triazol-5-yl]-1,3-benzoxazole (compound **7**), and evaluated its antibacterial efficacy. The compound was tested against Gram-negative (*Escherichia coli*) and Gram-positive (*Staphylococcus aureus*) bacterial strains to assess its antimicrobial potential. The study's results demonstrated that compound **7** exhibited significant antibacterial activity against both bacterial strains. This finding suggests that the benzoxazole derivative has broad-spectrum antibacterial properties, making it a promising candidate for further development as an antimicrobial agent. The potent activity observed against both Gram-positive and Gram-negative bacteria highlights its potential utility in treating various bacterial infections (Srivastava et al., 2015).

Vodela et al. synthesized a series of 2-(5-substituted-[1,3,4]oxadiazol-2-yl)benzoxazole derivatives and evaluated their antimicrobial activity against a variety of bacterial and fungal strains using the disk diffusion method. The study focused on four Gram-positive bacterial strains (Staphylococcus aureus, Staphylococcus albus, Streptococcus faecalis, and Klebsiella pneumoniae), four Gram-negative bacterial strains (Escherichia coli, Pseudomonas aeruginosa, Proteus mirabilis, and Salmonella typhi), and two fungal strains (Candida albicans and Aspergillus fumigatus). The results revealed that compound **8** exhibited good activity against S. faecalis but was almost inactive against E. coli. It showed moderate antimicrobial potential against the other tested organisms. Compound **9** displayed mild to moderate activity against both Gram-positive and Gram-negative bacteria. Interestingly, compound **10**, with an ethyl substituent, was inactive against E. coli compared to other molecules in the series. Compound **11**, which featured a para-nitro phenyl derivative, demonstrated significant antimicrobial activity against S. aureus, S. albus, S. faecalis, K. pneumoniae, and P. aeruginosa, surpassing the standard drug amikacin in these cases. However, it only showed moderate activity against E. coli and P. mirabilis. Notably, this compound also exhibited high antifungal activity against both C. albicans and A. fumigatus, achieving a marked activity index. The introduction of a nitro group in compound **11** was associated with enhanced antimicrobial activity against various organisms. Additionally, compounds **12** and **13**, with related substituents, demonstrated the highest antifungal activity against C. albicans and A. fumigatus, outperforming the standard antifungal drug fluconazole. This study highlights the potential of specific substitutions, particularly nitro groups, to improve the antimicrobial efficacy of benzoxazole derivatives (Vodela et al., 2013).

Jayanna et al. synthesized a series of (5,7-dichloro-1,3-benzoxazol-2-yl)-3-phenyl-1H-pyrazole-4-carbaldehyde derivatives (compounds **14-18**) and assessed their antimicrobial activity. The compounds were tested against a selection of microorganisms, including two Gram-positive bacterial strains (Staphylococcus aureus and Bacillus subtilis), two Gram-negative bacterial strains (Escherichia coli and Salmonella typhi), and two fungal strains (Aspergillus niger and Candida albicans). The antimicrobial testing was conducted using concentrations ranging from 10 to 50 µg. Ciprofloxacin and fluconazole served as the positive controls for bacterial and fungal comparisons, respectively. Among the synthesized compounds, 43b demonstrated the highest antimicrobial activity across all tested bacterial and fungal strains. This indicates that compound **15** has broad-spectrum antimicrobial potential, making it a promising candidate for further development and study in antimicrobial applications (Jayanna et al., 2013).

Figure 1. Compounds of antimicrobial activity

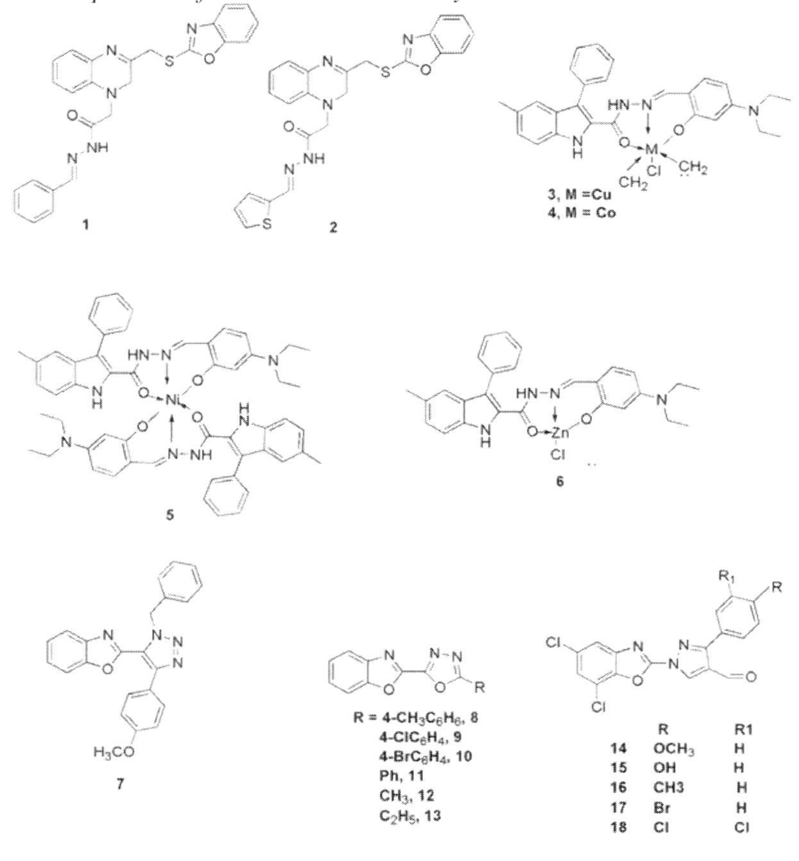

Anticancer Activity

Numerous benzoxazole derivatives exhibit significant anticancer properties. These compounds can induce apoptosis, inhibit cell proliferation, and interfere with cancer cell signaling pathways. Benzoxazole-based drugs are being explored for the treatment of various cancers, including breast, lung, and colon cancer.

Omer et al. synthesized a series of 2-substituted benzoxazole derivatives and investigated their in vitro cytotoxic effects on two breast cancer cell lines: MDA-MB-231 and MCF-7. The study aimed to identify compounds with significant antiproliferative activity against these cancer cell lines. The results revealed that certain compounds exhibited selective cytotoxicity towards the different cell lines. Specifically, compounds **19a**, **19b**, and **20** demonstrated a higher cytotoxic potential against the MCF-7 cell line, indicating their effectiveness in inhibiting the proliferation of these cells. On the other hand, compounds **19c**, **21**, and **22** showed pronounced cytotoxic activity against the MDA-MB-231 cell line, suggesting their capability to target and kill these more aggressive breast cancer cells. These findings highlight the potential of these 2-substituted benzoxazole derivatives as selective anticancer agents, with different compounds showing efficacy against specific breast cancer cell lines. The study's detailed analysis underscores the importance of structural variations in determining the cytotoxicity and selectivity of benzoxazole derivatives in cancer treatment (Omar et al., 2020).

Metal complexes exhibit excellent anticancer potency when paired with specific organic ligands such as indole, benzimidazole, benzothiazole, and benzoxazole. In a noteworthy example, novel Pd^{+2} complexes **(23-29)** incorporating indole-3-carbaldehyde thiosemicarbazones were synthesized and thoroughly characterized. These compounds underwent extensive analysis using techniques including 1H–13C HSQC; 31P NMR; 1H–31P HMBC; 1H NMR; 1H–13C HMBC; 13C NMR; 1H–1H COSY; DEPT-NMR; FT-IR; mass spectrometry; elemental analyses; single-crystal XRD and UV–visible. Crystallographic and spectroscopic data confirmed the co-ordination of the N- and S-atoms of thiosemicarbazone with Pd^{+2}, forming a stable five-membered ring. The DNA and bovine serum albumin (BSA) binding activities of these metal complexes were assessed, revealing that complexes MLC49 and MLC50 exhibited the highest DNA binding affinity, while MLC49 also showed significant BSA binding affinity. The cytotoxic effects of these complexes were tested against three cancer cell lines: lung cancer (A549); breast cancer (MCF7); and liver cancer (HepG-2). The complexes demonstrated moderate cytotoxicity towards MCF7 and A549 cell lines but showed outstanding cytotoxic activity against the HepG-2 cell line. The mode of cell death in HepG-2 cells was confirmed to be apoptosis, as evidenced by morphological changes observed through staining and DNA fragmentation methods. The presence of triphenylphosphine and the bulkiness of the substituent at the terminal N-atom of thiosemicarbazone positively influenced the activity of these complexes. Among the tested compounds, **26** and **27** emerged as the most promising, showing significant anticancer potential across the tested cell lines. These findings suggest that the specific structural features of **26** and **27** contribute to their enhanced biological activity, making them strong candidates for further development as anticancer agents (Haribabu et al., 2018).

Philoppes and Lamiet synthesized a series of benzoxazole derivatives featuring a phthalimide core and evaluated their anticancer activity against liver cancer (HepG2) and breast cancer (MCF7) cell lines. Through rigorous screening, the researchers identified compound **30** as having particularly potent anticancer effects. This compound demonstrated significant cytotoxicity, with IC50 values of 0.011μM against HepG2 cells and 0.006μM against MCF7 cells. These low IC50 values indicate a high level of effectiveness in inhibiting cancer cell growth, suggesting that compound 20 holds great promise as a potential anticancer agent for further development and clinical investigation (Philoppes & Lamie, 2019).

El-Hady et al. synthesized a series of benzoxazole derivatives and assessed their antitumor activity against two cancer cell lines: breast cancer (MCF7) and liver cancer (HepG2). The study revealed that among the synthesized compounds, 4-{2-(1,3-benzoxazol-2-yl)-2-[(phenylcarbonyl)amino]ethyl}-2-bromophenyl acetate (compound **31**) exhibited significant cytotoxicity against the MCF7 cell line. Meanwhile, another derivative, 4-{2-(1,3-benzoxazol-2-yl)-2-[(phenylcarbonyl) amino]ethyl}-2-bromophenyl chloroacetate (compound **32**), showed higher antitumor potential against the HepG2 cell line. Compound **31** demonstrated potent activity specifically towards MCF7 cells, highlighting its potential as a therapeutic agent for breast cancer. Compound **32** exhibited notable efficacy against HepG2 cells, with IC50 values ranging from 6.7μg/mL to 6.9μg/mL, indicating its strong antitumor potential for liver cancer treatment. These findings underscore the importance of structural variations in benzoxazole derivatives for targeting specific cancer types. The potent antitumor activities of compounds **31** and **32** make them promising candidates for further development as targeted cancer therapies (El-Hady & Abubshait, 2017).

Two novel ruthenium (+2/+3) complexes, designated as **33** and **34**, incorporating a benzimidazole ligand, were designed and evaluated for their chemotherapeutic potential. These complexes were screened for cytotoxic activity against two cancer cell lines, breast cancer (MCF7) and colorectal cancer (Caco2), as well as one normal cell line, liver epithelial cells (THLE-2). Among the two, complex **34** demonstrated significant anticancer activity against both MCF7 and Caco2 cell lines. The cytotoxic effects of **34** were associated with the induction of cell apoptosis and the arrest of the cell cycle at the G2/M phase. Notably, both ruthenium complexes were inactive against the normal THLE-2 cell line, indicating their selective toxicity towards cancer cells. Further investigation into the anticancer efficacy of **34** was conducted using an Ehrlich Ascites Carcinoma (EAC) mouse model. The in vivo studies revealed that **34** exhibited pronounced anticancer potency, effectively reducing the cancer burden at low doses, which resulted in minimal nephrotoxicity and hepatotoxicity. **34** was found to reduce oxidative stress and increase the levels of antioxidant enzymes, particularly superoxide dismutase (SOD). This enhancement of antioxidant activity reflects the complex's ability to improve normal cell repair mechanisms, contributing to its overall therapeutic potential. These findings highlight the promise of **34** as a potent anticancer agent with selective cytotoxicity, minimal side effects, and the ability to enhance oxidative stress resistance in normal cells (Elsayed et al., n.d.).

Belal and Abdelgawed synthesized a series of ten benzoxazole-pyrazole hybrid compounds and evaluated their anticancer potential against three human cancer cell lines: lung cancer (A549); breast cancer (MCF7), and liver cancer (HeP3B). Among these hybrids, compound **35** demonstrated the highest anticancer activity, specifically against the A549 cell line. The study involved comprehensive screening to assess the cytotoxic effects of these hybrids on the selected cancer cell lines. The results indicated that compound **35** exhibited significant inhibitory effects on the proliferation of A549 cells, suggesting its strong potential as a therapeutic agent for lung cancer. This enhanced anticancer activity of compound **35** highlights the promise of benzoxazole-pyrazole hybrids in developing effective cancer treatments, with specific efficacy against lung cancer (Belal & Abdelgawad, 2017).

Jauhari et al. synthesized a series of 2-substituted benzoxazole derivatives and evaluated their anticancer activity against four human cancer cell lines: cervical cancer (HeLa), colorectal cancer (WiDr), liver cancer (HepG2), and breast cancer (MCF7). The study revealed that among the synthesized compounds, compounds **36** and **37** demonstrated notably higher anticancer activity across all four tested cancer cell lines. The comprehensive screening of these compounds involved assessing their cytotoxic effects on the selected cancer cell lines. The results indicated that compounds **36** and **37** were particularly effective, showing strong inhibitory effects on the proliferation of WiDr, HeLa, MCF7 and HepG2 cells. This significant anti-cancer activity suggests that these benzoxazole derivatives possess broad-spectrum potential in cancer treatment, making compounds **36** and **37** promising candidates for further development as anticancer agents (Jauhari et al., 2008).

Figure 2. Compounds of anticancer activity

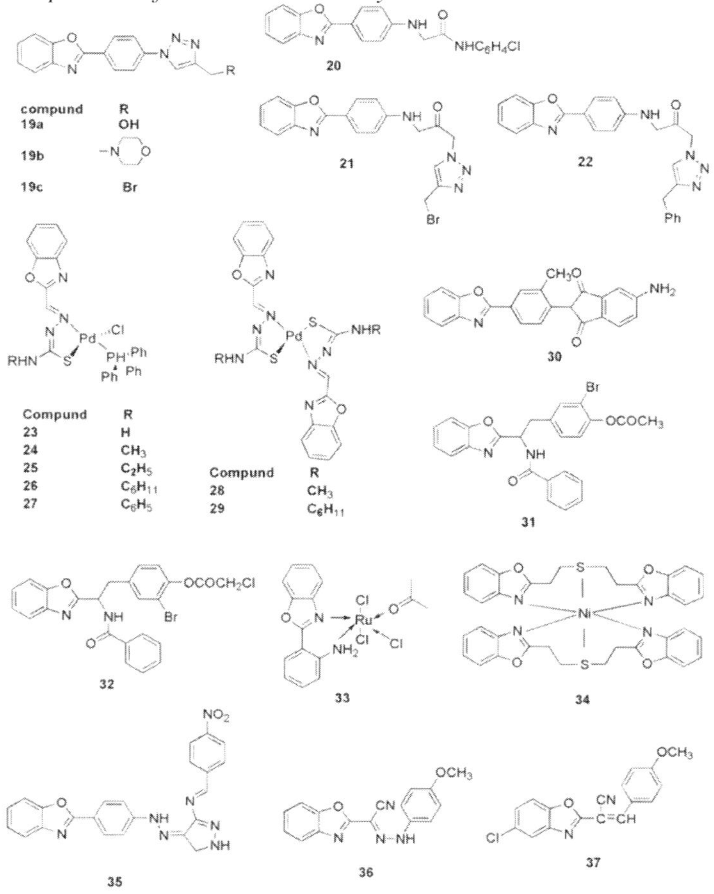

Anti-inflammatory Activity

Benzoxazoles possess anti-inflammatory properties, making them potential candidates for the treatment of inflammatory diseases. They can inhibit the production of pro-inflammatory cytokines and modulate the activity of inflammatory enzymes such as cyclooxygenase (COX) and lipoxygenase (LOX).

Angajala and Subashini synthesized a series of 2-substituted benzoxazole derivatives and evaluated their anti-inflammatory properties using two distinct methods: membrane stabilization and proteinase inhibition. These methods are well-regarded for assessing the potential of compounds to alleviate inflammation by preventing cell lysis and inhibiting proteolytic enzymes that contribute to inflammatory processes.

The study's findings highlighted that compound **38, 39,** and **40** exhibited significant anti-inflammatory activity. Specifically, these compounds demonstrated notable membrane stabilization activity with percentage inhibitions of 74.26 ± 1.04%; 80.16 ± 0.24%; and 70.24 ± 0.68%; respectively, at a concentration of 100µg/mL. Additionally, their proteinase inhibitory efficacy was recorded at 80.19±0.05%; 85.30±1.04%; and 75.68 ±1.28%; respectively, at the same concentration. These results indicate that compounds **38, 39,** and **40** are effective in stabilizing cell membranes and inhibiting proteinase activity, thereby reducing inflammation. The robust anti-inflammatory potential of these benzoxazole derivatives suggests their promise as therapeutic agents for treating inflammatory conditions, warranting further investigation and development (Angajala and Subashini (2020).

Two Cu(II) complexes, designated as **41** and **42**, were synthesized, thoroughly characterized, and evaluated for their anti-inflammatory, analgesic, and antipyretic activities using in vivo models involving mice and albino rats. These complexes underwent various characterization techniques to confirm their structural properties before biological testing. The in vivo studies revealed that complex **42** exhibited significant dose-dependent analgesic and anti-inflammatory activities, even at lower concentrations. This suggests that **42** is effective in reducing pain and inflammation in the animal models used for the study. The robust performance of **42** at lower doses highlights its potential as a therapeutic agent with strong analgesic and anti-inflammatory properties, meriting further investigation for potential clinical applications (Hussain et al. (2019).

Figure 3. Compounds of anti-inflammatory activity

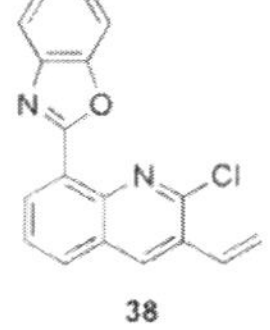

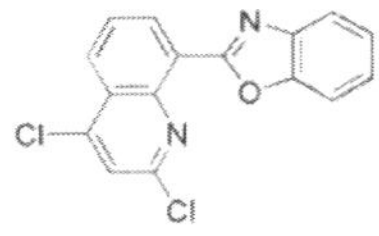

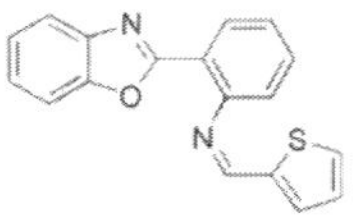

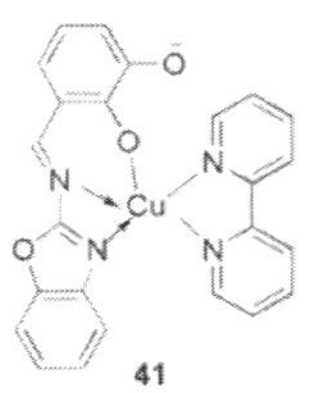

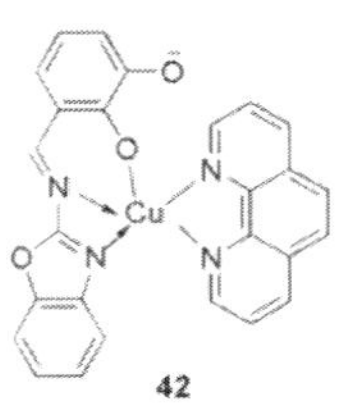

Antiviral Activity

Research has shown that benzoxazoles can act as antiviral agents against a range of viruses, including HIV, hepatitis C, and influenza. These compounds often target viral replication enzymes or viral entry into host cells, thereby preventing the spread of the infection.

Arulmurugan et al. (2013) synthesized a series of benzoxazole-based compounds and evaluated their cytotoxic effects on cancer cell lines, specifically MCF-7 (breast cancer) and HT-29 (colon cancer). Among the synthesized compounds, 3-(4-((2-(benzo[d]oxazol-2-yl)ethyl)amino)phenyl)-2-methylquinazolin-4(3H)-one (compound **43**) demonstrated notable cytotoxic activity. Compound **43** exhibited significant cytotoxicity against MCF-7 and HT-29 cell lines, with IC50 values of 22.5 µg/mL and 21.7 µg/mL, respectively. These values indicate the concentration at which the compound inhibits 50% of cell viability, showing its effectiveness in reducing cancer cell growth. For comparison, the standard chemotherapy drug 5-fluorouracil was used as a benchmark. The results underscore the potential of compound **43** as a promising anticancer agent, with efficacy comparable to established treatments. This study highlights the importance of benzoxazole moieties in developing new therapeutic agents for cancer treatment, warranting further investigation into their mechanisms of action and potential clinical applications (Arulmurugan and Kavitha (2013).

Compounds with halo group substitutions at the C-5 position exhibit high biological activity and lipid affinity, comparable to other synthesized derivatives. Specifically, the compounds

(Z)-N-(benzoxazol-2-yl)-2-(5-fluoro-2-oxoindolin-3-ylidene)hydrazinecarboxamide (**44**),
(Z)-N-(benzoxazol-2-yl)-2-(5-chloro-2-oxoindolin-3-ylidene)hydrazinecarboxamide (**45**), and
(Z)-N-(benzoxazol-2-yl)-2-(5-bromo-2-oxoindolin-3-ylidene)hydrazinecarboxamide (**46**)

demonstrated more potent activity in biological assays. These compounds exhibited IC50 values ranging from 13.71 to 21.21µM, indicating their effectiveness in inhibiting 50% of cell viability at these concentrations. The high activity of these compounds can be attributed to the presence of the halo substituents, which enhance their interaction with lipid membranes and biological targets. Compound **44**, with a fluorine atom at the C-5 position, compound **45**, with a chlorine atom, and compound **46**, with a bromine atom, all showed enhanced potency compared to other derivatives without these substitutions. The significant cytotoxic activity of these halo-substituted benzoxazole derivatives underscores their potential as effective therapeutic agents. Their favourable IC50 values suggest they are promising candidates for further development in cancer treatment, warranting additional studies to explore their mechanisms of action and potential clinical applications **(Gudipati, Reddy Anreddy, and Manda (2011).**

Figure 4. Compounds of antiviral activity

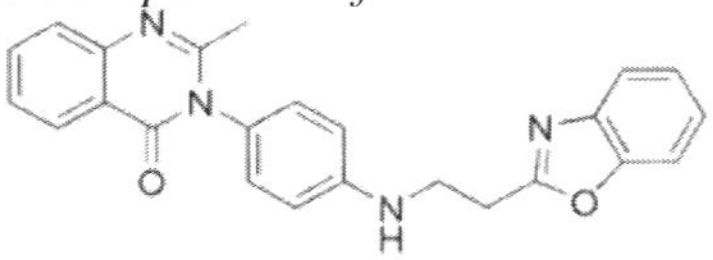

43

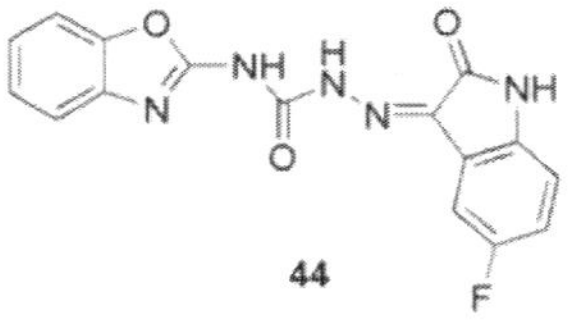

44

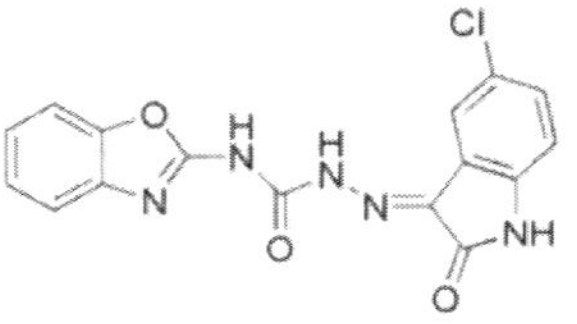

45

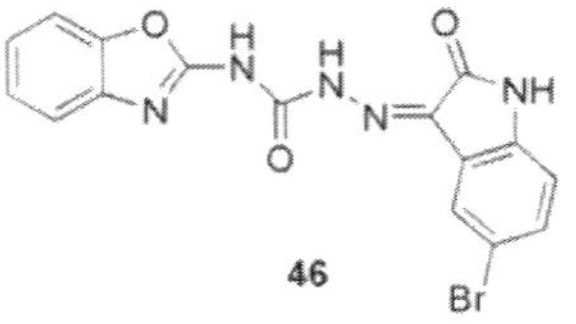

46

Antioxidant Activity

Benzoxazoles exhibit antioxidant properties, which can be beneficial in the treatment of diseases associated with oxidative stress. These compounds can scavenge free radicals and enhance the body's antioxidant defenses, reducing cellular damage.

Aichaoui et al. synthesized a series of 2(3H)-benzoxazolone derivatives and evaluated their in vitro antioxidant potential. The study aimed to determine the ability of these compounds to prevent human LDL (low-density lipoprotein) from undergoing copper-induced oxidation, with Cu^{2+} serving as the oxidizing agent. At a concentration of 10μM, the antioxidant efficacy of these derivatives was assessed. Among the synthesized compounds, compound **47** demonstrated superior antioxidant activity. This compound effectively inhibited both the initiation and propagation phases of copper-mediated LDL oxidation. The assessment was conducted in a time- and dose-dependent manner, highlighting compound **47**'s robust potential to protect LDL from oxidative damage. The findings indicate that compound **47** can significantly prevent the oxidative modification of LDL, suggesting its promise as a potent antioxidant agent. This characteristic makes it a valuable candidate for further development in the prevention of oxidative stress-related diseases (Aichaoui et al. (2009).

The antioxidant activities of two newly synthesized Cu(II) complexes, **48** and **49**, featuring benzoxazole ligands were thoroughly evaluated. Both complexes demonstrated significant antioxidant properties, evidenced by their remarkably low IC50 values of 0.112 µM and 0.191 µM, respectively. These low IC50 values indicate that only minimal concentrations of **48** and **49** are needed to inhibit 50% of the free radicals, underscoring their potent antioxidant activities. The exceptional antioxidant capabilities of these complexes are attributed to their geometric flexibility, particularly upon the approach of superoxide anions (O_2^-). This flexibility allows for the rapid exchange of the weakly bound water molecule at the axial position with the incoming superoxide anion. Such dynamic coordination chemistry enhances the ability of these complexes to effectively scavenge free radicals, thereby contributing to their high antioxidant efficacy. The study highlights the impressive antioxidant potential of Cu(II) complexes **48** and **49** with benzoxazole ligands. Their low IC50 values and ability to adapt their geometry in response to reactive oxygen species make them promising candidates for further development in antioxidant therapies, potentially offering new avenues for combating oxidative stress-related diseases (Gan et al. (2017).

Figure 5. Compounds of antioxidant activity

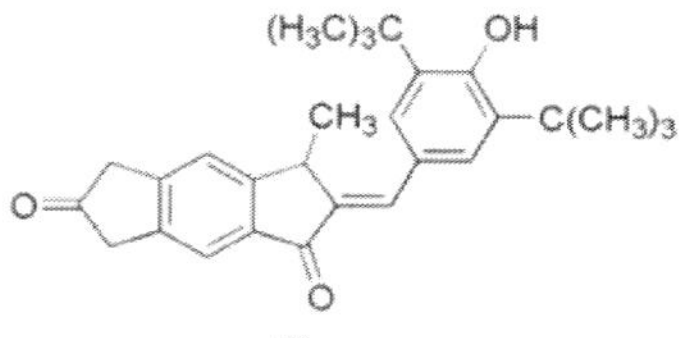

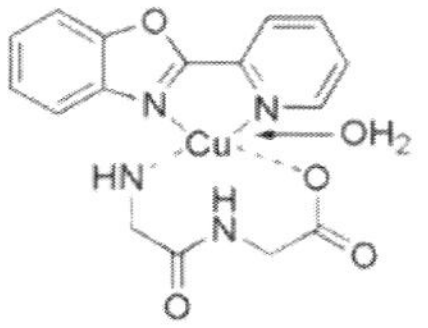

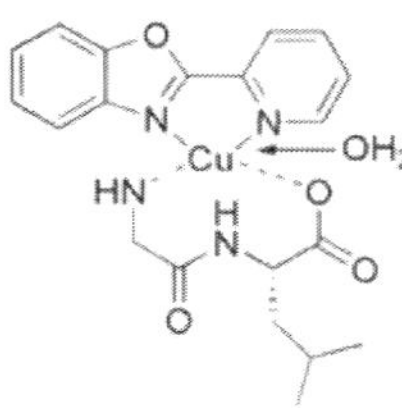

Neuroprotective Activity

Benzoxazoles have been investigated for their neuroprotective effects in neurodegenerative diseases such as Alzheimer's and Parkinson's disease. They can modulate neurotransmitter levels, inhibit neuroinflammation, and protect against neuronal damage.

Luisa et al. synthesized a series of 2-amino-6-(trifluoromethoxy)benzoxazole derivatives (**50-54**) to investigate their neuroprotective potential in the context of amyotrophic lateral sclerosis (ALS). The neuroprotective effects of these compounds were assessed by evaluating their ability to block voltage-dependent Na+

currents using the patch clamp technique in primary cultures of cerebellar and cortical neurons. Riluzole, a well-known neuroprotective agent, served as the positive control in this study. The results revealed that compounds **54** and **55** exhibited exceptional voltage-dependent Na+ current blocking activity, achieving inhibition rates of 97±2% and 98±2%, respectively. This high level of inhibition indicates a strong capacity to block Na+ currents, which is crucial in mitigating the excitotoxic damage associated with ALS. In contrast, compounds **52, 53,** and **56** demonstrated moderate blocking activity, reducing the Na+ currents by 50-70%. These findings highlight the potential of compounds **55** and **56** as promising neuroprotective agents due to their superior ability to inhibit voltage-dependent Na+ currents. The study underscores the importance of benzoxazole derivatives in developing new therapeutic strategies for ALS and potentially other neurodegenerative disorders where excitotoxicity plays a key role Calabrò et al. (2013).

Figure 6. Compounds of neuroprotective activity

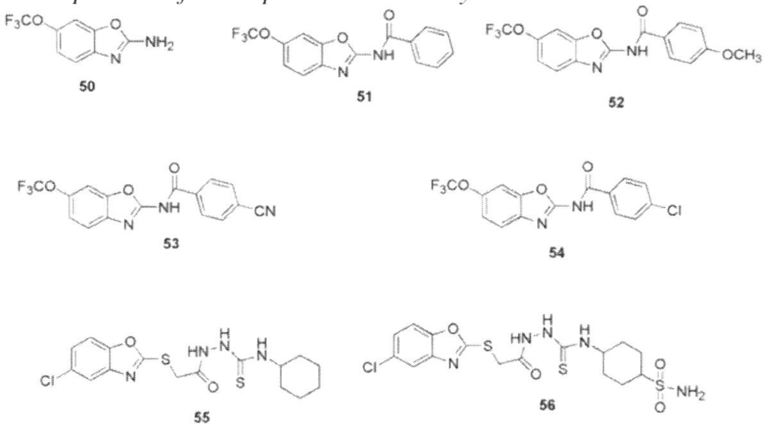

CHALLENGES AND FUTURE SCOPE OF BENZOXAZOLES IN DRUG DELIVERY

Benzoxazoles hold significant promise in drug delivery, yet their application is fraught with challenges that need to be addressed for their full potential to be realized. One of the primary obstacles is the synthetic complexity associated with benzoxazole derivatives. These compounds often require intricate and costly synthesis processes, which can hinder large-scale production and limit their availability for extensive research and development. Additionally, while benzoxazoles exhibit promising biological activities, some derivatives may present toxicity or undesirable side effects, necessitating a careful determination of their therapeutic window to ensure safety. Another challenge is the development of drug resistance over time, which can reduce the efficacy of benzoxazole-based therapies, particularly in the context of cancer and infectious diseases. Stability issues also pose a significant hurdle; certain benzoxazole derivatives may degrade prematurely under physiological conditions, leading to reduced efficacy. Moreover, navigating the regulatory approval process for new drug delivery systems involving benzoxazoles can be lengthy and complex, requiring extensive preclinical and clinical testing to meet stringent safety and efficacy standards. Furthermore, despite their demonstrated biological activities, a comprehensive understanding of the molecular mechanisms underlying the action of benzoxazoles remains limited, impeding the rational design of more effective derivatives.

Looking to the future, there is substantial scope for the advancement of benzoxazoles in drug delivery. Researchers can focus on synthesizing novel benzoxazole derivatives with enhanced pharmacokinetic and pharmacodynamic properties. Combining benzoxazoles with other therapeutic agents or nanoparticles could also improve their efficacy and targeting capabilities. Development of more sophisticated drug delivery systems, such as stimuli-responsive carriers or prodrugs, can offer improved targeted delivery and controlled release, responding to specific triggers within the diseased microenvironment. Emphasizing biocompatible and biodegradable formulations will minimize toxicity and environmental impact, making these therapies safer for patients and the environment.

The future of benzoxazole-based drug delivery also lies in the realm of multifunctional therapeutics, where compounds can be engineered to combine various therapeutic functions, such as anticancer and anti-inflammatory properties, providing comprehensive treatment for complex diseases like cancer. Advanced computational methods and high-throughput screening techniques will accelerate the discovery and optimization of benzoxazole derivatives, predicting their biological activities efficiently. The advent of personalized medicine presents another exciting avenue, where benzoxazole-based drug delivery systems can be tailored to the genetic and molecular profiles of individual patients, enhancing treatment efficacy and safety. Increased clinical trials and translational research are essential to evaluate the safety and efficacy of benzoxazole-based drug delivery systems in humans, facilitating their transition into clinical practice. Collaborative efforts between researchers, clinicians, and pharmaceutical companies can expedite this process, ultimately revolutionizing drug delivery and offering more effective treatments for a range of diseases. Despite the challenges, the future of benzoxazoles in medicine looks promising, with the potential to make significant advancements in therapeutic delivery systems.

CONCLUSION

Benzoxazoles, characterized by their unique heterocyclic structure, have emerged as versatile and promising compounds in the field of medicinal chemistry. Their diverse biological activities-including antibacterial, antifungal, antiviral, anticancer, anti-inflammatory, and neuroprotective properties—demonstrate their significant therapeutic potential. This chapter has highlighted the various medical applications of benzoxazoles, showcasing their effectiveness in treating a range of diseases from infections to cancer and neurodegenerative disorders. The ability of benzoxazoles to interact with critical biological targets, such as enzymes and receptors, underpins their efficacy in these therapeutic roles. The development of novel benzoxazole derivatives with optimized pharmacological profiles is paving the way for more effective treatments and innovative drug delivery systems. Despite the progress, several challenges remain, including synthetic complexity, potential toxicity, drug resistance, and stability issues. Addressing these challenges is crucial for realizing the full potential of benzoxazoles in clinical settings. Future research should focus on overcoming these hurdles by improving synthetic methods, enhancing drug stability, and minimizing side effects. Advances in drug delivery systems, including the use of nanoparticles and stimuli-responsive carriers, hold promise for better targeting and controlled release. The integration of benzoxazoles into multifunctional therapeutics and personalized medicine approaches could further enhance their clinical applications. With ongoing research and collaborative efforts, benzoxazoles are poised to make significant contributions to modern medicine, offering new solutions for complex diseases and advancing the field of drug discovery and development. Their continued exploration and development hold the potential to revolutionize therapeutic strategies and improve patient outcomes across a broad spectrum of medical conditions.

REFERENCES

Abdullahi, S. H., Moin, A. T., Uzairu, A., Umar, A. B., Ibrahim, M. T., Usman, M. T., Nawal, N., Bayil, I., & Zubair, T. (2024). Molecular docking studies of some benzoxazole and benzothiazole derivatives as VEGFR-2 target inhibitors: In silico design, MD simulation, pharmacokinetics and DFT studies. *Intelligent Pharmacy*, 2(2), 232–250. DOI: 10.1016/j.ipha.2023.11.010

Aichaoui, H., Guenadil, F., Kapanda, C. N., Lambert, D. M., McCurdy, C. R., & Poupaert, J. H. (2009). Synthesis and pharmacological evaluation of antioxidant chalcone derivatives of 2 (3 H)-benzoxazolones. *Medicinal Chemistry Research*, 18(6), 467–476. DOI: 10.1007/s00044-008-9143-y

Al-Saidi, H. M., & Khan, S. (2022). A review on organic fluorimetric and colorimetric chemosensors for the detection of Ag (I) ions. *Critical Reviews in Analytical Chemistry*, ●●●, 1–27. PMID: 36251012

Al-Saidi, H. M., & Khan, S. (2024). Recent advances in thiourea based colorimetric and fluorescent chemosensors for detection of anions and neutral analytes: A review. *Critical Reviews in Analytical Chemistry*, 54(1), 93–109. DOI: 10.1080/10408347.2022.2063017 PMID: 35417281

Aljaar, N., Gujjarappa, R., Al-Refai, M., Shtaiwi, M., & Malakar, C. C. (2019). Overview on Recent Approaches towards Synthesis of 2-Keto-annulated Oxazole Derivatives. *Journal of Heterocyclic Chemistry*, 56(10), 2730–2743. DOI: 10.1002/jhet.3673

Aljaar, N., Ibrahim, M. M., Younes, E. A., Al-Noaimi, M., Abu-Safieh, K. A., Ali, B. F., Kant, K., Al-Zaqri, N., Sengupta, R., & Malakar, C. C. (2023). Strategies towards the Synthesis of 2-Ketoaryl Azole Derivatives using C-H Functionalization Approach and 1, 2-Bis-Nucleophile Precursors. *Asian Journal of Organic Chemistry*, 12(4), e202300036. DOI: 10.1002/ajoc.202300036

Aljaar, N., Malakar, C. C., Conrad, J., & Beifuss, U. (2015). Base-promoted domino reaction of 5-substituted 2-nitrosophenols with bromomethyl aryl ketones: A transition-metal-free approach to 2-aroylbenzoxazoles. *The Journal of Organic Chemistry*, 80(21), 10829–10837. DOI: 10.1021/acs.joc.5b02000 PMID: 26399156

Aljaar, N., Malakar, C. C., Conrad, J., Frey, W., & Beifuss, U. (2013). Reaction of 1-nitroso-2-naphthols with α-functionalized ketones and related compounds: The unexpected formation of decarbonylated 2-substituted naphtho [1, 2-d][1, 3] oxazoles. *The Journal of Organic Chemistry*, 78(1), 154–166. DOI: 10.1021/jo3022956 PMID: 23181942

Alrooqi, M., Khan, S., Alhumaydhi, F.A., Asiri, S.A., Alshamrani, M., Mashraqi, M.M., Alzamami, A., Alshahrani, A.M. and Aldahish, A.A., 2022. A therapeutic journey of pyridine-based heterocyclic compounds as potent anticancer agents: a review (from 2017 to 2021). *Anti-Cancer Agents in Medicinal Chemistry (Formerly Current Medicinal Chemistry-Anti-Cancer Agents), 22*(15), pp.2775-2787.

Angajala, G., & Subashini, R. (2020). Synthesis, molecular modeling, and pharmacological evaluation of new 2-substituted benzoxazole derivatives as potent anti-inflammatory agents. *Structural Chemistry*, 31(1), 263–273. DOI: 10.1007/s11224-019-01374-1

Arulmurugan, S., & Kavitha, H. P. (2013). Synthesis and potential cytotoxic activity of some new benzoxazoles, imidazoles, benzimidazoles and tetrazoles. *Acta Pharmaceutica (Zagreb, Croatia)*, 63(2), 253–264. DOI: 10.2478/acph-2013-0018 PMID: 23846147

Balaghi, S. E., Safaei, E., Chiang, L., Wong, E. W., Savard, D., Clarke, R. M., & Storr, T. (2013). Synthesis, characterization and catalytic activity of copper (II) complexes containing a redox-active benzoxazole iminosemiquinone ligand. *Dalton Transactions (Cambridge, England)*, 42(19), 6829–6839. DOI: 10.1039/c3dt00004d PMID: 23487254

Belal, A., & Abdelgawad, M. A. (2017). New benzothiazole/benzoxazole-pyrazole hybrids with potential as COX inhibitors: Design, synthesis and anticancer activity evaluation. *Research on Chemical Intermediates*, 43(7), 3859–3872. DOI: 10.1007/s11164-016-2851-x

Calabrò, M. L., Caputo, R., Ettari, R., Puia, G., Ravazzini, F., Zappalà, M., & Micale, N. (2013). Synthesis and biological evaluation of new 2-amino-6-(trifluoromethoxy) benzoxazole derivatives, analogues of riluzole. *Medicinal Chemistry Research*, 22(12), 6089–6095. DOI: 10.1007/s00044-013-0594-4

Cornforth, J. W., & Cornforth, R. H. (1947). *24. A new synthesis of oxazoles and iminazoles including its application to the preparation of oxazole. Journal of the Chemical Society*. Resumed.

Dinakaran, V. S., Bomma, B., & Srinivasan, K. K. (2012). Fused pyrimidines: The heterocycle of diverse biological and pharmacological significance. *Der Pharma Chem*, 4(1), 255–265.

El-Hady, H. A., & Abubshait, S. A. (2017). Synthesis and anticancer evaluation of imidazolinone and benzoxazole derivatives. *Arabian Journal of Chemistry*, 10, S3725–S3731. DOI: 10.1016/j.arabjc.2014.05.006

Elsayed, S. A., Harrypersad, S., Sahyon, H. A., El-Magd, M. A., & Walsby, C. J. (2020). Ruthenium (II)/(III) DMSO-based complexes of 2-aminophenyl benzimidazole with in vitro and in vivo anticancer activity. *Molecules (Basel, Switzerland)*, 25(18), 4284. DOI: 10.3390/molecules25184284 PMID: 32962014

Faydalı, N., & Arpacı, Ö. T. (2024). Benzimidazole and Benzoxazole Derivatives Against Alzheimer's Disease. *Chemistry & Biodiversity*, 21(6), e202400123. DOI: 10.1002/cbdv.202400123 PMID: 38494443

Fu, L., Huang, D., Peng, J., Li, N., Liu, Z., Shen, Y., Zhao, X., Gu, Y., & Xiang, Y. (2023). Charge redistribution in covalent organic frameworks via linkage conversion enables enhanced selective reduction of oxygen to H 2 O 2. *Journal of Materials Chemistry. A, Materials for Energy and Sustainability*, 11(35), 18945–18952. DOI: 10.1039/D3TA03382A

Gan, Q., Qi, Y., Xiong, Y., Fu, Y., & Le, X. (2017). Two new mononuclear copper (II)-dipeptide complexes of 2-(2′-Pyridyl) benzoxazole: DNA interaction, antioxidation and in vitro cytotoxicity studies. *Journal of Fluorescence*, 27(2), 701–714. DOI: 10.1007/s10895-016-1999-5 PMID: 27981404

Gemili, M., Nural, Y., Keleş, E., Aydıner, B., Seferoğlu, N., Ülger, M., Şahin, E., Erat, S., & Seferoğlu, Z. (2019). Novel highly functionalized 1, 4-naphthoquinone 2-iminothiazole hybrids: Synthesis, photophysical properties, crystal structure, DFT studies, and anti (myco) bacterial/antifungal activity. *Journal of Molecular Structure*, 1196, 536–546. DOI: 10.1016/j.molstruc.2019.06.087

Gudipati, R., Reddy Anreddy, R. N., & Manda, S. (2011). Synthesis, anticancer and antioxidant activities of some novel N-(benzo [d] oxazol-2-yl)-2-(7-or 5-substituted-2-oxoindolin-3-ylidene) hydrazinecarboxamide derivatives. *Journal of Enzyme Inhibition and Medicinal Chemistry*, 26(6), 813–818. DOI: 10.3109/14756366.2011.556630 PMID: 21476831

Gujjarappa, R., Kabi, A. K., Sravani, S., Garg, A., Vodnala, N., Tyagi, U., Kaldhi, D., Gupta, S., & Malakar, C. C. (2022). Overview on Biological Activities of Thiazole Derivatives. In *Nanostructured Biomaterials: Basic Structures and Applications* (pp. 101–134). Springer Singapore. DOI: 10.1007/978-981-16-8399-2_5

Gujjarappa, R., Vodnala, N., Reddy, V. G., & Malakar, C. C. (2020). A Facile C-H Insertion Strategy using Combination of HFIP and Isocyanides: Metal-Free Access to Azole Derivatives. *Asian Journal of Organic Chemistry*, 9(11), 1793–1797. DOI: 10.1002/ajoc.202000481

Gul, Z., Khan, S., Ullah, S., Ullah, H., Khan, M. U., Ullah, M., & Altaf, A. A. (2024). Recent development in coordination compounds as a sensor for cyanide ions in biological and environmental segments. *Critical Reviews in Analytical Chemistry*, 54(3), 508–528. DOI: 10.1080/10408347.2022.2085027 PMID: 35671238

Han, J. H., Lee, E. J., Park, W., Ha, K. T., & Chung, H. S. (2023). Natural compounds as lactate dehydrogenase inhibitors: Potential therapeutics for lactate dehydrogenase inhibitors-related diseases. *Frontiers in Pharmacology*, 14, 1275000. DOI: 10.3389/fphar.2023.1275000 PMID: 37915411

Haribabu, J., Tamizh, M. M., Balachandran, C., Arun, Y., Bhuvanesh, N. S., Endo, A., & Karvembu, R. (2018). Synthesis, structures and mechanistic pathways of anticancer activity of palladium (II) complexes with indole-3-carbaldehyde thiosemicarbazones. *New Journal of Chemistry*, 42(13), 10818–10832. DOI: 10.1039/C7NJ03743K

Hussain, A., AlAjmi, M. F., Rehman, M. T., Amir, S., Husain, F. M., Alsalme, A., Siddiqui, M. A., AlKhedhairy, A. A., & Khan, R. A. (2019). Copper (II) complexes as potential anticancer and Nonsteroidal anti-inflammatory agents: In vitro and in vivo studies. *Scientific Reports*, 9(1), 5237. DOI: 10.1038/s41598-019-41063-x PMID: 30918270

Jarrar, Q., Ayoub, R., Jarrar, Y., Jaffal, H., Goh, K. W., Ming, L. C., Moshawih, S., & Sirhan, A. (2024). Unveiling the antinociceptive mechanisms of Methyl-2-(4-chloro-phenyl)-5-benzoxazoleacetate: Insights from nociceptive assays in mice. *European Review for Medical and Pharmacological Sciences*, 28(5). PMID: 38497888

Jauhari, P. K., Bhavani, A., Varalwar, S., Singhal, K., & Raj, P. (2008). Synthesis of some novel 2-substituted benzoxazoles as anticancer, antifungal, and antimicrobial agents. *Medicinal Chemistry Research*, 17(2-7), 412–424. DOI: 10.1007/s00044-007-9076-x

Jayanna, N. D., Vagdevi, H. M., Dharshan, J. C., Raghavendra, R., & Telkar, S. B. (2013). Synthesis, antimicrobial, analgesic activity, and molecular docking studies of novel 1-(5, 7-dichloro-1, 3-benzoxazol-2-yl)-3-phenyl-1 H-pyrazole-4-carbaldehyde derivatives. *Medicinal Chemistry Research*, 22(12), 5814–5822. DOI: 10.1007/s00044-013-0565-9

Jyothi, M. (2017). An update on the synthesis of benzoxazoles. *Asian Journal of Pharmaceutical and Clinical Research*, 10(10), 48–56. DOI: 10.22159/ajpcr.2017.v10i10.19457

Kabi, A. K., Sravani, S., Gujjarappa, R., Garg, A., Vodnala, N., Tyagi, U., Kaldhi, D., Singh, V., Gupta, S., & Malakar, C. C. 2022. An overview on biological activity of benzimidazole derivatives. *Nanostructured Biomaterials: Basic Structures and Applications*, pp.351-378.

Kabi, A. K., Sravani, S., Gujjarappa, R., Garg, A., Vodnala, N., Tyagi, U., Kaldhi, D., Velayutham, R., Gupta, S., & Malakar, C. C. (2022). An introduction on evolution of azole derivatives in medicinal chemistry. In *Nanostructured Biomaterials: Basic Structures and Applications* (pp. 79–99). Springer Singapore. DOI: 10.1007/978-981-16-8399-2_4

Kakkar, S., Tahlan, S., Lim, S. M., Ramasamy, K., Mani, V., Shah, S. A. A., & Narasimhan, B. (2018). Benzoxazole derivatives: Design, synthesis and biological evaluation. *Chemistry Central Journal*, 12(1), 1–16. DOI: 10.1186/s13065-018-0459-5 PMID: 30101384

Kaldhi, D., Vodnala, N., Gujjarappa, R., Nayak, S., Ravichandiran, V., Gupta, S., Hazra, C. K., & Malakar, C. C. (2019). Organocatalytic oxidative synthesis of C2-functionalized benzoxazoles, naphthoxazoles, benzothiazoles and benzimidazoles. *Tetrahedron Letters*, 60(3), 223–229. DOI: 10.1016/j.tetlet.2018.12.017

Katritzky, A. R., & Weeds, S. M. (1967). The literature of heterocyclic chemistry. [). Academic Press.]. *Advances in Heterocyclic Chemistry*, 7, 225–299. DOI: 10.1016/S0065-2725(08)60592-9

Kaur, A., Wakode, S. and Pathak, D.P., 2015. Benzoxazole: The molecule of diverse pharmacological importance. *International Journal of Pharmacy and Pharmaceutical Sciences, ISSN-0975-1491*, pp.17-23.

Khan, E., Khan, S., Gul, Z., & Muhammad, M. (2021). Medicinal importance, coordination chemistry with selected metals (Cu, Ag, Au) and chemosensing of thiourea derivatives. A review. *Critical Reviews in Analytical Chemistry*, 51(8), 812–834. PMID: 32571090

Khan, S., Alhumaydhi, F.A., Ibrahim, M.M., Alqahtani, A., Alshamrani, M., Alruwaili, A.S., Hassanian, A.A. and Khan, S., (2022). Recent advances and therapeutic journey of Schiff base complexes with selected metals (Pt, Pd, Ag, Au) as potent anticancer agents: A review. *Anti-Cancer Agents in Medicinal Chemistry (Formerly Current Medicinal Chemistry-Anti-Cancer Agents), 22*(18), pp.3086-3096.

Khan, S., Muhammad, M., Al-Saidi, H. M., Hassanian, A. A., Alharbi, W., & Alharbi, K. H. (2022). Synthesis, characterization and applications of schiff base chemosensor for determination of Cu2+ ions. *Journal of Saudi Chemical Society*, 26(4), 101503. DOI: 10.1016/j.jscs.2022.101503

Kumar, R., Saha, P., Keshamma, E., Sachitanadam, P., & Subramanian, M. (2022). Docking studies of some novel Hetrocyclic compound as Acat inhibitors: A meta analysis. *Journal for Research in Applied Sciences and Biotechnology*, 1(3), 33–41. DOI: 10.55544/jrasb.1.3.5

Liu, M., Zhang, J., Li, X., & Wang, Y. (2024). Research progress of DDR1 inhibitors in the treatment of multiple human diseases. *European Journal of Medicinal Chemistry*, 268, 116291. DOI: 10.1016/j.ejmech.2024.116291 PMID: 38452728

Mohammad Abu-Taweel, G., Ibrahim, M. M., Khan, S., Al-Saidi, H. M., Alshamrani, M., Alhumaydhi, F. A., & Alharthi, S. S. (2024). Medicinal importance and chemosensing applications of pyridine derivatives: A review. *Critical Reviews in Analytical Chemistry*, 54(3), 599–616. DOI: 10.1080/10408347.2022.2089839 PMID: 35724248

Muhammad, M., Khan, S., Shehzadi, S. A., Gul, Z., Al-Saidi, H. M., Kamran, A. W., & Alhumaydhi, F. A. (2022). Recent advances in colorimetric and fluorescent chemosensors based on thiourea derivatives for metallic cations: A review. *Dyes and Pigments*, 205, 110477. DOI: 10.1016/j.dyepig.2022.110477

Muhammed, M. T., Kuyucuklu, G., Kaynak-Onurdag, F., & Aki-Yalcin, E. (2022). Synthesis, antimicrobial activity, and molecular modeling studies of some benzoxazole derivatives. *Letters in Drug Design & Discovery*, 19(8), 757–768. DOI: 10.2174/1570180819666220408133643

Omar, A. M. M., AboulWafa, O. M., El-Shoukrofy, M. S., & Amr, M. E. (2020). Benzoxazole derivatives as new generation of anti-breast cancer agents. *Bioorganic Chemistry*, 96, 103593. DOI: 10.1016/j.bioorg.2020.103593 PMID: 32004897

Omar, A. M. M., AboulWafa, O. M., El-Shoukrofy, M. S., & Amr, M. E. (2020). Benzoxazole derivatives as new generation of anti-breast cancer agents. *Bioorganic Chemistry*, 96, 103593. DOI: 10.1016/j.bioorg.2020.103593 PMID: 32004897

Parvathi, J., Swetha, S., Pavithra, P., Sruthi, M., Divya, E., & Seetaramswamy, S. (2013). Synthesis and pharmacological screening of novel substituted benzoxazole derivatives as an anti-inflammatory agents. *International Journal of Current Pharmaceutical Research*, 6, 4–7.

Philoppes, J. N., & Lamie, P. F. (2019). Design and synthesis of new benzoxazole/benzothiazole-phthalimide hybrids as antitumor-apoptotic agents. *Bioorganic Chemistry*, 89, 102978. DOI: 10.1016/j.bioorg.2019.102978 PMID: 31136900

Shafiei, M., Peyton, L., Hashemzadeh, M., & Foroumadi, A. (2020). History of the development of antifungal azoles: A review on structures, SAR, and mechanism of action. *Bioorganic Chemistry*, 104, 104240. DOI: 10.1016/j.bioorg.2020.104240 PMID: 32906036

Sharma, P. C., Sharma, D., Sharma, A., Bansal, K. K., Rajak, H., Sharma, S., & Thakur, V. K. (2020). New horizons in benzothiazole scaffold for cancer therapy: Advances in bioactivity, functionality, and chemistry. *Applied Materials Today*, 20, 100783. DOI: 10.1016/j.apmt.2020.100783

Singh, L. P., Chawla, V., Chawla, P., & Saraf, S. K. (2010). Synthesis and antimicrobial activity of some 2-phenyl-benzoxazole derivatives. *Der Pharma Chemica*, 2(4), 206–212.

Srivastava, A., Aggarwal, L., & Jain, N. (2015). One-pot sequential alkynylation and cycloaddition: Regioselective construction and biological evaluation of novel benzoxazole–triazole derivatives. *ACS Combinatorial Science*, 17(1), 39–48. DOI: 10.1021/co500135z PMID: 25396730

Sun, Y., Xu, S., You, F., & Shi, X. (2022). Synthesis and characterization of the titanium catalysts supported by pyrrolide-benzoxazole ligands and their application in ethylene polymerization. *Polyhedron*, 219, 115791. DOI: 10.1016/j.poly.2022.115791

Tian, S., Zhong, K., Yang, Z., Fu, J., Cai, Y., & Xiao, M. (2024). Investigating the mechanism of tricyclic decyl benzoxazole-induced apoptosis in liver Cancer cells through p300-mediated FOXO3 activation. *Cellular Signalling*, 121, 111280. DOI: 10.1016/j.cellsig.2024.111280 PMID: 38960058

Vodela, S., Mekala, R. V. R., Danda, R. R., & Kodhati, V. (2013). Design, synthesis and screening of some novel benzoxazole based 1, 3, 4-oxadiazoles as potential antimicrobial agents. *Chinese Chemical Letters*, 24(7), 625–628. DOI: 10.1016/j.cclet.2013.04.005

Vodnala, N., Gujjarappa, R., Kabi, A. K., Kumar, M., Beifuss, U., & Malakar, C. C. (2018). Facile Protocols towards C2-Arylated Benzoxazoles using Fe (III)-Catalyzed C (sp2-H) Functionalization and Metal-Free Domino Approach. *Synlett*, 29(11), 1469–1478. DOI: 10.1055/s-0037-1609718

Xu, Y., Xiao, L., Sun, S., Pei, Z., Pei, Y., & Pang, Y. (2014). Switchable and selective detection of Zn 2+ or Cd 2+ in living cells based on 3′-O-substituted arrangement of benzoxazole-derived fluorescent probes. *Chemical Communications*, 50(56), 7514–7516. DOI: 10.1039/C4CC02335H PMID: 24890988

Yang, Z., Bai, X., Ma, S., Liu, X., Zhao, S., & Yang, Z. (2017). A benzoxazole functionalized fluorescent probe for selective Fe 3+ detection and intracellular imaging in living cells. *Analytical Methods*, 9(1), 18–22. DOI: 10.1039/C6AY02660E

Yernale, N. G., Matada, B. S., Vibhutimath, G. B., Biradar, V. D., Karekal, M. R., Udayagiri, M. D., & Mathada, M. H. (2022). Indole core-based Copper (II), Cobalt (II), Nickel (II) and Zinc (II) complexes: Synthesis, spectral and biological study. *Journal of Molecular Structure*, 1248, 131410. DOI: 10.1016/j.molstruc.2021.131410

Yin, X., Wang, J., Ge, M., Feng, X., & Zhang, G. (2024). Designing Small Molecule PI3Kγ Inhibitors: A Review of Structure-Based Methods and Computational Approaches. *Journal of Medicinal Chemistry*, 67(13), 10530–10547. DOI: 10.1021/acs.jmedchem.4c00347

Chapter 9
Synthetic Approaches and Medicinal Attributes of Benzotriazoles

Amol Arjun Nagargoje
https://orcid.org/0000-0001-6689-3436
K.M.C. College, Khopoli, India

Sharad Pandit Panchgalle
https://orcid.org/0000-0001-9706-7567
K.M.C. College, Khopoli, India

Mubarak Hanif Shaikh
https://orcid.org/0000-0002-1190-2371
Radhabai Kale Mahila Mahavidyalaya, Ahmednagar, India

Bapurao B. Shingate
Dr. Babasaheb Ambedkar Marathwada University, India

ABSTRACT

Benzotriazole represents a vital scaffold in the design of new pharmacologically active compounds. Its diverse biological activities and the potential for structural modification make it a promising candidate for future drug development endeavours. Continued research into its synthesis and SAR will further enhance its therapeutic potential. This chapter aims to provide a comprehensive overview of benzotriazole, focusing on its synthesis, biological activities, and potential applications in drug design. By exploring the historical development, modern synthetic methods, and the diverse biological activities of benzotriazole and its derivatives, we will highlight its significance in medicinal chemistry. The environmental and safety aspects of benzotriazole are also discussed, emphasizing the importance of sustainable and

DOI: 10.4018/979-8-3693-7267-8.ch009

safe practices in its use and development.

1. INTRODUCTION TO BENZOTRIAZOLE

Benzotriazole (BTA) is a fascinating five-membered heterocyclic compound comprising three nitrogen atoms within its ring structure, fused to a benzene ring. This bicyclic compound can be viewed as fused structure of aromatic benzene and triazole rings. This unique arrangement endows benzotriazole with a set of distinct chemical and biological properties, making it a subject of extensive research in various scientific fields, including medicinal chemistry, materials science, and environmental chemistry.[1] Chemically, benzotriazole exhibits exceptional stability, largely due to the delocalization of electrons across the aromatic system. This stability is complemented by its reactivity, allowing it to participate in various chemical reactions and form numerous derivatives. These derivatives often display enhanced biological activities compared to the parent compound, broadening the potential applications of benzotriazole in therapeutic contexts.[2] Biologically, benzotriazole has garnered attention for its broad spectrum of activities. It serves as a lead compound in the development of antimicrobial, antiviral, anticancer, and anti-inflammatory agents. Its ability to interact with multiple biological targets, coupled with its relative ease of synthesis and modification, makes benzotriazole a versatile and valuable scaffold in drug design. In addition, benzotriazole derivatives have found profound applications as corrosion inhibitors, UV filters, and materials for solar and photovoltaic cells. Recently, Bajaj *et al* reviewed the applications of benzotriazoles and its derivatives in medicinal chemistry and material chemistry.[3]

2. HISTORICAL PERSPECTIVE AND SIGNIFICANCE

The significance of benzotriazole in medicinal chemistry can be traced back to its initial discovery and subsequent studies that highlighted its potential as a bioactive molecule. Early research focused on its synthesis and basic chemical properties, laying the groundwork for later investigations into its biological activities. Over the years, advances in synthetic methods have enabled the production of a wide range of benzotriazole derivatives, each with distinct and potent bioactivities. The role of benzotriazole in drug development is underscored by its presence in various pharmaceuticals and its ongoing investigation in preclinical and clinical studies. Its structural simplicity, combined with the rich chemistry it offers, makes it an attractive starting point for the development of new therapeutic agents. Additionally, benzotriazole's environmental and industrial applications, such as corrosion inhibition and UV stabilization, further highlight its versatility and importance.[4]

3. CHEMICAL STRUCTURE AND PROPERTIES OF BENZOTRIAZOLE

3.1 Molecular Structure and Isomers

Benzotriazole is a heterocyclic compound with the molecular formula $C_6H_5N_3$. Its structure consists of a fused ring system where a triazole ring (a five-membered ring containing three nitrogen atoms) is fused to a benzene ring. The triazole ring can exist in two tautomeric forms: $1H$-benzotriazole (A) and $2H$-benzotriazole (B) **(Fig. 1)**.

Figure 1. Benzotriazole in its tautomeric forms A and B

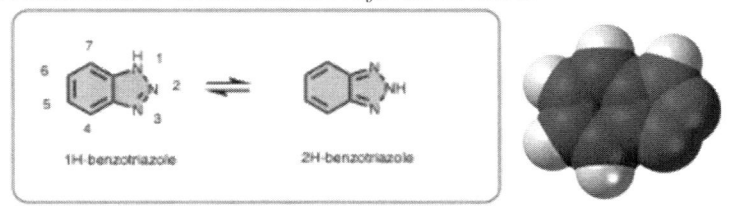

In the $1H$-tautomer, the hydrogen atom is attached to the nitrogen atom at the 1-position, whereas in the $2H$-tautomer, it is attached to the nitrogen at the 2-position. This tautomerism can influence the reactivity and interaction of benzotriazole with various biological targets.

3.2 Physical and Chemical Properties

Benzotriazole is a white to off-white crystalline powder with a melting point of around 97-99 °C. It is moderately soluble in water but highly soluble in organic solvents such as ethanol, acetone, and dimethyl sulfoxide (DMSO). The aromaticity of the benzene ring and the triazole ring's nitrogen atoms contribute to its chemical stability and reactivity. The presence of three nitrogen atoms in the triazole ring confers basic properties to benzotriazole, allowing it to act as a ligand in coordination chemistry. It can form stable complexes with metal ions, which is a basis for its applications in corrosion inhibition and other industrial uses.

3.3 Stability and Reactivity

Benzotriazole is known for its exceptional thermal and chemical stability, which is attributed to the delocalization of electrons across its aromatic ring system. This stability is crucial for its use in various applications, including as a corrosion inhibitor in industrial settings and as a stabilizer in photographic chemicals. Despite its stability, benzotriazole is also reactive under appropriate conditions, enabling the synthesis of a wide range of derivatives. The nitrogen atoms in the triazole ring are nucleophilic and can participate in various substitution reactions. These reactions allow for the introduction of different functional groups, which can significantly alter the biological activity of benzotriazole derivatives.

The utility of 1H-benzotriazole as a halogen surrogate in acylation, aroylation, and alkylation reactions is reviewed together by Hall and coworkers with its application to the synthesis of heterocycles, peptides, isopeptides, cyclic peptides, peptidomimetics, and peptide conjugates.[5]

3.4 Acid-Base Behaviour

Benzotriazole exhibits weakly acidic properties due to the presence of nitrogen atoms in the triazole ring. It can act as a proton donor, forming salts with bases. The *pKa* of benzotriazole is approximately 8.2, indicating that it can ionize under mildly basic conditions. This acid-base behaviour is important in its application as a corrosion inhibitor, where it can adsorb onto metal surfaces, forming a protective layer that prevents corrosion. It can bind to other species by utilizing lone pair of electrons to form stable coordination compounds on the surface of metals like copper and acts as corrosion inhibitor.[6]

3.5 Spectral Characteristics

Various structural analyses with UV, IR and ¹H-NMR spectra indicate that tautomer A is dominant. Benzotriazole exhibits distinct spectral characteristics that facilitate its identification and analysis. In infrared (IR) spectroscopy, benzotriazole shows characteristic absorption bands corresponding to the stretching vibrations of the C-H and N-H bonds, as well as the aromatic C=C bonds. FTIR spectrum of benzotriazole shows a strong absorption band at 3445 cm^{-1} which is attributed to -OH stretching. The -NH stretching vibration of 1H- and 2H-benzotriazole (BT) in the gas phase are in the region between 3440 cm^{-1} and 3540 cm^{-1}. In nuclear magnetic resonance (NMR) spectroscopy, the protons of the benzene ring and the triazole ring exhibit distinct chemical shifts, providing detailed information about its structure.

^{1}H NMR (DMSO-d_6), δ (ppm): C4–H, 8.00; C5–H, 7.44; C6–H, 7.44; C7–H, 8.0.

^{13}C NMR (DMSO-d_6), δ (ppm): C4, 130.3; C5, 130.3; C6, 130.3; C7, 130.3.

Ultraviolet-visible (UV-Vis) spectroscopy of benzotriazole reveals strong absorption in the UV region due to the π-π* transitions in the aromatic ring system. This property is utilized in analytical chemistry to detect and quantify benzotriazole in various samples.

3.6 Implications for Biological Activity

The unique chemical structure and properties of benzotriazole significantly influence its biological activity. Its ability to interact with metal ions and form stable complexes is crucial for its role in inhibiting microbial growth and enzyme activities. The aromaticity and stability of benzotriazole also enable it to intercalate into DNA, disrupting biological processes in pathogens and cancer cells.

The presence of multiple nitrogen atoms allows for hydrogen bonding and other interactions with biological targets, enhancing its efficacy as an antimicrobial, antiviral, and anticancer agent. These properties, combined with its chemical versatility, make benzotriazole a valuable scaffold in the design and development of new bioactive compounds.

4. SYNTHESIS OF BENZOTRIAZOLE

The synthesis of benzotriazole has evolved considerably since its initial discovery. The classical method involves cyclization of o-phenylenediamine with nitrous acid, a process still relevant for large-scale production (**Fig. 2**). However, recent advances have introduced greener approaches, such as microwave-assisted synthesis, which offer higher yields and reduced environmental impact.

Figure 2. Classical synthesis of Benzotriazole

O-phenylene diamine   Acetic acid   Sodium nitrate   Benzotriazole

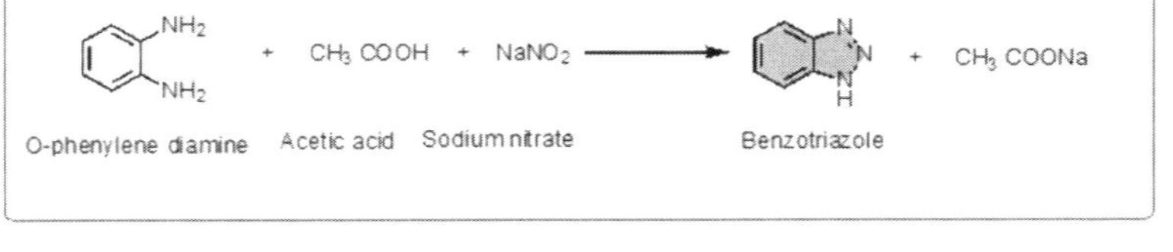

Larock[7] and others,[8-12] described the reaction of orthotrimethylsilyl triflates with fluoride ions generates arynes which undergo [3+2] cycloaddition with azides **(Fig. 3A)**. The resulting benzotriazoles were obtained after reaction times of up to 24 hours. To accelerate the reaction, a microwave protocol was developed for the synthesis of benzyl substituted benzotriazoles but required a reaction temperature of 125 °C **(Fig. 3B)**.[13] As heating of organic azides and highly reactive aryne intermediates poses the danger of an explosion, upscaling is problematic.[14-16] In recent years, flow chemistry has evolved as an alternative to overcome these limitations.[17-25] Kleoff and coworkers reported metal free synthesis of benzotriazoles using flow chemistry via [3+2] cycloaddition of azides and arynes in a good yield with improved safety and sustainability **(Fig. 3C)**.[26]

Figure 3. A) Synthesis of benzotriazoles via [3+2] cycloaddition of arynes and azides in batch at room temperature. B) Microwave-assisted reaction of arynes with in situ formed benzyl azides. C) Expedient and scalable synthesis of benzotriazoles in flow. TBAT: tetrabutylammonium triphenyldifluorosilicate; Tf: trifluoromethanesulfonyl; TMS: trimethylsilyl.

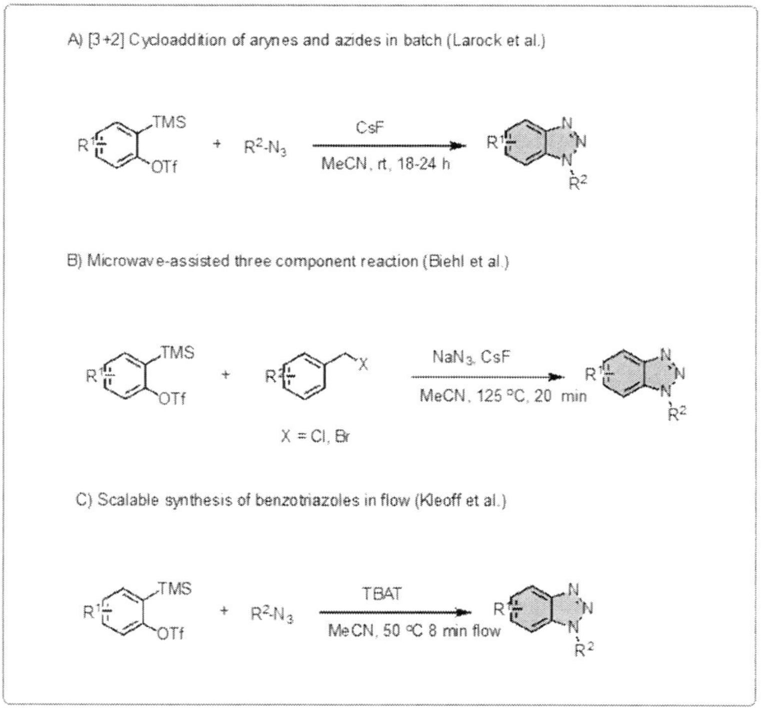

Zhou et al developed a 1,7-palladium migration-cyclization-dealkylation sequence for the regioselective synthesis of benzotriazoles in excellent yields with high regioselectivities. The mechanism of the reaction has also been investigated **(Fig. 4A)**.[27] Kumar et al developed a protocol in which C-H activation of aryl triazene was done followed by intramolecular amination in the presence of a catalytic amount of Pd(OAc)$_2$ to yield 1-aryl-1H-benzotriazoles at moderate temperature **(Fig. 4B)**.[28]

Figure 4. Pd catalysed C-H Activation of aryl triazines to regioselective synthesis of benzotriazoles

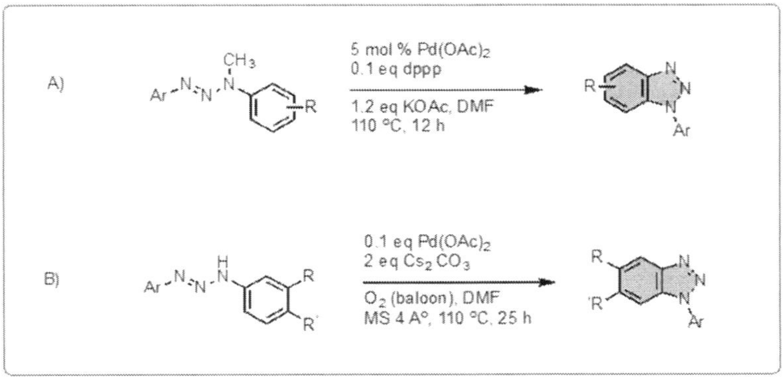

A cyclocondensation of 2-(arylamino)aryliminophosphoranes enables the synthesis of 1-aryl-1,2,3-benzotriazoles under mild conditions. The reaction involves a three-step, halogen-free route starting from simple nitroarenes and arylamines **(Fig. 5)**.[29]

Figure 5. Synthesis of 1-aryl-1,2,3-benzotriazoles under classical mild conditions

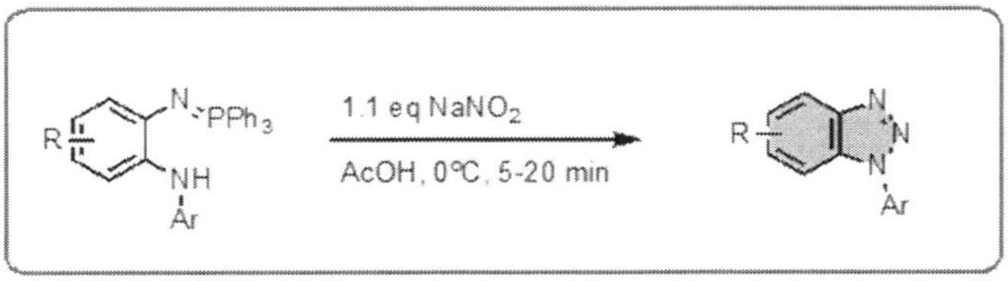

Faggyas and coworkers developed a mild method for the general synthesis of benzotriazoles from readily available 1,2-aryldiamines. The reaction uses a polymer-supported nitrite reagent and p-tosylic acid. It proceeds via a diazotization, followed by cyclization **(Fig. 6)**.[30]

Figure 6. Synthesis of benzotriazoles using polymer supported nitrite reagent & p-TSA

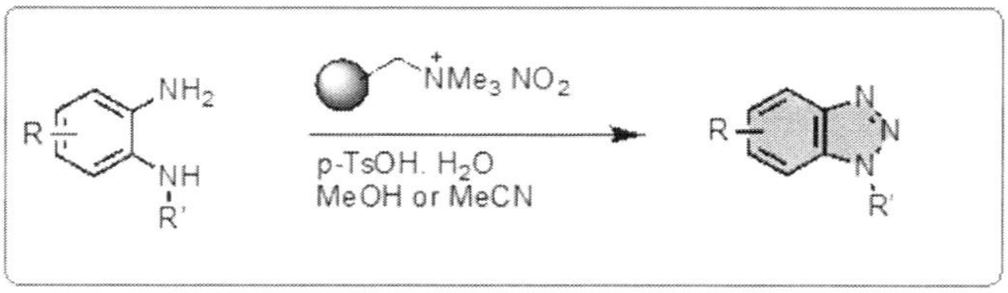

Shang et al developed a protocol for the synthesis of $2H$-benzotriazoles via copper-catalyzed C-N coupling of 2-haloaryltriazenes or 2-haloazo compounds with sodium azide and the intramolecular addition of a nitrene species to N-N bonds (TMEDA=N,N,N',N'-tetramethylethylenediamine; CTAB=hexadecyltrimethylammonium bromide). This approach allows the synthesis of various N-amino- and N-aryl-$2H$-benzotriazoles in water, in good to excellent yields **(Fig. 7)**.[31]

Figure 7. Synthesis of 2H-benzotriazoles via copper-catalyzed C-N coupling

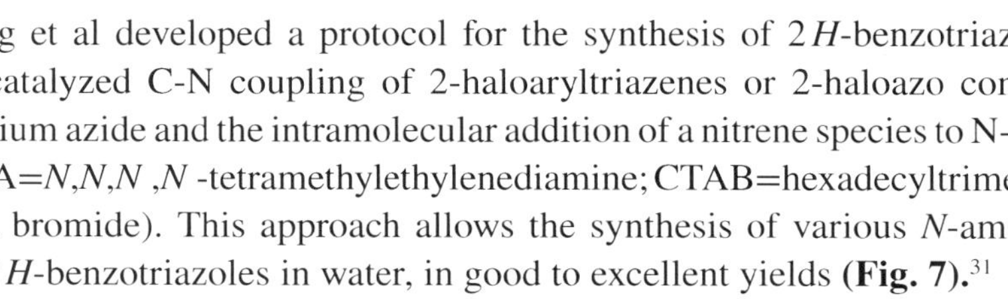

The synthesis of benzotriazole has evolved from classical methods to sophisticated modern techniques, reflecting advancements in organic chemistry and materials science. Classical cyclization methods provide a robust foundation, while metal-catalyzed, microwave-assisted, and green chemistry approaches offer enhanced efficiency, selectivity, and sustainability. These synthetic methods enable the production of benzotriazole and its derivatives, facilitating their exploration and application in various fields, including medicinal chemistry, materials science, and environmental protection.

5. BIOLOGICAL ACTIVITIES OF BENZOTRIAZOLE

Benzotriazole and its derivatives exhibit a broad spectrum of biological activities. As antimicrobial agents, they disrupt the integrity of bacterial cell walls, while their antiviral properties stem from inhibiting viral replication processes. Notably, benzotriazole derivatives have shown promise as anticancer agents by inducing apoptosis in malignant cells. Benzotriazole and its derivatives are renowned for their wide spectrum of biological activities, making them valuable in medicinal chemistry. The biological activities of benzotriazole include antimicrobial, antiviral, anticancer, anti-inflammatory, antioxidant properties, and enzyme inhibition. These activities are attributed to the compound's ability to interact with various biological targets, such as enzymes, receptors, and nucleic acids.

5.1 Antimicrobial Activity

Benzotriazole exhibits antimicrobial activity by interfering with critical biological processes in microbes. It can disrupt cell wall synthesis, inhibit protein synthesis, and interfere with nucleic acid synthesis. Benzotriazole shows broad-spectrum antimicrobial activity against both Gram-positive and Gram-negative bacteria. It is also effective against various fungi, making it a potential antifungal agent. Studies have identified several benzotriazole derivatives with potent antibacterial activity. For instance, benzotriazole derivatives substituted with halogens or alkyl groups have shown enhanced activity against strains such as *Staphylococcus aureus* and *Escherichia coli*.

Liu *et al.* synthesized[32] chalcone-benzotriazole derivatives and tested them for their antibacterial and antifungal properties. Both *N*-1 derived benzotriazole **1** and *N*-2 derived benzotriazole **2** showed significant inhibitory effects against certain tested strains. Notably, derivative **2** demonstrated superior antifungal activity against *C. utilis*, *S. cerevisiae*, and *A. flavus* (MIC $= 0.01, 0.02, 0.02\ \mu$mol/mL, respectively) compared to Fluconazole. When compound **1** or **2** was combined with antibacterial agents Chloromycin, Norfloxacin, or the antifungal Fluconazole, the combination displayed enhanced antimicrobial efficacy with lower doses and a broader antimicrobial spectrum than when used individually **(Fig. 8)**.

Figure 8. Chalcone-benzotriazole derivatives as potential antimicrobial agents

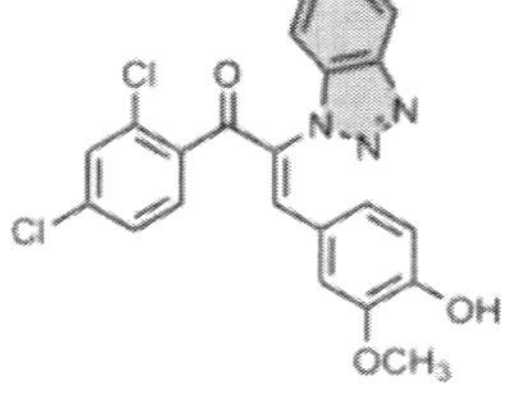

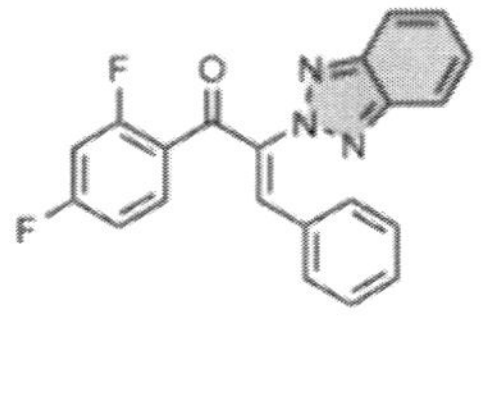

Ren *et al.* synthesized[33] a series of benzotriazole derivatives and evaluated their antimicrobial properties. Compound **3** showed significant antibacterial activity against MRSA with a MIC value of 4 µg/mL, which was twice as potent as Chloromycin. It also exhibited three times stronger antifungal activity (MIC = 4 µg/mL) than fluconazole (MIC = 16 µg/mL) against beer yeast. Initial studies on the interaction of compound **3** with calf thymus DNA indicated that compound **3** could effectively intercalate into DNA, forming a compound 3-DNA complex that might hinder DNA replication and thereby exert antimicrobial effects. Molecular docking studies suggested that compound **3** fits into the base pairs of a DNA hexamer duplex, forming two hydrogen bonds with guanine. Theoretical calculations supported these experimental findings (**Fig. 9**).

Figure 9. Benzotriazole derivative of Chloromycin derivative as antimicrobial agents

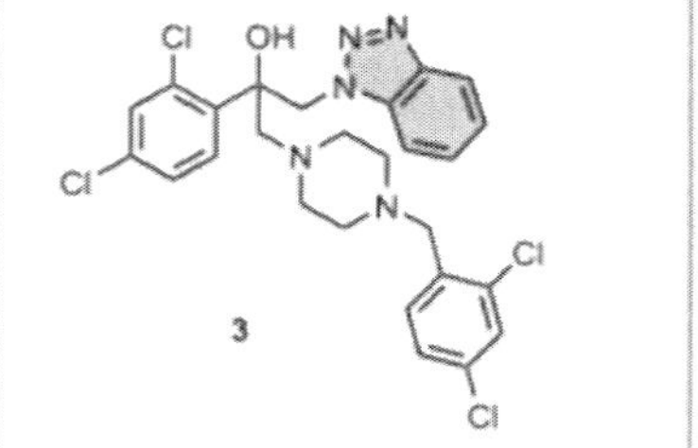

Shah *et al.* synthesized[34] 5-substituted benzotriazole derivatives and tested them as inhibitors of fungal cytochrome P450 lanosterol 14-alpha demethylase. Derivative **6** inhibited ergosterol biosynthesis at a level comparable to fluconazole, while derivatives **4** and **7** demonstrated even better activity, with nearly 50% inhibition (IC_{50}). Additionally, derivatives **4** and **7** exhibited greater antifungal activity, while derivatives **5** and **6** showed similar antifungal effects, when tested at a concentration of 50 µg/mL, compared to the standard fluconazole in both antifungal assays (**Fig. 10**).

Figure 10. 5-substituted benzotriazole derivatives as potential antifungal agents

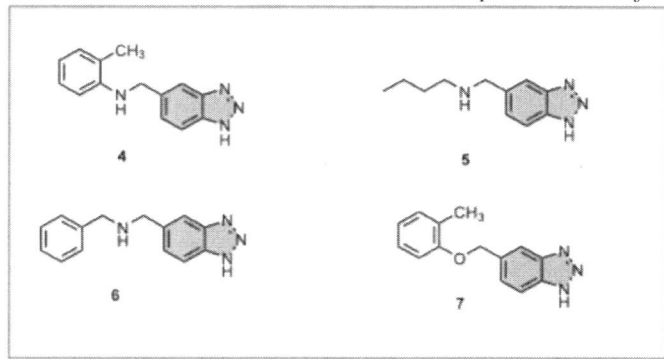

Singh *et al.* synthesized[35] benzotriazole-based β-amino alcohols and oxazolidine heterocyclic derivatives and evaluated their *in vitro* antibacterial activity. The results showed that compounds **8**, **9**, and **17** exhibited activity against *Staphylococcus aureus* (ATCC-25923) with minimum inhibitory concentrations (MICs) of 32, 8, and 64 μM, respectively. Additionally, compounds **8-19** demonstrated effective activity against *Bacillus subtilis* (ATCC 6633) with MICs of 64, 16, 16, 16, 64, 16, 64, 64, 32, 64, 8, and 16 μM, respectively. A biological investigation, including molecular docking studies, was performed on two compounds with several receptors to identify and confirm the most favorable ligand-protein interactions **(Fig. 11)**.

Figure 11. Benzotriazole-based β-amino alcohols and benzotriazole-based oxazo-lidine heterocyclic derivatives as potential antibacterial agents

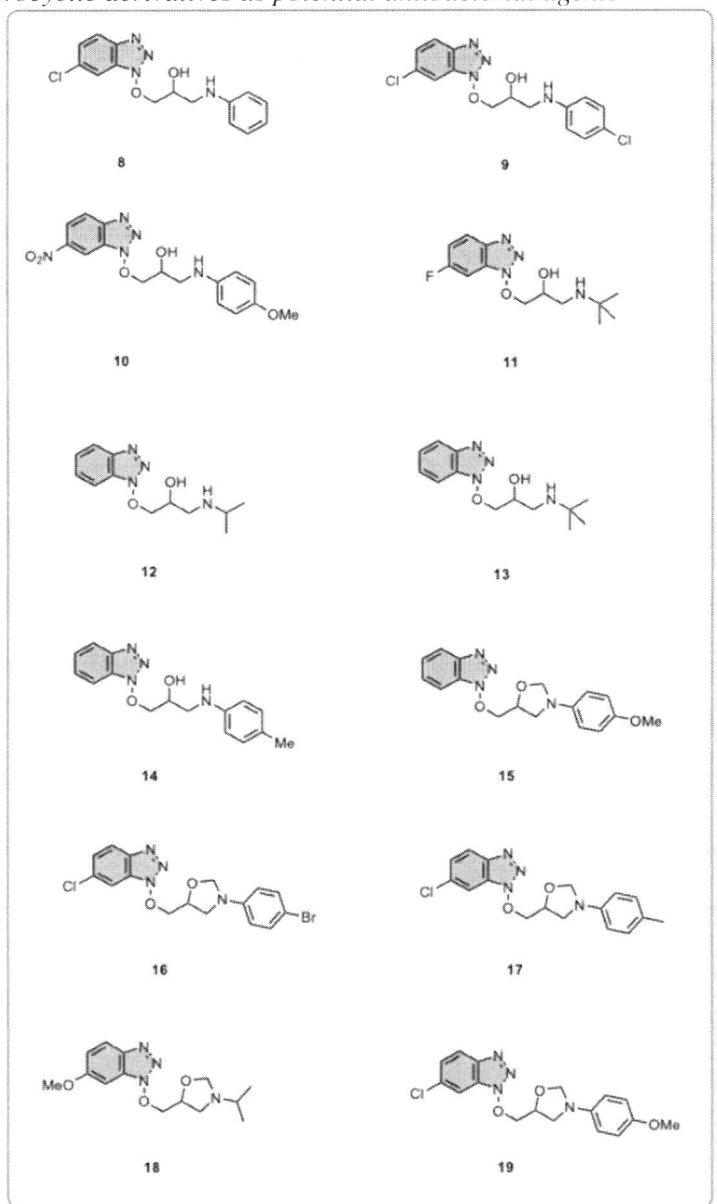

Chand *et al.* synthesized[36] 1,2,4-triazolylbenzotriazoles and evaluated their *in vitro* antimicrobial activity against various bacterial and fungal strains. Compounds **20-24** were found to have antibacterial activity comparable to ciprofloxacin against a *Klebsiella pneumoniae* strain **(Fig. 12)**.

Figure 12. 1,2,4-triazolylbenzotriazoles as potential antimicrobial agents

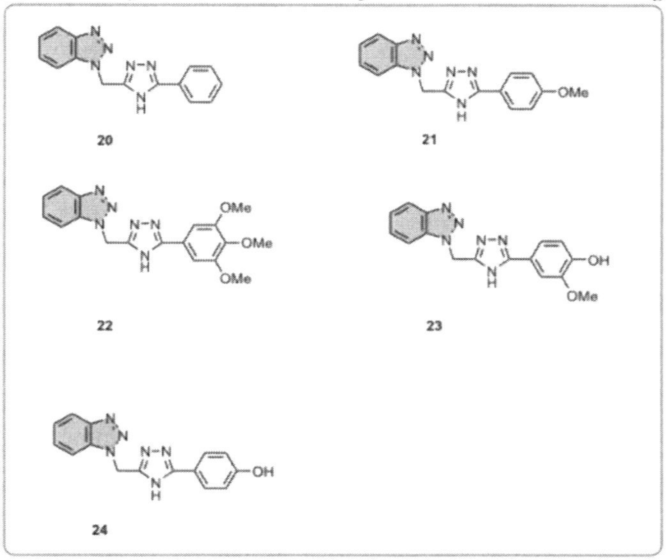

Navneet Singh *et al.* synthesized[37] benzotriazoles **25-28** using both conventional and microwave-assisted methods and evaluated their antimicrobial properties. These benzotriazoles demonstrated promising antimicrobial activity, along with significant ligand pose energy values and efficient docking run times, making them strong candidates for the antimicrobial drug discovery process **(Fig. 13)**.

*Figure 13. Benzotriazoles **25-28** as potential antimicrobial agents*

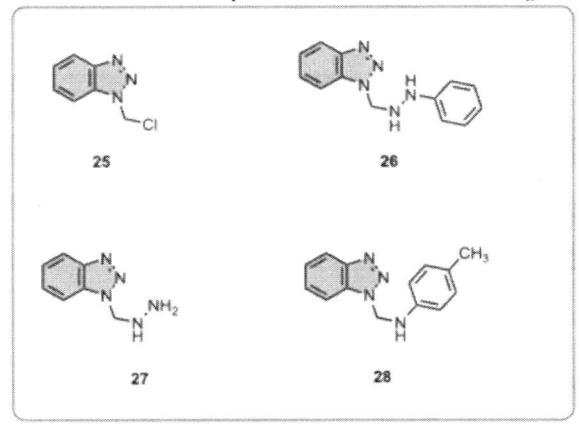

Chinh and colleagues synthesized[38] a chalcone derivatives containing benzotri-azole moieties and assessed their antimicrobial properties. Among the synthesized compounds, three, specifically compounds **29**, **30**, and **31**, exhibited significant antibacterial activity against Gram-positive bacteria, including *Bacillus subtilis* and *Staphylococcus aureus*. The minimum inhibitory concentration (MIC) values for these compounds were determined, revealing potent activity. Compound **29** showed MIC values of 25 μg/mL against *Bacillus subtilis* and 12.5 μg/mL against *Staphylococcus aureus*. Both compounds **30** and **31** demonstrated even stronger antibacterial effects, with MIC values of 12.5 μg/mL against both bacterial strains. These findings highlight the potential of these benzotriazole-containing chalcones as effective antibacterial agents, particularly against Gram-positive bacteria **(Fig. 14)**.

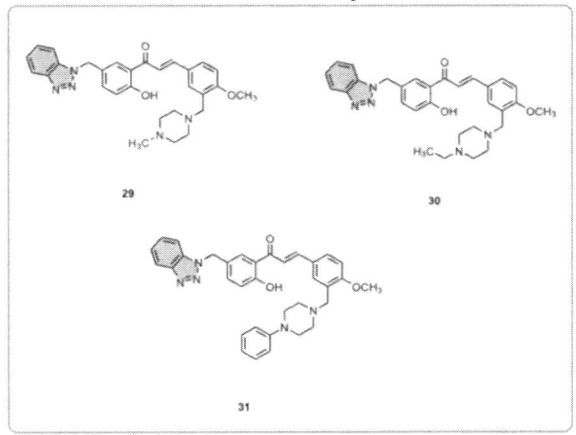

Lv and colleagues synthesized[39] benzotriazole-azo-phenol/aniline derivatives and evaluated their antifungal activity against six phytopathogenic fungi: *Fusarium graminearum*, *Fusarium solani*, *Alternaria alternata*, *Valsa mali*, *Botrytis cinerea*, and *Curvularia lunata*. Among these, compounds **32, 34**, and 35 exhibited broad-spectrum antifungal potency, being 3.5 to 10.8 times more effective than the standard fungicide carbendazim against *Alternaria alternata* and *Curvularia lunata*. Additionally, compounds **31, 32**, and **35** demonstrated significant protective and therapeutic effects against *Botrytis cinerea* at a concentration of 200 µg/mL. These results suggest that these benzotriazole-azo derivatives have strong potential as antifungal agents for agricultural use **(Fig. 15)**.

Figure 15. Benzotriazole-azo-phenol/aniline derivatives **31-35** *as potential anti-fungal agents*

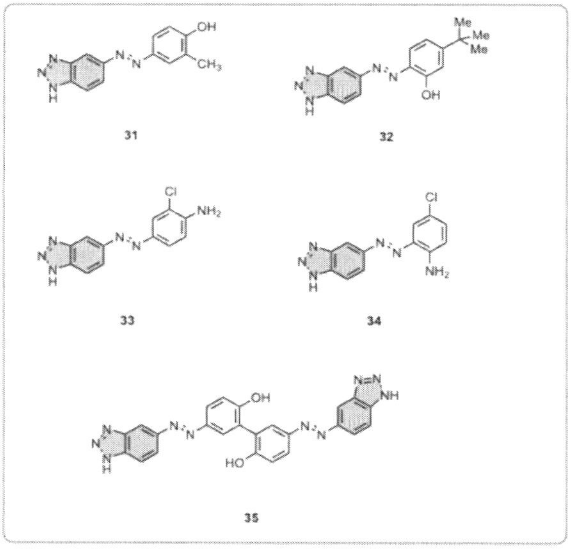

Atmajit Singh and colleagues developed[40] triazole-linked isatin-benzotriazole derivatives and evaluated their antifungal efficacy against Candida strains. Among these derivatives, compound **36** demonstrated significant activity against fluconazole-resistant *Candida albicans*, with minimum inhibitory concentration (MIC) and minimum fungicidal concentration (MFC) values of 3.9 and 7.8 μM, respectively. This compound shows promise as a lead for the development of more effective and safer antifungal agents targeting Candida infections **(Fig. 16)**.

Figure 16. Triazole tethered isatin-benzotriazole derivative **36** *as potential anti-fungal agent*

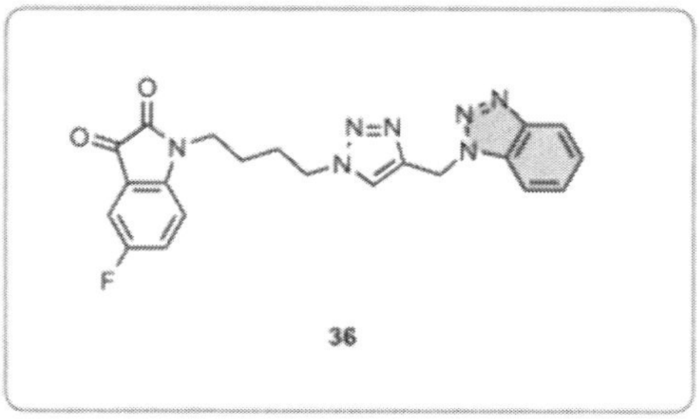

Ochal and colleagues synthesized[41] a series of 5-halogenomethylsulfonyl benzotriazole derivatives and evaluated their antibacterial activity through *in vitro* testing. Among the derivatives, compound **37** demonstrated the highest potency, indicating it was the most effective in combating bacterial strains in their study. This suggests that compound **37** holds significant promise as a potent antibacterial agent and could be a valuable lead for further research and development in the field of antimicrobial agents **(Fig. 17)**.

*Figure 17. Benzotriazole derivative **37** as potential antifungal agent*

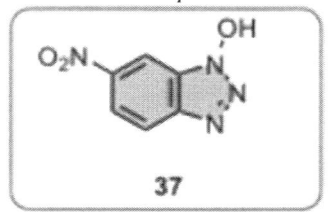

Muvvala and colleagues synthesized[42] a series of 1,2,3-benzotriazole derivatives and evaluated their antibacterial activity. Compounds **38** to **53** showed significant antibacterial effects against various bacterial strains. Specifically, these compounds were effective against gram-negative bacteria such as *Pseudomonas aeruginosa* (MTCC – 1035) and gram-positive bacteria including *Bacillus cereus* (MTCC – 430), *Bacillus subtilis* (MTCC – 441), *Staphylococcus aureus* (MTCC – 737), and *Staphylococcus epidermidis* (MTCC – 3086). The screening was conducted using the paper disc diffusion method, highlighting the potential of these derivatives as effective antibacterial agents **(Fig. 18)**.

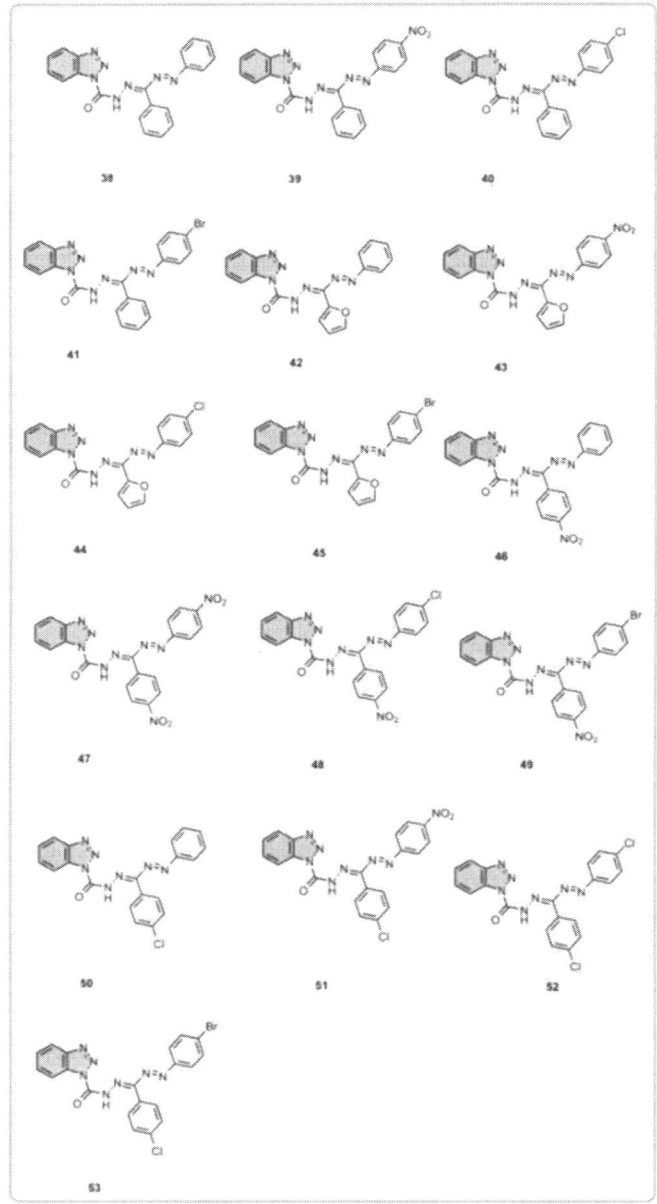

Patel and colleagues synthesized[43] a series of substituted benzotriazoles and assessed their *in vitro* antifungal activity against *Candida* and *Aspergillus* species. Compounds **54-56** demonstrated notable antifungal efficacy. Specifically, these compounds exhibited minimum inhibitory concentrations (MICs) ranging from 1.6 µg/mL to 25 µg/mL against *Candida* species and from 12.5 µg/mL to 25 µg/mL against *Aspergillus niger*. These results underscore the potential of these benzotriazole derivatives as effective antifungal agents **(Fig. 19)**.

Figure 19. Benzotriazole derivatives **54-56** *as potential antifungal agent*

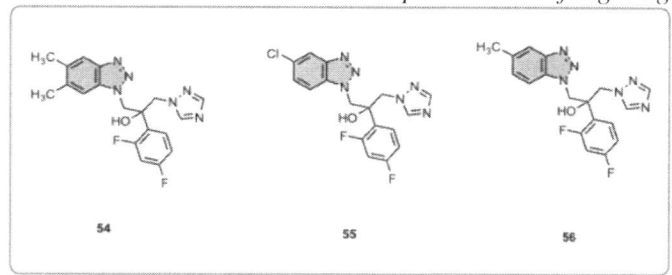

Gaikwad and colleagues synthesized[44] a series of thiazole-substituted benzotriazole derivatives and evaluated their antifungal activity. Compounds **57** to **64** showed varying degrees of effectiveness, with minimum inhibitory concentrations (MICs) ranging from 16 to 32 µg/mL against *Aspergillus niger* and *Candida albicans*. These findings indicate that the derivatives exhibit good to moderate antifungal activity **(Fig. 20)**.

Figure 20. Benzotriazole derivatives **57-64** *as potential antifungal agent*

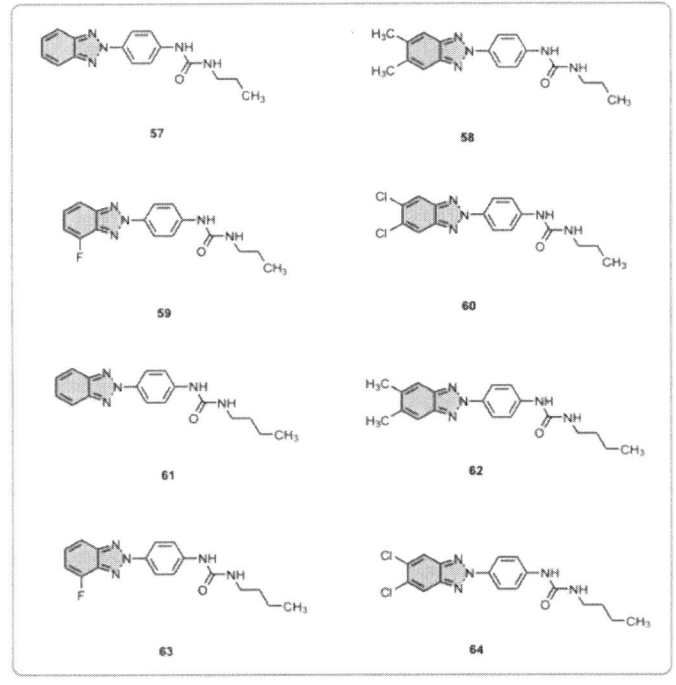

Ibrahim and colleagues synthesized[45] a range of benzotriazole derivatives and assessed their antifungal and antibacterial potential. Compounds **65** to **67** exhibited significant antimicrobial activity, highlighting their potential as effective agents against microbial infections **(Fig. 21)**.

Figure 21. Benzotriazole derivatives **65-67** *as potential antimicrobial agent*

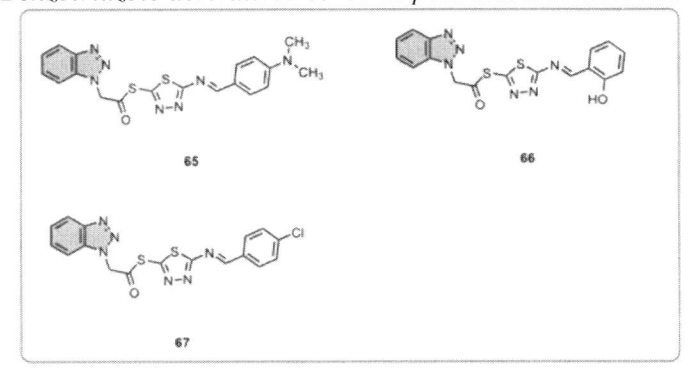

5.2 Antiviral Activity

Benzotriazoles have gained attention in recent years for their potential antiviral activities. These compounds have been investigated for their ability to inhibit the replication of various viruses, including human immunodeficiency virus (HIV), herpes simplex virus (HSV), and influenza virus. The antiviral mechanism of benzotriazoles is often linked to their capacity to interfere with viral enzyme function or to inhibit the fusion of the virus with host cells. For example, certain benzotriazole derivatives have been shown to inhibit the reverse transcriptase enzyme in HIV, thereby preventing the synthesis of viral DNA from RNA. Similarly, studies have demonstrated that benzotriazoles can disrupt the viral replication process in HSV by inhibiting key enzymes necessary for viral DNA synthesis. Recent research has also explored the structure-activity relationship (SAR) of benzotriazoles, leading to the development of more potent derivatives with enhanced antiviral efficacy and lower cytotoxicity. These findings suggest that benzotriazoles hold promise as scaffolds for the design of new antiviral agents, potentially offering a novel approach to combat viral infections[46]

Ibba and colleagues synthesized[47] *bis*-benzotriazole-dicarboxamide derivatives, which were evaluated for their ability to inhibit viral helicase activity of poliovirus and screened for *in vitro* antiviral activity. Derivatives **68-70** demonstrated notable antiviral activity against Picornaviruses, specifically *Coxsackievirus B5* and *Poliovirus-1*. These compounds exhibited EC_{50} values ranging from 9 to 13 µM against *Coxsackievirus B5*. Additionally, derivative **71** emerged as the most active compound among the bis-benzotriazole-dicarboxamides, showing significant antiviral effects against both tested Picornaviruses with EC_{50} values of 23 µM against *Coxsackievirus B5* and 43 µM against *Poliovirus-1* (**Fig. 22**).

Figure 22. Bis- benzotriazole-dicarboxamide derivatives **68-71** *as potential antiviral agent*

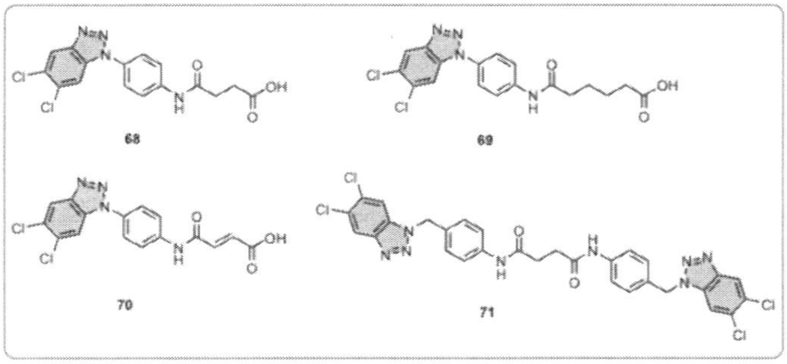

In a further study, Ibba and coworkers synthesized[48] additional benzotriazole derivatives and tested their *in vitro* antiviral activity against a broad spectrum of RNA positive- and negative-sense viruses. Compounds **72** through **76** exhibited selective antiviral activity specifically against *Coxsackievirus B5* (CVB5), a member of the *Picornaviridae* family, with EC_{50} values ranging from 6 to 18.5 µM. Among these, compounds **73** and **75** showed significant activity against CVB5 and were further evaluated for their safety profile on cell monolayers using the transepithelial electrical resistance (TEER) test **(Fig. 23)**.

*Figure 23. Benzotriazole derivatives **72-76** as potential antiviral agent*

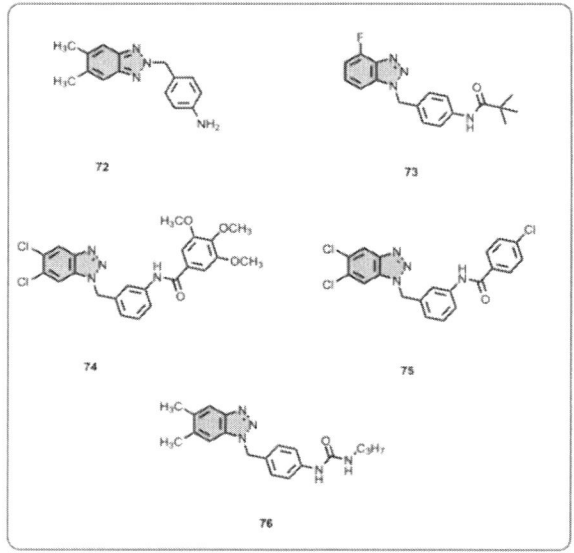

Loddo and colleagues synthesized[49] a 5-[4-(Benzotriazol-2-yl)phenoxy]-2,2-dimethylpentanoic acid derivatives and evaluated their antiviral activity. Compound **77** showed promising antiviral activity against Sb-1, with an EC_{50} of 7 µM and a selectivity index (SI) of 13. The 5,6-dichloroderivative, compound **78**, exhibited a broad spectrum of antiviral activity, including efficacy against BVDV, YFV, DENV-2, WNV, CVB-5, and RSV, with EC_{50} values ranging from 6 to 25 µM (mean EC50 = 12.5 µM). Further modifications of the structure led to compounds **79-82** and **83-90**, which featured dimethylalkanoic side chains of varying lengths. These compounds mainly targeted CVB-5 and RSV, with compounds **88** and **89** standing out. Compound **88** demonstrated high activity against both CVB-5 (EC_{50} = 0.7 µM, SI = 21.4) and RSV (EC_{50} = 2 µM, SI = 7.5). Meanwhile, compound **89** showed potent and selective activity against CVB-5, with an impressive EC_{50} of 0.15 μM and an SI of 100 (**Fig. 24**).

*Figure 24. Benzotriazole- dimethylalkanoic acid derivatives **77-90** as antiviral agent*

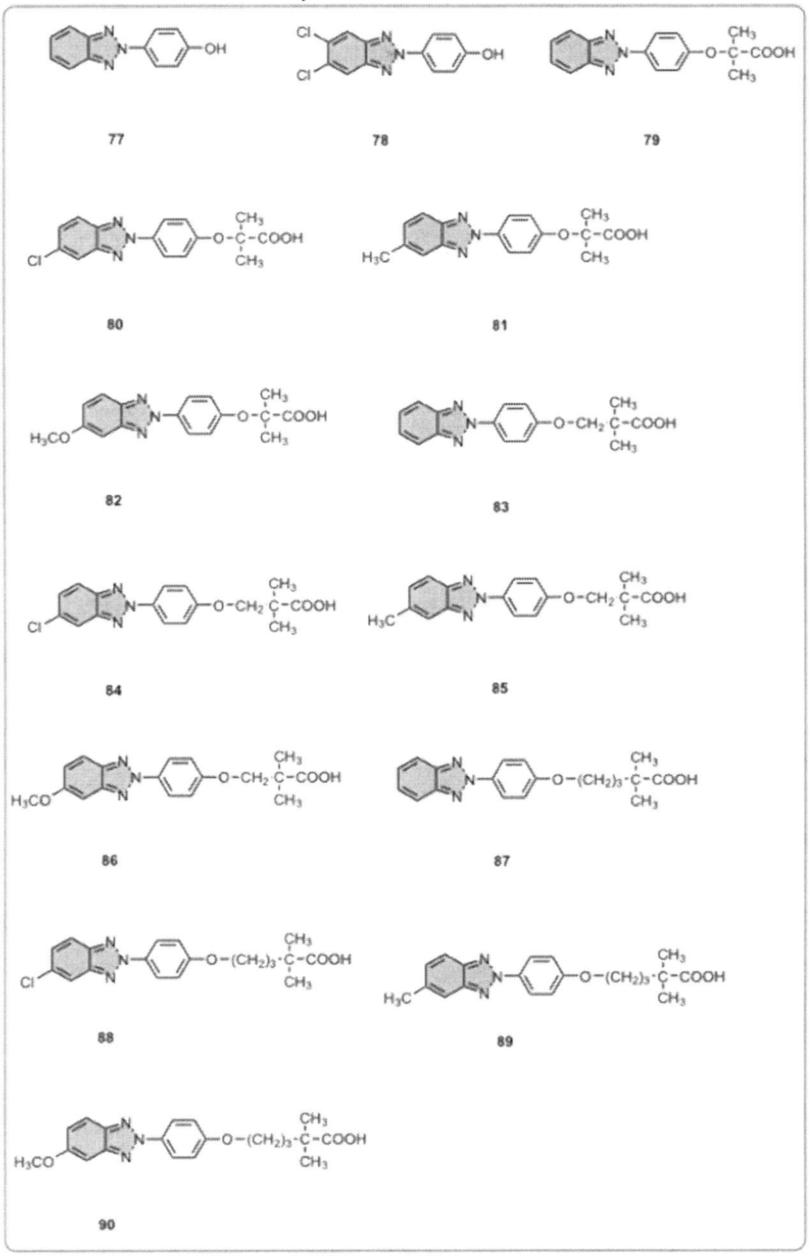

Piras and colleagues synthesized[50] dichlorophenyl-benzotriazole derivatives and investigated their potential to inhibit Human Respiratory Syncytial Virus (RSV), a leading cause of bronchopneumonia in infants and children. Among the synthesized compounds, derivative **91** emerged as a promising lead, demonstrating potent anti-RSV activity with an EC_{50} of 2 μM. This compound also showed high selectivity against RSV and exhibited a favorable safety profile across various continuous cell lines from different organs and species, making it a strong candidate for further development as an antiviral agent **(Fig. 25)**.

*Figure 25. Dichlorophenyl-benzotriazole **91** as antiviral agent*

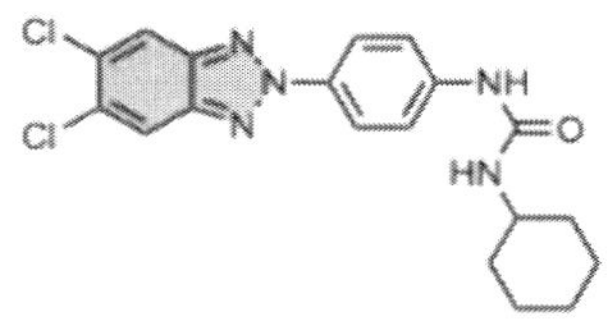

Corona and colleagues synthesized[51] a benzotriazole derivatives and evaluated their antiviral activity. Among the tested compounds, **92** and **93** were identified as the most active. Compound **92** exhibited an EC_{50} of 6.9 μM against *Coxsackievirus B5* (CV-B5) and 20.5 μM against *Poliovirus (Sb-1)*. Similarly, compound **93** showed an EC_{50} of 5.5 μM against CV-B5 and 17.5 μM against Sb-1. These results suggest that both compounds have strong potential as antiviral agents, particularly against these viruses **(Fig. 26)**.

*Figure 26. Benzotriazoles **92-93** as antiviral agent*

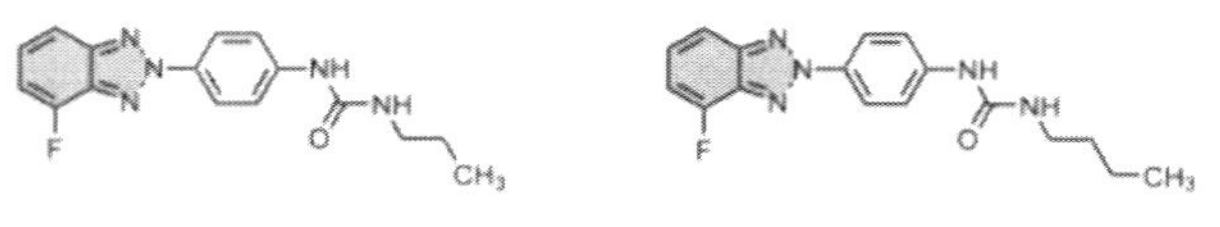

5.3 Anticancer Activity

Benzotriazole and its derivatives have garnered attention in the field of oncology due to their multifaceted mechanisms of action against cancer. The compound exerts its anticancer effects primarily through the induction of apoptosis, the programmed cell death essential for eliminating cancerous cells. By promoting apoptosis, benzotriazole helps in reducing the survival of cancer cells. Additionally, it plays a crucial role in inhibiting cell proliferation, which is the rapid multiplication of cancer cells that leads to tumor growth. One of the significant ways benzotriazole disrupts cancer progression is by interfering with the cell cycle. The cell cycle is a tightly regulated process that controls cell division and replication. Benzotriazole disrupts this cycle, effectively halting the division of cancer cells, which is critical in preventing the spread of cancer within the body. Beyond these mechanisms, benzotriazole also has the ability to inhibit angiogenesis. Angiogenesis is the formation of new blood vessels from pre-existing ones, a process that tumors exploit to secure a blood supply that supports their rapid growth and metastasis. By inhibiting angiogenesis, benzotriazole can starve tumors of the necessary nutrients and oxygen, thereby limiting their growth and potential to spread.

The anticancer potential of benzotriazole is further enhanced when specific functional groups, such as nitro or sulfonamide, are introduced into its structure. These modifications have been shown to significantly increase the cytotoxicity of benzotriazole derivatives against various cancer cell lines, including breast, lung, and colon cancers. This suggests that the chemical structure of benzotriazole can be fine-tuned to target specific types of cancer more effectively. Moreover, some benzotriazole derivatives have been identified as multi-targeted anticancer agents. These compounds do not just act on a single pathway but instead inhibit multiple signaling pathways that are critical for cancer cell survival and proliferation. This multi-targeted approach is particularly valuable because it reduces the likelihood of cancer cells developing resistance to treatment, which is a common challenge in cancer therapy.

Currently, several benzotriazole derivatives are in the preclinical stage of development. These studies, often conducted in animal models, have shown promising results, with these compounds effectively reducing tumor growth and spreading in vivo. The success of these preclinical studies has paved the way for clinical trials, where the safety and efficacy of these compounds are being evaluated in human patients.

If these clinical studies are successful, benzotriazole derivatives could become a valuable addition to the arsenal of anticancer therapies, offering new hope for patients with various types of cancer. These compounds have the potential to be used alone or in combination with existing treatments, providing a more comprehensive approach to cancer therapy.

Onar *et al* synthesized[52] silver complexes of the $[Ag(NHC)_2]+[AgCl_2]$- type by reacting the respective carbene precursor with Ag_2O, resulting in Benzotriazole-functionalized *N*-heterocyclic carbene-silver(I). The cytotoxic effects of complex **94** were tested against human breast cancer (MCF-7), colorectal cancer (Caco-2), and non-cancerous L-929 cell lines. The findings revealed that complex **94** exhibited significantly higher cytotoxicity towards cancer cell lines compared to cisplatin, showing approximately 25-fold greater potency against Caco-2 cells and 4-fold higher potency against MCF-7 cells. Additionally, complex **94** demonstrated lower toxicity towards non-cancerous cells. Molecular docking studies predicted that complex **94** effectively binds to DNA, indicating its potential as a drug candidate. This DNA binding ability was further validated through agarose gel electrophoresis, which showed altered migration patterns of pBR322 plasmid DNA upon interaction with complex **94**. Stability tests in aqueous medium over two weeks, using ^1H NMR spectroscopy, confirmed that complex **94** retained over 90% stability, suggesting its suitability for biological applications **(Fig. 27)**.

Table 1.

Compounds	Caco-2	MCF-7	L-929
94	11.65 ±0.21	34.43 ±0.61	109.82 ±9.51
Cisplatin	310.63 ±19.36	139.72 ±5.54	112.93 ±1.89

Figure 27. Benzotriazoles **94** *as anticancer agent*

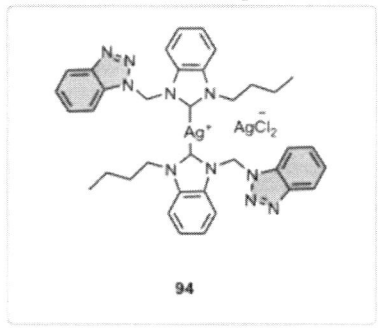

Edyta Lukowska-Chojnacka and coworkers synthesized[53] benzimidazole and benzotriazole compounds featuring a tetrazole moiety and evaluated their anti-cancer properties. These compounds were produced through *N*-alkylation. Their cytotoxicity was tested against human T-cell lymphoblast (CCRF-CEM) and breast adenocarcinoma (MCF-7) cell lines. The ability of these compounds to inhibit human recombinant casein kinase 2 alpha subunit (rhCK2a) was also assessed. The compounds **95** and **96** exhibited the highest cytotoxicity against the CCRF-CEM cell line, reducing cell viability by approximately 40% at a concentration of 50 μM, regardless of incubation time. Additionally, analogs **95** and **96** showed the strongest inhibitory effects on MCF-7 cells, with cytotoxicity observed across all tested concentrations in a time-dependent manner **(Fig. 28)**.

*Figure 28. Benzotriazoles **95** and **96** as anticancer agent*

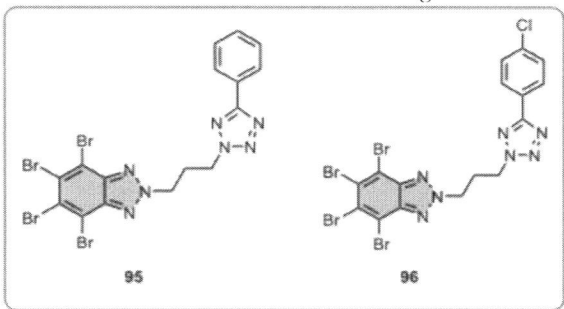

Li and colleagues synthesized[54] benzotriazole derivatives as tyrosine-protein kinase inhibitors using a fragment-based design approach. These compounds were synthesized by reacting benzotriazole with chloro-acetonitrile and various aromatic aldehydes, utilizing ultrasonic-microwave irradiation. The antiproliferative effects of these compounds were evaluated against several cancer cell lines, including VX2, A549, MGC, and MKN45, with Gefitinib serving as a reference standard for comparison. The inhibitory potency of the compounds was measured through their IC_{50} values using the CCK-8 assay. Among them, analogue **97** showed superior inhibitory activity across all four cancer cell lines compared to Gefitinib. Specifically, compound **97** exhibited strong antiproliferative effects, with IC_{50} values of 5.47 ± 0.41 μM for VX2, 5.47 ± 1.11 μM for A549, 4.59 ± 0.14 μM for MGC, and 3.04 ± 0.02 μM for MKN45 cell lines. With IC_{50} values ranging from 5.47 to 3.04 μM, compound **97** stands out as a promising lead for developing new chemotherapeutic agents, highlighting its potential as a foundational structure for further optimization and drug discovery initiatives **(Fig. 29)**.

*Figure 29. Benzotriazole **97** as tyrosine-protein kinase inhibitors*

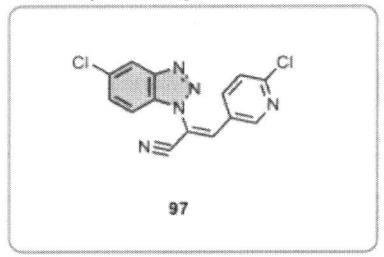

Kassab et al synthesized[55] novel benzotriazole *N*-acylarylhydrazone hybrids with potent anticancer activity. Analogue **98** showed exceptional efficacy against leukemia (CCRF-CEM and HL-60(TB)) and ovarian cancer (OVCAR-3) cell lines, with IC$_{50}$ values of 25-29 nM, significantly outperforming doxorubicin. It effectively inhibited FAK and Pyk2 enzymes, leading to apoptosis through caspase-3 activation and cell cycle arrest. Analogue **99** displayed broad-spectrum anticancer activity across 34 tumor cell lines, particularly against colon cancer (HT-29) and melanoma (MDA-MB-435), making it a strong candidate for further cancer therapy development **(Fig. 30)**.

*Figure 30. Benzotriazole N-acylarylhydrazone hybrids **98-99** with potent anticancer activity*

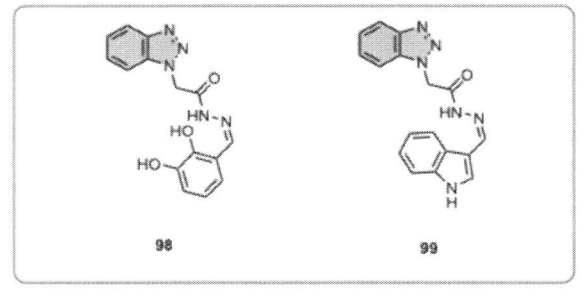

Gulnihan Onar and team synthesized[56] palladium and ruthenium complexes with benzotriazole-substituted *N*-heterocyclic carbene ligands. The ruthenium complexes, particularly compounds **100** and **101**, exhibited strong anticancer activity against colorectal (Caco-2) and breast (MCF-7) cancer cells, comparable to cisplatin, while sparing non-cancerous cells. These findings suggest that ruthenium-NHC complexes could be a promising alternative to traditional platinum-based therapies, offering selective and effective anticancer treatment **(Fig. 31)**.

*Figure 31. Palladium and ruthenium complexes **100-101** with benzotriazole-substituted N-heterocyclic carbene ligands with potent anticancer activity*

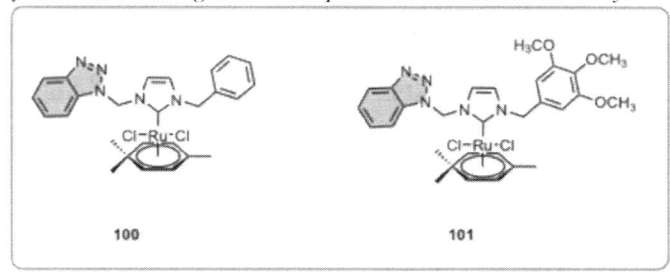

Roberta Ibba and colleagues synthesized[47] a series of benzotriazole-dicarboxamide derivatives, identifying Analogue **71** as a promising antitumor candidate due to its broad-spectrum antiproliferative activity. Analogue **71** effectively inhibited the growth of six out of seven tested cancer cell lines, with low micromolar IC_{50} values. It showed significant activity, particularly against breast, prostate, ovarian, CNS, and non-small cell lung cancers, with a notable impact on ovarian cancer (OVCAR-4) where it achieved over 100% growth inhibition. These findings suggest Analogue **71**'s potential as a potent therapeutic agent *(See the**figure 16**for structure of compound**71**)*.

Alraqa and co-workers synthesized[57] benzotriazole-1,2,3-triazole hybrids using a Cu(I)-catalyzed *click* reaction, significantly reducing reaction times with microwave-assisted synthesis. The compounds were tested for anticancer activity against A549 and H1299 lung cancer cell lines. Compounds **102** and **103** emerged as the most potent, reducing cell viability to as low as 10-11% in both cell lines. These compounds also demonstrated strong groove binding with DNA, as indicated by significant hyperchromic effects. Compound **102** showed a shift in λmax from 265 nm to 266 nm, while compound **103** shifted from 212 nm to 214 nm, further confirming their binding affinity **(Fig. 32)**.

Figure 32. Benzotriazole-1,2,3-triazole hybrids **100-101** *with potent anticancer activity*

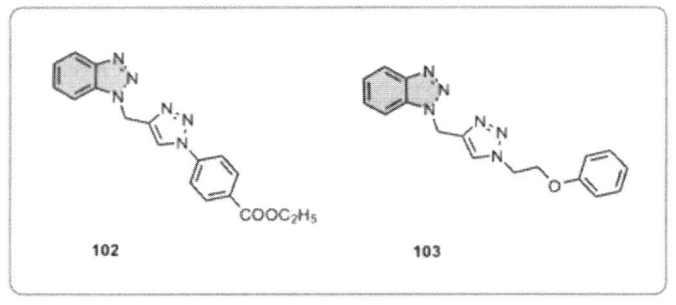

Matheswari and colleagues synthesized[58] compound **104**, a biologically important macromolecule, and evaluated its cytotoxic potential against MCF-7 breast cancer cells. Using the MTT assay, they found that compound **104** exhibited superior anticancer activity compared to doxorubicin, with a lower IC_{50} value of 83.45 μg/ml. The compound's efficacy peaked at 100 μg/ml, surpassing doxorubicin in potency. Compound **104** employs a receptor-mediated targeting strategy, binding specifically to receptors like ER, PR, and EGFR on tumor cells, leading to targeted drug delivery. This approach enhances therapeutic concentration within cancer cells while minimizing damage to healthy tissues, making compound **104** a promising candidate for further development as a targeted anticancer agent **(Fig. 33)**.

Figure 33. Benzotriazole-macromolecule **104** *active against MCF-7 breast cancer*

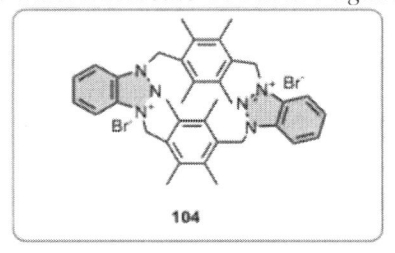

Mermer and co-workers synthesized[59] benzotriazole-oxadiazole hybrids using conventional and ultrasonic methods, offering an eco-friendly approach. The compounds were tested for anticancer activity against the PANC-1 pancreatic cancer cell line. Analogue **106** emerged as the most potent, with an IC_{50} value of 87.82 μg/ml, outperforming the standard drug gemcitabine (IC_{50}: 300 μg/ml). Analogue **105** also showed strong activity (IC_{50}: 117.50 μg/ml). Molecular docking revealed that Analogue **106** had high binding affinities with key pancreatic cancer proteins, supporting its potential as a promising anticancer agent **(Fig. 34)**.

Figure 34. benzotriazole-oxadiazole hybrids **105-106** *active against PANC-1 pancreatic cancer cell line*

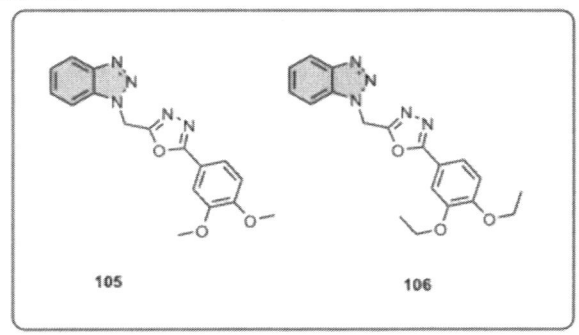

105   106

5.4 Enzyme Inhibition

Benzotriazole and its derivatives can inhibit a range of enzymes, including proteases, kinases, and oxidoreductases. Benzotriazole derivatives can act as competitive or non-competitive inhibitors, binding to the active site or allosteric sites of enzymes. This inhibition can disrupt critical biochemical pathways, leading to therapeutic effects. Specific benzotriazole derivatives have been identified as potent inhibitors of enzymes such as tyrosine kinases, making them potential candidates for targeted cancer therapy. Other derivatives inhibit matrix metalloproteinases (MMPs), enzymes involved in tissue remodeling and inflammation, highlighting their potential in treating conditions like arthritis and fibrosis.

Khan and colleagues synthesized[60] a benzotriazole-based *bis*-Schiff bases and conducted an *in vitro* evaluation of their α-glucosidase inhibitory activity, which is crucial for managing conditions like diabetes by slowing carbohydrate absorption and reducing postprandial blood glucose levels. Among the synthesized compounds, **107-122** showed the most promising inhibition profiles, with IC_{50} values ranging from 1.10 ± 0.05 μM to 28.30 ± 0.60 μM. These values indicate the concentration of each compound needed to inhibit 50% of the enzyme's activity. Notably, several of these compounds demonstrated superior inhibitory effects compared to acarbose, a standard α-glucosidase inhibitor used clinically, which has an IC_{50} of 10.30 ± 0.20 μM. The results suggest that these benzotriazole-based bis-Schiff bases could be strong candidates for the development of new antidiabetic drugs, offering potentially more effective alternatives to existing therapies **(Fig. 35)**.

Figure 35. Benzotriazole-based bis-Schiff bases **107-122** *with α-glucosidase inhibitory potential*

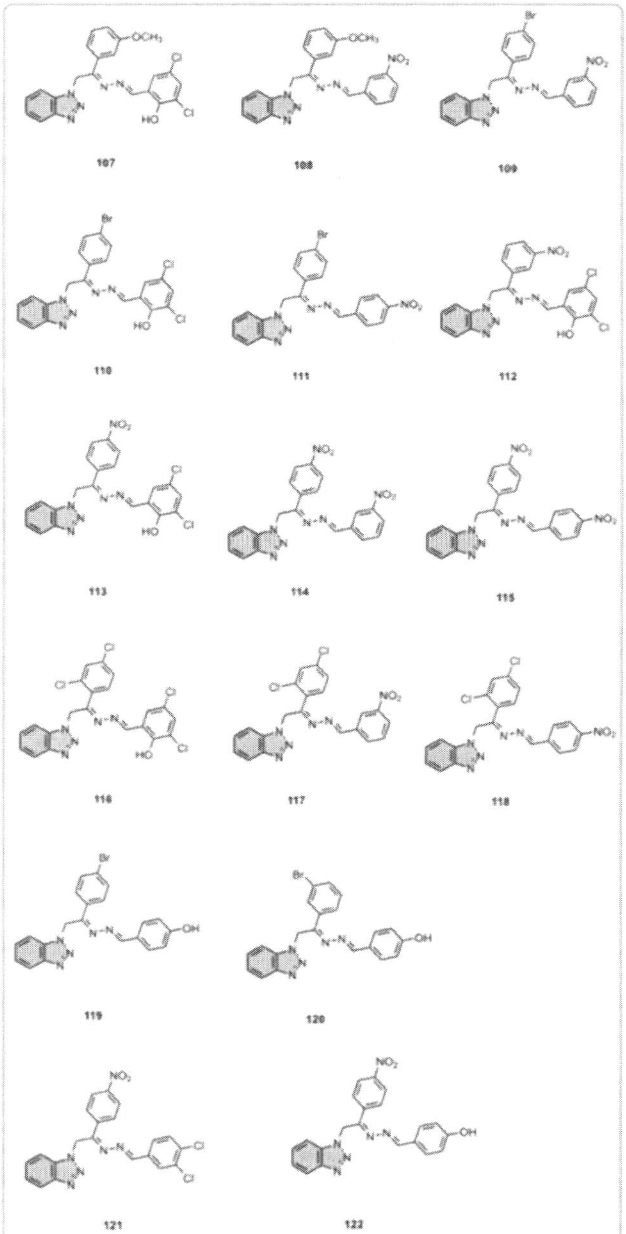

Hameed and colleagues synthesized[61] a benzotriazole derivatives and evaluated their inhibitory activity against two key enzymes involved in carbohydrate metabolism: α-glucosidase and α-amylase. These enzymes play crucial roles in the breakdown of carbohydrates, and their inhibition is a valuable strategy in managing diabetes by controlling blood sugar levels. The synthesized compounds exhibited moderate to good inhibitory activity, with IC_{50} values ranging from 2.00 to 5.6 µM for α-glucosidase and 2.04 to 5.72 µM for α-amylase. Among the tested molecules, compounds **123-125** emerged as the most potent inhibitors. Interestingly, the study also revealed that these compounds exhibited different modes of inhibition depending on the enzyme. They acted as competitive inhibitors of α-amylase, meaning they directly competed with the substrate for the enzyme's active site. In contrast, they exhibited non-competitive inhibition against α-glucosidase, indicating that they bind to a different site on the enzyme, altering its activity without directly competing with the substrate. This dual-mode of inhibition suggests that these benzotriazole derivatives have a versatile mechanism of action, making them promising candidates for the development of new therapeutic agents to manage diabetes (**Fig. 36**).

Figure 36. Benzotriazole-derivatives **123-125** *with α-glucosidase inhibitory potential*

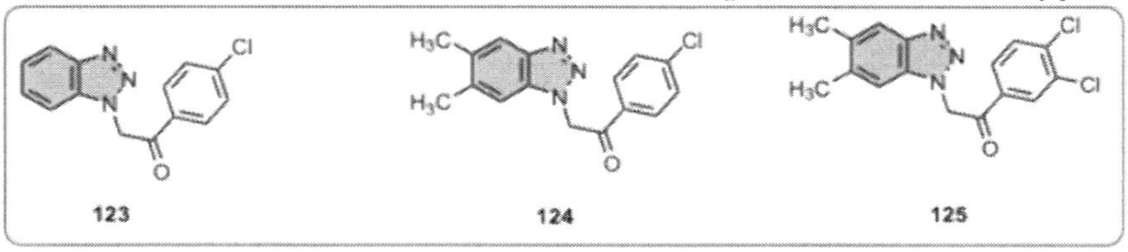

5.5 Environmental and Industrial Applications

Benzotriazoles, a class of heterocyclic compounds, have been widely used in various environmental and industrial applications due to their excellent corrosion inhibition properties, UV stabilization capabilities, and other chemical functionalities. Since 2015, significant advancements have been made in understanding and applying benzotriazoles, with a focus on improving their efficiency, environmental safety, and exploring novel applications.

One of the primary industrial uses of benzotriazoles is as corrosion inhibitors, particularly for protecting copper and its alloys. These compounds are essential in cooling systems, antifreeze, and lubricants due to their ability to form protective films on metal surfaces. Recent research has focused on enhancing the efficiency of benzotriazole-based inhibitors while minimizing their environmental impact.

For instance, studies have explored modifying benzotriazoles to reduce toxicity and combining them with other inhibitors to achieve synergistic effects that offer better protection with lower chemical concentrations[62]

Gopi and colleagues synthesized[63] benzotriazole derivatives, specifically compounds **126** and **127**, and conducted a study on their effectiveness in inhibiting the corrosion of mild steel in groundwater environments at various temperatures. The investigation revealed that these benzotriazole derivatives provided significant inhibition efficiency, protecting the mild steel from corrosion in the groundwater medium. This suggests that compounds **126** and **127** could be promising candidates for corrosion prevention in such environments **Fig. 37**).

*Figure 37. Benzotriazole-derivatives **126-127** as corrosion inhibitor*

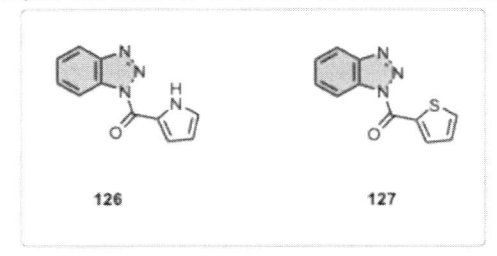

In addition to corrosion inhibition, benzotriazoles are commonly used as UV stabilizers in polymers and coatings. They help prevent material degradation by absorbing ultraviolet radiation and dissipating it as heat. Advances in this area have been driven by the demand for more sustainable materials. For example, researchers have developed bio-based polymers incorporating environmentally friendly benzotriazole derivatives. Moreover, nanotechnology has enabled the creation of nano-structured benzotriazoles, which provide enhanced UV protection and extended product lifespans.[64]

However, the persistence and environmental impact of benzotriazoles have raised significant concerns. These compounds are resistant to biodegradation, leading to their accumulation in aquatic environments, which can harm aquatic life. As a result, there has been increased regulatory scrutiny, particularly in Europe and North America, leading to stricter regulations on their use. Concurrently, research has focused on developing more efficient methods for removing benzotriazoles from wastewater, such as advanced oxidation processes (AOPs) and biofiltration.[65] The U.S. Environmental Protection Agency (EPA)[66] and the European Chemicals Agency (ECHA)[67] have been actively monitoring these compounds, with updated guidelines and restrictions being implemented in recent years.

Beyond traditional applications, benzotriazoles have found new uses in various industries. In electronics, they are used in manufacturing semiconductors and printed circuit boards (PCBs) to prevent corrosion, ensuring the longevity and reliability of electronic components. Additionally, benzotriazole derivatives have been explored in pharmaceuticals as scaffolds for drug development, leveraging their stability and ability to form strong bonds with biological targets. In the field of photovoltaics, research has shown promise in using benzotriazoles in organic photovoltaic (OPV) cells, where they serve as electron-donating materials, potentially contributing to more efficient and durable solar energy devices.[68]

Given the environmental concerns associated with benzotriazoles, significant research has been directed toward developing mitigation strategies. Techniques such as advanced oxidation processes, biodegradation using specialized microbial communities, and adsorption using activated carbon have been explored to reduce the environmental footprint of these compounds. Additionally, green chemistry approaches are being applied to design less toxic, more biodegradable benzotriazole derivatives that still perform effectively in industrial applications.[69] Chung and colleagues investigated the persistence and photostabilizing properties of benzotriazole and its derivative, 5-methyl-1H-benzotriazole. Their study revealed that these compounds are effective in reducing the photochemical activity of common photosensitizers and organic compounds in aqueous environments. This suggests that benzotriazole and its derivatives can play a crucial role in mitigating the degradation of substances exposed to light, thereby enhancing their stability and longevity in environmental contexts.[70]

5.6 Anti-Inflammatory Activity

Benzotriazole exhibits significant anti-inflammatory properties, making it a potential candidate for developing anti-inflammatory drugs. Its effects stem from inhibiting key enzymes like cyclooxygenase (COX) and lipoxygenase (LOX), which are crucial in producing pro-inflammatory mediators. By targeting COX-2 and LOX, benzotriazole reduces the synthesis of prostaglandins and leukotrienes, thus alleviating inflammation. Additionally, it modulates the production of inflammatory cytokines (TNF-α, IL-1β, IL-6) and reactive oxygen species (ROS), further minimizing tissue damage. Benzotriazole derivatives have shown effectiveness in various inflammation models, such as carrageenan-induced paw edema and LPS-induced inflammation. Notably, derivatives with specific substituents offer selective COX-2 inhibition, reducing inflammation while minimizing gastrointestinal side effects associated with traditional NSAIDs. Overall, benzotriazole and its derivatives hold promise as safer, more effective anti-inflammatory agents.

Patil and colleagues synthesized[71] a series of benzotriazole derivatives using microwave-assisted synthesis, a technique known for its efficiency in speeding up chemical reactions. These synthesized compounds were then evaluated for their potential as COX-2 inhibitors, which are enzymes involved in the inflammatory response and are key targets for pain relief drugs. Among the compounds tested, **128-131** stood out for their significant analgesic activity, indicating their effectiveness in reducing pain, potentially through the inhibition of COX-2. This suggests that these derivatives could be promising candidates for the development of new analgesic medications **(Fig. 38)**.

Figure 38. Benzotriazole-derivatives **128-131** *as COX-2 inhibitor*

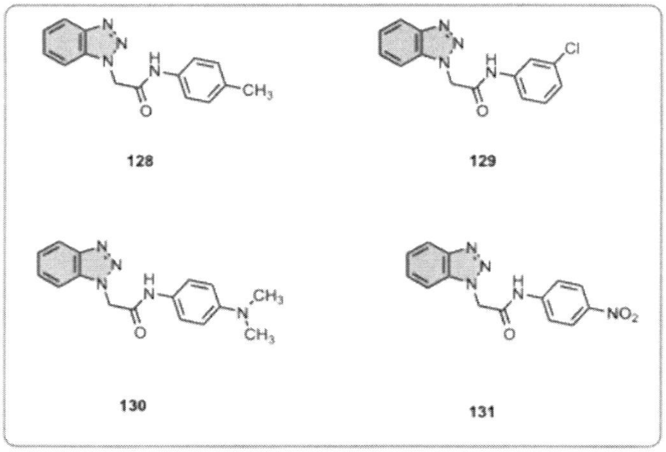

Chen et al synthesized[72] benzotriazole derivatives and screened for analgesic and anti-inflammatory activity. The compounds **132** and **133** exhibited moderate activity **(Fig. 39)**.

Figure 39. Benzotriazole-derivatives **132-133** *as anti-inflammatory agents*

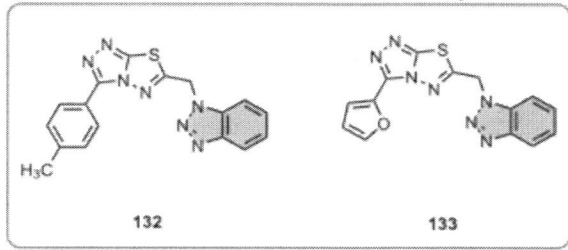

Boido et al. synthesized[73] a series of benzotriazol-1/2-ylalkanoic acids and evaluated them for their anti-inflammatory properties. Among the compounds tested, **134** and **135** exhibited significant anti-inflammatory activity, highlighting their potential as effective agents in the treatment of inflammatory conditions **(Fig. 40)**.

Figure 40. Benzotriazol-1/2-ylalkanoic acids **134-135** *as anti-inflammatory agents*

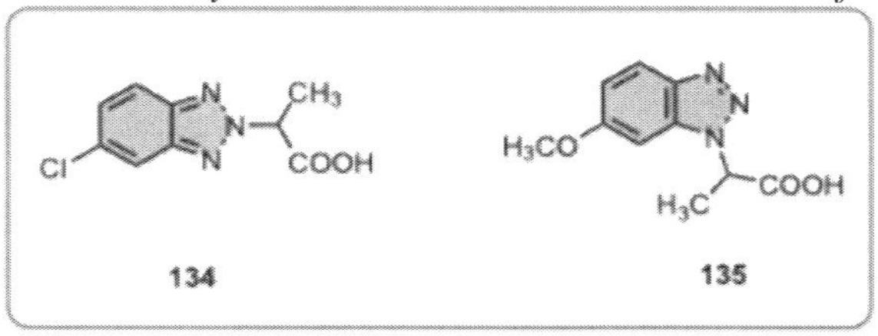

Rajasekaran and coworkers synthesized[74] a series of benzotriazole derivatives and assessed their anti-inflammatory potential. Among the compounds tested, **136** and **137** stood out, displaying promising anti-inflammatory activity. These results suggest that these particular derivatives may serve as potential candidates for the development of new anti-inflammatory agents **(Fig. 41)**.

*Figure 41. Benzotriazole derivatives **136-137** as anti-inflammatory agents*

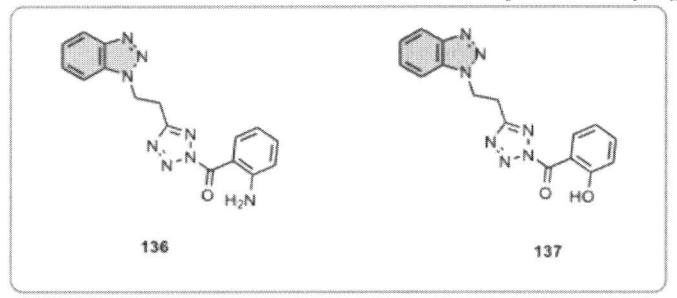

136   137

Dawood et al. synthesized[75] a series of benzotriazole derivatives and assessed their anti-inflammatory properties. Among the tested compounds, **138** demonstrated significant anti-inflammatory activity, highlighting its potential as a strong candidate for further development in anti-inflammatory therapy **(Fig. 42)**.

*Figure 42. Benzotriazole derivative **138** as anti-inflammatory agents*

138

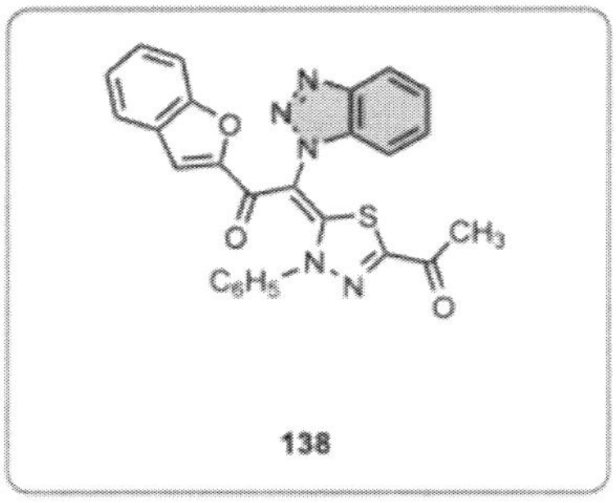

5.7 Antioxidant Properties

Benzotriazole demonstrates strong antioxidant properties by scavenging free radicals and protecting cells from oxidative damage. It chelates metal ions like iron and copper, preventing them from catalyzing the formation of reactive oxygen species (ROS). Benzotriazole derivatives with hydroxyl or methoxy groups show potent antioxidant activity in assays like DPPH radical scavenging. These derivatives also protect against oxidative stress in both cellular and animal models, reducing damage in tissues such as the liver, brain, and heart. This makes benzotriazole a promising candidate for developing therapies to combat oxidative stress-related diseases.

Jamkhandi et al synthesized[76] benzotriazole substituted with *N*-phenylacetamide and acetylcarbamic acid derivatives and evaluated for antioxidant activity. The Benzotriazole derivatives **139-141** showed comparable percentage of nitric oxide scavenging activity with standards reference ascorbic acid **(Fig. 43)**.

Figure 43. Benzotriazole substituted with N-phenylacetamide and acetylcarbamic acid **139-141** *as antioxidant agents*

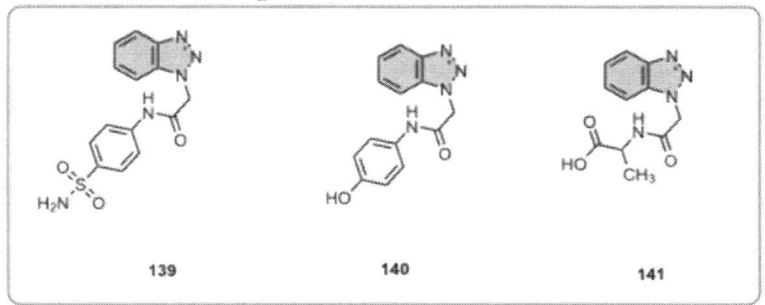

George et al. synthesized[77] Mannich bases of benzotriazolyl triazoles and evaluated their antioxidant properties. The DPPH assay indicated that compounds **142-145** displayed moderate free radical scavenging activity. Structure-activity relationship analysis revealed that compounds with a fluoro substitution on the benzene ring, specifically **142** and **144**, exhibited both excellent antioxidant activity **(Fig. 44)**.

Figure 44. Mannich bases of benzotriazolyl triazoles **142-145** *as antioxidant agents*

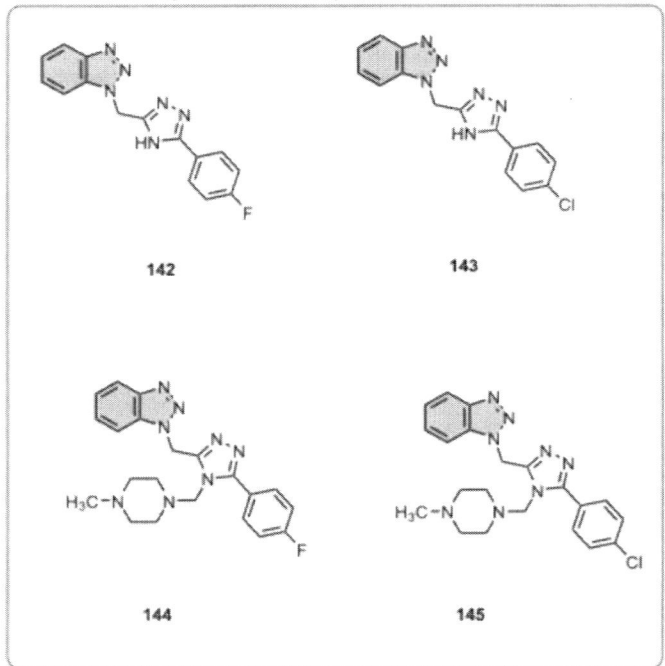

Ling et al. synthesized[78] a series of benzotriazole-based Zn(II) complexes. Among them, complex **3** exhibited moderate antioxidant activity in the DPPH radical scavenging assay, indicating its potential as an antioxidant agent **(Fig. 45)**.

Figure 45. Benzotriazole-based Zn(II) complex **146** *as antioxidant agents*

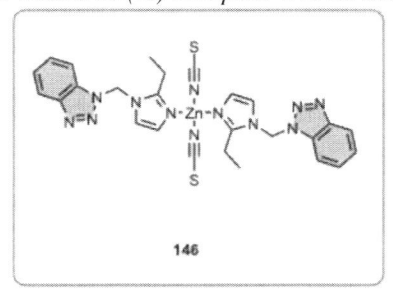

5.8 Miscellaneous

Besides, antibacterial, antifungal, antiviral, anticancer, antioxidant, anti-inflammatory and environmental and industrial applications, benzotriazoles have wide range of other activities like, antidiabetic, antitubercular, anthelmintic activity etc.

Singh et al synthesized[79] a triazole-coumarin-benzotriazole hybrids and screened for Alzheimer's disease (AD) treatment. The compound **147** showed the most potent acetylcholinesterase (AChE) inhibition (IC_{50} = 0.059 µM) with selective AChE inhibition and negligible activity against butyrylcholinesterase. **147** also inhibited copper-induced Aβ1-42 aggregation, chelated metal ions involved in AD, and protected DNA from oxidative damage. Molecular modeling confirmed its ability to block key AChE sites. This makes **147** a promising lead for developing multifunctional anti-Alzheimer's agents **(Fig. 46)**.

*Figure 46. Triazole-coumarin-benzotriazole hybrid **147** against Alzheimer's disease*

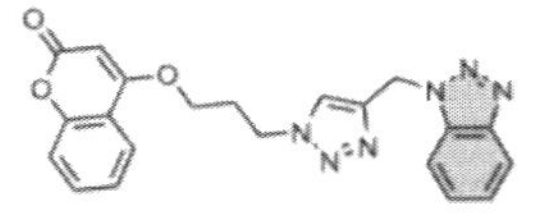

147

Yuan et al synthesized[80] benzotriazole derivatives and evaluated as potential p90 ribosomal S6 protein kinase 2 (RSK2) inhibitors. The compounds **148-151** exhibited notable in vitro RSK2 inhibitory activity, with IC_{50} values of 3.19, 3.05, 4.49, and 2.09 µmol/L, respectively. Molecular docking studies, along with structure-activity relationship (SAR) analysis, suggested that these active compounds interact with the ATP binding site at the *N*-terminal kinase domain (NTKD) of RSK2. The presence of electron-donating groups at the 4-position of the phenyl ring was identified as a key factor for enhancing inhibitory activity **(Fig. 47)**.

Figure 47. Benzotriazole derivatives **148-151** *as p90 ribosomal S6 protein kinase 2 (RSK2) inhibitors.*

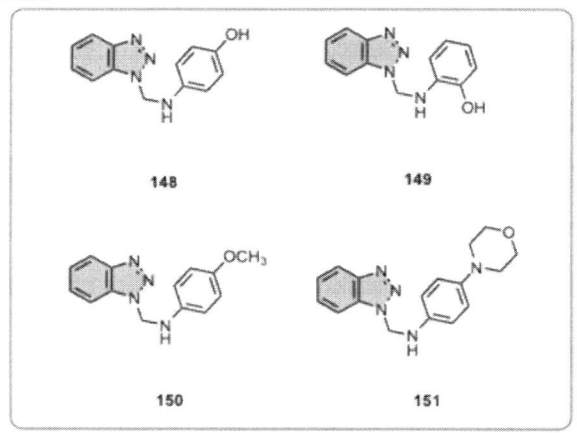

Ambekar and coworkers synthesized[81] coumarin-benzotriazole hybrids and evaluated for their anti-mycobacterial activity against H37Rv strain of *M. tuberculosis*. The compound **152** displayed good antimycobacterial activity towards *M. tuberculosis* with an MIC value of 15.5 μM **(Fig. 48)**.

Figure 48. Coumarin-benzotriazole hybrid **152** *as anti-mycobacterial agent*

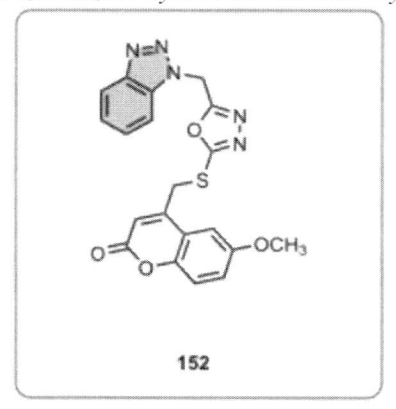

Sudhir et al. synthesized[82] a series of 1,2,3-benzotriazole derivatives and evaluated them for anthelmintic activity. Among these, compounds **153-156** which feature a *p*-nitrophenyl substituent attached to the azo group of the benzotriazole moiety, demonstrated anthelmintic activity comparable to that of albendazole. Notably, compound **155** exhibited superior activity, likely due to the presence of an additional p-nitrophenyl substituent attached to the cyano group **(Fig. 49)**.

*Figure 49. 1,2,3-benzotriazole derivatives **153-156** as anthelmintic agent*

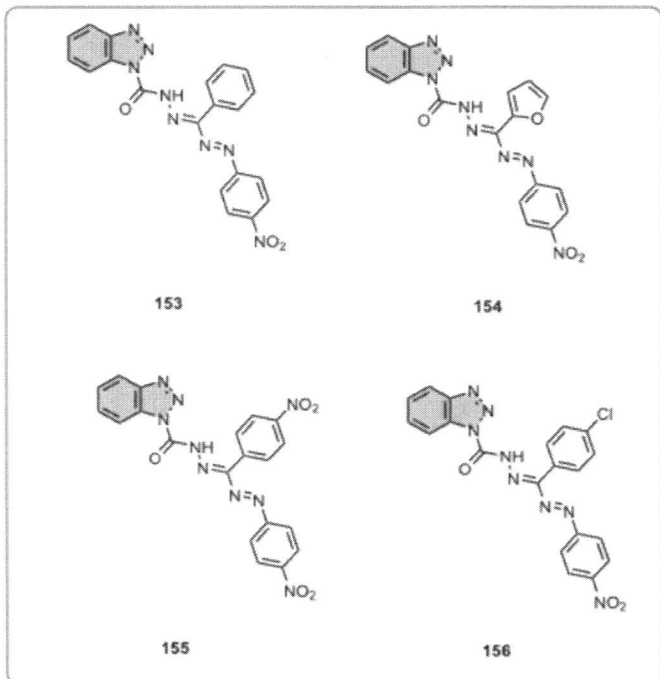

Ling et al synthesized[78] benzotriazole-based Zn(II) complexes and evaluated for the antidiabetic activities against alpha-amylase. The complexes **157, 159** and **146** displayed promising antidiabetic activity, particularly complex **146** shows excellent activity with IC_{50} value of 1.69 mg/mL than that of standard acarbose (2.29 mg/mL) **(Fig. 50)**.

*Figure 50. Benzotriazole-based Zn(II) complexes **157-158, 146** as antidiabetic activity*

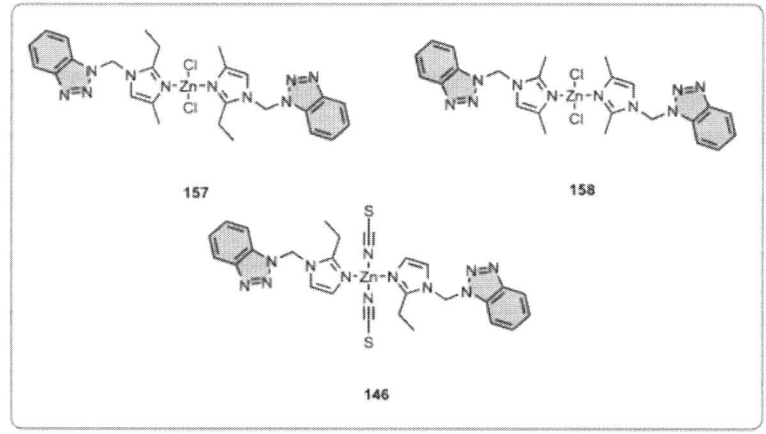

6. CONCLUSION

The current topic delves into the foundational aspects of benzotriazoles, covering their structure and reactivity, acidity and basicity, spectral and physical properties, historical development, and synthetic methods. Benzotriazole and its derivatives are recognized for their broad spectrum of biological activities, making them highly valuable in the field of medicinal chemistry. They have demonstrated significant potential as antimicrobial, antiviral, anticancer, anti-inflammatory, and antioxidant agents, as well as enzyme inhibitors, highlighting their versatility as therapeutic agents. In addition to their medicinal applications, benzotriazoles play a vital role in environmental safety and various industrial processes. Their widespread utility in these areas underscores their importance beyond pharmacology. The ongoing research into the structure-activity relationship (SAR) of benzotriazole derivatives is continuously revealing new opportunities to enhance their biological efficacy and expand their therapeutic potential. This research is crucial for developing more potent and selective benzotriazole-based compounds for a range of applications.

6.1 Abbreviations

AChE - Acetylcholinesterase
AOPs - Advanced Oxidation Processes
BT - Benzotriazole
BTA - Benzotriazole

COX - Cyclooxygenase

Caco-2 - Human Colorectal Cancer Cell Line

CCRF-CEM - Human T-cell Lymphoblast

DMSO - Dimethyl Sulfoxide

DNA - Deoxyribonucleic Acid

DPPH - 2,2-Diphenyl-1-picrylhydrazyl

EC50 - Half-Maximal Effective Concentration

ECHA - European Chemicals Agency

EPA - Environmental Protection Agency

EGFR - Epidermal Growth Factor Receptor

ER - Estrogen Receptor

FTIR - Fourier Transform Infrared

HSV - Herpes Simplex Virus

HIV - Human Immunodeficiency Virus

IC50 - Half-Maximal Inhibitory Concentration

IL-1β - Interleukin-1 beta

IL-6 - Interleukin-6

LOX - Lipoxygenase

MFC - Minimum Fungicidal Concentration

MIC - Minimum Inhibitory Concentration

MMPs - Matrix Metalloproteinases

MRSA - Methicillin-Resistant Staphylococcus Aureus

MKN45 - Human Gastric Cancer Cell Line

N-1 - N-1 Derived Benzotriazole

N-2 - N-2 Derived Benzotriazole

NHC - N-Heterocyclic Carbene

NSAIDs - Non-Steroidal Anti-Inflammatory Drugs

NMR - Nuclear Magnetic Resonance

OPV - Organic Photovoltaic

PCBs - Printed Circuit Boards

P450 - Cytochrome P450

PANC-1 - Human Pancreatic Cancer Cell Line

PR - Progesterone Receptor

RSV - Respiratory Syncytial Virus

RSK2 - p90 Ribosomal S6 Kinase 2

ROS - Reactive Oxygen Species

SAR - Structure-Activity Relationship

SI - Selectivity Index

TBAT - Tetrabutylammonium Triphenyldifluorosilicate

TEER - Transepithelial Electrical Resistance

Tf - Trifluoromethanesulfonyl
TMS - Trimethylsilyl
TMEDA - N, N, N′, N′-Tetramethylethylenediamine
TNF-α - Tumor Necrosis Factor-alpha
UV-Vis - Ultraviolet-Visible
VX2 - Rabbit Squamous Cell Carcinoma Cell Line
Zn(II) - Zinc (II)

REFERENCES

Alraqa, S. Y., Alharbi, K., Aljuhani, A., Rezki, N., Aouad, M. R., & Ali, I. (2021). Design, click conventional and microwave syntheses, DNA binding, docking and anticancer studies of benzotriazole-1,2,3-triazole molecular hybrids with different pharmacophores. *Journal of Molecular Structure*, 1225, 129192. DOI: 10.1016/j.molstruc.2020.129192

Ambekar, S. P., Mohan, C. D., Shirahatti, A., Kumar, M. K., Rangappa, S., & Mohan, S., Basappa, Kotresh, O., Rangappa, K. S. (2018). Synthesis of Coumarin-benzotriazole Hybrids and Evaluation of their Anti-tubercular Activity. *Letters in Organic Chemistry*, 15, 23.

Ankati, H., & Biehl, E. (2009). Microwave-assisted benzyne-click chemistry: Preparation of 1*H*-benzo[*d*][1,2,3]triazoles. *Tetrahedron Letters*, 50(32), 4677–4682. DOI: 10.1016/j.tetlet.2009.06.004

Atkinson, D. J., Sperry, J., & Brimble, M. A. (2011). Synthesis of benzotriazole analogues of the helicobactericidal agents CJ-13,015, CJ-13,102, CJ-13,108, and CJ-13,104 using a regioselective 1,3-dipolar cycloaddition. *Synlett*, 2011(1), 99–103. DOI: 10.1055/s-0030-1259081

Bajaj, K., & Sakhuja, R. (2015). Benzotriazole: Much More Than Just Synthetic Heterocyclic Chemistry. In Monbaliu, J.-C. M. (Ed.), *The Chemistry of Benzotriazole Derivatives* (pp. 235–283). Springer. DOI: 10.1007/7081_2015_198

Boido, A., Vazzana, I., Mattioli, F., & Sparatore, F. (2003). Antiinflammatory and antinociceptive activities of some benzotriazolylalkanoic acids. *Il Farmaco*, 58(1), 33–44. DOI: 10.1016/S0014-827X(02)00003-4 PMID: 12595035

Bräse, S., & Banert, K. (Eds.). (2010). *Organic Azides: Synthesis and Applications*. Wiley.

Bräse, S., Gil, C., Knepper, K., & Zimmermann, V. (2005). Organic Azides: An Exploding Diversity of a Unique Class of Compounds. *Angewandte Chemie International Edition*, 44(33), 5188–5240. DOI: 10.1002/anie.200400657 PMID: 16100733

Briguglio, I., Piras, S., Corona, P., Gavini, E., Nieddu, M., Boatto, G., & Carta, A. (2015). Benzotriazole: An overview on its versatile biological behaviour. *European Journal of Medicinal Chemistry*, 97, 612–648. DOI: 10.1016/j.ejmech.2014.09.089 PMID: 25293580

Cambié, D., Bottecchia, C., Straathof, N. J. W., Hessel, V., & Noël, T. (2016). Applications of Continuous-Flow Photochemistry in Organic Synthesis, Material Science, and Water Treatment. *Chemical Reviews*, 116(17), 10276–10341. DOI: 10.1021/acs.chemrev.5b00707 PMID: 26935706

Campbell-Verduyn, L., Elsinga, P. H., Mirfeizi, L., Dierckx, R. A., & Feringa, B. L. (2008). Copper-free 'click': 1,3-dipolar cycloaddition of azides and arynes. *Organic & Biomolecular Chemistry*, 6(19), 3461. DOI: 10.1039/b812403e PMID: 19082144

Cantwell, M. G., Sullivan, J. C., & Burgess, R. M. (2015). Benzotriazoles: History, Environmental Distribution, and Potential Ecological Effects. In: *Comprehensive Analytical Chemistry* (ed. E. Y. Zeng) 67: 513-545. Netherlands: Elsevier.

Chand, M., Kaushik, R., & Jain, S. C. (2018). Synthesis and antimicrobial and antioxidant activities of hybrid molecules containing benzotriazole and 1,2,4-triazole. *Turkish Journal of Chemistry*, 42(6), 1663–1677. DOI: 10.3906/kim-1803-61

Chandrasekhar, S., Seenaiah, M., Rao, C. L., & Reddy, C. R. S. (2008). A smooth access to benzotriazoles via azide-benzyne cycloaddition. *Tetrahedron*, 64(49), 11325–11327. DOI: 10.1016/j.tet.2008.08.115

Chen, X., Liu, C., Wang, J., & Li, Y. (2010). Synthesis of Some Novel 3-Alkyl/aryl-6-((1Hbenzo[d][1,2,3]triazol-1-yl)methyl)-[1,2,4]triazolo[3,4-b][1,3,4]thiadiazoles. *Journal of Heterocyclic Chemistry*, 47(5), 1225–1229. DOI: 10.1002/jhet.178

Chinh, L. V., Hung, T. N., Nga, N. T., Hang, T. T. N., Mai, T. T. N., & Tarasevich, V. A. (2014). Synthesis and Antimicrobial Activity of Chalcones Containing Benzotriazolylmethyl and Imidazolylmethyl Substituents. *Russian Journal of Organic Chemistry*, 50(12), 1767–1774. DOI: 10.1134/S1070428014120094

Chung, K. H.-Y., Lin, Y. C., & Lin, A. Y.-C. (2018). The persistence and photostabilizing characteristics of benzotriazole and 5-methyl-1H-benzotriazole reduce the photochemical behavior of common photosensitizers and organic compounds in aqueous environments. *Environmental Science and Pollution Research International*, 25(6), 5911–5920. DOI: 10.1007/s11356-017-0900-7 PMID: 29235031

Cohen, S., Haham, H., Pellach, M., & Margel, S. (2017). Design of UV-Absorbing Polypropylene Films with Polymeric Benzotriazole Based Nano- and Microparticle Coatings. *ACS Applied Materials & Interfaces*, 9(1), 868–875. DOI: 10.1021/acsami.6b12821 PMID: 28005334

Corona, P., Piras, S., Ibba, R., Riu, F., Murineddu, G., Sanna, G., Madeddu, S., Delogu, I., Loddo, R., & Carta, A. (2020). Antiviral Activity of Benzotriazole Based Derivatives. *The Open Medicinal Chemistry Journal*, 14(1), 83–98. DOI: 10.2174/1874104502014010083

Dawood, K. M., Abdel-Gawad, H., Rageb, E. A., Ellithey, M., & Mohamed, H. A. (2006). Synthesis, anticonvulsant, and anti-inflammatory evaluation of some new benzotriazole and benzofuran-based heterocycles. *Bioorganic & Medicinal Chemistry*, 14(11), 3672–3680. DOI: 10.1016/j.bmc.2006.01.033 PMID: 16464601

Dennis Hall, C., & Panda, S. S. (2016). The Benzotriazole Story. In: *Advances in Heterocyclic Chemistry* (ed. C. A. Ramsden, E. F. V. Scriven) 119: 1-23. London: Academic Press. DOI: 10.1016/bs.aihch.2016.01.001

Elsherbini, M., & Wirth, T. (2019). Electroorganic synthesis under flow conditions. *Accounts of Chemical Research*, 52(12), 3287–3296. DOI: 10.1021/acs.accounts.9b00497 PMID: 31693339

Empel, C., & Koenigs, R. M. (2020). Continuous-flow photochemical carbene transfer reactions. *Journal of Flow Chemistry*, 10(1), 157–160. DOI: 10.1007/s41981-019-00054-9

European Chemicals Agency (ECHA) (2023). Restriction of Benzotriazoles in Environmental Applications.

Faggyas, R., Sloan, N., Buijs, N., & Sutherland, A. (2019). Synthesis of Structurally Diverse Benzotriazoles via Rapid Diazotization and Intramolecular Cyclization of 1,2-Aryldiamines. *European Journal of Organic Chemistry*, 31(31-32), 5344–5353. DOI: 10.1002/ejoc.201900463

Finsgar, M., & Milosev, I. (2010). Inhibition of copper corrosion by 1,2,3-benzotriazole: A review. *Corrosion Science*, 52(9), 2737–2749. DOI: 10.1016/j.corsci.2010.05.002

Gaikwad, N. D., Patil, S. V., & Bobade, V. D. (2012). Synthesis and biological evaluation of some novel thiazole substituted benzotriazole derivatives. *Bioorganic & Medicinal Chemistry Letters*, 22(10), 3449–3454. DOI: 10.1016/j.bmcl.2012.03.094 PMID: 22520260

George, S., Chakraborty, R., Parameswaran, M., Rajan, A., & Ravi, T. K. (2015). Synthesis and Biological Activity Studies of Some Novel Mannich Bases of Benzotriazolyl Triazoles. *Journal of Heterocyclic Chemistry*, 52(1), 211–214. DOI: 10.1002/jhet.2029

Gopi, D., Sherif, E. S. M., Manivannan, V., Rajeswari, D., Surendiran, M., & Kavitha, L. (2014). Corrosion and Corrosion Inhibition of Mild Steel in Groundwater at Different Temperatures by Newly Synthesized Benzotriazole and Phosphono Derivatives. *Industrial & Engineering Chemistry Research*, 53(11), 4286–4294. DOI: 10.1021/ie4039357

Gutmann, B., Cantillo, D., & Kappe, C. O. (2015). Continuous-Flow Technology-A Tool for the Safe Manufacturing of Active Pharmaceutical Ingredients. *Angewandte Chemie International Edition*, 54(23), 6688–6728. DOI: 10.1002/anie.201409318 PMID: 25989203

Hameed, S., Kanwal, , Seraj, F., Rafique, R., Chigurupati, S., Wadood, A., Rehman, A. U., Venugopal, V., Salar, U., Taha, M., & Khan, K. M. (2019). Synthesis of benzotriazoles derivatives and their dual potential as α-amylase and α-glucosidase inhibitors in vitro: Structure-activity relationship, molecular docking, and kinetic studies. *European Journal of Medicinal Chemistry*, 183, 111677. DOI: 10.1016/j.ejmech.2019.111677 PMID: 31514061

Ibba, R., Corona, P., Nonne, F., Caria, P., Serreli, G., Palmas, V., Riu, F., Sestito, S., Nieddu, M., Loddo, R., Sanna, G., Piras, S., & Carta, A. (2023). Design, Synthesis, and Antiviral Activities of New Benzotriazole-Based Derivatives. *Pharmaceuticals (Basel, Switzerland)*, 16(3), 429. DOI: 10.3390/ph16030429 PMID: 36986528

Ibba, R., Piras, S., Corona, P., Riu, F., Loddo, R., Delogu, I., Collu, G., Sanna, G., Caria, P., Dettori, T., & Carta, A. (2021). Synthesis, antitumor and antiviral *in vitro* activities of new benzotriazole-dicarboxamide derivatives. *Frontiers in Chemistry*, 9, 660424. DOI: 10.3389/fchem.2021.660424 PMID: 34017818

Ibrahim, A. A., Khalaf, S. D., Ahmed, N. A. A.-S., & Dalaf, A. H. (2021). Synthesis, characterization and biological evaluation (antifungal and antibacterial) of new derivatives of indole, benzotriazole and thioacetyl chloride. *Materials Today: Proceedings*, 47, 6201–6210. DOI: 10.1016/j.matpr.2021.05.160

Jamkhandi, C. M., & Disouza, J. I. (2013). Evaluation of antioxidant activity for some benzotriazole substituted with N-phenylacetamide and acetylcarbamic acid derivatives. *International Journal of Pharmacy and Pharmaceutical Sciences*, 5, 249.

Kang, I.-J., Wang, L.-W., Yeh, T.-K., Lee, C., Lee, Y.-C., Hsu, S.-J., Wu, Y.-S., Wang, J.-C., Chao, Y.-S., Yueh, A., & Chern, J.-H. (2010). Synthesis, activity, and pharmacokinetic properties of a series of conformationally-restricted thiourea analogs as novel hepatitis C virus inhibitors. *Bioorganic & Medicinal Chemistry*, 18(17), 6414–6421. DOI: 10.1016/j.bmc.2010.07.002 PMID: 20675142

Kassab, A., & Hassan, R. (2018). Novel benzotriazole N-acylarylhydrazone hybrids: Design, synthesis, anticancer activity, effects on cell cycle profile, caspase-3 mediated apoptosis and FAK inhibition. *Bioorganic Chemistry*, 80, 531–544. DOI: 10.1016/j.bioorg.2018.07.008 PMID: 30014921

Khan, I., Rehman, W., Rahim, F., Hussain, R., Khan, S., Fazil, S., Rasheed, L., Taha, M., Shah, S. A. A., Abdellattif, M. H., & Farghaly, T. A. (2023). Synthesis, In Vitro α-Glucosidase Inhibitory Activity and Molecular Docking Study of New Benzotriazole-Based Bis-Schiff Base Derivatives. *Pharmaceuticals (Basel, Switzerland)*, 16(1), 17. DOI: 10.3390/ph16010017 PMID: 36678514

Kim, J.-H., Kim, H. U., Song, C. E., Kang, I.-N., Lee, J.-K., Shin, W. S., & Hwang, D.-H. (2013). Benzotriazole-based donor–acceptor type semiconducting polymers with different alkyl side chains for photovoltaic devices. *Solar Energy Materials and Solar Cells*, 108, 113–125. DOI: 10.1016/j.solmat.2012.09.019

Klahn, P., Erhardt, H., Kotthaus, A., & Kirsch, S. F. (2014). The Synthesis of α-Azidoesters and Geminal Triazides. *Angewandte Chemie International Edition*, 53(30), 7913–7917. DOI: 10.1002/anie.201402433 PMID: 24895221

Kleoff, M., Boeser, L., Baranyi, L., & Heretsch, P. (2021). Scalable Synthesis of Benzotriazoles via [3+2] Cycloaddition of Azides and Arynes in Flow. *European Journal of Organic Chemistry*, 2021(6), 979–982. DOI: 10.1002/ejoc.202001543

Kumar, R. K., Ali, M. A., & Punniyamurthy, T. (2011). Pd-Catalyzed C−H Activation/C−N Bond Formation: A New Route to 1-Aryl-1H-benzotriazoles. *Organic Letters*, 13(8), 2102–2105. DOI: 10.1021/ol200523a PMID: 21391666

Li, Q., Liu, G., Wang, N., Yin, H., & Li, Z. (2019). Synthesis and anticancer activity of benzotriazole derivatives. *Journal of Heterocyclic Chemistry*, 57(3), 1220–1227. DOI: 10.1002/jhet.3859

Ling, N., Wang, X., Zeng, D., Zhang, Y.-W., Fang, X., & Yang, H.-X. (2020). Synthesis, characterization and biological assay of three new benzotriazole-based Zn(II) complexes. *Journal of Molecular Structure*, 1206, 127641. DOI: 10.1016/j. molstruc.2019.127641

Liu, H., Gopala, L., Avula, S. R., Jeyakkumar, P., Peng, X., Zhou, C., & Geng, R. (2017). Chalcone-Benzotriazole Conjugates as New Potential 2017Antimicrobial Agents: Design, Synthesis, Biological Evaluation and Synergism with Clinical Drugs. *Chinese Journal of Chemistry*, 35(4), 483–496. DOI: 10.1002/cjoc.201600639

Liu, Y.-S., Ying, G.-G., Shareef, A., & Kookana, R. S. (2012). Occurrence and removal of benzotriazoles and ultraviolet filters in a municipal wastewater treatment plant. *Environmental Pollution*, 165, 225–232. DOI: 10.1016/j.envpol.2011.10.009 PMID: 22019204

Loddo, R., Novelli, F., Sparatore, A., Tasso, B., Tonelli, M., Boido, V., Sparatore, F., Collu, G., Delogu, I., Giliberti, G., & Colla, P. L. (2015). Antiviral activity of benzotriazole derivatives. 5-[4-(Benzotriazol-2-yl)phenoxy]-2,2-dimethylpentanoic acids potently and selectively inhibit Coxsackie Virus B5. *Bioorganic & Medicinal Chemistry*, 23(21), 7024–7034. DOI: 10.1016/j.bmc.2015.09.035 PMID: 26443549

Łukasik, E., & Wróbel, Z. (2014). 2-(Arylamino)aryliminophosphoranes as Easily Available and Convenient Starting Materials in the Synthesis of 1,2,3-Benzotriazoles. *Synlett*, 25(14), 1987–1990. DOI: 10.1055/s-0034-1378448

Lukowska-Chojnacka, E., Winska, P., Wielechowska, M., & Bretner, M. (2016). Synthesis of polybrominated benzimidazole and benzotriazole derivatives containing a tetrazole ring and their cytotoxic activity. *Monatshefte für Chemie*, 147(10), 1789–1796. DOI: 10.1007/s00706-016-1785-8 PMID: 27729714

Lv, M., Ma, J., Li, Q., & Xu, H. (2018). Discovery of benzotriazole-azo-phenol/aniline derivatives as antifungal agents. *Bioorganic & Medicinal Chemistry Letters*, 28(2), 181–187. DOI: 10.1016/j.bmcl.2017.11.032 PMID: 29191555

Mallia, C. J., & Baxendale, I. R. (2016). The use of gases in flow synthesis. *Organic Process Research & Development*, 20(2), 327–360. DOI: 10.1021/acs.oprd.5b00222

Matheswari, P. P., Jeyamalar, I. J., & Iruthayaraj, A. (2024). Synthesis, structural, multitargeted molecular docking analysis of anti-cancer, anti-tubercular, DNA interactions of benzotriazole based macrocyclic ligand. *Bioorganic Chemistry*, 147, 107361. DOI: 10.1016/j.bioorg.2024.107361 PMID: 38613924

Mermer, A., Bulbul, M. V., Kalender, S. M., Keskin, I., Tuzun, B., & Eyupoglu, O. E. (2022). Benzotriazole-oxadiazole hybrid Compounds: Synthesis, anticancer Activity, molecular docking and ADME profiling studies. *Journal of Molecular Liquids*, 359, 119264. DOI: 10.1016/j.molliq.2022.119264

Moral, A., Borrull, F., Fourton, K. G., Kabir, A., Marce, R. M., & Fontanals, N. (2023). Extraction of selected benzothiazoles, benzotriazoles and benzenesulfonamides from environmental water samples using a home-made sol-gel silica-based mixed-mode zwitterionic sorbent modified with graphene. *Talanta*, 256, 124315. DOI: 10.1016/j.talanta.2023.124315 PMID: 36739742

Movsisyan, M., Delbeke, E. I. P., Berton, J. K. E. T., Battilocchio, C., Ley, S. V., & Stevens, C. V. (2016). Taming hazardous chemistry by continuous flow technology. *Chemical Society Reviews*, 45(18), 4892–4928. DOI: 10.1039/C5CS00902B PMID: 27453961

Müller, S. T. R., & Wirth, T. (2015). Diazo compounds in continuous-flow technology. *ChemSusChem*, 8(2), 245–250. DOI: 10.1002/cssc.201402874 PMID: 25488620

Muvvala, S. S., & Ratnakaram, V. N. (2014). Antibacterial activity of some newer 1,2,3 – benzotriazole derivatives synthesized by ultrasonication in solvent – free conditions. *Izvestiia po Himiia*, 46, 25.

Noël, T., & Buchwald, S. L. (2011). Cross-coupling in flow. *Chemical Society Reviews*, 40(10), 5010. DOI: 10.1039/c1cs15075h PMID: 21826351

Ochal, Z., Bretner, M., Wolinowska, R., & Tyski, S. (2013). Synthesis and in vitro Antibacterial Activity of 5-HalogenomethylsulfonylBenzimidazole and Benzotriazole Derivatives. *Medicinal Chemistry (Shariqah, United Arab Emirates)*, 9(8), 1129–1136. DOI: 10.2174/1573406411309080015 PMID: 23339322

Onar, G., Gurses, C., Karatas, M. O., Balcioglu, S., Akbay, N., Ozdemir, N., Ates, B., & Alici, B. (2019). Palladium(II) and ruthenium(II) complexes of benzotriazole functionalized N-heterocyclic carbenes: Cytotoxicity, antimicrobial, and DNA interaction studies. *Journal of Organometallic Chemistry*, 886, 48–56. DOI: 10.1016/j.jorganchem.2019.02.013

Onar, G., Karatas, M. O., Balcioglu, S., Tok, T. T., Gurses, C., Kılıç-Cıkla, I., Özdemir, N., Ateş, B., & Alıcı, B. (2018). Benzotriazole functionalized N-heterocyclic carbene-silver(I) complexes: Synthesis, cytotoxicity, antimicrobial, DNA binding, and molecular docking study. *Polyhedron*, 153, 31–40. DOI: 10.1016/j.poly.2018.06.052

Patel, P. D., Patel, M. R., Kocsis, B., Kocsis, E., Graham, S. M., Warren, A. R., Nicholson, S., Billack, B., Fronczek, F. R., & Talele, T. T. (2010). Design, synthesis and determination of antifungal activity of 5(6)-substituted benzotriazoles. *European Journal of Medicinal Chemistry*, 45(6), 2214–2222. DOI: 10.1016/j.ejmech.2010.01.062 PMID: 20181413

Patil, A., & Chaurasia, G. (2013). Synthesis of N-(alkyl or aryl)-2-(1H-benzotriazol-1-yl)-acetamides as selective Cox-2 inhibitor. *International Journal of Pharmaceutical Sciences and Research*, 4, 4371.

Piras, S., Sanna, G., Carta, A., Corona, P., Ibba, R., Loddo, R., Madeddu, S., Caria, P., Aulic, S., Laurini, E., Fermeglia, M., & Prici, S. (2019). Dicloro-phenyl-benzotriazoles: A New Selective Class of Human Respiratory Syncytial Virus Entry Inhibitors. *Frontiers in Chemistry*, 7, 247. DOI: 10.3389/fchem.2019.00247 PMID: 31041309

Plutschack, M. B., Pieber, B., Gilmore, K., & Seeberger, P. H. (2017). The Hitch-hiker's Guide to Flow Chemistry. *Chemical Reviews*, 117(18), 11796–11893. DOI: 10.1021/acs.chemrev.7b00183 PMID: 28570059

Rajasekaran, A., & Rajagopal, K. A. (2009). Synthesis of some novel triazole derivatives as anti-nociceptive and anti-inflammatory agents. *Acta Pharmaceutica (Zagreb, Croatia)*, 59(3), 355. DOI: 10.2478/v10007-009-0026-7 PMID: 19819831

Ren, Y., Zhang, H. Z., Zhang, S. L., Luo, Y. L., Zhang, L., Zhou, C. H., & Geng, R. X. (2015). Synthesis and bioactive evaluations of novel benzotriazole compounds as potential antimicrobial agents and the interaction with calf thymus DNA. *Journal of Chemical Sciences*, 127(12), 2251–2260. DOI: 10.1007/s12039-015-0991-y

Shah, J. J., Khedkar, V., Coutinho, E. C., & Mohanraj, K. (2015). Design, synthesis and evaluation of benzotriazole derivatives as novel antifungal agents. *Bioorganic & Medicinal Chemistry Letters*, 25(17), 3730–3737. DOI: 10.1016/j.bmcl.2015.06.025 PMID: 26117563

Shang, X., Zhao, S., Chen, W., Chen, C., & Qiu, H. (2014). Copper-Catalyzed Cascade Cyclization Reaction of 2-Haloaryltriazenes and Sodium Azide: Selective Synthesis of 2*H*-Benzotriazoles in Water. *Chemistry (Weinheim an der Bergstrasse, Germany)*, 20(7), 1825–1828. DOI: 10.1002/chem.201303712 PMID: 24488951

Shi, F., Waldo, J. P., Chen, Y., & Larock, R. C. (2008). Benzyne Click Chemistry: Synthesis of Benzotriazoles from Benzynes and Azides. *Organic Letters*, 10(12), 2409–2412. DOI: 10.1021/ol800675u PMID: 18476707

Singh, A., Kaur, K., Kaur, H., Mohana, P., Arora, S., Bedi, N., Chadha, R., & Bedi, P. M. S. (2023). Design, synthesis and biological evaluation of isatin-benzotriazole hybrids as new class of anti-Candida agents. *Journal of Molecular Structure*, 1274, 134456. DOI: 10.1016/j.molstruc.2022.134456

Singh, A., Sharma, S., Arora, S., Attri, S., Kaur, P., Gulati, H. K., Bhagat, K., Kumar, N., Singh, H., Singh, J. V., & Bedi, P. M. S. (2020). New coumarin-benzotriazole based hybrid molecules as inhibitors of acetylcholinesterase and amyloid aggregation. *Bioorganic & Medicinal Chemistry Letters*, 30(20), 127477. DOI: 10.1016/j.bmcl.2020.127477 PMID: 32781220

Singh, G., Kumar, R., Swett, J., & Zajc, B. (2013). Modular Synthesis of *N*-Vinyl Benzotriazoles. *Organic Letters*, 15(16), 4086–4089. DOI: 10.1021/ol401661j PMID: 23915255

Singh, N., Abrol, V., Parihar, S., Kumar, S., Khanum, G., Mir, J. M., Dar, A. A., Jaglan, S., Sillanpaa, M., & Al-Farraj, S. (2023). Design, Synthesis, Molecular Docking, and In Vitro Antibacterial Evaluation of Benzotriazole-Based β-Amino Alcohols and Their Corresponding 1,3-Oxazolidines. *ACS Omega*, 8(44), 41960–41968. DOI: 10.1021/acsomega.3c07315 PMID: 37969976

Singh, N., Mahant, V., Bhasin, R., Verma, K., Kumar, A., & Vyas, A. (2024). Antimicrobial and Computational Studies of Newly Synthesized Benzotriazoles. *Indian Journal of Microbiology*, 64(3), 1339–1346. Advance online publication. DOI: 10.1007/s12088-024-01344-0 PMID: 39282187

Sudhir, M. S., & Nadh, R. V. (2013). Evaluation of in vitro anthelmintic activities of novel 1,2,3-benzotriazole derivatives synthesized in ultrasonic and solvent free conditions. *Journal of Pharmacy Research*, 7(1), 47–52. DOI: 10.1016/j.jopr.2013.01.004

Suma, B. V., Natesh, N. N., & Madhavan, V. (2011). Benzotriazole in medicinal chemistry: An overview. *Journal of Chemical and Pharmaceutical Research*, 3(6), 375.

U.S. Environmental Protection Agency (EPA) (2022). Benzotriazole: Environmental Monitoring and Risk Assessment.

Xiang, T., Zhang, Y., Cui, L., Wang, J., Chen, D., Zheng, S., & Qiang, Y. (2023). Synergistic inhibition of benzotriazole and sodium D-gluconate on steel corrosion in simulated concrete pore solution. *Colloids and Surfaces. A, Physicochemical and Engineering Aspects*, 661, 130918. DOI: 10.1016/j.colsurfa.2023.130918

Yuan, J., Zhong, Y., Li, S., Zhao, X., Luan, G., Zhao, Z., Huang, J., Li, H., & Xu, Y. (2013). Triazole and Benzotriazole Derivatives as Novel Inhibitors for p90 Ribosomal S6 Protein Kinase 2: Synthesis, Molecular Docking and SAR Analysis. *Chinese Journal of Chemistry*, 31(9), 1192–1198. DOI: 10.1002/cjoc.201300443

Zhang, F., & Moses, J. E. (2009). Benzyne click chemistry with in situ generated aromatic azides. *Organic Letters*, 11(7), 1587–1590. DOI: 10.1021/ol9002338 PMID: 19267454

Zhou, J., He, J., Wang, B., Yang, W., & Ren, H. (2011). 1,7-Palladium Migration via C−H Activation, Followed by Intramolecular Amination: Regioselective Synthesis of Benzotriazoles. *Journal of the American Chemical Society*, 133(18), 6868–6870. DOI: 10.1021/ja2007438 PMID: 21495635

Chapter 10
Synthetic Strategies and Biomedical Applications of Tetrazoles

H. S. Lalithamba

Siddaganga Institute of Technology, Tumakuru, India

G. K. Prashanth

https://orcid.org/0000-0001-6691-4030

Sir M. Visvesvaraya Institute of Technology, India

Aisha Siddekha

https://orcid.org/0000-0002-0349-8508

Government First Grade College, India

M. Ramya

REVA University, Bangalore, India

G. Nagendra

https://orcid.org/0000-0002-5099-3384

REVA University, Bangalore, India

ABSTRACT

The present chapter Tetrazoles: synthetic strategies and Biomedical applications reveals different synthetic protocols for the preparation of tetrazoles. A standard aside, nitrile condensation route to metal-catalyzed condensation route, including the greener approaches and ionic liquid mediated approaches were discussed further these tetrazoles are considered on bio isosteres of carboxylic and found various applications in pharma and medicine and exhibit catalytic activities in organic Synthesis. Various methods limit the role of NaN_3 and emissions the alter-

DOI: 10.4018/979-8-3693-7267-8.ch010

nate protocol for the construction of tetrazole. Tetrazole alone and in conjunction with the organic groups including heterocycles exhibit various pharmacological activities that were covered in the chapter. Finally, the chapter allows a reader to choose the best possible pathway to construct the tetrazole ring and provides all necessary references.

INTRODUCTION

Peptides, the shorter chains of amino acids, exhibit a high degree of flexibility in their structure (Hersh, Broyles, Capcha, Dikici, Shehadeh, Daunert, & Deo, 2021) (Ganguly, Sharma, & Majumder, 2020) owing to the unique features of the amino acids that make them up, as well as the surrounding environment. However, while this flexibility is advantageous in certain contexts, it can hinder the ability of peptides to effectively bind to specific biological targets, such as receptors or enzymes. In order to overcome this restraint, researchers have sought to introduce structural motifs that impose conformational restrictions on peptides. By doing so, the binding affinity and therapeutic potential of the peptides can be enhanced. Non-natural amino acids play an essential role in this process, as they enable the design and synthesis of pharmacologically relevant molecules, analogs of bioactive peptides (Freidinger, 2003) (Zaky, Simal-Gandara, Eun, Shim, & Abd El-Aty, 2022) (Giannis & Kolter, 1993), and peptidomimetics (Bhandari, Rafiq, Gat, Gat, Waghmare, & Kumar, 2020) (Jimmidi, 2023). Of particular significance are heterocyclic α-amino acids (Wang, 2023) (Poulie & Bunch, 2013), which have garnered attention in various scientific domains, such as biochemistry, enzymology, and pharmacology (Kabir & Uzzaman, 2022). Their exceptional properties make them valuable building units in the development of novel therapeutic agents and biologically active compounds.

Heterocyclic substances, which are distinguished by the presence of at least one non-carbon atom in the ring structure, are ubiquitous in nature and also a prominent area of focus in organic chemistry research (Saini, Aran, Jaya, & Rakesh, 2013). These compounds are integral to biochemical processes as they form the core structure of essential components in living cells, including the nucleic acids (Cyranski, Miroslaw, Mariusz, & Tadeusz, 2003) (Dua, Suman, Sonwane, & Srivastava, 2011).

Numerous heterocyclic compounds and their derivates are very well known in the literature with a vast array of pharmacologically active heterocyclic compounds finding widespread use in medicinal chemistry. Notably, natural products containing heterocyclic units, act as antibiotics-penicillin and cephalosporin, alkaloids-vinblastine, ellipticine, morphine, reserpine and cardiac glycosides, all of which have significant therapeutic implications (Ghisalberti, Marcello, & Elizabeth, 1998).

Heterocyclic chemistry is currently undergoing a significant transformation, driven by the growing adoption of organometallic processes for both the synthesis of heterocycles and carbon substitution. Within the realm of heterocycles, based on their structure, substances with one or more heteroatoms in a ring, such as nitrogen (N), oxygen (O), or sulfur (S), these molecules are categorized. Particular attention is focussed on unsaturated derivatives in the study of heterocyclic chemistry, with an emphasis on unstrained five- and six-membered rings. One group of compounds which have garnered significant interest is tetrazoles, which are five-membered heterocyclic compounds featuring poly-nitrogen electron-rich planar structures. Notably, derivatives of tetrazole have found use as valuable drugs, including as antifungal, antineoplastic, antiviral and anti-inflammatory agents (Angelica, Barreto, Veronica, & Carlos, 2017)

(Rajasekaran & Thampi, 2004). Additionally, tetrazoles have diverse usage in fields such as medicine, material sciences, agriculture (Santhosh, Sagar, Vishwanatha, & Sureshbabu, 2017), and photography (Maleki & Afshin, 2015).

The study heterocyclic chemistry has been a captivating area of research. Developing new synthetic protocols for tetrazoles, including the enhancement of established preparation methods, has been accomplished through two primary tactics: First being, heterocyclization of existing nitrogen-containing substrates, and the other is, the functionalization of heterocyclics with substitutions. The direct method for tetrazole formation involves the ubiquitous [2 + 3] cycloaddition of nitriles and azides. Tetrazoles were conventionally created by adding azide ions to organic nitriles or cyanamides (Modarresi, Ali, & Mahmoud, 2009). By employing addition of sodium azide to nitriles in stoichiometric ratios of salts in water, Sharpless *et al.* have presented a novel method for synthesizing tetrazole (Himo, Zachary, Louis, & Barry, 2002).

Herbst proposed that substituting a carboxyl group with a 5-tetrazolyl group in biologically active carboxylic acids may yield compounds with noteworthy properties (Smissman, Terada, & El-antably, 1976). Subsequently, tetrazole analogs of physiologically and pharmacologically active carboxylic acids were synthesized and assessed for their biological effects. Tetrazoles and their derivatives were investigated for their medicinal potential in diverse areas such as antibacterials, cancer treatment, cardiovascular disease, and neurodegenerative disorders, among others (McGuire, Russell, Bolanowska, Freitag, Jones, & Kalman, 1990) (Morley, 1969). They function as precursors to various nitrogen-containing heterocycles and serve as convenient lipophilic spacers in pharmaceutical formulations.

In the field of peptidomimetic chemistry, it is well known that the tetrazole group can serve as a substitute for carboxylic acid residues (Zabrocki, Smith, Dunbar, Iijima, & Marshall, 1988). Furthermore, the 1,5-disubstituted tetrazole ring is recognized as an excellent mimic of the *cis*-amide bond **(Figure 1)** (Larsen, King, Chen, Corley,

Foster, Roberts, Yang, Lieberman, Reamer, Tschaen, Verhoeven, Reider, Young, Rossano, Brookes, Meloni, Moore, Arnett, 1994). Research has demonstrated that the tetrazole analog is capable of accessing nearly 88% of the conformations accessible to the cis-isomer of a dipeptide, indicating significant conformational mimicry. In numerous instances, these analogs are equally as active as the native bonds but offer improved metabolic stability.

Figure 1. Tetrazole as an acid analog and amide bond isostere.

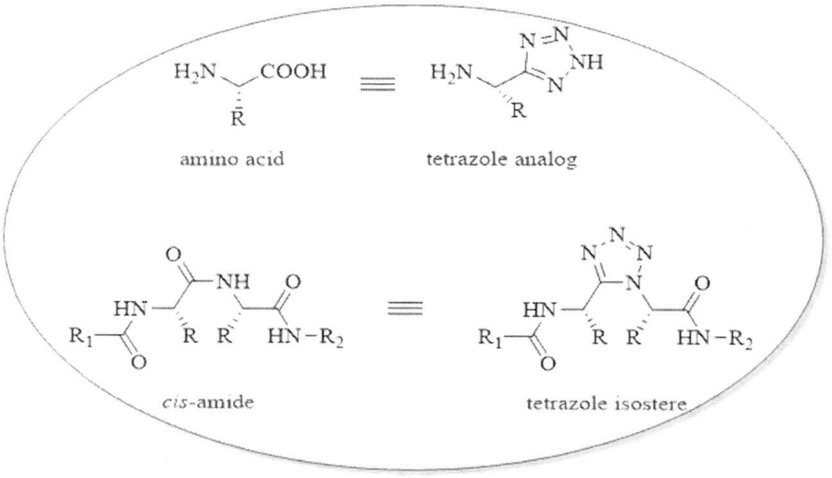

Several tetrazole analogous amino acids were prepared and compared the dissociation constants between 5-aminoalkyl tetrazoles and their corresponding amino acids, which reveal closely aligned pKa values. Additionally, 5-substituted tetrazoles containing a free –NH bond, known as tetrazolic acids, exist in near 1:1 ratios of $1H$- and $2H$-tautomeric forms.

Figure 2. Tetrazole containing bioactive molecules.

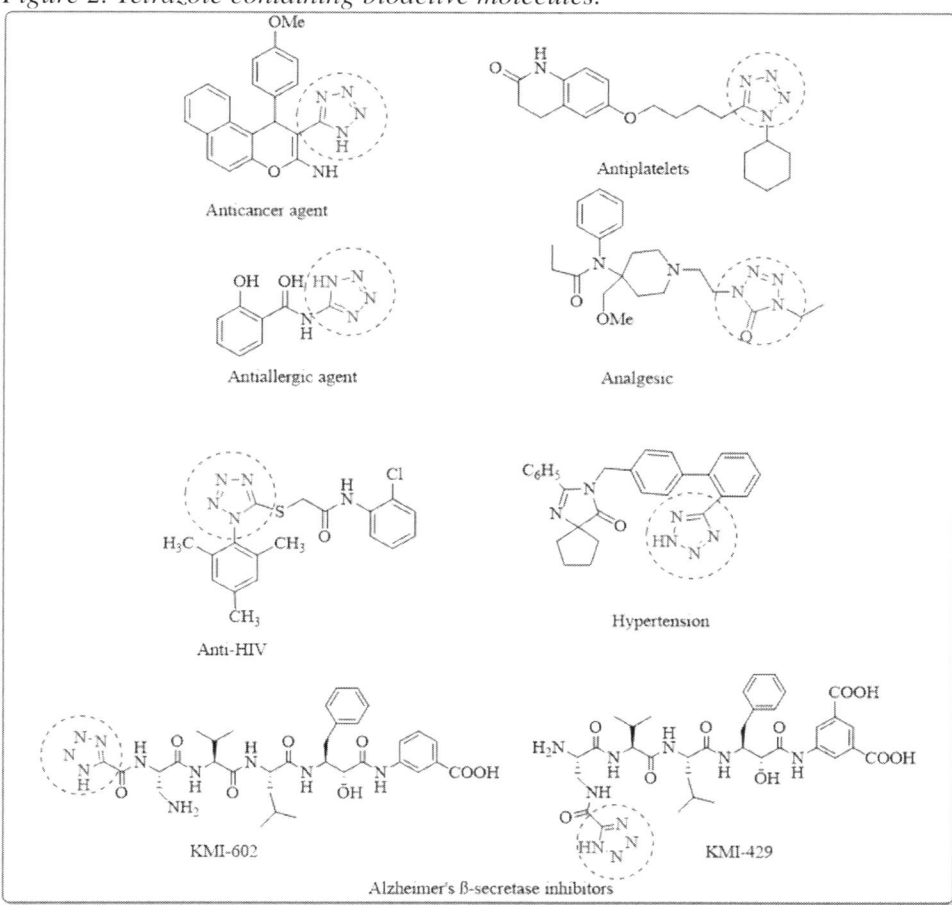

Tetrazole acids surpass carboxylic acids in therapeutic molecular design because of their resistance to biological degradation mechanisms. The tetrazole ring has played an important role in the development of many very successful medications. Losartan, an angiotensin II receptor antagonist used for hypertension treatment, and KMI-602 and KMI-429, renowned for their function in Alzheimer's β-secretase (BACE1) inhibitors (May & bell, 2002)

(Kimura, Shuto, Hamada, Igawa, Kasai, Liu, Hidaka, & Hamada, 2005), have extensively emphasized the tetrazole structure. (**Figure 2**).

The simple synthesis of *1H*-tetrazoles, by adding sodium azide to nitriles is facilitated by zinc salts as catalysts in water (**Figure 1.1**). This methodology demonstrates a broad reaction scope, accommodating various substrates such as cyanamides, nitriles- aromatic, activated and inactivated alkyl nitriles, substituted vinyl nitriles, thiocyanates and thus offering a versatile route for tetrazole formation (Demko & Barry, 2001).

Figure 3. Synthesis of tetrazole.

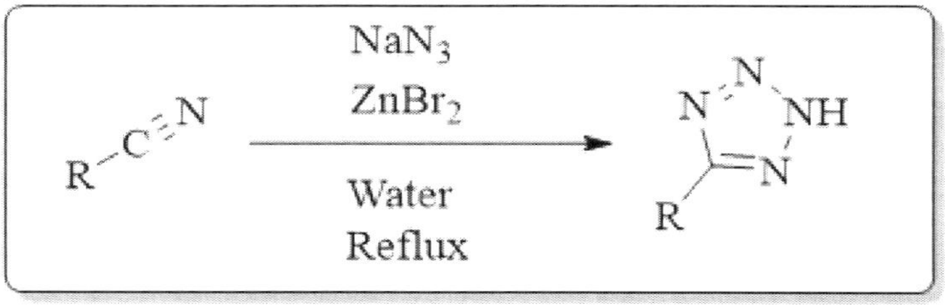

Matthews *et al.* have devised a route for the conversion of chloroalkyl nitriles (having low reactivity) to chloroalkyl tetrazoles using a combination of $AlCl_3$ and NaN_3 and refluxing in THF. The possibilities of chloroalkyl tetrazoles as alkylating agents and precursors for parallel synthesis with trityl chloride resin were then investigated. The alkylation process proceeded readily following an *insitu* Finkelstein *trans* halogenation of *N*-methylpyrrolidinone and K_2CO_3 at 60 °C, in the presence of KI, leading to the conversion of the unreactive chloroalkyl to the iodo derivative in a single step (Matthews, Jonathan, & Anthony, 2000) **(Figure 1.2)**.

Figure 4. Synthesis of chloroalkyl tetrazoles.

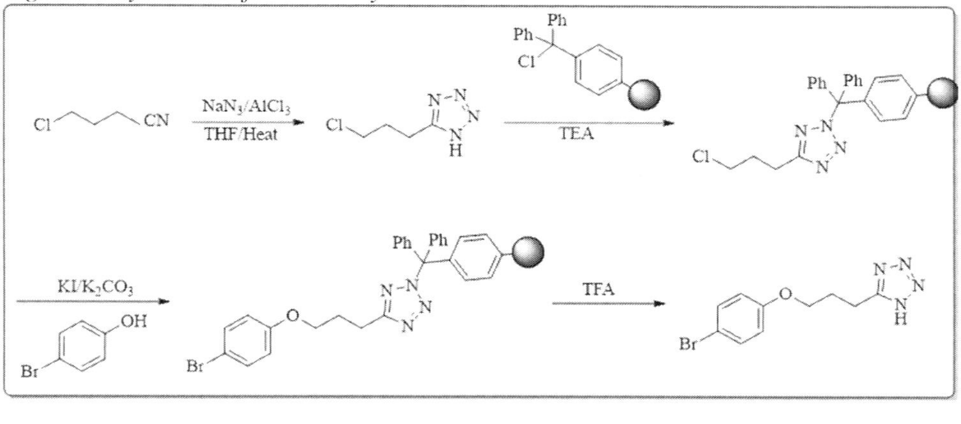

Darvish *et al.* successfully conducted $FeCl_3$-catalyzed synthesis of 1-substituted 1*H*-1,2,3,4-tetrazoles in a single step. This process involved a solvent free reaction of, a diverse range of 1° amines in conjunction with trimethylsilyl azide and trimethyl orthoformate (Darvish & Shima, 2015) **(Figure 1.3)**.

Figure 5. Synthesis of 1H-1,2,3,4-tetrazoles.

Tetrabutylammonium fluoride (TBAF) facilitates the reaction as a catalyst in a reaction that combines organic nitriles and trimethylsilyl azide ($TMSN_3$) under solvent-free environment (Amantini, Romina, Francesco, Ferdinando, & Luigi, 2004). This method, as reported by *David et al.***(Figure 1.4),** successfully converted benzonitrile into 5-phenyl-*1H*-tetrazole. This TBAF catalyst was found to be effective for the construction of tetrazoles.

Figure 6. Synthesis of 5-phenyl-1H-tetrazole.

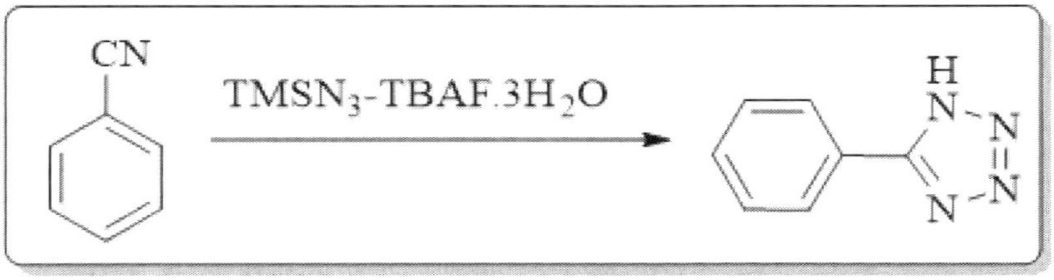

N-bromosuccinimide (NBS), nitriles, $TMSN_3$ and alkenes can be reacted in a single pot with a metal triflate catalyst to create 1,5-disubstituted tetrazoles, according to a technique established by Hajra *et al.* They used $Zn(OTf)_2$ among other metal triflates to produce a library of 1,5-disubstituted tetrazoles from various combinations of alkenes and nitriles (Hajra, Debarshi, & Manishabrata, 2007) **(Figure 1.5).**

Figure 7. disubstituted tetrazoles synthesis.

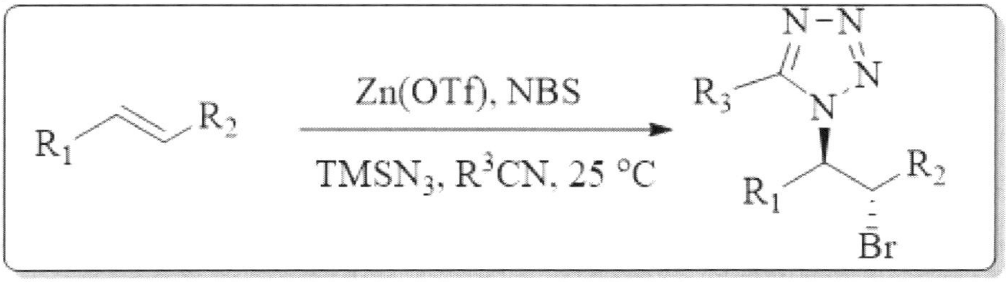

Tetrazole derivatives were made using $FeCl_3$, BF_3 (Kumar, Ramamurthi, & Harold, 1996) $Pd(PPh_3)_4$ and $Yb(OTf)_3$ as catalysts. Specifically, 5-(hydroxyphenyl) tetrazoles were efficiently produced by refluxinhg cyanophenols, sodium azide, and boron trifluoride in dimethylformamide **(Figure 1.6).** When these tetrazoles were heated at 620-650 °C, they produced tropone, benzaldehyde, *o*-cresol, 4-cyanophenol, hydrogen azide and nitrogen.

Figure 8. Synthesis of 5-(hydroxyphenyl) tetrazoles.

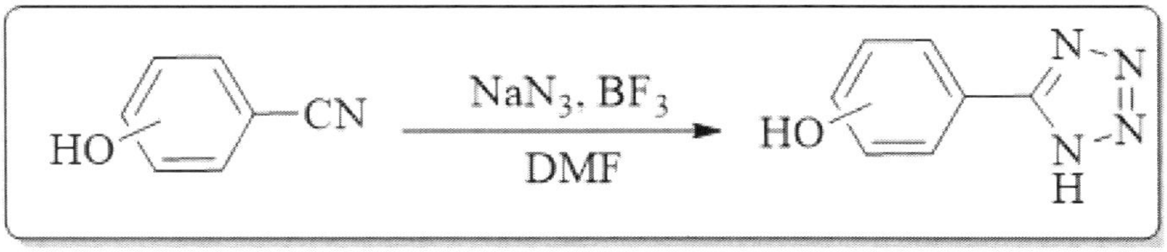

A palladium-catalyzed three-component coupling (TCC) reaction was used to synthesize Tetrazoles, wherein, derivatives of malononitrile, allyl acetate, and trimethylsilyl azide were brought together in the presence of catalytic dose $Pd(PPh_3)_4$, which produced 2-allyl tetrazoles as shown in **(Figure 1.7).** Additionally, activated cyano compounds, and allyl methyl carbonate underwent the TCC reaction to yield 2-allyltetrazoles with trimethylsilyl azide. The π-allylpalladium azide complex as a key intermediate in the catalytic cycle played a pivotal role. The tetrazole frameworks were notably formed via the [3+2] cycloaddition between cyano compounds and the π-allylpalladium azide complex (Shin, Tienan, Zhibao, Young, Jae, & Yoshinori, 2003).

Figure 9. Synthesis of 2-allyltetrazoles.

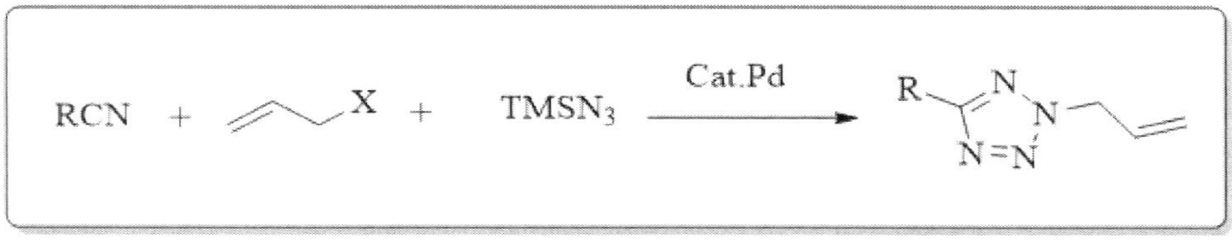

The effectiveness of lanthanide triflates as Lewis acid catalysts in hetero-cycloaddition processes was investigated by Wei-Ke Su *et al* **(Figure 1.8)**. This study demonstrated the effectiveness of $Yb(OTf)_3.H_2O$ as a catalyst, in the cyclization of amines, triethyl orthoformate and sodium azide, leading to 1-substituted 1*H*-1,2,3,4-tetrazoles (Su, Zhi, Wei, & Xing, 2006).

Figure 10. Synthetic route for 1-substituted 1H-1,2,3,4-tetrazoles.

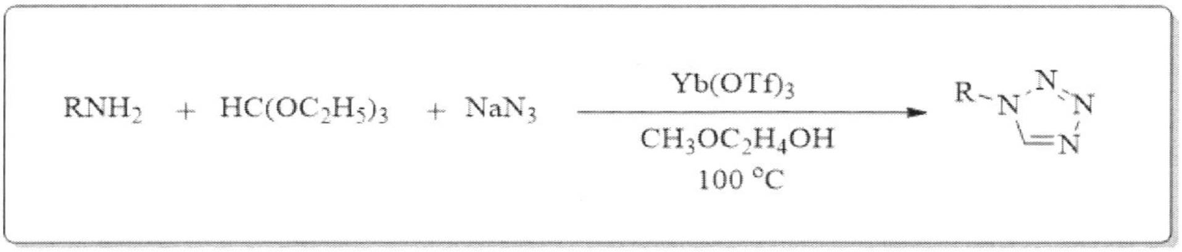

Sharpless et al. reported the transformation of α-aminonitriles to the α-amino acid tetrazole analogues. The transformation involved the refluxing the starting material and sodium azide, in water and 2-propanol mixture at 80 °C using catalytic zinc bromide. This method offers a convenient route for the preparation of tetrazole analogues of α-amino acids, showcasing the potential for further exploration in peptide and peptidomimetic (Demko & Sharpless, 2002) **(Figure 1.9)**.

Figure 11. Synthesis of tetrazole.

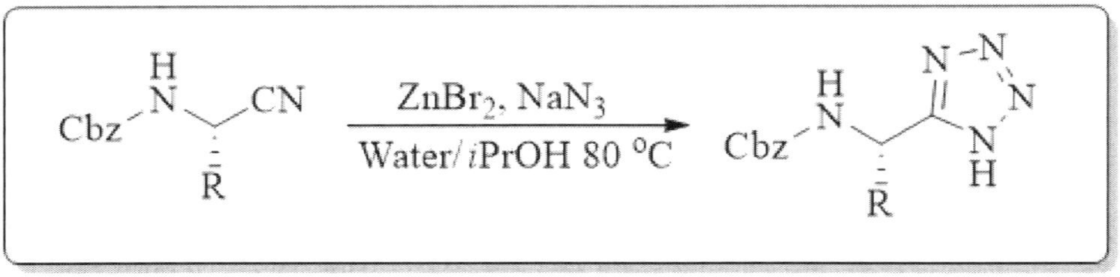

Carsten *et al.* presented a means of producing N-(*1H*)-tetrazole sulfoximines. The procedure entails mixing of sodium azide to the corresponding *N*-cyano derivatives, in a solvent mixture of water and ethanol, in presence of $ZnBr_2$ (Olga & Carsten, 2007) **(Figure 2.0)**.

Figure 12. Synthesis of N-(1H)-tetrazole sulfoximines.

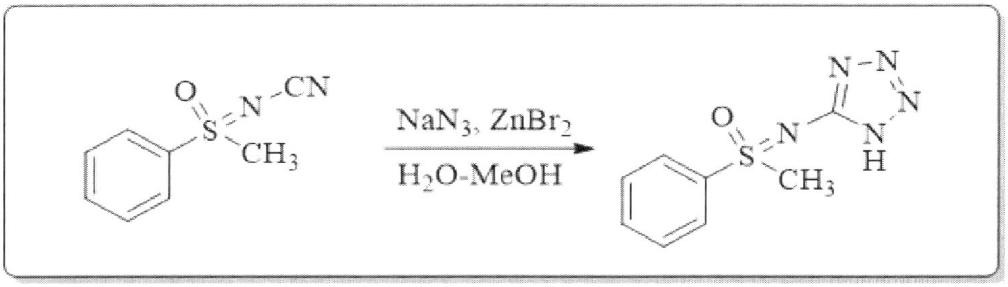

The tetrazole group's acidity is similar to that of a carboxylic acid. Substituting the C-terminal amino acid residue in biologically active peptides with a tetrazole analog can lead to a new class of compounds for structure-function relationship studies. Replacing the C-terminus carboxylic acid with a tetrazole often maintains or enhances biological activity. A two-component synthesis of $1H$-tetrazoles from aldehydes and azides, using Co@g-C_3N_4 as a photocatalyst, was reported as a green alternative method by Fooleswar Verma (**Figure 2.1**). This method involves NaN_3 acting as a three-nitrogen donor for the tetrazole ring and transforming the aldehyde into isocyanide as a one-nitrogen supplier, resulting in an efficient protocol with an excellent 95% yield (Verma, Sahu, Singh, Rai, & Singh, 2018).

Figure 13. Synthesis of 1H-tetrazoles.

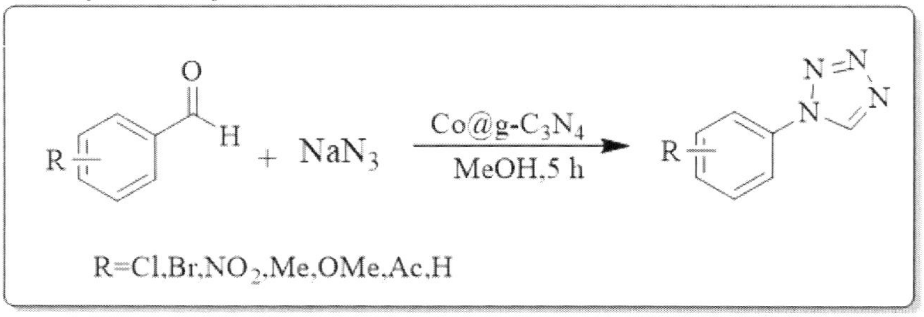

$$R=Cl, Br, NO_2, Me, OMe, Ac, H$$

5-Substituted tetrazoles can be synthesized from aldehydes using water as a green solvent (**Figure 2.2**). This is achieved by treating the aldehydes with hydroxylamine hydrochloride and sodium azide in the presence of *humic acid* as a nontoxic and effective catalyst (Wang, Wang, Han, Zhao, & Wang, 2020).

Figure 14. Synthesis of 5-substituted tetrazoles.

$$R = aryl$$

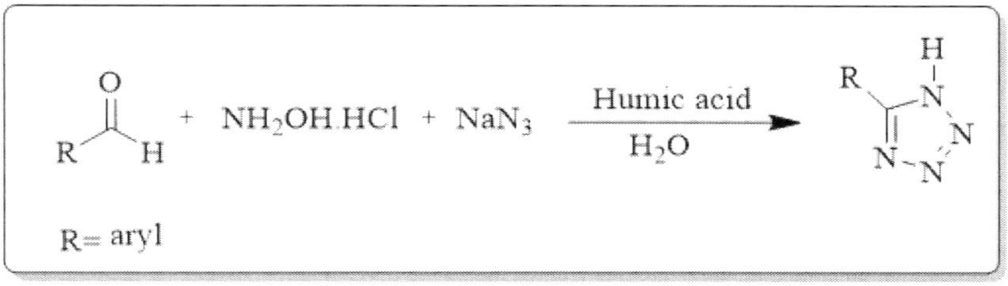

In a single step, tetrazolopiperidinones were formed using a Ugi multicomponent reaction. This reaction involves α-amino acid derived isocyanides, carbonyl components, NaN_3, aqueous ammonia, and base in MeOH/H_2O (**Figure 2.3**). The reaction mixture was kept at RT for 18 h, resulting in a 59% yield of the product (Patil, Kurpiewska, Kalinowska-Tłuścik, & Dömling, 2017).

Figure 15. Synthesis of 5-substituted tetrazolopiperidinone.

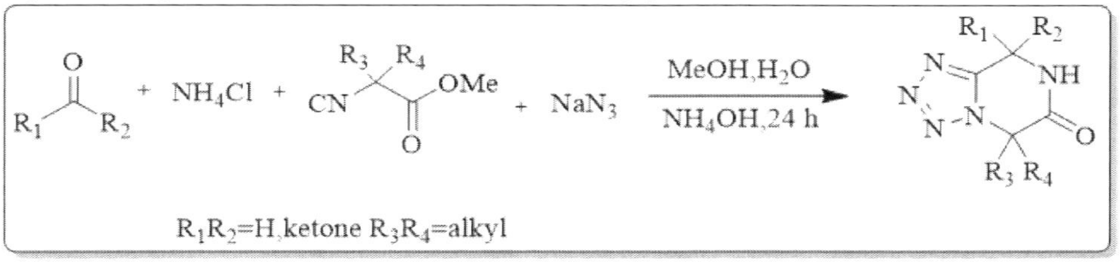

The synthesis of 5-substituted-1*H*-tetrazole can be catalyzed by $AgNO_3$ through [3+2] cycloaddition of nitriles and sodium azide at around 120 °C. Another study describes tetrazole preparation from various nitriles and NaN_3 (sodium azide) using $CuFe_2O_4$ nanoparticles, producing good yields of products under mild reaction conditions (Mani, Ashawani, & Satish, 2014) (**Figure 2.4**).

Figure 16. Synthesis of 5-substituted tetrazole.

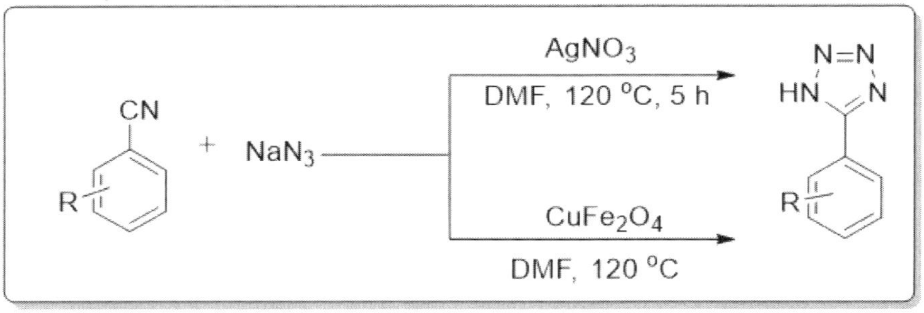

Tetrazolopyrazinone and tetrazolotriazepinone synthesized from *N*-Boc-protected hydrazine and α-amino acid derived isocyanides (**Figure 2.5**). The reactions led to the formation of two cyclic products, 7-aminotetrazolopyrazinone and tetrazolotriazepinone, under acidic conditions. Under basic conditions, Boc-protected 7-aminotetrazolopyrazinone was obtained in good yield (Wang, Patil, Kurpiewska, Kalinowska-Tluscik, & Dömling, 2017).

Figure 17. Synthesis of 7-Boc amino/amino tetrazolopyrazinone and tetrazolotri-azepinone

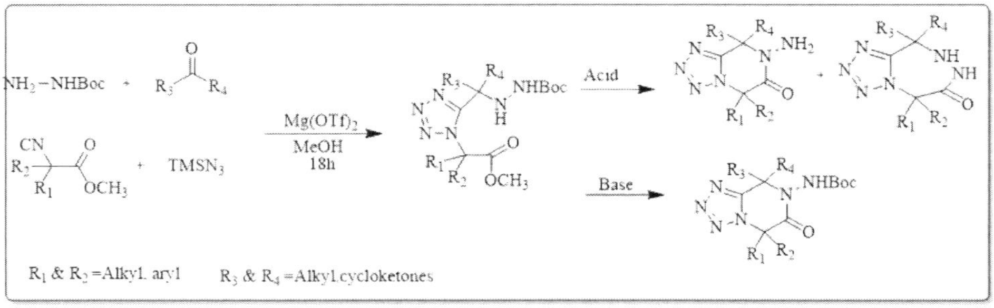

In 2022, various 1,5-disubstituted, 1-substituted, and 5-substituted $1H$ tetrazoles were reported. This was accomplished by utilizing Diphenylphosphoryl azide (DPPA) or p-nitro diphenylphosphoryl azide (p-NO$_2$DPPA) (**Figure 2.6**). The process involved a 16 h cycloaddition reaction of amides with phosphorazidates in the presence of pyridine or 4-methyl pyridine, resulting in the production of tetrazole derivatives with a high yield (Ishihara, Ishihara, Tanaka, Shioiri, & Matsugi, 2022).

Figure 18. Synthesis of 1-5 disubstituted, 1-substituted, and 5-substitued 1H-tetrazoles

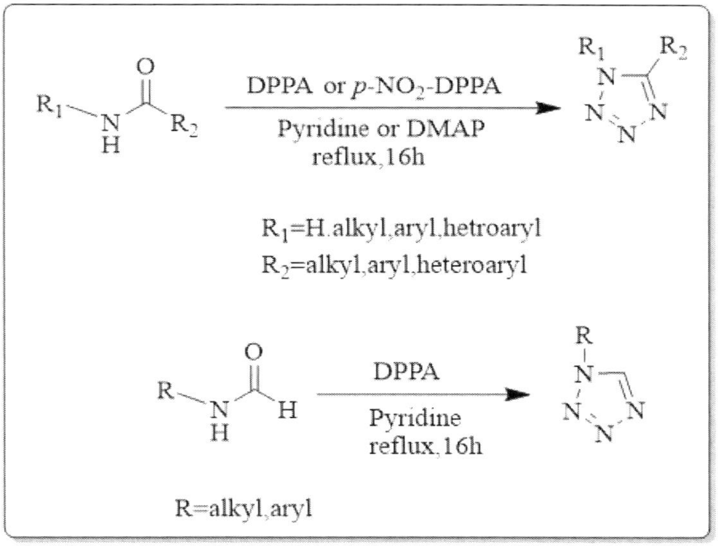

The synthesis of 1-(aryl)-5-(allylthio)-1*H*-tetrazoles involves several steps. Firstly, aniline is reacted with carbon disulfide (CS$_2$) in the presence of benzene and triethylamine (TEA) as a base. The resulting intermediate, triethylammonium dithiocarbamate salt, is subsequently dissolved in ethyl chloroformate at 0 °C for 10 min to yield isothiocyanate (**Figure 2.7**). This isothiocyanate is then treated with NaN$_3$ to produce 1*H*-tetrazole-5-thiol. Finally, the thiol was reacted with allyl bromide in basic condition to obtain 1-(aryl)-5-(allylthio)-1*H*-tetrazoles (Slyvka, Goreshnik, Veryasov, Morozov, Andrii, & Nazariy, 2019).

Figure 19. Synthesis of 1-(aryl)-5-(allylthio)-1H-tetrazoles.

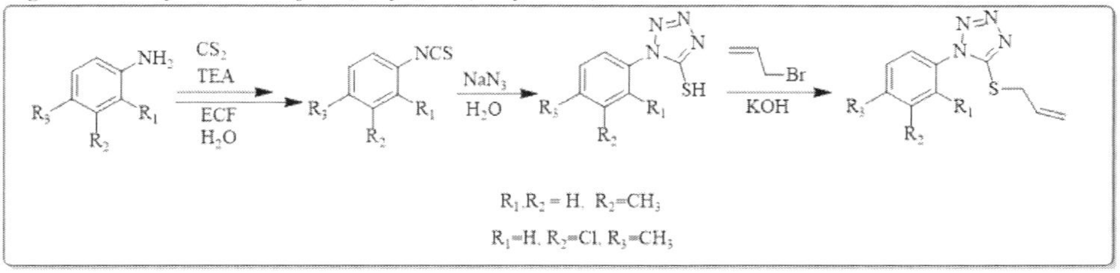

There have been several documented techniques for creating functionalized tetrazoles. *Bolm and others* demonstrated the tetrazoles formation using substituted benzonitriles with TMS-N$_3$. The reaction was carried out at 80 °C for 24 h using Fe(OAc)$_2$ in water, and DMF as the solvent, resulting in high yields of tetrazoles. Additionally, the literature mentions the use of CdCl$_2$ and Sb$_2$O$_3$ catalyst in DMF solvent. Zheng *et al.* reported the use of FeCl$_3$-SiO$_2$ catalyst at 120 °C for synthesizing tetrazoles from benzonitriles and sodium azide in DMF. Furthermore, they also mentioned the use of mesoporous ZnS nanospheres and metal tungstate as catalysts for tetrazole synthesis. Lastly, the successful synthesis of tetrazoles has been achieved using Zn-HAP (zinc hydroxyapatite) and Zn/Al hydrotalcite (**Figure 2.8**) as shown in (**Table 1**).

Figure 20. Synthesis of tetrazole using metal catalysts.

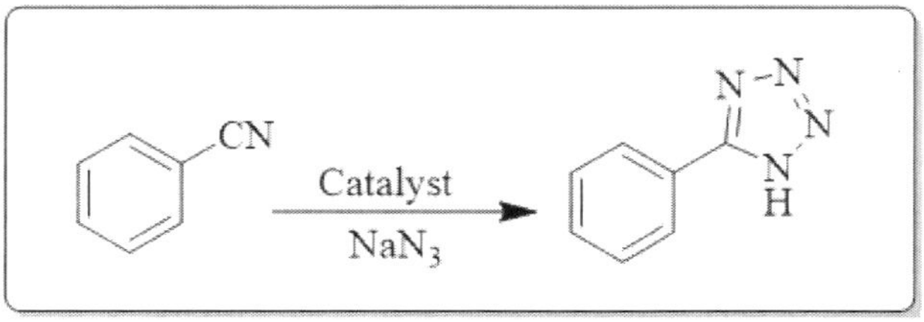

Table 1. Synthesis of tetrazole using metal catalysts.

Catalyst	References
$Fe(OAc)_2$	(Bonnamour & Carsten, 2009)
$CdCl_2$	(Venkateshwarlu, Premalatha, Rajanna, & Saiprakash, 2009)
Sb_2O_3	(Venkateshwarlu, Rajanna, & Saiprakash, 2009)
$FeCl_3$-SiO_2	(Nasrollahzadeh, Yadollah, Davood, & Saeed, 2009)
ZnS	(Lang, Baojun, Wei, Li, Zheng, & Gui, 2010)
Tungstates	(Jinghui, Baojun, Fasheng, Zheng, & Gui, 2009)
Zn-HAP	(Lakshmi Kantam, Balasubrahmanyam, & Kumar, 2006)
Zn/Al Hydrotalcite	(Lakshmi Kantam, Kumar, & Phani Raja, 2006)

Satish N. Dighe et al. have developed an innovative and efficient method for creating 1-substituted-1H-1,2,3,4-tetrazoles (Figure 2.9). By integrating amines, triethyl orthoformate, and sodium azide, and using ionic liquid (IL) and DMSO as a solvent and promoter, they have devised a gentle and straightforward synthesis process. This approach also enables the synthesis of 1-aryl tetrazoles by leveraging the combined effects of DMSO and an ionic liquid as a solvent system at room temperature Satish N. Dighe and colleagues have developed an innovative and efficient method for creating 1-substituted-1H-1,2,3,4-tetrazoles. By integrating amines, triethyl orthoformate, and sodium azide, and using ionic liquid and DMSO as a solvent and promoter, they have devised a gentle and straightforward synthesis process. This approach also enables the synthesis of 1-aryl tetrazoles by leveraging the combined effects of DMSO and an ionic liquid as a solvent system at room temperature. (Dighe, Jain, & Srinivasan, 2009).

Figure 21. Synthesis of 1-aryl tetrazole.

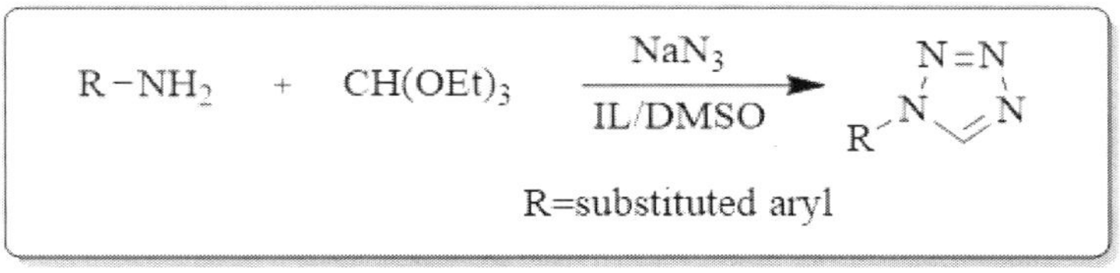

In a single-step process, Mahmoud Nasrollahzadeh's group successfully prepared 5-substituted 1H-tetrazoles from aryl halides using Pd/MnO_2 as a heterogeneous catalyst (**Figure 3.0**). This innovative method utilized $K_4[Fe(CN)_6]$ as a non-toxic cyanide source and sodium azide, showcasing a sustainable approach to chemical synthesis (Sardarian, Eslahi, & Esmaeilpour, 2018).

Figure 22. Synthesis of 5-substituted 1H-tetrazole

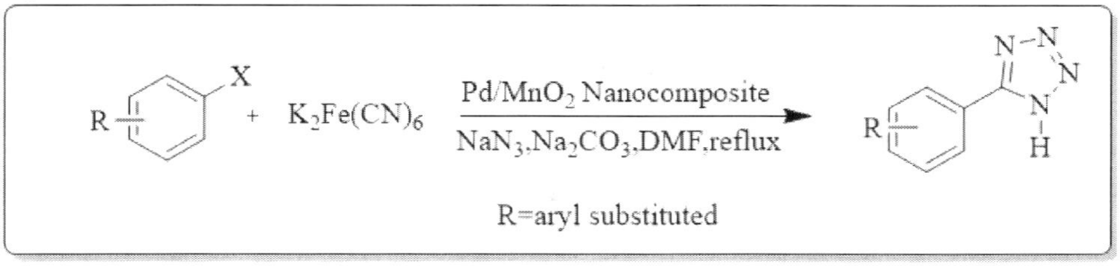

Since zirconyl chloride is largely non-toxic and not-sensitive to air, and it has been reported to be utilized as a catalyst in organic synthesis. A versatile method for preparing 5-substituted-1H-tetrazoles from aryl nitriles using $ZrOCl_2 \cdot 8H_2O$ as a catalyst at 100°C has been shown to yield the desired products (Madhusudana, Muthukur, & Mohamed, 2011) (**Figure 3.1**).

Figure 23. Synthesis of 5-substituted-1H-tetrazoles

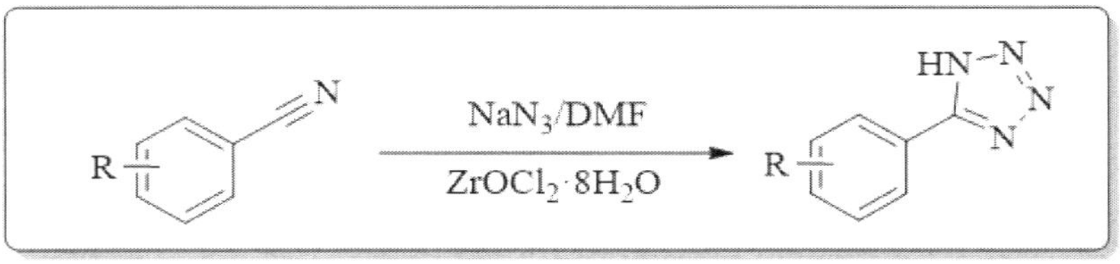

Synthesis of 5-substituted tetrazoles *via* one-pot synthesis hydroxylamine and aldehyde with azide in the presence of different nano metal catalysts ($Fe_3O_4@SiO_2$-PVA/Cu(II) (Sardarian, Eslahi, & Esmaeilpour, 2018), $Cu(OAc)_2$ (Patil, Kumthekar, & Nagarkar, 2012), $Fe_3O_4@NFC@NSaloph(Cu)CO_2H$ (Ghamari kargar & Bagherzade, 2021) (**Figure 3.2**)).

Figure 24. Synthesis of 5-substituted tetrazoles using different metal catalysts

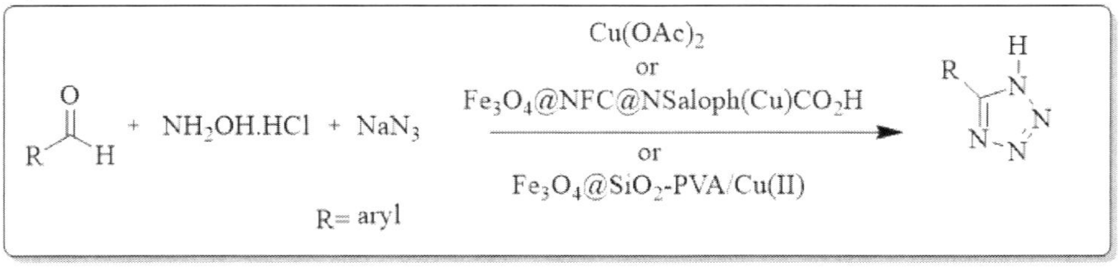

413

Lakshmi Kantam and group achieved the synthesis of 5-substituted 1*H*-tetrazoles with high yields by employing nanocrystalline ZnO as a catalyst for the cycloaddition of sodium azide with nitriles. Additionally, various other heterogeneous catalysts, including nano-sized Cu-MCM-41, nano TiO_2–H_3BO_3, and ZnO/Co_3O_4 (**Table 2**), were efficiently used in the production of tetrazoles **(Figure 3.3)**.

Figure 25. Synthesis of 5-substituted tetrazoles.

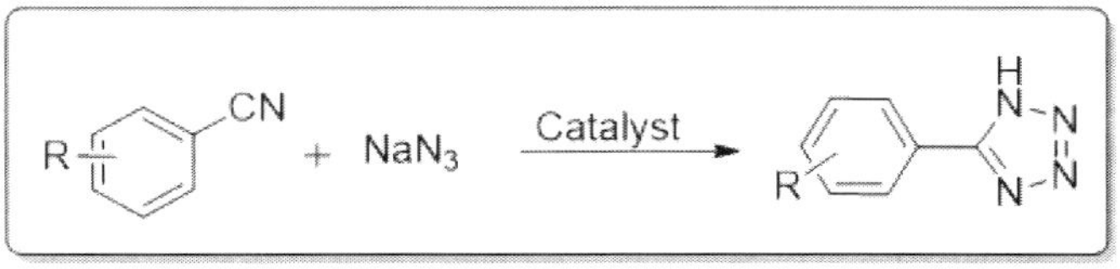

Table 2. Synthesis of 5-substituted tetrazoles using different metal catalyst

Catalyst	References
ZnO	(Lakshmi Kantam, Kumar, & Sridhar, 2005)
Cu-MCM-41	(Abdollahi & Ali, 2016)
TiO_2-H_3BO_3	(Abrishami, Maryam, & Fatemeh, 2018)
ZnO/Co_3O_4	(Agawane & Jayashree, 2012)

When 4-arylthio-2-chloroquinazolines react with NaN_3, the resulting product is 5-azidotetrazolo[1,5-a]quinazoline. The synthesis of 5-(arylthiol)tetrazolo[1,5-c]-quinazolines involves treating 4-arylthio-2-chloroquinazolines with aryl thiol **(Figure 3.4)**. 5-(alkyl/arylthio)tetrazolo[1,5-a]quinazolines were formed, when the compounds engage in regioselective nucleophilic substitution reactions with thiols at the C-5 position in both scenarios (Jeminejs, Goliškina, Novosjolova, Stepanovs, Bizdēna, & Turks, 2021).

Figure 26. Synthesis of 5-(alkyl/arylthio)tetrazolo[1,5-a]quinazoline

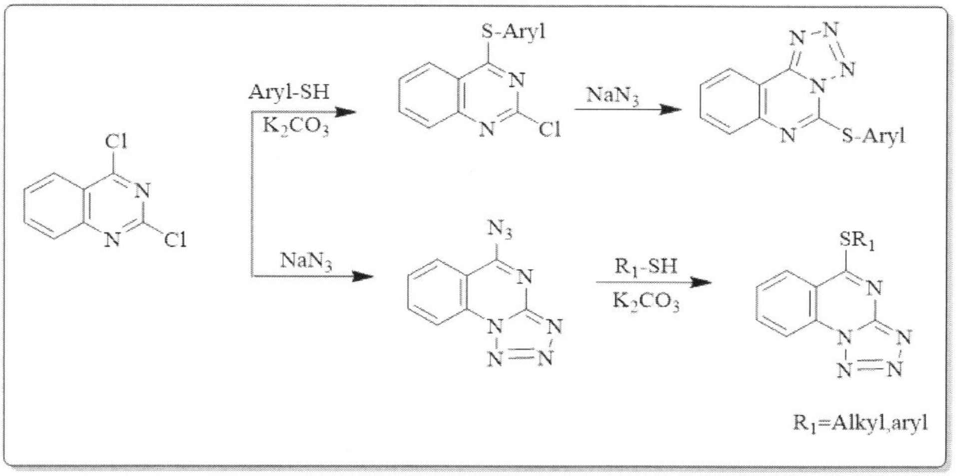

Browsing through the literature, we come across various methodologies for the synthesis of 5-substituted-1*H*-tetrazoles. The cycloaddition technique has been successfully used to create a series of tetrazoles of diverse aryl and alkyl nitriles with sodium azide in DMSO, employing $CuSO_4.5H_2O$ as a catalyst **(Figure 3.5).** A wide array of aryl nitriles, upon [3+2] cycloaddition, yielded tetrazoles in excellent yields (Akhlaghinia & Soodabeh, 2012) under mild conditions.

Figure 27. Synthesis of 5-substituted-1H-tetrazoles

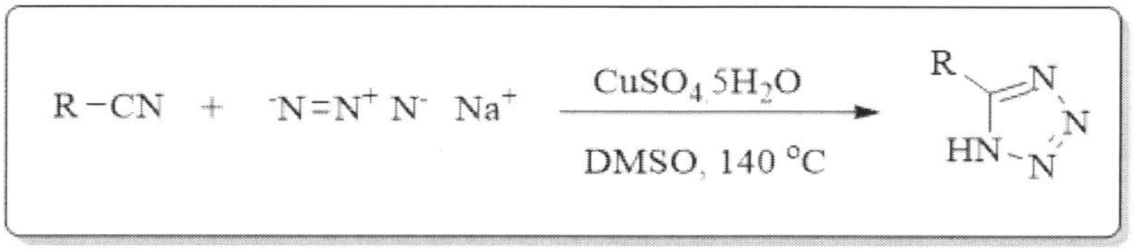

Zhenting and group (Zhenting, Changmei, Youqiang, Yin, & Jing, 2012) describes a process in which a silica-supported sulfuric acid catalyst is used to facilitate a [3+2] cycloaddition of nitriles and sodium azide, resulting in the formation of tetrazoles. Furthermore, Nagarkar and group successfully prepared tetrazole derivatives using Amberlyst-15 (Shelkar, Abhilash, & Jayashree, 2013) as a heterogeneous catalyst in a separate study **(Figure 3.6).**

Figure 28. Synthesis of tetrazoles.

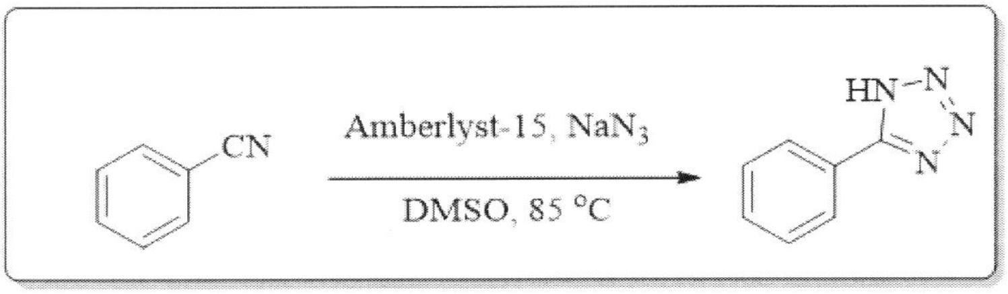

The reaction of 3-amino-1,2,4-triazol-5-carbonitrile with sodium nitrate in the presence of sulphuric acid yields nitro triazole derivatives. Subsequent treatment of these derivatives with NaN_3 and hydroxylamine results in the formation of 1-hydroxy-tetrazoles (Bauer, Benz, Klapötke, Pignot, & Stierstorfer, 2022) (**Figure 3.7**).

Figure 29. Synthesis of 3-amino-1,2,4-tetrazole-5-carbonitrile

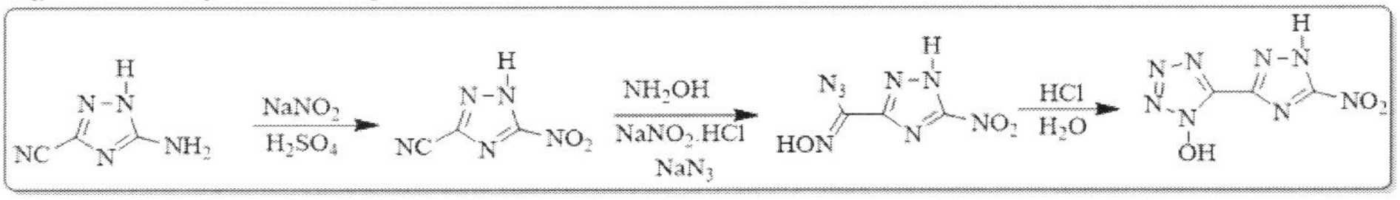

In 1959, McManus and Herbst reported the synthesis of tetrazole analogs of amino acids, replacing the carboxyl group with the 5-tetrazolyl group (McManus & Herbst, 1959). They prepared analogs of various amino acids using two different routes (**Figure 3.8**).

Figure 30. Herbst approach to the synthesis of tetrazole analogs of amino acids.

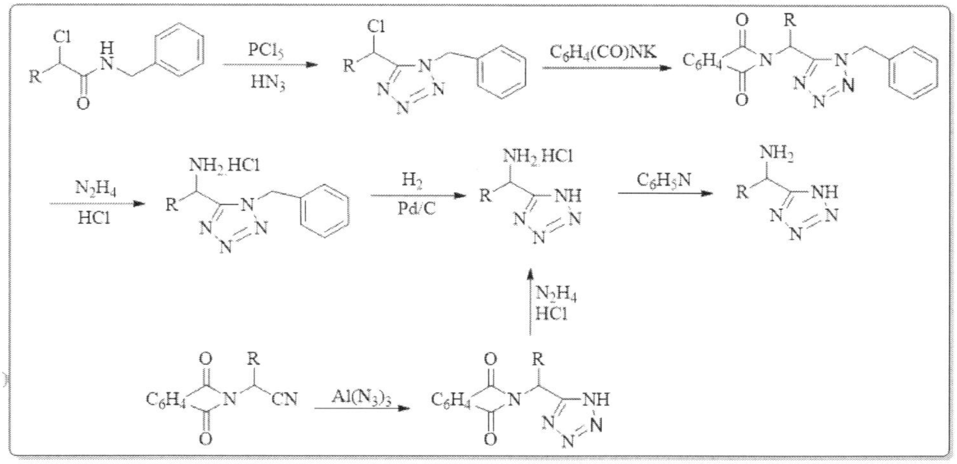

Grzonka *et al.* synthesized amino acid tetrazole analogs from amino acids, protecting the amino groups with benzyloxycarbonyl, phthaloyl, or benzoyl groups. Nitriles were converted into 5-substituted tetrazoles by treating with aluminum azide or using ammonium azide. (**Figure 3.9**) When compared to equivalent *N*-protected amino acids, the tetrazole analogs had higher melting temperatures and were produced as crystalline solids in good yields. Further, different techniques were employed to eliminate the amine-protecting groups (Grzonka & Liberek, 1971).

Figure 31. Synthesis of tetrazole analogs of amino acids

Pg = Benzyloxycarbonyl, Phthaloyl or Benzoyl; X = HCl, HBr
R or *S* or *R* and *S* at *

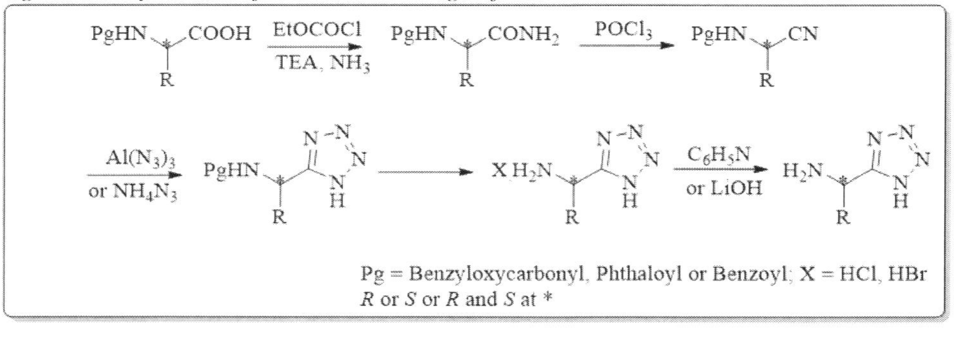

Z-Protected amino alcohols derived from α- and β-amino acids were oxidized using NaOCl and a catalytic amount of AcNH-TEMPO radical to yield aldehydes (**Figure 4.0**). The aldehydes were then converted to unsaturated nitriles with an increased number of carbon atoms. Subsequently, the nitriles were treated to produce the desired tetrazoles, and finally, through catalytic hydrogenation, unprotected and saturated γ- and δ-amino tetrazoles were obtained (Moutevelis-Minakakis, Filippakou, Sinanoglou, & Kokotos, 2006).

Figure 32. Synthesis of tetrazole analogs of γ- and δ-amino acids.

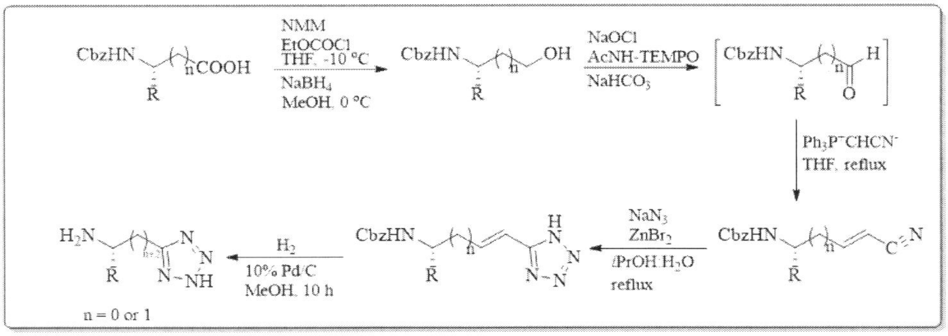

Suresh babu et al. reported a new method that involves the reaction of amino acid derived isocyanides, amines, aldehydes, and TMS-N_3 in a Ugi-azide multi-component reaction. The result is the creation of a new category of tetrazole linked peptidomimetics (**Figure 4.1**) By utilizing enantiopure N and C-terminal isocyanides, the process can generate both symmetrical and non-symmetrical tetrazole peptidomimetics. The Ugi-azide method provides various synthesis benefits and is effective in producing a wide range of novel peptidomimetics (Santhosh, Nagamangala, Thimmalapura, & Vommina, 2017).

Figure 33. Synthesis of tetrazole-linked peptidomimetics.

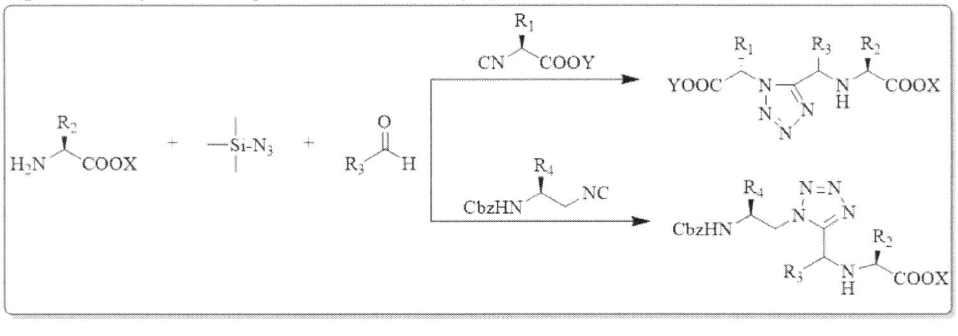

N-Protected amino acids were converted to acyl azide and then to isocyanates, which were transferred into corresponding triazoles using TMS-N$_3$ (**Figure 4.2**). Fmoc-protecting group yielded more tetrazole derivatives compared to Boc and Cbz-protecting groups (Nagendra, Narendra, & Vommina, 2012).

The same group also modified the side chains of glutamic and aspartic acid by adding a tetrazolone ring using the same protocol (**Figure 4.2**).

Figure 34. Synthesis of amino acid-derived tetrazolones.

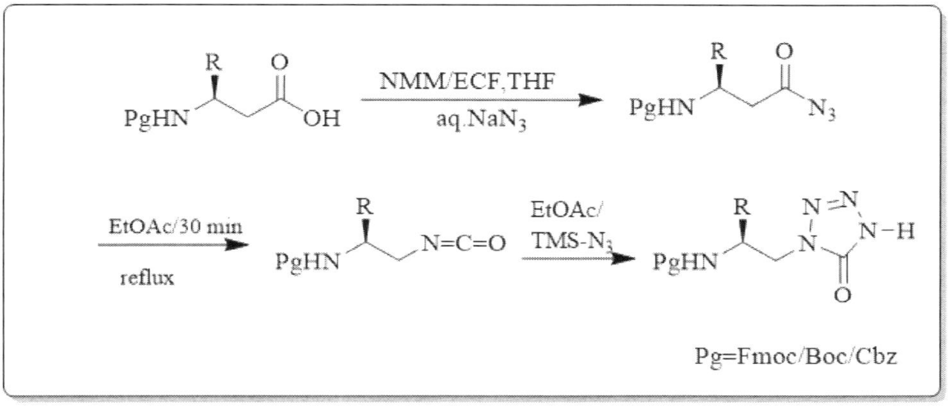

Figure 35. Synthesis of side chain modified tetrazolones.

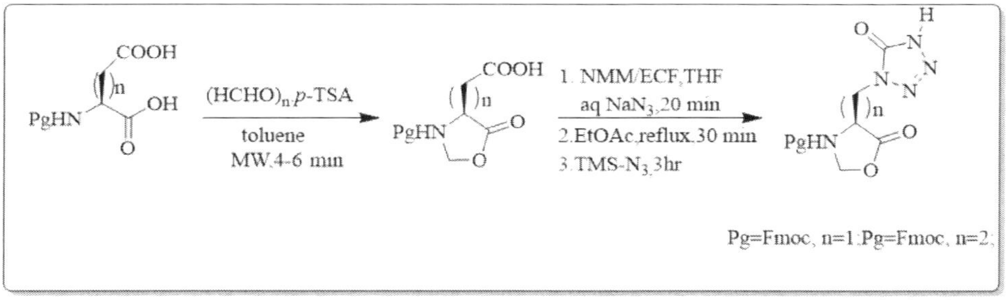

Girish *et al.* have suggested a simple strategy for inserting a 1,5-disubstituted tetrazole molecule into the peptide backbone starting with the thiopeptide. The formation of *cis*-amid bond isostere 1,5-disubstituted tetrazole peptidomimetics was achieved using HgCl$_2$, NaN$_3$, and TEA, in a good yield compared to other thiophilic reagents (**Figure 4.3**). This approach offers the benefit of inserting the tetrazole moiety into the peptide-thiopeptide hybrid, which improves upon previous methods (Prabhu, Nagendra, Sagar, Pal, GuruRow, & Sureshbabu, 2016).

Figure 36. Synthesis of derived 1,5-disubstituted tetrazoles.

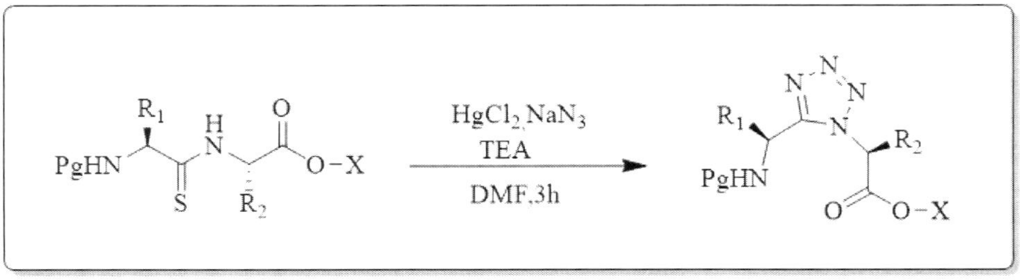

Peptidic-amino tetrazole compounds with *N*-terminal moieties were synthesized *via* a solid-phase methodology. This entailed the *N*-terminal amino group engaging in a reaction with an aryl isothiocyanate to generate a thiourea intermediate, subsequently undergoing dehydrothiolation to yield a peptidic carbodiimide. The 5-amino tetrazole moiety was then obtained through electrocyclization followed by nucleophilic entrapment with sodium azide (Gavrilyuk, Evindar, Chen, & Batey, 2007) (**Figure 4.4**).

Figure 37. Formation of substituted 5-aminotetrazoles from thioureas.

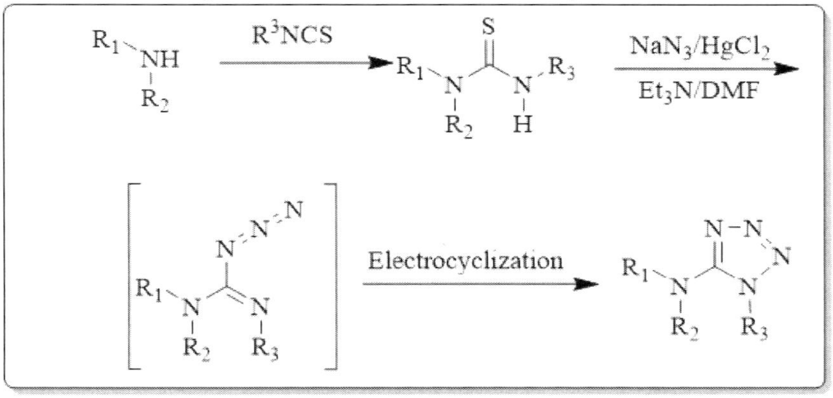

Figure 38. Formation of peptide N-terminal 5-aminotetrazoles.

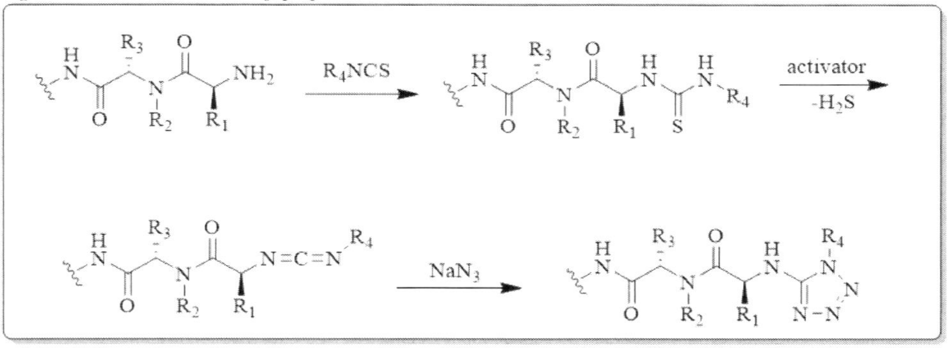

Yanira Mendez *et al.* have introduced a novel protocol for realizing the tetrazole peptidomimetics using a Ugi-azide-4-component reaction with resin-bound amino acids. This involved the paraformaldehyde and resin-bound amino acids reaction in a mixed solvent of THF and MeOH (1:1) and subsequent sequential addition of the piperidinium ion, the isocyanide component, and TMS-N$_3$ to the same reaction vessel (**Figure 4.5**). The synthesized compounds underwent antibacterial testing against *Escherichia coli*, demonstrating notable activity. Additional investigation revealed that two of the synthesized compounds exhibited moderate cytotoxicity against murine myeloma P3X63Ag cells. (Méndez, De Armas, Pérez, Rojas, Valdés-Tresanco, Izquierdo, Alonso del Rivero, Álvarez-Ginarte, Valiente, Soto, de León, Vasco, Scott, Westermann, González-Bacerio, Rivera, 2019).

Figure 39. Solid phase synthesis of tetrazole-peptidomimetics by on resin Ugi-azide –4CR

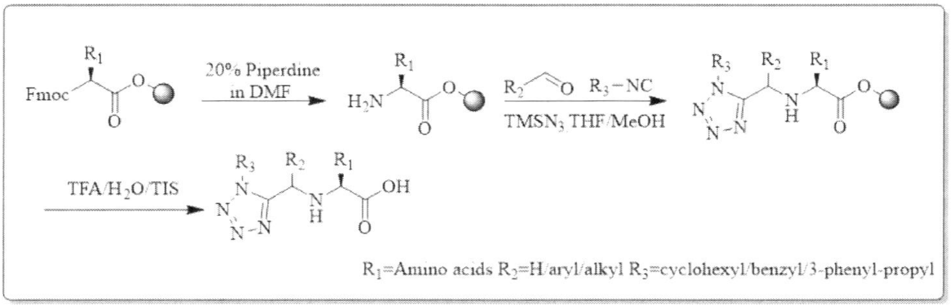

Vasantha *et al.* described the synthesis of 5-substituted S/Se-linked tetrazoles using *N*-Fmoc aminoalkyl iodides in conjunction with potassium thiocyanate in the presence of tetrabutylammonium bromide (**Figure 4.6**). The resulting thiocyanate underwent a cycloaddition reaction with NaN_3 and zinc bromide (Sureshbabu, Vasantha, & Hemanth, 2011).

Figure 40. Synthesis of 5-substituted S/se-linked tetrazole

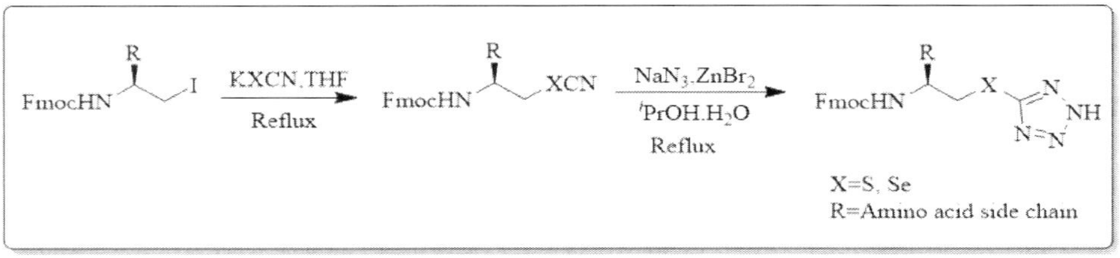

Saad Shaaban conducted a study on the synthesis of tetrazole-based diselenide and selenoquinones using the azido-Ugi and SN (nucleophilic substitution) method. The process involved the addition of two equivalents of aldehyde, TMS-N_3, and isocyanide to synthesize tetrazole-based symmetrical diselenides. Subsequently, tetrazole-based selenoquinones were obtained by reducing diselenides with sodium hydroxide to produce the corresponding sodium selenolate (Figure 4.7). This sodium selenolate then underwent an addition/elimination sequence or SN mechanism with 2-bromo-3-methyl-1,4-naphthoquinone. The study also included in silico antioxidant activity assessments of the newly formed tetrazole moieties and the investigation of the cytotoxicity of these compounds against HepG2 cell and MCF-7 cell lines by conventional MTT assays (Shaaban, Negm, Ashmawy, Ahmed, Wessjohann, 2016).

Figure 41. Synthesis of tetrazole-based diselenides and selenoquinones 2016).

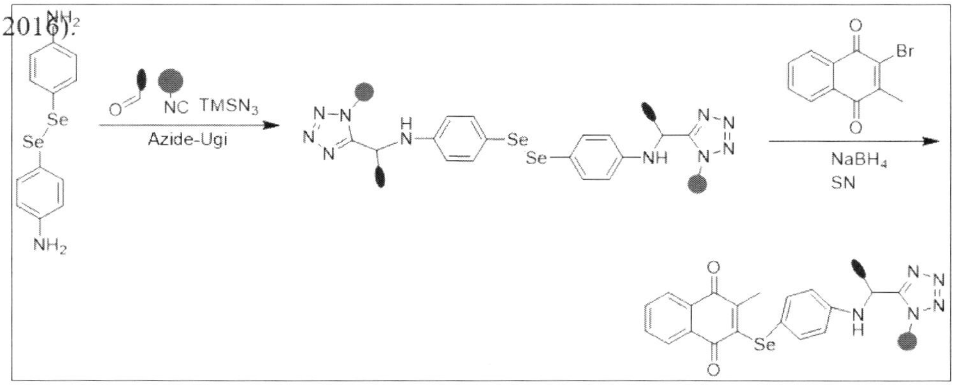

A series of compounds known as azabicyclo[3.2.2]nonane derivatives containing terminal tetrazoles were executed through a multistep process. Initially, 3-azabicyclo[3.2.2]nonanes were refluxed with 2-chloroacetamide, followed by hydrogenation using LiAlH$_4$ (**Figure 4.8**). The resulting *N*-(2-Aminoalkyl) analogs were then subjected to the Ugi-azide strategy to produce the desired tetrazole compounds. Evaluation of their antiprotozoal activities *in vitro* against strains of *Plasmodium falciparum and Trypanosoma brucei rhodesiense (STIB 900)* (Dolensky, Hinteregger, Leitner, Seebacher, Saf, Belaj, Mäser, Kaiser, Weis, 2022) was described.

Table 3.

R$_3$	R$_4$	
n	m	R$_1$+R$_2$
0	2	CH3
0	2	(CH2)3
1	1	(CH2)4

Figure 42. Synthesis of terminal tetrazole derivatives with azabicyclo[3.2.2]nonane

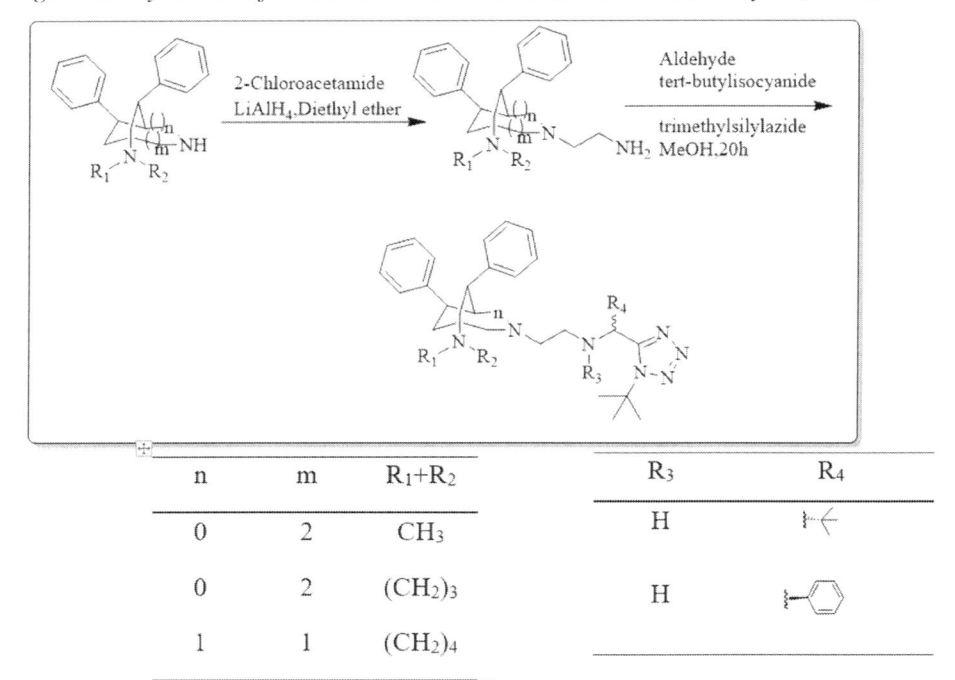

n	m	R$_1$+R$_2$	R$_3$	R$_4$
0	2	CH$_3$	H	
0	2	(CH$_2$)$_3$	H	
1	1	(CH$_2$)$_4$		

Eva S. Schaffert's group synthesized 1,5-disubstituted and 5-monosubstituted aminomethyl tetrazole compounds through the modification of glycine using TMSN$_3$ in the Ugi reaction (**Figure 4.9**). Testing of these molecules for their ability to inhibit and selectively target the four murine GABA transporter subtypes mGAT1–mGAT4, indicated that the 5-monosubstituted tetrazoles were effective in significantly inhibiting [3H] GABA uptake, while the 1,5-disubstituted tetrazole derivatives exhibited distinct activity specifically targeting the GABA transport proteins mGAT2–mGAT4 (Schaffert & Höfner, 2011).

Figure 43. Synthesis of aminomethyl tetrazoles.

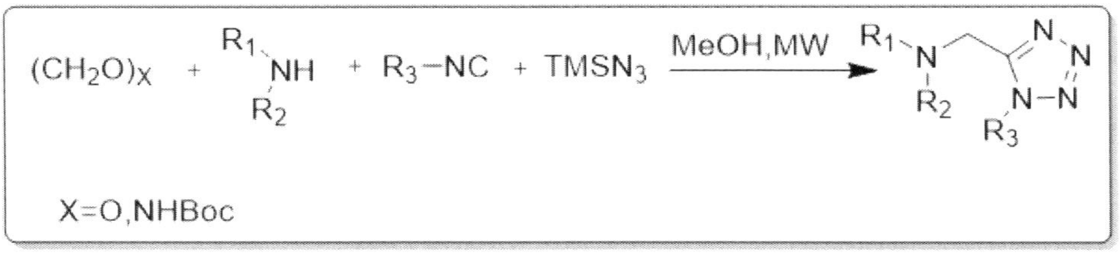

1,5-diaryl-substituted tetrazole from amides by treating with NaN$_3$, was reported by Al-Hourani *et al.* in the presence of SiCl$_4$ in dry CH$_3$CN (**Figure 5.0**). For the synthesized tetrazole compounds *in vitro* studies demonstrated inhibitory potency against the COX-2 enzyme, whereas no inhibition activities were identified against the COX-1 enzyme. Tetrazoles with 4-(sulfonamide) phenyl at the fifth position (C$_5$) inhibited COX-2, more effectively compared to those with 4-(sulfonamide) phenyl at the first position (Jawabrah Al-Hourani, Sharma, Kaur, & Wuest, 2015).

Figure 44. Synthesis of 1,5-diaryl-substituted tetrazole

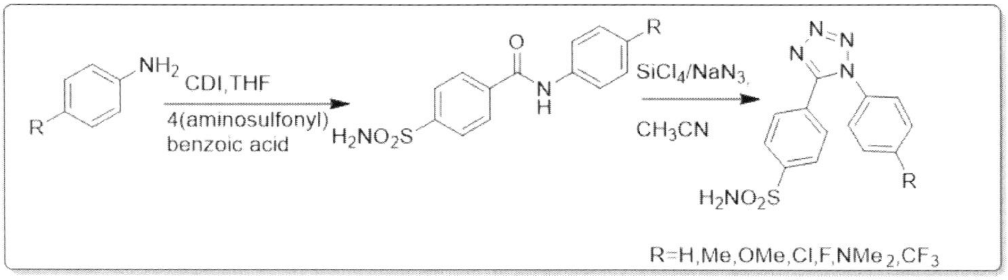

Firoz A. Kalam et al. produced biphenyl tetrazole-thiazolidinediones with good yields. The synthesized molecules were tested for *in vitro* PDF inhibition of enzymes and antibacterial activities (**Figure 5.1**). Docking studies carried out on these molecules against the PDF enzyme of *E. coli* and showed effective binding characteristics (Khan, Jadhav, Patil, Shinde, Arote, & Sangshetti, 2016)

Figure 45. Synthesis of biphenyl tetrazole-thiazolidinediones

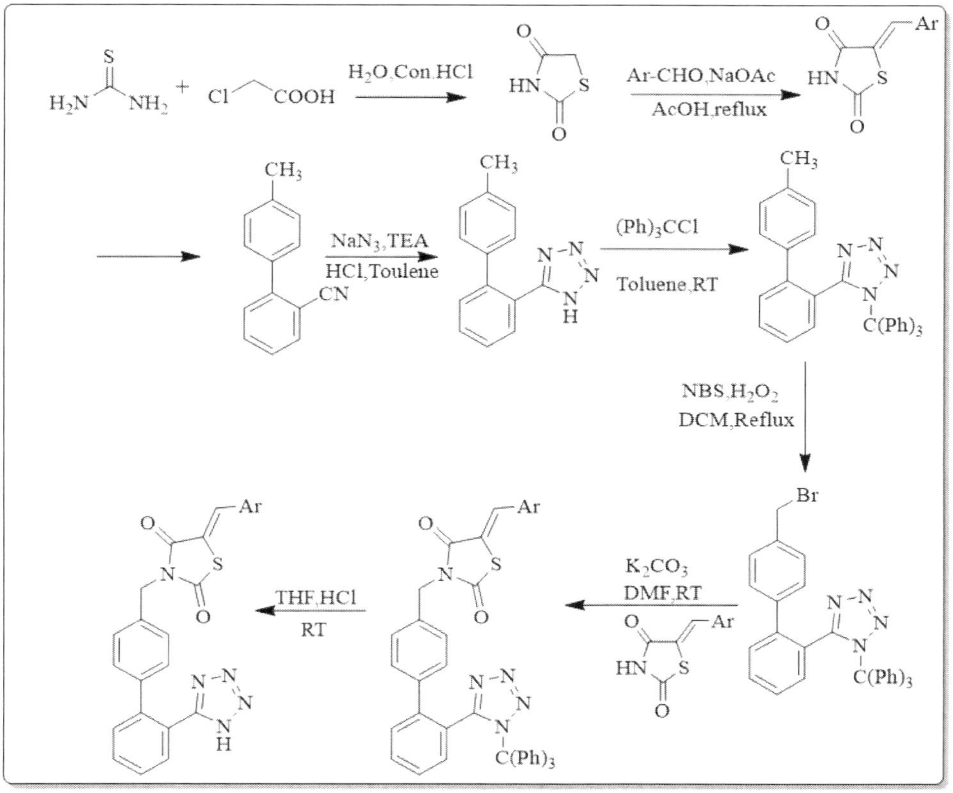

A novel approach to synthesizing tetrazole from Fmoc/Cbz protected amin acid derived nitrile, sodium azide, and micro MgO as a catalyst in the presence of methanol as a solvent was described by Lalithamba *et al*. (**Figure 5.2**). It was demonstrated that nano MgO was an effective heterogeneous catalyst. Using fluconazole as a reference, the synthesized compounds were tested against Aspergillus niger (A.Niger-ATCC No. 36607) for antifungal activity.

Figure 46. Synthesis of peptidyl tetrazole using nano MgO.

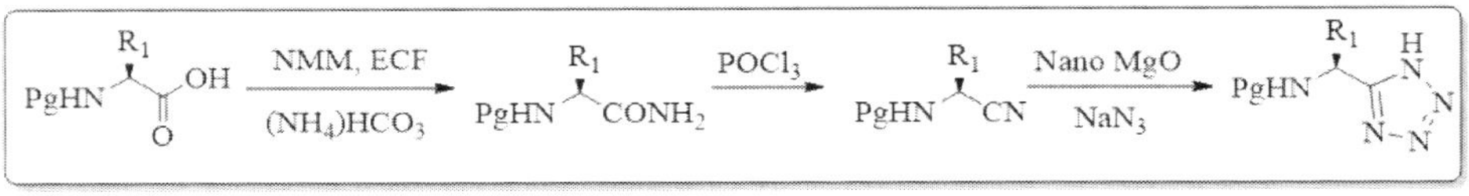

Figure 47. Different derivatives of tetrazoles.

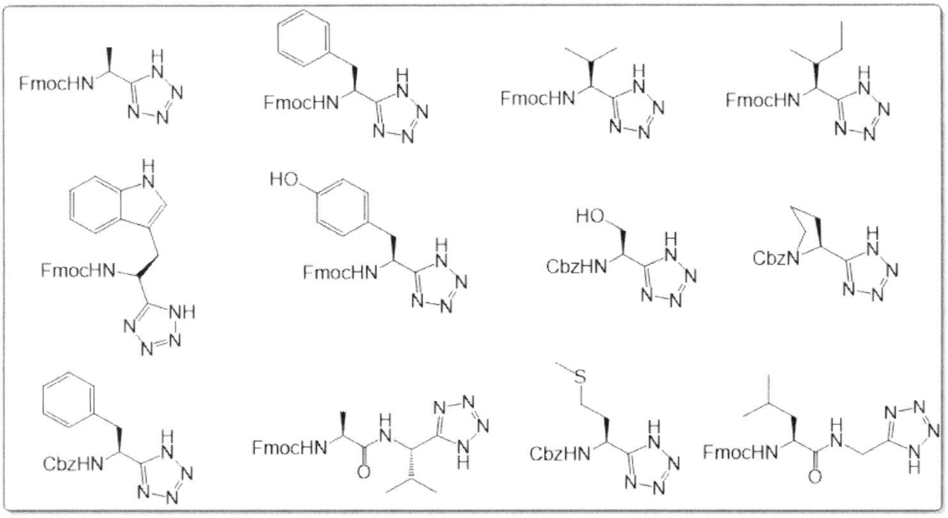

CONCLUSION

In conclusion, the authors demonstrate the protocols available for the construction of five-membered heterocycle tetrazole. Various possible approaches and different building blocks that can be used for the synthesis are highlighted in the chapter. From the classical azide-nitrile condensation route to most recent catalytic, it is most useful for the organic chemists who are leading to construct the tetrazoles and also who work on the biological applications of *N*-heterocycles.

REFERENCES

Abdollahi, A. M., & Ali, M. (2016). Cu-MCM-41 nanoparticles: An efficient catalyst for the synthesis of 5-substituted 1*H*-tetrazoles *via* [3+ 2] cycloaddition reaction of nitriles and sodium azide. *Journal of Chemical Sciences*, 128(1), 93–99. DOI: 10.1007/s12039-015-1005-9

Abrishami, F., Maryam, E., & Fatemeh, R. (2018). Nano TiO_2–H_3BO_3 as an efficient and recyclable catalyst for the synthesis of 5-substituted-1*H*-tetrazoles. *Iranian Journal of Catalysis*, 8(2), 103–111.

Agawane, S. M., & Jayashree, M. N. (2012). Synthesis of 5-substituted 1*H*-tetrazoles using a nano ZnO/Co_3O_4 catalyst. *Catalysis Science & Technology*, 2(7), 1324–1327. DOI: 10.1039/c2cy20094e

Akhlaghinia, B., & Soodabeh, R. (2012). A novel approach for the synthesis of 5-substituted-1*H*-tetrazoles. *Journal of the Brazilian Chemical Society*, 23(12), 2197–2203. DOI: 10.1590/S0103-50532013005000005

Amantini, D., Romina, B., Francesco, F., Ferdinando, P., & Luigi, V. (2004). TBAF-catalyzed synthesis of 5-substituted 1*H*-tetrazoles under solventless conditions. *The Journal of Organic Chemistry*, 69(8), 2896–2898. DOI: 10.1021/jo0499468 PMID: 15074950

Angelica, D. F., Barreto, S., Veronica, A. D. S., & Carlos, K. Z. A. (2017). Consecutive hydrazino-Ugi-azide reactions: Synthesis of acylhydrazines bearing 1, 5-disubstituted tetrazoles. *Beilstein Journal of Organic Chemistry*, 13(1), 2596–2602. PMID: 29259669

Bauer, L., Benz, M., Klapötke, T. M., Pignot, C., & Stierstorfer, J. (2022). Combining the most suitable energetic tetrazole and triazole moieties: Synthesis and characterization of 5-(1-hydroxy-3-nitro-1,2,4-triazol-5-yl)-1-hydroxy-tetrazole and its nitrogen-rich ionic derivatives. *Materials Advances*, 3(9), 3945–3951. DOI: 10.1039/D2MA00135G

Bhandari, D., Rafiq, S., Gat, Y., Gat, P., Waghmare, R., & Kumar, V. (2020). A Review on Bioactive Peptides: Physiological Functions, Bioavailability and Safety. *International Journal of Peptide Research and Therapeutics*, 26(1), 139–150. DOI: 10.1007/s10989-019-09823-5

Bonnamour, J., & Carsten, B. (2009). Iron salts in the catalyzed synthesis of 5-Substituted 1*H*-Tetrazoles. *Chemistry (Weinheim an der Bergstrasse, Germany)*, 15(18), 4543–4545. DOI: 10.1002/chem.200900169 PMID: 19296482

Cyranski, M. K., Miroslaw, G., Mariusz, J., & Tadeusz, M. K. (2003). On the Aromatic Character of the Heterocyclic Bases of DNA and RNA. *The Journal of Organic Chemistry*, 68(22), 8607–8613. DOI: 10.1021/jo034760e PMID: 14575493

Darvish, F., & Shima, K. (2015). FeCl$_3$ Catalyzed One pot synthesis of 1-Substituted 1*H*-1, 2, 3, 4-tetrazoles under solvent-free conditions. *International Journal of Organic Chemistry*, 5(2), 75–75. DOI: 10.4236/ijoc.2015.52009

Demko, Z. P., & Barry, K. S. (2001). Preparation of 5-substituted 1*H*-tetrazoles from nitriles in water. *The Journal of Organic Chemistry*, 66(24), 7945–7950. DOI: 10.1021/jo010635w PMID: 11722189

Demko, Z. P., & Sharpless, B. K. (2002). An expedient route to the tetrazole analogues of α-amino acids. *Organic Letters*, 4(15), 2525–2527. DOI: 10.1021/ol020096x PMID: 12123367

Dighe, S. N., Jain, K. S., & Srinivasan, K. V. (2009). Tetrahedron Lett 50(45):6139-6142.Sardarian AR, Eslahi H, Esmaeilpour M (2018). *ChemistrySelect*, 3(5), 1499–1511.

Dolensky, J., Hinteregger, C., Leitner, A., Seebacher, W., Saf, R., Belaj, F., Mäser, P., Kaiser, M., & Weis, R. (2022). Antiprotozoal Activity of Azabicyclo-Nonanes Linked to Tetrazole or Sulfonamide Cores. *Molecules (Basel, Switzerland)*, 27(19), 6217–6231. DOI: 10.3390/molecules27196217 PMID: 36234752

Dua, R., Suman, S., Sonwane, S. K., & Srivastava, S. K. (2011). Pharmacological significance of synthetic heterocycles scaffold: A review. *Advances in Biological Research (Faisalabad)*, 5(3), 120–144.

Freidinger, R. M. (2003). Design and Synthesis of Novel Bioactive Peptides and Peptidomimetics. *Journal of Medicinal Chemistry*, 46(26), 5553–5566. DOI: 10.1021/jm030484k PMID: 14667208

Ganguly, A., Sharma, K., & Majumder, K. (2020). Biopolymer-Based Formulations. *Biomedical and Food Applications*, 10, 87–104.

Gavrilyuk, J. I., Evindar, G., Chen, J. Y., & Batey, R. A. (2007). Peptide-Heterocycle Hybrid Molecules: Solid-Phase-Supported Synthesis of Substituted *N*-Terminal 5-Aminotetrazole Peptides via Electrocyclization of Peptidic Imidoylazides. *Journal of Combinatorial Chemistry*, 9(4), 644–651. DOI: 10.1021/cc060119p PMID: 17580974

. Ghamari kargar P, Bagherzade G (2021) RSC Adv 11(31):19203-23206.

Ghisalberti, E. L., Marcello, P., & Elizabeth, A. (1998). Survey of secondary plant metabolites with cardiovascular activity. *Pharmaceutical Biology*, 36(4), 237–279. DOI: 10.1076/phbi.36.4.237.4583

Giannis, A., & Kolter, T. (1993). Peptidomimetics for Receptor Ligands—Discovery, Development, and Medical Perspectives. *Angewandte Chemie International Edition in English*, 32(9), 1244–1267. DOI: 10.1002/anie.199312441

Grzonka, Z., & Liberek, B. (1971).. . *Roczniki Chem*, 45, 967–980.

Hajra, S., Debarshi, S., & Manishabrata, B. (2007). Metal triflate catalyzed reactions of alkenes, NBS, nitriles, and $TMSN_3$: Synthesis of 1, 5-disubstituted tetrazoles. *The Journal of Organic Chemistry*, 72(5), 1852–1855. DOI: 10.1021/jo062432j PMID: 17266377

Hersh, J., Broyles, D., Capcha, J. M. C., Dikici, E., Shehadeh, L. A., Daunert, S., & Deo, S. (2021). Peptide-Modified Biopolymers for Biomedical Applications. *ACS Applied Bio Materials*, 4(1), 229–251. DOI: 10.1021/acsabm.0c01145 PMID: 34250454

Himo, F., Zachary, P. D., Louis, N., & Barry, K. S. (2002). Mechanisms of tetrazole formation by addition of azide to nitriles. *Journal of the American Chemical Society*, 124(41), 12210–12216. DOI: 10.1021/ja0206644 PMID: 12371861

Ishihara, K., Ishihara, K., Tanaka, Y., Shioiri, T., & Matsugi, M. (2022). Practical synthesis of tetrazoles from amides and phosphorazidates in the presence of aromatic bases. *Tetrahedron*, 108, 132642–132651. DOI: 10.1016/j.tet.2022.132642

Jawabrah Al-Hourani, B., Sharma, S. K., Kaur, J., & Wuest, F. (2015). Synthesis, bioassay studies, and molecular docking of novel 5-substituted 1H tetrazoles as cyclooxygenase-2 (COX-2) inhibitors. *Medicinal Chemistry Research*, 24(1), 78–75. DOI: 10.1007/s00044-014-1102-1

Jeminejs, A., Goliškina, S. M., Novosjolova, I., Stepanovs, D., Bizdēna, Ē., & Turks, M. (2021).. . *Synthesis*, 53, 1379–1530. DOI: 10.1055/a-1348-9122

Jimmidi, R. (2023). Synthesis and Applications of Peptides and Peptidomimetics in Drug Discovery. *European Journal of Organic Chemistry*, 26(18), e202300028. DOI: 10.1002/ejoc.202300028

Jinghui, H., Baojun, L., Fasheng, C., Zheng, X., & Gui, Y. (2009). Tungstates: Novel heterogeneous catalysts for the synthesis of 5-substituted 1*H*-tetrazoles. *Journal of Molecular Catalysis A Chemical*, 304(1-2), 135–138. DOI: 10.1016/j.molcata.2009.01.037

Kabir, E., & Uzzaman, M. (2022). A review on biological and medicinal impact of heterocyclic compounds. *Results in Chemistry*, 4, 100606–100617. DOI: 10.1016/j.rechem.2022.100606

Khan, K., Jadhav, K. S., Patil, R. H., Shinde, D. B., Arote, R. B., & Sangshetti, J. N. (2016). Biphenyl tetrazole-thiazolidinediones as novel bacterial peptide deformylase inhibitors: Synthesis, biological evaluations and molecular docking study. *Biomedicine and Pharmacotherapy*, 83, 1146–1153. DOI: 10.1016/j.biopha.2016.08.036 PMID: 27551762

Kimura, T., Shuto, D., Hamada, Y., Igawa, N., Kasai, S., Liu, P., Hidaka, K., Hamada, T., Hayashi, Y., & Kiso, Y. (2005). Design and synthesis of highly active Alzheimer's β-secretase (BACE1) inhibitors, KMI-420 and KMI-429, with enhanced chemical stability. *Bioorganic & Medicinal Chemistry Letters*, 15(1), 211–215. DOI: 10.1016/j.bmcl.2004.09.090 PMID: 15582441

Kumar, A., Ramamurthi, N., & Harold, S. (1996). Rearrangement reactions of (hydroxyphenyl) carbenes. *The Journal of Organic Chemistry*, 61(13), 4462–4465. DOI: 10.1021/jo952269k PMID: 11667354

Lakshmi Kantam, M., Balasubrahmanyam, V., & Kumar, K. B. S. (2006). Zinc hydroxyapatite-catalyzed efficient synthesis of 5-substituted 1*H*-tetrazoles. *Synthetic Communications*, 36(12), 1809–1814. DOI: 10.1080/00397910600619630

Lakshmi Kantam, M., Kumar, K. B. S., & Phani Raja, K. (2006). An efficient synthesis of 5-substituted 1*H*-tetrazoles using Zn/Al hydrotalcite catalyst. *Journal of Molecular Catalysis A Chemical*, 247(1-2), 186–188. DOI: 10.1016/j.molcata.2005.11.046

Lakshmi Kantam, M., Kumar, K. B. S., & Sridhar, C. (2005). Nanocrystalline ZnO as an Efficient Heterogeneous Catalyst for the Synthesis of 5-Substituted 1*H*-Tetrazoles. *Advanced Synthesis & Catalysis*, 347(9), 1212–1214. DOI: 10.1002/adsc.200505011

Lang, L., Baojun, L., Wei, L., Li, J., Zheng, X., & Gui, Y. (2010). Mesoporous ZnS nanospheres: A high activity heterogeneous catalyst for synthesis of 5-substituted 1*H*-tetrazoles from nitriles and sodium azide. *Chemical Communications*, 46(3), 448–450. DOI: 10.1039/B912284B PMID: 20066321

Larsen, R. D., King, A. O., Chen, C. Y., Corley, E. G., Foster, B. S., Roberts, F. E., Yang, C., Lieberman, D. R., Reamer, R. A., Tschaen, D. M., Verhoeven, T. R., Reider, P. J., Young, S. L., Rossano, L. T., Brookes, S., Meloni, D., Moore, J. R., & Arnett, J. F. (1994). Efficient Synthesis of Losartan, A Nonpeptide Angiotensin II Receptor Antagonist. *The Journal of Organic Chemistry*, 59(21), 6391–6394. DOI: 10.1021/jo00100a048

Madhusudana, R., Muthukur, B. G., & Mohamed, A. P. (2011). A versatile and an efficient synthesis of 5-substituted-1*H*-tetrazoles. *Journal of Chemical Sciences*, 123(1), 75–79. DOI: 10.1007/s12039-011-0065-8

Maleki, A., & Afshin, S. (2015). Synthesis of tetrazoles *via* isocyanide-based reactions. *RSC Advances*, 5(75), 60938–60955. DOI: 10.1039/C5RA11531K

Mani, P., Ashawani, K. S., & Satish, K. A. (2014). $AgNO_3$ catalyzed synthesis of 5-substituted-1*H*-tetrazole *via* [3+ 2] cycloaddition of nitriles and sodium azide. *Tetrahedron Letters*, 55(11), 1879–1882. DOI: 10.1016/j.tetlet.2014.01.117

Matthews, D. P., Jonathan, E. G., & Anthony, J. S. (2000). Parallel synthesis of alkyl tetrazole derivatives using solid support chemistry. *Journal of Combinatorial Chemistry*, 2(1), 19–23. DOI: 10.1021/cc990035z PMID: 10813880

McGuire, J. J., Russell, C. A., Bolanowska, W. E., Freitag, C. M., Jones, C. S., & Kalman, T. I. (1990).. . *Cancer Research*, 50, 1726–1731. PMID: 2306727

McManus, J. M., & Herbst, R. M. (1959). Tetrazole Analogs of Amino Acids. *The Journal of Organic Chemistry*, 24(11), 1643–1649. DOI: 10.1021/jo01093a006

Méndez, Y., De Armas, G., Pérez, I., Rojas, T., Valdés-Tresanco, M. E., Izquierdo, M., Alonso del Rivero, M., Álvarez-Ginarte, Y. M., Valiente, P. A., Soto, C., de León, L., Vasco, A. V., Scott, W. L., Westermann, B., González-Bacerio, J., & Rivera, D. G. (2019). Discovery of potent and selective inhibitors of the Escherichia coli M1-aminopeptidase via multicomponent solid-phase synthesis of tetrazole-peptidomimetics. *European Journal of Medicinal Chemistry*, 163, 481–499. DOI: 10.1016/j.ejmech.2018.11.074 PMID: 30544037

Modarresi, A., Ali, R., & Mahmoud, N. (2009). Synthesis of 5-arylamino-1*H* (2*H*)-tetrazoles and 5-amino-1-aryl-1*H*-tetrazoles from secondary arylcyanamides in glacial acetic acid: A simple and efficient method. *Turkish Journal of Chemistry*, 33(2), 267–280. DOI: 10.3906/kim-0808-44

Morley, J. S. (1969). Polypeptides. Part XI. Tetrazole analogues of the C-terminal tetrapeptide amide sequence of the gastrins. *Journal of the Chemical Society, Section C: Organic*, 5(5), 809–813. DOI: 10.1039/j39690000809

Moutevelis-Minakakis, P., Filippakou, M., Sinanoglou, C., & Kokotos, G. (2006). Synthesis of tetrazole analogs of γ- and δ-amino acids. *Journal of Peptide Science*, 12(6), 377–382. DOI: 10.1002/psc.737 PMID: 16432805

Nagendra, G., Narendra, N., & Vommina Sureshbabu, V. (2012).. . *Indian Journal of Chemistry*, 51, 486–492.

Nasrollahzadeh, M., Yadollah, B., Davood, H., & Saeed, M. (2009). $FeCl_3$–SiO_2 as a reusable heterogeneous catalyst for the synthesis of 5-substituted 1*H*-tetrazoles *via* [2+3] cycloaddition of nitriles and sodium azide. *Tetrahedron Letters*, 50(31), 4435–4438. DOI: 10.1016/j.tetlet.2009.05.048

Nasrollahzadeh, N., Ghorbannezhad, F., & Sajadi, S. M. (2019). Biosynthesis of Pd/MnO_2 nanocomposite using *Solanum melongena* plant extract and its application for the one-pot synthesis of 5-substituted 1 *H* -tetrazoles from aryl halides. *Applied Organometallic Chemistry*, 33(1), e4698. DOI: 10.1002/aoc.4698

Olga, G. M., & Carsten, B. (2007). Synthesis of *N*-(1*H*)-tetrazole sulfoximines. *Organic Letters*, 9(15), 2951–2954. DOI: 10.1021/ol071302+ PMID: 17595098

Patil, P., Kurpiewska, K., Kalinowska-Tłuścik, J., & Dömling, A. (2017). Ammonia-Promoted One-Pot Tetrazolopiperidinone Synthesis by Ugi Reaction. *ACS Combinatorial Science*, 19(5), 343–350. DOI: 10.1021/acscombsci.7b00033 PMID: 28240545

Patil, U. B., Kumthekar, K. R., & Nagarkar, J. M. (2012). A novel method for the synthesis of 5-substituted 1H-tetrazole from oxime and sodium azide. *Tetrahedron Letters*, 53(29), 3706–3709. DOI: 10.1016/j.tetlet.2012.04.093

Poulie, C. B. M., & Bunch, L. (2013). Heterocycles as Nonclassical Bioisosteres of α-Amino Acids. *ChemMedChem*, 8(2), 205–215. DOI: 10.1002/cmdc.201200436 PMID: 23322633

Prabhu, G., Nagendra, G., Sagar, N. R., Pal, R., Guru Row, T. N., & Sureshbabu, V. V. (2016). A Facile Synthesis of 1,5-Disubstituted Tetrazole Peptidomimetics by Desulfurization/Electrocyclization of Thiopeptides. *Asian Journal of Organic Chemistry*, 5(1), 127–137. DOI: 10.1002/ajoc.201500384

Rajasekaran, A., & Thampi, P. P. (2004). Synthesis and analgesic evaluation of some 5-[*β*-(10-phenothiazinyl) ethyl]-1-(acyl)-1, 2, 3, 4-tetrazoles. *European Journal of Medicinal Chemistry*, 39(3), 273–279. DOI: 10.1016/j.ejmech.2003.11.016 PMID: 15051176

Saini, M. S., Aran, K., Jaya, D., & Rakesh, S. (2013). A review: Biological significances of heterocyclic compounds. *International Journal of Pharmaceutical Sciences and Research*, 4(3), 66–77.

Santhosh, L., Nagamangala, S. R., Thimmalapura, V. M., & Vommina, S. V. (2017). Synthesis of 1, 5- Disubstituted Tetrazole *via* Ugi Azide Reaction: An Asymmetric Induction Approach. *ChemistrySelect*, 2(20), 5497–5500. DOI: 10.1002/slct.201701032

Santhosh, L., Sagar, R. N., Vishwanatha, M. T., & Sureshbabu, V. V. (2017). Synthesis of 1, 5-disubstituted tetrazole *via* Ugi-azide reaction: An asymmetric induction approach. *ChemistrySelect*, 2(20), 5497–5500. DOI: 10.1002/slct.201701032

Sardarian, A. R., Eslahi, H., & Esmaeilpour, M. (2018). Copper(II) Complex Supported on Fe_3O_4 @SiO_2 Coated by Polyvinyl Alcohol as Reusable Nanocatalyst in *N* -Arylation of Amines and *N(H)* - Heterocycles and Green Synthesis of 1 *H* -Tetrazoles. *ChemistrySelect*, 3(5), 1499–1511. DOI: 10.1002/slct.201702452

Schaffert, E. S., Höfner, G., & Wanner, K. T. (2011). Aminomethyltetrazoles as potential inhibitors of the γ-aminobutyric acid transporters mGAT1–mGAT4: Synthesis and biological evaluation. *Bioorganic & Medicinal Chemistry*, 19(21), 6492–6504. DOI: 10.1016/j.bmc.2011.08.039 PMID: 21940173

. Shaaban S, Negm A, Ashmawy A, Ahmed DM, Wessjohann LA (2016) Elsevier 122:55-71.

Shelkar, R., Abhilash, S., & Jayashree, N. (2013). Amberlyst-15 catalyzed synthesis of 5-substituted 1-*H*-tetrazole *via* [3+ 2] cycloaddition of nitriles and sodium azide. *Tetrahedron Letters*, 54(1), 106–109. DOI: 10.1016/j.tetlet.2012.10.116

Shin, K., Tienan, J., Zhibao, H., Young, S. G., Jae, G. S., & Yoshinori, Y. (2003). Tetrazole synthesis *via* the palladium-catalyzed three component coupling reaction. *Molecular Diversity*, 6, 181–192. PMID: 15068080

Slyvka, Y., Goreshnik, E., Veryasov, G., Morozov, D., Andrii, A. F., & Nazariy, P. (2019). The novel copper(I) π,σ-complexes with 1-(aryl)-5-(allylthio)-1 *H* -tetrazoles: Synthesis, structure characterization, DFT-calculation and third-order nonlinear optics. *Journal of Coordination Chemistry*, 72(5-7), 1049–1058. DOI: 10.1080/00958972.2019.1580699

Smissman, E. E., Terada, A., & El-antably, S. (1976). Synthesis of inhibitors of bacterial cell wall biogenesis. Analogs of D-alanyl-D-alanine. *Journal of Medicinal Chemistry*, 19(1), 165–167. DOI: 10.1021/jm00223a030 PMID: 812991

Su, W. K., Zhi, H., Wei, G. S., & Xing, X. Z. (2006). A facile synthesis of 1-Substituted-1*H*-1, 2, 3, 4-tetrazoles catalyzed by Ytterbium triflate hydrate. *European Journal of Organic Chemistry*, 2006(12), 2723–2726. DOI: 10.1002/ejoc.200600007

Sureshbabu, V. V., Vasantha, B., & Hemanth, H. P. (2011). Synthesis of N-Fmoc-Protected Amino Alkyl Thiocyanates/Selenocyanates and Their Application in the Preparation of 5-Substituted S/Se-Linked Tetrazoles. *Synthesis*, 9(9), 1447–1455. DOI: 10.1055/s-0030-1259972

Venkateshwarlu, G., Premalatha, A., Rajanna, K. C., & Saiprakash, P. K. (2009). Cadmium chloride as an efficient catalyst for neat synthesis of 5-substituted 1H-tetrazoles. *Synthetic Communications*, 39(24), 4479–4485. DOI: 10.1080/00397910902917682

Venkateshwarlu, G., Rajanna, K. C., & Saiprakash, P. K. (2009). Antimony Trioxide as an Efficient Lewis Acid Catalyst for the Synthesis of 5-Substituted 1H-Tetrazoles. *Synthetic Communications*, 39(3), 426–432. DOI: 10.1080/00397910802378381

Verma, F., Sahu, A., Singh, P., Rai, A., Singh, M., & Rai, V. K. (2018). Visible-light driven regioselective synthesis of 1 H -tetrazoles from aldehydes through isocyanide-based [3 + 2] cycloaddition. *Green Chemistry*, 20(16), 3783–3789. DOI: 10.1039/C8GC01321G

. Wang H, Wang Y, Han Y, Zhao W, Wang X (2020) Green chm. 10:784-789.

Wang, Y., Patil, P., Kurpiewska, K., Kalinowska-Tluscik, J., & Dömling, A. (2017). Two Cycles with One Catch: Hydrazine in Ugi 4-CR and Its Postcyclizations. *ACS Combinatorial Science*, 19(3), 193–198. DOI: 10.1021/acscombsci.7b00009 PMID: 28181791

Wang, Z. (2023). Amino Acids. *Insights and Roles in Heterocyclic Chemistry*, 1, 414–432.

Zabrocki, J., Smith, G. D., Dunbar, J. B., Iijima, H., & Marshall, G. (1988). Conformational mimicry. 1. 1,5-Disubstituted tetrazole ring as a surrogate for the cis amide bond. *Journal of the American Chemical Society*, 110(17), 5875–5880. DOI: 10.1021/ja00225a045

Zaky, A. A., Simal-Gandara, J., Eun, J. B., Shim, J. H., & Abd El-Aty, A. M. (2022). Bioactivities, Applications, Safety, and Health Benefits of Bioactive Peptides From Food and By-Products: A Review. *Frontiers in Nutrition*, 8, 1–18. DOI: 10.3389/fnut.2021.815640

Zhenting, D., Changmei, S., Youqiang, L., Yin, W., & Jing, L. (2012). Improved synthesis of 5-substituted 1H-tetrazoles *via* the [3+2] cycloaddition of nitriles and sodium azide catalyzed by silica sulfuric acid. *International Journal of Molecular Sciences*, 13(4), 4696–4703. DOI: 10.3390/ijms13044696 PMID: 22606004

Chapter 11
Imidazo–Pyridines:
A Hybrid N–Heterocycles for Their Sustainable Synthetic Approaches and Significant Clinical Diversity

Kartik Sanghavi

RK University, India

Bonny Patel

RK University, India

Vijay Khedkar

Vishvkarma University, India

Khushal M. Kapadiya

https://orcid.org/0000-0001-9025-1177

RK University, India

ABSTRACT

The structural resemblance between the fused imidazopyridine heterocyclic ring system (a purine system) has prompted biological investigations to assess their potential therapeutic significance. They are known to play a crucial role in numerous disease conditions. In recent years, new preparative methods for the synthesis of imidazopyridines using various catalysts or non-catalytic systems have been described. In the present chapter, we summarise the recent approaches adopted for the synthesis of functionalized imidazo-pyridines over the last two decades along with their clinical advancement and applications. The key points adopted here including, traditional cyclo-condensation, reaction with nitro olefins, reaction with alkynes, 3-CCR (3-Component Condensation Reaction) based MCRs and miscellaneous aspects of imidazo-pyridine along with its biological importance have been

DOI: 10.4018/979-8-3693-7267-8.ch011

presented and discussed. This chapter will provide new initiatives to the chemists towards the synthesis of imidazo-pyridines and possible medicinal applications with the reported methods.

1.0 INTRODUCTION

Imidazopyridine the imidazole moiety fused with the pyridine ring' is an important class of biologically active nitrogen-containing heterocycle (Sanapalli *et al.,* 2022). Among the various imidazopyridine derivatives, imidazo[1,2-*a*] pyridine moiety is the most important in the field of natural products and pharmaceuticals (Pawar *et al.,* 2023). These derivatives show a wide range of biological activities such as antifungal, anti-inflammatory, antitumor, antiviral, antibacterial, antiprotozoal, antipyretic, analgesic, antiapoptotic, hypnoselective, and anxioselective activities (Geedkar *et al.,* 2022). There are several drugs such as zolpidem (used in the treatment of insomnia), alpidem (as an anxiolytic agent), olprinone (for the treatment of acute heart failure), zolimidine (used for the treatment of peptic ulcer), necopidem and saripidem (both work as an anxiolytic agent) are available in the market which contain imidazo[1,2-*a*] pyridine moiety (Panda *et al.,* 2022; Vanya *et al.,* 2021). The optically active **GSK812397** is a drug for the treatment of HIV infection (Jenkinson *et al.,* 2010). The antibiotic drug Rifaximin also contains this fused heterocyclic moiety **(Figure 1)** (Giorgio *et al.,* 2024). In addition, some abnormal N-heterocyclic carbenes are also prepared based on imidazo[1,2-*a*] pyridines (Hendriks *et al.,* 2015).

Figure 1. Drug candidates based on Imidazo-pyridine

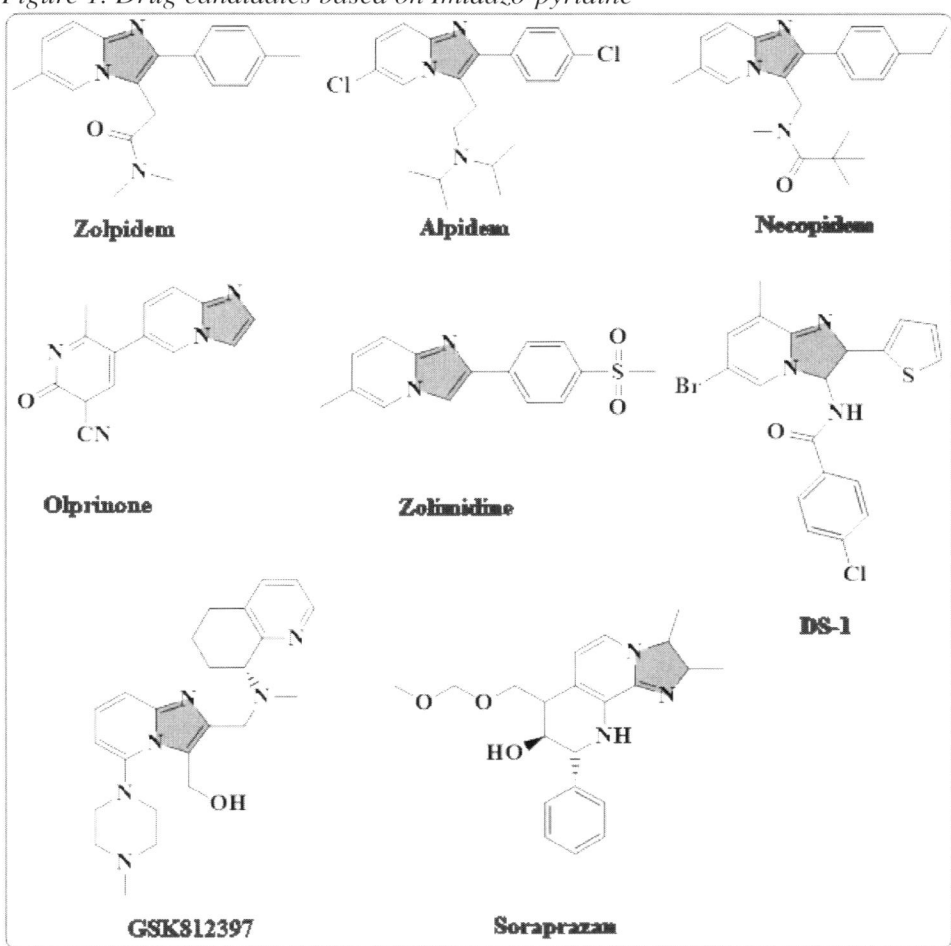

Several synthetic routes have been developed for imidazo[1,2-*a*] pyridines. Particularly, the last decade has witnessed a remarkable advancement in the synthesis of imidazo[1,2-*a*] pyridines by employing several interesting approaches, such as multicomponent reactions, tandem sequences, and transition-metal-catalysed C–H functionalization. These methods offer easy access to imidazo[1,2-*a*] pyridines from simple and readily available precursors (Olomiiets *et al.,* 2023). The present review is complimentary to that review and systematically summarizes various synthetic protocols for the synthesis of imidazo[1,2-*a*] pyridines and their biological importance.

2.0 SYNTHESIS OF IMIDAZO[1,2-*A*]PYRIDINES

2.1 Traditional Cyclo-Condensation Reactions

A new rouate has been developed by a catalyst-free method to synthesize imidazo[1,2-*a*] pyridines, pyrazines, and pyrimidines under microwave irradiation in green solvent. Studying with various solvents and mixtures, among them a mixture of water-IPA as a solvent, provides excellent yield. Figure 2. This method was efficient in tolerating different functional groups (Rao *et al.*, 2018).

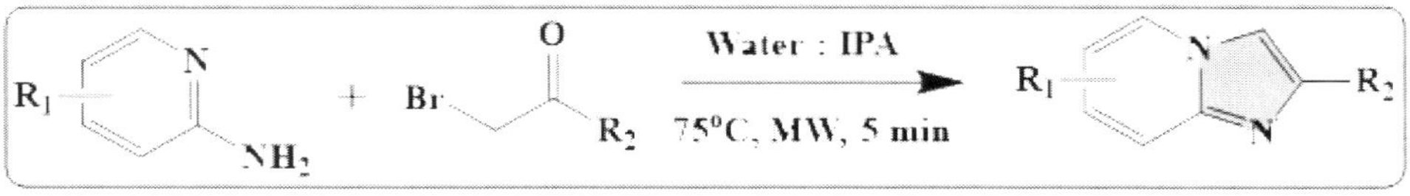

A group of researchers prepared analogues of Imidazo[1,2-*a*]pyridines by condensation of the substituted 2-aminopyridine with ethyl 2-chloro -acetoacetate in ethanol as a solvent at reflux temperature to give corresponding imidazo[1,2-*a*] pyridines (Ku *et al.*, 2012).

Figure 3. These were condensation with hydrazine hydrate to give respective hydrazone followed by reductive amination with 3,4-dimethoxy benzaldehyde in ethanol to obtain desired compounds in a novel structural framework.

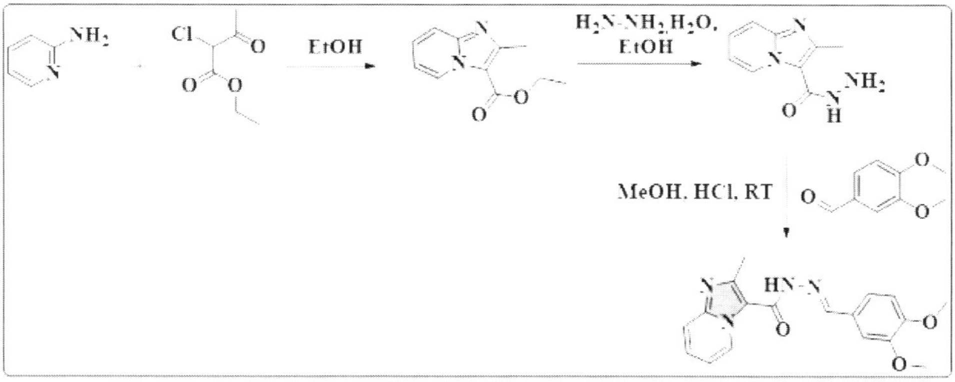

A library of 2,3-diaryl substituted imidazo[1,2-*a*]pyridines were prepared by one pot condensation and C-H Arylation. The reaction of 2-amino pyridines with appropriate alpha bromo ketone in the presence of bases like $NaHCO_3$ and PEG-400

(polyethylene glycol) as a solvent at 120°C by microwave irradiation for 10 mins, followed by arylation using Aryl halide, Palladium catalyst and Base like KOAc at 220°C for 2 hr by microwave irradiation to obtained derivatives of 2,3-diaryl substituted imidazo[1,2-a]pyridine (Hiebel *et al.*, 2014).

Figure 4. A method was devloped to synthesise imidazo[1,2-a]pyridine benzylic ethers using BCl₃ by formation of C-N, C-O and C-S bond (Singh et al., 2019).

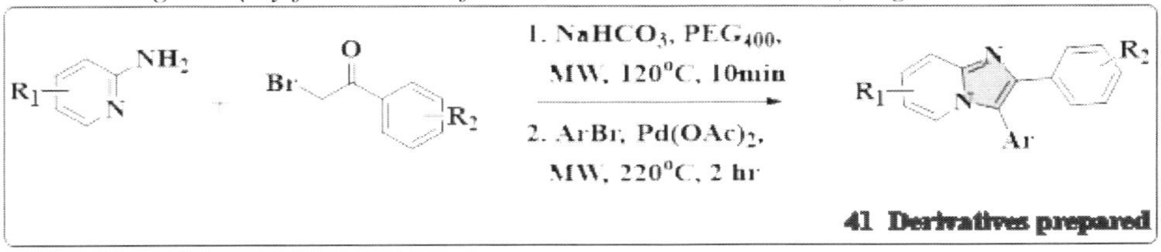

Figure 5. The Developed process is applicable for the synthesis of a wide variety of ((3-amino / thio/ alkoxy) -methyl)-imidazo[1,2-a]pyridines. The salient feature of this method corresponds to the substitution of different nucleophiles via in situ unconventional debenzylation.

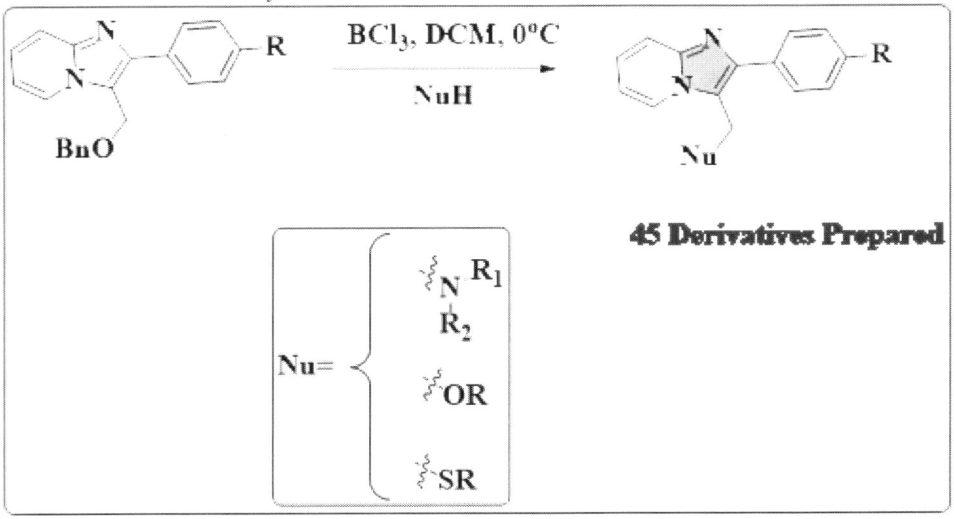

A series was synthesised on a novel imidazo[1,2-a]pyridine-2-carboxamide derivatives and their evaluation as an anti-mycobacterial (Jose *et al.*, 2015).

Figure 6. The imidazo[1,2-a]pyridine ring system was accomplished by the condensation of 2-amino-5-iodopyridine with ethyl bromopyruvate (α-halo carbonyl compound) in refluxing ethanol to give ethyl-6-iodo-H-imidazo[1,2-a]pyridine-2-carboxylate.

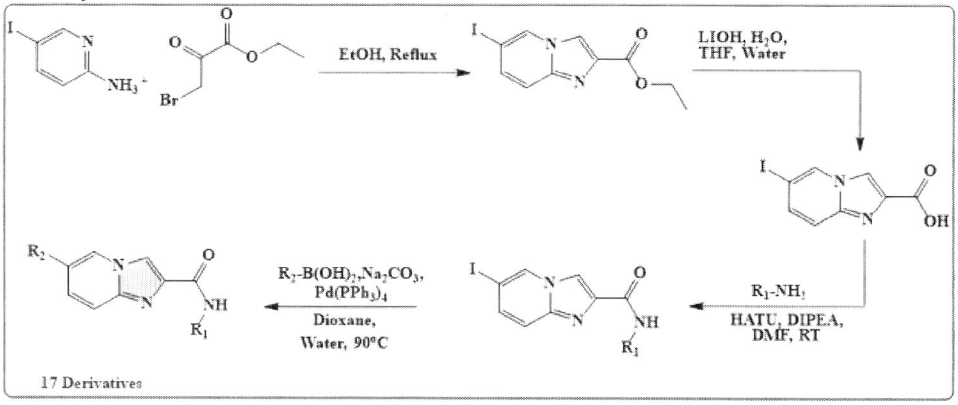

It was hydrolysed with Lithium hydroxide mono hydrate and followed by acid-amine coupling with different aromatic or aliphatic amines (R_1-NH_2) by using HATU, DIPEA in DMF solvent. The iodo was replaced with a different boronic acid or pinacol ester in the presence of Pd(0) catalyst, Na_2CO_3 in Dioxane solvent to yield imidazo[1,2-a]pyridine-2-carboxamide derivatives.

A diverse and straight forward synthesis of 2-phenyl imidazo[1,2-a]pyridine by condensation of 2-amino pyridine and 2-bromo acetophenone, sodium bicarbonate in acetonitrile as a solvent at reflux temperature.

Figure 7. They also developed their purification technique using crystallization and mapping the risk hazard of the reaction (Santaniello et al., 2017).

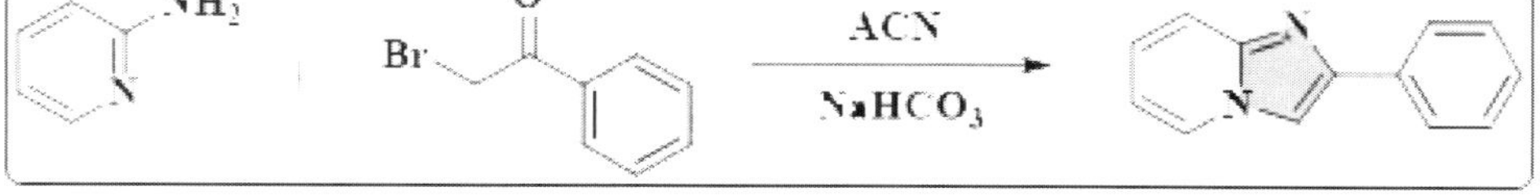

A process of solvent and metal-free synthesis of 2,3-substituted imidazo[1,2-a] pyridines and zolimidine under high-speed ball milling was developed. Under solvent-free high-speed ball milling, an I_2-promoted condensation/cyclization of substituted methyl ketone or 1,3-dicarbonyl compounds with substituted 2-amino pyridines in the presence of DMAP, which allows the quick assembly of 2,3,-substituted imidazo[1,2-*a*]pyridines. The advantages of high yields, different functional group compatibility, short time reaction, solvent, metal-free and heating-free (Wang *et al.*, 2016).

Figure 8. 22 and 7 derivatives

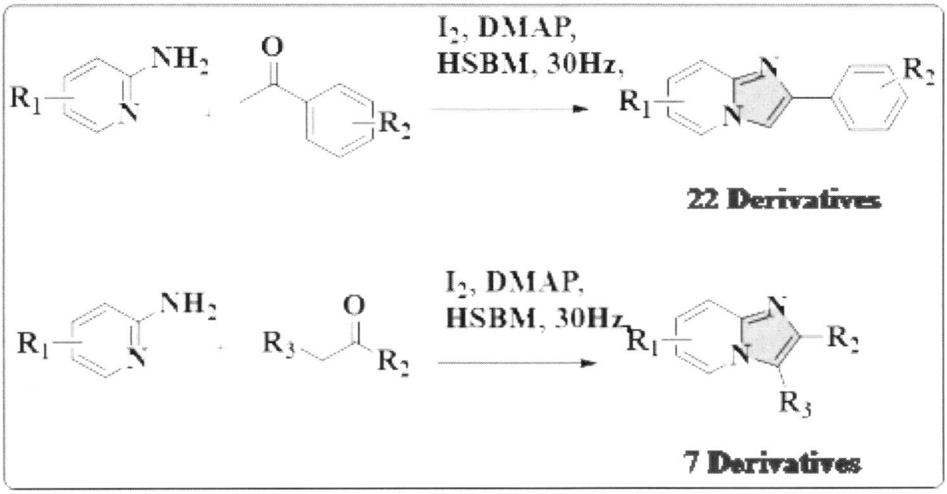

A metal-free synthesis of 3-formyl substituted imidazo[1,2-*a*] pyridines using substituted 2-amino pyridine and bromo malonaldehyde in ethanol: water (1:1) mixture at 110-100ºC by microwave irradiation has been put forwarded.

Figure 9. They studied different solvent and their ratio, the moderated yield and reduced reaction time found with ethanol: water 1:1 ratio (Damian et al., 2019).

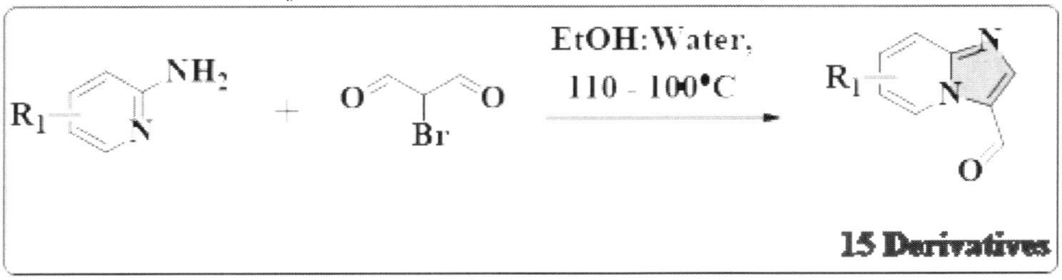

An imidazo[1,2-*a*] pyridine derivative by catalyst-free and solvent-free grindstone chemistry was developed by environment-friendly, timesaving (mostly within 3-7 mins) and high-yielding process.

Figure 10. This process favoured most of the functional groups (Godugu et al., 2021).

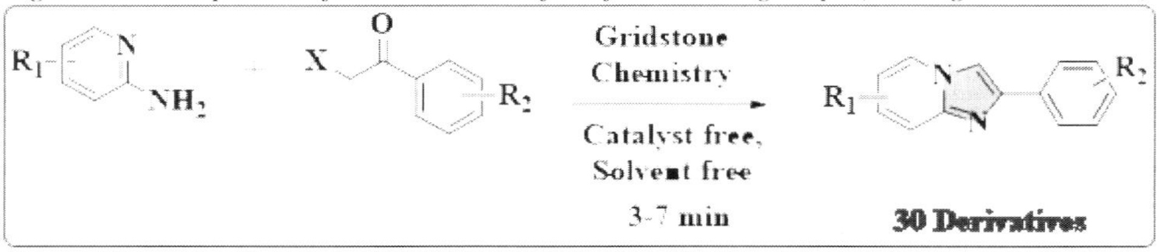

A green synthesis of imidazo[1,2-*a*]pyridines in the aqueous medium. 2-Amino pyridine and phenacyl bromide derivatives in aqueous medium by microwave irradiation have resulted in bridged azaheterocycles in good to excellent yields.

Figure 11. They also examined it as an antibacterial agent (Gui et al., 2021).

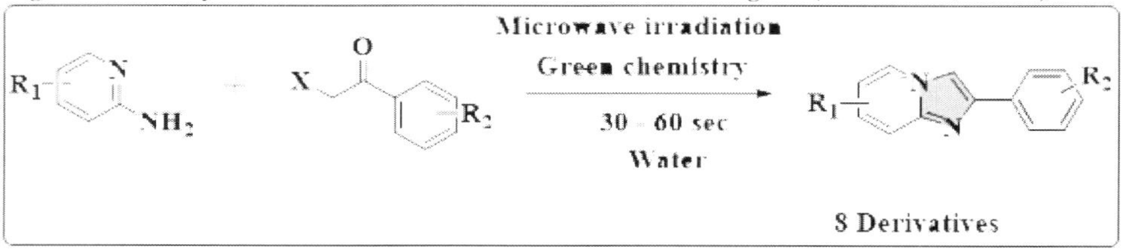

A new imidazo[1,2-*a*]pyridine derivatives have been synthesized using the traditional method of 2-amino pyridine and substituted α-halo acetoacetate followed by hydrolysis and amide coupling and exhibited bactericidal activity in the mouse *M. tuberculosis* infection model (Yong at al., 2014).

Figure 12.

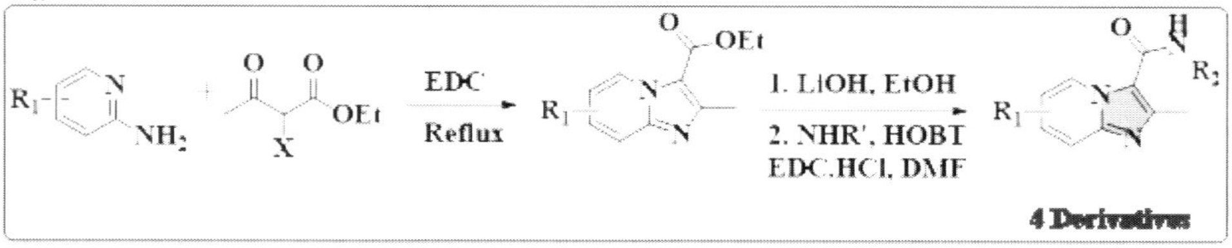

A group have discovered the synthesis of some imidazo[1,2-*a*]pyridine derivatives by using substituted 2-amino pyridine and α-bromo pyruvate in ethanol as solvent at reflux temperature. It was hydrolysed and followed by acid-amine coupling to obtain amide derivatives and evaluate their biological activity as an antiviral drug (Rani *et al.*, 2020).

Figure 13.

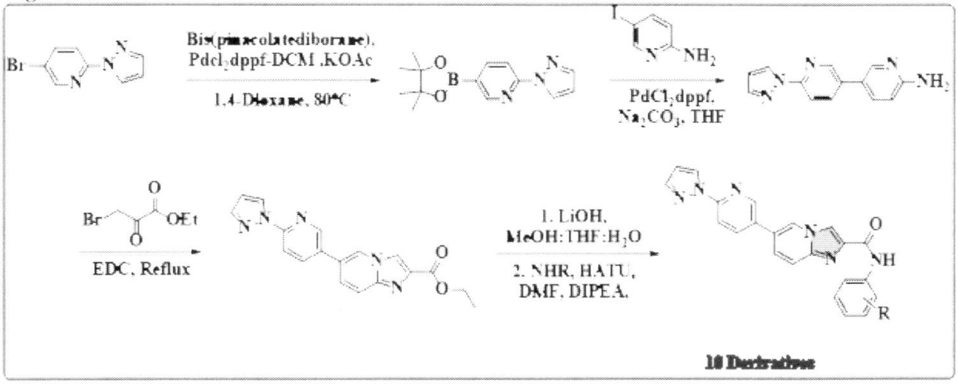

Newer 2-arylimidazo[1,2-*a*]pyridines were synthesized using substituted 2-amino pyridine and substituted acetophenone and iodine as a promoter in aqueous medium. The whole reaction was carried out in Micellar Media (which was prepared from Anionic/ Cationic surfactants added in water and followed by vigorous stirring) to reduce the solubility problem. Advantages of this reaction (water approach), moderated yield compared to in water approach, metal-free synthesis, very mild condition, wide substrate scope, functional group tolerance, co-effective and ease of scale-up preparation. Here, commonly used NH_4Cl as a surfactant and I_2 (Iodine) is a catalyst (Bhutia *et al.*, 2020).

Figure 14.

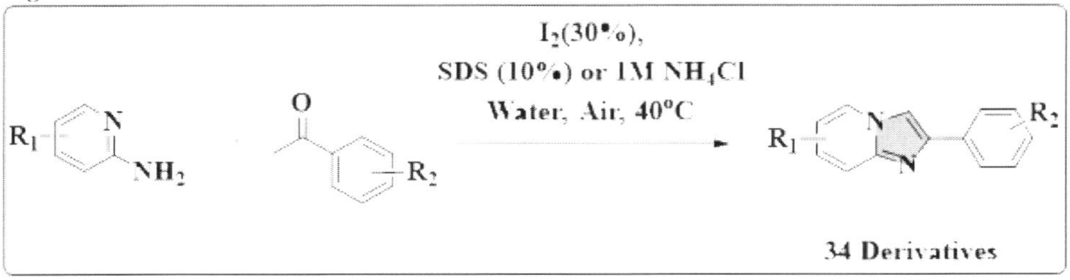

A method was developed for the synthesis of a marketed drug zolimidine and developed a green and efficient route to synthesise 2-aryl imidazo[1,2-*a*]pyridines using an electrical grinder. I_2 (Iodine) was used as the catalyst and mechanochemical grinding facilitates the cyclo-condensation reaction between various aryl methyl ketones and 2-amino pyridines to afford 2-aryl imidazo[1,2-a]pyridine in moderated yield at ambient temperature (Das *et al.,* 2020).

Figure 15.

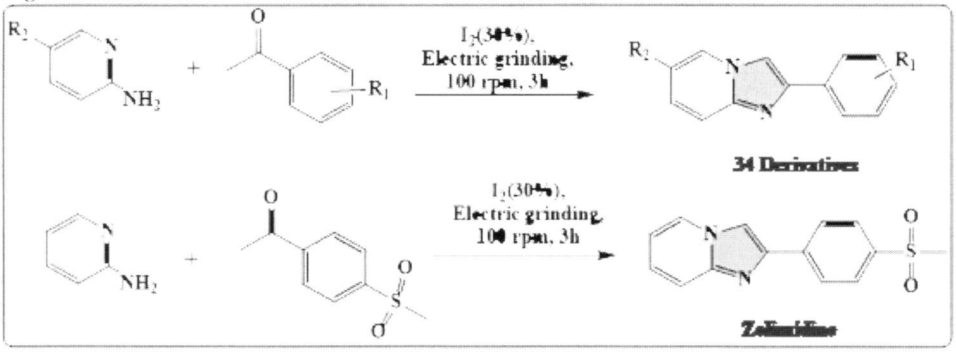

A process to synthesise imidazo[1,2-a]pyridines derivatives by cyclisation of ketones with 2-amino pyridines, via electrochemically initiated intermolecular C-N bond formation was introduced (Feng *et al.,* 2019). Hydroiodic acid is used as a catalyst, and it creates a redox mediator. The advantages of this reaction were A variety of ketones including acetophenones, unsaturated and alkyl ketones are amenable to this reaction, affording products in moderate to excellent yields. A three-component tandem reaction realizing C-N, C-S/C-Se bond formation can also be achieved under standard conditions. They studied various catalysts (like, NaI, TBAI, $BF_3.OEt$ and HI) and conditions, among them 25% HI with 2.0 V at 50°C temperature obtained the best yield.

Figure 16.

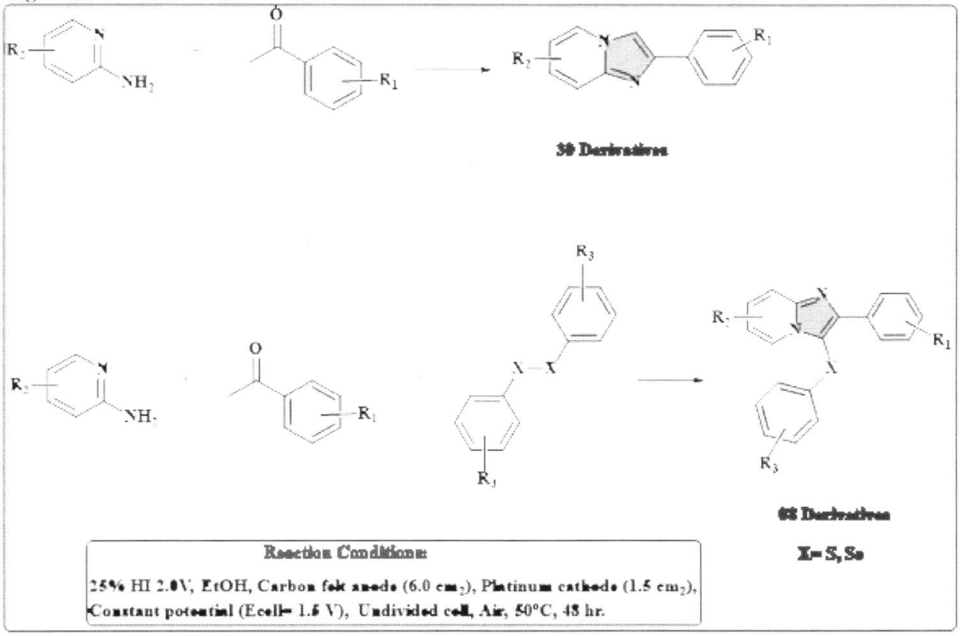

Figure 16.

2.2 Reaction With Nitro Olefins

A group discovered new process for one pot synthesis of imidazo[1,2-*a*] pyridines from aminopyridines and nitro Using Air as an Oxidant in presence of Cu(I) catalyst agent. Generally, CuI, CuBr, Cu(OTf)$_2$, CuCl used as a catalyst and DMSO, CAN, NMP, DMF and Me-OH were used in reaction. CuBr showed highest activity and high conversion rate (Yan *et al.*, 2012).

Figure 17.

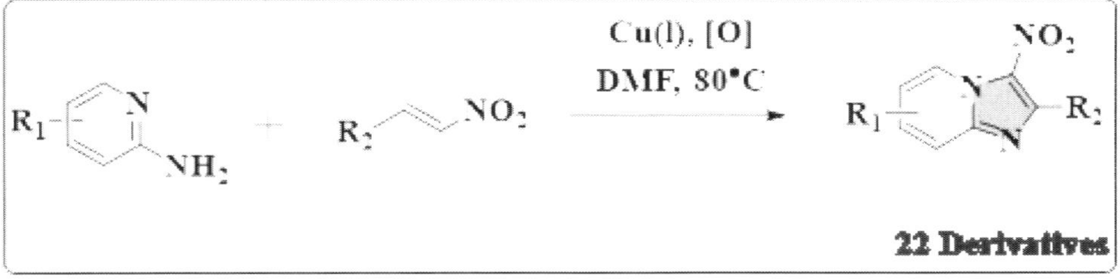

Figure 17.

A easy process for synthesised imidazo[1,2-*a*]pyridines derivatives by reaction between 2-amino pyridine and substituted nitro olefines in presence of Iron(III) as a catalyst was generated. The reaction proceeds through Michael addition followed by intramolecular cyclization and *in situ* de-nitration (Santra at al., 2013).

Figure 18.

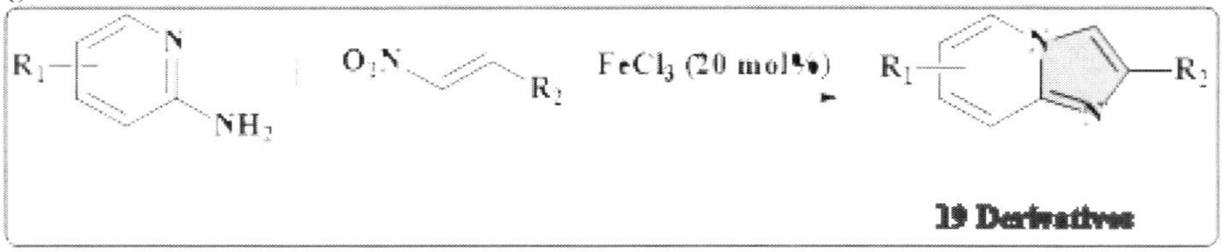

A metal free synthesis of imidazo-pyridine from nitroalkene and 2-aminopyridine in the presence of a catalytic amount of iodine and aqueous hydrogen peroxide was developed. DMSO was preferred solvent for this reaction (Tachikawa *et al.*, 2015).

Figure 19.

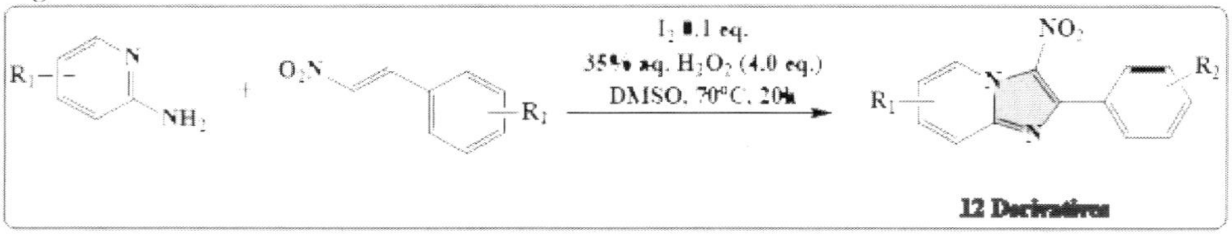

The imidazopyridines were obtained in good to high yield regardless of an electron-donating or -withdrawing group at the para- and meta positions of the aromatic nucleus of nitroalkenes. On the other hand, ortho-substituted nitroalkene gave low yield due to steric hindrance. 3, 4 or 5 substituted 2-aminopyridines, including methyl, chloro, and ester groups were good substrates in this reaction.

Oxidative coupling of amino pyridines with nitro-olefins was used to synthesized imidazo[1,2-*a*] pyridine derivatives using TBAI as the catalyst and TBHP as oxidation agent (Xu *et al.*, 2014).

Figure 20.

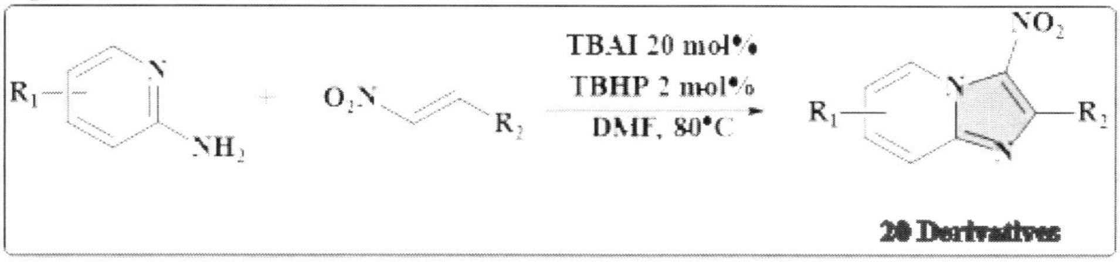

Also investigate the steric hindrance and electronic nature of the substituents on the aminopyridines ring had a pronounced effect on the efficiency of the oxidative coupling. e.g. aminopyridine with 6-methyl and electron-withdrawing chloride substituents are less effective. Reactions of aminopyridine with nitro-olefin were screened. Both electron-donating -Me, -i-Pr, -OMe substituents and electron withdrawing halide, $-CF_3$, $-CO_2Me$ substituents at para and meta position were compatible with the optimal conditions, whereas ortho-substituted nitroolefin did not undergo the cyclization oxidative coupling due to the steric effect.

A advanced synthesis of 3-nitro-2-arylimidazo[1,2-*a*]pyridines by reaction of 2-aminopyridines and nitro-styrene in the presence of di-chloro iodide ($NaICl_2$) was developed (Jagadhane *et al.,* 2014). Also performed out using different solvents such as DMSO, MeOH, EtOH, and MeCN, but very low yields were observed in each case and DMF was deemed the most suitable solvent for this reaction.

Figure 21.

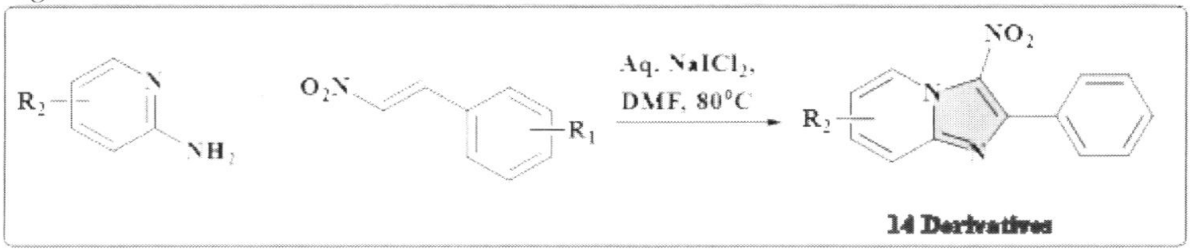

The process of super magnetic nanoparticle-catalysed coupling of 2-amino pyridines with trans-chalcones or nitro styrene reaction proceed in the presence of $CuFe_2O_4$ Super magnetic nano particle catalyst (commonly copper or iron complexes), Iodine, Oxygen as oxidant and 1,4-dioxane as preferred solvent was scrutinized (Nguyen *et al.,* 2019).

Figure 22.

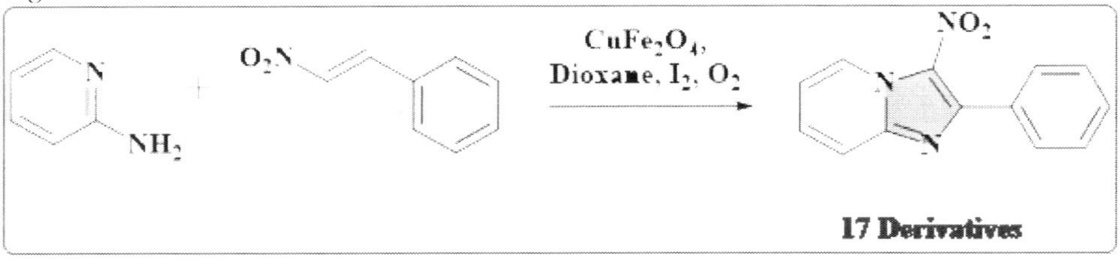

Strongly coordinated solvents such as DMF or DMSO are not suitable for this reaction, while chlorobenzene is inferior to 1,4-dioxane. No product was obtained if I_2 was omitted. Low yield was obtained if the amount of $CuFe_2O_4$ and using less than 2 equivalents of I_2. The coupling was not successful under argon atmosphere, which proving that oxygen oxidant is crucial.

A new approach to synthesis of imidazo[1,2-*a*]pyridine from phenyl acrylic acids. First to convert phenyl acrylic acid to nitro styrene using $Cu(NO_3)_2$ in acetonitrile. Then react with 2-amino pyridine derivatives to obtained derivatives of imidazo[1,2-*a*]pyridine (Mutkule *et al.*, 2020). They were also notified presence of $Cu(NO_3)_2$, where in absence of $Cu(NO_3)_2$ the reaction did not yield the desired product.

Figure 23.

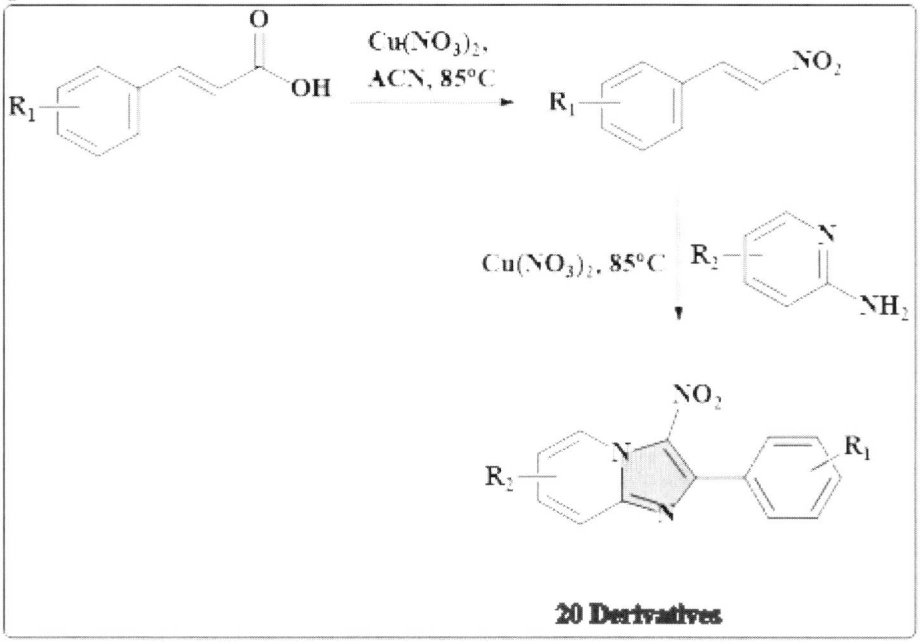

2.3 Reaction With Alkynes

An oxidative cross-coupling/cyclization to prepare heteroaromatic imidazo[1,2-*a*] pyridines derivatives was reported by group of scientists. They prepare novel silver-mediated highly selective C-H/N-H oxidative cross coupling/cyclization between 2-aminopyridines and terminal alkynes and Ag_2CO_3 (Yu *et al.*, 2012).

Figure 24.

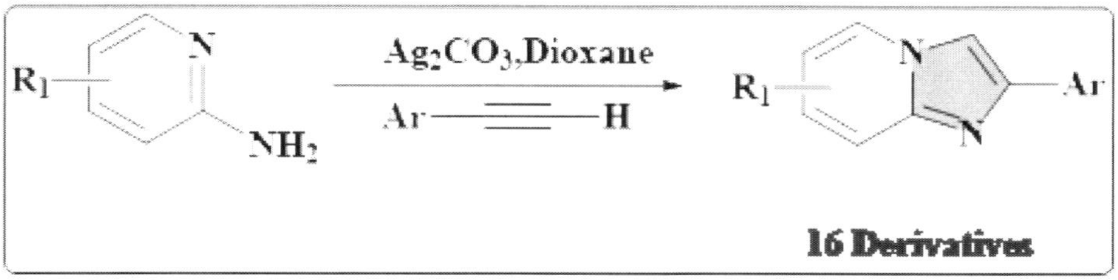

A library of 3-aryl-2-phosphinoimidazo[1,2-*a*]pyridine ligands were prepared, which was used in palladium – catalysed cross-coupling reactions (Tran *et al.*, 2019).

Figure 25.

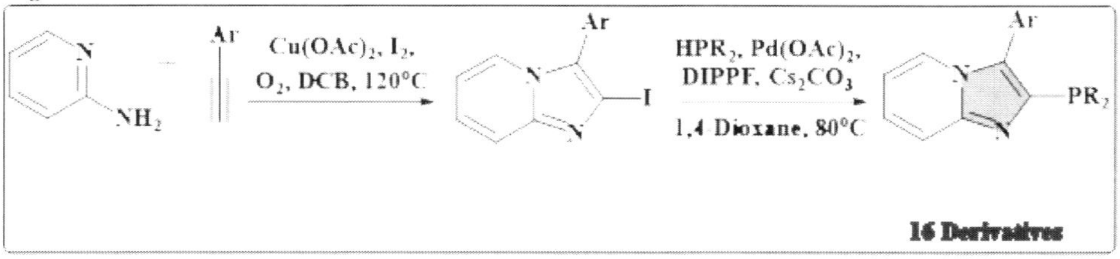

From the studies, among those some derivative of phosphorus ligand were gave excellent reaction conversion (>99%) with such bases like K_3PO_4, K_2CO_3, KOtBu, and NaOtBu. Ligand works better with $Pd(OAc)_2$, 1,4-dioxane and K_3PO_4 as a reagent.

A diverse method to synthesize imidazo[1,2-*a*]pyridine-thiones by lewis acid catalysed intermolecular annulation/cyclization by using three components like, substituted 2-amino pyridine, ynals and Sulphur were reported and optimized. Reaction obtained moderated yielded in $SC(OTf)_3$ used as lewis catalyst among of listed lewis acid and DMF as a solvent. Required more mol equivalents in use of other Lewis's acid catalyst like TEMPO, DPE and BHT. Advantages of this reaction are one pot synthesis, functional group tolerance, easy for scale-up preparation (Chen *et al.*, 2019).

Figure 26.

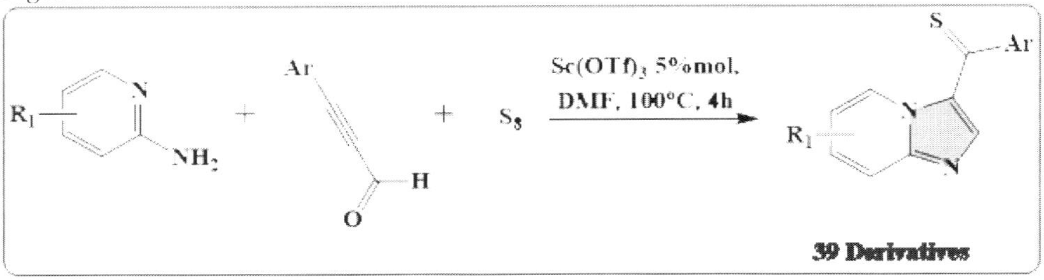

3.0 MULTI-COMPONENT REACTIONS

3.1 Three-Component Reaction of Aldehyde, 2-Aminopyridine and Isocyanide. (Groebke-Blackburn-Bienaymé Reaction)

Groebke and co-researchers were established novel synthesis method of 3-amino imidazo[1,2-*a*]pyridine. They discover highly versatile and efficient synthetic methods for a broad variety of chemical structures by condensation reaction of three component, 2-amino pyridine derivative, aldehyde and corresponding isonitrile in protic solvent. They examined and found that these 3-amino imidazo[1,2-*a*]pyridine derivatives were pharmacologically active compounds such as benzodiazepine receptor agonists, anti-inflammatory agents, inhibitors of gastric acid secretion, calcium channel blockers and antibacterial (Groebke *et al.,* 1998).

Figure 27.

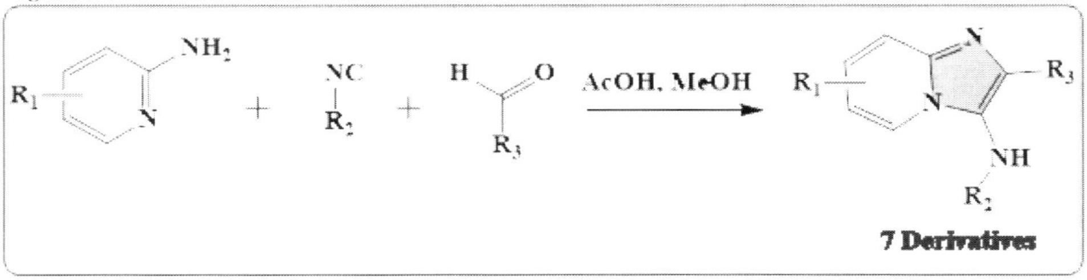

A process to synthesized imidazo[1,2-*a*]pyridines derivatives by using multicomponent with catalytic Zinc Chloride was prepared. Before this discovery, reaction was carried out in the presence of an acid catalyst, usually scandium (III) triflate, ammonium chloride, acetic acid, perchloric acid Montmorillonite clay K1013 to catalyse the reaction (Rousseau *et al.*, 2007).

Figure 28.

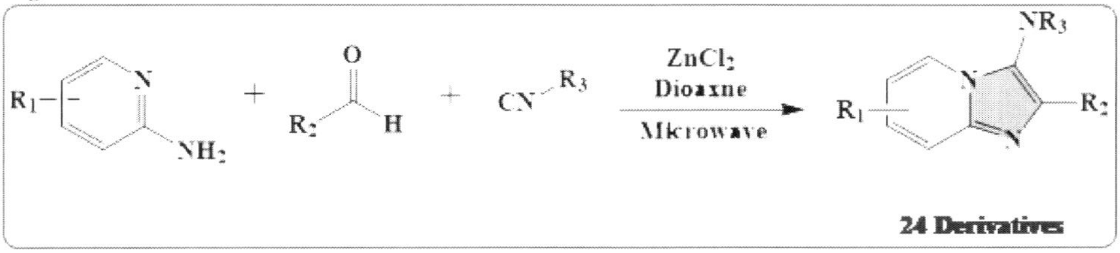

They have described the novel application of zinc chloride, a cheap catalyst, for the one-pot preparation of imidazo[1,2-*a*]pyridines using either conventional heating or microwave irradiation. In which 2-amino pyridine was treated with aldehydes, Zinc Chloride and derivative of isocyanates to yielded analogues of imidazo[1,2-*a*] pyridines.

A one pot synthesis of imidazo[1,2-*a*]pyridines derivatives from aryl halides or aryl tosylates with substituted 2-aminopyridines and isocyanides was developed and was examined as in the reaction benzyl halides or benzyl tosylates are oxidized to aldehydes under mild Kornblum conditions which then undergo a three-component reaction with various 2-aminopyridines and isocyanides to afford the imidazo[1,2-*a*] pyridines in excellent yields (Adib *et al.*, 2011).

Figure 29.

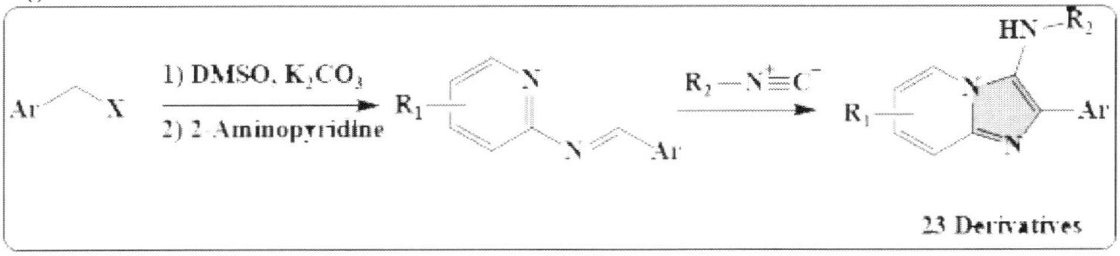

A new class of novel imidazo[1,2-*a*]pyridines and their *in vitro* evaluation as a antimicrobial agents was prepared. A substituted 2-aminopyridine or substituted 2-aminobenzothiazole condensed with substituted aldehyde in presence of $Sc(OTf)_3$ followed by the addition of the different phenyl isocyanide, yielded analogous of imidazo[1,2-*a*]pyridine compounds (Taleb *et al.*, 2011).

Figure 30.

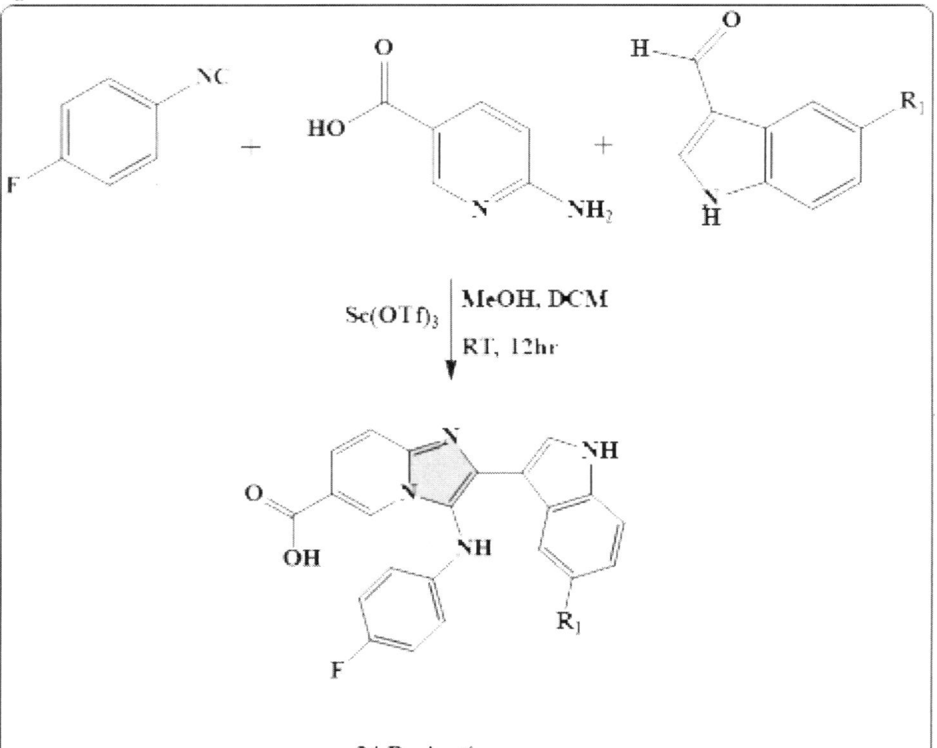

24 Derivatives

A different procedure to prepare imidazo[1,2-*a*]pyridine derivatives by GBB (Groebke-Blackburn-Bienayme) reaction using Biocatalytic Lipase enzyme was introduced by group of researchers. Among different lipase enzymes, *candida antarctica lipase B* (CALB) and *Aspergillus niger lipase* were found best to obtained moderated yield. MeOH and Ethanol were suitable solvent for the reaction. DMF, THF and Hexane were no product conversion, while H_2O and DMSO were showed minor product conversion. Donating group on 4th position of 2-amino pyridine favours the high conversion (Budhiraja *et al.,* 2020).

Figure 31.

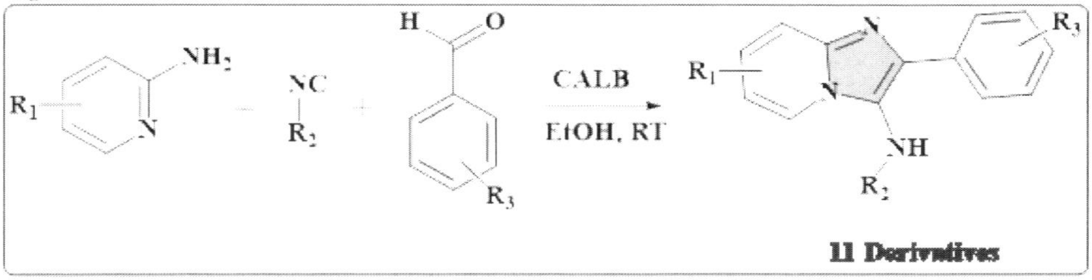

Two different sustainable and efficient one pot synthetic methods to synthesize various imidazo[1,2-*a*]pyridine derivatives were put forwarded by team of researchers. Earlier process has some disadvantage in water mediated organic synthesis was the limited solubility of hydrophobic organic reactants in water. This limitation was addressed using surfactants in water as it enhances the solubility and reactivity by the formation of micellar or vesicular system. These surfactants provide a hydrophobic micellar core for the water insoluble reactants and thereby promotes the micellar catalysis in water. In method-A, micellar condition using SDS (Sodium dodecyl sulphate) surfactant in water and in Method-B, catalysed by the organo-catalyst thiamine hydrochloride (VB1) under solvent-free condition to produce diverse imidazopyridine heterocycles (Mathavan *et al.,* 2020).

Figure 32.

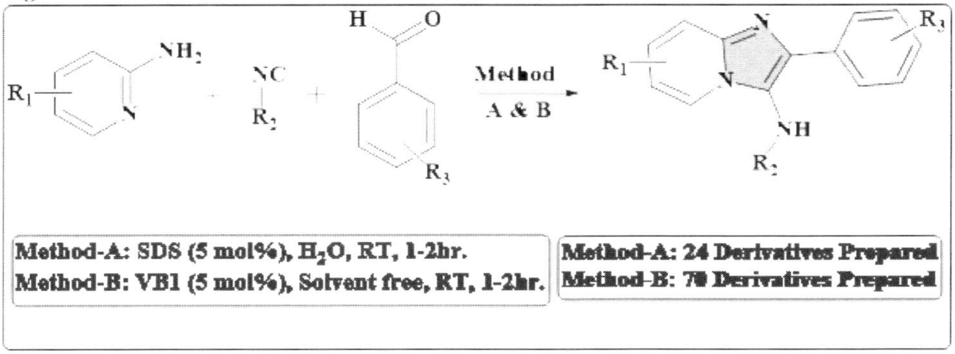

A library of 3-aminoimidazole[1,2-*a*]pyridines through the multicomponent reaction of benzylamine, 2-aminopyridine, and *t*-butyl isocyanide under visible light using eosin Y as a photocatalyst was prepared and optimized. This process also known as HAT (Hydrogen Atom Transfer) process. In photocatalysis, there are usually three modes of the HAT procins. In the first mode, the activated photocatalyst abstracts a hydrogen atom from the substrate. The catalytic cycle is then turned over to a newly produced intermediate via a reverse HAT (RHAT). In Second mode, one more catalyst was activated by the photocatalyst, which was already excited. After activation, one more catalyst stimulates the reaction through the hydrogen atom transfer pathway. In 3rd mode, the proton-coupled electron transfer (PCET) process, which involves the coordinated electron transfer and proton transfer from the reagent. Eosin Y possibly will be the best hydrogen atom transfer photocatalyst and may abstract proton from benzylic C−H from benzylamine. The process has many advantages like, one pot synthesis, metal free and environment friendly, mild reaction condition, enumerating tolerance of an extensive range of electron-donating and electron-withdrawing groups (Singh *et al.*, 2020).

Figure 33.

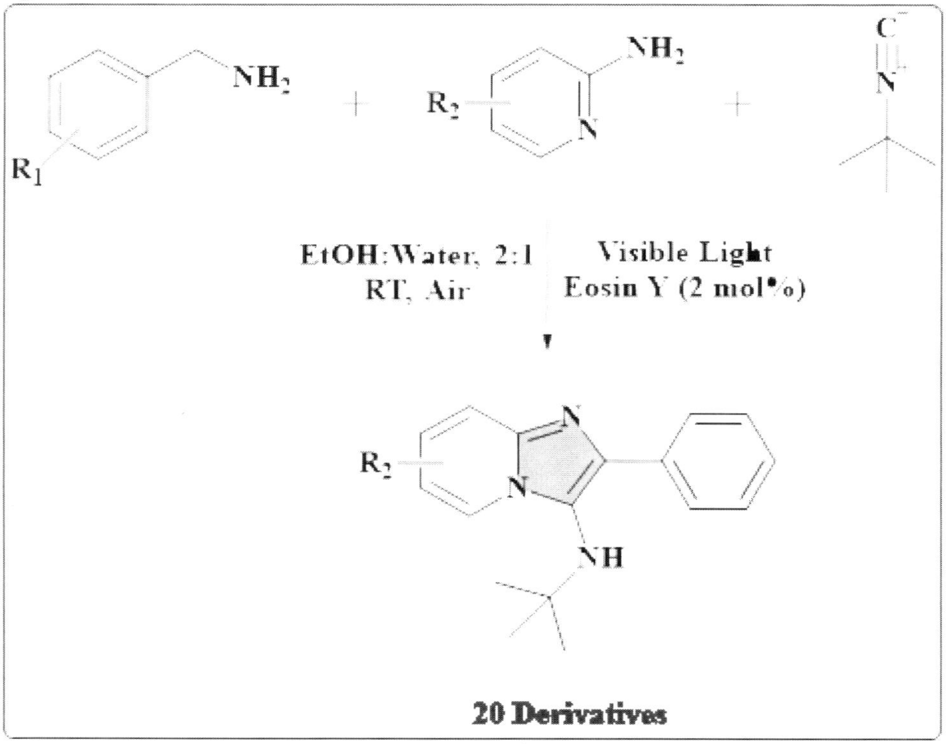

20 Derivatives

The Groebke–Blackburn–Bienayme reaction (GBBR) using aldehydes, isocyanides, and substituted 2-amino pyridine under catalyst and solvent-free conditions at ambient temperature was employed to synthesized imidazo[1,2-*a*]pyridines under visible light. This transformation proceeds via formation of C-N and C-C bond which were activated with visible light. They studied with different Visible light like, absence of light, 20W, 24W, 32W CFL and 32W LED, among all of them, the best yield found with 32W CFL at room temperature for 3.0 hr (Shivhare *et al.*, 2018).

Figure 34.

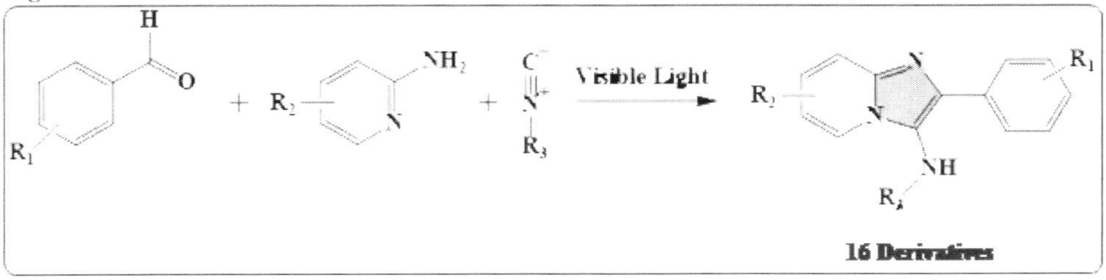

3.2 Three-Component Reaction of 2-Aminopyridine, Aldehyde, and Alkyne

Diverse imidazo[1,2-*a*]pyridine- fused isoquinolines through the post GBB transformation strategy was successfully isolated. In this methodology, GBB reaction product, treated with cyclisation using Au(Johnphos)Cl to yielded desired analogous (Shao *et al.*, 2017).

Figure 35.

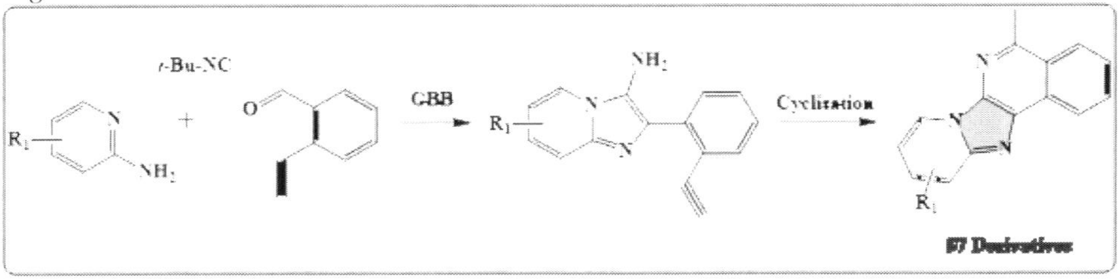

An efficient and green synthetic route to prepare imidazo[1,2-*a*]pyridine derivatives by domino A^3-coupling reaction catalysed by Cu(II)-ascorbate in aqueous micellar media in the presence of sodium dodecyl sulphate (SDS) have been employed. The catalyst, a dynamic combination of Cu(II)/Cu(I), was generated *in situ* in the reaction mixture by mixing $CuSO_4$ with sodium ascorbate and aided a facile 5-exo-dig cyclo-isomerization of alkynes with the condensation products of

2-aminopyridines and aldehydes to afford a variety of imidazo[1,2-a]pyridines in good overall yields (Bhutia et al., 2019).

Figure 36.

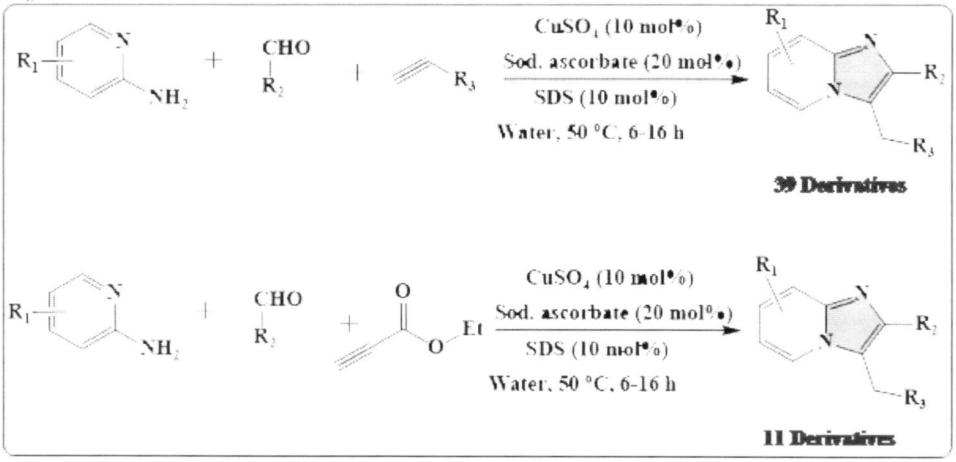

The synthesis imidazo[1,2-a]pyridines using CuI/Fe_3O_4 NPs@Biimidazole IL-KCC-1 as a catalyst in aqueous medium was developed. an innovative leach proof nano-catalyst based on dendritic fibrous nano silica (DFNS) modified with ionic liquid loaded Fe_3O_4 NPs and CuI salts were applied for the rapid synthesis of imidazo[1,2-a]pyridines from the reaction of phenyl acetylene, 2-aminopyridine, and aldehydes in aqueous medium. Higher catalytic activity of CuI/Fe_3O_4NPs@IL -KCC-1 is due to exceptional dendritic fibrous structure of KCC-1 and the ionic liquid groups that perform as strong anchors to the loaded magnetic nanoparticles (MNPs) and avoid leaching them from the pore of the nano-catalyst. The advantages of this synthetic method were green reaction media, shorter reaction times, higher yields, easy workup, and no need to use the chromatographic columns for purification (Azizi et al., 2021).

Figure 37.

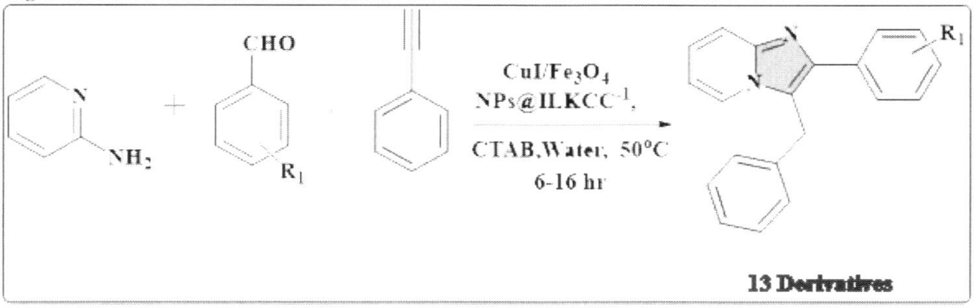

3.3 Miscellaneous Multicomponent Procedures

An acyclo-α-nucleosides imidazo pyridines and pyrimidines derivatives were synthesized and examined biological activity as an antiviral agent. The VilsMeier-Haack formylation on Substituted imidazo[1,2-*a*]pyridines gave the corresponding aldehyde which was reduced to the corresponding alcohol. It was alkylated with alkyl halide to yielded acyclo-α-nucleosides (Gueiffier *et al.*, 1996).

Figure 38.

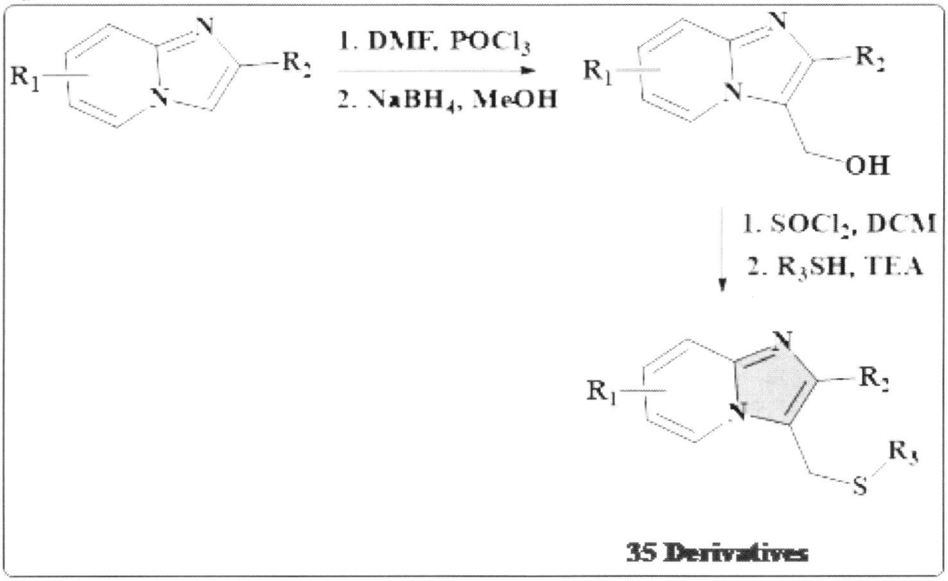

A novel class of inhibitors of human Rhinovirus was synthesized in a stereospecific manner of 2-amino-3-alkyl substituted imidazo[1,2-*a*]pyridine derivatives and examined their pharmacological activity as an anti-rhinovirus agent. The imidazo ring in this class of compounds was constructed starting from the aminopyridine after tosylation and subsequent treatment with the appropriate acetamides. The key steps in the synthesis include the development and use of a new Horner-Emmons reagent for the direct incorporation of methyl vinyl carboxamide. The reaction was stereospecific and leading exclusively to the desired E-isomer (Hamdouchi *et al.*, 1999).

Figure 39.

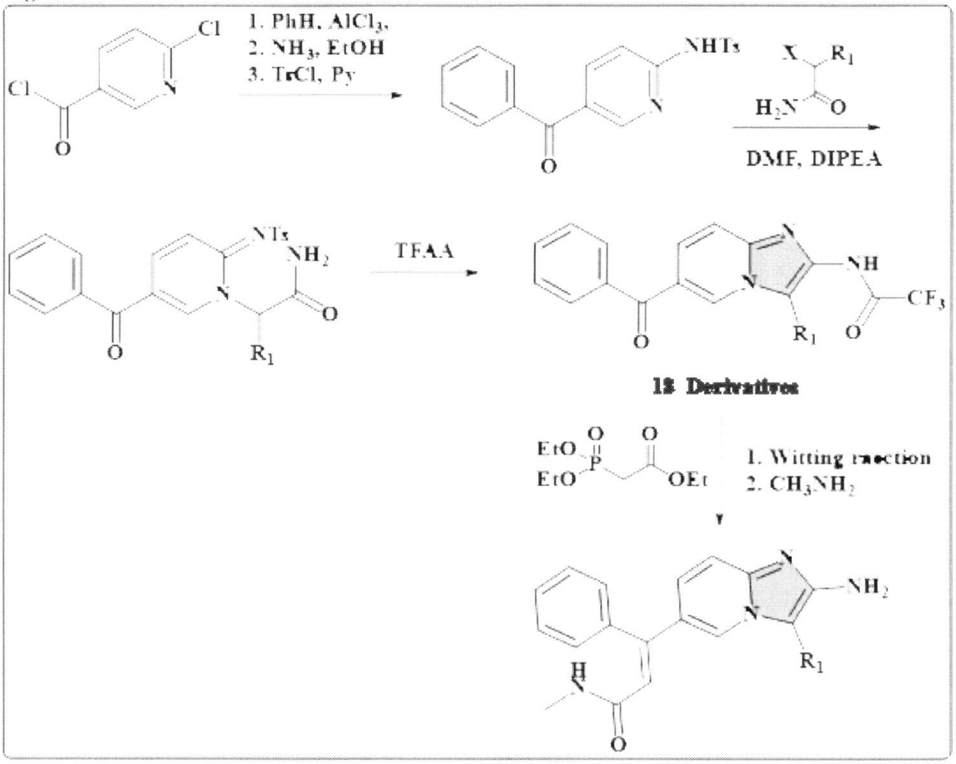

A group of researchers were synthesized imidazo[1,2-*a*]pyridines bearing a 3-(dithiolane, dioxolan or oxathiolan-2-yl) substituent were synthesized from the corresponding aldehydes. The dithiolanyl derivatives proved active against cytomegalovirus (Hassani *et al.*, 1999). Also, novel 3-sulfonyl pyrazole imidazo[1,2-*a*] pyridine derivatives and examined as PI3 Kinase p110α inhibitors by another group (Hayakawa *et al.*, 2007).

Figure 40.

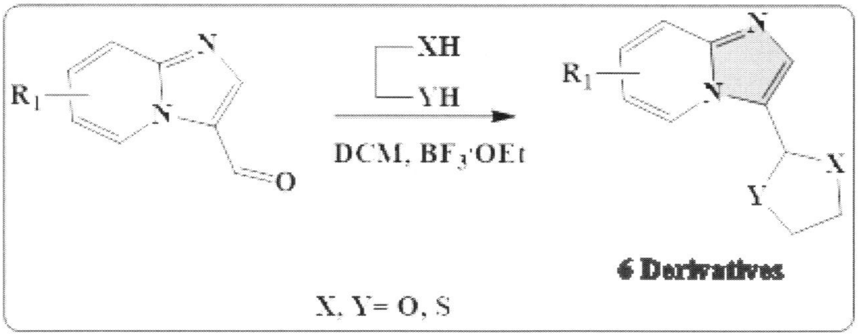

X, Y= O, S

Figure 41.

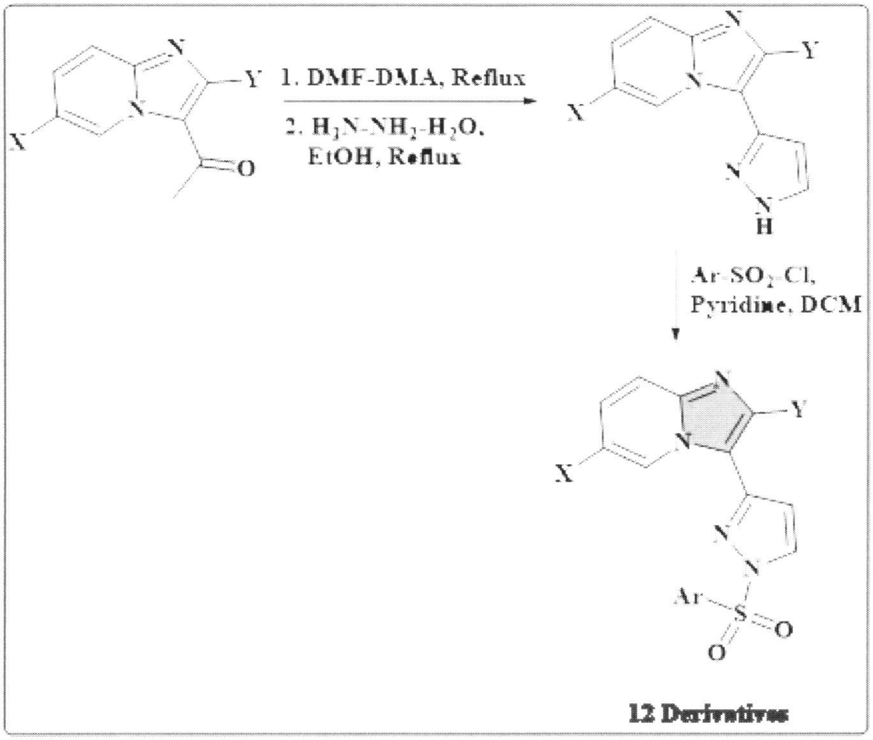

Substituted acetophenone, condensation with 1,1-dimethoxy-*N*,*N*-dimethylmeth-anamine (DMF-DMA) to obtained enamines intermediate, which was cyclized by using hydrazine hydrate to give pyrazole derivatives. It's sulphonation with the appropriate aryl sulfonyl chlorides. Using the similar approach and idea, for the exploration of SAR and optimisation of the imidazo[1,2-*a*]pyridine, CDK inhibitors has led to the discovery of novel, potent and selective inhibitors of the cyclin-dependent kinase CDK2. The important contribution of the aniline and pyrimidine groups to the binding and activity of these compounds.

3-Acetyl imidazo[1,2-*a*]pyridines were prepared by reaction of 2-amino pyridine with 3-chloro-2,4-pentanedione. It condensed with DMF-DMA to give dimethyl amino propanone, followed by cyclisation using thiourea and methylated to get key intermediate. Later, thio methyl group will be displaced with substituted aniline to get desired compounds (Byth *et al.,* 2004).

Figure 42.

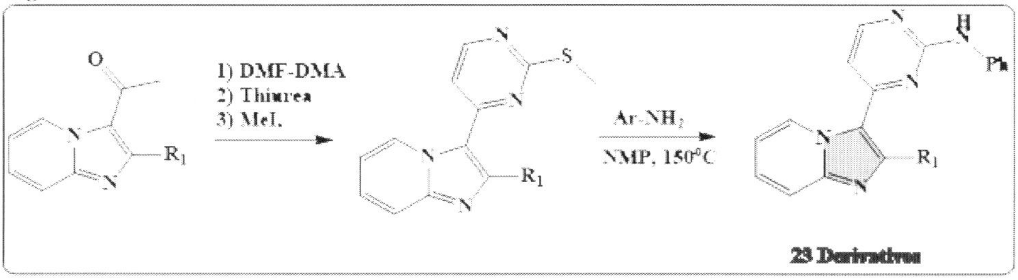

2-Amino pyridines react with substituted 1,3-dione or (β-keto esters) in presence of trifluoro borane diethyl ether, catalysed by TBAI (Tetra butyl Ammonium Iodide) and TBHP (Tert-butyl hydroperoxide) or H_2O_2 in aprotic solvent given extraordinary framework (Ma *et al.,* 2011). And the bio-potent, imidazo[1,2-*a*]pyridines derivatives were articulated in the presence of $Cu(OTf)_2$ as catalyst (Yadav *et al.,* 2007).

Figure 43.

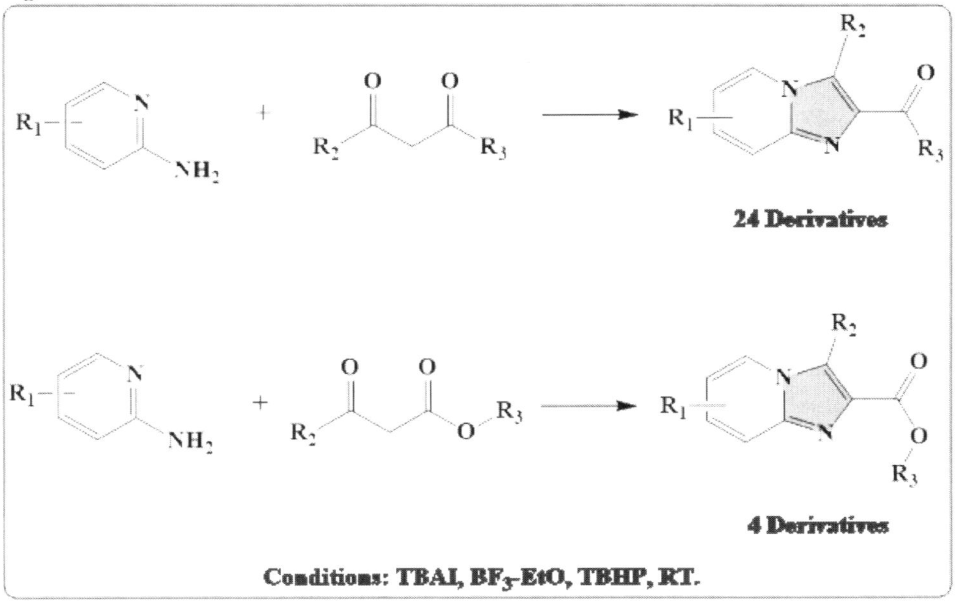

24 Derivatives

4 Derivatives

Conditions: TBAI, BF₃-EtO, TBHP, RT.

Figure 44.

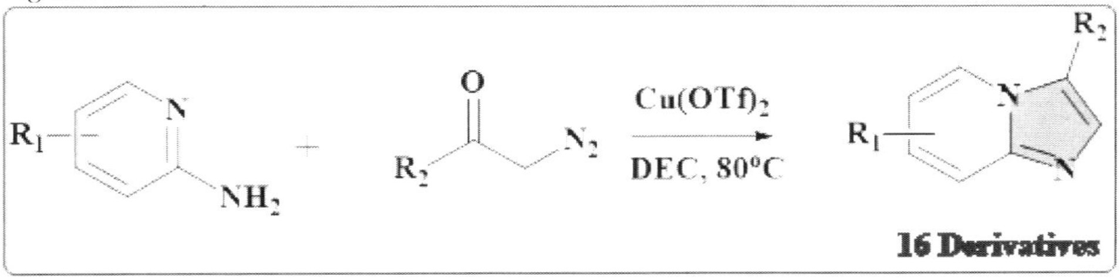

16 Derivatives

Substituted 2-amino pyridine was treated with derivatives of alkyl or aryl α-diazo ketone and copper triflate in DEC at reflux temperature yielded analogues of imidazo[1,2-*a*]pyridines. $Rh_2(OAc)_4$ is also found to be an equally effective catalyst for this transformation.

A copper catalysed approach was used to generate imidazo[1,2-*a*]pyridine derivative by reaction with pyridine and substituted oxime esters. Generally, pyridines derivatives synthesised Cu-catalysed cyclization with a variety of ketone oxime acetates. A number of oximes derived from aryl methyl ketones smoothly participated in the reaction, affording the corresponding imidazo[1,2-*a*] pyridines in moderate to good yields, with tolerance of functional groups, including fluoro, chloro, bromo, trifluoromethyl, methoxyl, and methyl sulfonyl groups (Huang *et al.*, 2013).

Figure 45.

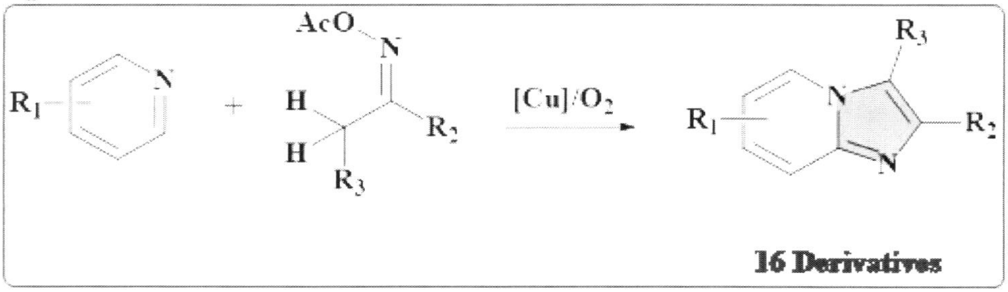

A novel process for regiospecific synthesis of 3-substituted imidazo[1,2-*a*] pyridines derivatives, by reaction between 2-amino pyridine and derivatives of 1,2-Bis(benzotriazolyl)-1,2-(dialkylamino)ethanes was synthesized via double condensation of glyoxal, benzotriazole, and amines (Katritzky *et al.*, 2003).

Figure 46.

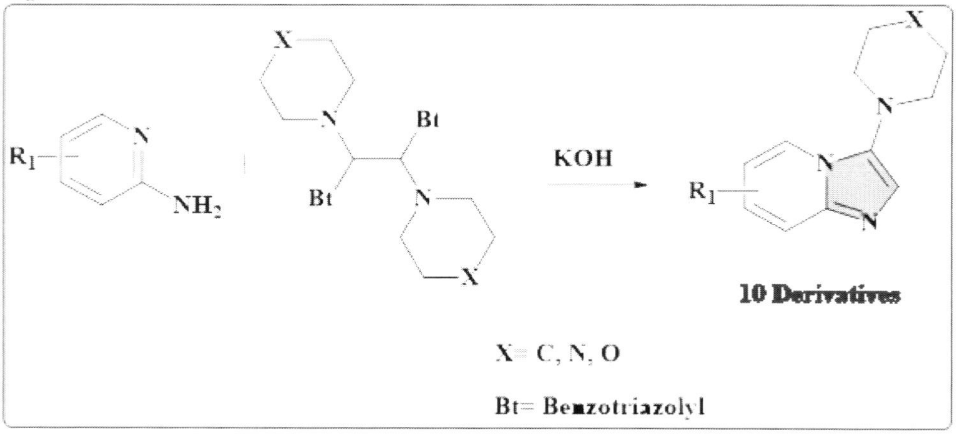

X= C, N, O

Bt= Benzotriazolyl

Wang and co-researchers established process to synthesised 2,3-diarylimidazo[1,2-*a*]pyridines analogues via one pot, ligand free, palladium-catalysed three-component reaction under microwave irradiation. This methodology is superior to the existing procedure and great efficiency in expanding molecules diversity (Wang *et al.*, 2014).

Figure 47.

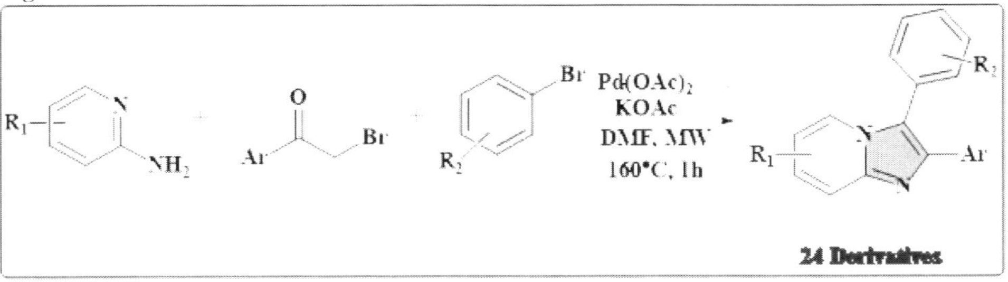

To studied the process to catalyst free synthesis *N*-arylidene-2-arylimidazo[1,2-*a*]pyridine-3-ylamine derivative via Strecker reaction under microwave irradiation, condensation between 2-amino pyridine and aromatic aldehyde to give schiff's base which was react with benzoyl cyanide or cyanamide followed by aromatic aldehyde to yielded desired analogues (Mohamed *et al.*, 2017).

Figure 48.

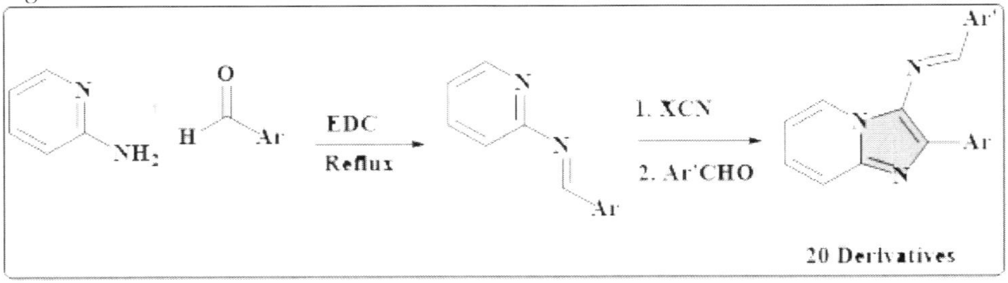

A newer scaffolds were prepared as imidazo[1,2-*a*]-*N*-heterocyclic derivatives with gem-difluorinated side chain. In this methodology, 2-amino pyridine react with α,β-unsaturated ketone in presence of $AlCl_3$, Iodine at reflux temperature to yielded desired new analogues of imidazo[1,2-*a*]pyridine. $Cu(OAc)_2$ also used as catalyst under oxygen atmosphere (Hariss *et al.*, 2017).

Figure 49.

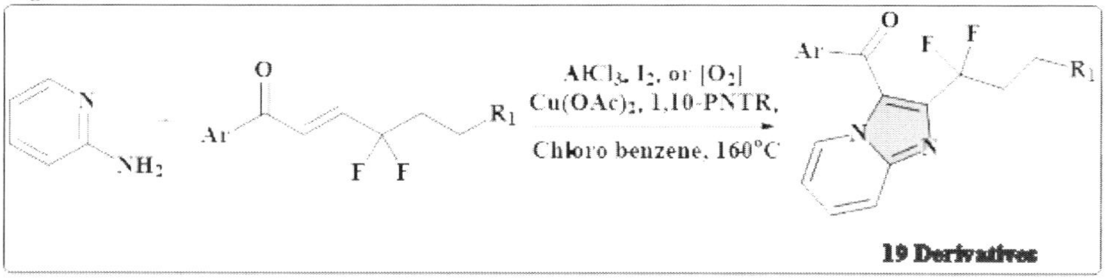

To synthesised novel imidazo[1,2-*a*]pyridine derivatives and examined as their anti-bacterial activity, substituted 2-amino pyridine was react with acetophenone with bromine and AlCl$_3$ to yielded imidazo[1,2-*a*]pyridine derivatives, which was followed by formylation, cyanation, hydrolysis, amide coupling using HATU to prepare desired product, those were studied against gram-positive and gram-negative bacteria. Among the all-synthesized compounds *N*-benzyl-4-((2-(6-methyl-2-(*p*-tolyl)imidazo[1,2-*a*] pyridin-3-yl)acetamido)methyl)benzamide are possessing high activity against *bacillus subtilis* (Budumuru at al., 2018).

Figure 50.

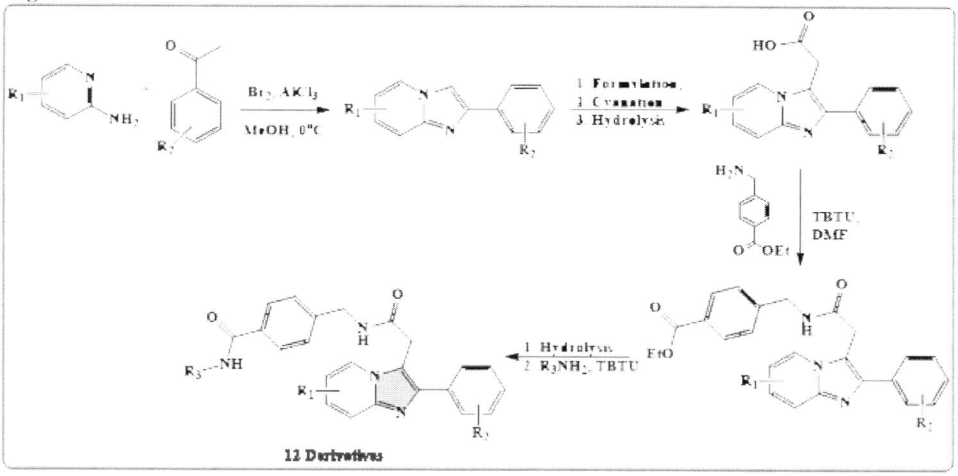

Saini and his group were developed highly efficient copper-catalysed one pot tandem protocol has been developed for the synthesis of naphtho-fused imidazo[1,2-*a*]pyridine derivatives. The transformation involves a Knoevenagel condensation followed by a chemo selective cross-coupling reaction along with a carbon-carbon bond cleavage. This protocol can tolerate a variety of functional groups and provided naphtho[1',2':4,5]imidazo[1,2-*a*]pyridines in moderate to excellent yields. They were also studied for different base like, KOtBu. K_2CO_3, KOH, Cs_2CO_3 and different Solvents. Among of them, K_2CO_3, L-Proline, $CuCl_2$ in DMF obtained moderated yield (Saini *et al.*, 2015).

Figure 51.

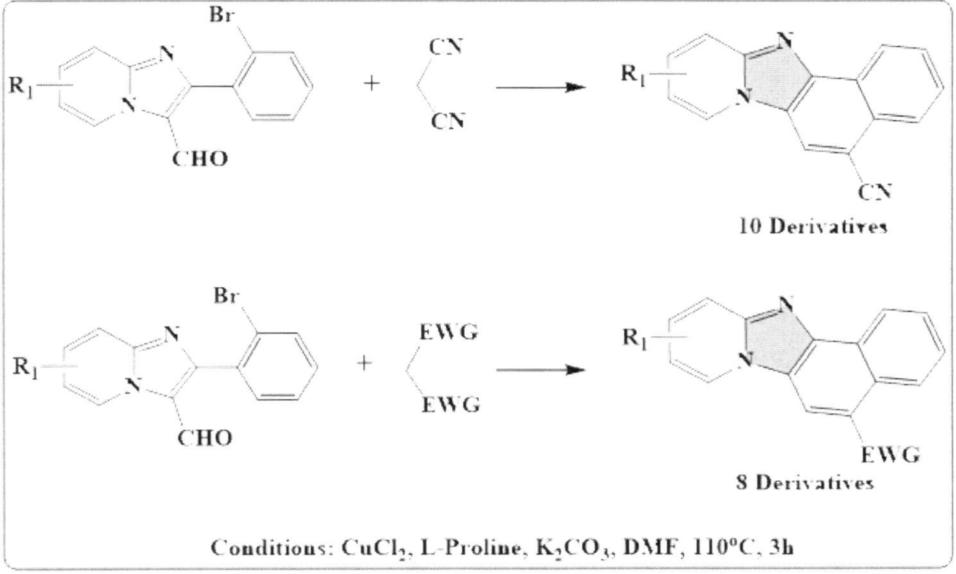

Conditions: $CuCl_2$, L-Proline, K_2CO_3, DMF, 110°C, 3h

A untraditional method was developed to synthesis of α-amino acid and α-amino ketones derivatives of imidazo[1,2-*a*]pyridines using substituted imidazo[1,2-*a*]pyridine and α-amino carbonyl compounds via visible-light-promoted dehydrogenative cross-coupling reaction using Eosin Y (catalyst), citric acid monohydrate, ethanol as a solvent in open air under blue LED light at room temperature. The synthetic method has the advantages of wide substrate scope, good functional tolerance, and mild reaction conditions, which make this transformation more practical and sustainable (Zhu *et al.*, 2020).

Figure 52.

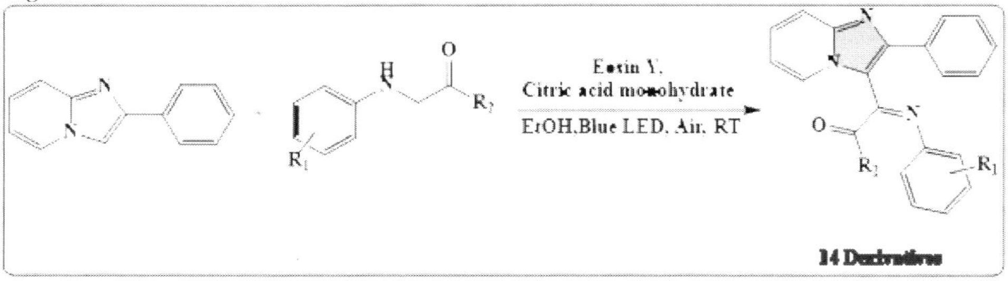

A newly developed method to synthesize aryl and phosphoryl analogues of imidazo[1,2-*a*]pyridines by using substituted 6-bromo imidazo[1,2-*a*]pyridines and phosphorus derivative in the presence of Pd_2dba_3 (catalyst), Xantphos (ligand), triethylamine (Base) in toluene at 110°C to obtained 6-phosphoryl imidazopyridine analogues was examined and optimized. In other method, they optimized the condition to prepared 3-aryl imidazo[1,2-*a*]pyridine via C-H bond activation using imidazo pyridine, aryl halide, $Pd(OAc)_2$ (catalyst), KOAc (Base) in DMA (dimethyl acetamide) as a solvent to obtained moderated yield (Gernet *et al.,* 2020).

Figure 53.

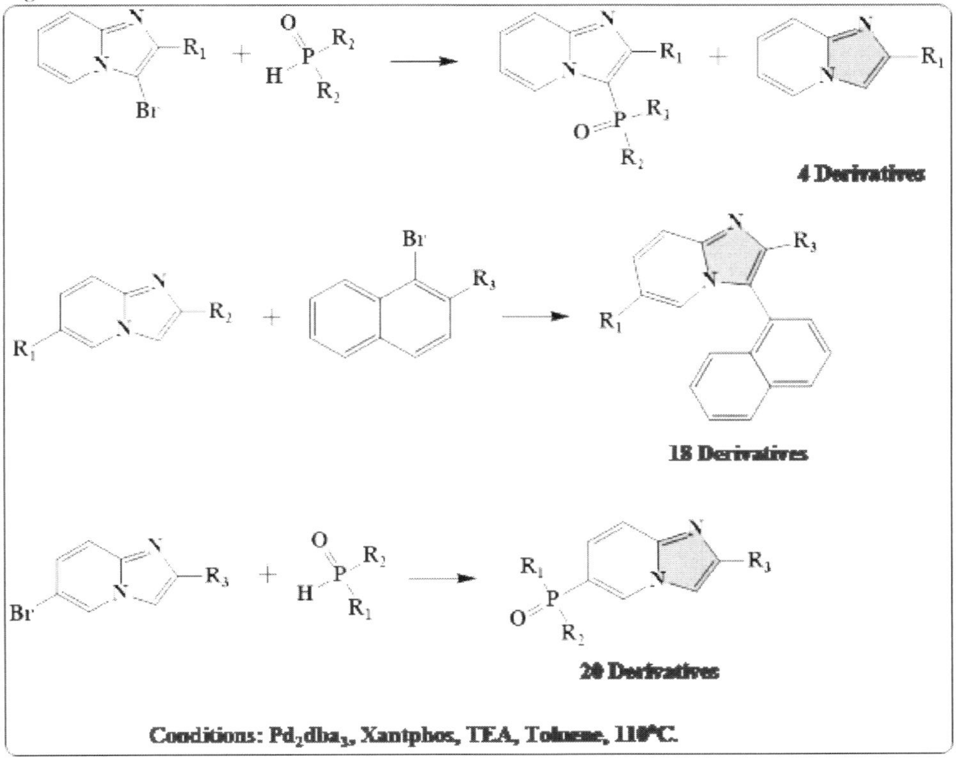

One pot synthesis of 3-alkyl/aryl imidazo[1,2-*a*]pyridine imidazo[1,2-*a*]pyridine and 3-thio imidazo[1,2-*a*]pyridine derivatives by aerobic oxidative formation of C-N and C-S bonds using multi component like, 2-substituted amino pyridine, carbonyl compound, and thiols in presence of flavin and iodine as a catalyst was introduced by team of researchers. This was the first metal free aerobic system in which used coupled flavin-iodine (TFO/I_2) catalyst instead of copper, zinc, indium and boron as a catalyst (Okai *et al.*, 2020).

Figure 54.

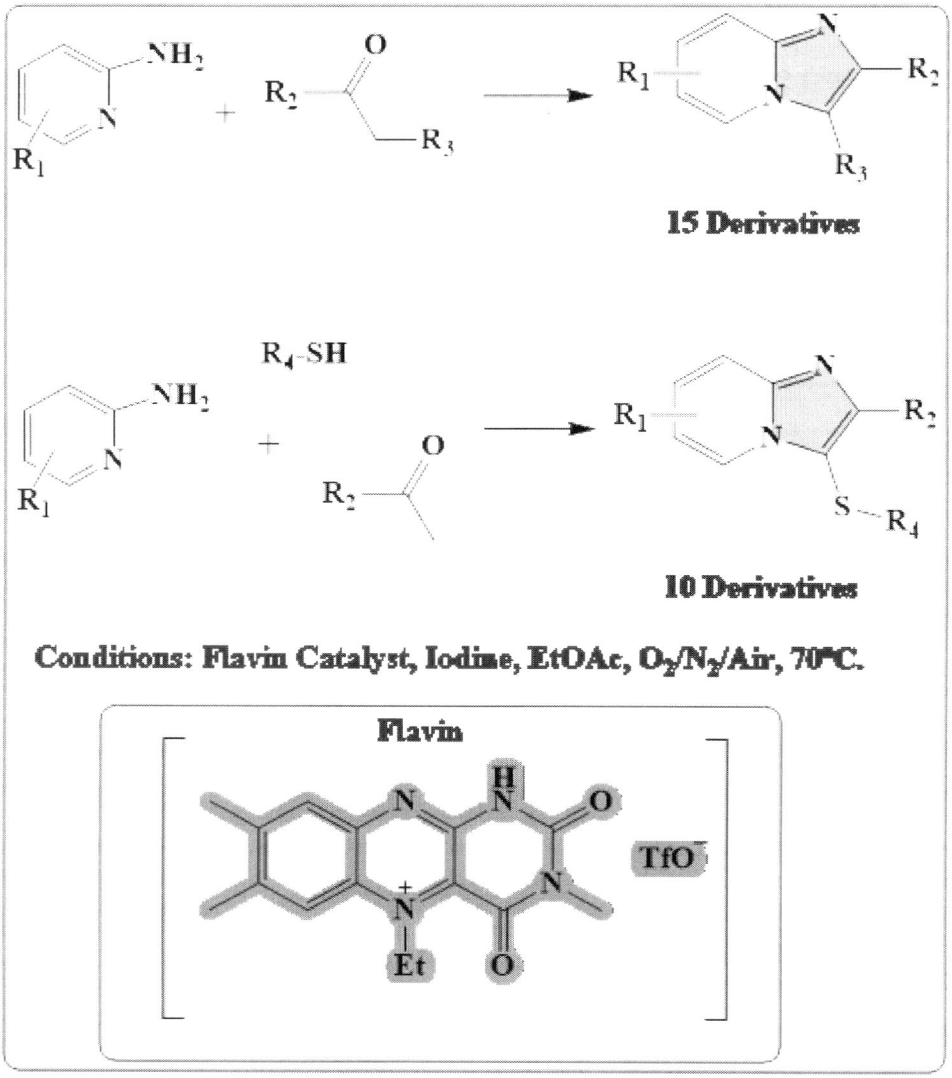

15 Derivatives

10 Derivatives

Conditions: Flavin Catalyst, Iodine, EtOAc, O₂/N₂/Air, 70°C.

Flavin

Adiyala and co-researchers were developed novel strategy to synthesized imidazo[1,2-*a*]pyridine derivatives via efficient catalyst/metal free annulation of α-keto vinyl azides and 2-amino pyridine derivatives. This was one pot synthesis in which first condensation of vinyl azides with 2-amino pyridines would lead to the formation of an imine intermediate, which should undergo annulation to yield desired substituted imidazo[1,2-*a*]pyridine derivatives. They also studied with different solvent and various temperature. The best yield found in THF rather than EtOH, ACN, Chloroform, 1,2-DEC and 1,4-dioxane at 70ºC (Adiyala *et al.*, 2015).

Figure 55.

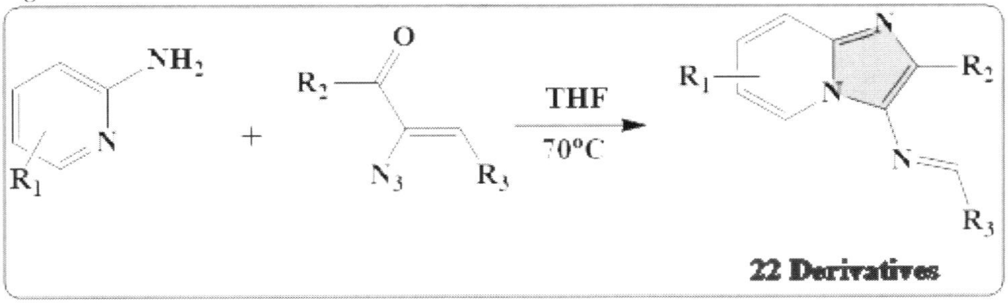

This *N*-(imidazo[1,2-*a*]pyridin-3-yl)-1-arylmethanimines was reduced by Groebke reaction using NaBH$_4$ in methanol to obtain their 3-amino substituted imidazo[1,2-*a*]pyridine derivatives.

Figure 56.

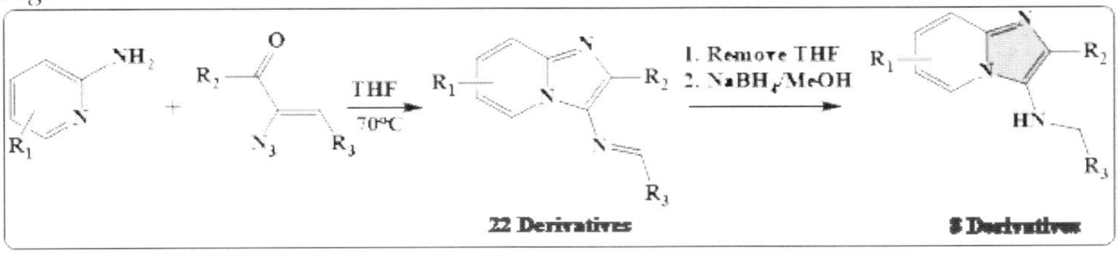

A new process to synthesize 3-thio imidazo[1,2-*a*]pyridine derivatives via Tandem Flavin-Iodin catalysed aerobic oxidation sulfenylation of imidazo[1,2-*a*] pyridines and various thiols was developed. In this process, reaction beginning with the aerobic oxidation of a thiols to afford a disulfide that is utilized in the oxidative sulfenylation of the imidazo[1,2-*a*]pyridine (Lida *et al.*, 2018).

Figure 57.

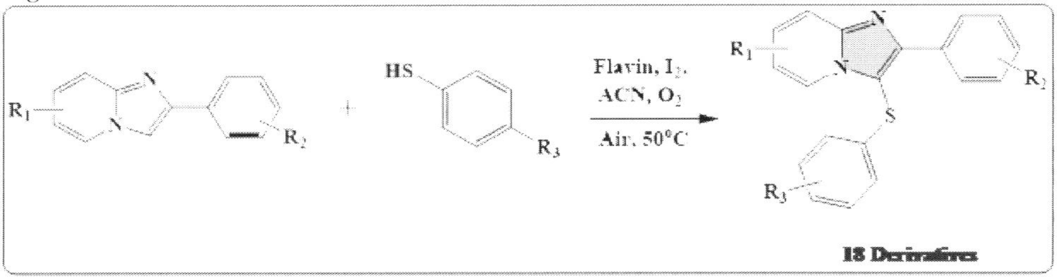

Commercially available riboflavin (**X**) and its tetraacetate (**Y**) failed to promote the sulfenylation reaction. The cationic flavin catalysts 5-ethylisoalloxazinium (**P·TfO and Q·TfO**), 5-ethylalloxazinium (**6·TfO**), and 1,10-ethylene-bridged alloxazinium salts (**R·TfO** and **S·TfO**), which can be derived from **X**. They also investigated, as these new flavin derivatives display different redox and catalytic properties which dependent upon their π-conjugation of the ring structure. The results showed that the success of the reaction was likely dependent upon the electrophilicity of the catalyst, as the most electrophilic of the cationic catalysts **X·TfO** smoothly cata-lysed the sulfenylation of imidazo[1,2-*a*]pyridine to give the desired product 3-thio imidazo[1,2-*a*]pyridine analogous.

Figure 58.

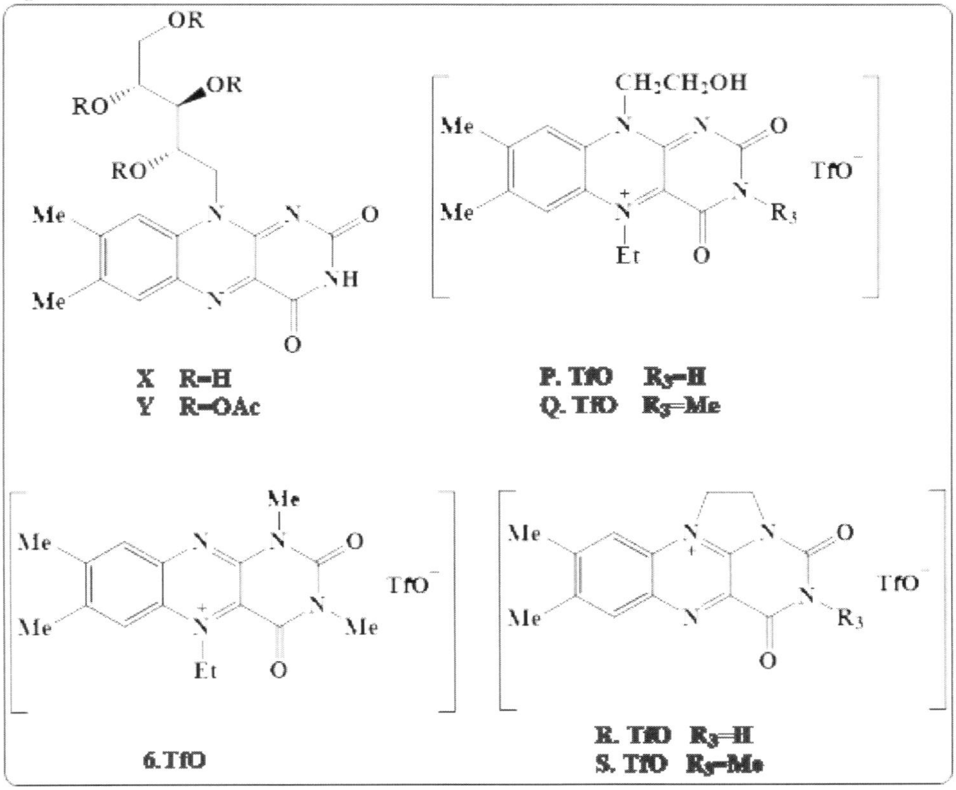

A metal free process was developed to prepare quinoline-fused imidazopyridines via NH_4I catalysed cross-dehydrogenative coupling (CDC) of ethers with 2-(2-amino aryl)imidazo[1,2-*a*]pyridines cascade cyclization. Aqueous H_2O_2 was used as an oxidizing agent. They studied for different catalyst (like Bu_4NI, KI, NH_4I, I_2 and CuI) and oxidizing agent like, TBHP, DTBP, BPO, $K_2S_2O_8$, and H_2O_2). Best yield obtained by NH_4I as a catalyst and H_2O_2 as an oxidizing agent. Advantages of this reaction, various functional group tolerance, easy for scale-up preparation, cheap and commercially available reagents, moderate to good yields (Lian *et al.*, 2018).

Figure 59.

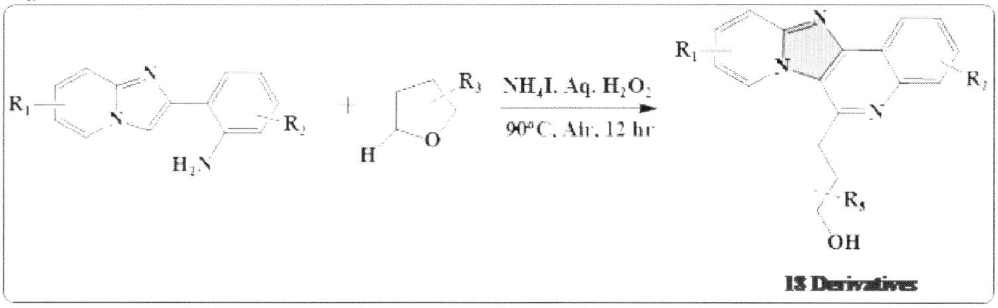

Yuanjiu Xiao *et al.* were developed process to synthesise C3-ethoxy carbonyal methylation of Imidazo[1,2-*a*]pyridines with ethyl diazoacetate via visible light promoted C-H bond functionalization. The advantages of this method were very mild and efficient, very good yield, various functional group tolerance, economic reagent. They studied on various photo-catalyst (like, Rose B, Eosin Y, Ru(bpy)$_3$Cl$_2$.6H$_2$O and solvent (like ACN, Toluene, DMF, 1.4dioxane, MeOH, EtOH, H$_2$O and mixture of MeOH: Water (10:1). Among from those Ru(bpy)$_3$Cl$_2$.6H$_2$O and mixture of MeOH: Water (10:1) under blue LED light obtained desired product with moderate yield (Xiao *et al.*, 2020).

Figure 60.

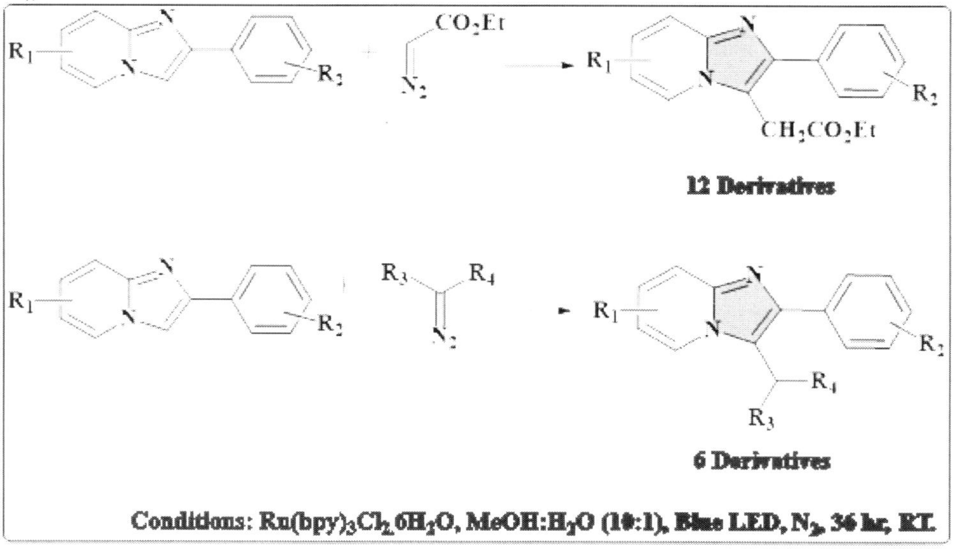

The metal free process to synthesise 3-thio imidazo[1,2-*a*]pyridine derivatives via sulfenylation of imidazo[1,2-*a*]pyridine which promote by using *N*-chloro succinimide was introduced and optimized. They studied on *N*-halo succinimide (like, *N*-chloro, *N*-bromo and *N*-iodo succinimide) among from them 1.5 eq. NCS and DCM solvent obtained good yield. This method is very mild and effective (Ravi *et al.*, 2014).

Figure 61.

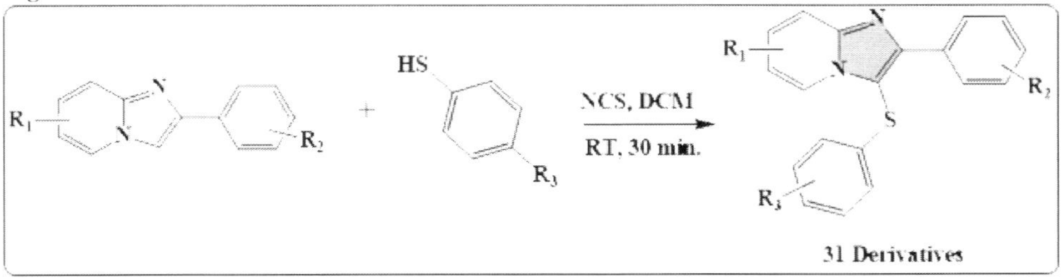

This sulfonylation is regio-selective reaction, it's don't work on 2nd position of imidazo[1,2-*a*]pyridine, it's only forwarded on 3rd position.

Figure 62.

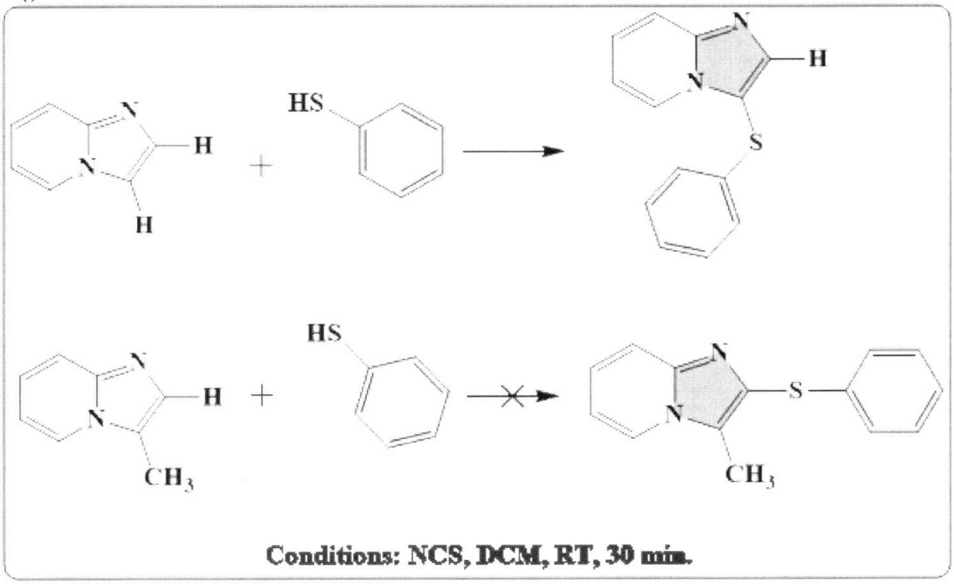

Conditions: NCS, DCM, RT, 30 min.

An Ullmann type C-N bond coupling reaction to synthesised 1,2,3-triazole-fused imidazo[1,2-*a*]pyridines via copper-catalysed tandem azide-alkyne cycloaddition (CuAAC) and intramolecular direct arylation was put forwarded by group of scientists. 3-Bromo-2-(2-bromophenyl) imidazo [1,2-a] pyridine, phenylacetylene and sodium azide were in the presence of $CuCl_2 2H_2O$ and K_2CO_3 in DMF at 150ºC for 24 h to obtained desired product 1-phenylpyrido [2',1':2,3]imidazo[4,5-*c*][1,2,3] triazolo[1,5-*a*]quinoline. Advantages of this process were one pot synthesis, easy to scale-up and cheap catalyst and reagent and various function group tolerance (Pericherla *et al.*, 2013).

Figure 63.

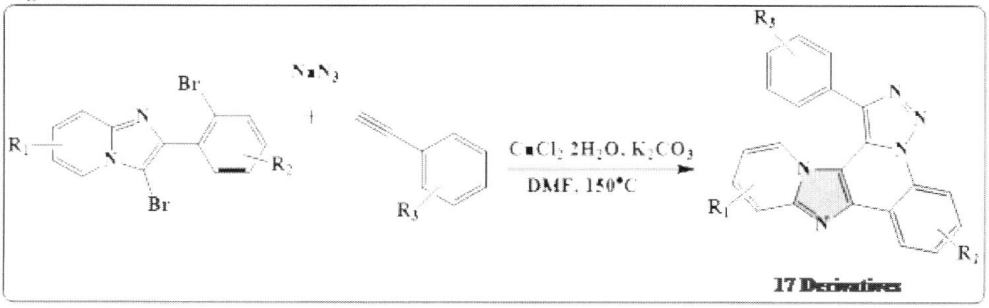

To synthesise 3-phenoxy imidazo pyridine derivatives via copper catalysed oxidative cyclization of 2-amino-azaarenes with lignin models was given by Zhang and his group. A catalytic oxidative cyclization of 2-aminopyridines or 2-aminobenzothiazole with 2-phenoxyacetophenones (a kind of lignin platform compound) was developed, efficiently providing valuable 3-phenoxy imidazo[1,2-*a*]pyridines or 3-phenoxy benzo[*d*]imidazo[2,1-*b*]thiazoles. The reaction was realized under oxygen by simply using inexpensive CuI as the catalyst (Zhang *et al.*, 2017).

Figure 64.

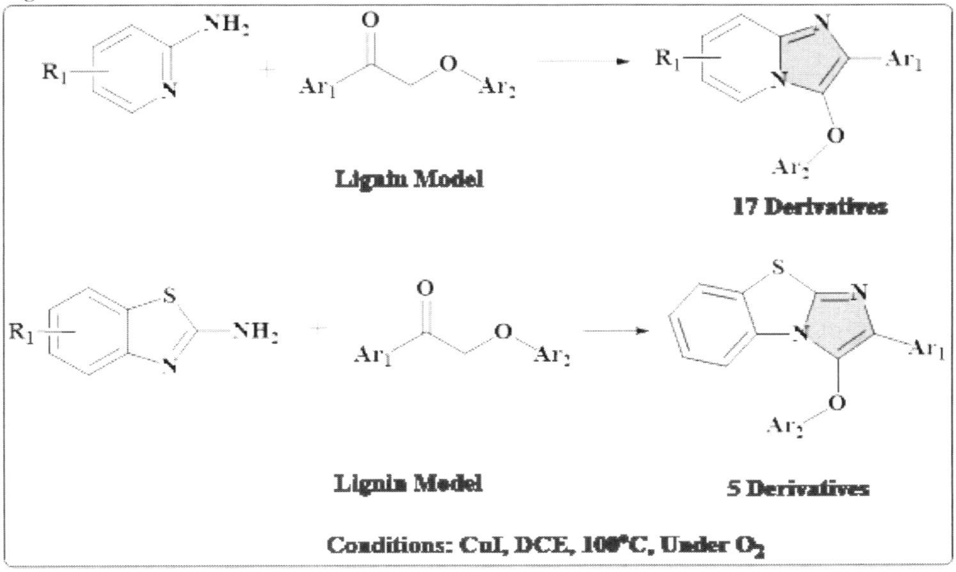

Lignin Model

17 Derivatives

Lignin Model

5 Derivatives

Conditions: CuI, DCE, 100°C, Under O$_2$

4.0 BIOLOGICAL IMPORTANCE

A library of versatile imidazo[1,2-*a*]pyridine derivative was prepared and examined for analgesic, anti-inflammatory, anticonvulsant and muscle relaxant activity ([Almirante *et al.,* 1995).

Table 1. Analgesic effect of imidazo[1,2-a]pyridine derivatives.

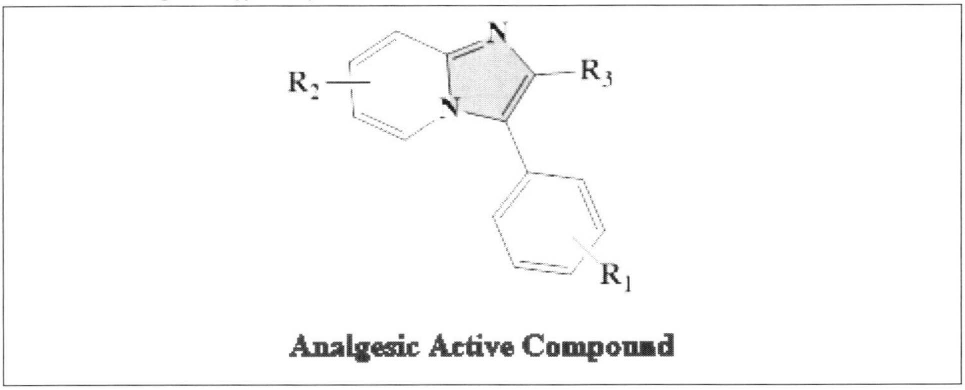

Analgesic Active Compound

R_1 replace with methyl, methyl-thio and methane sulfonyl groups enhanced analgesic activity.

Table 2. Anti-convulsant activity of imidazo[1,2-a]pyridine derivatives.

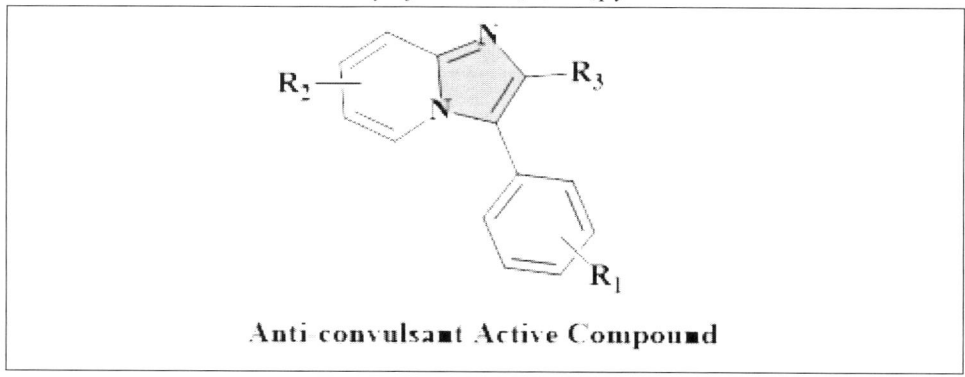

Anti convulsant Active Compound

R_1 and R_2 replace with *p*-methoxy and *p*-methane sulfonyl groups enhanced anti-convulsant activity.

Table 3. Anti-inflammatory activity of imidazo[1,2-a]pyridine derivatives.

Comp. ID	R_1	R_2	R_3
3	4-NO_2	-H	-H

James and team have been identified Gastric Antisecretory and Cyto- protective properties of substituted Imidazo[1,2-a]pyridines. The presence of an alkyl group at the 2-position of the imidazo[1,2-a]pyridine ring system appeared to be necessary to maintain oral antisecretory potency. The 2-methyl-substituted analogue **27** and the 2-ethyl-substituted analogue **29** exhibited comparable oral antisecretory activity in the dog (Palmer *et al.*, 2007).

Table 4. Anti-secretory activity of imidazo[1,2-a]pyridine derivatives.

Comp. ID	R_1	R_2	R_3	R_4
27	-H	-OCH_2Ph	-CH_3	-CH_2CN
28	-H	-OCH_2Ph	-CH_2CH_3	-CH_2CN

The gastric antisecretory and cytoprotective properties of a new class of antiulcer agents, the substituted imidazo[1,2-a]pyridines, has been described. Structure-activity studies in this series led to the identification of three compounds, **27, 51** and **57,** that exhibited novel and potentially promising antiulcer profiles.

Table 5. Anti-secretory and cytoprotective study of imidazo[1,2-a]pyridine derivatives.

Comp. ID	R_1	R_2	R_3	R_4
27	-H	$-OCH_2Ph$	$-CH_3$	$-CH_2CN$
51	-H	$-OCH_2$-Thionyl	$-CH_3$	$-CH_2CN$
57	-H	$-NHCH_2Ph$	$-CH_3$	$-CH_2CN$

A compound **27** was represent of a novel class of antiulcer agents that are neither competitive histamine (H_2) receptor antagonists nor prostaglandin analogues yet exhibit both gastric antisecretory and cytoprotective properties.

Fisher and Aino Lusi were prepared 2-substitued imidazo[1,2-a]pyridines by reaction of 2-amino pyridines with appropriate halo ketones and the biological activities results shown anthelmintic and anti-fungal activity against six fungi, *Aspergillus niger, Pullularia pullulans, Penicillium luteum, Chaetomium globosum, Trichoderma viride*, and *Rhizoctonia solaniI* (Fisher *et al.,* 1972).

Table 6. Anti-hementic activity of imidazo[1,2-a]pyridine derivatives.

Comp. ID	R_1	R_2
40	$-NHCH_2OCH_3$	$-OCH_2Ph$
47	$-NHCH_2OCH_3$	$-OCH_2Ph$

In vivo anthelmintic results show the lowest oral dose in mg/kg which demonstrated activity against trichostrongyles in a laboratory animal model assay or inactivity at 400 mg/kg. Compound **40** and **47** was active against a broad spectrum.

Figure 65.

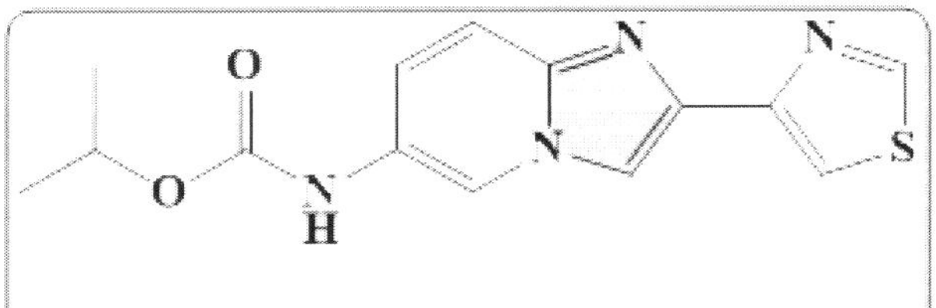

Compound 48

Compound **48** possess anti-fungal activity on *Aspergillus niger* (An), *Pullularia pullulans* (Pp), *Penicillium luteum* (Pl) at 10 mg/Kg.

A diverse and bioactive, 2-(4'-dimethylaminophenyl)-6-Iodoimidazo[1,2-*a*] pyridine and their related derivatives were derived in 2003. They performed Structure Activity Relationship of Imidazo[1,2-*a*]pyridines as Ligands for Detecting B-Amyloid Plaques in the Brain (Zhuang *et al.,* 2003).

Figure 66.

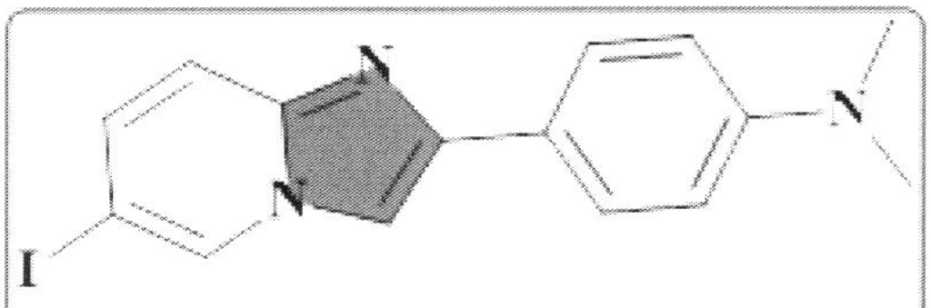

Compound 16 (IMPY)

Formation of β-amyloid (Aβ) plaques in the brain is a pivotal event in the pathology of Alzheimer's disease (AD). As on *in vivo* study in mice, lower values of φ exhibited a higher binding affinity toward the aggregates and the partition coefficient (PC) of **16** is 100 (1-octanol/buffer). A relatively high lipophilicity may be critical for the initial brain penetration by a simple diffusion mechanism. The most important feature of this novel ligand, **16** (IMPY), is the faster rate of washout from the normal brain.

A group was discovered novel Dicatonic imidazo[1,2-*a*]pyridine derivatives. Among them 5,6,7,8-Tetrahydro-imidazo[1,2-*a*]pyridines were a anti-protozoal agents (Ismail *et al.,* 2004).

Figure 67.

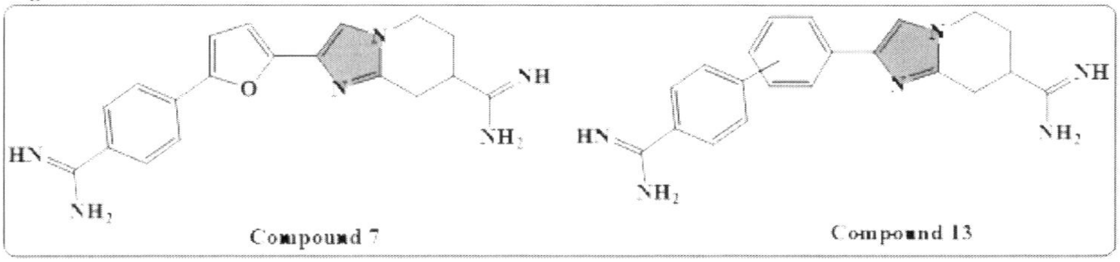

Compound 7   Compound 13

Aromatic diamidines exhibit broad-spectrum anti-microbial activity including effectiveness against the protozoan diseases caused by *Trypanosoma sp.* and *Plasmodium sp.* Several prodrugs of these aza-analogues show excellent oral activity *in vivo* which are superior to that of their respective diamidines. A new class of diaryl diamidines Compound **7** and Compound **13**, which show strong DNA affinity and high *in vitro* activity ag0ainst *T. b. rhodesiense* and *P. f.* The diamidines show excellent in vivo activity against *T. b. rhodesiense* on ip dosage.

Sheela Gopal and Anitha I. were examined as an antibacterial activity against *Escherichia coli* (NCIM 2343), *Klebsiella pneumonia* (NCIM 2707) and *Staphylococcus aureus* (NCIM 2127) (Sheela *et al.*, 2016).

Figure 68.

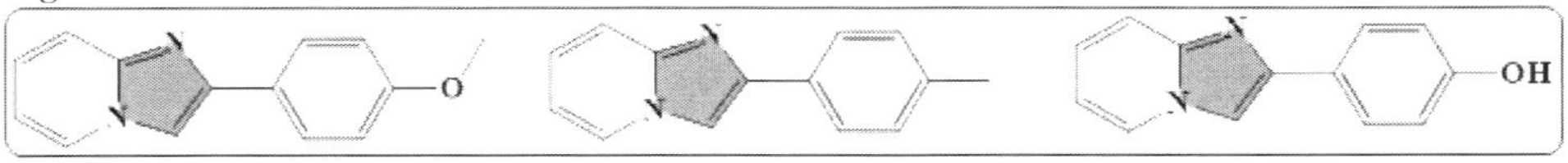

Compounds with methoxy, methyl and hydroxy group in imidazo[1,2-*a*]pyridine were screened, and the MIC values shows antibacterial activity against *Escherichia coli*, *Klebsiella pneumoniae* and *Staphylococcus aureus*.

A tubercular antagonist was developed as imidazo[1,2-*a*]pyridine derivative and identified their bactericidal activity on mouse *M. tuberculosis* infection model. A compound ND-09759 exhibit potent activity against *M. tuberculosis* (Cheng *et al.*, 2014).

Figure 69.

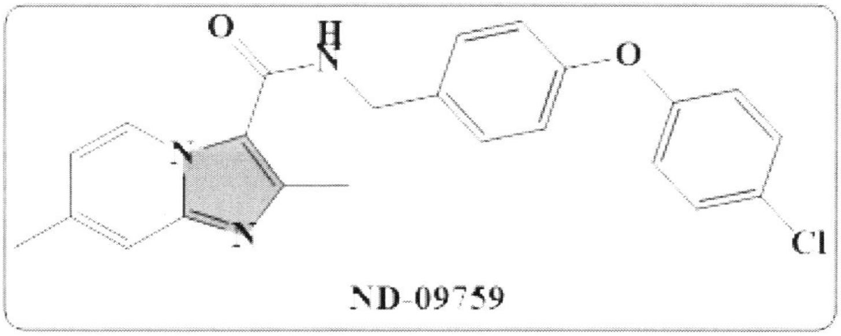

ND-09759

ND-09759 belongs to a novel class of drug compounds, imidazo[1,2-*a*]pyridines, that exhibit potent activity against *M. tuberculosis*. Its *in-vivo* activity combined with its synthetically tractability and druggable properties suggests that this family of compounds could be developed into a new class of anti-TB drugs.

Gómez and team were examined and found that 3-amino imidazo[1,2-*a*]pyridine derivatives like, compounds **A**, **B** and **C** were pharmacologically active compounds such as benzodiazepine receptor agonists, anti-inflammatory agents, inhibitors of gastric acid secretion, calcium channel blockers and antibacterial (Gómez *et al.*, 2019).

Figure 70.

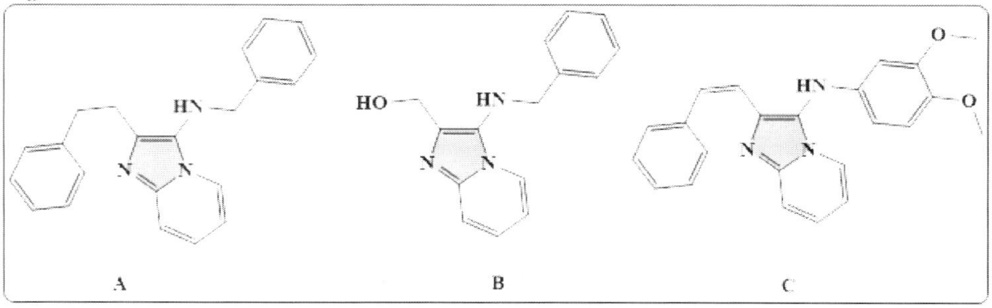

Idealized acyclo-nucleosides imidazo pyridines and pyrimidines were derivatives and examined biological activity as an antiviral agent (Gueiffier *et al.*, 1996).

Figure 71.

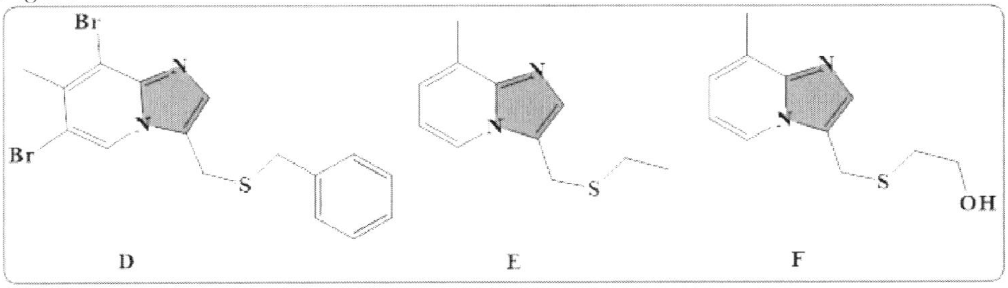

D   E   F

Compounds **D**, **E**, and **F** were proved herpes simplex virus (HSV), varicellazoster virus (VZV), cytomegalovirus (CMV).

Novel class of inhibitors of human Rhinovirus were synthesized stereo-specifically active 2-amino-3-Alkyl substituted imidazo[1,2-*a*]pyridine derivatives and examined their pharmacological activity as an anti-Rhinovirus agent (Hamdouchi *et al.*, 1999).

Figure 72.

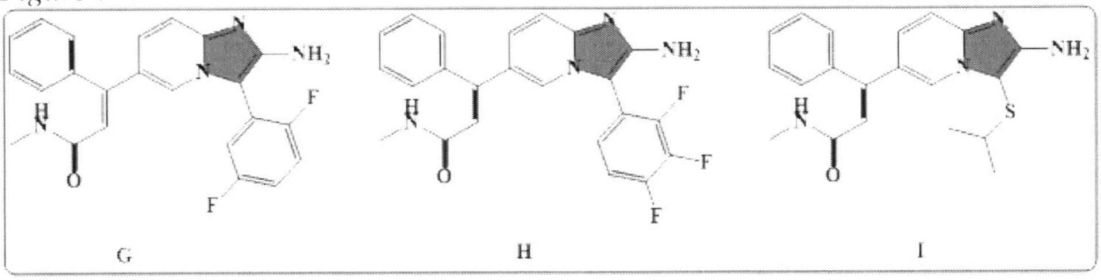

G   H   I

Compounds **G**, **H** and **I** exhibited a strong anti-rhino virus activity, and no apparent cellular toxicity was found.

Followed to a novel imidazo[1,2-*a*]pyridines bearing a 3-(dithiolane, dioxolan- or oxathiolan-2-yl) substituent, it was proved active against cytomegalovirus (Hassani *et al.*, 1999).

Figure 73.

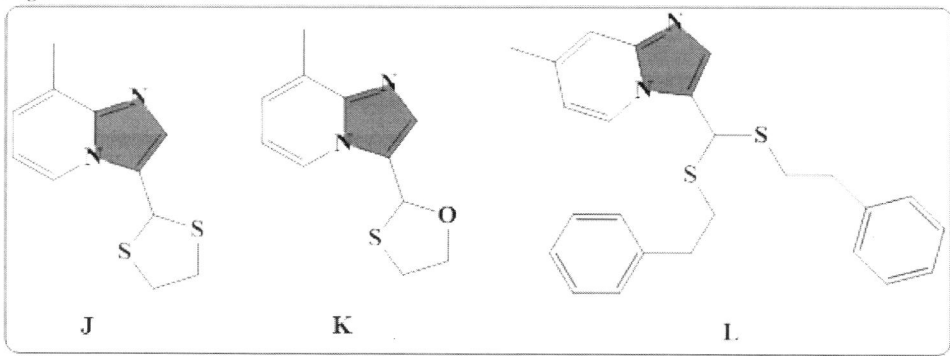

The 8-methyl dithiolanyl derivative **J** showed activity against CMV (strains AD-169 and davis). The thioacetal **K** was shown to be the most potent anti-CMV agent. In addition, the 8-methyl oxathiolanyl derivative **L** exhibited activity against respiratory syncytial virus (RSV).

The exploration of SAR and optimisation of the imidazo[1,2-*a*]pyridine, CDK inhibitors has led to the discovery of novel, potent and selective inhibitors of the cyclin-dependent kinase CDK2 (Byth *et al.*, 2004).

Figure 74.

biological profiles of compounds **M** and **N** were more fully characterised. These results show that compounds **M** and **N** inhibit CDK1 with similar potency to CDK2 though show selectivity with respect to CDK4.

Al-tel and his group were synthesised some novel imidazo[1,2-*a*]pyridines and was carried out *in vitro* evaluation as an antimicrobial (Al-tel *et al.*, 2011).

Figure 75.

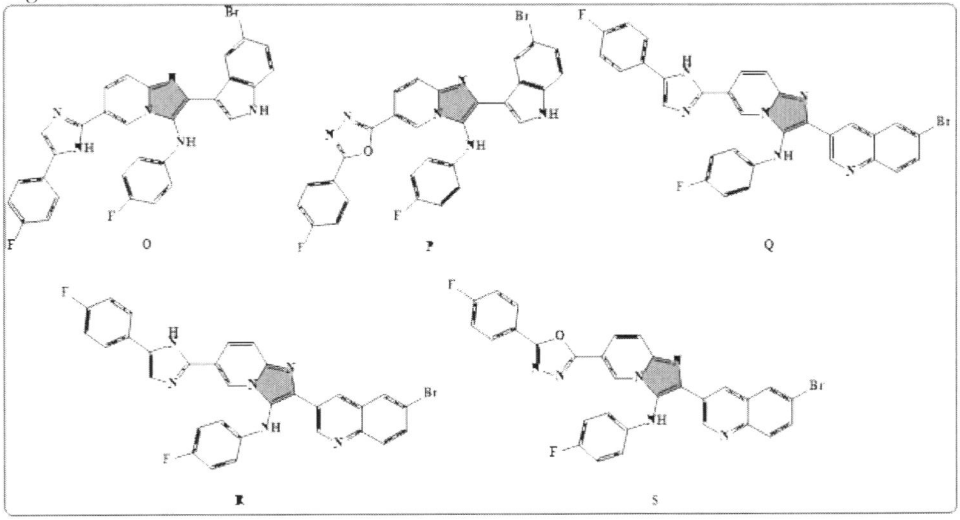

Study of antimicrobial activities were conducted of these analogues against various Gram-positive, Gram-negative bacteria and fungi. Compounds **O, P, Q, R** and **S** exerted strong inhibition of the investigated bacterial and fungal strains compared to control antibiotics amoxicillin and cefixime and the antifungal agent fluconazole.

The synthesis of novel imidazo[1,2-*a*]pyridine derivatives were examined as their anti-bacterial activity against Gram-positive and Gram-negative bacteria.

Figure 76.

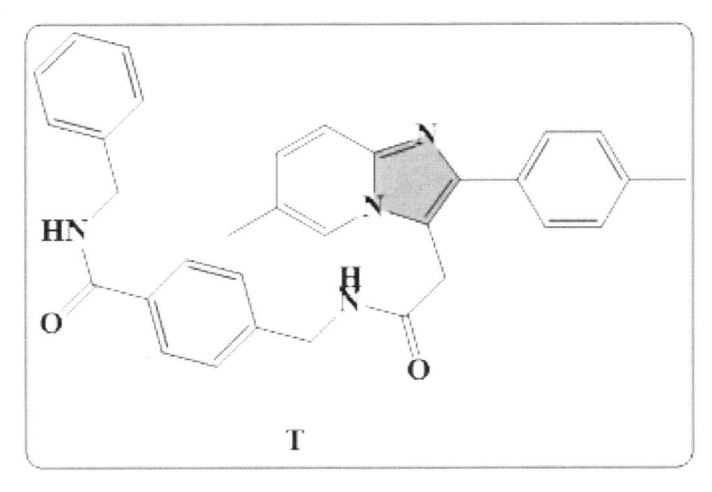

483

Among the synthesized compounds *N*-benzyl-4-((2-(6-methyl-2-(*p*-tolyl) imidazo[1,2-*a*]pyridin-3-yl)acetamido)methyl)benzamide (**T**) are possessing high activity against *Bacillus subtilis*. The zone of inhibition produced by the compound **T** is wider than that of remaining compounds used in targeted study (Pushpalatha *et al.*, 2018).

CONCLUSIONS

Imidazo[1,2-*a*]pyridine analogues have taken a lead role in the recent literature because of their wide variety of applications in various disciplines like medicinal chemistry, organometallics, and material science. Several novel synthetic routes have been developed to achieve these skeletons. These newer synthetic routes are based on the combination of several interesting strategies such as multicomponent reactions, tandem sequences, and C–H activation. As is evident from the discussion in this review, these synthetic procedures offered easy access to imidazo[1,2-*a*] pyridines from simple and readily available precursors without the need of any pre-functionality. The development of these synthetic procedures is very useful especially for medicinal and material chemists.

REFERENCES

Adib, M., Sheikhi, E., & Rezaei, N. (2011). One-pot synthesis of imidazo [1, 2- a] pyridines from benzyl halides or benzyl tosylates, 2-aminopyridines and isocyanides. *Tetrahedron Letters*, 52(25), 3191–3194. DOI: 10.1016/j.tetlet.2011.04.002

Adiyala, R., Mani, S., Nanubolu, B., Shekar, C., & Maurya, A. (2015). Access to Imidazo[1,2 - a]pyridines via Annulation of α - Keto Vinyl Azides and 2 - Aminopyridines. *Organic Letters*, 17(17), 4308–4311. DOI: 10.1021/acs.orglett.5b02124 PMID: 26308984

Al-tel, H., Al-qawasmeh, A., & Zaarour, R. (2011). European Journal of Medicinal Chemistry Design, synthesis and in vitro antimicrobial evaluation of novel Imidazo [1, 2- a] pyridine and imidazo [2, 1- b][1, 3] benzothiazole motifs. *European Journal of Medicinal Chemistry*, 46(5), 1874–1881. DOI: 10.1016/j.ejmech.2011.02.051 PMID: 21414694

Almirante, L., Polo, L., Mugnaini, A., Provinciali, E., Rugarli, P., Biancotti, A., & Murmann, W. (1965). Derivatives of imidazole. I. Synthesis and reactions of imidazo [1, 2-α] pyridines with analgesic, antiinflammatory, antipyretic, and anticonvulsant activity. *Journal of Medicinal Chemistry*, 8(3), 305–312. DOI: 10.1021/jm00327a007 PMID: 14329509

Azizi, S., Shadjou, N., & Soleymani, J. (2021). CuI/Fe3O4 NPs@Biimidazole IL-KCC-1 as a leach proof nanocatalyst for the synthesis of imidazo[1,2-a]pyridines in aqueous medium. *Applied Organometallic Chemistry*, 35(1), 1–13. DOI: 10.1002/aoc.6031

Bhutia, T., Das, D., Chatterjee, A., & Banerjee, M. (2019). Efficient and "green" synthetic route to imidazo[1,2- a]pyridine by Cu(II)-Ascorbate-Catalyzed A 3 - coupling in aqueous micellar media. *ACS Omega*, 4(3), 4481–4490. DOI: 10.1021/acsomega.8b03581 PMID: 31459643

Bhutia, T., Panjikar, C., Panjikar, C., Iyer, S., Chatterjee, A., & Banerjee, M. (2020). Iodine Promoted Efficient Synthesis of 2-Arylimidazo[1,2-a]pyridines in Aqueous Media: A Comparative Study between Micellar Catalysis and an "on-Water" Platform. *ACS Omega*, 5(22), 13333–13343. DOI: 10.1021/acsomega.0c01478 PMID: 32548520

Budhiraja, M., Kondabala, R., Ali, A., & Tyagi, V. (2020). First biocatalytic Groebke-Blackburn-Bienaymé reaction to synthesize imidazo[1,2-a]pyridine derivatives using lipase enzyme. *Tetrahedron*, 76(47), 131643. DOI: 10.1016/j.tet.2020.131643

Budumuru, P., Golagani, S., Siva, V., & Kantamreddi, S. (2018). Design and synthesis of novel imidazo [1, 2-a] pyridine derivatives and their anti-bacterial activity. *Asian Journal of Pharmaceutical and Clinical Research*, 11(8), 252. DOI: 10.22159/ajpcr.2018.v11i8.26241

Byth, F., Culshaw, D., Green, S., Oakes, S. E., & Thomas, A. P. (2004). Imidazo [1, 2- a] pyridines. Part 2 : SAR and optimisation of a potent and selective class of cyclin-dependent kinase inhibitors. *Bioorganic & Medicinal Chemistry Letters*, 14(9), 2245–2248. DOI: 10.1016/j.bmcl.2004.02.015 PMID: 15081017

Byth, F., Culshaw, D., Green, S., Oakes, S. E., & Thomas, A. P. (2004). Imidazo [1, 2- a] pyridines. Part 2 : SAR and optimisation of a potent and selective class of cyclin-dependent kinase inhibitors. *Bioorganic & Medicinal Chemistry Letters*, 14(9), 2245–2248. DOI: 10.1016/j.bmcl.2004.02.015 PMID: 15081017

Chen, Z., Liang, P., Xu, F., Qiu, R., Tan, Q., Long, L., & Ye, M. (2019). Lewis Acid-Catalyzed Intermolecular Annulation: Three-Component Reaction toward Imidazo[1,2- a]pyridine Thiones. *The Journal of Organic Chemistry*, 84(14), 9369–9377. DOI: 10.1021/acs.joc.9b01188 PMID: 31274309

Cheng, Y., Moraski, C., Cramer, J., Miller, J., & Schorey, S. (2014). Bactericidal Activity of an Imidazo [1,2-a] pyridine Using a Mouse M tuberculosis Infection Model. *PLoS One*, 9(1), 1–8. DOI: 10.1371/journal.pone.0087483 PMID: 24498115

Cheng, Y., Moraski, C., Cramer, J., Miller, J., Schorey, S. (2014). Bactericidal Activity of an Imidazo [1, 2-a] pyridine Using a Mouse M . tuberculosis Infection Model. *journal.pone, 9*, 1-8. DOI: 10.1371/journal.pone.0087483

Das, D., Bhutia, T., Panjikar, C., Chatterjee, A., & Banerjee, M. (2020). A simple and efficient route to 2-arylimidazo[1,2-a]pyridines and zolimidine using automated grindstone chemistry. *Journal of Heterocyclic Chemistry*, 57(11), 4099–4107. DOI: 10.1002/jhet.4106

Deepika, G., Ashok, K., & Pratibha, S. (2022). Molecular Iodine-Catalyzed Synthesis of Imidazo[1,2-a]Pyridines: Screening of Their In Silico Selectivity, Binding Affinity to Biological Targets, and Density Functional Theory Studies Insight. *ACS Omega*, 7(26), 22421–22439. DOI: 10.1021/acsomega.2c01570 PMID: 35811892

Feng, L., Li, Q., He, Z., Xi, Y., Chen, Y., & Yu, Q. (2019). Electrochemically initiated intermolecular C-N formation/cyclization of ketones with 2-aminopyridines: An efficient method for the synthesis of imidazo[1,2-: A] pyridines. *Green Chemistry*, 21(7), 1619–1624. DOI: 10.1039/C8GC03622E

Fisher, M. H., & Lusi, A. (1972). Imidazo(1,2-a)pyridine anthelmintic and anti-fungal agents. *Journal of Medicinal Chemistry*, 15(9), 982–985. DOI: 10.1021/jm00279a026 PMID: 5065787

Gaurav, P., Shaik, G., Swanand, J., Naiyaz, M., Arunava, D., Madhavi, V., Siddharth, C., & Srinivas, N. (2023). Microwave-assisted Cu(i)-catalyzed one-pot tandem synthesis of pyridoimidazole-fused quinolines as new antimycobacterial agents: DFT and ESI-HRMS study. *New Journal of Chemistry*, 47(12), 5961–5969. DOI: 10.1039/D2NJ06165A

Gernet, A., Sevrain, N., Volle, J., Ayad, T., Pirat, J., & Virieux, D. (2020). Diversity-Oriented Synthesis toward Aryl- and Phosphoryl- Functionalized Imidazo[1,2 - a] pyridineš. *The Journal of Organic Chemistry*, 85(22), 14730–14743. DOI: 10.1021/acs.joc.0c02059 PMID: 33166470

Giorgio, V., Enzo, L., Roberto, R. (2024). Imidazopyridine Family: Versatile and Promising Heterocyclic Skeletons for Different Applications. *Molecules. 29*. 2668. doi; .DOI: 10.3390/molecules29112668

Godugu, K., & Nallagondu, R. (2021). Solvent and catalyst-free synthesis of imidazo[1,2-a]pyridines by grindstone chemistry. *Journal of Heterocyclic Chemistry*, 58(1), 250–259. DOI: 10.1002/jhet.4164

Gómez, M. A. R., Islas-Jácome, A., & Gámez-Montaño, R. (2019). Synthesis of Imidazo[1,2-*a*]pyridines via Multicomponent GBBR Using α-isocyanoacetamides. *Proceedings.*, 9, 53. DOI: 10.3390/ecsoc-22-05692

Groebke, K., Weber, L., & Mehlin, F. (1998). Synthesis of Imidazo [1, 2-a] annulated Pyridines, Pyrazines and Pyrimidines by a Novel Three-Component Condensation. *ChemInform*, 29(39), 661–663. DOI: 10.1002/chin.199839120

Gueiffier, A., Hassani, M., & Elhakmaoui, A. (1996). Synthesis of Acyclo- C - nucleosides the Imidazo [1, 2- a] pyridine and Pyrimidine Series as Antiviral Agents. *Journal of Medicinal Chemistry*, 39(14), 2856–2859. DOI: 10.1021/jm9507901 PMID: 8709116

Gueiffier, A., Hassani, M., & Elhakmaoui, A. (1996). Synthesis of Acyclo- C - nucleosides in the Imidazo [1, 2- a] pyridine and Pyrimidine Series as Antiviral Agents. *Journal of Medicinal Chemistry*, 2623(14), 2856–2859. DOI: 10.1021/jm9507901 PMID: 8709116

Gui, Q., Wang, B., Zhu, S., Li, F., Zhu, M., Yi, M., Yu, J., Wu, Z., & He, W. (2021). Four-component synthesis of 3-aminomethylated imidazoheterocycles in EtOH under catalyst-free, oxidant-free and mild conditions. *Green Chemistry*, 23(12), 4430–4434. DOI: 10.1039/D1GC01017D

Hamdouchi, C., de Blas, J., del Prado, M., Gruber, J., Heinz, B. A., & Vance, L. (1999). 2-Amino-3-substituted-6- [(E) -1-phenyl-2- (N -methylcarbamoyl) vinyl] imidazo- [1, 2- a] pyridines as a Novel Class of Inhibitors of Human Rhinovirus : Stereospecific Synthesis and Antiviral Activity. *Journal of Medicinal Chemistry*, 42(1), 50–59. DOI: 10.1021/jm9810405 PMID: 9888832

Hamdouchi, C., de Blas, J., del Prado, M., Gruber, J., Heinz, B. A., & Vance, L. (1999). 2-Amino-3-substituted-6-[(E) -1-phenyl-2- (N -methylcarbamoyl) vinyl] imidazo- [1, 2- a] pyridines as a Novel Class of Inhibitors of Human Rhino virus : Stereospecific Synthesis and Antiviral Activity. *Journal of Medicinal Chemistry*, 42(1), 50–59. DOI: 10.1021/jm9810405 PMID: 9888832

Hariss, L., Hadir, B., El-masri, M., Roisnel, T., Grée, R., & Hachem, A. (2017). Grée R, Hachem A. Preparation of imidazo [1, 2- a] - N -heterocyclic derivatives with gem -difluorinated side chains. *Beilstein Journal of Organic Chemistry*, 13, 2115–2121. DOI: 10.3762/bjoc.13.208 PMID: 29062431

Hassani, M., Chavignon, O., & Chezal, J. (1999). New products Synthesis and antiviral activity of imidazo [1, 2- a] pyridines. *European Journal of Medicinal Chemistry*, 34(3), 271–274. DOI: 10.1016/S0223-5234(99)80061-0

Hassani, M., Chavignon, O., & Chezal, J. (1999). New products Synthesis and antiviral activity of imidazo [1, 2- a] pyridines. *European Journal of Medicinal Chemistry*, 34(3), 271–274. DOI: 10.1016/S0223-5234(99)80061-0

Hayakawa, M., Kaizawa, H., Kawaguchi, K., Ishikawa, N., Koizumi, T., Ohishi, T., Yamano, M., Okada, M., Ohta, M., & Tsukamoto, S. (2007). Synthesis and biological evaluation of imidazo [1, 2- a] pyridine derivatives as novel PI3 kinase p110 a inhibitors. *Bioorganic & Medicinal Chemistry*, 15(1), 403–412. DOI: 10.1016/j.bmc.2006.09.047 PMID: 17049248

Hendriks, C., Nürnberg, P., & Bolm, C. (2015). Zolimidine Analogues: The Synthesis of Imidazo[1,2-α]pyridine-Based Sulfilimines and Sulfoximines. *Synthesis*, 47(8), 1190–1194. DOI: 10.1055/s-0034-1380109

Hiebel, A., Fall, Y., Scherrmann, C., & Berteina-Raboin, S. (2014). Straightforward synthesis of various 2,3-diarylimidazo[1,2-a]pyridines in peg400 medium through one-pot condensation and C-H arylation. *European Journal of Organic Chemistry*, 2014(21), 4643–4650. DOI: 10.1002/ejoc.201402079

Huang, H., Ji, X., Tang, X., Zhang, M., Li, X., & Jiang, H. (2013). Conversion of Pyridine to Imidazo [1, 2 - a] pyridines by Copper-Catalyzed Aerobic Dehydrogenative Cyclization with Oxime Esters. *Organic Letters*, 15(24), 6254–6257. DOI: 10.1021/ol403105p PMID: 24261576

Ismail, A., Brun, R., Wenzler, T., Tanious, A., Wilson, D., & Boykin, W. (2004). Novel Dicationic Imidazo [1, 2- a] pyridines and 5, 6, 7, 8-Tetrahydro-imidazo [1, 2- a] pyridines as Antiprotozoal Agents. *Journal of Medicinal Chemistry*, 47(14), 3658–3664. DOI: 10.1021/jm0400092 PMID: 15214792

Jagadhane, B., & Telvekar, N. (2014). Synthesis of 3-Nitro-2-arylimidazo[1,2- a]pyridines Using Sodium DichloroÂ-iodide. *Synlett*, 25(18), 2636–2638. DOI: 10.1055/s-0034-1379185

Jenkinson, S., Thomson, M., McCoy, D., Edelstein, M., Danehower, S., Lawrence, W., Wheelan, P., Spaltenstein, A., & Gudmundsson, K. (2010). Blockade of X4-tropic HIV-1 cellular entry by GSK812397, a potent noncompetitive CXCR4 receptor antagonist. *Antimicrobial Agents and Chemotherapy*, 54(2), 817–824. DOI: 10.1128/AAC.01293-09 PMID: 19949058

Jose, G., Kumara, T., & Nagendrappa, G. (2015). Synthesis, molecular docking and anti-mycobacterial evaluation of new imidazo[1,2-a]pyridine-2-carboxamide derivatives. *European Journal of Medicinal Chemistry*, 89, 616–627. DOI: 10.1016/j.ejmech.2014.10.079 PMID: 25462270

Katritzky, R., Xu, Y., & Tu, H. (2003). Regiospecific Synthesis of 3-Substituted Imidazo [1, 2-c] pyrimidine. *The Journal of Organic Chemistry*, 68, 4935–4937. DOI: 10.1021/jo026797p PMID: 12790603

Ku, E., Schmitt, M., Cardozo, V., Lugnier, C., Villa, P., Lopes, B., Romeiro, N., Justiniano, H., Martins, A., Fraga, A., Bourguignon, J., & Barreiro, J. (2012). Design, Synthesis, and Pharmacological Evaluation of N - Acylhydrazones and Novel Conformationally Constrained Compounds as Selective and Potent Orally Active Phosphodiesterase - 4 Inhibitors. *Journal of Medicinal Chemistry*, 55(17), 7525–7545. DOI: 10.1021/jm300514y PMID: 22891752

Kusy, D., Maniukiewicz, W., & Błażewska, M. (2019). Microwave-assisted synthesis of 3-formyl substituted imidazo[1,2-a]pyridines. *Tetrahedron Letters*, 60(45), 2–6. DOI: 10.1016/j.tetlet.2019.151244

Lian, G., Li, J., Liu, P., & Sun, P. (2019). An Approach to Quinoline-Fused Imidazopyridines via CDC of Ethers with Imidazopyridines under Metal-Free Conditions. *The Journal of Organic Chemistry*, 84(24), 16346–16354. DOI: 10.1021/acs.joc.9b02897 PMID: 31773963

Lida, H., Demizu, R., & Ohkado, R. (2018). Tandem Flavin-Iodine-Catalyzed Aerobic Oxidative Sulfenylation of Imidazo[1,2-a]Pyridines with Thiols. *The Journal of Organic Chemistry*, 83(19), 12291–12296. DOI: 10.1021/acs.joc.8b01878 PMID: 30165018

Ma, L., Wang, X., Yu, W., & Han, B. (2011). ChemComm TBAI-catalyzed oxidative coupling of aminopyridines with b -keto esters. *Chemical Communications (Cambridge)*, 47, 11333–11335. DOI: 10.1039/c1cc13568f PMID: 21879040

Mathavan, S., & Yamajala, R. (2020). Sustainable Synthetic Approaches for 3-Aminoimidazo-fused Heterocycles via Groebke-Blackburn-Bienaymé Process. *ChemistrySelect*, 5, 10637–10642. DOI: 10.1002/slct.202002894

Mohamed, A., Hameed, A., Moustafa, S., & Al-mousawi, M. (2017). An efficient and catalyst-free synthesis of ylamine derivatives via Strecker reaction under controlled microwave heating. *Green Process Synth.*, 6, 371–375. DOI: 10.1515/gps-2017-0019

Mutkule, N., Bugad, N., Mokale, S., Choudhari, V., & Ubale, M. (2020). Novel approach in the synthesis of imidazo [1, 2-a] pyridine from phenyl acrylic acids. *Journal of Heterocyclic Chemistry*, 57(8), 3186–3192. DOI: 10.1002/jhet.4026

Nguyen, O., Ha, T., Dang, V., Vo, Y. H., Nguyen, T. T., Le, N. T. H., & Phan, N. T. S. (2019). Superparamagnetic nanoparticle-catalyzed coupling of 2-amino pyridines/pyrimidines with trans-chalcones. *RSC Advances*, 9(10), 5501–5511. DOI: 10.1039/C9RA00097F PMID: 35515937

Okai, H., Tanimoto, K., Ohkado, R., & Iida, H. (2020). Multicomponent Synthesis of Imidazo[1,2 - a]pyridines: Aerobic Oxidative Formation of C − N and C − S Bonds by Flavin − Iodine- Coupled Organocatalysis. *Organic Letters*, 22(20), 8002–8006. DOI: 10.1021/acs.orglett.0c02929 PMID: 33006477

Olomiiets, V., Tsygankov, V., Kornet, N., Brazhko, A., Musatov, I., Chebanov, A., & Beilstein, J. (2023). Synthesis of imidazo[1,2-a]pyridine-containing peptidomimetics by tandem of Groebke–Blackburn–Bienaymé and Ugi reactions. *Beilstein Journal of Organic Chemistry*, 19, 727–735. DOI: 10.3762/bjoc.19.53 PMID: 37284590

Palmer, M., Grobbel, B., Jecke, C., Brehm, C., Zimmermann, J., Buhr, W., & Kromer, W. (2007). Synthesis and Evaluation of 7 H-8, 9-Dihydropyrano [2, 3-c] imidazo [1, 2-a] pyridines as Potassium-Competitive Acid Blockers. *Journal of Medicinal Chemistry*, 50(24), 6240–6264. DOI: 10.1021/jm7010063 PMID: 17975907

Panda, J., Raiguru, B., Mishra, M., Mohapatra, S., & Nayak, S. (2022). Recent Advances in the Synthesis of Imidazo[1,2-a]pyridines: A Brief Review. *ChemistrySelect*, 7(3), e202103987. Advance online publication. DOI: 10.1002/slct.202103987

Pericherla, K., Jha, A., Khungar, B., & Kumar, A. (2013). Copper-Catalyzed Tandem Azide À Alkyne Cycloaddition, Ullmann Type C À N Coupling, and Intramolecular Direct Arylation. *Organic Letters*, 15(17), 4304–4307. DOI: 10.1021/ol401655r PMID: 23947761

Pushpalatha, B., Srinivasarao, G., & Venkata, K. (2018). Design and synthesis of novel imidazo[1,2-a]pyridine derivatives and their anti-bacterial activity. *Asian Journal of Pharmaceutical and Clinical Research*, 11(8), 252. DOI: 10.22159/ajpcr.2018.v11i8.26241

Rani, S., Reddy, G., Susithra, E., Mak, K.-K., Pichika, M. R., Reddymasu, S., & Rao, M. V. B. (2020). Synthesis and anticancer evaluation of amide derivatives of imidazo-pyridines. *Medicinal Chemistry Research*, 30(1), 74–83. DOI: 10.1007/s00044-020-02638-w

Rao, N., Balamurali, M., Maiti, B., Thakuria, R., & Chanda, K. (2018). Efficient Access to Imidazo[1,2- a] pyridines/pyrazines/pyrimidines via Catalyst-Free Annulation Reaction under Microwave Irradiation in Green Solvent. *ACS Combinatorial Science*, 20(3), 164–171. DOI: 10.1021/acscombsci.7b00173 PMID: 29373013

Ravi, C., Mohan, C., & Adimurthy, S. (2014). N - Chlorosuccinimide-Promoted Regioselective Sulfenylation of Imidazoheterocycles at Room Temperature. *Organic Letters*, 16(11), 2978–2981. DOI: 10.1021/ol501117z PMID: 24838116

Rousseau, L., Matlaba, P., & Parkinson, J. (2007). Multicomponent synthesis of imidazo [1, 2- a] pyridines using catalytic zinc chloride. *Tetrahedron Letters*, 48(23), 4079–4082. DOI: 10.1016/j.tetlet.2007.04.008

Saini, K., Kaswan, P., Pericherla, K., & Kumar, A. (2015). Synthesis of Naphtho-Fused Imidazo [1, 2- a] pyridines through Copper-Catalyzed Cascade Reactions. *Asian Journal of Organic Chemistry*, 4(12), 1380–1385. DOI: 10.1002/ajoc.201500297

Sanapalli, B., Ashames, A., Sigalapalli, D., Shaik, A., Bhandare, R., & Yele, V. (2022). Synthetic Imidazopyridine-Based Derivatives as Potential Inhibitors against Multi-Drug-Resistant Bacterial Infections: A Review. *Antibiotics (Basel, Switzerland)*, 11(12), 1680. DOI: 10.3390/antibiotics11121680 PMID: 36551338

Santaniello, S., Price, J., & Murray, K.Jr. (2017). Synthesis and Characterization of 2-Phenylimidazo[1,2-A]pyridine: A Privileged Structure for Medicinal Chemistry. *Journal of Chemical Education*, 94(3), 388–391. DOI: 10.1021/acs.jchemed.6b00286

Santra, S., Bagdi, K., Majee, A., & Hajra, A. (2013). Iron ACHTUNG RE (III) - Catalyzed Cascade Reaction between Nitroolefins and 2-Aminopyridines : Synthesis of Imidazo [1,2- a] pyridines and Easy Access towards Zolimidine. *Advanced Synthesis & Catalysis*, 355(6), 1065–1070. DOI: 10.1002/adsc.201201112

Shao, T., Gong, Z., Su, T., Hao, W., & Che, C. (2017). A practical and efficient approach to imidazo [1, 2- a] pyridine- fused isoquinolines through the post-GBB transformation strategy. *Beilstein Journal of Organic Chemistry*, 13, 817–824. DOI: 10.3762/bjoc.13.82 PMID: 28546839

Sheela, M., & Anitha, I. (2016). Green synthesis of imidazo [1, 2-a] pyridines in aqueous medium. *J. Appl. Chem*, 9, 1–5. DOI: 10.9790/5736-0904010105

Shivhare, N., Jaiswal, K., Srivastava, A., Tiwari, K., & Siddiqui, R. (2018). Visible-light-activated C-C and C-N bond formation in the synthesis of imidazo[1,2-: A] pyridines and imidazo[2,1- b] thiazoles under catalyst and solvent-free conditions. *New Journal of Chemistry*, 42(20), 16591–16601. DOI: 10.1039/C8NJ03339K

Singh, D., Kumar, G., Dheer, D., Jyoti, K., Manoj, A., & Qazi Naveed, S. (2019). BCl 3 -Mediated C-N, C-S, and C-O Bond Formation of Imidazo[1,2- a]pyridine Benzylic Ethers. *ACS Omega*, 4(3), 4530–4539. DOI: 10.1021/acsomega.9b00035 PMID: 31459646

Singh, K., Kamal, A., Kumari, S., Kumar, D., Maury, S. K., Srivastava, V., & Singh, S. (2020). Eosin Y-catalyzed synthesis of 3-aminoimidazo[1,2-a]pyridines via the hat process under visible light through formation of the C−N bond. *ACS Omega*, 5(46), 29854–29863. DOI: 10.1021/acsomega.0c03941 PMID: 33251420

Tachikawa, Y., Nagasawa, Y., Furuhashi, S., Cui, L., Yamaguchi, E., Tada, N., Miura, T., & Itoh, A. (2015). Metal-free synthesis of imidazopyridine from nitroalkene and 2-aminopyridine in the presence of a catalytic amount of iodine and aqueous hydrogen peroxide. *RSC Advances*, 5(13), 9591–9593. DOI: 10.1039/C4RA14970J

Taleb, H., Al-qawasmeh, A., & Zaarour, R. (2011). European Journal of Medicinal Chemistry Design, synthesis and in vitro antimicrobial evaluation of novel Imidazo [1, 2- a] pyridine and imidazo [2, 1- b][1, 3] benzothiazole motifs. *European Journal of Medicinal Chemistry*, 46(5), 1874–1881. DOI: 10.1016/j.ejmech.2011.02.051 PMID: 21414694

Tran, Q., Jacoby, A., Roberts, E., Swann, W. A., Harris, N. W., Dinh, L. P., Denison, E. L., & Yet, L. (2019). Synthesis of 3-aryl-2-phosphinoimidazo[1,2-a]pyridine ligands for use in palladium-catalyzed cross-coupling reactions. *RSC Advances*, 9(31), 17778–17782. DOI: 10.1039/C9RA02200G PMID: 35520553

Vanya, K. (2021). Recent Progress in Metal-Free Direct Synthesis of Imidazo[1,2-a]pyridines. *ACS Omega*, 6(51), 35173–35185. DOI: 10.1021/acsomega.1c03476 PMID: 34984250

Wang, J., Xu, H., Xin, M., & Zhang, Z. (2016). I2 -mediated amination/cyclization of ketones with 2-aminopyridines under high-speed ball milling: Solvent- and metal-free synthesis of 2,3-substituted imidazo[1,2-a]pyridines and zolimidine. *Molecular Diversity*, 20(3), 659–666. DOI: 10.1007/s11030-016-9666-y PMID: 26975201

Wang, Y., Frett, B., & Li, H. (2014). E ffi cient Access to 2,3-Diarylimidazo[1,2 - a]pyridines via a One-Pot, Ligand-Free, Palladium-Catalyzed Three-Component Reaction under Microwave Irradiation. *Organic Letters*, 16(11), 3016–3019. DOI: 10.1021/ol501136e PMID: 24854606

Xiao, Y., Yu, L., Yu, Y., Tan, Z., & Deng, W. (2020). Visible-light-mediated C3-ethoxycarbonylmethylation of imidazo[1,2-a]pyridines and convenient access to Zolpidem. *Tetrahedron Letters*, 152606(51), 152606. Advance online publication. DOI: 10.1016/j.tetlet.2020.152606

Xu, X., Hu, P., & Yu, W. (2014). Bu4NI-catalyzed synthesis of imidazo[1,2- a] pyridines via oxidative coupling of aminopyridines with nitroolefins. *Synlett*, 25(5), 718–720. DOI: 10.1055/s-0033-1340485

Yadav, S., Reddy, S., Rao, G., Srinivas, M., & Narsaiah, V. (2007). Cu (OTf) 2 -catalyzed synthesis of imidazo [1, 2- a] pyridines from a -diazoketones and 2-aminopyridines. *Tetrahedron Letters*, 48(43), 7717–7720. DOI: 10.1016/j.tetlet.2007.08.090

Yan, R., Yan, H., Ren, Z., Gao, X., Huang, G., & Liang, Y. (2012). Cu(I)-Catalyzed Synthesis of Imidazo[1,2-a]pyridines from Aminopyridines and Nitroolefins Using Air as the Oxidant. *The Journal of Organic Chemistry*, 77(4), 2024–2028. DOI: 10.1021/jo202447p PMID: 22239920

Yu, L., Bhaswati, G., & Kwok-Kong, M. (2012). In situ formation of β-glycosyl imidinium triflate from participating thioglycosyl donors: Elaboration to disarmed–armed iterative glycosylation. *Chemical Communications*, 48, 2–5. DOI: 10.1039/c2cc35927h PMID: 23023321

Zhang, J., Lu, X., Li, T., Wang, S., & Zhong, G. (2017). Copper-Catalyzed Oxidative Cyclization of 2 - Amino-azaarenes with Lignin Models: Synthesis of 3 - Phenoxy Imidazo Heterocycles. *The Journal of Organic Chemistry*, 82(10), 5222–5229. DOI: 10.1021/acs.joc.7b00480 PMID: 28429945

Zhu, Z., Guo, D., Ji, J., Zhu, X., Tang, J., Xie, Z.-B., & Le, Z.-G. (2020). Visible-Light-Induced Dehydrogenative Imidoylation of Imidazo[1,2 - a]pyridines with α - Amino Acid Derivatives and α - Amino Ketones. *The Journal of Organic Chemistry*, 85(23), 15062–15071. DOI: 10.1021/acs.joc.0c01940 PMID: 33135893

Zhuang, Z., Kung, M., Wilson, A., Lee, C.-W., Plössl, K., Hou, C., Holtzman, D. M., & Kung, H. F. (2003). Structure - Activity Relationship of Imidazo [1, 2- a] pyridines as Ligands for Detecting -Amyloid Plaques in the Brain. *Journal of Medicinal Chemistry*, 46(2), 237–243. DOI: 10.1021/jm020351j PMID: 12519062

Chapter 12
Aqua Mediated Multicomponent Synthesis of N, O– Heterocycles and Biological Activity of Fused Isoxazole Derivatives

Shalini Jaiswal
https://orcid.org/0000-0003-0137-7734
Amity University, Greater Noida, India

ABSTRACT

In the literature, several Drugs substituted with isoxazole ring shows various biological activities. The various drugs available in the market like sulfamethoxazole A, muscimol-B, ibotenic acid C, parecoxib-D, and leounomide contain the core structure of isoxazole. The use of water and phase transfer catalyst combined acted as a green attribute for the current approach. The current methodology has several benefits, such as a single-pot reaction, environmental friendliness, economic viability, wide substrate range, ease of operation, quick reaction time, simple workup process, and elevated yields. This chapter concentrates on the many antimicrobial characteristics of isoxazole derivatives, such as their analgesic, antitubercular, anticancer, and antibacterial qualities. We also report the synthesis of isoxazole derivatives using the aqua-mediated reaction of Chalcone hydroxylamine hydrochloride, and p-toluene sulfonic acid as catalyst.

DOI: 10.4018/979-8-3693-7267-8.ch012

1. INTRODUCTION

A heterocyclic compound is a cyclic compound whose ring(s) contains at least two different elements. (Moss et al., 1995). All nucleic acids, most medications, most biomass (cellulose and related materials), and a wide variety of artificial and natural pigments are examples of heterocyclic compounds. Nitrogen heterocycles are included in 59% of US FDA-approved medications. (Vitaku et al., 2014).

The electronic structure of heterocyclic organic molecules can be used to classify them effectively. The acyclic derivatives are similar in behaviour to the saturated organic heterocycles. Heterocycles can be carbon-free. Tetra sulphur tetranitride S_4N_4, borazine (B_3N_3 ring), and Hexachloro-phosphazenes (P_3N_3 rings) are a few examples. Inorganic ring systems are mostly of theoretical interest and have many practical uses.

There is also a sizable class of compounds with five members having 1,2,3 or more heteroatoms (Tables 1 and 2). The dithiazole class, which has one nitrogen and two sulphur atoms, is one example. Imidazole, isoxazole, pyrazole, oxazole, tetrazole, thiazole, triazole, and other aliphatic heterocycles are important subclasses of these chemicals that serve as the building blocks of numerous medications and bioactive substances. Azoles are a group of five-membered rings that have two heteroatoms, at least one of which contains nitrogen (Jaiswal,2019, pp. 36-39). Both nitrogen and sulphur atoms can be found in the rings of thiazoles and Isothiazoles. There are two sulphur atoms in dithiolanes.

Table 1. Five-membered rings with one heteroatom

Heteroatom	Saturated	Unsaturated
Antimony	Stibolane	Stibole
Arsenic	Arsolane	Arsole
Bismuth	Bismolane	Bismole
Boron	Borolane	Borole
Nitrogen	Pyrrolidine	Pyrrole
Oxygen	Tetrahydrofuran	Furan
Phosphorus	Phospholane	Phosphole
Selenium	Selenolane	Selenophene
Silicon	Silacyclopentane	Silole
Sulfur	Tetrahydrothiophene	Thiophene
Tellurium		Tellurophene
Tin	Stannolane	Stannole

Table 2. Five-membered rings with two heteroatoms

Heteroatoms	Saturated	Unsaturated & partially unsaturated
2 Nitrogen atom	Imidazolidine and Pyrazolidine	Imidazole (Imidazoline) Pyrazole (Pyrazoline)
1Oxygen & 1 Sulphur atom	1,3-Oxathiolane and 1,2-Oxathiolane	Oxathiole (Oxathioline) Isoxathiole
1Nitrogen &1 Oxygen atom	Oxazolidine and Isoxazolidine	Oxazole Isoxazole
1Nitrogen & 1 Sulphur atom	Thiazolidine and Isothiazolidine	Thiazole Isothiazole
2 Oxygen atom	Dioxolane	
2 Sulphur atom	Dithiolane	Dithiole

These Heterocyclic compounds are currently the subject of extensive modern research globally. As a part of our research, we also synthesised various heterocyclic compound with antimicrobial activity (Jaiswal & Dwivedi, 2017;). Particularly regarded as a "privileged" structure for the synthesis and development of novel pharmaceuticals are heterocycles containing nitrogen (Sheldon, 1997; Dabholkar & Ansari, 2010). The Connecting oxygen and nitrogen atoms in a five-membered heteroaromatic ring known as an isoxazole. It serves as both a crucial pharmacophore and foundation for biological activity. Because its two adjacent electronegative heteroatoms facilitate hydrogen donor-acceptor interactions with a variety of target enzymes and receptors that other ring systems are unable to access, the isoxazole ring exhibits desirable pharmacological activity. One significant pharmacophore that is essential to biological function is isoxazole. Of the 351 ring systems discovered in medications that are sold, the isoxazole ring is the 33rd most frequently occurring, indicating that the synthesis of functional isoxazoles is of major importance.

Different methods for the synthesis and functionalization of isoxazoles have been developed recently. The most recent synthetic methods for functionalized isoxazoles are outlined in this thorough review, which focuses especially on the previous three years about the following reaction types: Direct functionalization, condensation, cyclo-isomerization, and 1,3-dipolar cycloaddition are the first four processes. (Morita et al., 2018) An essential family of five-membered heterocycles containing one oxygen atom and one nitrogen atom in proximity is called isoxazole.

Organic chemistry currently lacks research in areas such as the functionalization of easily accessible or commercial core structures, new methods for synthesizing substituted five-membered heterocycles, and material and pharmacological studies of these molecules. The creation of new five-membered heterocycles will be aided by the application of cutting-edge functionalization techniques and synthetic procedures, such as the domino protocol and green chemistry techniques. The application side of the challenge will be addressed with the aid of biological and material research of these new chemicals.

A review of the literature showed that several substituted isoxazoles had been made using various synthetic methods. By synthesizing the first compound in the family, isoxazole, by oxidizing propargyl aldehyde acetal, Claisen made the first contribution to the chemistry of isoxazoles in 1903. (Claisen, 1903, pp.3664–3673).

Different routes can be used to synthesis isoxazoles, utilizing both homogeneous and heterogeneous catalysts. A useful method for creating pharmacologically active five-membered isoxazoles is 1,3-Dipolar Cycloaddition. It has been observed that 1,3-Dipolar cyclization of nitrile oxides with alkynes provides good yields of isoxazoles (Tornoe et al., 2002; Violette et al.,2005; Matthew et al., 2006). There are two proposed processes known for the 1,3-dipolar cycloaddition reaction. first, a coordinated mechanism for the pericyclic cycloaddition reaction and, secondly, via a gradual process involving the production of diradical intermediates.(Firestone, 1968, pp. 2285–2290 ; Huisgen, 1968, pp. 2291–2297).A variety of physiologically dynamic compounds have been synthesized using substituted oxazole, pyrazole, isoxazole, and their analogues as precursors (Gothelf & Jorgensen, 1998 ;Jose et al.,2003) These substances sparked interest in the realms of medicine and synthetics (Hill et al., 2001).

In the contemporary chemical business, going green with chemical processes has become a contentious issue. A green solvent has developed as a novel research technique in the combination of multicomponent reactions (MCRs) with water, allowing both MCRs and green solvents to grow simultaneously toward an ideal organic synthesis.

In addition to minimizing time, labor, expense, and waste production, designing organic reactions in aqueous media offers an inherent flexibility for producing molecular complexity and diversity (Hulme & Gore, 2001; Ugi, 2001; Weber, 2002; Lieby-Muller et al., 2006; Evdokimov et al., 2007; Tejedor & Garcia-Tellado, 2007).

Some reports state that a specific reaction with a negative activation volume proceeds more quickly in water than it does in organic solvents (Kljin and Engberts, 2005; Narayan et al., 2005; Kanizsai et al., 2007). A negative activation of volume is proposed for MCRs (Pirrung and Das Sarma, 2004; Hailes, 2007). Because of this, the design of novel MCRs combined with environmentally friendly techniques has received a lot of attention, particularly in the fields of material science, organic synthesis, and drug development, because of its exceptional ability to reduce the possibility of adverse effects.

2. GREEN CHEMISTRY AND MICROWAVE-ASSISTED ORGANIC SYNTHESIS OF HETEROCYCLIC COMPOUNDS

Compared to conventional heating procedures, microwave irradiation offers several benefits, such as fast and uniform heating, remarkable response accelerations due to the high heating rate, and selective heating (Figure 1). The primary hypothesis underlying the microwave heating mechanism is the contact of the charged atom of the reaction material with the electromagnetic wavelength of a certain frequency. Heat output from electromagnetic radiation occurs either by transmission, impact, or both. The polarity of the wave energy shifts from positive to negative. Rapid molecular orientation and reorientation result from this, and collision heating is the result.

Figure 1. Comparison of conventional heating with microwave heating mechanism with a water sample

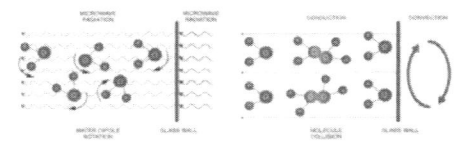

When particles that are charged with a substance are allowed to flow through a material (such as an electron in a carbon sample), current is generated that can move in phase with the field. A portion of the electric field causes charging particles within material areas to travel before the opposite is completed. As a fundamental component of Green Chemistry, MW use leads to improved reaction rates, quicker initial heating, greater reaction kinetics, clean reaction products with quicker starting material consumption, and higher yields. (Lancaster,2002; Grewal et al., 2013) The science of using microwave radiation in chemical reactions is known as Green chemistry and it has various applications. (Anastas & Warner, 1998; Kidwai et al, 2001;Lidstrom et al., 2001; Loupy,2002; Ravichandran, 2001; Kumar et al.,2010, Jaiswal et al., 2017) The most popular and extensively studied use of microwaves in chemical reactions are N-Alkylation, R-Alkylation, Oxidation (Figure 2), Reduction, Knoevenagel Condensation, Hydrolysis (Figure 3), Esterification, Decarboxylation, Deacetylation, Mannich reaction (Figure 4). (Deng et al., 1994; Gabriel et al., 1998 ; Bogda, et al., 1999; Chakraborty, & Bordoloi,1999; Jiang et al., 2001; Gupta & Wakhloo, 2007; Sreeram et al., 2008; Ravichandran & Karthikeyan,2011; Boscencu,2012; Jacob, 2012; Petersen, 2012)

Figure 2. Microwaves irradiation in N-alkylation, R-alkylation and oxidation

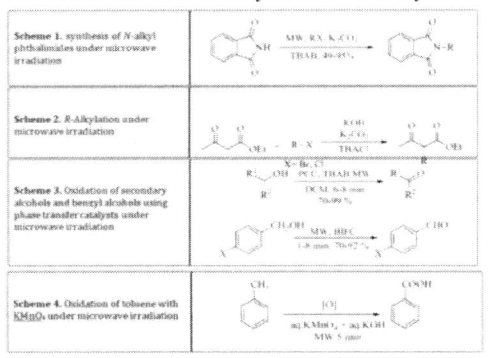

Figure 3. Microwaves irradiation in reduction, Knoevenagel condensation, hydrolysis

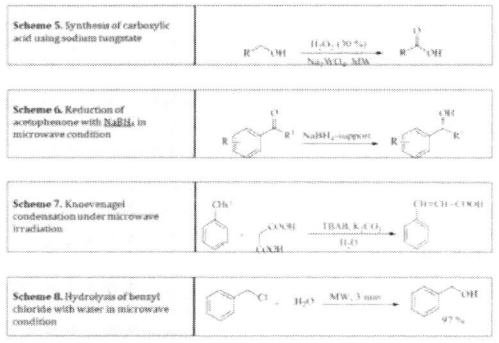

Figure 4. Microwaves irradiation in reduction, Knoevenagel condensation, hydrolysis

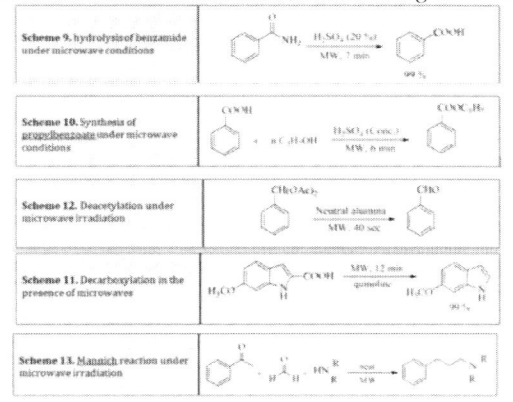

As per literature, a variety of organic reactions have been successfully carried out by scientists. (Langa et al., 1997; Larhed & Hallberg, 2001; Lew et al., 2002; Ley & Baxendale, 2002; Lidström et al., 2002; Wilson et al.,2004; Mont et al., 2006; Algul et al., 2008; Collins, 2010; Gaba & Dhingra, 2011;).

The traditional approach is insufficient to prepare the heterocyclic molecule due to its low yield, lengthy reaction durations, and usage of dangerous and unique chemical solvents. Green chemistry is one of the most appealing concepts in chemistry for environmental sustainability. Green Chemistry reduces or eliminates the use or production of hazardous elements, production, and use of chemical products. Multicomponent coupling reactions (MCRs) are a cost-effective, efficient, and waste-free method of synthesizing heterocyclic compounds as compared to conventional synthetic methods. (Dömling et al., 2012; Gawande et al., 2013)

Constructing N-heterocycles(Polshettiwar & Varma, 2008) through bond formation in one-pot MCR synthesis opens the door to a sustainable and environmentally friendly method of discovering novel molecules.(Brauch et al., 2013).The synthesis of thiazolo[3,2-α]pyridines and similar fused heterocyclic compounds has been accomplished by a MW-assisted three-component reaction involving malononitrile, aromatic aldehydes, and 2-mercaptoacetic acid in water (Figure 5).

Figure 5. Synthesis of Thiazolo [3, 2-α] pyridines

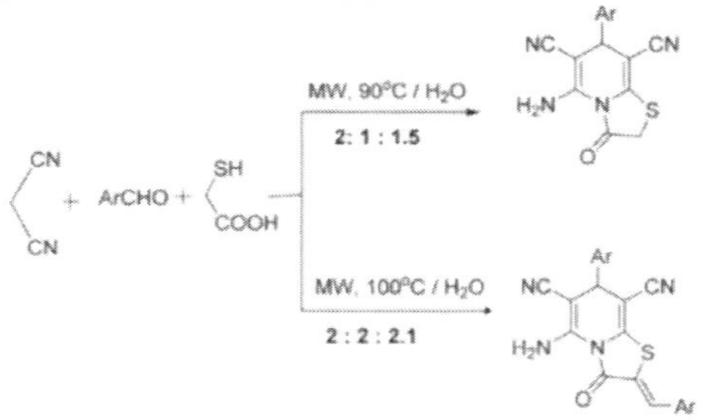

These compounds have demonstrated noteworthy biological activity. (Shi et al., 2009)

The MW-assisted synthesis of novel isoniazid (INH) analogues45 has been reported by Sriram et al. utilizing dimedone and benzaldehydes in water with a catalytic quantity of DBS. Pyrido[3,2,1-jk] carbazoles are produced when 2-(3-oxo-1,3-diarylpropyl)-1-cyclo-hexanones react domino-style with phenylhydrazine hydrochloride in water (Figure 6). (Chitra et al., 2011)

Figure 6. Synthesis of Pyridocarbazoles

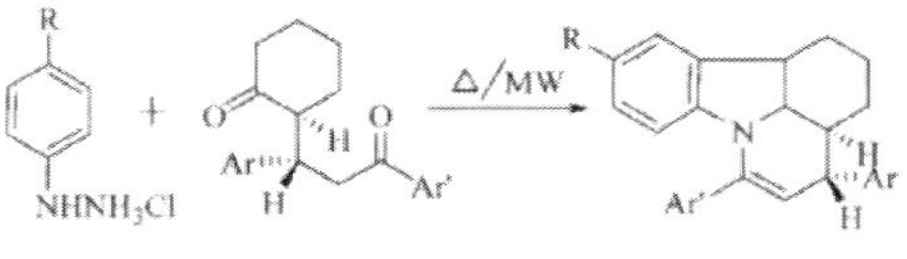

Using MW, pyrazoles and diazepines have been produced without the need for solvents or catalysts (Vaddula et al., 2013) a complete conversion takes place at 120 °C in 5–15 minutes (Figure 7), and the technique can be applied to various fused pyrazoles and diazepines.

Figure 7. Synthesis of Pyrazoles and Diazepines

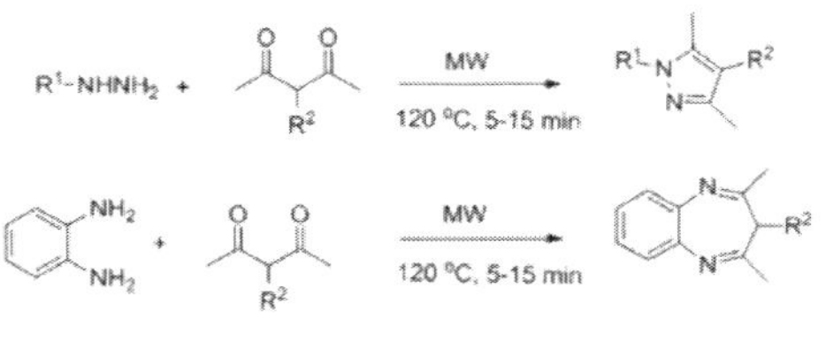

The Suzuki-Miyaura reaction between 6-chloropurines, a nucleoside, and sodium tetraarylborate, an arylation reagent, was described by Qu and their colleagues in plain water (Figure 8). (Qu et al., 2011)

Figure 8. Synthesis of 6-Arylpurines

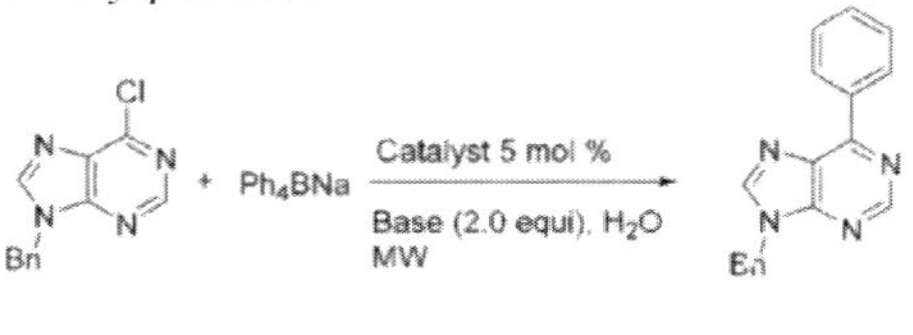

Using magnetite-Glu-Cu (Cu on glutathione modified ferrites), multicomponent processes under MW irradiation have been used to synthesize 1,2,3-triazoles in aqueous media (Figure 9) (Baig & Varma, 2013).

Figure 9. Magnetite-Glu-Cu catalysed synthesis of 1, 2, 3-Triazoles

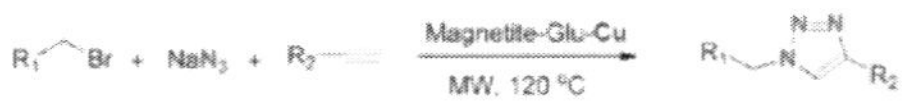

Therefore, the microwave-aided approach is considered a big step toward green chemistry. This area is greatly impacted by green innovation: drug development; medicinal chemistry; testing fields; and combinatorial chemistry. Green chemistry with chemical processes has become a big issue in the modern chemical industry. The multicomponent reactions (MCRs) with water, a green solvent have emerged as a novel research technique that permits both MCRs and green solvents to grow simultaneously toward an ideal organic synthesis.

Designing organic processes in aqueous medium with low molecular complexity and variation while minimizing labor, time, money, and waste creation is naturally adaptable. Thus, because of its exceptional decreasing the chance of side reactions, the design of novel MCRs combined with eco-advantageous strategies has received attention in various field like drug development, organic synthesis, and material research. Compared to the conventional method, Microwave Assisted Organic Synthesis2 is superior. The synthesis of several organic substances became simple, quick, clean, economical, and efficient thanks to the MAOS.

3. POTENTIAL BIOLOGICAL ACTIVITY OF HETEROCYCLIC DRUGS HAVING ISOXAZOLE NUCLEUS.

Because they contain physiologically active compounds, heterocyclic ring structures have earned great attention. A brief overview of the most potent pharmacophores indicates that heterocycles based on nitrogen and oxygen are the most prevalent class of physiologically significant small compounds. N-, O-heterocycles continue to be employed as scaffolds for molecules exhibiting intriguing biological activity in pharmacological domains.

The literature showed that distinct activities are imparted when different groups are substituted on the isoxazole ring. The synthesis of a broad range of functionalized heterocyclic scaffolds is important to medicinal chemists because it allows them to increase the amount of drug-like chemical space for binding to biological targets due to their chemical diversity.(Swinney & Anthony, 2011; Azzarito et al., 2013) To further expedite the drug discovery procedure, it is very desirable to create robust synthetic methods for the synthesis of a varied array of heterocyclic compounds.

Natural products containing isoxazole rings can treat microbial activity, inflammation, and malaria (Maguire et al., 1994; Dube et al., 1998). Certain isoxazole compounds have also been shown to treat viral, allergy, and cholesterol-related disorders (Graeve et al., 1971; Jiang & Isaacson, 1987; Raghavendra et al., 2006). Enzymatic activity breaks nucleotide bonds because of chemicals intercalating with DNA. The anticancer activity was discovered to demonstrate the effects of topoisomerase II cleaving and releasing DNA (Gatto et al., 1999). The chemical interactions with heterocyclic compounds were documented in the 1980s (Downey et al., 1980; Yang et al., 2004; Song et al., 2015;). Synthetic compounds were designed and employed as therapeutic moieties in basic research on DNA cleavage activity in the past year (Mokle et al., 2010). Due to study findings indicating a connection to the creation of novel DNA reagents for biotechnology and medicine, biochemists have focused a great deal of interest on the interaction of isoxazoles with DNA. (Sigel & Sigel,2002).

A key component in drug discovery research is a five-membered heterocyclic isoxazole pharmacophore which is utilized extensively. (Zhu et al.,2018) Functionalized isoxazole scaffolds exhibit a variety of biological activities (Figure 10), including antioxidant, antibacterial, antimicrobial, and maybe HDAC inhibitory properties. (Kang et al.,2000; Shen et al.,2019; Garella et al.,2020; Kumari et al.,2020)

Figure 10. Different activity of Isoxazole derivatives

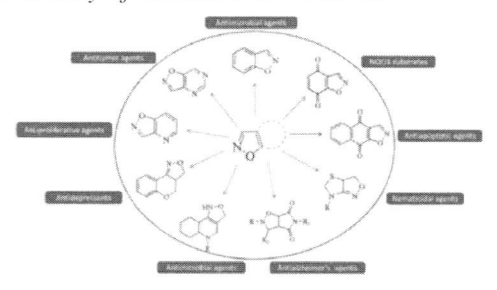

Furthermore, isoxazole derivatives play a significant role in the tiny chemical entities found in everyday synthetic goods. The talks that follow demonstrate how isoxazole is used pharmacologically to treat different biological characteristics.

a. Anti-Inflammatory and Analgesic Activity

Rajanarendar synthesized a few 6-methyl isoxazolo[5,4-d] isoxazol-3-yl aryl methadone's and evaluated their solubility, drug-likeness, lipophilicity, and molecular properties. They also tested the compounds' in vitro COX inhibitory activity and used the carrageenan-induced paw edema method to screen for anti-inflammatory activity. Significant anti-inflammatory activity was shown by the compounds with chloro or bromo substitutions on the phenyl ring (Figure 11, molecule 1) which were also more selective for the COX-2 enzyme. (Rajanarendar et al., 2015)

Figure 11. Isoxazole derivatives show anti-inflammatory and analgesic activity (1-12)

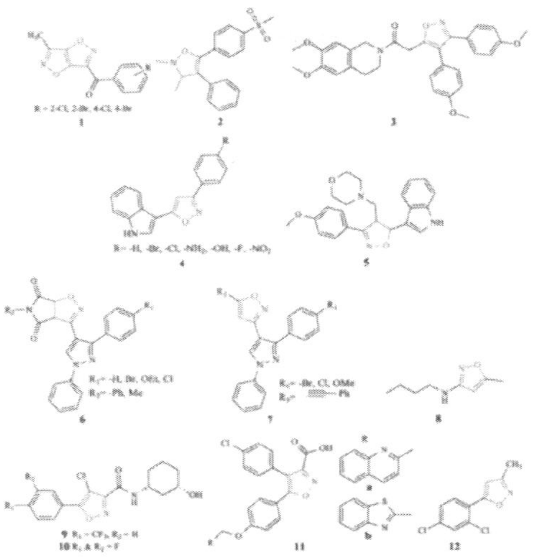

The synthesis of 4,5-diphenyl-4-isoxazolines with various substituents at the para position of one phenyl ring was described by Habeeb, who also assessed the compounds' analgesic and selective COX-2 anti-inflammatory properties. The substances with the C-3 Me substituent on the core isoxazolines ring exhibited strong analgesic properties and were specifically directed towards COX-2. 4-(4-methylsulfonylphenyl)-5-diphenyl-5-phenyl-4-isoxazoline and tested for anti-inflammatory properties. The compound 2 (Figure 11) with the sulfonylmethyl group in the para position of the phenyl ring was the most effective anti-inflammatory and analgesic of these. Studies using molecular modelling on Compound 2 revealed that the C-3 Me group and the S atom of sulfonylmethyl were essential for the specific inhibition of COX-2. (Habeeb et al.,2001)

A novel series of isoxazoles was synthesized by Perrone, who then assessed the compounds' selectivity and COX inhibitory action. It was discovered that compound 3 (Figure 11) is a sub-micromolar selective COX-2 inhibitor (IC50 0.95 μM). (Perrone et al., 2016)

Panda synthesized a few indolyl-isoxazoles (Figure 11, compound 4) and used the carrageenan-induced rat paw edema method to assess the compounds' acute anti-inflammatory efficacy. According to, every chemical showed good anti-inflammatory action, with edema reduction ranging from 36.6 to 73.7%. (Panda et al.,2009)

Amir synthesized a different series of indole derivatives with an isoxazolines moiety and used the carrageenan-induced rat paw edema method to test the compounds' anti-inflammatory properties in vivo. According to these investigations, compound 5 (Figure 11) demonstrated the greatest reduction in lipid peroxidation and ulcerogenic activity as well as the greatest anti-inflammatory effect. (Amir et al.,2010)

Karthikeyan reported employing 1,3-dipolar cycloaddition of pyrazole generated nitrile oxides with different dipolarophiles to synthesize a range of pyrazolyl isoxazolines 6 and isoxazoles 7 (Figure 11). The chemicals that were produced were assessed for their ability to inhibit pain. All the substances showed efficacy that was on par with aspirin and pentazocine, two common medications. Chemical-cooling compounds such as menthol work by activating a Ca2+-permeable ligand-gated cation channel called TRPM8 channels, which in turn cause analgesia. (Karthikeyan et al.,2009)

Ostacolo created some derivatives based on amino-isoxazoles that can control TRPM8, which is useful for cold-evoked analgesia. Using an in vivo model of cold allodynia and [Ca2+]-imaging tests in sensory neurons, the compounds' capacity to function as TRPM8 agonists was investigated. While some of the chemicals had potencies up to 200 times greater than menthol, none of them demonstrated increased effectiveness. The most promising was compound 8 (Figure 11). Primary afferent neurons produce TRPV1 channels, which are non-selective Ca2+-permeable cation channels that integrate nociceptive responses to noxious and chemical inputs. (Ostacolo et al.,2013)

Palin developed an isoxazole-3-carboxamide series and assessed its potential to regulate the TRPV1 channel following the HTS. Both potency and solubility were improved by substituting the 1S, 3R-3-aminocyclohexanol motif for isoxazole-3-carboxamide (Figure 11, compounds 9 and 10). These substances underwent additional testing on animals, where they both reduced the rat complete Freund's adjuvant assay's acute anti-inflammatory response. (Palin et al.,2011)

Banoglu developed a series of 4,5-diaryloisoxazol-3-carboxylic acids as inhibitors of leukotriene biosynthesis that target FLAP. Leukotrienes play a significant part in several anti-inflammatory illnesses. FLAP is involved in leukotriene biosynthesis at the very initial stage. Compounds 11a and 11b (Figure 11) were found to be powerful anti-inflammatory drugs with strong inhibitory effect against the formation of cellular 5-Lipoxygenase products (IC50) of 0.24 µM each. (Banoglu et al.,2016)

Isomexazole compounds were created by Silva as nAChR ligands for the purpose of developing analgesics. The positive benefits of nicotine in CNS disorders like Parkinson's disease, Alzheimer's disease, and pain relief have drawn a lot of attention to nAChR. The compound with the best analgesic performance in the series was Compound 12 (Figure 11) Silva et al.,2002).

b. Anticancer Activity

Vitale designed and synthesized new isoxazoles taking [3-(5-chlorofuran-2-yl)-5-methyl-4-phenylisoxazole], a highly selective COX-1 inhibitor as a lead. COX-1 inhibition is considered as a promising therapeutic strategy for cancer and neuro-inflammation-derived neurodegenerative diseases. All the compounds were evaluated for their in vitro COX inhibitory activity and selectivity. The substitution of methyl in the lead compound **16** with an amino group (Figure 12) improved the COX-1 selectivity and inhibitory activity. Instead low or no activity was observed for its N, N-dialkyl and N, N-diacetyl derivatives. The inhibitory activity of compound **16** was also assessed in ovarian cancer cell line (OVCAR-3) where it was found to be 20-fold more potent than the lead compound.(Vitale et al.,2014)

Figure 12. Isoxazole derivatives showing anticancer activity (16-26)

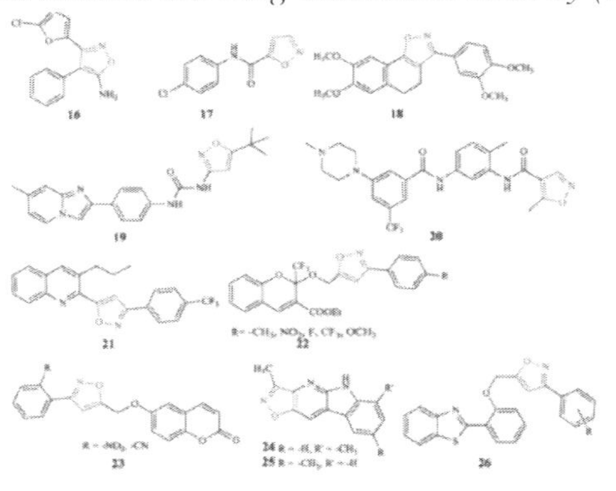

N-phenyl-5-carboxamidyl isoxazoles were synthesized by Shaw and tested for their ability to kill colon 38 and CT-26 murine colon carcinoma cells. In both human and mouse colon cancer cells, the most potent and solid tumor compound **17** (Figure 12) (IC50 2.5 μg/mL) dramatically reduced the production of phosphorylated-STAT3, a novel target for chemotherapeutic medicines. (Shaw et al., 2012)

Human cervical cancer (HeLa) and human breast carcinoma (MCF-7) cells were the two tumor cell lines, and the two endothelial cell lines (HMEC-1 and MBEC) were used for evaluation of tetralone-isoxazole fused compounds. molecule **18** (Figure 12) showed encouraging cytostatic activity; demethylation produced a molecule that was ten times less active. One important factor in the onset and progression of acute myeloid leukaemia is FLT3 overexpression. (Tzanetou et al.,2014)

Xu synthesized several new N-(5-(tert-butyl) isoxazol-3-yl)-N -phenyl urea derivatives. Compound 19 (Figure 12) was the most produced of the compounds and caused total tumor regression in the MV4-11 xenograft model by inhibiting the phosphorylation of FLT3. Compound **19** demonstrated favourable pharmacokinetic characteristics, good water solubility, and strong cytotoxic selectivity against MV4-11 cells. The antiproliferative effect was shown to be enhanced by an electron-rich fused ring (R) at the phenyl, according to SAR investigations. (Xu et al.,2015)

A series of 4-arylamido 3-methylisoxazoles were assessed their antiproliferative effects against the hematopoietic cell line U937 and the melanoma A375P. While the majority of the compounds showed poor cytotoxic action against the A375P and human normal cell line, they were superior to the reference medication Sorafenib in their selective antiproliferative activity against the U937 cell line. With an IC50 of 9.95 nM, it was discovered that the most potent inhibitor, **20**, (Figure 12), was also a potent inhibitor of FMS kinase. FMS kinase inhibitors show considerable promise in treating inflammation, osteolysis of the bone, and malignancies that are aided by macrophages. (Im et al.,015)

The 5-(3-alkylquinolin-2-yl)-3-aryl isoxazole derivatives were synthesized and tested against four human cancer cell lines (A549, COLO 205, MDA-MB 231 and PC-3). At an IC50 of less than 12 µM, compound **21** (Figure 12) demonstrated strong cytotoxicity against every tested cell line. According to the SAR, the cytotoxicity is enhanced by the fluorine or trifluoromethyl group located at the fourth position of the phenyl in the isoxazole ring. (Sambasiva et al.,2014)

Using 5-Fluorouracil as a positive control by MTT assay, isoxazole functionalized chromene derivatives (Figure 12, compound 22) were screened against four human cancer cell lines: A549-Lung Cancer (CCL-185), MCF7-Breast cancer (HTB-22), DU145-Prostate cancer (HTB-81), and HeLa-Cervical cancer (CCL-2). Every component showed mediocre to above average activity. (Ratnakar Reddy et al.,2014)

6-hydroxycoumarins were coupled to isoxazole and then tested for cytotoxic activity against five distinct human cancer cell lines: lung (A-549), prostate (PC-3), colon (HCT-116 and Colo-205), leukaemia (HL-60), and colon (HCT-116). The 6-hydroxy coumarin ortho substituted isoxazole derivatives (Figure 12, compound 23) are potent cytotoxic agents against the PC-3 cancer cell line. (Shakeel-u-Rehman et al.,2014)

A series of novel isoxazolo[5′,4′:5,6] pyrido[2,3-b] indoles and assessed them against three human cancer cell lines: Hela, MCF-7, and NCI-H460 (lung cancer) using the MTT test method. Regarding all the examined cancer cell lines, compounds **24** and **25** (Figure 12) showed strong anticancer activity equivalent to that of the reference medication. The remaining drugs demonstrated limited selectivity and moderate to good anticancer efficacy. (Rajanarendar et al.,2015). The isoxazoles linked 2-phenylbenzothiazole **26** (Figure 12) exhibited noteworthy anticancer activity against MCF-7 cells (IC50 26–43 µM), A549 cells (human lung adenocarcinoma; IC50 11–24 µM), and Colo-205 cells (human colon cancer; IC50 11–21 µM) and could induce apoptosis. (Kumbhare et al.,2012).

Isoxazolo(aza)naphthoquinones were assessed for their cytotoxic Hsp90 inhibitory activity in a range of human tumor cell lines. Most compounds showed strong affinity and antiproliferative action, primarily 3-pyridyl derivatives (Figure 13, compound 30). (Bargiotti et al.,2012).

The HDAC6 and Hsp90 inhibitory action of 4,5,6,7-tetrahydro-isoxazole-[4,5-c] pyridine compounds were assessed. The N-5 substitution using a 2,4-resorcinol carboxamide was essential for inhibitory action. Hsp90 and HDAC6 were both suppressed by compound **31** (Figure 13), which has a long-chain hydroxamic residue attached to the C-3 amide region. (Baruchello et al.,2014).

Figure 13. Isoxazole derivatives showing anticancer activity (30-38)

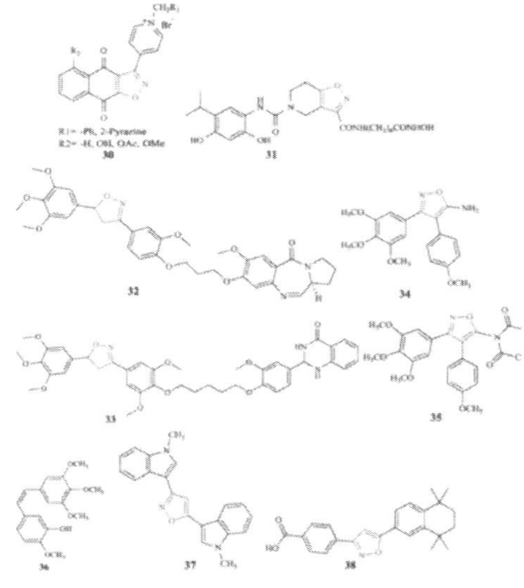

To assess their anticancer efficacy, Kamal synthesized a variety of 2,3-dihydroquinazolinones and 3,5-diaryl isoxazoline/isoxazole-linked pyrrolobenzodiazepine. Compound **32** (Figure 13) from the previous series showed G0/G1 arrest, caspase-3 activation, and ultimately apoptotic cell death. The later series, compound **33** (Figure 13) shows nuclei fragmentation and microtubule disruption. It also proved to be an excellent cyclin B1 inhibitor and CDK1 inhibitor, and this hybrid molecule was effective against eighteen human cancer cell lines. (Kamal et al.,2010, 2011)

Human myeloid leukaemia cells K562, human oesophageal carcinoma cells ECA-109, human non-small lung cancer cells A549, human hepatocellular carcinoma cells SMMC-7721, and human prostate carcinoma cells PC-3 where the five human cancer cell lines were evaluated against a series of cis-restricted 3,4-diaryl-5-aminoisoxazoles. Compared to other cell lines, the investigated drugs exhibited more effective cytotoxic activity against leukemia cells K562. But the compounds **34** (Figure 13) and its N, N-diacetyl derivative (Figure 13, compound 35) showed strong in vitro cytotoxic activities against all tested cancer lines, with IC50 values in the 0.04–12.00 µM range, (Figure 13, compound 36). (Liu et al.,2009)

The antiproliferative activity of a novel series of 2,5-bis(3′-indolyl) furans and 3,5-bis(3′-indolyl) isoxazoles were evaluated against a variety of human tumor cell lines. The compound **37** (Figure 13) was the most active and had high selectivity towards 29 cell lines among the isoxazoles (Diana et al.,2010).

Strong apoptotic retinoic acid receptor ligands were synthesized. A large range of normal and malignant cell types have their proliferation and differentiation significantly regulated by retinoids. Compound **38** showed excellent affinity for the retinoid receptors (Figure 13). Furthermore, the fact that it exhibited effects on K562 and HL60R cell lines indicated that it may have significant therapeutic applications for various leukaemia (Simoni et al.,2001)

c. Anti-Microbial Activity

5-(heteroaryl)isoxazoles synthesized by and tested for antibacterial efficacy against P. aeruginosa, S. aureus, and E. coli. Significant activity was shown by the isoxazoles substituted with 2-thienyl (Figure 14, compound 48) or 5-bromo-2-thienyl moieties (Figure 14, compound 49) at position 5. (Rama Rao et al.,2011).

Figure 14. Isoxazole derivatives showing antimicrobial activity (48-60)

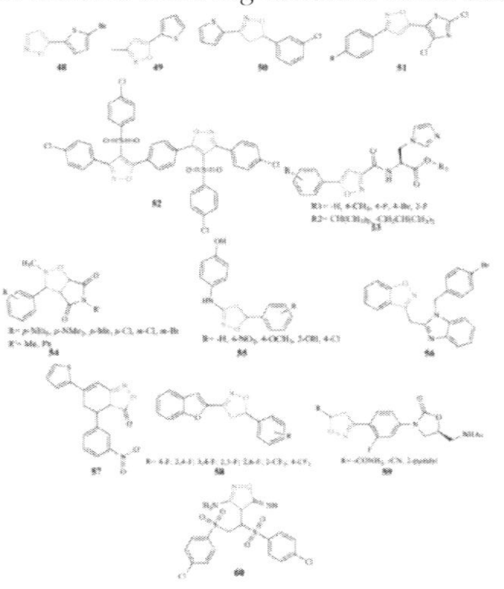

A series of 4,5-dihydro-5-(phenyl)-3-(thiophene-2-yl) isoxazoles and tested their antibacterial and antifungal properties in vitro against S. aureus, B. subtilis, E. coli, and P. aeruginosa and A. niger and C. albicans, respectively. Owing to the presence of chloride on the aromatic ring, compound **50** (Figure 14) was shown to be very active against P. aeruginosa, C. albicans, and E. coli. This finding may explain its higher lipid solubility. It was discovered that the remaining substances were moderately to extremely active. (Gautam & Singh.,2013)

The thiazolyl isoxazoles were investigated for their antifungal and antibacterial properties against P. chrysogenum, A. niger, S. aureus, B. subtilis, K. pneumoniae, and P. aeruginosa. Phenyl ring-containing thiazolyl isoxazoles (Figure 14, compound 51) that have been substituted with nitro and chloro were discovered to possess antibacterial properties against A. niger and S. aureus. (Basha et al., 2015)

The molecule replaced with chloro on the aromatic ring in (Lavanya et al. (2014)'s (1,4-phenylene) bis (arylsulfonyl isoxazoles) series (Figure 14, compound 52) While the remaining compounds reduced the germination of spores against P. chrysogenum and A. niger, compound 11 shown remarkable antibacterial activity (38 mm at 100 µg/mL) against B. subtilis.

Several aromatic heterocyclic compounds were assessed for their in vitro antifungal efficacy. In particular, the compounds with the isoxazole nucleus (Figure 14, compound 53) enhanced the efficacy against Aspergillus spp. Additionally, the compounds with the fluoro substituent at the 2-position of the aryl moiety showed excellent blood plasma stability and weak inhibition profiles for different isoforms of human cytP450, as well as remarkable activity against Candida spp., C. neoformans, A. fumigatus, and fluconazole-resistant C. albicans strain. (Zhao et al.,2017). Agirbas produced the 2,3,5-substituted perhydropyrrolo[3,4-d] isoxazole-4,6-diones (Figure 14, compound 54) by reacting N-methyl-C-arylnitrones with N-substituted maleimides, then screening the result for antibacterial activity. Most of them shown efficacy against S. aureus and E. faecalis. (Agirbas et al.,2007).

A series of novel 4-(5′-substituted-aryl-4′,5′-dihydro-isoxazole-3′-yl-amino) phenols were synthesized and evaluated for antifungal and antibacterial activity against C. albicans and A. niger as well as in vitro against S. aureus and S. typhi. The most effective antibacterial and antifungal agent in the series was discovered to be a compound with a 4-Cl phenyl substitution at the 5-position of isoxazole. (Sahu et al.,2009)

A series of N-substituted-2-(benzo[d]isoxazol-3-ylmethyl)-1H-benzimidazoles were synthesized and tested for antibacterial efficacy against S. aureus and S. typhimurium. Compound **56** (Figure 14) was the only one that showed strong action against S. aureus even at 1 µg/mL. All other compounds showed low activity against S. typhimurium. Fused ring isoxazolinones demonstrated a broad spectrum of antibacterial action in a different series. (Vaidya et al.,2007). The compound **57**

(Figure 14) with a thiophene ring was the most effective against B. subtilis and C. albicans. (Mazimba et al.,2014)

Some fluorine-containing heterocycles were synthesized by combining benzofuran with dihydroisoxazoles (Figure 14, compound 58). The compounds were assessed for their in vitro antibacterial and antifungal activity against the fungus Candida albicans, A. niger, and A. clavatus, as well as the Gram-positive bacteria S. aureus, B. subtilis, S. pneumoniae, and S. pyogenes, and the Gram-negative bacteria E. coli, P. aeruginosa, and S. typhi. When tested against all tested bacterial strains and Candida albicans, all of the compounds exhibited moderate to high activity. (Chundawat et al.,2014)

A series of 5-substituted isoxazol-3-yl coupled with oxazolidinone was synthesized and tested for in vitro antibacterial efficacy against susceptible and resistant Gram-positive pathogens. Several analogues (Figure 14, compound 59) from this series showed in vivo action in a deadly mouse infection model, albeit not as much as linezolid. They were likewise as potent, if not more so, than linezolid in vitro. (Weidner-Wells et al.,2004)

By Michael condensation of 2-(1′,2′-diaroylethyl) malononitrile and 2-(1,2-diarylsulfonylethyl) malononitrile with hydroxylamine hydrochloride, produced some amino-imino isoxazole derivatives. The compounds were shown to have superior antibacterial activity against Gram-positive bacteria and good action against Gram-negative bacteria. The compound **60** (Figure 14) with a Cl group in the aryl moiety exhibited the highest level of activity. (Padmaja et al.,2009)

d. Antitubercular Activity

A series of 1-[p-(3′-chloro-2′-benzo(b)thiophenoylamino)-phenyl]-5-aryl-isoxazoles was synthesized and evaluated for antitubercular activity. Significant action was demonstrated by compounds **70a, 70b,** and **70c** (Figure 15) against MTB H37Rv. (Kachhadia et al.,2004) Using the Lownstein-Jensen MIC approach, coumarin-based isoxazoles and assessed for their antimycobacterial activity against MTB H37Rv. With a MIC of 62.5 µg/mL and 99% inhibition, compound **71** (Figure 15) with a methoxy group demonstrated strong antimycobacterial action, while the remaining compounds displayed MICs in the range of 100–1000 µg/mL. (Patel et al.,2014)

Figure 15. Isoxazole derivatives showing antitubercular activity (70-77)

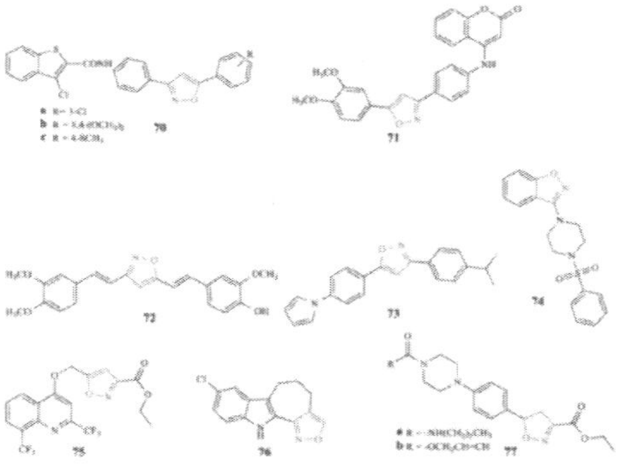

Curcumin's isoxazole analogues were produced and tested for antimycobacterial activity and activity was increased by p-methoxy and p-hydroxy groups on the aromatic ring. It was discovered that Compound **72** (Figure 15) containing mono-O-methyl curcumin isoxazole (MIC 0.09 µg/mL) was roughly 18 and 2 times more active than common medications isoniazid and kanamycin, respectively, and 1131 times more active than curcumin. Compound 72 (Fig. 13.13) likewise showed strong efficacy against clinical isolates of MTB that were resistant to multiple drugs (MIC 0.195–3.125 µg/mL).(Changtam et al.,2010)

The isoxazole-modified pyrrolyl derivatives were assessed for their anti-MTB efficacy. When compound **73** (Figure 15) was examined for mammalian cell toxicity using the A549 cancer cell line, it was shown to be non-toxic and to have substantial action against the H37Rv strain. To assess the 3-(4-(substituted sulfonyl) piperazine-1-yl) benzo[d]isoxazole analogs' in vitro antitubercular efficacy against the MTB H37Rv strain, Naidu et al. (2014) synthesized a series of them. The MIC for each chemical ranged from 3.125 to 50 µg/mL. With a selectivity of >130 and a dose of 3.125 µg/mL, compound 74 (Fig. (Fig.13)13) showed itself to be the most promising candidate for the series, blocking 99% development of the MTB H37Rv strain. (Joshi et al.,2016)

Mefloquine-isoxazole carboxylic esters were produced and their antitubercular action was assessed. It was discovered that Compound **75** (Figure 15) has outstanding intracellular and extracellular activity and selectivity against MTB H37Rv. Compound 75's ester bio-isosteres were ineffective as an antitubercular agent, indicating that the ester functioning as a prodrug. (Mao et al.,010). The isoxazolo cyclohepta[b] indoles were tested against MTB H37Rv for in vitro antimycobacterial activity. Compound **76** (Figure 15) with a chloro group in the cyclohepta[b]indole moiety was determined to have the highest activity. It was discovered that this substance is not mutagenic, making it safe to consume biologically. (Yamuna et al.,2012)

The isoxazoline esters that are 3,5-disubstituted were assessed for their anti-tuberculosis efficacy. The anti-tubercular action was enhanced by substituting piperazyl-urea (77a; Fig. Fig.13)13) and piperazyl-carbamate (Figure 15, compound 77b) for isoxazoline at the C-5 position. Nevertheless, total loss of activity resulted from substituting bioisosteric groups for the ester group at position C-3. (Rakesh et al.,2009)

e. Market Drugs Containing Isoxazole Nucleus

Numerous commercially available medications containing the isoxazole nucleus have various therapeutic effects, which led to numerous methods for synthesizing this important component. It appears that many Drugs, including sulfamethoxazole A (Rostamizadeh et al., 2019), which functions as an antibiotic, muscimol B(Johnston,2014) which functions as GABA, Ibotenic acid C(Obermeier & Muller,2020) which functions as a neurotoxin, parecoxib D(Noveck & Hubbard,2004) which functions as a COX2 inhibitor, and leounomide E(Prakash & Jarvis,1999) which functions as an immunosuppressive agent, contain the core structure of isoxazole (Table 3).

Table 3. Commercial drugs containing isoxazole nucleus

Name of the drug	Chemical structure	Antimicrobial activity
Sulfamethoxazole		Antibacterial (Majewsky et al., 2014)
Acetylsulfisoxazole		Antibacterial (Cronin et al., 2002)
Acivicin		Antitumour, Antileishmania (Conti et al.,2003)
Cycloserine		Antitubercular, Antileprotic (Desjardins et al.,2016)
Broxaterol		Bronchodilatory agent (Giustina et al., 1995)
Oxacillin		Antibacterial (Lainson et al., 2017)
Isoxautole		Pesticide (Zhao et al.,2017a)
Drazoxolon		Antifungal (Lehtonen et al., 1972)
Isocarboxazid		Antidepressant (Shader &Greenblatt 1999)
Leunomide		Antirheumatic (Golicki et al., 2012)
Danazol		Androgenic (McKinney et al., 2013)
Valdecoxib		COX-2 inhibitor (Bartzatt 2014)
Risperidone		Antpsychotic (Schoretsanitis et al., 2017)

4. MATERIALS AND METHODS

Based on the solubility of the reactants during chemical reactions, reactions in aqueous media are classified as "in-water" and "on-water" reactions (Gawande et al., 2012). The organic chemistry with solvent-free and water-mediated processes, the MW heating approach has become indispensable. (Polshettiwaret et al., 2007; Polshettiwar & Varma,2008; Gawande & Branco, 2011; Gawande et al., 2014). The MW heating in an aqueous media is beneficial for C−C coupling processes like the Sonogashira cross-coupling reaction. (Negishi & Anastasia, 2003; Baig et al., 2012, Kou et al., 2012). As part of our research previously we synthesized various Heterocyclic compounds using microwave radiation in with and without solvent (Jaiswal & Siddiqui, 2014a; Jaiswal & Sharma, 2014b; Jaiswal & Sigh, 2014c; Jaiswal, & Diwedi, 2016;Jaiswal & Dwivedi, 2017)

Melting point measurements were made in open capillaries without correction. TLC kept an eye on the product formation described. The methanol and chloroform (8:2) were used as a solvent mixture in TLC. A standard household microwave oven with a power output of 300 W and running at 2250 MHz was used in all the studies. In the Perkin-Elmer BX series FT-IR spectrophotometer, infrared spectra were computed. Tetra Methyl Silane was used as a reference chemical while calculating 1H-Nuclear Magnetic Resonance (CDCl3) spectra on Bruker AMX 500 MHz and AC 200 MHz spectrometers. Documentation of mass spectra was done using a Shimadzu-GCMS 50508. All the solvents and chemicals used in this experiment are analytical grade. The named compound's improved yield and melting point are displayed in Table 4.

Table 4. Physical and ^{1}HNMR Spectra of Compound 2(a-f)

Compd.	R	Ring Proton	Protons in the substituent	Yield (%)	M.P. (°C)	Reaction
2a	H	3.22, 4.5	7.1-7.6(m,10H, Ar-H)	77	237	60
2b	2-Cl	3.18,4.7	7.2-7.4(m,9H, Ar-H)	58	155	80
2c	4-OCH$_3$	3.28,4.6	3.7(s,3H, Ar-OCH$_3$)6.8-7.2(m,9H, Ar-H)	70	220	70
2d	3-NO$_2$	3.33,4.4	7.1-8.6(m,9H, Ar-H)	68	260	60
2e	4-N(CH$_3$)$_2$	3.30,4.2	2.8(S,6H, Ar-N, N(CH$_3$), 6.6-.4(m,9H, Ar-H)	82	168	90
2f	3-OH,4-OCH$_3$	3.33,4.6	3.9(s,3H, Ar-OCH$_3$) 5.0(S,1H, Ar-OH)6.6-7.0(m,8H, Ar-H)	79	217	90

Green Synthesis of Titled Compound 4,5-dihydro-3,5-diphenylisoxazole derivatives (a-f)

Chalcone1, hydroxylamine hydrochloride, p-toluene sulfonic acid as catalyst, and water as solvent were all added to a beaker in an equivalent molar ratio for catalysis. The compound was thoroughly mixed for two to three minutes, and then heated for the two to three minutes indicated in Table 6 at 250 watts in the microwave oven. Thin Thin-layer chromatography was used to verify the creation of the compound with the given name, and the mixture was heated until product formation was accomplished. After the reaction mixture reached room temperature, it was decanted over crushed ice. To get $2_{(a-f)}$, the separated material was filtered, dried, and then recrystallized from Benzene. (Figure 16, Scheme 1)

Figure 16. Scheme 1

5. RESULTS AND DISCUSSION

A review of the literature showed that several substituted isoxazoles had been made using various synthetic methods. When Claisen created the first compound of the series, isoxazole, in 1903 by oxidating propargyl aldehyde acetal, to the chemistry of isoxazoles (Claisen, 1903). There have been several reports on how to make isoxazoles with different substituents at positions 3, 4, and 5. Alkynyl dimethyl silyl ethers and aryl and alkyl nitrile oxides underwent a [3+2] cycloaddition reaction to create the first isoxazolyl silanols, as reported by (Denmark and Kallemeyn, 2005). When these heterocyclic silanols react cross-coupled with different aryl iodides, different 3,4,5-trisubstituted isoxazoles are formed.

6. FUTURE PERSPECTIVES

The literature evaluation notes that there has been a sufficient expansion of isoxazole research in recent years. Compounds containing isoxazoles have been successfully developed as possible inhibitors of various targets, such as the inhibition of aromatase, induction of apoptosis, inhibition of tubulin polymerization, inhibition of tyrosine kinase, inhibition of ERα, inhibition of thymidylate, inhibition of PLA2, inhibition of HDAC, inhibition of HER2, and inhibition of HSP90. Even though isoxazole derivatives have advanced remarkably, certain investigations are still necessary. These include a thorough analysis of the biological activity and mechanism of action (MOA), especially in laboratory animals, clinical trials, and investigation of the pharmacodynamic and pharmacokinetic properties of the identified powerful anticancer compounds. To determine which chemicals are the strongest, extensive research is needed. The analysis of the side effects linked to isoxazole derivatives will be a field of further work. The popular analogues that are more effective and have fewer negative effects would be a wonderful addition to the advancement of humanity.

7. CONCLUSION

Owing to their various biological functions, heterocyclic compounds and their derivatives that have sulphur and nitrogen in their skeleton are the most fascinating substances. In heterocyclic chemistry, isoxazole has a unique place, and its derivatives are very important because of their many applications in pharmacology and chemistry. The electromagnetic spectrum's microwave region is located between radio and infrared frequencies. Three main fields of drug research employ microwave instruments: peptide synthesis, DNA amplification, and organic drug screening. Five members make up the heterocyclic molecule isoxazole, which has a variety of pharmacological effects Numerous search groups in the literature have already demonstrated the growing interest in isoxazole moiety research. Because of how versatile they are as synthetic building blocks, isoxazoles and their derivatives are of tremendous interest.

REFERENCES

Agirbas, H., Guner, S., Budak, F., Keceli, S., Kandemirli, F., Shvets, N., Kovalishyn, V., & Dimoglo, A. (2007). Synthesis and structure–antibacterial activity relationship investigation of isomeric 2,3,5-substituted perhydropyrrolo[3,4-d]isoxazole-4,6-diones. *Bioorganic & Medicinal Chemistry*, 15(6), 2322–2333. DOI: 10.1016/j.bmc.2007.01.029 PMID: 17276071

Algul, O., Kaessler, A., Apcin, Y., Yilmaz, A., & Jose, J. (2008). Comparative Studies on Conventional and Microwave Synthesis of Some Benzimidazole, Benzothiazole and Indole Derivatives and Testing on Inhibition of Hyaluronidase. *Molecules (Basel, Switzerland)*, 13(4), 736–748. DOI: 10.3390/molecules13040736 PMID: 18463575

Amir, M., Javed, S. A., & Kumar, H. (2010). Design and synthesis of 3-[3-(substituted phenyl)-4-piperidin-1-ylmethyl/-4-morpholin-4-ylmethyl-4,5-dihydro-isoxazol-5-yl]-1H-indoles as potent anti-inflammatory agents. *Medicinal Chemistry Research*, 19(3), 299–310. DOI: 10.1007/s00044-009-9194-8

Anastas, P. T., & Warner, J. C. (1998). *Green Chemistry: Theory and Practice*. Oxford University Press.

Azzarito, V., Long, K., Murphy, N. S., & Wilson, A. J. (2013). Inhibition of α-helix-mediated protein–protein interactions using designed molecules. *Nature Chemistry*, 5(3), 161–173. DOI: 10.1038/nchem.1568 PMID: 23422557

Baig, R. B. N., & Varma, R. S. (2012). A highly active and magnetically retrievable nanoferrite-DOPA-copper catalyst for the coupling of thiophenols with aryl halides. *Chemical Communications (Cambridge)*, 48(20), 2582–2584. DOI: 10.1039/c2cc17283f PMID: 22293995

Baig, R. B. N., & Varma, R. S. (2013). Copper on chitosan: A recyclable heterogeneous catalyst for azide-alkyne cycloaddition reactions in water. *Green Chemistry*, 15(7), 1839–1843. DOI: 10.1039/c3gc40401c

Banoglu, E., Çelikoglu, E., Volker, S., Olgac, A., Gerstmeier, J., Garscha, U., Caliskan, B., Schubert, U. S., Carotti, A., Macchiarulo, A., & Werz, O. (2016). 4,5-Diarylisoxazol-3-carboxylic acids: A new class of leukotriene biosynthesis inhibitors potentially targeting 5-lipoxygenase-activating protein (FLAP). *European Journal of Medicinal Chemistry*, 113, 1–10. DOI: 10.1016/j.ejmech.2016.02.027 PMID: 26922224

Bargiotti, A., Musso, L., Dallavalle, S., Merlini, L., Gallo, G., Ciacci, A., Giannini, G., Cabri, W., Penco, S., Vesci, L., Castorina, M., Milazzo, F. M., Cervoni, M. L., Barbarino, M., Pisano, C., Giommarelli, C., Zuco, V., De Cesare, M., & Zunino, F. (2012). Isoxazolo(aza)naphthoquinones: A new class of cytotoxic Hsp90 inhibitors. *European Journal of Medicinal Chemistry*, 53, 64–75. DOI: 10.1016/j.ejmech.2012.03.036 PMID: 22538015

Bartzatt, R. (2014). Drug analogs of COX-2 selective inhibitors lumiracoxib and valdecoxib derived from in silico search and optimization. *Anti-Inflammatory & Anti-Allergy Agents in Medicinal Chemistry*, 13(1), 17–28. DOI: 10.2174/18715230113129990019 PMID: 23984829

Baruchello, R., Simoni, D., Marchetti, P., Rondanin, R., Mangiola, S., Costantini, C., Meli, M., Giannini, G., Vesci, L., Carollo, V., Brunetti, T., Battistuzzi, G., Tolomeo, M., & Cabri, W. (2014). 4,5,6,7-Tetrahydro-isoxazolo-[4,5-c]-pyridines as a new class of cytotoxic Hsp90 inhibitors. *European Journal of Medicinal Chemistry*, 76, 53–60. DOI: 10.1016/j.ejmech.2014.01.056 PMID: 24565573

Basha, S. S., Divya, K., Padmaja, A., & Padmavathi, V. (2015). Synthesis and antimicrobial activity of thiazolyl pyrazoles and isoxazoles. *Research on Chemical Intermediates*, 41(12), 10067–10083. DOI: 10.1007/s11164-015-2013-6

Boscencu, R. (2012). Microwave Synthesis Under Solvent-Free Conditions and Spectral Studies of Some Mesoporphyrinic Complexes. *Molecules (Basel, Switzerland)*, 17(5), 5592–5603. DOI: 10.3390/molecules17055592 PMID: 22576229

Brauch, S., van Berkel, S. S., & Westermann, B. (2013). Higher-order multicomponent reactions: Beyond four reactants. *Chemical Society Reviews*, 42(12), 4948–4962. DOI: 10.1039/c3cs35505e PMID: 23426583

Chakraborty, V., & Bordoloi, M. (1999). Microwave-assisted Oxidation of Alcohols by Pyridinium Chlorochromate. *Journal of Chemical Research Synopses*, 2, 118–122.

Changtam, C., Hongmanee, P., & Suksamrarn, A. (2010). Isoxazole analogs of curcuminoids with highly potent multidrug-resistant antimycobacterial activity. *European Journal of Medicinal Chemistry*, 45(10), 4446–4457. DOI: 10.1016/j.ejmech.2010.07.003 PMID: 20691508

Chitra, S., Paul, N., Muthusubramanian, S., & Manisankar, P. (2011). A facile, water mediated, microwave-assisted synthesis of 4,6-diaryl- 2,3,3a,4-tetrahydro-1H-pyrido 3,2,1-jk carbazoles by a domino Fischer indole reaction-intramolecular cyclization sequence. *Green Chemistry*, 13(10), 2777–2785. DOI: 10.1039/c1gc15483d

Chundawat, T. S., Sharma, N., & Bhagat, S. (2014). Synthesis and in vitro antimicrobial evaluation of novel fluorine-containing 3-benzofuran-2-yl-5-phenyl-4,5-dihydro-1H-pyrazoles and 3-benzofuran-2-yl-5-phenyl-4,5-dihydro-isoxazoles. *Medicinal Chemistry Research*, 23(3), 1350–1359. DOI: 10.1007/s00044-013-0735-9

Claisen, L. (1903). Zur Kenntniss des Propargylaldehyds und des Phenylpropargylaldehyds. *Berichte der Deutschen Chemischen Gesellschaft*, 36(3), 3664–3673. DOI: 10.1002/cber.190303603168

Collins, M. J.Jr. (2010). Future trends in microwave synthesis. *Future Medicinal Chemistry*, 2(2), 151–155. DOI: 10.4155/fmc.09.133 PMID: 21426181

Conti, P., Roda, G., Stabile, H., Vanoni, M. A., Curti, B., & De Amici, M. (2003). Synthesis and biological evaluation of new amino acids structurally related to the antitumoragent acivicin. *Il Farmaco*, 58(9), 683–690. DOI: 10.1016/S0014-827X(03)00107-1 PMID: 13679161

Cronin, M. T. D., Aptula, A. O., Dearden, J. C., Duffy, J. C., Netzeva, T. I., Patel, H., Rowe, P. H., Schultz, T. W., Worth, A. P., Voutzoulidis, K., & Schüürmann, G. (2002). Structure-based classification of antibacterial activity. *Journal of Chemical Information and Computer Sciences*, 42(4), 869–878. DOI: 10.1021/ci025501d PMID: 12132888

Deng, R., Wang, Y., & Jiang, Y. (1994). Solid-Liquid Phase Transfer Catalytic Synthesis X: The Rapid Alkylation of Ethyl Acetoacetate Under Microwave Irradiation. *Synthetic Communications*, 24(1), 111–115. DOI: 10.1080/00397919408012633

Denmark, S. E., & Kallemeyn, J. M. (2005). Synthesis of 3,4,5-trisubstituted isoxazoles via sequential [3 + 2] cycloaddition/silicon-based cross-coupling reactions. *The Journal of Organic Chemistry*, 70(7), 2839–2842. DOI: 10.1021/jo047755z PMID: 15787583

Desjardins, C. A., Cohen, K. A., Munsamy, V., Abeel, T., Maharaj, K., Walker, B. J., Shea, T. P., Almeida, D. V., Manson, A. L., Salazar, A., Padayatchi, N., O'Donnell, M. R., Mlisana, K. P., Wortman, J., Birren, B. W., Grosset, J., Earl, A. M., & Pym, A. S. (2016). Genomic and functional analyses of Mycobacterium tuberculosis strains implicate ald in D-cycloserine resistance. *Nature Genetics*, 48(5), 544–551. DOI: 10.1038/ng.3548 PMID: 27064254

Diana, P., Carbone, A., Barraja, P., Kelter, G., Fiebig, H. H., Cirrincione, G. S. N. C. V. L., & Bhadra, M. P. (2010). Synthesis and antitumor activity of 2,5-bis(3'-indolyl)-furans and 3,5-bis(3'-indolyl)-isoxazoles, nortopsentin analogues. *Bioorganic & Medicinal Chemistry*, 18(12), 4524–4529. DOI: 10.1016/j.bmc.2010.04.061 PMID: 20472437

Dömling, A., Wang, W., & Wang, K. (2012). Chemistry and biology of multicomponent reactions. *Chemical Reviews*, 112(6), 3083–3135. DOI: 10.1021/cr100233r PMID: 22435608

Downey, V. M., Que, B. R., & So, A. G. (1980). Synthesis, characterization, theoretical study and biological activities of oxovanadium (IV) complexes with 2-thiophene carboxylic acid hydrazide. *Biochemical and Biophysical Research Communications*, 93, 264–270. DOI: 10.1016/S0006-291X(80)80275-0 PMID: 6445734

Dube, D., Blouin, M., Brideau, C., Chan, C. C., Desmarais, S., Ethier, D., Falgueyret, J.-P., Friesen, R. W., Girard, M., Girard, Y., Guay, J., Riendeau, D., Tagari, P., & Young, R. N. (1998). Quinolines as potent 5-lipoxygenase inhibitors: Synthesis and biological profile of L-746,530. *Bioorganic & Medicinal Chemistry Letters*, 8(10), 1255–1260. DOI: 10.1016/S0960-894X(98)00201-7 PMID: 9871745

Evdokimov, N. M., Kireev, A. S., Yakovenko, A. A., Antipin, M. Y., Magedov, I. V., & Kornienko, A. (2007). One-Step Synthesis of Heterocyclic Privileged Medicinal Scaffolds by a Multicomponent Reaction of Malononitrile with Aldehydes and Thiols. *The Journal of Organic Chemistry*, 2(9), 3443–3453. DOI: 10.1021/jo070114u PMID: 17408286

Gaba, M., & Dhingra, N. (2011). Microwave Chemistry: General Features and Applications. *Indian Journal of Pharmaceutical Education and Research*, 45, 175–183.

Gabriel, C. S., Gabriel, S., Grant, E. H., Halstead, B. S. J., & Mingos, P. M. (1998). Dielectric parameters relevant to microwave dielectric heating. *Chemical Society Reviews*, 27, 213–224. DOI: 10.1039/a827213z

Garella, D., Borretto, E., Di Stilo, A., Martina, K., Cravotto, G., & Cintas, P. (2013). Microwave-assisted synthesis of N-heterocycles in medicinal chemistry. *MedChemComm*, 4(10), 1323–1343. DOI: 10.1039/c3md00152k

Gatto, B., Capranico, G., & Palumbo, M. (1999). Drugs acting on DNA topoisomerases: Recent advances and future perspectives. *Current Pharmaceutical Design*, 5(3), 195–215. DOI: 10.2174/1381612805666230109215114 PMID: 10066890

Gautam, K. C., & Singh, D. P. (2013). Synthesis and antimicrobial activity of some isoxazole derivatives of thiophene. *Chemical Science Transactions*, 2, 992–996.

Gawande, M. B., Bonifacio, V. D. B., Luque, R., Branco, P. S., & Varma, R. S. (2012). Benign by design: Catalyst-free in-water, on-water green chemical methodologies in organic synthesis. *Chemical Society Reviews*, 42(12), 5522–5551. DOI: 10.1039/c3cs60025d PMID: 23529409

Gawande, M. B., Bonifacio, V. D. B., Luque, R., Branco, P. S., & Varma, R. S. (2014). Solvent-free and catalysts-free chemistry: A benign pathway to sustainability. *ChemSusChem*, 7(1), 24–44. DOI: 10.1002/cssc.201300485 PMID: 24357535

Gawande, M. B., Bonifacio, V. D. B., Varma, R. S., Nogueira, I. D., Bundaleski, N., Ghumman, C. A. A., Teodoro, O. M. N. D., & Branco, P. S. (2013). Magnetically recyclable magnetite-ceria (Nanocat-Fe-Ce) nanocatalyst - applications in multicomponent reactions under benign conditions. *Green Chemistry*, 15(5), 1226–1231. DOI: 10.1039/c3gc40375k

Gawande, M. B., & Branco, P. S. (2017). An efficient and expeditious Fmoc protection of amines and amino acids in aqueous media. *Green Chemistry*, 13(12), 3355–3359. DOI: 10.1039/c1gc15868f

Gawande, M. B., Rathi, A. K., Nogueira, I. D., Varma, R. S., & Branco, P. S. (2008). Magnetite-supported sulfonic acid: A retrievable nanocatalyst for the Ritter reaction and multicomponent reactions. *Green Chemistry*, 15(7), 1895–1899. DOI: 10.1039/c3gc40457a

Giustina, A., Malerba, M., Bresciani, E., Desenzani, P., Licini, M., Zaltieri, G., & Grassi, V. (1995). Effect of two β2-agonist drugs, salbutamol and broxaterol, on the growth hormone response to exercise in adult patients with asthmatic bronchitis. *Journal of Endocrinological Investigation*, 18(11), 847–852. DOI: 10.1007/BF03349831 PMID: 8778156

Golicki, D., Newada, M., Lis, J., Pol, K., Hermanowski, T., & Tłustochowicz, M. (2012). Leflunomide in monotherapy of rheumatoid arthritis: Meta-analysis of randomized trials. *Polish Archives of Internal Medicine*, 122(1-2), 22–32. DOI: 10.20452/pamw.1131 PMID: 22353705

Gothelf, K. V., & Jorgensen, K. A. (1998). Asymmetric 1, 3-dipolar cycloaddition reactions. *Chemical Reviews*, 98(2), 863–909. DOI: 10.1021/cr970324e PMID: 11848917

Graeve, R. E., Pociask, J. R., & Stein, R. G. (1971). *US Patent* 3, 600,393.

Grewal, A. S., Kumar, K., Redhu, S., & Bhardwaj, S. (2013).. . *International Research Journal of Pharmaceutical and Applied Sciences*, 3, 278–285.

Gupta, M., & Wakhloo, B. P. (2007). Tetrabutylammoniumbromide mediated Knoevenagel condensation in water: Synthesis of cinnamic acids. *ARKIVOC*, 15(1), 94–98. DOI: 10.3998/ark.5550190.0008.110

Habeeb, A. G., Praveen Rao, P. N., & Knaus, E. E. (2001). Design and synthesis of 4,5-diphenyl-4-isoxazolines: Novel inhibitors of cyclooxygenase-2 with analgesic and antiinflammatory activity. *Journal of Medicinal Chemistry*, 44(18), 2921–2927. DOI: 10.1021/jm0101287 PMID: 11520200

Hailes, H. C. (2007). Reaction Solvent Selection: The Potential of Water as a Solvent for Organic Transformations. *Organic Process Research & Development*, 11(1), 114–120. DOI: 10.1021/op060157x

Hill, D. J., Mio, M. J., Prince, R. B., Hughes, T. S., & Moore, J. S. (2001). A Field Guide to Foldamers. *Chemical Reviews*, 101(12), 3893–4011. DOI: 10.1021/cr990120t PMID: 11740924

Huisgen, R. (1968). *Journal Organic Chemistry*, 33, 2291–2297. Firestone, R. A. (1968). *The Journal of Organic Chemistry*, 33, 2285–2290.

Hulme, C., & Gore, V. (2001). "Multi-component Reactions : Emerging Chemistry in Drug Discovery" 'From Xylocain to Crixivan'. *Current Medicinal Chemistry*, 10(1), 51–80. DOI: 10.2174/0929867033368600 PMID: 12570721

Im, D., Jung, K., Yang, S., Aman, W., & Hah, J. M. (2015). Discovery of 4-arylamido 3-methyl isoxazole derivatives as novel FMS kinase inhibitors. *European Journal of Medicinal Chemistry*, 102, 600–610. DOI: 10.1016/j.ejmech.2015.08.031 PMID: 26318067

Jacob, J. (2012). Microwave-assisted reactions in organic chemistry: A review of recent advances. *International Journal of Chemistry*, 4(6), 29–43. DOI: 10.5539/ijc.v4n6p29

Jaiswal, S. (2019). Five and Six Membered Heterocyclic Compound with Antimicrobial Activity. *Journal for Modern Trends in Science and Technology*, 5(06), 36–39.

Jaiswal, S., & Diwedi, S. (2016). One-pot synthesis of chalcone derivatives by using anhydrous AlCl3 as condensing agents. *Int. J. Eng. Res. Generic Sci*, 4, 204–209.

Jaiswal, S., & Dwivedi, S. (2017). Alumina Supported Solvent-free Synthesis and Structural Elucidation of New 3-aryl-5-benzylidene-2-thioxohydantoin Derivatives. *Indian Journal of Heterocyclic Chemistry*, 27(3), 265–268.

Jaiswal, S., Kapoor, D., Kumar, A., & Sharma, K. (2017). Applications of green chemistry. *Int. J. Cybern. Informatics*, 6, 127–133.

Jaiswal, S., & Sharma, A. (2014b). Green Synthesis of Some Novel Substituted and Unsubstituted Benzimidazole Derivatives by Using Microwave Energy. *International Journal of Pharmaceutical Sciences Review and Research*, 27, 153–156.

Jaiswal, S. & Siddiqui, I. R. (2014a). Montmorillonite K-10 Supported Solvent-Free Synthesis of Novel 2-thioxoimidazole-4-one N-nucleosides. *chemistry*, Int. J. Pharm. Sci. Rev. Res.,26(1),*16*, 18.

Jaiswal, S., & Sigh, S. (2014c). A novel POCl3 catalysed expeditious synthesis and antimicrobial activities of 5-subtituted-2-arylbenzalamino-1, 3, 4-thiadiazole. *International Journal of Engineering Research and General Science*, 2(6), 166–169.

Jiang, J. B., & Isaacson, D. (1987). *US Patent*. 4, 656, 274.

Jiang, Z. L., Feng, Z. W., & Shen, X. C. (2001).. . *Chinese Chemical Letters*, 12, 551–554.

Johnston, G. A. R. (2014). Muscimol as an Ionotropic GABA Receptor Agonist. *Neurochemical Research*, 39(10), 1942–1947. DOI: 10.1007/s11064-014-1245-y PMID: 24473816

Jose, A., Arno, S. M., & Luis, R. D. (2003). A DFT study for the regioselective 1,3- dipolar cycloadditions of nitrile N-oxides towardalkynylboronates. *Tetrahedron*, 59(46), 9167–9171. DOI: 10.1016/j.tet.2003.09.050

Joshi, S. D., Dixit, S. R., Kirankumar, M. N., Aminabhavi, T. M., Raju, K. V. S. N., Narayan, R., Lherbet, C., & Yang, K. S. (2016). Synthesis, antimycobacterial screening and ligand-based molecular docking studies on novel pyrrole derivatives bearing pyrazoline, isoxazole and phenyl thiourea moieties. *European Journal of Medicinal Chemistry*, 107, 133–152. DOI: 10.1016/j.ejmech.2015.10.047 PMID: 26580979

Kachhadia, V. V., Patel, M. R., & Joshi, H. S. (2004). Synthesis of isoxazoles and cyanopyridines bearing benzo (b) thiophene nucleus as potential antitubercular and antimicrobial agents. *Journal of Sciences Islamic Republic of Iran*, 15, 47–51.

Kamal, A., Bharathi, E. V., Reddy, J. S., Ramaiah, M. J., Dastagiri, D., Reddy, M. K., Viswanath, A., Reddy, T. L., Shaik, T. B., Pushpavalli, S. N. C. V. L., & Bhadra, M. P. (2011). Synthesis and biological evaluation of 3,5-diaryl isoxazoline/isoxazole linked 2,3-dihydroquinazolinone hybrids as anticancer agents. *European Journal of Medicinal Chemistry*, 46(2), 691–703. DOI: 10.1016/j.ejmech.2010.12.004 PMID: 21194809

Kamal, A., Surendranadha Reddy, J., Janaki Ramaiah, M., Dastagiri, D., Vijaya Bharathi, E., Ameruddin Azhar, M., Sultana, F., Pushpavalli, S. N. C. V. L., Pal-Bhadra, M., Juvekar, A. S. N. C. V. L., & Bhadra, M. P. (2010). Design, synthesis and biological evaluation of 3,5-diaryl-isoxazoline/isoxazole-pyrrolobenzodiazepine conjugates as potential anticancer agents. *European Journal of Medicinal Chemistry*, 45(9), 3924–3937. DOI: 10.1016/j.ejmech.2010.05.047 PMID: 20557981

Kang, Y. K., Shin, K. J., Yoo, K. H., Seo, K. J., Hong, C. Y., Lee, C. S., Park, S. Y., Kim, D. J., & Park, S. W. (2000). Synthesis and antibacterial activity of new carbapenems containing isoxazole moiety. *Bioorganic & Medicinal Chemistry Letters*, 10(2), 95–99. DOI: 10.1016/S0960-894X(99)00646-0 PMID: 10673088

Kanizsai, I., Gyónfalvi, S., Szakonyi, Z., Sillanpää, R., & Fülöp, F. (2007). Synthesis of bi- and tricyclic β-lactam libraries in aqueous medium. *Green Chemistry*, 9(4), 357–360. DOI: 10.1039/B613117D

Karthikeyan, K., Veenus Seelan, T., Lalitha, K. G., & Perumal, P. T. (2009). Synthesis and antinociceptive activity of pyrazolyl isoxazolines and pyrazolyl isoxazoles. *Bioorganic & Medicinal Chemistry Letters*, 19(13), 3370–3373. DOI: 10.1016/j.bmcl.2009.05.055 PMID: 19481931

Kidwai, M., Venkataraman, R., & Dave, B. (2001).. . *Green Chemistry*, 3(6), 278–279. DOI: 10.1039/b106034c

Kljin, J. E., & Engberts, J. B. N. (2005).. . *Nature*, 435, 746. DOI: 10.1038/435746a PMID: 15944683

Kou, J. H., Saha, A., Bennett-Stamper, C., & Varma, R. S. (2012). Inside-out core-shell architecture: Controllable fabrication of Cu2O@Cu with high activity for the Sonogashira coupling reaction. *Chemical Communications (Cambridge)*, 48(47), 5862–5864. DOI: 10.1039/c2cc31577g PMID: 22572855

Kumar, R. A., Subramani, K., & Ravichandran, S. (2010).. . *International Journal of Chemtech Research*, 2, 278–281.

Kumari, P., Mishra, V. S., Narayana, C., & Chakrabarty, A. K. (2020). Design and efficient synthesis of pyrazoline and isoxazole bridged indole C-glycoside hybrids as potential anticancer agents. *Scientific Reports*, 10(1), 6660. DOI: 10.1038/s41598-020-63377-x

Kumbhare, R. M., Kosurkar, U. B., Janaki Ramaiah, M., Dadmal, T. L., Pushpavalli, S. N. C. V. L., & Pal-Bhadra, M. (2012). Synthesis and biological evaluation of novel triazoles and isoxazoles linked 2-phenyl benzothiazole as potential anticancer agents. *Bioorganic & Medicinal Chemistry Letters*, 22(17), 5424–5427. DOI: 10.1016/j.bmcl.2012.07.041 PMID: 22858144

Lainson, J. C., Daly, S. M., Triplett, K., Johnston, S. A., Hall, P. R., & Diehnelt, C. W. (2017). Synthetic Antibacterial Peptide Exhibits Synergy with Oxacillin against MRSA. *ACS Medicinal Chemistry Letters*, 8(8), 853–857. DOI: 10.1021/acsmedchemlett.7b00200 PMID: 28835801

Lancaster, M. (2002). Principles of sustainable and green chemistry. *Handbook of green chemistry and technology*, 10-27.

Langa, F., Cruz, P., Hoz, A., Diaz-Ortiz, A., & Diez-Barra, E. (1997). Microwave irradiation: More than just a method for accelerating reactions. *Contemporary Organic Synthesis*, 4(5), 373–386. DOI: 10.1039/CO9970400373

Larhed, M., & Hallberg, A. (2001). Microwave-assisted high-speed chemistry: A new technique in drug discovery. *Drug Discovery Today*, 6(8), 406–416. DOI: 10.1016/S1359-6446(01)01735-4 PMID: 11301285

Lavanya, G., Mallikarjuna Reddy, L., Padmavathi, V., & Padmaja, A. (2014). Synthesis and antimicrobial activity of (1,4-phenylene)bis(arylsulfonylpyrazoles and isoxazoles). *European Journal of Medicinal Chemistry*, 73, 187–194. DOI: 10.1016/j.ejmech.2013.11.041 PMID: 24398288

Lehtonen, K., Summers, L. A., & Carter, G. A. (1972). Fungitoxicity of acid and alkali hydrolysis products of Drazoxolon and related arylhydrazonoisoxazolones. *Pesticide Science*, 3(3), 357–364. DOI: 10.1002/ps.2780030313

Lew, A., Krutzik, P. O., Hart, M. E., & Chamberlin, A. R. (2002). Increasing rates of reaction: Microwave-assisted organic synthesis for combinatorial chemistry. *Journal of Combinatorial Chemistry*, 4(2), 95–105. DOI: 10.1021/cc010048o PMID: 11886281

Ley, S. V., & Baxendale, I. R. (2002). New tools and concepts for modern organic synthesis. *Nature Reviews. Drug Discovery*, 1(8), 573–586. DOI: 10.1038/nrd871 PMID: 12402498

Lidstrom, P., Tierney, J., Wathey, B., & Westman, J. (2001). Microwave assisted organic synthesis—A review. *Tetrahedron*, 57(45), 9225–9283. DOI: 10.1016/S0040-4020(01)00906-1

Lidstrom, P., Tierney, J., Watheyb, B., & Westmana, J. (2001). Microwave assisted organic synthesis—A review. *Tetrahedron*, 37(45), 9225–9283. DOI: 10.1016/S0040-4020(01)00906-1

Lidström, P., Westman, J., & Lewis, A. (2002). Enhancement of Combinatorial Chemistry by Microwave-Assisted Organic Synthesis. *Combinatorial Chemistry & High Throughput Screening*, 5(6), 441–458. DOI: 10.2174/1386207023330147 PMID: 12470274

Lie'by-Muller, F., Simon, C., Constantieux, T., & Rodriguez, J. (2006). Current Developments in Michael Addition-Based Multicomponent Domino Reactions Involving 1,3-Dicarbonyls and Derivatives. *QSAR & Combinatorial Science*, 25(5-6), 432–438. DOI: 10.1002/qsar.200540201

Liu, T., Dong, X., Xue, N., Wu, R., He, Q., Yang, B., Hu, Y. S. N. C. V. L., & Bhadra, M. P. (2009). Synthesis and biological evaluation of 3,4-diaryl-5-aminoisoxazole derivatives. *Bioorganic & Medicinal Chemistry*, 17(17), 6279–6285. DOI: 10.1016/j.bmc.2009.07.040 PMID: 19665898

Loupy, A. (2002). *Microwaves in Organic Synthesis*. Wiley-VCH. DOI: 10.1002/3527601775

Maguire, M. P., Sheets, K. R., McVety, K. T., Spada, A. P., & Zilberstein, A. (1994). A New Series of PDGF Receptor Tyrosine Kinase Inhibitors: 3-Substituted Quinoline Derivatives. *Journal of Medicinal Chemistry*, 37(14), 2129–2137. DOI: 10.1021/jm00040a003 PMID: 8035419

Majewsky, M., Wagner, D., Delay, M., Bräse, S., Yargeau, V., & Horn, H. (2014). Antibacterial activity of sulfamethoxazole transformation products (TPs): General relevance for sulfonamide TPs modified at the para position. *Chemical Research in Toxicology*, 27(10), 1821–1828. DOI: 10.1021/tx500267x PMID: 25211553

Mao, J., Yuan, H., Wang, Y., Wan, B., Pak, D., He, R., & Franzblau, S. G. (2010). Synthesis and antituberculosis activity of novel mefloquine-isoxazole carboxylic esters as prodrugs. *Bioorganic & Medicinal Chemistry Letters*, 20(3), 1263–1268. DOI: 10.1016/j.bmcl.2009.11.105 PMID: 20022500

Matthew, P., Bourbeau, James, T.R. (2006). Copper-Catalyzed Alkylation of Nitroalkanes with α-Bromonitriles: Synthesis of β-Cyanonitroalkanes. *Organic Letters*, 8, 3679–3680.

Mazimba, O., Wale, K., Loeto, D., & Kwape, T. (2014). Antioxidant and antimicrobial studies on fused-ring pyrazolones and isoxazolones. *Bioorganic & Medicinal Chemistry*, 22(23), 6564–6569. DOI: 10.1016/j.bmc.2014.10.015 PMID: 25456077

McKinney, A. R., Cawley, A. T., Young, E. B., Kerwick, C. M., Cunnington, K., Stewart, R. T., Ambrus, J. I., Willis, A. C., & McLeod, M. D. (2013). The Metabolism of Anabolic-Androgenic Steroids in The Greyhound. *Bioanalysis*, 5(7), 769–781. DOI: 10.4155/bio.13.40 PMID: 23534422

Mokle, S. S., Khansole, S. V., Patil, R. B., & Vibhute, Y. B. (2010). Synthesis and antibacterial activity of some new chalcones and flavones having 2-chloro-8-methoxyquinolinyl moiety. *International Journal of Pharma and Bio Sciences*, 1, 1–7M.

Montes, I., Sanabria, D., García, M., & Fajardo, J. (2006). A Greener Approach to Aspirin Synthesis Using Microwave Irradiation. *Journal of Chemical Education*, 83(4), 628–631. DOI: 10.1021/ed083p628

Morita, T., Yugandar, S., Fuse, S., & Nakamura, H. (2018). Recent progresses in the synthesis of functionalized isoxazoles. *Tetrahedron Letters*, 59(13), 1159–1171. DOI: 10.1016/j.tetlet.2018.02.020

Moss, G. P., Smith, P. A. S., & Tavernier, D. (1995). Glossary of class names of organic compounds and reactivity intermediates based on structure (IUPAC Recommendations 1995). *Pure and Applied Chemistry*, 67(8-9), 1307–1375. DOI: 10.1351/pac199567081307

Narayan, S., Fokin, M. G., Kolb, H. C., & Sharpless, K. B. (2005). "On Water": Unique Reactivity of Organic Compounds in Aqueous Suspension. *Angewandte Chemie International Edition*, 44(21), 3275–3279. DOI: 10.1002/anie.200462883 PMID: 15844112

Negishi, E., & Anastasia, L. (2003). Palladium-catalyzed alkynylation. *Chemical Reviews*, 103(5), 1979–2017. DOI: 10.1021/cr020377i PMID: 12744698

Noveck, R. J., & Hubbard, R. C. (2004). Parecoxib Sodium, an Injectable COX-2-Specific Inhibitor, Does Not Affect Unfractionated Heparin-Regulated Blood Coagulation Parameters. *Journal of Clinical Pharmacology*, 44(5), 474–480. DOI: 10.1177/0091270004264166 PMID: 15102867

Obermeier, S., & Muller, M. (2020). Angewantde Chemie *International Edition*, 59, 12432–12435.

Ostacolo, C., Ambrosino, P., Russo, R., Lo Monte, M., Soldovieri, M. V., Laneri, S., Sacchi, A., Vistoli, G., Taglialatela, M., & Calignano, A. (2013). Isoxazole derivatives as potent transient receptor potential melastatin type 8 (TRPM8) agonists. *European Journal of Medicinal Chemistry*, 69, 659–669. DOI: 10.1016/j.ejmech.2013.08.056 PMID: 24095758

Padmaja, A., Payani, T., Reddy, G. D., & Padmavathi, V. (2009). Synthesis, antimicrobial and antioxidant activities of substituted pyrazoles, isoxazoles, pyrimidine and thioxopyrimidine derivatives. *European Journal of Medicinal Chemistry*, 44(11), 4557–4566. DOI: 10.1016/j.ejmech.2009.06.024 PMID: 19631423

Palin, R., Abernethy, L., Ansari, N., Cameron, K., Clarkson, T., Dempster, M., Dunn, D., Easson, A. M., Edwards, D., MacLean, J., Everett, K., Feilden, H., Ho, K. K., Kultgen, S., Littlewood, P., McArthur, D., McGregor, D., McLuskey, H., Neagu, I., & Walker, G. (2011). Structure-activity studies of a novel series of isoxazole-3-carboxamide derivatives as TRPV1 antagonists. *Bioorganic & Medicinal Chemistry*, 21(3), 892–898. DOI: 10.1016/j.bmcl.2010.12.092 PMID: 21236666

Panda, S., Chowdary, P., & Jayashree, B. (2009). Synthesis, antiinflammatory and antibacterial activity of novel indolyl-isoxazoles. *Indian Journal of Pharmaceutical Sciences*, 71(6), 684–687. DOI: 10.4103/0250-474X.59554 PMID: 20376225

Patel, D., Kumari, P., & Patel, N. B. (2014). Synthesis and biological evaluation of coumarin based isoxazoles, pyrimidinthiones and pyrimidin-2-ones. *Arabian Journal of Chemistry*, 10, 3990–4001. DOI: 10.1016/j.arabjc.2014.06.010

Perrone, M. G., Vitale, P., Panella, A., Ferorelli, S., Contino, M., Lavecchia, A., & Scilimati, A. (2016). Isoxazole-based-scaffold inhibitors targeting cyclooxygenases (COXs). *ChemMedChem*, 11(11), 1172–1187. DOI: 10.1002/cmdc.201500439 PMID: 27136372

Petersen, S. L., Tofteng, A. P., Malik, L., & Jensen, K. J. (2012). Microwave heating in solid-phase peptide synthesis. *Chemical Society Reviews*, 41(5), 1826–1844. DOI: 10.1039/C1CS15214A PMID: 22012213

Pirrung, M. C., & Das Sarma, K. (2004). Multicomponent Reactions Are Accelerated in Water. *Journal of the American Chemical Society*, 126(2), 444–445. DOI: 10.1021/ja038583a PMID: 14719923

Polshettiwar, P., Nadagouda, M. N., & Varma, R. S. (2001). Microwave-Assisted Chemistry: A Rapid and Sustainable Route to Synthesis of Organics and Nanomaterials. *Australian Journal of Chemistry*, 62(1), 16–26. DOI: 10.1071/CH08404

Polshettiwar, V., & Varma, R. S. (2007). Tandem bis-aldol reaction of ketones: A facile one-pot synthesis of 1,3-dioxanes in aqueous medium. *The Journal of Organic Chemistry*, 72(19), 7420–7422. DOI: 10.1021/jo701337j PMID: 17696550

Polshettiwar, V., & Varma, R. S. (2008). Greener and rapid access to bio-active heterocycles: Room temperature synthesis of pyrazoles and diazepines in aqueous medium. *Tetrahedron Letters*, 49(2), 397–400. DOI: 10.1016/j.tetlet.2007.11.017

Polshettiwar, V., & Varma, R. S. (2008). Aqueous microwave chemistry: A clean and green synthetic tool for rapid drug discovery. *Chemical Society Reviews*, 37(8), 1546–1557. DOI: 10.1039/b716534j PMID: 18648680

Prakash, A., & Jarvis, B. (1999). Leflunomide. *Drugs*, 58(6), 1137–1164. DOI: 10.2165/00003495-199958060-00010 PMID: 10651393

Qu, G. R., Xin, P. Y., Niu, H. Y., Jin, X., Guo, X. T., Yang, X. N., & Guo, H. M. (2011). Microwave promoted palladium-catalyzed Suzuki-Miyaura cross-coupling reactions of 6-chloropurines with sodium tetraarylborate in water. *Tetrahedron*, 67(47), 9099–9103. DOI: 10.1016/j.tet.2011.09.082

Raghavendra, M., Naik, H. S. B., & Sherigara, B. S. (2006). One pot synthesis of some new 2-hydrazino-[1,3,4]thiadiazepino[7,6-b] quinolines under microwave irradiation conditions. *ARKIVOC*, 15, 153–159.

Rajanarendar, E., Rama Krishna, S., Nagaraju, D., Govardhan Reddy, K., Kishore, B., & Reddy, Y. N. (2015). Environmentally benign synthesis, molecular properties prediction and anti-inflammatory activity of novel isoxazolo[5,4-d] isoxazol-3-yl-aryl-methanones via vinylogous Henry nitroaldol adducts as synthons. *Bioorganic & Medicinal Chemistry Letters*, 25(7), 1630–1634. DOI: 10.1016/j.bmcl.2015.01.041 PMID: 25708616

Rakesh, S. D., Lee, R. B., Tangallapally, R. P., & Lee, R. E. (2009). Synthesis, optimization and structure-activity relationships of 3,5-disubstituted isoxazolines as new anti-tuberculosis agents. *European Journal of Medicinal Chemistry*, 44(2), 460–472. DOI: 10.1016/j.ejmech.2008.04.007 PMID: 18524421

RamaRao, R. J., Rao, A. K. S., Sreenivas, N., Kumar, B. S., & Murthy, Y. L. N. (2011). Synthesis and antibacterial activity of novel 5-(heteroaryl) isoxazole Derivatives. *Journal of the Korean Chemical Society*, 55(2), 243–250.

Ratnakar Reddy, K., Sambasiva Rao, P., Jitender Dev, G., Poornachandra, Y., Ganesh Kumar, C., Shanthan Rao, P., & Narsaiah, B. (2014). Synthesis of novel 1,2,3-triazole/isoxazole functionalized 2H-Chromene derivatives and their cytotoxic activity. *Bioorganic & Medicinal Chemistry Letters*, 24(7), 1661–1663. DOI: 10.1016/j.bmcl.2014.02.069 PMID: 24641975

Ravichandran, S. (2001). FACILE STEREOSELECTIVE SYNTHESIS OF (*E*)- AND (*Z*)-ALLYL BROMIDES FROM THE BAYLIS-HILLMAN ADDUCTS USING MgBr $_2$. *Synthetic Communications*, 31(13), 2059–2062. DOI: 10.1081/SCC-100104426

Ravichandran, S., & Karthikeyan, E. (2011)... *International Journal of Chemtech Research*, 3, 466–470.

Rees, C. W. (1992). Polysulfur-Nitrogen Heterocyclic Chemistry. *Journal of Heterocyclic Chemistry*, 29(3), 639–651. DOI: 10.1002/jhet.5570290306

Rostamizadeh, S., Daneshfar, Z., & Moghimi, H. (2019). Synthesis of sulfamethoxazole and sulfabenzamide metal complexes; evaluation of their antibacterial activity. *European Journal of Medicinal Chemistry*, 171, 364–371. DOI: 10.1016/j.ejmech.2019.03.002 PMID: 30928708

Sahu, S., Banerjee, M., Sahu, D., Behera, C., Pradhan, G., & Azam, M. A. (2009). Synthesis, analgesic and antimicrobial activities of some novel isoxazole derivatives. *Dhaka University Journal of Pharmaceutical Science*, 7(2), 113–118. DOI: 10.3329/dujps.v7i2.2165

Sambasiva Rao, P., Kurumurthy, C., Veeraswamy, B., Poornachandra, Y., Ganesh Kumar, C., & Narsaiah, B. (2014). Synthesis of novel 5-(3-alkylquinolin-2-yl)-3-aryl isoxazole derivatives and their cytotoxic activity. *Bioorganic & Medicinal Chemistry Letters*, 24(5), 1349–1351. DOI: 10.1016/j.bmcl.2014.01.038 PMID: 24507927

Schoretsanitis, G., Spina, E., Hiemke, C., & de Leon, J. (2017). A systematic review and combined analysis of therapeutic drug monitoring studies for long-acting risperidone. *Expert Review of Clinical Pharmacology*, 10(9), 965–981. DOI: 10.1080/17512433.2017.1345623 PMID: 28699847

Shader, R. I., & Greenblatt, D. J. (1999). The reappearance of a monamine oxidase inhibitor (isocarboxazid). *Journal of Clinical Psychopharmacology*, 19(2), 105–106. DOI: 10.1097/00004714-199904000-00001 PMID: 10211910

Shakeel-u-Rehman, R., Tripathi, V. K., Singh, J., Ara, T., Koul, S., Farooq, S., & Kaul, A. (2014). Synthesis and biological evaluation of novel isoxazoles and triazoles linked 6-hydroxycoumarin as potent cytotoxic agents. *Bioorganic & Medicinal Chemistry Letters*, 24(17), 4243–4246. DOI: 10.1016/j.bmcl.2014.07.031 PMID: 25088398

Shaw, J., Chen, B., Bourgault, J. P., Jiang, H., Kumar, N., Mishra, J., Valeriote, F. A., Media, J., Bobbitt, K., Pietraszkiewicz, H., Edelstein, M., & Andreana, P. R. (2012). Synthesis and biological evaluation of novel N-phenyl-5-carboxamidyl isoxazoles as potential chemotherapeutic agents for colon cancer. *American Journal of Biomedical Sciences*, 4, 14–25. DOI: 10.5099/aj120100014 PMID: 25285182

Sheldon, R. A. (1997). *J. Chem. Technol. Biotechno*logy, 68, 381. Dabholkar V.V., & Ansari, F.Y. (2010). *Green Chemistry Letters and Reviews*, 3, 245.

Shen, S., Hadley, M., Ustinova, K., Pavlicek, J., Knox, T., Noonepalle, S., Tavares, M. T., Zimprich, C. A., Zhang, G., Robers, M. B., Bařinka, C., Kozikowski, A. P., & Villagra, A. (2019). Discovery of a New Isoxazole-3-hydroxamate-Based Histone Deacetylase 6 Inhibitor SS-208 with Antitumor Activity in Syngeneic Melanoma Mouse Models. *Journal of Medicinal Chemistry*, 62(18), 8557–8577. DOI: 10.1021/acs.jmedchem.9b00946 PMID: 31414801

Shi, F., Li, C. M., Xia, M., Miao, K. J., Zhao, Y. X., Tu, S. J., Zheng, W. F., Zhang, G., & Ma, N. (2009). Green chemoselective synthesis of thiazolo 3,2- a pyridine derivatives and evaluation of their antioxidant and cytotoxic activities. *Bioorganic & Medicinal Chemistry Letters*, 19(19), 5565–5568. DOI: 10.1016/j.bmcl.2009.08.046 PMID: 19729303

Sigel, H., & Sigel, A. (2002). *Metal ions in biological systems*. Marcel Dekker.

Silva, N. M., Tributino, J. L. M., Miranda, A. L. P., Barreiro, E. J., & Fraga, C. A. M. (2002). New isoxazole derivatives designed as nicotinic acetylcholine receptor ligand candidates. *European Journal of Medicinal Chemistry*, 37(2), 163–170. DOI: 10.1016/S0223-5234(01)01327-7 PMID: 11858848

Simoni, D., Roberti, M., Invidiata, F. P., Rondanin, R., Baruchello, R., Malagutti, C., Mazzali, A., Rossi, M., Grimaudo, S., Capone, F., Dusonchet, L., Meli, M., Raimondi, M. V., Landino, M., D'Alessandro, N., Tolomeo, M., Arindam, D., Lu, S., Benbrook, D. M. S. N. C. V. L., & Bhadra, M. P. (2001). Heterocycle-Containing Retinoids. Discovery of a novel isoxazole arotinoid possessing potent apoptotic activity in multidrug and drug-induced apoptosis-resistant cells. *Journal of Medicinal Chemistry*, 44(14), 2308–2318. DOI: 10.1021/jm0010320 PMID: 11428925

Song, Y. L., Li, Y. T., & Wu, Z. Y. (2015). Synthesis and Biological Studies of Some Lanthanide Complexes of Schiff Base. *Journal of Inorganic Biochemistry*, 45, 1617–1626.

Sreeram, K. J., Nidhin, K. J., & Nair, B. U. (2008). Microwave assisted template synthesis of silver nanoparticles. *Bulletin of Materials Science*, 31(7), 937–942. DOI: 10.1007/s12034-008-0149-3

Swinney, D. C., & Anthony, J. (●●●). (20111). *Nature Reviews. Drug Discovery*, 10, 507–519. DOI: 10.1038/nrd3480 PMID: 21701501Tejedor, D., & Garcia-Tellado, F. (2007). Chemo-differentiating ABB′ multicomponent reactions. Privileged building blocks. *Chemical Society Reviews*, 36(3), 484–491. DOI: 10.1039/B608164A PMID: 17325787

Tornoe, C. W., Christensen, C., & Meldal, M. (2002). Peptidotriazoles on solid phase: [1,2,3]-triazoles by regiospecific copper(i)- catalyzed 1,3-dipolar cycloadditions of terminal alkynes to azides. *The Journal of Organic Chemistry*, 67(9), 3057–3064. DOI: 10.1021/jo011148j PMID: 11975567

Tzanetou, E., Liekens, S., Kasiotis, K. M., Melagraki, G., Afantitis, A., Fokialakis, N., & Haroutounian, S. A. (2014). Antiproliferative novel isoxazoles: Modeling, virtual screening, synthesis, and bioactivity evaluation. *European Journal of Medicinal Chemistry*, 81, 139–149. DOI: 10.1016/j.ejmech.2014.05.011 PMID: 24836066

Ugi, I. (2001). Recent progress in the chemistry of multicomponent reactions. *Pure and Applied Chemistry*, 73(1), 187–191. DOI: 10.1351/pac200173010187

Vaddula, B. R., Varma, R. S., & Leazer, J. (2013). Mixing with microwaves: Solvent-free and catalyst-free synthesis of pyrazoles and diazepines. *Tetrahedron Letters*, 54(12), 1538–1541. DOI: 10.1016/j.tetlet.2013.01.029

Vaidya, S. D., Kumar, B. V. S., Kumar, R. V., Bhise, U. N., & Mashelkar, U. C. (2007). Synthesis, anti-bacterial, anti-asthmatic and anti-diabetic activities of novel N-substituted-2-(benzo[d]isoxazol-3-ylmethyl)-1H-benzimidazoles. *Journal of Heterocyclic Chemistry*, 44(3), 685–691. DOI: 10.1002/jhet.5570440327

Violette, A, Averlant-P.etit, M.C., Semetey, V., Hemmerlin, C., Casimir, R., et al. (. (2005). N, N'-Linked oligoureas with proteinogenic side chains are peptide backbone mimetics belonging to the γ-peptide lineage. *Journal of the American Chemical Society*, 127, 2156–2164. DOI: 10.1021/ja044392b PMID: 15713093

Vitaku, E., Smith, D. T., & Njardarson, J. T. (2014). Analysis of the Structural Diversity, Substitution Patterns, and Frequency of Nitrogen Heterocycles among U.S. FDA Approved Pharmaceuticals. *Journal of Medicinal Chemistry*, 57(24), 10257–10274. DOI: 10.1021/jm501100b PMID: 25255204

Weber, L. (2002). Multi-component reactions and evolutionary chemistry. *Drug Discovery Today*, 7(2), 143–147. DOI: 10.1016/S1359-6446(01)02090-6 PMID: 11790626

Weidner-Wells, M. A., Werblood, H. M., Goldschmidt, R., Bush, K., Foleno, B. D., Hilliard, J. J., Melton, J., Wira, E., & Macielag, M. J. (2004). The synthesis and antimicrobial evaluation of a new series of isoxazolinyl oxazolidinones. *Bioorganic & Medicinal Chemistry Letters*, 14(12), 3069–3072. DOI: 10.1016/j.bmcl.2004.04.037 PMID: 15149646

Wilson, N. S., Sarko, C. R., & Roth, G. P. (2004). Development and Applications of a Practical Continuous Flow Microwave Cell. *Organic Process Research & Development*, 8(3), 535–538. DOI: 10.1021/op034181b

Xu, Y., Wang, N. Y., Song, X. J., Lei, Q., Ye, T. H., You, X. Y., Zuo, W. Q., Xia, Y., Zhang, L. D., & Yu, L. T. (2015). Discovery of novel N-(5-(tert-butyl)isoxazol-3-yl)-N′-phenylurea analogs as potent FLT3 inhibitors and evaluation of their activity against acute myeloid leukemia in vitro and in vivo. *Bioorganic & Medicinal Chemistry*, 23(15), 4333–4343. DOI: 10.1016/j.bmc.2015.06.033 PMID: 26142317

Yamuna, E., Kumar, R. A., Zeller, M., & Rajendra Prasad, K. J. (2012). Synthesis, antimicrobial, antimycobacterial and structure-activity relationship of substituted pyrazolo-, isoxazolo-, pyrimido- and mercaptopyrimidocyclohepta[b] indoles. *European Journal of Medicinal Chemistry*, 47, 228–238. DOI: 10.1016/j.ejmech.2011.10.046 PMID: 22119150

Yang, Z. S., Wang, Y. L., & Zhao, G. C. (2004). Electrocatalytic Oxidation of Dopamine and Ascorbic Acid at Poly (Eriochrome Black-T) Modified Carbon Paste Electrode. *Analytical Sciences*, 20, 1127–1135. DOI: 10.2116/analsci.20.1127 PMID: 15352498

Zhao, N., Zuo, L., Li, W., Guo, W., Liu, W., & Wang, J. (2017). Greenhouse and field evaluation of isoxaflutole for weed control in maize in China. *Scientific Reports*, 7(1), 12690–12698. DOI: 10.1038/s41598-017-12696-7 PMID: 28978910

Zhao, S., Zhang, X., Wei, P., Su, X., Zhao, L., Wu, M., Hao, C., Liu, C., Zhao, D., & Cheng, M. (2017). Design, synthesis and evaluation of aromatic heterocyclic derivatives as potent antifungal agents. *European Journal of Medicinal Chemistry*, 137, 96–107. DOI: 10.1016/j.ejmech.2017.05.043 PMID: 28558334

Zhu, J., Mo, J., Lin, H. Z., Chen, Y., & Sun, H. P. (2018). The recent progress of isoxazole in medicinal chemistry. *Bioorganic & Medicinal Chemistry*, 23(12), 3065–3075. DOI: 10.1016/j.bmc.2018.05.013 PMID: 29853341

Chapter 13
Thiadiazoles:
Chemistry and Biological Activities

G. Nagendra
https://orcid.org/0000-0002-5099-3384
REVA University, Bangalore, India

D. N. Akshitha
REVA University, Bangalore, India

H. S. Lalithamba
Siddaganga Institute of Technology, Tumakuru, India

G. K. Prashanth
https://orcid.org/0000-0001-6691-4030
Sir M. Visvesvaraya Institute of Technology, Bengaluru, India

ABSTRACT

Heterocycles containing nitrogen, oxygen and sulfur have been under investigation for a long time because of their important medicinal properties. The literature survey has described that thiadiazole moieties serve as analgesic, anti-inflammatory, anti-microbial activities, antihypertensive, anticancer, antituberculosis and vasodilator and this heterocyclic nucleus still possess considerable characteristics to attract the chemists for designing of newer biologically active molecules. Among them 1,2,4-thiadiazole, 1,3,4-thiadiazole and their derivatives are recognized as heterocyclic nuclei of great value in the field of medicinal chemistry. As a heterocyclic unit in a peptidomimetics might add conformational limitations to the structure, influencing the structure-activity-relationship. In this chapter, we are focusing on the synthetic strategies of 1,2,3-thiadiazole, 1,2,5-thiadiazole, 1,2,4-thiadiazole, 1,3,4-thiadiazole and their application in the industrial and biomedical fields.

DOI: 10.4018/979-8-3693-7267-8.ch013

INTRODUCTION

Thiadiazole, a bioisotere of oxadiazole is a potent inhibitor of metalloproteases and aminopeptidases (Hu. Y, 2014). The most preferred heterocycles oxadiazole and thiadiazole's are found in numerous medications and natural compounds (Atmaram U.A, 2022).Thiadiazole has a toxophoric -N=C-S- moiety and is non-carcinogenic due to its strong polarity, its nucleus is linked to a wide range of biological activity (Siddiqui. N, 2009) . The development of sulfur-based medications and the subsequent identification of mesoionic chemicals significantly quickened the pace of advancement in the thiadiazole field (Sharma.B, 2013). Because of its mesoionic nature (Badami B.V) thiadiazole pharmacophore can readily penetrate biological membranes and interact with a variety of biological proteins. Thiadiazoles with amino, hydroxyl, and mercapto substituents can exist in a variety of tautomeric forms (Sandstrom. J,1969), this characteristic is the subject of extensive research using contemporary experimental techniques. Pharmacologically, thiadiazole has a wide range of actions, including diuretic (Ergena. A, 2022), antitubercular (Kolavi. G, 2006), anti-microbial (Farghaly. T, 2011), anti-inflammatory (Kadi. A, 2010), antifungal (Karaburun. A, 2018), anticonvulsant (Luszczki. A, 2015) and antitumor (Szeliga. M, 2020). The well-known commercial medications bearing thiadiazole moiety include Cefazolin (5) (Vergeron. M, 1973), Cefazedone (6) (Gao. L, 2015) (cell wall synthesis inhibitors), Methazolamide (7) (Maren T.H, 1977), Acetazolamide (8) (Reiss W.G,) (carbonic anhydrase inhibitors), Megazol (9) (Chauviere. G, 2003), (protein and DNA synthesis inhibitor), Sulphamethizole (10) (dihydropteroate synthase inhibitor) (Verzas Nevado J.J, 1991), Azetepa (11), (an alkylating agents) (Jchoy D. S, 1967), Furidiazine (12) (Wei . J, 2021), Besaglybuzole (13) (Otsuka .M, 1999) and Cefozopran (14) (Lizawa. Y, 1993). Over the last ten years, thiadiazole synthesis has advanced significantly. It is customary in these syntheses to create C–S and N–S bonds using disulfides, sulfonyl chlorides and sulfonyl hydrazides.

Figure 1. Structural isomers of oxadiazole

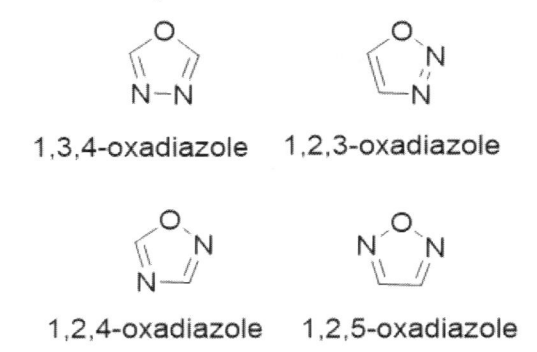

1,3,4-oxadiazole   1,2,3-oxadiazole

1,2,4-oxadiazole   1,2,5-oxadiazole

Figure 2. Structural isomers of thiadiazole

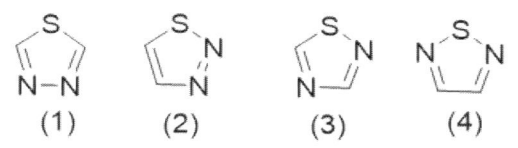

(1)   (2)   (3)   (4)

Lately, 1,2,3-thiadiazoles exhibited a wide range of applications in agrochemical-based industries (Kuckhoff. P, 2023). 1,2,5-thiadiazole derivatives are marked as an effective electron acceptor (Xia. D, 2016).

Figure 3. Examples of commercially available drugs bearing thiadiazole moiety

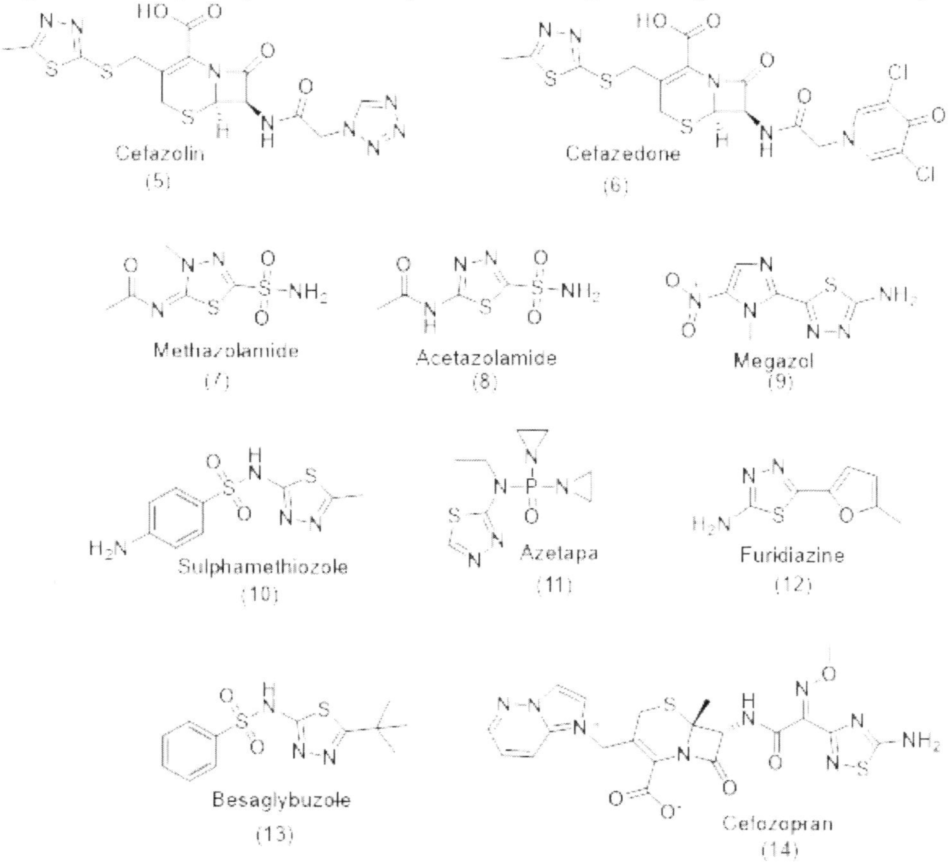

Cefazolin
(5)

Cefazedone
(6)

Methazolamide
(7)

Acetazolamide
(8)

Megazol
(9)

Sulphamethiozole
(10)

Azetapa
(11)

Furidiazine
(12)

Besaglybuzole
(13)

Cefozopran
(14)

1,2,3-Thiadiazole

1,2,3- Thiadiazole is one of the most promising bioactive moieties with wide applications in agrochemicals and polymer chemistry. They also act as antivirals (15) (Agouram. N, 2023), antipsychotics (17) (Irfan. A, 2021) neurodegenerative (20) (Masih. A, 2020), antibiotics (21) (Irfan. A, 2021). In plant cultivation, Tiadinil (19) a plant activator, a pesticide Methiadinil (18) (Fan. Z, 2011), Certain plants are protected from bacterial, fungal, and viral illnesses by the use of bion (16) as a plant inductor (Saitanis. C. J, 2015). The cotton-defoliant thidiazuron (22) (Guo. B, 2011) which encourages the growth of new hybrid and transgenic plant tissues and cellular cultures, shares properties with natural phytohormones known as cytokines.

Figure 4. Bioactive compounds containing 1,2,3-thiadiazole

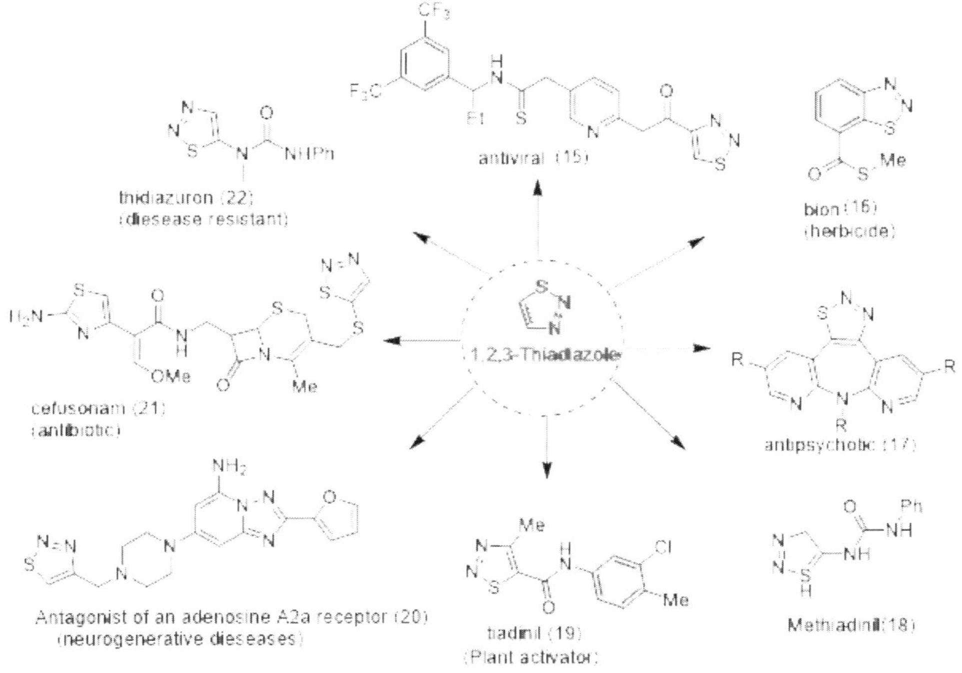

Traditionally 1,2,3-thiadiazole are synthesized by the widely accepted Hurd-Mori method by intramolecular cyclization of hydrazones using $SOCl_2$ or S_2Cl_2 (Figure 5). In Pechmann synthesis, isothiocyanates are reacted with diazomethane to give secondary amine substituted 1,2,3-thiadiazole (Figure 6), In Wolff synthesis α-diazo-1,3-dicarbonyl compounds on treatment with $(NH_4)_2S$ yielded substituted 1,2,3-thiadiazole (Figure 7).

Figure 5. Synthesis of 1,2,3-thiadiazole via Hurd-Mori method

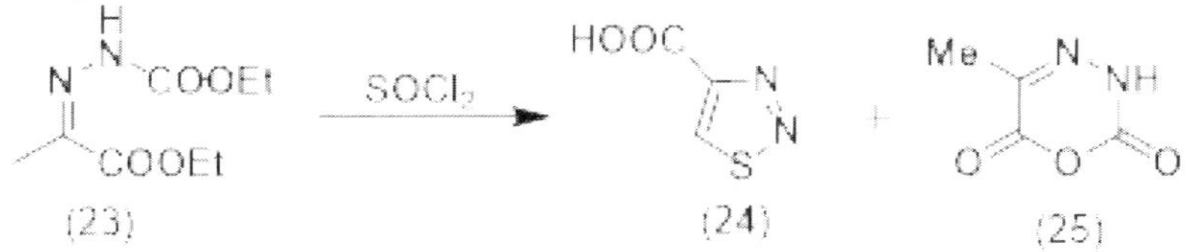

Figure 6. Synthesis of 1,2,3-thiadiazole via Pechmann method

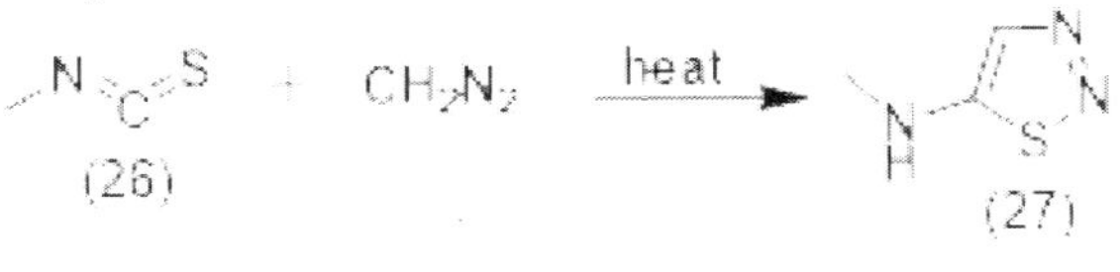

Figure 7. Synthesis of 1,2,3-thiadiazole via Wolff method

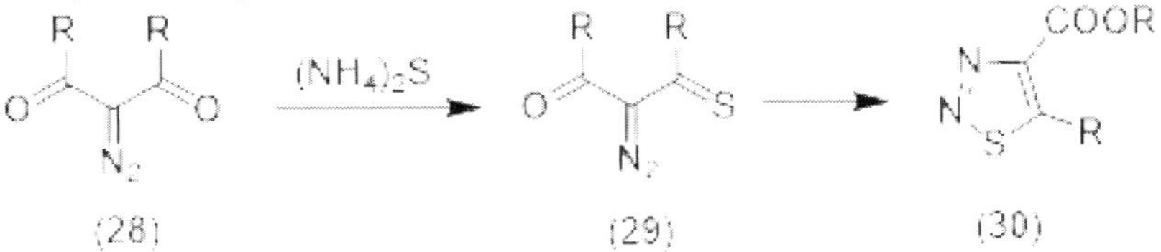

Filimonov and co-workers developed to promote the development of new hybrid and transgenic plant tissues and cellular cultures. Green synthesis of substituted 1,2,3-thiadiazole-4-carbimidamides (Figure 9) catalyzed by water/ alkali. The reaction between azides (32, 35) and 2-cyanothioacetamides (31, 34) in the presence of alkali in water yielded a final product monocyclic (Figure 8) and bicyclic 1,2,3-thiadiazole-4-carbimidamides (Figure 9) in a single step (Filimonov. V. O, 2019).

Figure 8. Synthesis of monocyclic 1,2,3-thiadiazole

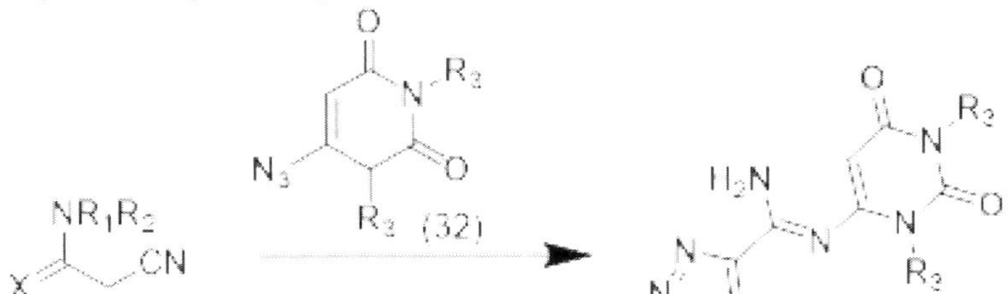

Figure 9. Synthesis of monocyclic and bicyclic 1,2,3-thiadiazole

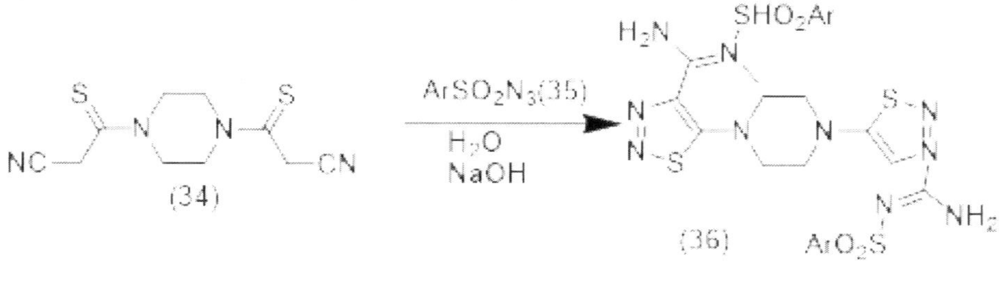

In a manner comparable to the Hurd-Mori reaction, the Chen group created a synthesis of substituted 1,2,3-thiadiazole without the use of metal catalyst (38). In the presence of $K_2S_2O_8$, the reaction of *N*-tosyl hydrazone (37) with elemental sulfur, catalyzed by TBAI, produced the product in excellent yield (Figure 10) (Chen. J, 2016).

Figure 10. Synthesis of 1,2,3-thiadiazole using TBAI/ $K_2S_2O_8$

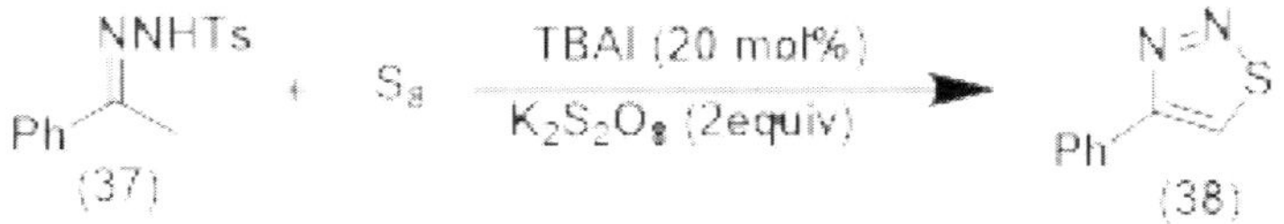

Employing an electron detosylation process, Liu *et al.*, constructed a 4-substituted 1,2,3-thiadiazole (41). A reaction between α-chloroketones (39) and toluene sulfonyl hydrazide (40) in-situ mediated by trisulfur radical anion (S_3^-) (Figure 11) (Liu. B. B, 2018).

Figure 11. Synthesis of 1,2,3-thiadiazole employing detosylation process

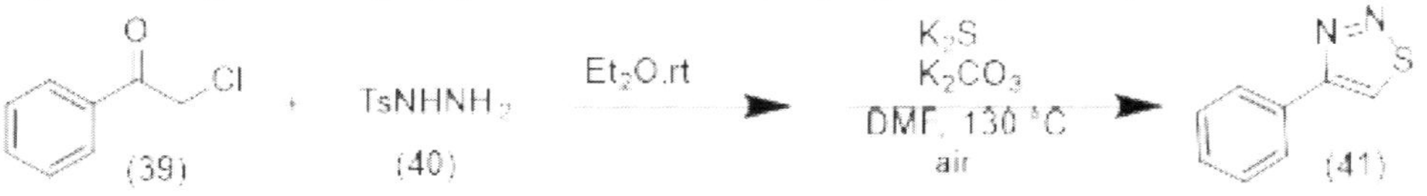

Bakulev group described the adaptable synthesis of 1,2,3-thiadiazole (45-47) from 2-cyanothioacetamides and azides. Benzenesulfonyl azide was used to annulate under diazo-transfer conditions at room temperature in the presence of pyridine 2-cyanoethanethioamides (43 Figure 12) (Filimonov. V. O, 2017).

Figure 12. Synthesis of 4,5-disubstituted-1,2,3-thiadiazole

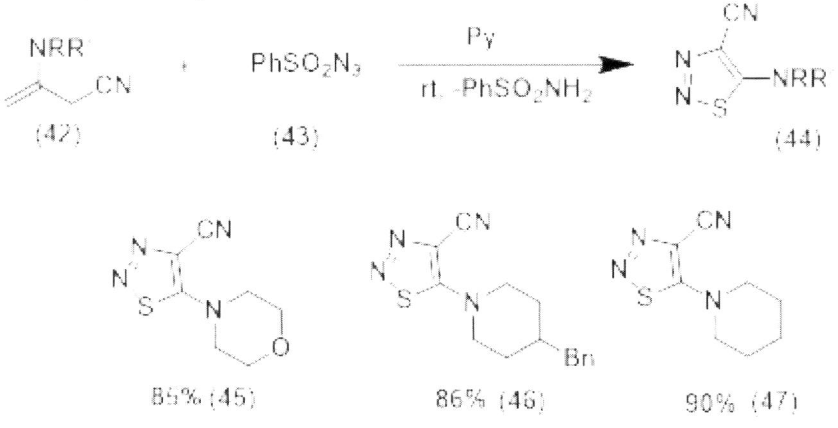

A metal-assisted one-pot three-component synthesis of substituted 1,2,3-thiadiazoles (51) was developed by Wang *et al.*, An aliphatic or aromatic substituted 1,2,3-thiadiazole was produced in good yield by the reaction of aromatic-substituted methyl ketones (48), *p*-toluenesulfonyl hydrazide (49), and potassium thiocyanate (50) in DMSO with the help of $I_2/CuCl_2$. (Figure 13) (Wang. C, 2019).

Figure 13. Synthesis of 1,2,3-thiadiazole from methyl ketones

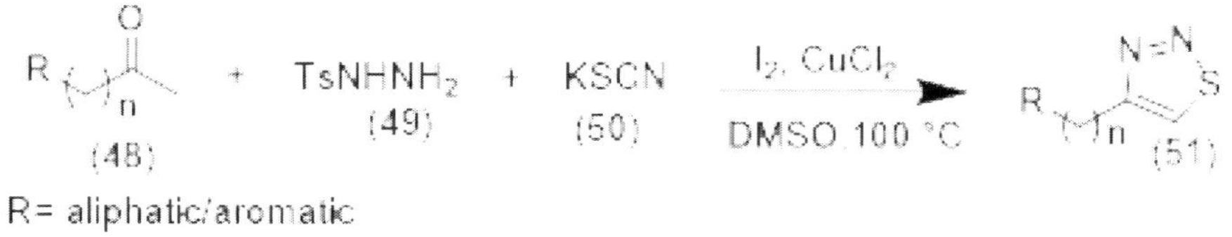

R= aliphatic/aromatic

Zhang's group published a versatile method to synthesize 1,2,3-thiadiazoles. 1,2-addition of carbon disulfide and benzyl bromide to α-diazo carbonyl compound (52) in the presence of KOH as a base in DCM at 50 C afforded 4,5- disubstituted-1,2,3-thiadiazoles (53) (Zhang. L, 2018).

Figure 14. Synthesis of ipso-1,2,3-thiadiazole

(52)   (53)

Dong *et al.*, designed a convenient base-promoted synthesis of 1,2,3-thiadiazoles (55). The transition-metal and oxidant-free intramolecular reaction of *N-tosyl* hydrazone-bearing thiocarbamates (54) in Cs_2CO_3 in THF produced a product (55) with good yield (Figure 15) (Dong C, 2022).

Figure 15. Synthesis of 3,5-substituted-1,2,3-thiadiazole

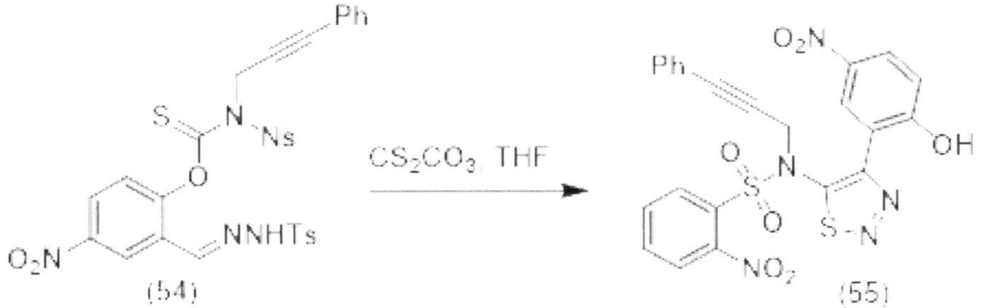

(54)   (55)

1,2,4-thiadiazole

1,2,4-thiadiazole is a special pharmacophore that exhibits a variety of biological characteristics., including anti-microbial, anti-inflammatory, anti-cancer (Kumar. D, 2011), Neuroprotector (58) (Grigoriev. V. V, 2017), cardiovascular (60) (Castro. A, 2006), Synthase kinase inhibitor (57) (Palomo. V, 2012), gramma secretase modulator (59) (Gurjar. A. S, 2014) and modulators of allostery (56) (Lanzafame. A, 2004).

Figure 16. Biologically active compounds containing 1,2,4-thiadiazole

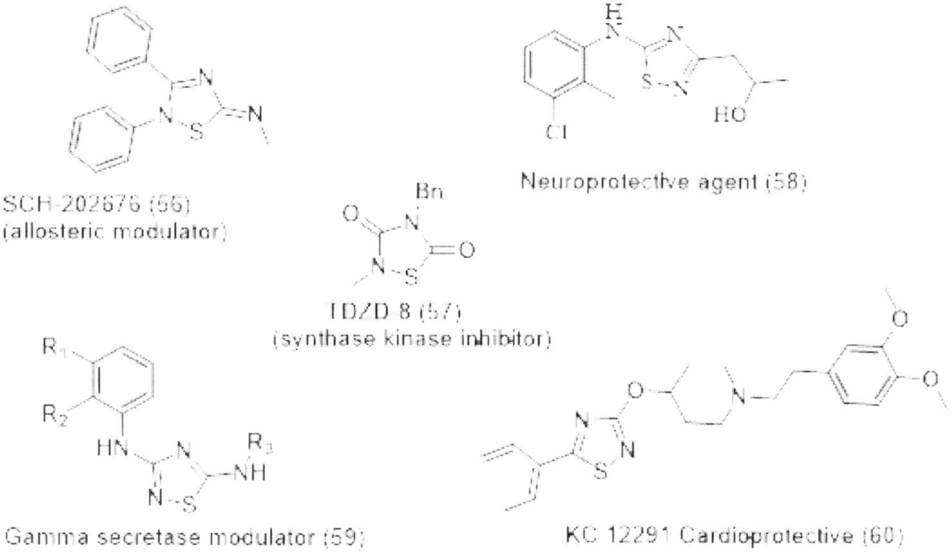
SCH-202676 (56)
(allosteric modulator)
TDZD-8 (57)
(synthase kinase inhibitor)
Neuroprotective agent (58)
Gamma secretase modulator (59)
KC 12291 Cardioprotective (60)

The earlier efficient method involves the production of 3,5-disubstituted-1,2,4-thiadiazoles (61) from aromatic thioamide by oxidative dimerization (Figure 17) in one synthesis (Yajima. K, 2015). Another familiar method utilizes the thioamide derivation and nitriles (62) to give 1,2,4-thiadiazoles (63, Figure 18) (Joana de A. e Silva, 2013).

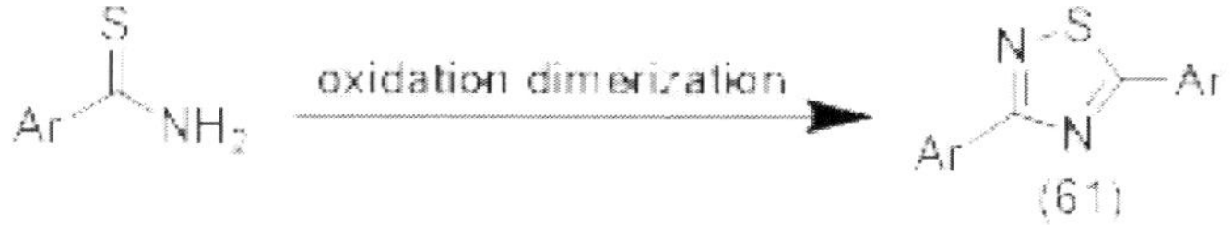
oxidation dimerization
(61)

Figure 18. Synthesis of 3,5-disubstituted-1,2,4-thiadiazoles

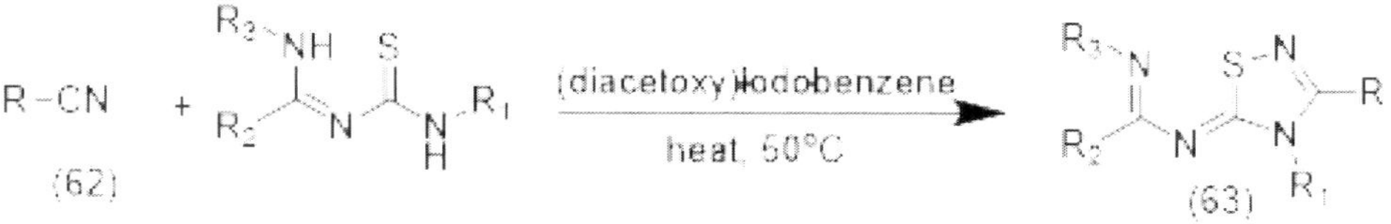
(diacetoxy)iodobenzene
heat, 50°C
(62)
(63)

Rajput *et al.,* presented a potentially effective one-pot two-step approach for the synthesis of 1,2,4-thiadiazoles, which is superior than a number of previously studied techniques involving the oxidative dimerization of thioamides. Conversion of primary amide to thioamide using Lawesson's reagent followed by reacting with TBHP to give 3,5-disubstituted-1,2,4-thiadiazole (64) in a solvent-free environment (Figure 19) (Rajput. K, 2024).

Figure 19. Electro-orgnaic synthesis of 3,5-disubstituted-1,2,4-thiadiazoles

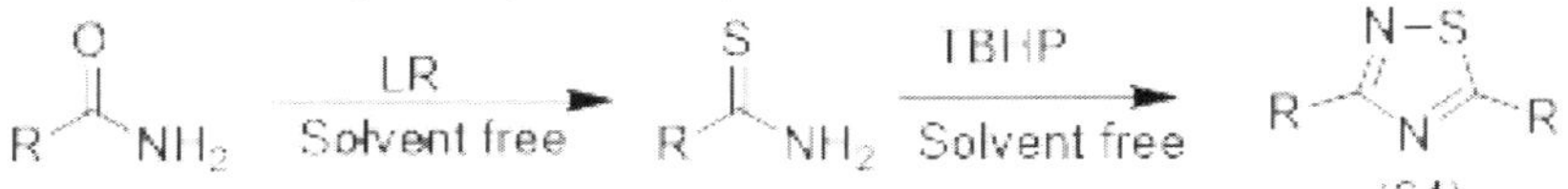

Using NH_4I as an electrolyte in $MeOH/CH_3CN$, Wang et al. devised the electro-chemical synthesis of disubstituted-1,2,4-thiadiazole (65), from substituted thioamides without using transition metals or oxidizing agents (Figure 20) (Wang. Z. Q, 2018).

Figure 20. Synthesis of disubstituted-1,2,4-thiadiazoles

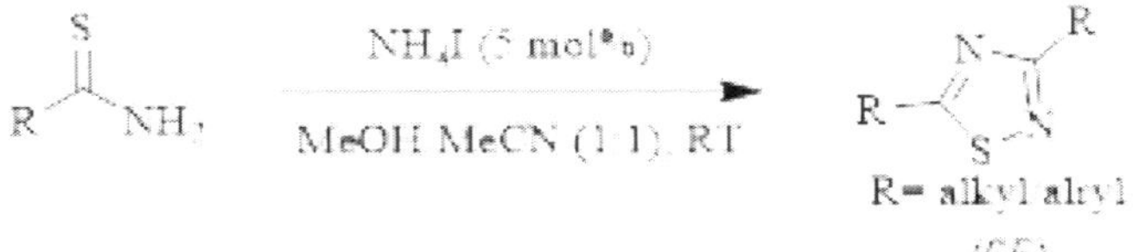

A single-step synthesis involving oxidative dimerization of thioamides has produced symmetrical diaryl substituted-1,2,4-thiadiazoles by the use of CAN as an oxidant, by combining ceric ammonium nitrate and thioamides in an equimolar ratio in acetonitrile (Figure 21) (Vanajatha. G, 2016).

Figure 21. Synthesis of diaryl-1,2,4-thiadiazoles using CAN

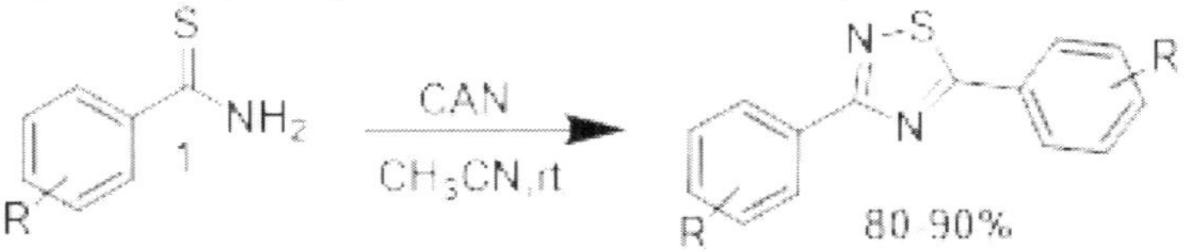

Halimehjani and co-workers reported the metal-free synthesis to access a 3,5-Bis-mercapto-1,2,4-thiadiazoles (67) by the oxidative dimerization of dithiocarbamates using an equimolar mixture of H_5IO_6 as an oxidant which yielded 85% at 0 C when the reaction is optimized in the DCM (Figure 22) (Halimehjani A. Z, 2019).

Figure 22. Synthesis of 3,5-Bis-mercapto-1,2,4-thiadiazoles

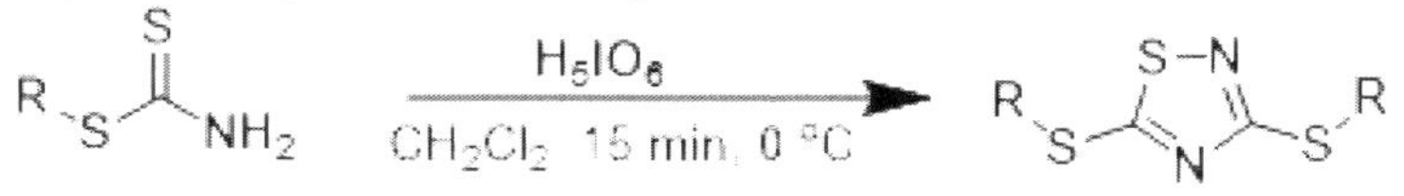

Tumula's group reported the synthesis of solvent-free 3,4-disubstituted 5-imino-1,2,4-thiadiazoles (68)by using by heating imidoyl thiourea and benzonitrile in the presence of PIDA (poly(diiododiacetylene) as a catalyst at 50 C (Figure 23) (Jatangi. N, 2018).

Figure 23. Synthesis of 1,2,4-thiadiazoles using PIDA

Tang Ma group demonstrated the selective synthesis of 1,2,4-thiadiazole (71) by using BAST (70) as a source of sulfur. The reaction of imidazo[1,5-a]pyridines (69) with BAST using NH_4I at room temperature. Here NH_4I acts a nitrogen source as well as a promoter (Ma. S. T, 2021).

Figure 24. Synthesis of 1,2,4-thiadiazoles using BAST

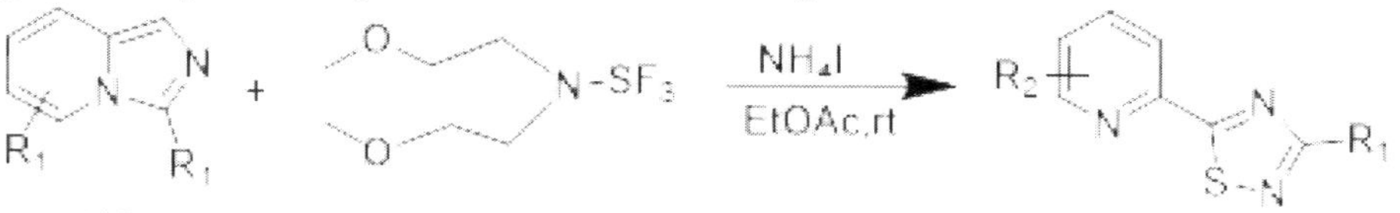

Bose *et al.,* reported a TCCA-H$_2$O-promoted one-spot synthesis of 3,5-diaryl-1,2,4-thiadiazoles (72). An oxidative dimerization of thioamides with aryl groups attached to the electron withdrawing and donating group using TCCA as a catalyst in the presence of water at room temperature afforded the product with good yield without formation of byproducts (Figure 25) (Subhas Bose. D, 2017).

Figure 25. Synthesis of 1,2,4-thiadiazoles using TCCA-H$_2$O

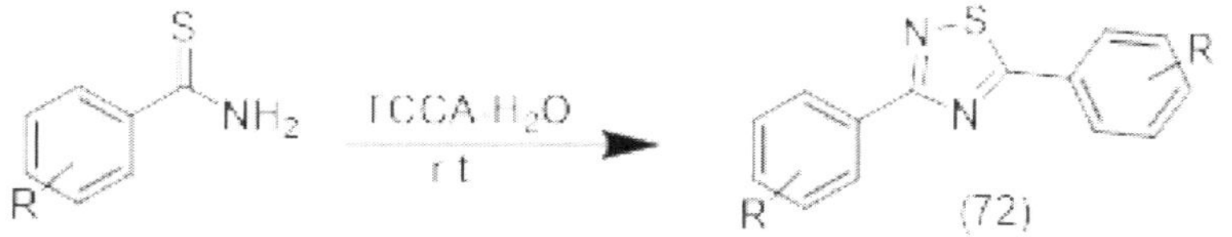

Chauhan's group published an ultrasound-accelerated green synthesis of 1,2,4-thiadiazole derivatives (73) using water as a solvent medium. Conversion of thio-amide to 1,2,4-thiadiazole by using chloranil in water at room temperature under ultrasound irradiation (Figure 26) (Chauhan. S, 2020).

Figure 26. Synthesis of 1,2,4-thiadiazoles using ultrasonication

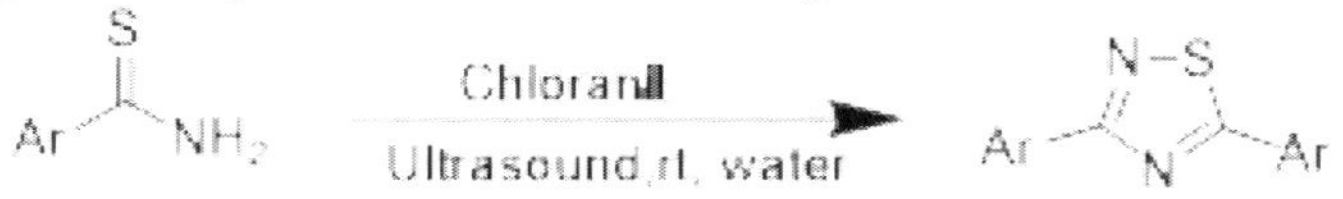

Halimehjani *et al.,* developed an environmentally friendly Protocol for the synthesis of 3,5-Bis (alkylthio)-1,2,4-thiadiazoles (74) via two methods. The first protocol describes an oxidative dimerization of S-alkyl dithiocarbamates catalyzed by I$_2$ in EtOH at room temperature. In the second method, they replaced I$_2$ by oxone as an oxidant in an aqueous medium at 50 C (Figure 27) (Halimehjani. A. Z, 2017).

Figure 27. Synthesis of 3,5-Bis (alkylthio)-1,2,4-thiadiazole

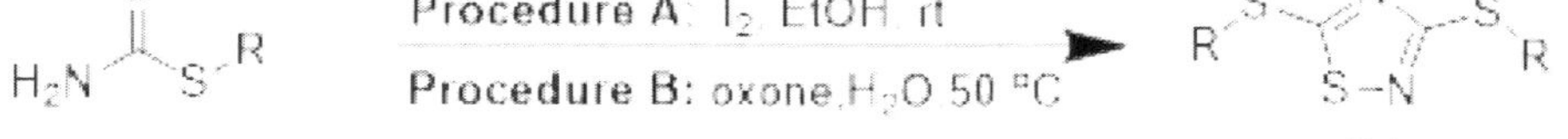

Coa reported the expedient green synthesis of 1,2,4-thiadiazole under metal-free conditions. Hydrogen peroxide assisted conversion of substituted thiourea into 1,2,4-thiadiazole in ethanol at ambient temperature which is a suitable procedure for the large-scale synthesis (Figure 28).

Figure 28.

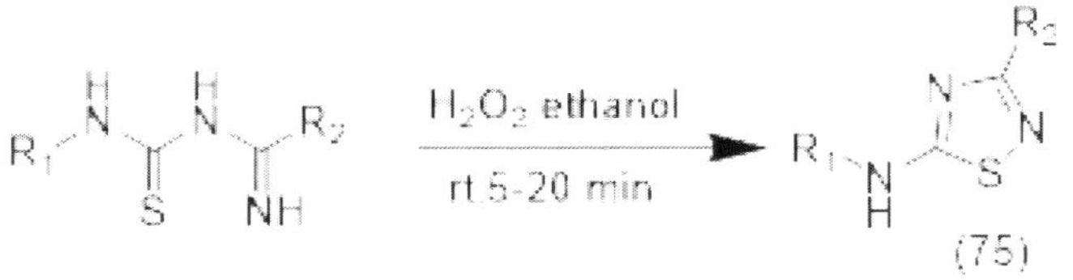

Jatangi group successfully designed metal-free facile synthesis of 3-substituted 5-amino-1,2,4-thiadiazole scaffolds (76). Substituted isothiocyanates and substituted amidoximes were reacted in the presence of I_2/K_2CO_3 at 60 °C in an aqueous condition (Figure 29). In a same fashion, Wang *et al.*, described a I_2 mediated simple method for the synthesis 5-amino and 3,5-diamino substituted 1,2,4-thiadiazoles. Here, different types of variety of imidoyl and guanyl thiourea substrates are reacted in the presence of molecular iodine alone (Figure 30) (Jatangi. N, 2018).

Figure 29.

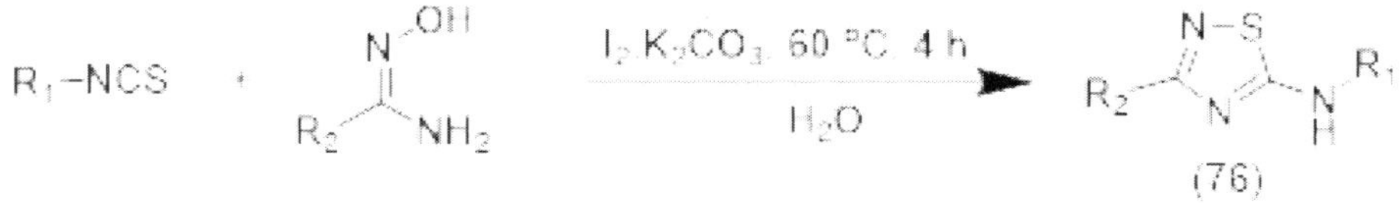

Figure 30. Synthesis of disubstituted-1,2,4-thiadiazoles

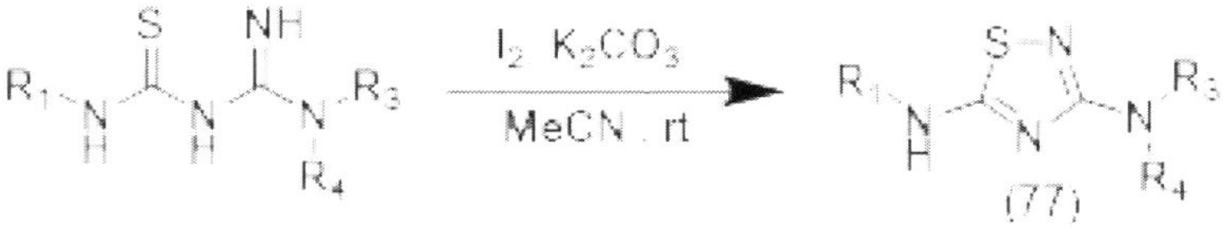

Yoshimura and co-workers utilized the 2-iodosylbenzoic acid activated by trifluoromethanesulfonic anhydride in the synthesis of 1,2,4-thiadiazole (78). IBA/Tf_2O was used to oxidatively dimerize thiobenzamide, which resulted in an excellent isolated yield of 3,5-diphenyl-1,2,4-thiadiazole at rt (Figure 31) (Yoshimura. A, 2021).

Figure 31. Synthesis of 1,2,4-thiadiazoles using IBA/Tf_2O

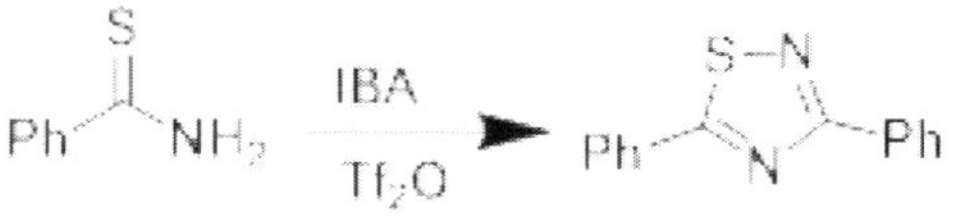

Mariappan group published an efficient method for the synthesis of 3-substituted 5-amino-1,2,4-thiadiazoles (79) via intramolecular oxidation. Conversion of imidoyl thioureas to 5-amino-1,2,4-thiadiazoles using PIFA (phenyliodine(III) bis(trifluoroacetate)) in the presence of DCE at room temperature as a cost-effective catalyst. This synthesis technique holds the benefit of a high product yield when the reactants contain an electron donating or withdrawing or both the groups (Figure 32) (Mariappan. A, 2016).

Figure 32. Synthesis of 3-substituted 5-amino-1,2,4-thiadiazoles

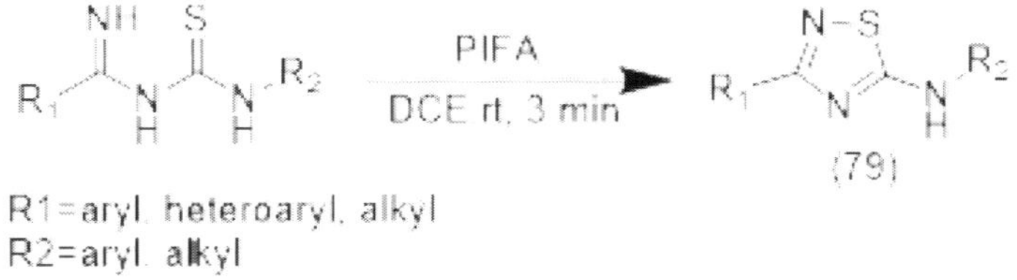

R1=aryl, heteroaryl, alkyl
R2=aryl, alkyl

Chauhan et al., successfully proposed a synthesis of 1,2,4-thiadiazole (80) via radical-induced dimerization. Dimerization of primary thioamides using *tert*-butyl nitrite as a radical initiator in DCM at room temperature in excellent yields (Figure 33) (Chauhan. S, 2018).

Figure 33. Synthesis of diaryl-substituted-1,2,4-thiadiazoles

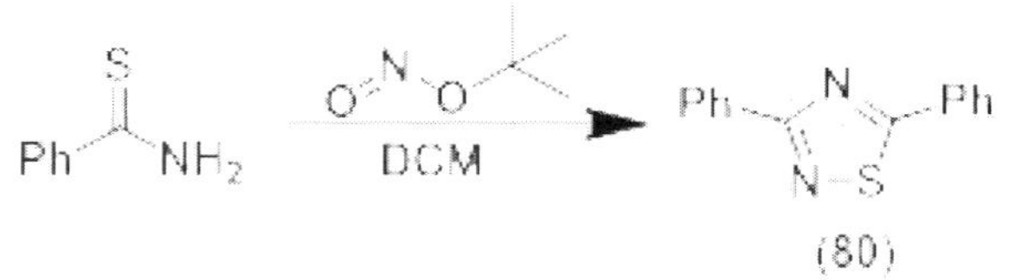

(80)

Chacko and Shivashankar reported the unprecedented synthesis of *N*-fused imi-nothiadiazolo isoquinoline derivatives (82). In this approach, 2-aminoisoquinolines (81) and phenylisothiocyanate in DMF at room temperature using montmorillonite K10 as a catalyst for the construction N-S bond (Figure 34) (Chacko. P, 2018).

Figure 34. Synthesis of 1,2,4-thiadiazoles bearing isoquinoline

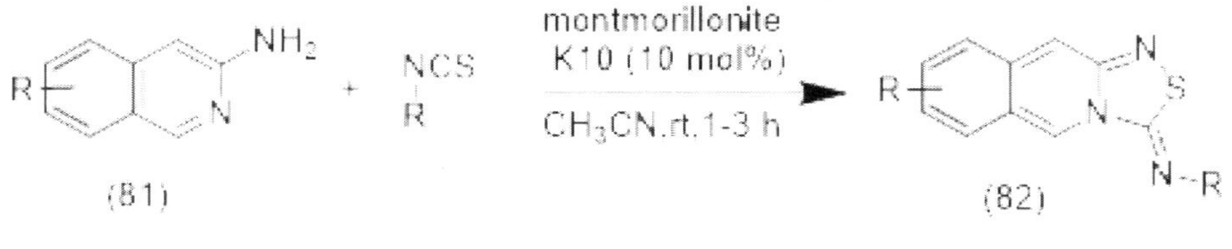

(81)   (82)

Yang and co-workers developed a regioselective green procedure for the synthesis of 5-amino- 1,2,4-thiadiazoles (83). The reaction between 2-aminopyridine/amidine and isothiocyanate catalyzed by TMEDA afforded metal-free synthesis of 1,2,4-thiadiazoles in MeCN and air as an oxidant at 90 C (Figure 35) (Yang. Z, 2019).

(83)

With the help of substituted aromatic acid chlorides and 3,4,5-trimethoxyben-zamidine in the presence of ($K_3PO_4 \cdot 3 H_2O$) as a base, substituted triazole reacted with 3,4,5-trimethoxybenzamidine at 130 C. Then, it coupled with sulfur in DMSO at rt and produced 1,2,4-thiadiazole-1,2,4-triazole derivatives (84) in good yields. With IC_{50} values ranging from 0.10±0.084 to 11.5±6.49 µM concerning the standard Etoposide, six compounds- 84a, 84b, 84c, 84d, 84e, and 84f exhibited more effective anti-cancer action for three distinct cancers: breast cancer (MCF-7, MDA MB-231), lung cancer (A549), and prostate cancer (DU-145) (Figure 36) (Pragathi. Y. J, 2021).

Figure 36. Synthesis of 3,5-substituted-1,2,4-thiadiazoles

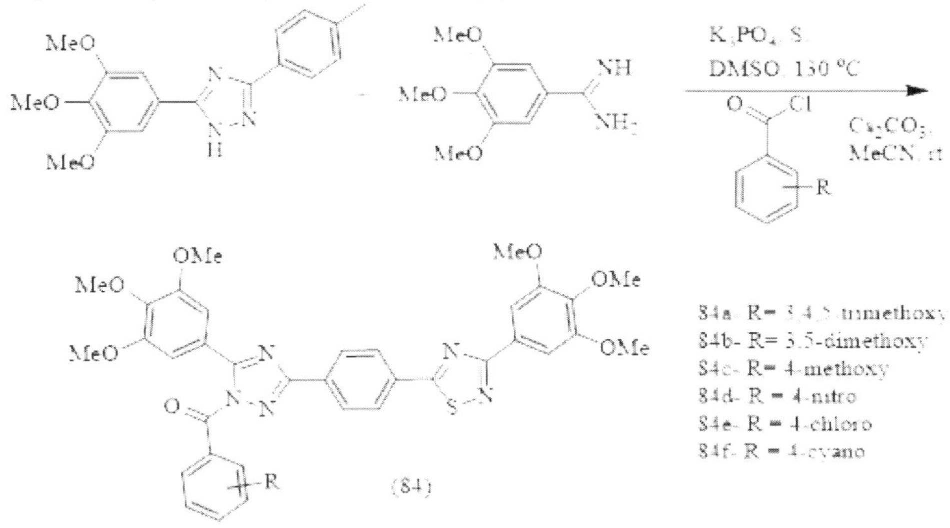

Volkova *et al.,* designed a 5-N-monoaminosubstituted-5-amino-3-(2-aminopropyl)- 1,2,4-thiadiazoles derivative as potent neuroprotectors. The compound (87) exhibited an excellent inhibition of the glutamate-stimulated Ca^{2+} uptake in vitro. By adding 3-amino-5-methylisoxazole to a solution of isothiocyanate in Acetonitrile followed by reacting with primary amine resulted in vinylamine 5-amino-3-(2-amino-1-propenyl)- [1,2,4]thiadiazole. Upon reduction of compound with $NaBH_4$, 5-amino-3- (2-hydroxypropyl)-[1,2,4]thiadiazole obtained in good yield (Figure 37) (Pragathi. Y. J, 2021).

Figure 37. Synthesis of 3,5-disubstituted-1,2,4-thiadiazoles

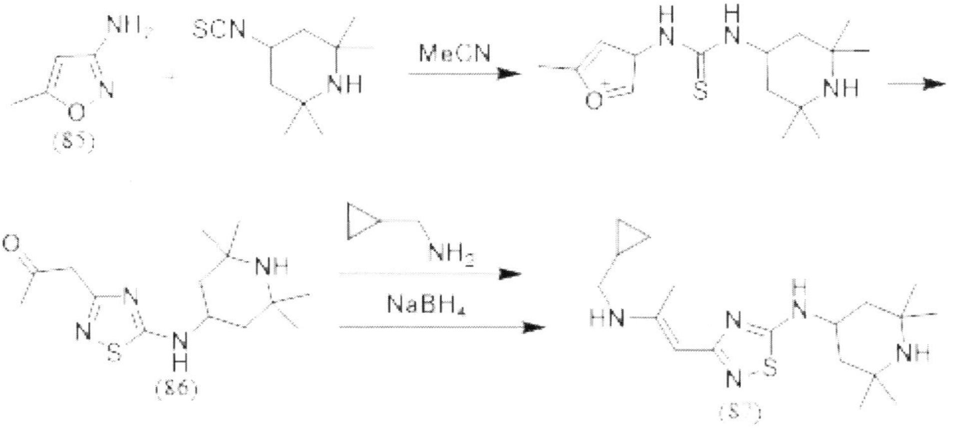

1,2,5-Thiadiazole

1,2,5-Thiadiazole has wide applications in material chemistry. Isomorphic fluorobenzo[c][1,2,5]thiadiazole based polymers (88) produced by Lee *et al.,* show an ultra-high mobility of 17.8 cm^2 $V^{-1}s^{-1}$, making them suitable for usage as superior optoelectronic materials (Lee. J, 2018).

Jin and coworkers developed novel narrow band gap π-conjugated polymers containing benzodithiophene or thiophene annulated with alkylthiophene as the electron-donating segment based on naphtho[1,2-c:5,6-c′]bis([1,2,5]thiadiazole) (89). When the 1,2,5-thiadiazole-based segment is coated on polymer solar cells (PSCs) exhibited excellent power conversion efficiencies (PCEs) greater than 11%. These conjugated polymers were promising scaffolds for the mass production of high-performance polymer solar cells (Jin. Y, 2017).

Ma's group utilized a Benzo[c][1,2,5]thiadiazole as third component electron acceptor for the for ternary polymer solar cells. 4,7-Bis(5-bromothiophen-2-yl)-5,6-difluorobenzo[c][1,2,5]thiadiazole (90) is used to fabricate the efficient ternary PSCs which resulted in the short-circuit current density, open-circuit voltage, increased light-harvesting with improved polymer crystallinity (Ma. Y, 2020).

There are many convenient methods reported for the synthesis of 1,2,5-thiadiazole among them use of 2-diamino derivatives is one of the most efficient methods for the synthesis of monocyclic 1,2,5-thiadiazoles. 3-hydroxy-1,2,5-thiadiazole was produced by the reaction of S_2Cl_2 and 2-amino3-methylbutanamide (Figure 38) (Rakitin. O. A, 2019).

Figure 38. Synthesis of substituted-1,2,5-thiadiazoles

(88)

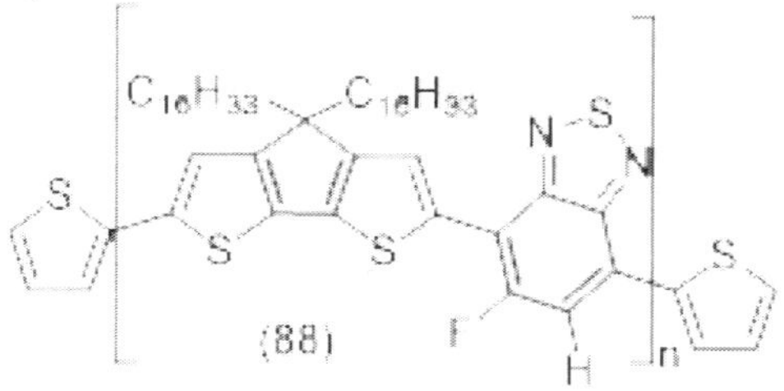

Taydakov et al., published a synthesis of [13,14-*c*][1,2,5]thiadiazole 14,14-dioxide by condensation of alpha-diketones with vicinal diamines. The reaction of 1,2-di(naphthalen-1-yl)ethane- 1,2-dione with sulfamide followed by oxidative ring fusion of 3,4-di(naphthalene-1-yl)-1,2,5-thiadiazole 1,1-dioxide in the presence of a catalyst $AlCl_3$ resulted in the compound (93, Figure 39) (Taydakov. I. V, 2017).

Figure 39. Synthesis of substituted-1,2,5-thiadiazoles from alpha-diketones

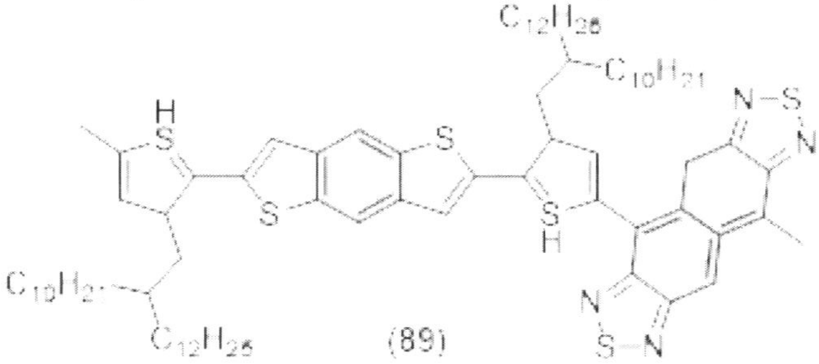

(89)

Kalogirou group designed a 3-aryl-4-(2-arylbenzo[*d*][1,3]dioxol-2-yl)-1,2,5-thiadiazoles via Wagner–Meerwein rearrangement. Thermal ring contraction of 3,5-diaryl-1,2,6-thiadiazine 4,4-catechol ketals in the presence of Lewis or bronsted acids resulted in excellent yield (Kalogirou. A. S, 2016).

Figure 40. Synthesis of 1,2,5-thiadiazole via Wagner–Meerwein rearrangement.

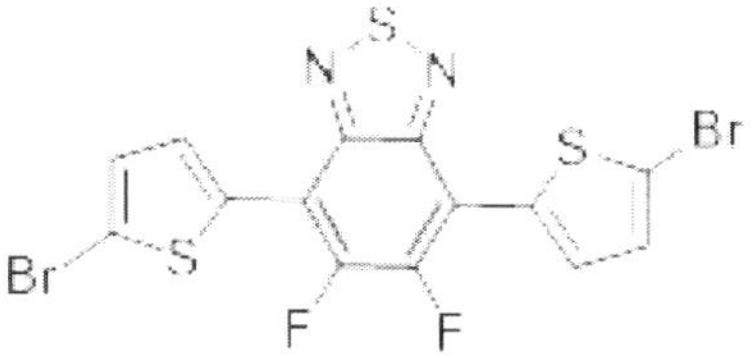

(90)

1,3,4-Thiadiazole

Among the four structural isomers of thiadiazole, the wide range of pharmaco-logical effects of 1,3,4-thiadiazole derivatives has led to their extensive research. Electron withdrawing groups and other heterocycles were substituted at the 1,3,4-thiadiazole ring to produce agents with increased potency and less toxicity (Jain. A. K, 2013). Nucleophiles can readily attack the 1,3,4-thiadiazole ring. However, not many methods are available for creating such frameworks. Access to 1,3,4-thiadiazoles from thiosemicarbazide derivatives is one of the most commonly used procedures (Linganna. N, 1998). Convenient literature reports for the synthesis of molecules containing a 1,3,4-thiadiazole (96) unit involve the reaction of either alkyl or aryl diacylhydrazines with LR or P_2S_5 at reflux (Figure 41).

Figure 41. Synthesis of 1,3,4-thiadiazole using Lawesson's reagent

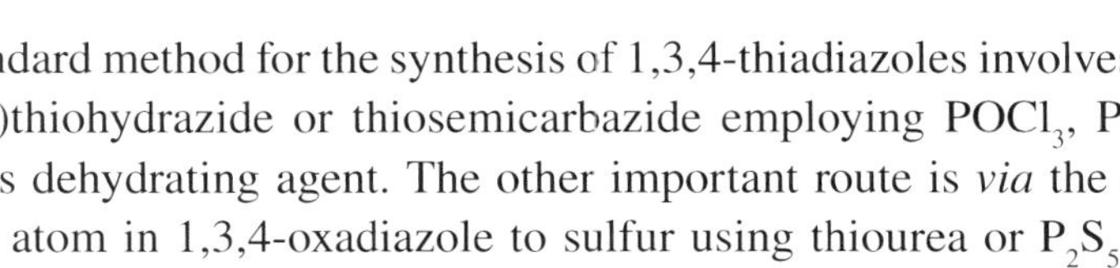

(91)

The standard method for the synthesis of 1,3,4-thiadiazoles involves cyclization of *N'*-(acyl)thiohydrazide or thiosemicarbazide employing $POCl_3$, PCl_5, MsOH, or H_2SO_4 as dehydrating agent. The other important route is *via* the exchange of the oxygen atom in 1,3,4-oxadiazole to sulfur using thiourea or P_2S_5 (Figure 42) (Linganna. N, 1998).

Figure 42. Synthesis of 1,3,4-thiadiazole from 1,3,4-oxadiazole

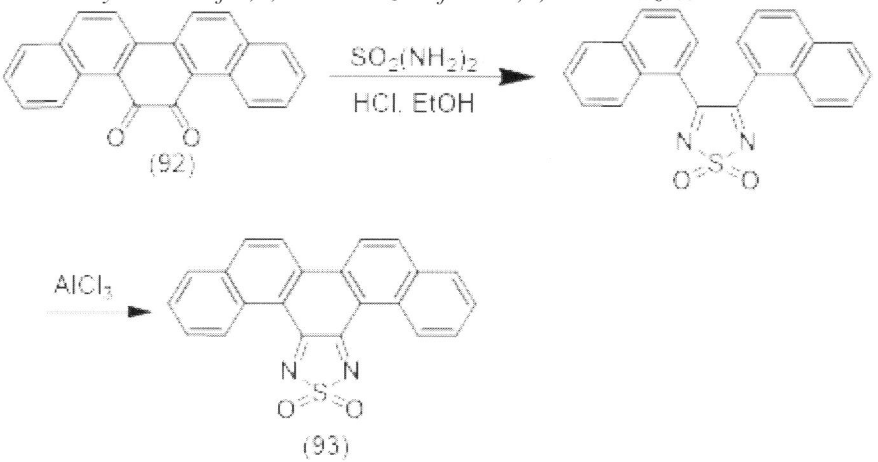

Ghomi's group developed a novel series of [1,2,4]triazolo[3,4-b][1,3,4]thiadiazole evaluated as urease inhibiting activity. According to the SARs analysis and kinetic study, the unsubstituted compound (99) exhibited excellent urease inhibition among the series with K_i of 1.37 µM. The reaction was initiated by the addition of substituted triazole and 4-amino-5-phenyl-4H pyrazole-3-thiol in MeOH, potassium hydroxide was added and refluxed at 70 °C to obtain 3-phenyl-[1,2,4]triazolo[3,4-b][1,3,4] thiadiazole-6-thiol, followed by the reaction with aryl bromide resulted in [1,2,4] triazolo[3,4-b][1,3,4]-thiadiazole derivatives (Figure 43) (Khalili Ghomi. M, 2023).

Figure 43. Synthesis of triazole linked 1,3,4-thiadiazole

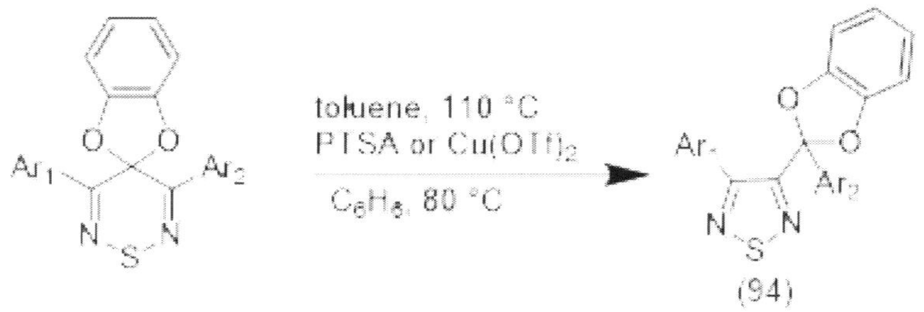

Dong and co-workers established a convenient route for the regioselective synthesis of 2-amino-1,3,4-thiadiazole. Nucleophilic addition reaction of acyl hydrazides with isothiocyanates resulted in the thiosemicarbazides which on cyclization with $POCl_3$ in chlorobenzene at 60 C resulted a product in excellent yields. Among the synthesized series N-(adamantan-1-yl)-5-(5-(azepan-1-yl)pyrazin-2-yl)-1,3,4-thiadiazol-2-amine (100) displayed excellent anti-influenza activity against influenza A virus strains A/HK/68 (H3N2) and A/PR/8/34 (H1N1) with EC_{50} values of 3.5 µM and 7.5 µM. This activity is due to the pyrazine or pyridine attached to the 1,3,4-thiadiazole moiety (Figure 44) (Dong. J, 2022).

Figure 44. regioselective synthesis of 2-amino-1,3,4-thiadiazole

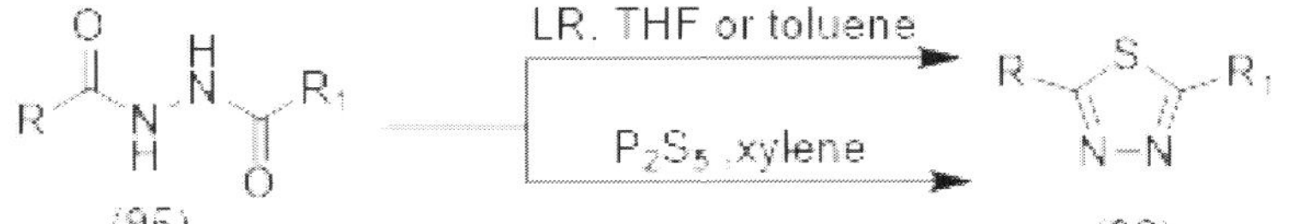

Freitas *et al.*, designed a synthesis of novel pyridinyl-1,3,4-thiadiazole derivatives from *N*-aminobenzyl or *N*-arylhydrazone and evaluated for trypanocidal activity. Compound (101) exhibited excellent trypanocidal activity against the trypomastigote form of *Trypanosoma cruzi*, showing IC_{50} value of 21.60 µM (Figure 45) (Freitas. R. H. C. N, 2020).

Figure 45. Synthesis of novel pyridinyl-1,3,4-thiadiazole derivatives

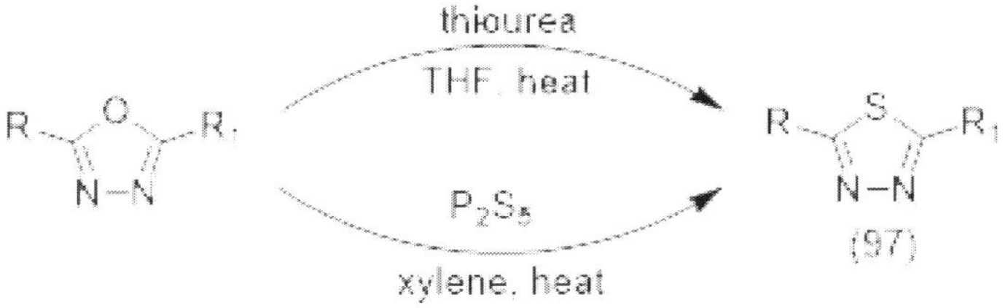

Jakovljevic *et al.*, designed the facile synthesis of 1,3,4-thiadiazoles from phenolic acids which are investigated for antioxidant and antiproliferative activity. The reaction between thiosemicarbazide and 3,4-dihydroxybenzoic 1 and 2,3-dihydroxy benzoic acid in the presence of phosphoryl chloride results in 5-substituted-1,3,4-thiadiazol-2-amines (102). In the synthesized analogs 1,3,4- thiadiazole with 4-chlorophenyl moiety exhibited excellent inhibition for lung carcinoma A549 cells. Excellent DPPH radical scavenging and good cytotoxic action against human acute promyelocytic leukemia were demonstrated by compound 3c, which has an adamantane ring. HL-60 cells (Figure 46) (Jakovljević. K, 2017).

Figure 46. Facile Synthesis of 1,3,4-thiadiazoles from phenolic acids

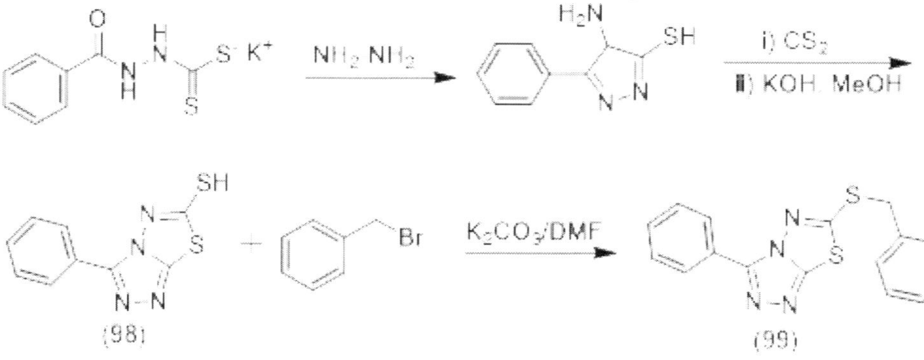

Zarei *et al.*, established a Vilsmeier reagent as a cyclo-dehydrating agent in the synthesis of 1,3,4-Thiadiazole derivatives (103) from hydrazine and carboxylic acid. This method offers a straightforward purifying synthesis because of the water-soluble by-products and ambient reaction conditions of Vilsmeier reagent (Figure 47) (Zarei. M, 2017).

Figure 47.

Muğlu and co-workers described a green protocol for the synthesis of 1,3,4-thiadiazole compounds which exhibited excellent antimicrobial activity. By reacting 4-phenoxybutyric acid with *N*-phenyl thiosemicarbazide derivatives followed by cyclization using $POCl_3$ afforded a product in excellent yields. The 5-[3-phenoxypro-pyl]-*N*-[4'-fluorophenyl]-1,3,4-thiadiazol-2-amine (104) displayed the highest anti-bacterial activity *S. aureus* among the synthesized series (Figure 48) (Muğlu. H, 2018).

Figure 48. Synthesis of 2,5-disubstututed-1,3,4-thiadiazole from

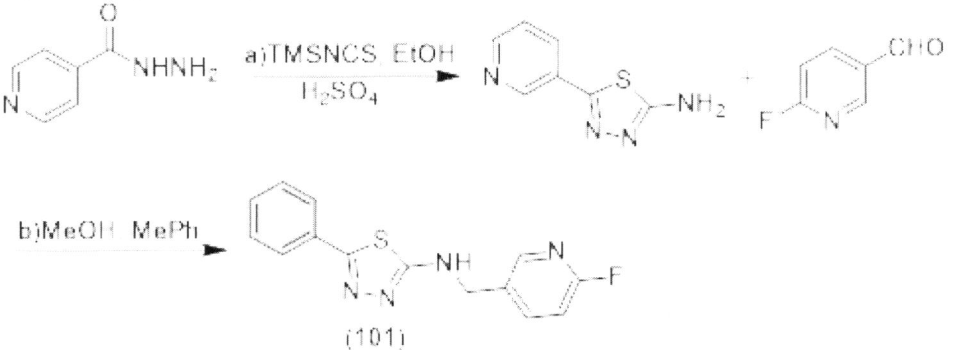

(101)

1,3,4-Thiadiazole (105) was synthesized by Zeebaree *et al.*, using copper nanoparticles that were isolated from *Trifolium resupinatum* plant leaves. Thiocarbohydrazide, acetophenone, and chalcone were added to ethanol in a multicomponent reaction followed by the addition of copper nanoparticles produced product in a good yield (Figure 49) (Sharaf Zeebaree. S. Y, 2019).

Figure 49. Synthesis of 1,3,4-thiadiazole using copper nanoparticles

(102)

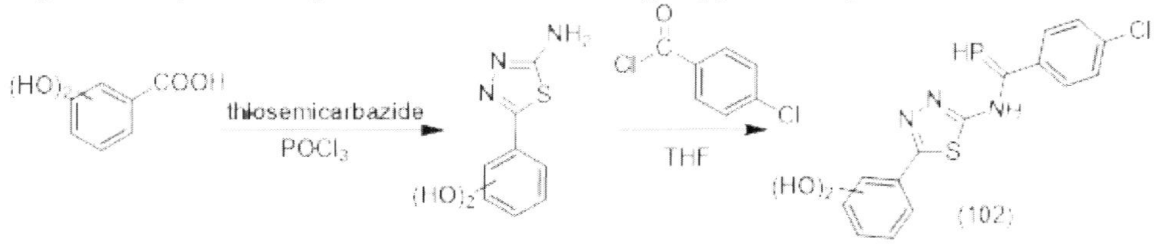

Lin's group synthesized 3,6-disubstituted-1,2,4-triazole-1,3,4-thiadiazole derivatives (105) and evaluated for antibacterial and antifungal. Twelve analogs were prepared in the presence of POCl$_3$ by reacting different aromatic carboxylic acids with 4-amino-5-substituted-4H- 1,2,4-triazole-3-thiols in good yield. Compound (105a) and (105b) exhibited excellent antibacterial activity against *E. coli, S. aureus, P. oryzae, R. solani* with the relative inhibition zone ranging 80% to 90%. The presence of fluorophenyl and nitrophenyl moieties are responsible for the high antibacterial activity (Figure 50) (Lin. L, 2017).

Figure 50. Synthesis of 3,6-disubstituted-1,2,4-triazole-1,3,4-thiadiazole derivatives

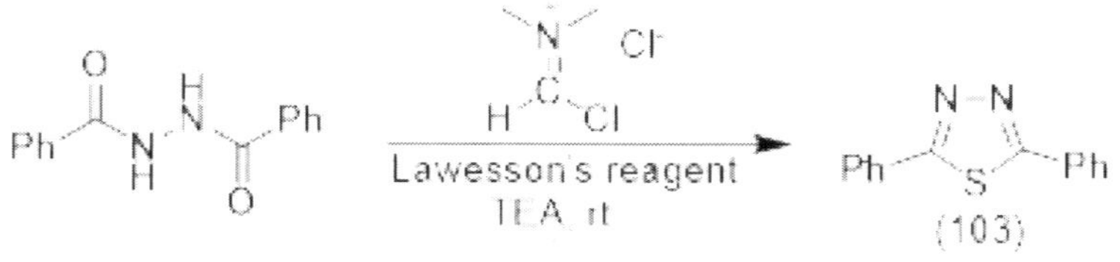

Narran *et al,*. developed 5-styryl-2-amino-1,3,4-thiadiazole by reacting 3-phenyl propenoic acid and thiosemicarbazide. The 5-styryl-2-amino-1,3,4-thiadiazoles (106) were treated with different anhydrides in presence of catalytic NaOAc to obtain the corresponding carbamates. The 5-styryl-2-amino-1,3,4-thiadiazole derivatives bearing styrl group and carbamate exhibit high biological activity against *Staphylococcus aureus* and *E. coli* (Figure 51) (Narran. S. F, 2022).

Figure 51. Synthesis 1,3,4-thiadiazole from cinnamic acid

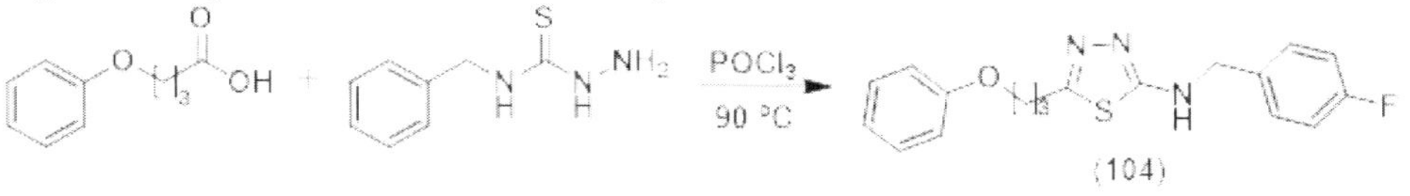

Chen *et al.,* reported 1,3,4-thiadiazole containing amide moiety from thiosemicarbazide followed by reacting with pyridyl chlorides give substituted 1,3,4 derivatives. compound (107) demonstrated strong nematocidal properties in vitro and in vivo against Meloidogyne incognita; the LC_{50} value and control effect were 6.5 mg/L and 83.3%, respectively as well as It had remarkable antibacterial properties against *Ralstonia solanacearum, Xanthomonas campestris pv. citri*, and *Xanthomonas oryzae* the EC50 values were 5.1, 6.7 and 0.4 mg/L (Figure 52) (Liu, B. B, 2018).

Figure 52. Synthesis of amide linked1,3,4-thiadiazole

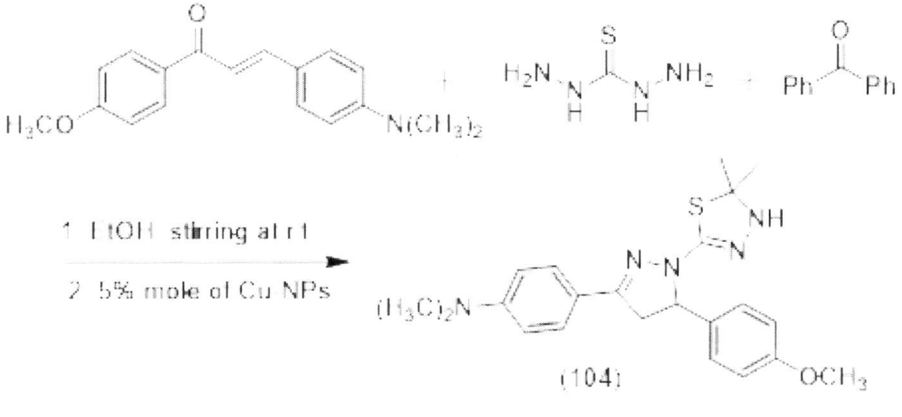

Ismail *et al.,* published their pioneer work by reacting thiosemicarbazide with 3-chlorobenzoic acid followed by the addition of acid chloride to afford 1,3,4-thiadiazole (108) and their derivatives in two steps via solvent-free protocol. The synthesized compounds exhibited potent insecticidal activity against cotton leafworm (*Spodoptera littoralis*). The most effective against cotton leafworm was Schiff base derivative, with an LC_{50} of 556.94 lg/mL (Figure 53) (Ismail. M. F, 2021).

Figure 53. Synthesis of 1,3,4-thiadiazolo[3,2-a]pyrimidine

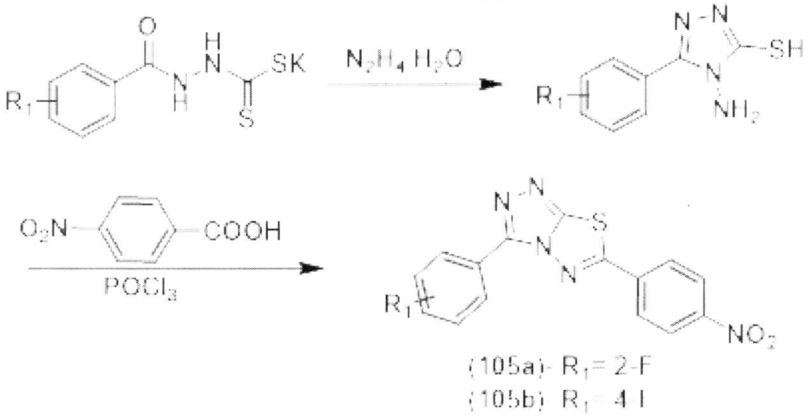

Ma and co-workers successfully established electrochemical synthesis of 1,3,4-thiadizole. This approach enables the oxidant and metal-free protocol by reacting isothiocyanates with hydrazone in the presence of electrolyte nBu$_4$NBF$_4$ and DDQ as a redox catalyst using graphite rod as an anode and Ni as cathode at 35 C in MeCN/H$_2$O afforded substituted 1,3,4-thiadiazole in good yield (109, Figure 54) (Ma. Z, 2021).

Figure 54.Electro-organic synthesis of 1,3,4-thiadiazole

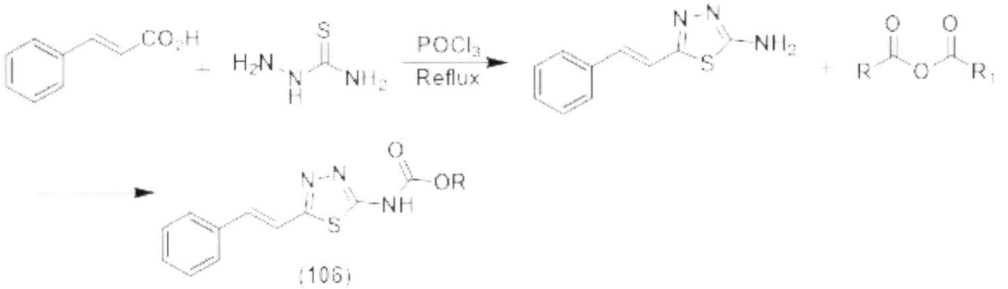

(106)

Karaburun group developed a efficient method for the synthesis of 1,3,4-thiadiazole derivative which exhibited excellent pharmokinetic profile as well as anti-fungal activity. Intermediate 5-((4-chlorophenyl)amino)-1,3,4-thiadiazole-2-thiol was synthesized by cyclizing substituted thiosemicarbazide which on reaction with 2-bromoacetophenone (0.83 mmol), and potassium carbonate enabled the targeted scaffold. The substance demonstrated exceptional antifungal properties against every type of fungus that was examined. Compound 110 proved to be the most successful derivative in combating *Candida albicans* ATCC 10231. Because of the fluoro and chloro groups at the second position of the phenyl molecule exhibited notable activity (Figure 55) (Karaburun. A. Ç, 2018).

Figure 55. Synthesis 2,5-substituted thiadiazole

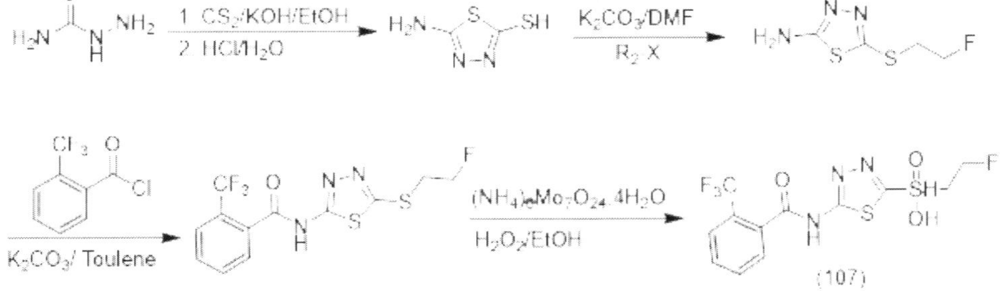

(107)

Kudelko *et al.*, successfully reported the synthesis of synthesis of azo dyes bearing a 1,3,4-thiadiazole. In this work 2-amino-5-aryl-1,3,4-thiadiazoles are synthesized from aniline, *N,N*-dimethylaniline and phenol and used as a precursors for the synthesis of three series of azodyes in excellent yield (Figure 56) (Kudelko. A, 2020).

Figure 56. Synthesis of 2-amino-5-aryl-1,3,4-thiadiazoles

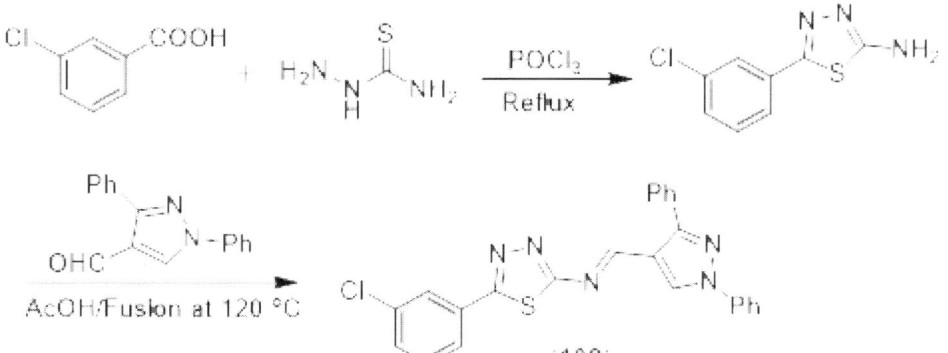

Kokovina and co-workers established a novel route for the synthesis of 1,3,4-thiadiazole-2-amine derivatives (112) from a thiosemicarbazide and carboxylic acid in the presence of polyphosphate ester (PPE) without the use of hazardous cyclizing agents like $SOCl_2$ or $POCl_3$, which offered a novel method for the one-pot, three-step synthesis of 2-amino-1,3,4-thiadiazoles. Using benzoic acid as a precursor yields great results (Figure 57) (Kokovina. T. S, 2021).

Figure 57. Synthesis of primary amine substituted 1,3,4-thiadiazole

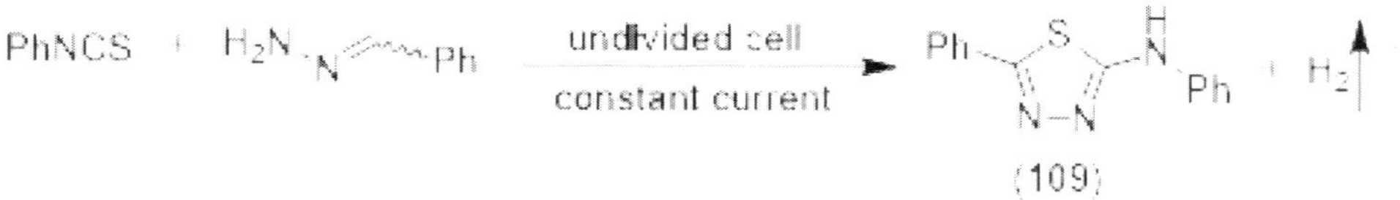

Patel group designed a versatile protocol for the synthesis of imidazo[2,1-b][1,3,4]thiadiazole derivatives from 1- methyl-1H-imidazole-2-carbonitrile which was obtained by reacting cyanogen bromide and 4-*N,N*-dimethylamiopyridine with 1- methyl-1H-imidazole in DMF. The compound were reacted with thiosemicarbazide and alpha-halo ketones in CF_3COOH to give compound (113). In vitro antitubercular activity against Mycobacterium Tuberculosis demonstrated the highest (98%) inhibitory activity among the investigated drugs with a minimum inhibitory concentration (MIC) of 3.14 lg/ml. The strong activity is caused by the existence of an electron-withdrawing nitro group (Figure 58) (Patel. H. M, 2017).

Figure 58. Synthesis of imidazole substituted 1,3,4-thiadiazole

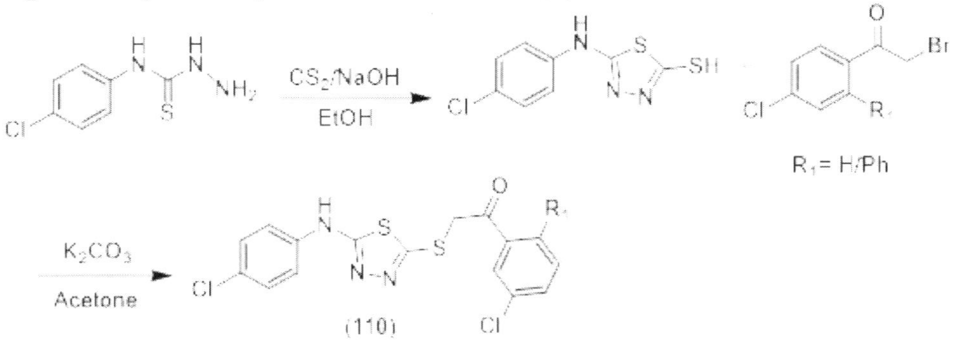

Kamel group on reacting methyl hydrazinecarbodithioate or hydrazinecarbothio-amide with 1-[3,5-dimethyl- 1-(4-nitrophenyl)-1*H*-pyrazol-4-yl]ethan-1-one rendered 2-[1-[5-methyl-1-(4-nitrophenyl)-1*H*-pyrazol-4-yl]ethylidene]hydrazine derivatives as a precursor for the synthesis of 1,3,4-thiadiazole derivatives which on treatment with hydrazonoyl chloride derivatives afforded a target molecule. The compound (114) were tested for anti-microbial activity against gram-positive bacteria (*B. mycoides*), whereas yeast (*C. albicans*) and gram-negative bacteria (*E. coli*). Compared to the positive control, the compound demonstrated superior antibacterial activity (Figure 59) (Kamel. M. G, 2022).

Figure 59. Synthesis of 2,3-disubstituted 1,3,4-thiadiazole

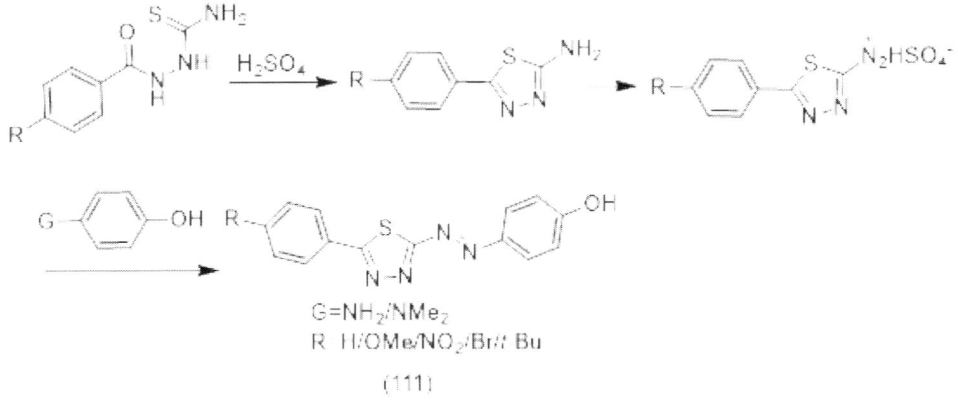

Bondock *et al.,* outlined the synthesis of 1,3,4-thiadiazole (115, 116) from hydrazonyl bromide and alkyl phenylcarbamodithioates. Two compounds were analyzed using DFT (B3LYP) to investigate the structural characterization by quantum methods with the values of 6-31 +g(d,p) basis set. According to studies on molecular electrostatic potential, the stronger the electron withdrawing group, such as the nitro group, the more powerful the site will be for an electrophilic attack (Figure 60) (Bondock. S, 2021).

Figure 60.

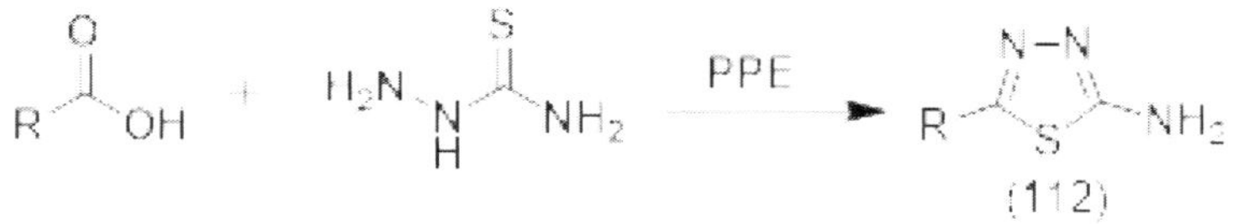

(112)

REFERENCES

Agouram, N. (2023). 1,2,3-Triazole Derivatives as Antiviral Agents. *Medicinal Chemistry Research*, 32(12), 2458–2472. DOI: 10.1007/s00044-023-03154-3

Atmaram, U. A., & Roopan, S. M. (2022). Biological Activity of Oxadiazole and Thiadiazole Derivatives. In *Applied Microbiology and Biotechnology* (pp. 3489–3505). Springer.

Badami, B. V. (2007). *Mesoionic Compounds: An Unconventional Class of Aromatic Heterocycles*.

Bergeron, M. G., Brusch, J. L., Barza, M., & Weinstein, L. (1973). Bactericidal Activity and Pharmacology of Cefazolin. *Antimicrobial Agents and Chemotherapy*, 4(4), 396–401.

Berzas Nevado, J. J., Flores, J. R., De La, M. L., & Parnj, M. (1991). Determination of sulphamethizole in the presence of nitrofurantoine by derivative spectrophotometry and ratio spectra derivative (Vol. 38).

Bondock, S., Albarqi, T., & Abboud, M. (2021). Advances in the Synthesis and Chemical Transformations of 5-Acetyl-1,3,4-Thiadiazolines. *Journal of Sulfur Chemistry*, 42(2), 202–240. DOI: 10.1080/17415993.2020.1843170

Castro, A., Castaño, T., Encinas, A., Porcal, W., & Gil, C. (2006). Advances in the Synthesis and Recent Therapeutic Applications of 1,2,4-Thiadiazole Heterocycles. *Bioorganic & Medicinal Chemistry*, 14(5), 1644–1652. DOI: 10.1016/j.bmc.2005.10.012 PMID: 16249092

Chacko, P., & Shivashankar, K. (2018). Montmorillonite K10-Catalyzed Synthesis of N-Fused Imino-1,2,4-Thiadiazolo Isoquinoline Derivatives. *Synthetic Communications*, 48(11), 1363–1376. DOI: 10.1080/00397911.2018.1448419

Chauhan, S., Chaudhary, P., Singh, A. K., Verma, P., Srivastava, V., & Kandasamy, J. (2018). Tert-Butyl Nitrite Induced Radical Dimerization of Primary Thioamides and Selenoamides at Room Temperature. *Tetrahedron Letters*, 59(3), 272–276. DOI: 10.1016/j.tetlet.2017.12.033

Chauhan, S., Verma, P., Mishra, A., & Srivastava, V. (2020). An Expeditious Ultrasound-Initiated Green Synthesis of 1,2,4-Thiadiazoles in Water. *Chemistry of Heterocyclic Compounds*, 56(1), 123–126. DOI: 10.1007/s10593-020-02632-5

Chauvière, G., Bouteille, B., Enanga, B., De Albuquerque, C., Croft, S. L., Dumas, M., & Périé, J. (2003). Synthesis and Biological Activity of Nitro Heterocycles Analogous to Megazol, a Trypanocidal Lead. *Journal of Medicinal Chemistry*, 46(3), 427–440. DOI: 10.1021/jm021030a PMID: 12540242

Chen, J., Jiang, Y., Yu, J. T., & Cheng, J. (2016). TBAI-Catalyzed Reaction between N-Tosylhydrazones and Sulfur: A Procedure toward 1,2,3-Thiadiazole. *The Journal of Organic Chemistry*, 81(1), 271–275. DOI: 10.1021/acs.joc.5b02280 PMID: 26675203

Choy, D. S. J., Arandia, J., & Rosenbaum, I. (1967). Clinical evaluation of a new alkylating agent, azetepa, in one hundred and twenty-five cases of malignant tumors. *International Journal of Cancer*, 2(2), 189–193.

Dong, C.; Mai, S.; Wang, S.; Li, X.; Song, Q. *Supporting Information for Base-Promoted Anaerobic Intramolecular Cyclization Synthesis of 4,5-Disubstituted 1,2,3-Thiadiazoles Electronic Supplementary Material (ESI) for Organic*; 2022.

Dong, J., Pei, Q., Wang, P., Ma, Q., & Hu, W. (2022). Optimized POCl3-Assisted Synthesis of 2-Amino-1,3,4-Thiadiazole/1,3,4-Oxadiazole Derivatives as Anti-Influenza Agents. *Arabian Journal of Chemistry*, 15(4), 103712. Advance online publication. DOI: 10.1016/j.arabjc.2022.103712

Ergena, A., Rajeshwar, Y., & Solomon, G. (2022). Synthesis and Diuretic Activity of Substituted 1,3,4-Thiadiazoles. *Scientifica*, 2022, 1–9. DOI: 10.1155/2022/3011531 PMID: 35433072

Fan, Z. (2011). Methiadinil, a Novel Elicitor Candidate for Crop Protection(New Chemistry,Poster,1) Pest Management, Crop Protection and Vector Control). *Journal of Pesticide Science*, 36(1), 162.

Farghaly, T. A., Abdallah, M. A., & Muhammad, Z. A. (2011). Synthesis and Evaluation of the Anti-Microbial Activity of New Heterocycles Containing the 1,3,4-Thiadiazole Moiety. *Molecules (Basel, Switzerland)*, 16(12), 10420–10432. DOI: 10.3390/molecules161210420 PMID: 22173335

Filimonov, V. O., Dianova, L. N., Beryozkina, T. V., Mazur, D., Beliaev, N. A., Volkova, N. N., Ilkin, V. G., Dehaen, W., Lebedev, A. T., & Bakulev, V. A. (2019). Water/Alkali-Catalyzed Reactions of Azides with 2-Cyanothioacetamides. Eco-Friendly Synthesis of Monocyclic and Bicyclic 1,2,3-Thiadiazole-4-Carbimidamides and 5-Amino-1,2,3-Triazole-4-Carbothioamides. *The Journal of Organic Chemistry*, 84(21), 13430–13446. DOI: 10.1021/acs.joc.9b01599 PMID: 31547663

Filimonov, V. O., Dianova, L. N., Galata, K. A., Beryozkina, T. V., Novikov, M. S., Berseneva, V. S., Eltsov, O. S., Lebedev, A. T., Slepukhin, P. A., & Bakulev, V. A. (2017). Switchable Synthesis of 4,5-Functionalized 1,2,3-Thiadiazoles and 1,2,3-Triazoles from 2-Cyanothioacetamides under Diazo Group Transfer Conditions. *The Journal of Organic Chemistry*, 82(8), 4056–4071. DOI: 10.1021/acs.joc.6b02736 PMID: 28328204

Freitas, R. H. C. N., Barbosa, J. M. C., Bernardino, P., Sueth-Santiago, V., Wardell, S. M. S. V., Wardell, J. L., Decoté-Ricardo, D., Melo, T. G., da Silva, E. F., Salomão, K., & Fraga, C. A. M. (2020). Synthesis and Trypanocidal Activity of Novel Pyridinyl-1,3,4-Thiadiazole Derivatives. *Biomedicine and Pharmacotherapy*, 127, 110162. Advance online publication. DOI: 10.1016/j.biopha.2020.110162 PMID: 32407986

Gao, L., Zhu, Y., Lyu, Y., Hao, F. L., Zhang, P., & Wei, M. J. (2015). A Pharmacokinetic and Pharmacodynamic Study on Intravenous Cefazedone Sodium in Patients with Community-Acquired Pneumonia. *Chinese Medical Journal*, 128(9), 1160–1164. DOI: 10.4103/0366-6999.156086 PMID: 25947397

Grigoriev, V. V., Makhaeva, G. F., Proshin, A. N., Kovaleva, N. V., Rudakova, E. V., Boltneva, N. P., Gabrel'yan, A. V., Lednev, B. V., & Bachurin, S. O. (2017). 1,2,4-Thiadiazole Derivatives as Effective NMDA Receptor Blockers with Anticholinesterase Activity and Antioxidant Properties. *Russian Chemical Bulletin*, 66(7), 1308–1313. DOI: 10.1007/s11172-017-1890-9

Guo, B., Abbasi, B. H., Zeb, A., Xu, L. L., & Wei, Y. H. (2011). Thidiazuron: A Multi-Dimensional Plant Growth Regulator. *African Journal of Biotechnology*, 10, 8984–9000. DOI: 10.5897/AJB11.636

Gurjar, A. S., Andrisano, V., Simone, A. D., & Velingkar, V. S. (2014). Design, Synthesis, in Silico and in Vitro Screening of 1,2,4-Thiadiazole Analogues as Non-Peptide Inhibitors of Beta-Secretase. *Bioorganic Chemistry*, 57, 90–98. DOI: 10.1016/j.bioorg.2014.09.002 PMID: 25303313

Halimehjani, A. Z., Nosood, Y. L., Didaran, S., & Aryanasab, F. (2017). Metal-Free Oxidative Dimerization of Dithiocarbamates: Direct Access to 3,5-Bis-Mercapto-1,2,4-Thiadiazoles. *SynOpen*, 1(1), 138–142. DOI: 10.1055/s-0036-1590964

Halimehjani, A. Z., Sharifi, A., & Rahimzadeh, H. (2019). Simple and Green Procedures for the Synthesis of 3,5-Bis(Alkylthio)-1,2,4-Thiadiazoles via Oxidative Dimerization of Dithiocarbamates. *ChemistrySelect*, 4(9), 2634–2638. DOI: 10.1002/slct.201803832

Hu, Y., Li, C. Y., Wang, X. M., Yang, Y. H., & Zhu, H. L. (2014). 1,3,4-Thiadiazole: Synthesis, Reactions, and Applications in Medicinal, Agricultural, and Materials Chemistry. *Chemical Reviews*, 114(10), 5572–5610. DOI: 10.1021/cr400131u

Hurd, C. D., & Mori, R. I. (1955). On Acylhydrazones and 1,2,3-Thiadiazoles. *Journal of the American Chemical Society*, 77(1), 5359–5364.

Iizawa, Y., Okonogi, K., Hayashi, R., Iwahi, T., Yamazaki, T., & Imada, A. *29th Interscience Confer-Ence On*; 1993; Vol. 17. https://journals.asm.org/journal/aac

Irfan, A.; Ullah, S.; Anum, A.; Jabeen, N.; Zahoor, A. F.; Kanwal, H.; Kotwica-Mojzych, K.; Mojzych, M. Synthetic Transformations and Medicinal Significance of 1,2,3-Thiadiazoles Derivatives: An Update. *Applied Sciences (Switzerland)*. MDPI AG June 2, 2021. .DOI: 10.3390/app11125742

Ismail, M. F., Madkour, H. M. F., Salem, M. S., Mohamed, A. M. M., & Aly, A. F. (2021). Design, Synthesis and Insecticidal Activity of New 1,3,4-Thiadiazole and 1,3,4-Thiadiazolo[3,2-a]Pyrimidine Derivatives under Solvent-Free Conditions. *Synthetic Communications*, 51(17), 2644–2660. DOI: 10.1080/00397911.2021.1945106

Jain, A. K., Sharma, S., Vaidya, A., Ravichandran, V., & Agrawal, R. K. (2013). 1,3,4-Thiadiazole and Its Derivatives: A Review on Recent Progress in Biological Activities. *Chemical Biology & Drug Design*, 81(May), 557–576. DOI: 10.1111/cbdd.12125 PMID: 23452185

Jakovljević, K., Matić, I. Z., Stanojković, T., Krivokuća, A., Marković, V., Joksović, M. D., Mihailović, N., Nićiforović, M., & Joksović, L. (2017). Synthesis, Antioxidant and Antiproliferative Activities of 1,3,4-Thiadiazoles Derived from Phenolic Acids. *Bioorganic & Medicinal Chemistry Letters*, 27(16), 3709–3715. DOI: 10.1016/j.bmcl.2017.07.003 PMID: 28709826

Jatangi, N., Tumula, N., Palakodety, R. K., & Nakka, M. (2018). I2-Mediated Oxidative C-N and N-S Bond Formation in Water: A Metal-Free Synthesis of 4,5-Disubstituted/N-Fused 3-Amino-1,2,4-Triazoles and 3-Substituted 5-Amino-1,2,4-Thiadiazoles. *The Journal of Organic Chemistry*, 83(10), 5715–5723. DOI: 10.1021/acs.joc.8b00753 PMID: 29717614

Jin, Y., Chen, Z., Xiao, M., Peng, J., Fan, B., Ying, L., Zhang, G., Jiang, X. F., Yin, Q., Liang, Z., Huang, F., & Cao, Y. (2017). Thick Film Polymer Solar Cells Based on Naphtho[1,2-c:5,6-c]Bis[1,2,5]Thiadiazole Conjugated Polymers with Efficiency over 11%. *Advanced Energy Materials*, 7(22), 1700944. Advance online publication. DOI: 10.1002/aenm.201700944

Kadi, A. A., Al-Abdullah, E. S., Shehata, I. A., Habib, E. E., Ibrahim, T. M., & El-Emam, A. A. (2010). Synthesis, Antimicrobial and Anti-Inflammatory Activities of Novel 5-(1-Adamantyl)-1,3,4-Thiadiazole Derivatives. *European Journal of Medicinal Chemistry*, 45(11), 5006–5011. DOI: 10.1016/j.ejmech.2010.08.007 PMID: 20801553

Kalogirou, A. S., Kourtellaris, A., & Koutentis, P. A. (2016). The Acid and/or Thermal Mediated Ring Contraction of 4H-1,2,6-Thiadiazines to Afford 1,2,5-Thiadiazoles. *Organic Letters*, 18(16), 4056–4059. DOI: 10.1021/acs.orglett.6b01929 PMID: 27483200

Kamel, M. G., Sroor, F. M., Othman, A. M., Hassaneen, H. M., Abdallah, T. A., Saleh, F. M., & Teleb, M. A. M. (2022). Synthesis and Biological Evaluation of New 1,3,4-Thiadiazole Derivatives as Potent Antimicrobial Agents. *Monatshefte für Chemie*, 153(10), 929–937. DOI: 10.1007/s00706-022-02967-z

Karaburun, A. Ç., Çevik, U. A., Osmaniye, D., Saglık, B. N., Çavuşoglu, B. K., Levent, S., Özkay, Y., Koparal, A. S., Behçet, M., & Asım Kaplancıklı, Z. (2018). Synthesis and Evaluation of New 1,3,4-Thiadiazole Derivatives as Potent Antifungal Agents. *Molecules (Basel, Switzerland)*, 23(12), 3129. Advance online publication. DOI: 10.3390/molecules23123129 PMID: 30501053

Khalili Ghomi, M., Noori, M., Nazari Montazer, M., Zomorodian, K., Dastyafteh, N., Yazdanpanah, S., Sayahi, M. H., Javanshir, S., Nouri, A., Asadi, M., Badali, H., Larijani, B., Irajie, C., Iraji, A., & Mahdavi, M. (2023). [1,2,4]Triazolo[3,4-b][1,3,4]Thiadiazole Derivatives as New Therapeutic Candidates against Urease Positive Microorganisms: Design, Synthesis, Pharmacological Evaluations, and in Silico Studies. *Scientific Reports*, 13(1), 10136. Advance online publication. DOI: 10.1038/s41598-023-37203-z PMID: 37349372

Kokovina, T. S., Gadomsky, S. Y., Terentiev, A. A., & Sanina, N. A. (2021). A Novel Approach to the Synthesis of 1,3,4-Thiadiazole-2-Amine Derivatives. *Molecules (Basel, Switzerland)*, 26(17), 5159. Advance online publication. DOI: 10.3390/molecules26175159 PMID: 34500593

Kolavi, G., Hegde, V., Khazi, I. A., & Gadad, P. (2006). Synthesis and Evaluation of Antitubercular Activity of Imidazo[2,1-b][1,3,4]Thiadiazole Derivatives. *Bioorganic & Medicinal Chemistry*, 14(9), 3069–3080. DOI: 10.1016/j.bmc.2005.12.020 PMID: 16406644

Kuckhoff, T. (2024). *Designing Photocatalytic Materials by Merging Macromolecules with Small Molecular Photocatalysts*. https://openscience.ub.uni-mainz.de/handle/20.500.12030/10047

Kudelko, A., Olesiejuk, M., Luczynski, M., Swiatkowski, M., Sieranski, T., & Kruszynski, R. (2020). 1,3,4-Thiadiazole-Containing Azo Dyes: Synthesis, Spectroscopic Properties and Molecular Structure. *Molecules (Basel, Switzerland)*, 25(12), 2822. Advance online publication. DOI: 10.3390/molecules25122822 PMID: 32570910

Kumar, D., Kumar, N. M., Chang, K. H., Gupta, R., & Shah, K. (2011). Synthesis and In-Vitro Anticancer Activity of 3,5-Bis(Indolyl)-1,2,4- Thiadiazoles. *Bioorganic & Medicinal Chemistry Letters*, 21(19), 5897–5900. DOI: 10.1016/j.bmcl.2011.07.089 PMID: 21873049

Lanzafame, A., & Christopoulos, A. (2004). Investigation of the Interaction of a Putative Allosteric Modulator, N-(2,3-Diphenyl-1,2,4-Thiadiazole-5-(2H)-Ylidene) Methanamine Hydrobromide (SCH-202676), with M1 Muscarinic Acetylcholine Receptors. *The Journal of Pharmacology and Experimental Therapeutics*, 308(3), 830–837. DOI: 10.1124/jpet.103.060590 PMID: 14617684

Lee, J., Kang, S. H., Lee, S. M., Lee, K. C., Yang, H., Cho, Y., Han, D., Li, Y., Lee, B. H., & Yang, C. (2018). An Ultrahigh Mobility in Isomorphic Fluorobenzo[c][1,2,5]Thiadiazole-Based Polymers. *Angewandte Chemie International Edition*, 57(41), 13629–13634. DOI: 10.1002/anie.201808098 PMID: 30133093

Lin, L., Liu, H., Wang, D. J., Hu, Y. J., & Wei, X. H. (2017). Synthesis and Biological Activities of 3,6-Disubstituted-1,2,4-Triazolo-1,3,4-Thiadiazole Derivatives. *Bulletin of the Chemical Society of Ethiopia*, 31(3), 481–489. DOI: 10.4314/bcse.v31i3.12

Linganna, N., & Lokanatha Rai, K. M. (1998). Transformation of 1,3,4-Oxadiazoles to 1,3,4-Thiadiazoles Using Thiourea. *Synthetic Communications*, 28(24), 4611–4617. DOI: 10.1080/00397919808004526

List, G. C. (2013). Graphical contents list. *Tetrahedron Letters*, 54(1), 1–7. DOI: 10.1016/S0040-4039(12)02016-3

Liu, B. B., Bai, H. W., Liu, H., Wang, S. Y., & Ji, S. J. (2018). Cascade Trisulfur Radical Anion (S3•-) Addition/Electron Detosylation Process for the Synthesis of 1,2,3-Thiadiazoles and Isothiazoles. *The Journal of Organic Chemistry*, 83(17), 10281–10288. DOI: 10.1021/acs.joc.8b01450 PMID: 30011993

Luszczki, J. J., Karpińska, M., Matysiak, J., & Niewiadomy, A. (2015). Characterization and Preliminary Anticonvulsant Assessment of Some 1,3,4-Thiadiazole Derivatives. *Pharmacological Reports*, 67(3), 588–592. DOI: 10.1016/j.pharep.2014.12.008 PMID: 25933973

Ma, S. T., Zhu, X. X., Xu, J. Y., Li, Y., Zhang, X. M., Feng, C. T., & Yan, Y. (2021). Iodide-Promoted Transformations of Imidazopyridines into Sulfur-Bridged Imidazopyridines or 1,2,4-Thiadiazoles. *Chemical Communications*, 57(43), 5338–5341. DOI: 10.1039/D1CC01044A PMID: 33928973

Ma, Y., Zhou, X., Cai, D., Tu, Q., Ma, W., & Zheng, Q. (2020). A Minimal Benzo[: C] [1,2,5]Thiadiazole-Based Electron Acceptor as a Third Component Material for Ternary Polymer Solar Cells with Efficiencies Exceeding 16.0%. *Materials Horizons*, 7(1), 117–124. DOI: 10.1039/C9MH00993K

Ma, Z., Hu, X., Li, Y., Liang, D., Dong, Y., Wang, B., & Li, W. (2021). Electrochemical Oxidative Synthesis of 1,3,4-Thiadiazoles from Isothiocyanates and Hydrazones. *Organic Chemistry Frontiers : An International Journal of Organic Chemistry / Royal Society of Chemistry*, 8(10), 2208–2214. DOI: 10.1039/D1QO00168J

Maren, T. H., Haywood, J. R., Chapman, S. K., & Zimmerman, T. J. (1977). The pharmacology of methazolamide in relation to the treatment of glaucoma. *Investigative Ophthalmology & Visual Science*, 16(8), 730–742.

Mariappan, A., Rajaguru, K., Merukan Chola, N., Muthusubramanian, S., & Bhuvanesh, N. (2016). Hypervalent Iodine(III) Mediated Synthesis of 3-Substituted 5-Amino-1,2,4-Thiadiazoles through Intramolecular Oxidative S-N Bond Formation. *The Journal of Organic Chemistry*, 81(15), 6573–6579. DOI: 10.1021/acs.joc.6b01199 PMID: 27379380

Masih, A., Singh, S., Agnihotri, A. K., Giri, S., Shrivastava, J. K., Pandey, N., Bhat, H. R., & Singh, U. P. (2020). Design and Development of 1,3,5-Triazine-Thiadiazole Hybrids as Potent Adenosine A2A Receptor (A2AR) Antagonist for Benefit in Parkinson's Disease. *Neuroscience Letters*, 735, 135222. Advance online publication. DOI: 10.1016/j.neulet.2020.135222 PMID: 32619652

Muğlu, H., Şener, N., Mohammad Emsaed, H. A., Özkınalı, S., Özkan, O. E., & Gür, M. (2018). Synthesis and Characterization of 1,3,4-Thiadiazole Compounds Derived from 4-Phenoxybutyric Acid for Antimicrobial Activities. *Journal of Molecular Structure*, 1174, 151–159. DOI: 10.1016/j.molstruc.2018.03.116

Narran, S. F., Mohammed, S. S., Omer, M. K., Hussein, I. A., Jawad, N. M., & Shweish, B. K. (2022). Synthesis and Biological Activities of Some New Derivatives Based on 5-Styryl-2-Amino-1,3,4-Thiadiazole. *Chemical Methodologies*, 6(2), 83–90. DOI: 10.22034/chemm.2022.2.1

Otsuka, M., Ofusa, T., & Matsuda, Y. (1999). Physicochemical characterization of glybuzole polymorphs and their pharmaceutical properties. *Drug Development and Industrial Pharmacy*, 25(2), 197–203.

Palomo, V., Perez, D. I., Perez, C., Morales-Garcia, J. A., Soteras, I., Alonso-Gil, S., Encinas, A., Castro, A., Campillo, N. E., Perez-Castillo, A., Gil, C., & Martinez, A. (2012). 5-Imino-1,2,4-Thiadiazoles: First Small Molecules As Substrate Competitive Inhibitors of Glycogen Synthase Kinase 3. *Journal of Medicinal Chemistry*, 55(4), 1645–1661. DOI: 10.1021/jm201463v PMID: 22257026

Patel, H. M., Noolvi, M. N., Sethi, N. S., Gadad, A. K., & Cameotra, S. S. (2017). Synthesis and Antitubercular Evaluation of Imidazo[2,1-b][1,3,4]Thiadiazole Derivatives. *Arabian Journal of Chemistry*, 10, S996–S1002. DOI: 10.1016/j.arabjc.2013.01.001

Pechmann, H. (1884). Neue Bildungsweise der Cumarine. Synthese des Daphnetins. *Berichte der Deutschen Chemischen Gesellschaft*, 17(1), 929–936. DOI: 10.1002/cber.188401701248

Pragathi, Y. J., Sreenivasulu, R., Veronica, D., & Raju, R. R. (2021). Design, Synthesis, and Biological Evaluation of 1,2,4-Thiadiazole-1,2,4-Triazole Derivatives Bearing Amide Functionality as Anticancer Agents. *Arabian Journal for Science and Engineering*, 46(1), 225–232. DOI: 10.1007/s13369-020-04626-z PMID: 32837812

Rajput, K., Singh, V., Singh, S., & Srivastava, V. (2024). A Chromatography-Free One-Pot, Two-Step Synthesis of 1,2,4-Thiadiazoles from Primary Amides via Thiolation and Oxidative Dimerization under Solvent-Free Conditions: A Greener Approach. *RSC Advances*, 14(31), 22480–22485. DOI: 10.1039/D4RA03993A PMID: 39015666

Rakitin, O. A. (2019). Recent Developments in the Synthesis of 1,2,5-Thiadiazoles and 2,1,3-Benzothiadiazoles. *Synthesis*, 51(23), 4338–4347. DOI: 10.1055/s-0039-1690679

Reiss, W. G., & Oles, K. S. (1996). Acetazolamide in the treatment of seizures. *The Annals of Pharmacotherapy*, 30(5), 514–519.

Saitanis, C. J., Lekkas, D. V., Agathokleous, E., & Flouri, F. (2015). Screening Agrochemicals as Potential Protectants of Plants against Ozone Phytotoxicity. *Environmental Pollution*, 197, 247–255. DOI: 10.1016/j.envpol.2014.11.013 PMID: 25432168

Sandstrom, J. (1969). Recent Advances in the Chemistry of I, 3,4-Thiadiazoles. *Advances in Heterocyclic Chemistry*, 9, 165–209.

Sharaf Zeebaree, S. Y., & Zeebaree, A. Y. S. (2019). Synthesis of Copper Nanoparticles as Oxidising Catalysts for Multi-Component Reactions for Synthesis of 1,3,4- Thiadiazole Derivatives at Ambient Temperature. *Sustainable Chemistry and Pharmacy*, 13, 100155. Advance online publication. DOI: 10.1016/j.scp.2019.100155

Sharma, B., Verma, A., Prajapati, S., & Sharma, U. K. (2013). Synthetic Methods, Chemistry, and the Anticonvulsant Activity of Thiadiazoles. *International Journal of Medicinal Chemistry*, 2013, 1–16. DOI: 10.1155/2013/348948 PMID: 25405032

Siddiqui, N., Ahuja, P., Ahsan, W., Pandeya, S. N., & Alam, S. (2009). Thiadiazoles: Progress Report on Biological Activities. *Journal of Chemical and Pharmaceutical Research*, 1(1), 19–30.

Subhas Bose, D., & Raghavender Reddy, K. (2017). A Simple and Convenient Method for the Synthesis of 3,5-Disubstituted 1,2,4-Thiadiazoles via Oxidative Dimerization of Primary Thioamides. *Journal of Heterocyclic Chemistry*, 54(1), 769–774. DOI: 10.1002/jhet.2627

Szeliga, M. (2020). Thiadiazole derivatives as anticancer agents. *Pharmacological Reports*, 72(5), 1079–1100.

Taydakov, I. V., Vashchenko, A. A., Lyssenko, K. A., Konstantinova, L. S., Knyazeva, E. A., & Obruchnikova, N. V. (2017). Synthesis, Crystal Structure and Electroluminescent Properties of Fac-Bromotricarbonyl([1,2,5]Oxadiazolo[3′,4′:5,6] Pyrazino-[2,3-f][1,10]Phenanthroline)Rhenium (I). *ARKIVOC*, 2017(3), 205–216. DOI: 10.24820/ark.5550190.p010.130

Vanajatha, G., & Reddy, V. P. (2016). High Yielding Protocol for Oxidative Dimerization of Primary Thioamides: A Strategy toward 3,5-Disubstituted 1,2,4-Thiadiazoles. *Tetrahedron Letters*, 57(22), 2356–2359. DOI: 10.1016/j.tetlet.2016.04.029

Wang, C., Geng, X., Zhao, P., Zhou, Y., Wu, Y. D., Cui, Y. F., & Wu, A. X. (2019). I2/CuCl2-Promoted One-Pot Three-Component Synthesis of Aliphatic or Aromatic Substituted 1,2,3-Thiadiazoles. *Chemical Communications*, 55(56), 8134–8137. DOI: 10.1039/C9CC04254G PMID: 31240291

Wang, Z. Q., Meng, X. J., Li, Q. Y., Tang, H. T., Wang, H. S., & Pan, Y. M. (2018). Electrochemical Synthesis of 3,5-Disubstituted-1,2,4-Thiadiazoles through NH4I-Mediated Dimerization of Thioamides. *Advanced Synthesis & Catalysis*, 360(21), 4043–4048. DOI: 10.1002/adsc.201800871

Wei, Z., Zhang, Q., Tang, M., Zhang, S., & Zhang, Q. (2021). Diversity-Oriented Synthesis of 1,2,4-Triazols, 1,3,4-Thiadiazols, and 1,3,4-Selenadiazoles from N-Tosylhydrazones. *Organic Letters*, 23(11), 4436–4440. DOI: 10.1021/acs.orglett.1c01379 PMID: 33988376

Wolff, L. (1902). Ueber Diazoanhydride. *Justus Liebigs Annalen der Chemie*, 325, 129–195. DOI: 10.1002/jlac.19023250202

Xia, D., Guo, X., Chen, L., Baumgarten, M., Keerthi, A., & Müllen, K. (2016). Layered Electron Acceptors by Dimerization of Acenes End-Capped with 1,2,5-Thiadiazoles. *Angewandte Chemie*, 128(3), 953–956. DOI: 10.1002/ange.201508361 PMID: 26592160

Yajima, K., Yamaguchi, K., & Mizuno, N. (2014). Facile access to 3,5-symmetrically disubstituted 1,2,4-thiadiazoles through phosphovanadomolybdic acid catalyzed aerobic oxidative dimerization of primary thioamides. *Chemical Communications*, 50, 6748–6750.

Yang, Z., Cao, T., Liu, S., Li, A., Liu, K., Yang, T., & Zhou, C. (2019). Transition-Metal-Free S-N Bond Formation: Synthesis of 5-Amino-1,2,4-Thiadiazoles from Isothiocyanates and Amidines. *New Journal of Chemistry*, 43(17), 6465–6468. DOI: 10.1039/C9NJ01419E

Yoshimura, A., Huss, C. D., Saito, A., Kitamura, T., & Zhdankin, V. V. (2021). 2-Iodosylbenzoic Acid Activated by Trifluoromethanesulfonic Anhydride: Efficient Oxidant and Electrophilic Reagent for Preparation of Iodonium Salts. *New Journal of Chemistry*, 45(36), 16434–16437. DOI: 10.1039/D1NJ03787K

Zarei, M. (2017). One-Pot Synthesis of 1,3,4-Thiadiazoles Using Vilsmeier Reagent as a Versatile Cyclodehydration Agent. *Tetrahedron*, 73(14), 1867–1872. DOI: 10.1016/j.tet.2017.02.042

Zhang, L., Sun, B., Liu, Q., & Mo, F. (2018). Addition of Diazo Compounds Ipso-C-H Bond to Carbon Disulfide: Synthesis of 1,2,3-Thiadiazoles under Mild Conditions. *The Journal of Organic Chemistry*, 83(7), 4275–4278. DOI: 10.1021/acs.joc.8b00383 PMID: 29552885

About the Contributors

Shrikaant Kulkarni, Ph.D., is currently working as an Adjunct Professor, the Faculty of Business, Victorian Institute of Technology, Melbourne, Australia.. Dr. Kulkarni has been a senior academician and researcher for more than four decades. He has delivered invited lectures and conducted sessions at national and international conferences as well as faculty development programs. He has guided many major and minor projects in subject areas like engineering chemistry, green chemistry, nanotechnology, analytical chemistry, catalysis, chemical engineering materials, industrial organization, and management. He has published over 100+ research papers in national and international journals and conferences of repute. He has authored 50+ book chapters in books published by CRC, Springer Nature, Apple Academic Press, Elsevier, Wiley, and IET. He has edited twenty five books published by Apple Academic Press/CRC Press and Springer Nature. Another ten books are in the offing. He authored four textbooks in engineering chemistry. Dr. Kulkarni possesses MSc, MPhil, and PhD degrees in Chemistry in addition to master's degrees in Economics, Business Management, and Political Science. He has published two Patents and two more are in the process for the grant. He has expertise in the fields of green chemistry and engineering, analytical chemistry, green nanoscience and nanotechnology, wastewater treatment, Green and Analytical Chemistry, and advanced areas in Chemical Engineering and Material Science.

Hemantkumar Akolkar, Ph.D., Working as Associate Professor, Department of Chemistry, Rayat Shikshan Sanstha's Abasaheb Marathe Arts and New Commerce, Science College, Rajapur, Dist- Ratnagiri, Maharashtra, India Dr. Hemantkumar N. Akolkar is an academician and researcher for last13 years at both undergraduate and postgraduate level. He has taught subjects like Heterocyclic Chemistry, Green Chemistry, Organic Spectroscopy, Organic Stereochemistry, Chemistry of Natural Products during his career. He possesses Masters in Organic chemistry and Ph.D. in Chemistry from Savitribai Phule Pune University, Pune. Hehas got 50+ research papers in national and international journals and conferences of repute. His areas of interest are Heterocyclic Chemistry, Synthetic Organic Chemistry and Green Chemistry. He is reviewer of several journals in Chemistry of international repute.

Bapurao B. Shingate, working as an Associate Professor in Organic Chemistry at Department of Chemistry, Dr. Babasaheb Ambedkar Marathwada Univeresity, Aurangabad, Maharashtra, India Dr. Shingate has more than fifteen years of teaching and twenty-one years of research experience. He is teaching various topic to the postgraduate students like Organic reaction mechanism, Synthetic organic chemistry, Reagents in organic synthesis, Organometallic compounds, Protective groups, Retrosynthetic analysis, Classics in total synthesis, Multistep organic transformations, Asymmetric synthesis. Dr. Shingate is working on various areas of synthetic organic and medicinal chemistry and published more than 130 research papers in quality journals with good impact factors. He has published a review article in the prestigious journal Chemical Reviews (Impact factor 72.087). On his credit two Indian patents also. His total citations are 3849, h-index 35 and i-10 index is 96. He has immensely contributed in the field of design and synthesis of bioactive compounds and drug analogs, development of new synthetic methodologies, and total synthesis of natural and unnatural products. His cumulative impact factors are 506. Dr. Bapurao's contribution to teaching and research already been recognized in the university circle and national level. He has received several awards like GTEA-Best Chemistry Professor of the Year (2020), Bentham Ambassador (2018--23), ISCB-Best Teacher Award (2018), Dr. BAMU-Research Professor Award (2017), Dr. BAMU-Ideal Teacher Award (2014) and Dr. BAMU-Shikshak Pratibha Puraskar. Dr. Shingate has guided 07 Ph.D and more than 200 for M. Sc Dissertation. Dr. Shingate has delivered 175+ invited talks and BOS member in chemistry for several institutions. Dr. Shingate is editorial board member for several journals and reviewer for 60+ journals.

Vijay M Khedkar working as Associate Professor in the Department of Pharmacy at Vishwakarma University, Pune, Maharashtra, India A highly optimistic and proactive computational medicinal chemist with Ph.D. degree in Pharmaceutical Chemistry (Computational Medicinal Chemistry) and Post-doctoral research experience. Successful track record of executing lead identification and optimization projects by applying molecular modeling techniques.

***

Khushbu Bagul has completed her graduation from SES's R. C. Patel Institute of Pharmaceutical Education and Research, Shirpur. Currently, she is pursuing

M. Tech in pharmaceutical chemistry and technology at the Institute of Chemical Technology, Mumbai, Marathwada Jalna campus. She is working on a project (Process Development of Mebendazole) under the guidance of Dr. Vikas N. Telvekar.

Pranjal K. Baruah completed his Master's in Chemistry, securing 1st Class and 2nd Rank from Dibrugarh University, India. He pursued his Ph.D. under the guidance of Professor G. J. Sanjayan at NCL, Pune, specializing in molecular recognition. His first postdoctoral research was with Professor Harold Kohn at the University of North Carolina at Chapel Hill, USA, where he focused on medicinal chemistry. After receiving the prestigious Marie Curie Fellowship, he moved to the University of Oxford, UK, working with Professor Martin D. Smith on the design and synthesis of foldamers and their application in asymmetric catalysis. In 2011, Dr. Baruah joined Gauhati University as an Assistant Professor, where he now serves as a Professor. He has published over 70 internationally recognized papers and several book chapters. His research interests include developing novel methodologies for heterocyclic molecules, C-H functionalization, multi-component reactions, catalysis, peptidomimetics, and foldamers.

Paran Jyoti Borpatra is currently working as D. S. Kothari Post-Doctoral Fellow (DSKPDF) under Prof. Satyendra Kumar Pandey at Banaras Hindu University, Varanasi, India. His is working in the field of synthetic methodology and C-H functionalization. He was born in Lakhimpur district of Assam, India. He obtained his Ph.D degree (2020) under the supervision of Dr. Pranjal Kumar Baruah at Gauhati University, Assam, working in the area of organic synthesis especially C-H activation reactions, multicomponent reactions. Dr. Borpatra completed a two years research project as JRF under 'DBT's Twinning Program for the NE' titled "De Novo Rationally Designed Curcumin Derived Analogues for Increased Bioavailability and Its Metal Analogues for Anticancer and Protective Studies" under the guidance of Dr. Pranjal Kumar Baruah. His current research area is C-H functionalization and development of novel methodologies for heterocyclic molecules. He has also co-authored a number of publications in journals with good impact.

Aditi Boruah received her MSc with a specialization in Organic Chemistry in 2019 from University of Science and Technology, Meghalaya. At present, she is pursuing a PhD as a DST-WISE Junior Research Fellow under the supervision of Prof. Pranjal Kumar Baruah in the Department of Applied Sciences (Chemical Science Division), Gauhati University, Guwahati, Assam. Her research interest focuses on C-H functionalization, synthesis of heterocyclic compounds using Multicomponent reactions.

Akshitha D. N. is presently working as a research assistant in the Department of Chemistry, REVA University, Rukmini Knowledge Park, Kattigena Halli, Yelahanka, Bangalore, Karnataka, India. She received her B.Sc from JSS College for Women's and M.Sc. degree from JSS Science and Technology University in the year 2021. Her research interests include the design and synthesis of biologically active peptides and peptidomimetics, synthesis of nano materials and their utility in organic synthesis.

Mintu Maan Dutta is currently working as Assistant Professor in Arya Vidyapeeth College (Autonomous), Guwahati. He was born in Nazira, Sibsagar, Assam, India. He obtained his B.Sc. degree from Cotton College in 2012 and M.Sc degree with specialization in Physical Chemistry in 2014, from Gauhati University, Assam. He obtained his Ph.D. degree in April 2020 from Department of Chemistry, Gauhati University under the supervision of Professor Prodeep Phukan. His PhD. thesis entitled "Surface functionalized magnetic nanoparticles as catalyst for organic transformations". He has co-authored number of publications in reputed journals that includes Dalton Transactions, Catalysis Communications, ChemistrySelect etc. His current research interest includes heterogeneous catalysis, magnetic nanoparticles, organic transformations.

Nagendra G received his B.Sc. and M.Sc. from Bangalore University, Bengaluru, India, in 2004 and 2006 respectively, and his Ph.D. in Peptides and Peptidomimetics from the Bangalore University in 2012. He is currently working as an Associate Professor in the Department of Chemistry, REVA University, Rukmini Knowledge Park, Kattigena Halli, Yelahanka, Bangalore. His research interests include the design and synthesis of biologically active peptides and peptidomimetics, synthesis of nano meterials and their utility in organic synthesis, total synthesis of natural products etc. He has published more than 33 scientific research papers in reputed journals.

Manoj M. Gadewar currently working as Professor at K. R. Mangalam University, Gurugram. An accomplished researcher, with expertise in nano-biotechnology, pharmacology and pre-clinical study design for execution of research work involved in the process of drug development, he has more than Seventeen years of experience in academia and research. He did his PhD from IIT, Guwahati and published more than 45 research papers in various journals of international repute, to his credit 6 book chapters and 9 patents. He has more than 1100 citations and cumulative impact factor of 100 with H-index of 13. He is GATE qualified in 2003 conducted by IIT Madras with all India rank of 763 and receiver of National Doctoral Fellowship offered by AICTE to pursue doctoral study. He received best researcher award from K R MANGALAM University for his extraordinary contribution to research. He has guided more than 30 PG students and 2 students completed their Ph. D under his supervision.

Mithun Kumar Ghosh currently serves as an Assistant Professor in the Department of Chemistry at Medi-Caps University, located in Indore, Madhya Pradesh, India. Previously, he held similar positions at Govt College Hatta, Damoh, and Pandit S.N Shukla University, Shahdol, both situated in Madhya Pradesh. He completed his doctoral studies in Chemistry at Indira Gandhi National Tribal University, Amarkantak, specializing in the synthesis of transition/rare earth metal clusters with applications in catalysis and biology. Dr. Ghosh's ongoing research endeavors focus on synthesizing homo/heterometallic clusters, coordination polymeric complexes, and investigating their potential in antibiotic degradation and Single Molecule Magnets. Dr. Ghosh has authored 43 publications in renowned international journals and holds a patent. Dr. Ghosh actively contributes to the academic community by serving as a reviewer for various international reputed journals. His diverse research interests encompass Metal Organic Frameworks, Catalysts, Density Functional Theory (DFT), photocatalysis, and an array of chemical properties such as dye adsorption, DNA binding, cytotoxicity, antioxidant activity, and sensing properties.

Prashanth G K, Associate Professor and Head of the Department of Chemistry at Sir M. Visvesvaraya Institute of Technology, Bengaluru-562 157, obtained his B.Sc. degree with 2nd place and his M.Sc. degree with 3rd rank and gold medals from the University of Mysore, Mysuru. He obtained M.Phil. and Ph.D. degrees in Chemistry. He has over 18 years of teaching experience. His research interests include the synthesis of nanomaterials, their characterization, and biological applications, crystallography and environmental Chemistry. He has several scientific research publications in various reputed international journals, as well as textbooks, book chapters, and patent publications to his credit.

Nandini Gour has completed her graduation from the Government College of Pharmacy, Sambhaji Nagar. Currently, she is pursuing M. Tech in pharmaceutical chemistry and technology at the Institute of Chemical Technology, Mumbai, Marathwada Jalna campus. She is working on a project (Process Development of Bupropion) under the guidance of Dr. Vikas N. Telvekar.

H.S. Lalithamba received her B.Sc. and M.Sc. degrees in Chemistry and Organic Chemistry from Bangalore University, Bengaluru, India, in 1994 and 1996 respectively, and her Ph.D. in Peptides and Peptidomimetics from the same university in 2012. She is currently working as Professor and Head in the Department of Chemistry at Siddaganga Institute of Technology, Tumakuru, India. Her research interests include the synthesis of biologically active peptides and peptidomimetics, biological activity, and metal oxides. She has actively participated in VGST, Government of Karnataka, and has published more than 60 scientific research papers in reputed journals.

Navnath Hatvate has completed his PG and PhD from the Department of Pharmaceutical Sciences and Technology, Institute of Chemical Technology, Mumbai. Currently, he is working as an Assistant Professor at ICT Mumbai, Marathwada Campus. He is actively involved in the development of NCE and its analogs, Modifications of Excipients and their applications in Drug Delivery, Process Chemistry of APIs and Intermediates, Total Synthesis of Natural Products and bio-active compounds, Process Developments of API and Intermediates.

Ariful Islam completed his MSc with a specialization in Organic Chemistry in 2023 at the University of Science and Technology, Meghalaya. He is currently pursuing a PhD in the Department of Applied Sciences (Chemical Science Division) at Gauhati University in Guwahati, Assam, under the supervision of Prof. Pranjal Kumar Baruah. His research interests include organic synthesis and C-H functionalization of heterocyclic compounds.

Shalini Jaiswal has more than 17 years' experience in academics and research and is the author of more than 65 scientific publications in Scopus\SCI\UGC and other reputed journal\Conference. Dr. Shalini Jaiswal received her D.Phil degree in Organic Chemistry from University of Allahabad, Allahabad India, in 2009. Dr. Shalini Jaiswal has14 month experience as a project fellow on UGC sponsored project entitled Green synthesis of multi- target nucleosides as a potential anti-viral agent. Her research interest includes different area of Chemistry, Nanotechnology, AI, Biotechnology and Environmental science. She is a Life Member of the Indian science congress association. She has reviewed more than 50 papers of various reputed journal and conferences. She is Editorial board member and Reviewer of various reputed journals like TPI (International Journal), IJATES, International journal of advance research, Advanced Scientific Research, IJEEE, Asian Journal of Research in Chemistry (AJRC) etc.Dr. Shalini Jaiswal Worked as Moderator in TOKYO SUMMIT-II (2020) and TOKYO SUMMIT-III (2021), Turkey (2022,2023,2024)

Ramya M is presently working as a research assistant in the Department of Chemistry, REVA University, Rukmini Knowledge Park, Kattigena Halli, Yelahanka, Bangalore, Karnataka, India. She received her B.Sc from Bangalore University and M.Sc. degree from Bangalore Central University in the year 2021. Her research interests include the design and synthesis of biologically active peptides and peptidomimetics, synthesis of nano materials and their utility in organic synthesis.

Amol Arjun Nagargoje has completed his Masters in organic chemistry(2010) and PhD in organic chemistry (2022) from Dr Babasaheb Ambedkar Marathwada University Aurangabad. He has qualified CSIR-NET examination with AIR-53 in June 2011. In January 2012 he joined as an Assistant Professor in department of Chemistry at KTSP Mandals K M C College Khopoli District Raigad. He has total 13 yrs of teaching experience at UG and PG courses. He has published 21 research papers of international repute with good impact factor. He has published 02 Indian Patents and also completed 02 Minor research Projects funded by BCUD, University of Mumbai. He has presented papers in 07 international and national conferences and received 03 best paper presentation awards. He is the recognised guide for the PhD in chemistry of University of Mumbai, presently 01 student is working for PhD under his supervision. Recently, he received the Summer Research fellowship by Indian Academy of Sciences and completed project at IISER Kolkata.

.P. Bhagya, Associate Professor & HOD, Department of Chemistry, Sai Vidya Institute of Technology, Bengaluru - 560064. She has made research contributions in the fields of photoluminescence, adsorption, and corrosion. She acquired her post-doctoral degree from Kuvempu University and her doctoral degree from Bharathiar University and. She has 15 years of experience in teaching and 8 years of experience in research. She has published research articles in reputed journals. She is a member of the IEEE society.

Sharad P. Panchgalle did his post-graduation in Chemistry from Swami Ramanand Teerth Marathwada University, Nanded in 2001 with second rank in merit order. After qualifying NET-JRF, he worked at National Chemical Laboratory, Pune and received Ph. D. degree from Savitribai Phule Pune University in 2010. In 2010-11, he had worked as post-doctoral researcher with Prof. Alfred Hassner at Bar-Ilan University, Ramat Gan, Israel. Since 2012, he is working as Assistant Professor in Department of Chemistry, KMC College Khopoli (affiliated to University of Mumbai). His areas of interest are organocatalysis, total synthesis, chemical biology and heterocycle synthesis through multi-component reactions. He had published 18 research articles in journals of international repute. He had successfully completed two minor research projects funded by University of Mumbai. He was awarded a Summer Research Fellowship twice in tandem by Indian Academy of Sciences Bengaluru, Indian National Science Academy New Delhi and The National Academy of Sciences India Prayagraj and under that he completed two projects one at IIT Mumbai (2019) and other at IISER Bhopal (2022).

Dattatraya Pansare has 14 years of research experience in organic synthesis and medicinal chemistry. He is published more than 85 research paper in International Journals and 16 book chapters in International books. He has granted One indian patent and filled Two indian patents. His educational background and experience from working in different research projects has enabled him to gain area of expertise in Asymmetric synthesis, Drug discovery in the area of Cancer, Tuberculosis, Medicinal Chemistry: Lead optimization studies, C-H activation, Total synthesis of bioactive molecules, Multistxep synthesis (milligram to gram scales). Presently he is working on development of novel thiazole heterocyclic methodologies. Specially focuses on the development of highly efficient synthesis. We have worked on synthesize the substituted thiazole, thiazolidinone, molecules and Its analoges. To generated focus synthetic library for biological evaluation.

Srilatha Rao, Professor, Department of Chemistry, Nitte Meenakshi Institute of Technology, Bengaluru-560 064. She has made her research contributions in the field of plant extracts and energy efficient water purification and desalination. She has published in scholarly journals of international repute. Dr. Rao acquired a doctoral degree and postgraduate degree in Chemistry from the Mangalore University, Karnataka. She has over 23 years of teaching and research experience. She was selected for the prestigious VGST Fellowship, from Vision Group Science & Technology, Government of Karnataka in the year 2018: A research project titled "Synthesis of proteins and its effect on nanomembranes" AICTE, Govt. of India. As an investigator she successfully guided a VTU funded research project. She is a member of the ECSI, IISc Bangalore, India.

Mubarak H. Shaikh completed his undergraduate and post-graduate studies in Chemistry from Dr. Babasaheb Ambedkar Marathwada University, Aurangabad (2005-10). He qualified prestigious NET-CSIR JRF with all India Ranking 27 in December 2010. In the 2011, he started his doctoral study at Department of Chemistry, Dr. Babasaheb Ambedkar Marathwada University, Aurangabad, which he completed under the supervision of Prof. Bapurao Shingate in August 2016. His doctoral study topic was "Design, Syntheses and Bio evaluation of New Triazoles, Tetrazoles, Pyrazoles and Naphthyridines." His post-doctoral research work was carried out at Shandong University, where he worked with Prof. Wenguang Wang. In December 2017, he joined Wockhardt research center Aurangabad as research scientist. In his two years of tenure at Wockhardt R & D, he successfully implemented improved process of Amidoxime, R-Nosylate and final API Nafithromycin core on plant scale up. Now, Dr. Mubarak H. Shaikh working as Assistant Professor at Department of Chemistry, Rayat Shikshan Sanstha's, Radhabai Kale Mahila Mahavidyalaya, Ahmednagar, Maharashtra, India. He is involved in conducting the research in the field of Organic synthesis with a focus on green catalytic synthesis and C-H functionalization, with an emphasis on the development of new synthetic methods that facilitate the construction of complex and bioactive molecules. After synthesis of the molecules, these molecules were tested for antimicrobial, antioxidant, anti-inflammatory, antitubercular, cytotoxic activity along with computational study. Till today, he had published more than 50 Research papers in the international repute along with seven Indian patents and 4 book chapters included in RSC, Springer and Taylor & Francis publication. His research papers total citations are 1111, h-index is 17 and i10-index is 22. Having the perfect blend of Academics, Administration and Research; Dr. Mubarak has a long way to achieve many milestones in his life.

Aisha Siddekha is presently working as Associate professor in the department of Chemistry, Government First Grade College, Tumkur, affiliated to Tumkur University, Karnataka, India. She has M.Sc. degree in Organic Chemistry (1996) from Bangalore University, and M.Phil. (2009) from Madurai Kamraj University, Madurai, India. She received the Ph.D. degree in the year 2018 in "Green chemistry" from Bangalore University. Her main research interests include synthesis of biologically active organic compounds, green chemistry, corrosion inhibition, vibrational studies, nanomaterials synthesis, characterization and applications. She has published research articles on these topics in Scopus indexed and peer reviewed international journals.

Kalyani Sonawane has completed her graduation from S V. P. M. College of Pharmacy Malegao, Baramati, Pune. Currently, she is pursuing M. Tech in pharmaceutical chemistry and technology at Institute of Chemical Technology, Mumbai, Marathwada Jalna campus. She is working on a project titled "Design and Synthesis of Antitubercular Agents" under the guidance of Dr. Yatin Gadkari.

Index

Symbols

1,2,3-thiadiazole 537, 540, 541, 542, 543, 544, 568

1,2,3-Triazole 129, 135, 146, 152, 153, 156, 157, 159, 160, 162, 163, 164, 165, 167, 169, 170, 171, 172, 173, 174, 175, 176, 177, 179, 180, 181, 182, 183, 184, 185, 186, 190, 191, 192, 193, 194, 195, 196, 197, 198, 199, 200, 201, 215, 216, 243, 269, 371, 389, 474, 531, 567, 568

1,2,4-thiadiazole 537, 545, 547, 548, 549, 550, 551, 552, 567, 569, 572, 574

1,2,5-thiadiazole 537, 539, 554, 555

1,3,4-thiadiazole 226, 291, 537, 556, 557, 558, 559, 560, 561, 563, 564, 565, 566, 568, 569, 570, 571, 572, 573

4,5-dihydro-3,5-diphenylisoxazole 517

A

Alkynes 14, 61, 101, 108, 135, 136, 138, 139, 142, 152, 161, 162, 163, 198, 200, 208, 246, 435, 449, 455, 498, 534

amino acids 177, 184, 194, 400, 402, 407, 416, 417, 419, 421, 428, 431, 432, 434, 521, 523

Analgesic 2, 17, 29, 32, 44, 57, 104, 110, 127, 203, 204, 223, 224, 242, 243, 247, 249, 294, 295, 304, 308, 318, 327, 337, 378, 432, 436, 475, 476, 485, 495, 504, 505, 506, 524, 532, 537

Anti-bacterial 19, 93, 104, 153, 256, 301, 464, 483, 486, 491, 534, 559

antibacterial activity 25, 26, 80, 81, 82, 93, 98, 115, 125, 180, 181, 182, 215, 216, 217, 250, 252, 290, 320, 321, 351, 352, 353, 355, 356, 359, 395, 479, 511, 512, 519, 521, 526, 528, 529, 530, 531, 532, 560, 565

anti-cancer 104, 113, 129, 134, 170, 171, 200, 203, 204, 213, 215, 219, 220, 242, 244, 248, 249, 256, 335, 338, 394, 545, 553

Anticancer 1, 2, 17, 21, 22, 23, 32, 34, 40, 45, 50, 55, 57, 70, 71, 73, 74, 91, 95, 96, 97, 99, 103, 113, 122, 124, 125, 127, 128, 146, 148, 155, 158, 159, 171, 172, 173, 187, 190, 191, 193, 196, 198, 200, 255, 286, 287, 288, 290, 300, 304, 305, 310, 313, 314, 317, 318, 322, 323, 324, 325, 328, 332, 333, 335, 336, 337, 338, 344, 347, 350, 351, 367, 369, 370, 371, 372, 373, 383, 386, 389, 393, 394, 491, 495, 503, 507, 508, 509, 518, 525, 526, 527, 537, 572, 574, 575

anti-cancer activity 170, 219

anticoagulant 128, 298, 299, 315

anticonvulsant 2, 50, 159, 194, 226, 245, 247, 286, 296, 297, 313, 391, 475, 485, 538, 572, 575

anti-diabetic 97, 118, 192, 204, 221, 251, 310, 534

Antidiabetic 28, 57, 86, 87, 123, 176, 187, 191, 255, 291, 313, 373, 383, 385

anti-diabetic activity 118, 221

antihypertensive activity 251

Anti-inflammatory 1, 18, 19, 29, 44, 55, 57, 66, 67, 68, 69, 94, 95, 96, 97, 98, 103, 104, 110, 111, 115, 118, 122, 123, 126, 127, 134, 155, 159, 177, 178, 181, 187, 197, 199, 203, 204, 215, 218, 220, 233, 242, 243, 246, 248, 249, 255, 256, 286, 294, 295, 304, 308, 318, 326, 327, 332, 333, 335, 337, 339, 344, 351, 377, 378, 379, 380, 383, 386, 387, 391, 396, 401, 436, 450, 475, 476, 480, 504, 505, 506, 519, 520, 531, 537, 538, 545, 571

Antimalarial 27, 32, 44, 46, 126, 318

anti-microbial 18, 24, 115, 134, 203, 204, 215, 244, 245, 248, 250, 479, 510, 537, 538, 545, 565, 568

Antimicrobial 18, 24, 25, 35, 39, 41, 48, 50, 51, 55, 57, 79, 81, 93, 94, 95, 98, 101, 120, 122, 124, 126, 127, 128, 146, 159, 180, 181, 182, 190, 191, 196, 197, 198, 251, 255, 266, 286,

H

heterocycle 28, 56, 103, 107, 335, 426, 428, 436, 533

Heterocycles 1, 2, 41, 43, 45, 49, 56, 63, 104, 129, 134, 197, 203, 204, 211, 255, 256, 257, 301, 304, 305, 307, 309, 310, 313, 346, 391, 400, 401, 426, 428, 432, 433, 435, 453, 490, 493, 495, 496, 497, 501, 503, 512, 522, 530, 534, 537, 538, 556, 567, 568

Heterocyclic 2, 3, 8, 31, 38, 42, 44, 51, 55, 56, 58, 65, 66, 70, 80, 85, 91, 94, 95, 96, 99, 103, 104, 105, 125, 126, 131, 134, 137, 156, 166, 194, 203, 204, 207, 215, 216, 246, 250, 255, 256, 264, 287, 299, 301, 302, 305, 308, 317, 318, 333, 334, 335, 338, 344, 345, 353, 368, 370, 375, 387, 389, 390, 391, 393, 395, 400, 401, 428, 430, 432, 434, 435, 436, 463, 486, 487, 488, 490, 496, 497, 499, 501, 503, 504, 511, 516, 517, 518, 522, 524, 532, 534, 535, 537, 567, 574, 575

Heterocyclic Compounds 51, 55, 65, 91, 94, 95, 104, 105, 126, 134, 194, 215, 216, 246, 255, 264, 287, 301, 308, 317, 318, 335, 375, 400, 401, 430, 432, 496, 497, 499, 501, 503, 504, 511, 516, 518, 567

I

Imidazole 41, 80, 101, 172, 184, 233, 255, 256, 292, 297, 304, 436, 485, 496, 497, 564

Imidazo pyridine 466, 474

Indazole 203, 204, 205, 206, 207, 211, 213, 214, 215, 216, 218, 219, 221, 222, 223, 224, 225, 226, 227, 228, 229, 230, 231, 232, 233, 234, 241, 242, 243, 244, 245, 246, 247, 248, 249, 250, 251, 252, 253

Isoxazole 23, 28, 29, 31, 50, 53, 103, 104, 105, 106, 107, 108, 109, 110, 111, 113, 114, 115, 116, 118, 120, 121, 122, 123, 124, 125, 126, 127, 128, 495,

496, 497, 498, 503, 504, 506, 507, 508, 509, 511, 512, 513, 514, 515, 517, 518, 519, 520, 522, 524, 525, 526, 528, 529, 530, 531, 532, 533, 535

M

Medicinal Chemistry 17, 37, 39, 40, 41, 42, 43, 44, 45, 46, 47, 48, 49, 50, 52, 53, 55, 58, 63, 66, 76, 93, 94, 95, 96, 97, 98, 99, 100, 101, 103, 104, 122, 123, 124, 126, 127, 128, 134, 148, 153, 155, 156, 187, 191, 193, 194, 197, 198, 199, 200, 201, 203, 216, 241, 242, 243, 245, 247, 248, 249, 250, 251, 252, 255, 294, 301, 302, 303, 304, 305, 306, 308, 310, 311, 313, 314, 317, 319, 333, 334, 335, 336, 337, 338, 339, 341, 343, 344, 350, 351, 386, 389, 391, 392, 394, 395, 396, 397, 400, 428, 429, 430, 431, 432, 433, 484, 485, 486, 487, 488, 489, 490, 491, 492, 494, 503, 519, 520, 521, 522, 524, 525, 526, 527, 528, 529, 530, 531, 532, 533, 534, 535, 537, 567, 568, 570, 571, 572, 574, 575

Multicomponent Reactions 46, 64, 437, 484, 498, 503, 520, 522, 523, 530, 533, 534

N

neuroprotective effect 222

O

o-phenylenediamines 258, 274, 304, 306, 316

Oxazoles 1, 2, 3, 4, 5, 6, 7, 10, 12, 13, 14, 15, 16, 17, 21, 22, 23, 24, 26, 27, 28, 29, 30, 31, 32, 34, 35, 36, 37, 41, 42, 43, 47, 49, 50, 51, 52, 53, 334, 335

P

peptides 346, 400, 408, 427, 428, 429, 434

Printed in the United States
by Baker & Taylor Publisher Services